Studies
in the History of Mathematics and Physical Sciences

15

Studies in the History of
Mathematics and Physical Sciences

continued after index

Jesper Lützen

Joseph Liouville 1809–1882:
Master of Pure and Applied Mathematics

With 96 Illustrations

Springer-Verlag
New York Berlin Heidelberg
London Paris Tokyo Hong Kong

Jesper Lützen
Department of Mathematics
University of Copenhagen
DK-2100 Copenhagen
Denmark

Mathematics Subject Classification: 01A55, 01A70, 26A33, 12H05, 34B25, 70-03 33A25, 12F10, 31-03, 51-03, 53-03, 80-03, 78-03, 11-03

Library of Congress Cataloging-in-Publication Data
Joseph Liouville, 1809–1882: master of pure and applied mathematics/
    Jesper Lützen
        p. cm.—(Studies in the history of mathematics and physical
sciences; 15)
    Includes bibliographical references.
    ISBN 0-387-97180-7 (alk. paper)
        1. Liouville, Joseph, 1809–1882. 2. Mathematicians—France—
Biography. I. Lützen, Jesper. II. Series.
QA29.L62J67   1990
510.92—dc20
    [B]   90-31345

Printed on acid-free paper

Camera-ready copy prepared using LaTeX.
Printed and bound by Edwards Brothers, Inc., Ann Arbor, Michigan.
Printed in the United States of America.

9 8 7 6 5 4 3 2 1

ISBN 0-387-97180-7 Springer-Verlag New York Berlin Heidelberg
ISBN 3-540-97180-7 Springer-Verlag Berlin Heidelberg New York

To my parents

Preface

Joseph Liouville was the most important French mathematician in the generation between Galois and Hermite. This is reflected in the fact that even today all mathematicians know at least one of the more than six theorems named after him and regularly study *Liouville's Journal*, as the *Journal de Mathématiques pures et appliquées* is usually nicknamed after its creator. However, few mathematicians are aware of the astonishing variety of Liouville's contributions to almost all areas of pure and applied mathematics. The reason is that these contributions have not been studied in their historical context. In the *Dictionary of Scientific Biography* 1973, Taton [1973] gave a rather sad but also true picture of the Liouville studies carried out up to that date:

> The few articles devoted to Liouville contain little biographical data. Thus the principal stages of his life must be reconstructed on the basis of original documentation. There is no exhaustive list of Liouville's works, which are dispersed in some 400 publications.... His work as a whole has been treated in only two original studies of limited scope those of G. Chrystal and G. Loria.

Since this was written, the situation has improved somewhat through the publications of Peiffer, Edwards, Neuenschwander, and myself. Moreover, C. Houzel and I have planned on publishing Liouville's collected works. However, considering Liouville's central position in French mathematical science in the middle of the 19th century, there is, I think, still a need for a more comprehensive study of his life and work. I hope that I have been able to remedy this neglect in part by bringing together the published evidence and the extensive archival material, in particular Liouville's *Nachlass* of 340 notebooks, consisting of more than 40,000 pages.

When I started this work in 1980, I had the ambition of writing *the* definitive biography of Liouville. Now, however, I know that the result is only a modest first approximation, which I hope will be extended by other historians and mathematicians. In particular, I have not gone as deeply into Liouville's private life, family matters, etc., as I had planned to do. The reason is that after the initial archival studies I discovered that Neuenschwander was doing the same job. This means, for example, that I have not used the Bordeaux archive that Neuenschwander had discovered

[Neuenschwander 1984a, note 5]. Neither have I studied the newspapers of the time.

The more limited aim of this book is to tell the story of Liouville's scientific career: his education and his work as a teacher, journal editor, politician, and academician, and not least to analyze his mathematical works and place them in a historical perspective.

The book consists of two parts. The first (Chapters I–VI) is a chronological account of Liouville's career. We follow him from his student days at the *École Polytechnique* and the *École des Ponts et Chaussées* to his glorious days as a celebrated academician and professor at the most prestigious schools in Paris. We follow his less successful and brief career as a politician during the Second Republic, his renewed scientific creativity during the 1850s, and finally his last years full of illness, pain, and misery.

In this connection, I have tried to give a picture of the rather complex institutional setting in Paris. This part also contains descriptions of Liouville's teachers, colleagues, and students, both French and foreign, and of his two scientific archenemies, Libri and Le Verrier (whose name, by the way, I shall spell the way Liouville usually did, i.e., with a capital V). Liouville's private life is of course also taken into account, in particular when it tells us something about his scientific career.

Liouville's mathematical and physical works also have a central place in the first part. Indeed, one can consider the first part as an explanation of how Liouville's various works fit into a global picture of his life. Yet, the description of the mathematics is not technical in the first part and is mostly without formulas.

In the second part, one can find a more thorough analysis of various aspects of Liouville's work. This part is divided into 10 chapters concentrating on different mathematical and physical fields, of which Chapters X and XI are slightly revised versions of two previously published papers [Lützen 1984a,b]. Thus, the chronology is broken. It has been my aim that these chapters can be read independently of each other and independently of Part I of the book. This means that mathematicians can concentrate on those particular parts of Liouville's production they like the best, but it also means that I have been forced to repeat certain things.

The mathematical and historical analysis of Liouville's work relies heavily on his notebooks; for not only do these reveal the genesis of many of his published results, they also contain many ingenious ideas, methods, and results that he did not publish. It is in fact remarkable that many of Liouville's most far-reaching ideas were never published.

The mathematical analysis of Liouville's notes has not only been the most challenging but also the most interesting and rewarding work involved in the research concerned with this biography. The bulk of the material is overwhelming, and the notes are usually written pell-mell among each other and often in a very sketchy style, which even made them hard for Liouville himself to understand. In a letter to Bertrand concerning a passage by Laplace on tautochrones, Liouville wrote:

Now I have found the note that I wrote on this subject at least twelve years ago; it is terrifying (you are too young to know that) to see that as time passes one forgets what one has known the best. My note which is part of a sort of commentary on the first two volumes of the *mécanique céleste* that I composed for my own use around 1834 and 1835 and which are full of abbreviations, which I now have some difficulty in understanding...

[Neuenschwander 1984a, III, 2]

Similarly, in a letter to Dirichlet [Tannery 1910, p. 40], Liouville admitted that his notebooks "are in complete disorder." These quotes have often comforted me when I was unable to understand the content, or even the subject, of a particular note. I have looked through the notebooks a few times in order to find interesting entries. Many of the notes, in particular in the last 200 notebooks, are clearly without much mathematical interest, but there may very well be hidden treasures that I have overlooked. Indeed, I suspect that one can find valuable information about Liouville's ideas on number theory. Since I have no knowledge of number theory, I have omitted an analysis of this perplexing chapter of Liouville's production, but otherwise I have tried to discuss all of Liouville's major contributions to mathematics and physics.

The published and, in particular, the unpublished materials have allowed me to make several observations concerning Liouville's work that I do not think have been made before. Here, I shall list a few of the more remarkable conclusions:

1.   Liouville began his career as a mathematical physicist writing on electrodynamics and the theory of heat.

2.   His works on electrodynamics probably motivated his research on the fractional calculus.

3.   His works on the theory of heat led to his first work on Sturm-Liouville theory.

4.   His generalization of Sturm-Liouville theory to a higher order partly failed because he tried to prove the completeness of a system of eigenfunctions that is not complete.

5.   Liouville's approach to integration in finite terms developed from being analytical to being algebraic.

6.   Liouville knew a theorem on integration in implicit finite terms, which was proved by Risch as late as 1976, and he even had a formally correct proof of it.

7.   In 1842, Liouville developed a theory of the stability of rotating masses of fluids that shares many methods and results in common with the published theories by Poincaré and Liapounoff from the mid-1880s.

8.   Liouville's proof of the existence of transcendental numbers was inspired by a correspondence between Goldbach and Daniel Bernoulli and was based on Lagrange's formulas for continued fractions.

9.   One can reconstruct how Liouville found Liouville's theorem on doubly periodic functions from theorems on trigonometric series.

10.   In his unpublished, notes Liouville succeeded in filling the most conspicuous holes in Galois's great memoir.

11.   Starting from potential theory, Liouville, in the 1840s, created a rather general theory of integral operators, which anticipated Hilbert's ideas. In particular, it was based on the Rayleigh-Ritz method of finding eigenvalues.

12.   This theory and Liouville's work on Sturm-Liouville theory show that he was the first great expert in spectral theory.

13.   Liouville wrote an unpublished memoir on a generalization of the Poisson and Lagrange brackets. It led him to a theory of transformations of the equations of mechanics.

14.   Liouville may have been the mathematician before Riemann who best understood the value of an intrinsic formulation of differential geometry.

15.   This insight allowed him to use geometric ideas in an interesting transformation of the principle of least action. In two variables, he saw that trajectories of a particle acted on by forces can be thought of as geodesics on another surface. His general transformation theorem stopped short of generalizing this to the observation that trajectories are nothing but geodesics in a suitable Riemannian metric.

These somewhat crudely formulated conclusions are amplified and documented in the chapters of Part II.

Generally, Liouville's production is characterized by a slow movement from applied to pure mathematics; here, Liouville followed a general tendency in mathematics of the time. Most of his life he continued to be inspired by physical problems, although Laboulay, in his speech at Liouville's funeral, argued that this was a self-imposed restraint from the most abstract spheres where he felt at home as well:

> His mind operated in these heights where few savants could follow him. He said jokingly that there had been a problem that could only be posed or understood by three initiates in the entire world: a Russian savant, an American lady and a third mathematician whose name he did not mention; but this was not the subject of the sciences, and he added that there were such problems that could only be understood by two people. It was Liouville himself who because of modesty renounced rising to this last summit of abstraction.
>
> [Laboulay 1882]

As a person, Liouville seems to have been helpful and kind, but also firm in his beliefs. Although a biographer should perhaps keep clear of such feelings, I cannot help liking him and feeling sorry for him when life became unendurable to him. I hope, however, that this has not made me too biased.

I have tried to write the first part in a narrative style, whereas the second part is more discursive. In an attempt to make the text easy to read, I have tried to avoid footnotes as much as possible. For the same reason, I have translated all the French quotes (except a few quotes in the appendices). This causes a problem of documentation, in particular concerning the unpublished material that the reader cannot easily find in the original. I contemplated appending all the original French quotes, but that would have made this already long book even longer. Only the excerpts from the *Procès Verbaux du Bureau des Longitudes* are quoted in the original as well, because the president of the Bureau has asked me to do that.

References to places in this book are included in parentheses, e.g., (Chapter XV, §8); the chapters carry Roman numerals, and the number of the section is preceded by a §. Only the chapters in the second part of the book are divided into numbered sections. Therefore, I refer to places in the first part by page number, e.g. (Chapter III, p. 10). Formulas are consecutively numbered in each chapter. A reference to a formula appears as (Chapter XV (6)) or just (XV (6)). When I refer to a formula or a section in the same chapter, I omit the number of the chapter, e.g., (6) or §8.

The formulas in some of the quotations are renumbered to fit into the consecutive numbering of the other formulas in the chapter. Such numbers are put in square brackets, just as all other additions to the original text.

Otherwise, square brackets are used for references to the works mentioned in the bibliography. These references are given by the author's name and year of publication. In some instances, the year of composition or of presentation to an academy is used. If the bibliography contains several publications by one author from the same year, these are labeled a, b, c,..., e.g., [Liouville 1836b]. In cases where it is clear which author is cited, only the year is given in square brackets. Three abbreviations have been used throughout: C. R. for the *Comptes Rendus des Séances de l'Académie des Sciences de Paris*, L. J. for the *Journal de Mathématiques pures et appliquées*, or Liouville's Journal as it is often called, and Ms for the manuscripts in Liouville's *Nachlass* at the *Bibliothèque de l'Institut de France*. I have paginated many of the first notebooks but not the last ones. Therefore, only some references to the Ms carry page numbers.

References to archival material are usually given by some easily recognizable abbreviation, e.g., P. V. Bur. Long. for the *Procès Verbaux du Bureau des Longitudes*. The manuscripts found at the *Archive de l'Académie des Sciences*, however, have been given a special code. This code refers to the list of manuscripts at the beginning of the bibliography. The bibliography continues with a list of Liouville's published works and ends with the publications by other authors.

Years of birth and death of the persons mentioned in the book can be found in the index, at least as far as I have been able to find them.

Acknowledgments: During the eight years of writing this book, I have held appointments at two departments: first at the Mathematical Department at the University of Odense, and then at the Mathematical Department at the University of Copenhagen. I am grateful to my colleagues and the staff at these institutions for their kind encouragement and help. In particular, I wish to thank Uffe Haagerup and Christian Berg for having cleared up some difficult mathematical points. All along I have been supported from my former teachers and colleagues at the Institute for the History of Exact Sciences at the University of Aarhus and in particular by Kirsti Andersen, whose usual kindness, constructive criticism, and encouragement have been of great help.

During my stays in Paris, which were partly funded by the Danish Royal Society and the French State, I was fortunate to get a room in the Fondation Danoise. My research there was also greatly facilitated by the helpfulness of the staff at the libraries and archives I visited, such as the *Archive de l'Académie des Sciences*, the *Bureau des Longitudes*, whose President Mr. Kergrohen kindly allowed me to read through their *Procès Verbaux*, the *Archive de l'École Polytechnique*, the *Archive de la Collège de France*, and not the least, the *Bibliotèque de l'Institut de France*, where I spent most of my time in Paris studying Liouville's notebooks. I am grateful to these institutions and individuals for their kind help.

I also wish to thank Bruno Belhoste for having helped me find my way in the Parisian archival jungle and Professor F. Jongmans and Jan-Erik Roos, who provided me with valuable material concerning Catalan, Bjerknes, and Holmgren; I am also grateful to many other colleagues from the generalized Oberwolfach circle, in particular, to Ivor Grattan-Guinness and Harold M. Edwards for their helpful suggestions concerning various points of this book.

Finally, my appreciation goes to Jeremy Gray who has made my broken English sound less broken, to Dita Andersen, who enthusiastically has struggled transforming my scrawl into a TₑX file, and to Springer-Verlag for their generous editorial care. It was their Mathematics Editor, W. Kaufmann-Buhler, who originally suggested, at a very early stage of my work, that I should contact him as soon as my plans had developed far enough. I am very sorry that he did not live to express his opinion on my book, for after having read his fine biography of Gauss, I am sure he would have suggested valuable changes.

Copenhagen   Jesper Lützen
June 1988

Contents

III. Professor, Academician, and Editor (1840–1848) 69

IV. The Second Republic (1848–1852) 149

V. The Last Flash of Genius (1852–1862) 175

VI. Old Age (1862–1882) 227

Part II. Mathematical Work

XI. Figures of Equilibrium of a Rotating Mass of Fluid.... 477

XII. Transcendental Numbers...........................513

XIII. Doubly Periodic Functions 527

Part I

The Career of a Mathematician

I

Youth (1809–1830)

Early Interests in Mathematics

Joseph Liouville lived through one of the most tumultuous periods of French history. Born at the height of the Napoleonic era, he experienced six violent dynastical changes before he died at a time when France was only slowly recovering from its humiliation in the Franco-Prussian war.

The political unrest affected his life from the start, as he was born the son of army captain Claude-Joseph Liouville (1772–1852). Claude-Joseph was a member of a distinguished Lorraine family, which probably originated in the small village of Liouville, close to Saint-Michel in the *arrondissement* of Commercy. His wife, Thérèse Balland, came from another upper middle class family from Lorraine [cf. Dumont 1843, Faye 1882, Taton 1973, Neuenschwander 1984a]. In 1803, she bore their first son Félix-Silvestre-Jean-Baptiste Liouville. At this time, they lived in the ancient Roman city of Toul, by the river Moselle, between Commercy and Nancy. However, soldiers in Napoléon's army had to move to the places where the restless emperor needed them, and when the second son Joseph, our mathematician, was born on March 24, 1809, the family was staying at Saint-Omer, Pas de Calais.

During some of the following years of war until Napoleon's final defeat, the two brothers stayed with an uncle in Vignot, 2 km from Commercy [Brochard 1902–7, Vol.13, p. 14]. Félix attended the *collège* in Commercy, but Joseph was probably too young to follow regular classes. Still, according to Dumont, it was in Commercy that he received his first education:

> and one should not forget that his teacher Mr. Rollin predicted several times that he would not go a long way because he was too fond of playing. He did not guess that the child who imperturbably played chess for an entire day to the exclusion of the pleasures of his age, would become a Pascal or a Newton.
> [Dumont 1843, Vol. III, pp. 433–434]

Thus, Joseph's mathematical gifts manifested themselves at an early age in play. However, as a refreshing contrast to the usual anecdotes of young mathematicians (Pascal, Gauss, etc.), his teacher did not discover his abilities and predicted that he would not go a long way. Liouville was not an infant prodigy.

Unlike the great majority of his fellow soldiers, Claude-Joseph Liouville survived the war. He retired and returned to Toul [Ec. Pol. Cours], where he settled with his family. Here, at the *collège* Joseph received his education in ancient languages and then went to Paris to study mathematics at the *Collège St. Louis* [Neuenschwander 1984a].

As a *Collège* student, Liouville became acquainted with the mathematical research of the day, studying the only French mathematical journal, Gergonne's *Annales de Mathématiques Pures et Appliquées*. He soon discovered that he was able to add observations of his own, and he began to draft coherent papers [Ms. 3615(1)]. With these drafts, he began a lifelong habit of sketching his mathematical ideas and correspondence in notebooks, of which 340 containing approximately 50,000 pages are still preserved at the *Bibliothèque de l'Institut de France*. They constitute a mine of information about the work of Liouville and the mathematical community surrounding him.

The first three articles bear the titles *Recherches sur les Caustiques* (24 pp.), *Considérations analytiques sur les figures rectilignes* (32 pp.), and *Théorèmes nouveaux sur les triangles et les quadrilatèrs* (6 pp.). In the latter, Liouville proved the following theorem:

> Theorem 1: Let two triangles ABC, abc be placed in a plane such that the directions of the sides of the first pass through the corners of the second, respectively, and moreover, such that the three straight lines joining the opposite corners cut each other in one point. Draw in the same plane an arbitrary line DD' cutting the sides BC, CA, AB of the first triangle in the three points A', B', C' and join these three points to the opposite corners of the same triangle by the lines AA', BB', CC'. Then these lines cut the opposite sides bc, ca, ab of the second triangle abc in three new points a', b', c', which lie on a straight line.
>
> [Ms. 3615(1)]

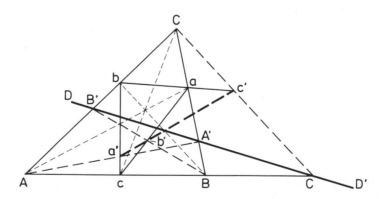

Liouville further formulated a converse of this theorem and generalized it to a quadrilateral. He mentioned certain modifications "giving rise to several corollaries which we shall pass over... we only remark that these corollaries comprise the theorem proved by Mr. Vecten in one of the last issues of the *Annales.*" It is this remark that has allowed me to date Liouville's first three notes to the winter of 1824–1825, because it clearly refers to Vecten's *Démonstration d'une propriété du quadrilatère complet,* which appeared in the November 1824 issue of Gergonne's *Annales* [Vecten 1824].

The note on triangles and quadrangles reveals that at the end of his *Collège* time, Liouville possessed a remarkable knowledge of and ingenuity in projective geometry, and the two other notes show a similar familiarity with analytic geometry and mechanics and knowledge of the basics of the calculus. He never published these notes, although their polished form indicates such a plan, and a desire to rehearse his mathematical style for future use. His second note began as follows:

> In this essay, we propose to study the relations existing between the lengths and the directions of the sides of an arbitrary rectilinear figure by a uniform and purely analytical method and to derive from this theory the solutions of several related questions. The results which we shall obtain are already known for the most part — we have tried to connect them to an attractively simple unity by relating them to a known principle. The general resolution of a plane polygon should have been a part of our work, but we have renounced this idea considering that Prof. Lhuilier is preparing a work in which this question will be treated in great detail (probably [Lhuilier 1828]).
>
> [Ms. 3615(1)]

Only a student who had decided to become a professional mathematician would write such a note.

Student at the École Polytechnique

It was therefore natural for Liouville to apply for admission to the *École Polytechnique.* To be sure, it was an engineering school, but it provided a solid basis in mathematics and mechanics and gave by far the highest education in these sciences in France, indeed in the whole world. The school had been founded during the great revolution in *an* III (1794), and its high scientific standards had from the very start been set by the teachers who were counted among the best scientific brains of the day; they included Monge, Lagrange, Prony, Fourcroy, Vauquelin, Berthollet, and Chaptal. Moreover, the choice of the *École Polytechnique* was particularly natural for the son of a soldier, for it was then — and still is — a military institution. [cf. Fourcy 1828, Marielle 1855].

The examiners of the school were also chosen from among the elite. Thus, in 1825, when Liouville applied for admission, the examiners were Poinsot, Dinet, and Reynaud, the latter being in charge of Liouville. [Ec. Pol. Corps Tome I, and registers]. The candidates were required to know about the metric system, the binomial theorem, the resolution of polynomial equations, logarithms, trigonometry and its application to geometry, conic sec-

PLATE 1. The entrance to the Ecole Polytechnique looks today as it did when Liouville studied at the school from 1825 to 1827. (Photographed by Jesper Lützen.)

tions, and the simple machines. Moreover, their knowledge of Latin and French and their ability to draw was tested [Ec. Pol. Programmes 1825]. The young Liouville, who passed this examination together with 110 other applicants, was described in the registers of the school as a 1.52-m-high young man with chestnut hair, a broad open face, grey-blue eyes, plumb nose and mouth, and a cleft chin. Like almost all his fellow students, he was declared to be of good health and in good condition [Ec. Pol. Santé]. These were important, because the schedule was hard. From 6 a.m. to 9 p.m., the students were occupied with lectures, seminars, study sessions, and free studies, except during their few hours of leisure on Wednesday afternoon and Sunday from 10 a.m. to 7.30 p.m. During the first year, Liouville was taught mathematical analysis (35 lectures), geometry (15), mechanics (35), descriptive geometry (72), applied analysis (i.e., differential geometry) (16), physics (33), chemistry (36), history, literature, etc. (34), topographic drawing (35), and landscape drawing (70). The second year most subjects, except for descriptive geometry, continued with almost the same number of lectures, and in addition the students were taught geodesy (28), social arithmetic (i.e., theory of probability and statistics) (6), and architecture (38) [Ec. Pol. Programmes 1825].

Of most importance to a student with mathematical interests was the *cours d'analyse et de mécanique*, which was taught in rotation by Cauchy

and Ampère, with Coriolis and Paul Binet as their respective *répétiteurs*.
Liouville's class got Ampère as their teacher, which may at first glance
seem to be bad luck, since at this time Cauchy was the young star in the
Parisian mathematical community. Of the old guard of mathematicians,
who had lent luster to the revolution, Lagrange, Monge, and Carnot were
dead. Legendre was applying his finishing touches to his treatise on elliptic
integrals, and Laplace and Fourier had almost stopped publishing. Poisson was at the height of his productivity, but he too was gradually being
eclipsed by Cauchy. As the young Norwegian mathematician Abel wrote
home with some exaggeration a year later:

> He [Cauchy] is the only one who at present works in pure mathe-
> matics; Poisson, Fourier, Ampère, etc. are exclusively occupied by
> magnetism and other physical theories.
>
> [Ore 1957, p. 147]

Since his nomination as a professor in 1815, Cauchy had drastically revised the analysis course at the *École Polytechnique*, replacing the old approach as expounded in the influential textbooks by Lacroix [1797–1800]
with his new, stricter standards of rigor. He declared in his lectures that
a divergent series had no sum, and that it was illegitimate to draw conclusions from the generality of algebra, for example, to transfer properties
of real functions to complex functions without proof. This greatly shocked
the old generation of mathematicians, who, following Euler and Lagrange,
had often violated these rules. Instead, Cauchy devised a theory of convergent series, introduced a new concept of continuity, and based it and
the definition of the derivative on the concept of limits. He further defined
the definite integral as the limit of a "Riemann" sum and proved that it
exists for continuous functions. Finally, he proved the existence of solutions
to certain differential equations, all revolutionary novelties not only in the
curriculum at the *École Polytechnique* but in mathematics as a whole.

Although Liouville missed Cauchy's lectures, he could still acquaint himself with their content, for Cauchy had published his lecture notes in [1821]
and [1823], and later he produced two more volumes in [1826] and [1829].
From Liouville's later writing, it is apparent that he soon read these masterpieces. Moreover, Ampère's lectures had some similarity with Cauchy's.
Already as a young man, Ampère had written a paper on the foundation of
the differential calculus [1806], and he had taught his ideas to his students
at the *École Polytechnique*, Cauchy among them. Cauchy was therefore
inspired by Ampère and later acknowledged that he had profited several
times from his observations (cf. [Grabiner 1981, pp. 127–132, and Guitard 1986]). Therefore, it is not surprising that Ampère actively supported
the young Cauchy in his attempts to reform the curriculum at the *École
Polytechnique*, and that he in turn adopted some of Cauchy's ideas in his
Précis de calcul différentiel et intégral, which was printed in part in 1824.
It was never finished, probably because when Liouville entered the school
in 1825 the curriculum was changed in a direction different from the one

Cauchy and Ampère had fought for: the introductory *analyse algébrique* was suppressed, and the lectures were to begin immediately with differential calculus [Belhoste 1985]. Still, the *Registre d'Instruction* of the year 1825–1826 reveals that Ampère stuck to the general outlines of his course. Thus, he probably continued to define a continuous function as did Cauchy, though with the additional requirement that it be piecewise monotonous. Moreover, he defined the derivative of $f(x)$ as the fraction $\frac{f(X)-f(x)}{X-x}$ when $X = x$, and he "proved" the "fundamental theorem" that if f is continuous this fraction cannot be zero or infinite except in "particular values" of x. This is a simplified version of his 1806 proof, and although basically wrong, it still contains several ideas similar to Cauchy's. Liouville's failure to attend Cauchy's lectures was therefore only a small loss, the more so since Cauchy was reportedly a rather bad lecturer and so occupied with his own work that on many occasions he neglected young mathematicians who sought his advice. Ampère, on the other hand, became a great help and support for Liouville's further development as a mathematician. He soon discovered the talents of his young student, and it was probably he who encouraged Liouville to follow his *Cours de Physique Mathématique* at the *Collège de France*.

Established around 1530 by François I, the *Collège Royale* - after the revolution called the *Collège de France* - occupied a place apart in the Parisian educational system. It offered no fixed syllabus and had no registered students or examinations. Its professors could freely choose the subject of their lectures within the field of their chair, and the science teachers, in particular, often benefited from this freedom which allowed them to present their own research. Therefore, the lectures drew many of the bright students from the *Faculté des Sciences* at the *Sorbonne* and from the specialized schools who wanted to acquaint themselves with the research front in their field.

Among the audience of Ampère's course of 1826–1827 was the young Liouville. The first two-thirds of the course cannot have excited him for it was an introduction to rational mechanics similar to the one he had heard at the *École Polytechnique*, but at the end of the course, Ampère developed his exciting new theory of electrodynamics. He had developed this theory over the preceding six years and had just published his definitive version of it in *Théorie des phénomènes électrodynamique, uniquement déduit de l'expérience* [Ampère 1826].

Liouville took careful notes on the lectures ([AC AL Ms. 1] (only the notes on rational mechanics are preserved)). He may have been invited to do so by Ampère, for Ampère later corrected them and saw to it that a fair copy was written ([AC AL Ms. 2] (also contains the notes on electrodynamics)). The purpose of the fair copy is unclear. In later years, it became common at the *École Polytechnique* for the teacher to approve the notes of one of the best students and for them to be lithographed so that all the students could use them when reading for their exams. Although there were no

PLATE 2. André-Marie Ampère (1775–1836). Liouville's teacher of analysis and mechanics at the École Polytechnique. Ampère's famous electrodynamical theory inspired Liouville's first original works. (Engraving by Ambroise Tarideu, 1825. Reproduced from Bern Dibner's "Ten Founding Fathers of Electrical Science", p. 24, Novo Connecticut, 1954. Courtesy Burndy Library.)

exams at the *Collège de France*, Ampère may similarly have planned to lithograph Liouville's notes, but this never happened, and the fair copy was left unfinished. Still, the contact with the master of electrodynamics alerted Liouville to an important branch of applied mathematics.

At the *École Polytechnique*, he became acquainted with other aspects of pure and applied mathematics. Thus, he was taught geodesy by another of his later supporters, Arago, and his répéteteur, Mathieu. The physics course was given by Dulong, and the descriptive geometry lectures by Leroy [Ec. Pol. Corps Tome I]. The efforts of these eminent scientists were not wasted on Liouville. He worked hard all through the two years at the school and obtained high marks in all disciplines except for the "literary essay" [Ec. Pol. Registre 1825–1826]. He passed from the *seconde division* (first year) to the *première division* (second year) as number 14 of 110 students, and at the final examination during the summer of 1827, the examiners Prony, Poisson, Demonferrand, and Chevreul ranked him as number 8.

The Ponts et Chaussées

The years at the *École Polytechnique* were only intended to give the future engineer a scientific basis to acquire the more practical skills at one of the specialized schools. Liouville wavered between the *École des Ponts et Chaussées*, the *École des Mines*, and the *École du Génie* [Neuenschwander 1984a, III, 1], but decided on the *Ponts et Chaussées*. In this respect, he followed many of the best savants of the time, for example, Cauchy. The only teacher at this school who made a great impression on Liouville was Navier, whom he described as "a very obliging and very active man." Still, when writing to LeVerrier in 1833, Liouville declared that "if I was to start over again, I would do what I have done, I would enter the *Ponts et Chaussées*" [Neuenschwander 1984a, III, 1]. Liouville was admitted to the corps of the school on November 20, 1827, as number 5 or 6 of 25 students [Neuenschwander 1984a, p. 120].

The *École des Ponts et Chaussées* was divided into three classes. Each year a competition was held to select the students who could pass to the next class. In each class, the students were arranged according to a system of degrees obtained both in the theoretical work at the school and in the practical work performed during the annual trainee period. During these periods, the students were sent out to assist the engineers of the *Ponts et Chaussées* with their constructions in various parts of France. Liouville regarded the trainee periods with mixed feelings:

> The trainee voyages of the *Ponts et Chaussées* have both their use-
> fulness and their inconvenience. They make you gad about the world
> a little but they also make you waste your time.
> [Neuenschwander 1984a, III, 1]

It is surprising to see that in spite of his more theoretical interests in mathematics and pure science, Liouville did very well in these more practical subjects. Thus, during his first year at the school, he twice obtained a "first honorable mention" in "style," and once in works on roads. The following two years he obtained a first prize and a first honorable mention, both in the literary test. He also devoted himself to the duties during the training periods, and with his great industry and talents he won the praise of his superiors. During his first year (approximately from May to November 1828), Liouville was sent on a mission to Chateauroux, south of Paris, to help with the work on the road from Poitiers through Chateauroux to Bourges and on other public works. The chief engineer of the *département* of Indre was highly satisfied with Liouville and wrote to the *Préfet*:

> ...holes in the Royal road N° 151. This work, which indirectly mat-
> ters to the Dep^t *de l'Indre*, has been entrusted to Mr. Liouville. Of
> a very weak constitution, and ill when he arrived at this Dep^t this
> student seemed only to be able to offer the sterile tribute of his good
> will. However, he is on the contrary the one for whom I must con-
> sider myself most lucky in all respects. And, in order to assure you,

Mr. *Baron* that I am infinitely satisfied with the services of this very interesting student, it is sufficient to me to have the honor of telling you that thanks to him I shall within a few days have accomplished 1° drafting the project of construction at the hole near Chitray, 2° drafting the project of the hole at Ciron, 3° sending to the Engineer of the *arrondissement* the probes, the plans, and the levelings etc. which will permit him to study the project of constructing a suspension bridge crossing the river Langlin replacing the ferry at Ingrande, ...

[Neuenschwander 1984a, Annexe 3]

When Liouville entered the *École Polytechnique*, his health and constitution had been characterized as good, but three years later (1828) it was debilitated and not up to the rigors of the life of an engineer in the field. Earlier in the century, Cauchy and Frenet had, for similar reasons, obtained permission to spend their training periods working on constructions in the neighborhood of Paris, but for Liouville, there was no circumventing fieldwork in the provinces. The following May 12 the school decided to send him to Avranches in Normandy, but this time he also fell ill, and a week later the doctor of the school recommended that he take one month's leave in Toul to recover from a "serious illness." The leave was granted by the head of the school, and Liouville went to stay with his family until July 4, when he took up his work at Avranches.

We do not know the nature of his illnesses. It may have been the early stages of the rheumatism, which racked him later in life, or it may have been psychologically conditioned by an aversion to leaving his well-known surroundings for the uncertain trainee jobs.

The following year there were further problems with Liouville's trainee place, this time, however, it was not due to an illness but to his family life. His mother had recently died, and he had decided to marry his 18-year-old maternal cousin, Marie-Louise Balland. Until the spring of 1830, Liouville had diligently followed the beaten track of an engineer in the French state, but now he began to kick over the traces. Having received an order on May 13 to report to the chief engineer in Grenoble on June 1, Liouville wrote to the *Directeur général des Ponts et Chaussées et des Mines, Conseiller d'Etat L. Becquey*, to ask for at least two months of leave:

Monsieur le directeur général,

I have received the order to travel to Grenoble, but a double motive compels me to ask you for a leave before I make my way to this residence.

First, I must ask you to give me permission to marry Miss Louise Balland, a daughter of a lawyer in Toul and grand daughter of a judge in the same town. Second, family matters, and the settlement of the inheritance left by my mother require my presence in Lorraine. Thus, much as I should like to perform the duties which have been assigned to me, these particular and urgent matters force me to apply for at least two months' leave from June 1 to August 1.

The proofs you have already given me of your kindness, Mr. *directeur général*, do not permit me to doubt the success of my request. Therefore, I respectfully attend your orders in Paris. The town of Grenoble (which I was sent to) is so far away and the matters I have told you about so essential that they excuse me from following the ordinary channels and dare to write this letter directly to you.

[Neuenschwander 1984a, Annexe 4]

This was a most irregular step, even for a student in his last year. It must be kept in mind that the *Ponts et Chaussées* was a half-civil and half-military institution, and as such, it was subject to strict regulations. It was a matter of course that orders were obeyed until new orders were given. Thus, it was very imprudent of Liouville to remain in Paris and wait for new orders. Moreover, it was most irregular to apply for a leave directly to the *directeur* of the school, for the chief engineer and the *Préfet* in the *arrondissement* involved had to give their consent. Finally, young men in those days were not supposed to marry before they had finished their education and acquired a living on which to support a family. The 1800 francs Liouville got in Avranches was not enough for such a purpose.

When the chief engineer of Isère, L. Crozet, heard in a letter from Liouville of the steps taken by the student that the school had promised to send him, he became very annoyed and wrote to the *chef de division* in the *Ministère des Travaux Publics, Mr. de Cheppe*:

Grenoble, June 3rd 1830

Sir and friend,

I have received from Mr. Liouville... a letter in which he tells me that he applies for three months of leave, so that he will arrive at the end of the campaign [the trainee period] and at a time when it will be too late to begin the works for which we have for a long time needed a fourth Engineer. If his demands are complied with, this help is gone; he is absolutely indispensable, and we would be obliged to send 3/4 of the money back to the government and the *département* who have granted us it for extraordinary works. I have asked the *Préfet* to oppose with all his power a request which is so inopportune and which violates all the rules, because a leave cannot be granted except on the recommendation of the Chief Engineer and the *Préfet*. Moreover, it is exceedingly absurd that a man who receives an order to go without delay to his job and to an urgent duty answers that he will come in three months; we really live at an insane time, duty and reason are unknown properties.

Mr. Liouville says that he will marry in Lorraine. To marry while still a student and to live in the mountains! These are two beautiful ideas. Moreover, when this student has married in Lorraine, he asks to be transferred or to continue the leave, so instead of having gained something for my project, I have had but a loss by the disorder and uncertainty all this has caused.

L. Crozet

[Neuenschwander 1984, Annexe 5]

The chief engineer might have saved himself the excitement, because the director of the school had only given Liouville one month's leave. However, by his clumsy procedure, Liouville had also annoyed the *Préfet* who asked the director to impress on Liouville that he be "punctual at the end of the leave." The director did so in a letter of June 19, but one day earlier Liouville had written to the director to ask for a prolongation of his leave. His grounds for asking was a delay of his wedding, which had only just taken place on June 15.

Apparently, Liouville went on a short honeymoon with his young bride, and so he did not receive the letter from the director until they returned to Toul on the June 23. He immediately sent a new application for prolongation of his leave [Neuenschwander 1984a, p. 123], but it was not granted. In his letter of June 23, Liouville had reassured the director that "if my request is refused, your Excellency has no reason to doubt the eagerness with which I shall obey his orders." Nevertheless, on July 1, we find Liouville in Paris applying for a few more days of leave and only then did he set out for the mountains with his young wife.

Thus, it happened that when the July revolution broke out on July 27, 1830 Liouville was in Saint-Marcellin, west of Grenoble, in the *département* Isère, where Fourier had previously been a *Préfet*. This revolution, which replaced the last of the Bourbons, Charles X, with the Citizen King, Louis-Philippe, had little immediate effect on Liouville's life. He had his own revolutions to fight that year. Next to his marriage, his most crucial step was to leave the engineering corps. As the letter from the chief engineer of Isère shows, there was a strained relationship between Liouville and his superiors even before he arrived at the mission in Saint-Marcellin. This may have influenced Liouville's decision. The inconveniences for his young wife, alluded to in the same letter, may also have been instrumental. Still, when Liouville wrote his letter of resignation, his only argument was his scientific isolation:

St Marcellin, October 1830

Monsieur le directeur général,

Since the study of the natural sciences, which is a necessity for me, hardly makes any sense outside of Paris and since I have no hope of returning to the capital if I stay in the *ponts et chaussées*, I have the honor of asking you to accept my resignation. However, if it is possible, I do not want to preclude myself from resuming my position if the circumstances change, as I have seen others do.

Whatever my new position will be, I beg you to receive the assurance of my zeal to serve, as much as I can, if a new situation should emerge, a corps which I leave with regret and which I shall always glory in having belonged to.

The works in my *arrondissement* are very urgent.... Therefore, if it is impossible to find a person to succeed me immediately I am willing to delay my departure from St Marcellin to the first half of December.

J. Liouville [Neuenschwander 1984, Annexe 7]

Thus, Liouville left his career as an engineer before he had formally finished his education in order to pursue an uncertain career as a scientist.

The Independent Researcher

Liouville was well prepared for his new career for, alongside his regular studies at the *École des Ponts et Chaussées*, he had begun to read the classics. From this time, or even earlier, stems an impressive list in his notebook of

Works to read:

Lacroix, probability — last edition Laplace, *analyt. des probabilités* — last ed. — to buy
Legendre, *théorie des nombres* — last edit.
Journal polytechnique, in particular the n° 19 [containing Poisson's theory of heat] — which I must buy
Lagrange *mécanique analyt.*
Laplace *mécanique céleste*

[Ms 3615 (1)]

The list continues with the geometrical works of Dupin, Monge, and Poncelet "to buy," Fourier's *Théorie analytique de la chaleur*, and Legendre's *Exercices de calcul intégral* (but not his *Traité des fonctions elliptiques*, which indicates that the list was written before 1825). In addition, Liouville listed the best journals of the time: the *Mémoires* of the Paris, Turin, Berlin, and St. Petersburg academies, the *Correspondance de l'École Polytechnique*, and the *Annales de Mathématique*.

Of course, Liouville did not read all these works intensively during his student years, but at the *École des Ponts et Chaussées*, he knew enough to embark on his own original research. During the first year, he tried to improve on the mathematical structure of Ampère's electrodynamics, as he had learned it in the course at the *Collège de France*. That these studies were conducted under expert guidance by the master himself can be seen from the many Liouville manuscripts, commented on by Ampère, which are preserved in Ampère's *Nachlass* at the *Académie des Sciences*.

Liouville's first work, *Théorie mathématique des phénomènes électrodynamiques* [AC AL Ms 3], was written as though Ampère was the author, and in fact, he was, in the sense that the memoir is only a slight modification of the lecture notes, and often follows them verbatim. Still, the memoir contains three new arguments, which are explicitly attributed to Liouville. One of these arguments, designed to show that the force between two elements of conductors acts along the line connecting their midpoints, was later published by Liouville. In the manuscript, Liouville/Ampère introduced this argument as follows:

> Based on this hypothesis, we have applied the calculus to the phenomena of electrodynamics, and the complete agreement between our results and those found by experiment has raised the fundamental principle beyond doubt. Nevertheless, it would be desirable to find a direct proof. This has been done by Mr. J. Liouville whose demonstration we quote below.
>
> [AC AL Ms 3, no. 2]

It is not clear whether Liouville wrote this manuscript on his own initiative or whether Ampère asked him to undertake the task. Neither is it clear why Ampère did not publish it. He may not have considered it as a sufficient improvement over his newly published book [1826] or he may have felt that the young Liouville should have the chance to present his own ideas in a paper of his own. In any case, the latter happened. Liouville composed a new mémoire written in his own name in which he presented slightly improved versions of his earlier arguments and a few more additions to Ampère's theory [AC L Ms 2]. On June 30, 1828, he sent the mémoire from Chateauroux to the *Académie des Sciences* in Paris. Submitting papers to the *Académie* was the official way of obtaining recognition and the first step for any young Frenchman with ambitions for a scientific career.

The *Académie Royale des Sciences* was founded by *le roi soleil* Louis XIV and his minister Colbert in 1666 [Hahn 1971]. Next to the Royal Society of London, it is the oldest scientific academy still in existence. During the revolution, it was reorganized as the first class of the Institut de France, which also included the other academies. With the return of the kings, the first class again took its old name, *Académie Royale des Sciences*. From its foundation through the 19th century and even till this day, the *Académie* has remained the highest scientific institution in the country and the first advisory assembly of the government in scientific matters. Its power, legally regulated by decree, was reinforced by the fact that it consisted of the best scientists in the country, whose advice was usually followed by the government.

During the 19th century, all the most important scientific discussions in France took place on Monday afternoons, when the *Académie des Sciences* had its weekly meetings. One of the important tasks of the *Académie* was to examine papers submitted by nonmembers. Many scientists used this facility, not so much to get their work reviewed, but in order to make their name and talents known to this learned society which, directly or indirectly through its individual members, governed most of the scientific life of Paris, including appointments to the chairs of the institutions for higher education and other scientific posts.

When the *Académie* had received a mémoire, it usually appointed a committee to evaluate it. Liouville's first mémoire was no exception, and it was refereed by Arago, Ampère, and Maurice. On September 15, 1828, they presented their report to the *Académie*. It was written by Maurice and signed by both him and Ampère [Maurice 1828]. The two academicians praised Liouville's proof that the elementary electric force is a central force, but they were rather critical toward the rest of the mémoire, which they considered superfluous and unnecessarily complicated. Still, they acknowledged its mathematical analytical elegance and the "sagacity in the details" and concluded:

> The *Académie* should make known to Mr. Liouville that although every part of his work does not obtain its approbation, nonetheless it encourages him to pursue similar studies on various points of mathematical physics whose theory still leaves something to be desired.
>
> [Maurice 1828]

The third member of the examining committee, Arago, who was a strong supporter of Ampère's electrodynamics, apparently agreed with the conclusions of the report, for he printed the approved part of Liouville's paper in the *Annales de Chimie et Physique*, of which he was one of the editors. This *Démonstration d'un théorème d'électricité dynamique* [1829] was Liouville's first published paper. It was not a great paper, made up as it was of a six-page circular argument, but it convinced the establishment; shortly afterward, Sturm published an abstract of it in Ferrusac's Bulletin.

The following year, on April 20, 1829, Liouville presented a second paper on electrodynamics to the *Académie*. Its title was *Mémoire sur la Théorie physique des phénomènes électrodynamiques*. I have not been able to find this paper, and the examining committee consisting of Arago, Ampère (replaced by Becquerel on August 3), and Dulong never wrote a report. Thus, the only thing known about the paper is that it treated, according to the *Procès-Verbaux* [April 20, 1829], "the hypothesis of the two electric fluids."

Two more notes in Liouville's hand are preserved among Ampère's papers [AC AL Ms 4 and 5]. The first and most important was later published in a revised form [1831]. In this, Liouville's second and last publication on electrodynamics, he clearly pointed out that it was, in principle, impossible to deduce the elementary force law between infinitesimal conductors from experiments of the kind used by Ampère. Ampère's force law was, therefore, only one of many possible laws. Although he had thus spotted a weakness in Ampère's theory, he was careful to express his full support of his patron.

Indeed, Liouville remained faithful to Ampère's main doctrine that all electromagnetic phenomena must be explained as a result of action at a distance between infinitesimally small conducting elements, and he severely criticized Biot's competing theory. Although in one case Liouville described

PLATE 3. Siméon-Denis Poisson (1781–1840). His works on heat conduction inspired the young Liouville (Delpech–Math Dept., University of Copenhagen.)

a new experiment, he was mainly interested in the mathematical perfection of the theory, just like Ampère's earlier student Savary, who had made essential contributions. However, when Liouville became interested in Ampère's electrodynamics, its most obvious mathematical problems had been solved, and Ampère himself was gradually losing his enthusiasm. This is probably why Liouville soon lost interest in the subject so that his name, unlike that of Savary, is not remembered in this connection.

During the beginning of his second class at the *École des Ponts et Chaussées*, in the interval between writing the two long memoirs on electrodynamics, Liouville also found time to work on pure mathematics. On December 1, 1828, he submitted three papers entitled *Premier (Seconde, Troisème) Mémoire sur le calcul aux différences partielles* [AC L Ms 3a,b,c] to the *Académie des Sciences*, in whose archives they are still preserved. In all three papers, Liouville discussed how the infinite series, obtained for the solution of a linear partial differential equation by separation of the variables, can be transformed into a multiple integral. Basic in these transformations was a theorem on infinite series established by Parseval [1806].

The mémoires were examined for the *Académie* by Poisson and Fourier, but a report never appeared. That was probably the best for Liouville, for his papers were rather unreadable and contained little of interest. The problem seems rather artificial and vague, its solution was complicated and the papers were loaded with many (overly) technical computations. Moreover, the papers were unnecessarily long because Liouville repeated much of the theory in all three of them.

These works bear witness to Liouville's developed but untamed analytical talent. In other juvenile papers, physics had posed the questions, but in these purely analytical investigations, Liouville clearly had difficulties in formulating relevant and concise problems, so his calculations tended to be mere formal manipulations. Still, some of the more or less implicit ideas point toward his later important works.

The third subject that Liouville took up as a student was the theory of heat. It had been a major breakthrough in the development of mathematical physics when in 1807 Fourier laid the foundations of the theory of heat conduction. At first, Fourier had difficulty in convincing the leading mathematicians about its importance, but when he became *secrétaire perpétuel* of the *Académie des Sciences*, he used all his power to promote the subject. He gathered around him a number of young talents and encouraged them to follow his track. By then Poisson had already extended Fourier's methods, and due to those two, the theory of heat became the standard field to which almost all young French mathematicians applied their skills.

When Liouville became interested in the subject, Fourier was very old and ill, and apparently Liouville did not belong to his circle of students. Instead, he learned the theory by reading Fourier's already classical book, *Théorie analytique de la chaleur* [1822] and Poisson's extensive *Mémoires sur la distribution de la chaleur dans les corps solides* [1823a]. His papers pursued ideas and problems indicated in these treatises. He presented three mémoires on the theory of heat to the *Académie des Sciences* during 1829 and 1830, of which only a part of the second one was published [1830]. The first, *Théorie analytique de la chaleur*, was presented on June 29, 1829. The examining committee consisting of Fourier and Cauchy never wrote a report, probably in part because Liouville soon decided to improve and extend its content and submitted it (February 15, 1830) under the new title *Recherches sur la théorie physico-mathématique de la chaleur* [AC L Ms 4] to the commission in charge of awarding the *Académie*'s annual *Grand prix de mathématique*.

Among the other competitors was Evariste Galois, who was two years younger than Liouville. Unfortunately for Liouville and Galois, the committee decided not to award the prize to any of the official competitors, but to honor the recently deceased Niels Henrik Abel (1802–1829) and Carl Gustav Jacob Jacobi (1804–1851) by awarding them the prize for their great work on elliptic functions. While Abel was still alive, the academicians had distinguished themselves by ignoring his work to the extent of mislaying one of his greatest papers [cf. Ore 1957]. This posthumous gesture should probably be seen as an act of atonement for their previous neglect.

Ironically, the *Académie* thereby committed another irreparable injustice, at least as fatal as the one they wanted to remedy, for they failed to acknowledge that Galois' work on the algebraic solution of equations was a masterpiece, in fact one of the most ingenious mathematical works of all time. Two years later Galois was dead, and it took another 14 years before a mathematician discovered the importance of Galois' works and had them published. That mathematician was Liouville (cf. Chapter XIV).

The decision of the *Académie* did not have such a fatal effect on Liouville, for he soon found other ways to promote his paper on heat conduction. He submitted it for publication in Gergonne's *Annales de mathématiques pures et appliquées*, and he presented it once more (August 16, 1830) to the *Académie des Sciences* as an ordinary memoir, hoping that the examining committee, consisting of Ampère, Cauchy, and Navier, would this time write a favorable review. Gergonne accepted to publish more than half of the memoir in the *Annales* [Liouville 1830], and perhaps as a result of that the committee did not feel obliged to write a report. Later, when Liouville presented an extended version of the part rejected by Gergonne to the *Académie* on November 2, 1830, under the title *Mémoire sur les questions primordiales de la théorie de la chaleur* [AC L Ms 5], Ampère finally wrote a favorable review [April 4, 1831].

Of these three mémoires presented to the *Académie*, only the last two are extant in the files of the respective meetings at the *Académie des Sciences*, but the contents of the first can, to a large extent, be determined from remarks in the latter papers. The mémoires deal with two distinct problems. The first is the solution of the heat equation for homogeneous or nonhomogeneous bodies that are unequally polished. This most mathematical of the questions was discussed in the first paper, considerably perfected in the second, and finally published with a few exceptions by Gergonne. The other, more physical part dealt with a molecular model of heat conduction. Liouville discussed it from two different points of view in the first two memoirs, and finally presented slightly extended versions of both methods in his last paper.

In the physical parts of the papers, Liouville placed himself in the research tradition that has been called Laplacian physics [Fox 1974]. In the spirit of Newtonian mechanics, Laplace suggested that all physical phenomena could be explained as the result of molecules acting upon each other at a distance. His application of this model to the conduction of heat was carried further by Poisson in sharp contrast to Fourier, who had claimed that he needed no molecular model in order to derive the heat equation. In this discussion, Liouville sided with Poisson, and in his memoirs he tried to find an explicit formula for the heat exchange between two molecules at a given distance and with given temperatures. This was the part of Liouville's work on which Ampère presented a favorable review to the *Académie* on April 4, 1831. It was also signed by Navier, but the third member of the committee, Cauchy, was unable to sign, since he had exiled himself after the 1830 revolution. The report is preserved in the file of the meeting of the *Académie*, but it was not included in the published version of the *Procès Verbaux*, because its conclusions met with opposition from an unexpected direction, namely, from Poisson. In a note which is also preserved in the files, Poisson severely criticized Liouville's derivation of the law, and therefore it was never officially appreciated by the *Académie*. Nevertheless, Poisson used almost the same arguments to obtain essentially the same law in his later book [1835a], and then he did not bother to refer to Liouville.

The more mathematical ideas that Gergonne published in his journal also received a mixed reception in spite of their indisputable ingenuity. In these papers, Liouville extended Fourier's and Poisson's investigations of heat conduction to unequally polished metal bars. That means that the coefficients of the heat equation are variable. Liouville found the "Fourier" series of the solution and showed that it is convergent. This result was remarkable, but its derivation was neither elegant nor rigorous. It heralded his later, more successful work on the Sturm-Liouville theory. The most interesting feature of Liouville's paper was his use of the method of successive approximations to obtain a stationary solution. It was not only the first application of this method to differential equations, but also the first published existence proof of a solution. To be sure, Liouville did not formulate the problem of existence as explicitly as Cauchy had done in his then unpublished lecture notes [Cauchy 1824–1981] and later in other publications. Probably, therefore, Liouville's use of successive approximations went unnoticed, so the method is often attributed to C. E. Picard [1890].

It must have been a great triumph for Liouville that Gergonne accepted this part of his paper for publication in the *Annales*. However, this triumph was spoiled by a long footnote at the end of the paper in which Gergonne severely criticized Liouville's style:

> I think I must apologize to the reader for having published a memoir which is so drearily, I may even say so incomprehensibly written. However, at the moment when I had planned to prepare this issue in peace and quiet, I regrettably had to take on administration which is always difficult, in particular at the start, and which has become much more difficult in the present situation. As a consequence of a move to a new address which I could not supervise, my papers were buried under piles of books that I did not have the time to remove. When this memoir cropped up I believed that the double qualification as engineer and former student at the *École Polytechnique* guaranteed the author's talent as a writer, and I immediately sent the work to the printer. Afterward the author sent me some corrections, but they would not have noticeably improved the memoir and in addition, they did not arrive until the memoir had been printed.
>
> I do not want to dispute the mathematical capacity of Mr. Liouville. But, what purpose does such a capacity serve if it is not

accompanied by the art of arranging the material and by the art of writing something that can be read, understood and enjoyed. Unfortunately, there are to-day too many young persons, otherwise of a great merit, who consider as an almost immaterial accessory this art which I regard as the essential merit, the merit par excellence for the lack of which all the rest is worthless.

I strongly hope that Mr. Liouville will soon have his revenge for this rather hard reproach, which I have regrettably felt obliged to make to-day, by publishing some memoirs which can be read almost as one reads a novel. However, to be honest I hope this more than I believe it. Long experience has shown me that the illness he suffers from is almost incurable.

[Gergonne in Liouville 1830]

It was kind of Gergonne to give advice to the young scholar, but it seems very unkind to do it publicly. One usually expects that when an editor accepts a paper, even when this is done under difficult circumstances, he will be loyal to the contributor. Gergonne's criticism seems unnecessarily harsh, for in fact Liouville's paper was not too badly written. To be sure, it was not a masterpiece of "literary style" (a subject in which Liouville always got low marks at the *École Polytechnique*, but distinction at the *École des Ponts et Chaussées*), and it was also too long, but the reason was in part that Liouville presented his main ideas in a special case before he solved the general problem. This pedagogical method made the paper easier to understand. Liouville also very explicitly formulated his problems, and as a whole the paper was considerably more readable than his previous memoirs.

It is my impression that Liouville's style did improve in his later production. Mathematically, his papers were always clear and didactically well written. As for the French style, I certainly read it more easily than a novel, but as a foreigner, I am a bad judge on this subject.

So, when Liouville left the engineering corps, he had already done promising work on partial differential equations, improved the mathematical theory of electricity, and made substantial innovations in the analytic theory of heat conduction. With these papers, he won recognition as a talented analyst, whereas the physical considerations received a mixed reception. The critics of his works in physics clearly had an element of personal likes, and dislikes, but with hindsight, one must agree that Liouville did show more mathematical than physical ingenuity. At any rate, the reviews of his work may have been instrumental in directing him away from Laplacian physics and toward purer mathematics. He never returned to hypothetical theorizing on the molecular structure of matter and, although he did not express any strong philosophical opinions, later in life he seems to have adopted Fourier's more positivist attitude.

Many of Liouville's most important mathematical ideas continued to be inspired by physics, and his first paper presented to the *Académie des*

Sciences after his resignation from the engineering corps was no exception. In his work on heat, he hit upon a special case of the Riccati differential equation, and on November 2, 1830, he presented a paper to the *Académie* in which he extended his earlier solution to the general case. As in his earlier papers on differential equations, his aim was to transform a series expansion of the solution into a multiple integral, but this time his goal was explicitly expressed and clearly reached. Parts of this paper were published in 1833 [1833a].

With Liouville's earlier productivity in mind, one would expect a stream of papers after he returned to Paris with the expressed purpose of becoming a full-time mathematician. However, during the first year, he only presented the one paper on the Riccati equation to the *Académie*, and only published one brief note on electrodynamics [1831]. Still, together with his previous papers, this proved to be sufficient to secure him his first academic job at the end of 1831.

Ampère resigned his post as a professor at the *École Polytechnique* after Liouville's class graduated, and his chair went to Claude-Louis Mathieu (not to be confused with the famous Emile Léonard Mathieu (1835–1890)). For the first few years, Mathieu automatically kept Paul Binet as his *répétiteur*, but at the meeting on November 3, 1831, in the *Conseil d'Instruction* of the school, he presented three candidates for the job. In addition to Binet, he suggested Chasles and Liouville.

Paul Binet's younger brother, the much more famous mathematician Jacques Binet, had been *Inspecteur des Études* at the *École Polytechnique* during the restoration, but due to his conservative and ultracatholic views, he had been dismissed after the 1830 revolution. This may have discredited his brother. However, the official reason for reconsidering Paul Binet's position was expressed during the following meeting in the *Conseil d'Instruction*, where the qualifications of the candidates were discussed. A member reminded the *Conseil* about:

> the negligence which the present *Répétiteur* [Paul Binet] has always exercised in his job and the trouble it has caused him without changing his behavior.
>
> [Ec. Pol. Conseil d'Instruction 11. novembre 1831]

Liouville, on the other hand, was strongly supported:

> Several members rose in order to recommend the election of Mr. Liouville, former student in the first rank at the *École Polytechnique* and at the *École des Ponts et Chaussées*, who has given up his career as an Engineer in order to devote himself exclusively to the cultivation of the sciences in Paris and whose capacity is testified by the highly remarkable memoirs he has presented to the *Académie*.
>
> [Ec. Pol. Conseil d'Instruction 11. novembre 1831]

After these statements, it was hardly surprising that the *Conseil* decided to let Liouville replace his own former *répétiteur*, Binet, on November 14, 1831. Liouville had taken the first step in a glorious academic career.

II

Climbing the Academic Ladder (1830–1840)

The Public School System

The academic establishment into which Liouville had decided to fight his way was a highly competitive world full of rivalries. The young Norwegian mathematician Abel described it quite explicitly when he visited Paris in 1826:

> On the whole, I do not like the French as well as the Germans; the French are extremely reserved towards strangers. It is very difficult to become more closely associated with them, and I dare not hope for it. Everybody works for himself without concern for others. All want to instruct, and nobody wants to learn. The most absolute egotism reigns everywhere.
>
> [Ore 1957, p. 147]

Even to a native Frenchman who was used to the system, it was a battle to survive. The first problem that faced the young Liouville after he had decided to quit the *École des Ponts et Chaussées* was to earn a living for himself and his wife.

As a student, he had received various grants the first year at the *École Polytechnique* only his sergeant uniform and equipment, but the second year a full grant from the funds of War. During the trainee periods at the *École des Ponts et Chaussées*, he made a little money, and the rest of his expenses were probably paid by his father. The father continued to support the young couple when they moved to Paris in the fall of 1830, but Liouville soon took to the traditional living of a mathematician: teaching.

In order to understand his institutional affiliation, we shall have to consider the French public educational system, which had since 1806–1808 been organized into the so-called *Université*— not to be confused with an ordinary university (they had been suppressed in France during the revolution) [Prost 1968]. It comprised the *lycées* and the *collèges*, but not the specialized schools, such as the *École Polytechnique* or the *Collège de France*. The *Université* was led by the *Conseil de savants* (Poinsot and Thénard representing mathematics and the sciences, respectively), which during the reign of Louis-Philippe had greater power than the Minister of Education. The university ran all the public schools and strictly regulated the private institutions. It was geographically divided into so-called academies (not to be confused with the *Académies* of the *Institut de France*). There was one for each appeal court, headed by a *Recteur*.

The Université had the right to confer the degrees of *baccalauréat, licence,* and *docteur*. The examinations were carried out by the professors belonging to the *Faculté des sciences* or the *Faculté des lettres* of the academies. In the provincial academies, the professors of the facultées were recruited from the local *lycée*, but in Paris they generally came from the *Collège de France*, the *École Polytechnique*, or one of the other specialized schools. Some *Faculté* professors also taught courses at the *Sorbonne*, but during the first decades of the 19th century this teaching was not among their formal duties.

In order to become a titular professor at the *lycées* of the *Université*, one had to pass an *agrégation*. In 1834, Liouville was "*agrégé* for the classes of science" [Neuenschwander 1984a, p. 119] (until 1840 this included mathematics as well), and the same year he began to teach a two-hour class one evening a week at the *Collège Louis le Grand*, in addition to classes at five private institutions (some of them may have been private students) [Ms 3615 (4), p. 93r]. However, after just one year of employment at the *Collège Louis le Grand*, he was fired for reasons we can implicitly infer from the testimonial written by the *vice-recteur*:

> Professor of a superior merit, I am very pleased with his zeal. Still,
> I fear that he is not suited for the teaching at the *Collège*.
> > [Neuenschwander 1984a, p. 58]

Apparently, Liouville could not adjust to the level of the college students and taught over the heads of his audience. He seems to have had more success at the private Institution Mayer, where in 1833 or earlier he began to teach simple arithmetic and analytic as well as synthetic geometry [Ms 3615 (3), p. 85r]. He kept this job for several years, and as late as 1852, when he had otherwise no connection with elementary education, he continued to give 15–20 nonoptional lectures on differential and integral calculus to the students of this institution, which was now called after its new owner Mr. Debain. However, 10 years later "it was a long time ago that I was employed by him [Mr. Debain]" [Ms 3615 (4), p. 92v, 93r; Ms 3620 (1); Neuenschwander 1984a, I (5); Ms 3629 (7); letter of March 12, 1863, to Barthelemy Saint Hilaire].

The École Centrale; Colladon and Sturm

By 1833 Liouville had been appointed professor at the *École Centrale des Arts et Manufacture*. This school had been founded in 1829 with the purpose of educating scientifically oriented engineers for private industry, in the same way that the *École Polytechnique* educated military engineers and civil servants (cf. [Pothier 1887], from which most of the following information stems). The school was located in the Marais quarter, indicating

PLATE 4a. Charles Francois Sturm (1803–1855). One of Louiville's best friends. Together they created the spectral theory of second order differential equations which is now named after them. (Pencil drawing by Daniel Colladon. Bibliothèque publique et Universitarie de Genève, Ms pr 3748, po 167)

thereby an intentional distance from the academia of the Latin quarter. This distance was also clearly displayed in the programmes of the three-year curriculum. Thus, the course on *Analyse géométrique et Mécanique rationnelle*, which was taught (from 1832) 60 hours in the first year and 10 hours in the second year, mainly provided the students with an understanding of simple machines. The first two teachers in this subject, Coriolis (1830–1831) and Colladon (1831–1833), taught in this spirit and with great success; Colladon even showed the students real commercially manufactured machines borrowed from Parisian factories.

It was through Colladon that Liouville made his way to the school. Born in Geneva, Switzerland, Daniel Colladon had moved in 1823 to France together with his townsman Charles-François Sturm, with whom he wrote a prizewinning essay on the compression of fluids [1827–1834]. In Paris, they were well received in the homes of Ampère, Arago, Fourier, and Dumas, and according to a very interesting series of recollections written in 1886 and 1887 by Colladon and printed in [Pothier, 1887, pp. 472–480], they often met Gay-Lussac, Dulong, and Mathieu there. Moreover, they entered a smaller group, including, among others, J. B. Say, Comte, Clément, Desormes, and "finally a younger society of physicists and mathematicians of which Sturm and I were members and which was composed of Joseph Liouville, Elie de Beaumont, Fresnel, Savary, etc." [Colladon in Pothier, 1887, p. 472]. In another place, Colladon, speaking about Coriolis, recalled:

> From 1827 we had a regular small gathering with him, Liouville, Elie de Beaumont, Fresnel, etc. to discuss questions of mathematical physics.
>
> [Colladon in Pothier 1887, p. 475]

Liouville later recalled [1855j] that he developed a close friendship with Colladon and Sturm during the early 1830s, but the late 1820s seems a more probable date. At any rate, Colladon asked Liouville to act as his substitute at the *École Centrale* during the last few lectures of 1832–1833. Liouville accepted and was paid so much by Colladon that he replied:

> What is the meaning of the 60 francs that you have sent Mr. Collet for me? You know the theory of proportions very badly when you pay me half a month's salary for one or two lectures.

This is a P.S. in a letter that Liouville sent to Colladon and that is reprinted in [Pothier, 1887, pp. 119–121]. I shall quote the entire letter because, in addition to explaining the circumstances of Liouville's nomination to the *École Centrale*, it gives evidence of the friendship between Liouville, Colladon, and Sturm and throws an interesting light on Liouville's view of applications. The letter was a result of Colladon's decision to return to his native Geneva to be married and become a professor at the university of that city. He suggested that Liouville should take over the chair of rational mechanics at the *École Centrale* which he was forced to leave while agreeing to return each May and June to teach the course on steam engines (which he actually did until 1836). Here follows Liouville's reaction, dated Toul, September 21, 1833:

> I hasten to answer the letter you have written to me on the 17th of this month. It is with a great pleasure that I receive the news of your nomination in Geneva. I agree with you that there is nothing like one's native air, and I shall get over the loss France will suffer because you do not leave us completely, and we can still look forward to seeing you some months each year. Would to heaven that the delights of your country may not corrupt you and that they may not make you forget the good habit of visiting us from time to time. The poor and excellent Sturm particularly needs to gather strength with you, and I fear that your departure will dishearten him.
>   Although the *École Centrale* is far away from my home and the way will be rough, in particular during winter, from the *rue d'Enfer* [now Avenue Denfert Rocheveau, rue Henri Barbusse, and part of Boul. Saint Michel] to the *rue Saint-Louis* [later rue des Métheirs and suppressed by the opening of the Boul. Sébastopol in 1854–1858], I accept with pleasure the place that you leave vacant, if it is offered to me. In addition, I give you a free hand to make whatever

arrangements in my name that you find convenient, if the case arises. However, I suppose that you will continue the course on steam engines, as you seem to say; it would be impossible for me to replace you since I completely lack the practical knowledge that is indispensable in such a matter. My taste and my work make industrial applications repugnant to me, just as much as you are attracted to them. Moreover, I even tremble to undertake the lectures of mechanics although they hardly deal with anything but theory, and almost everything deals with generalities that you can learn in depth without leaving your study. In fact, the less capable I am, the more I am indebted to you for the feeling of friendship that has made you think of me.

Supposing that nothing gets in the way, I shall stay in Toul until November 1. However, it is possible that my presence may be needed in Paris before that time. Therefore, if I am accepted by the *Conseil des professeurs*, please be so kind as to inform me and tell me the time when the courses begin.

Also, send me some news about Sturm. This year he has once more recoiled from the *agrégation* examination. I fear that he will one day be taken ill because he has lost his time in this way, leaving aside his memoirs because he says he wants to be agreggated and neglecting the *agrégation* under the pretext of working on his memoirs. Before leaving Paris, you must scold him rather severely. No one has such a great influence on his state of mind as you have, and if you do not succeed in curing him nobody will.

Since I shall probably not see you until next year, I shall here say goodbye to you and wish you all possible prosperity in Geneva.

All the best,
J. Liouville
[Pothier, 1887, p. 119–121]

It all went as Colladon had planned. In the middle of November 1833, the *Conseil d'Études* appointed Liouville professor of rational mechanics. It was surely a testimony of Colladon's friendship with Liouville that he suggested him as his successor, for he must have guessed the problems that would arise by letting Liouville teach at this very practically oriented school. To be sure, all through his life Liouville was a propagator of applied mathematics, but for him that meant the area of mutual inspiration between mathematics and theoretical physics and astronomy. What was called for at the *École Centrale*, on the other hand, was application of theoretical mechanics to actual machine constructions, which he openly declared that he detested; and indeed, although he taught on a lower level than at the *École Polytechnique*, Liouville's course was still very theoretical. This is the impression one gets from the notes of his lectures from 1837–1838, which are preserved in the *Bibliothèque Nationale* [V. 14275]. These notes, representing probably only the first half of the course (see below), are divided into three sections: (1) *Résumé des leçons de trigonométrie* (only elementary plane trigonometry), (2) *Résumé des leçons de géométrie analytique* (conic sections, the concept of function and introduction to three dimensions), and (3) *Notes pour le cours de statique* (conditions for equilibrium of forces, including parallel forces, centers of gravity, Guldin's theorem, equilibrium of some machines). The exposition is very traditional, large portions of the statics section being quoted directly from Poisson's *Traité de mécanique* [1811–1833].

It soon turned out that the students did not take to Liouville's teaching. Unlike the other subjects they got generally bad marks in mechanics, and at the beginning of 1836–1837, the students complained to the *Conseil des études* that Liouville taught them almost nothing but statics. The *Conseil* therefore decided to set up a committee to suggest modifications in the mechanics course. At the end of the year, and at the repeated request of the *Conseil*, the commission presented a report—a devastating criticism of the teaching, which was accused of lacking logical coherence and of failing to satisfy the programme. The chief author of this report was no doubt the new *Directeur des études*, Belanger, an industrially interested engineer educated at the *École Polytechnique*. When it was decided in the summer of 1837 to intensify the examinations of the classes, Belanger organized a special session where Liouville had to examine a selection of the students in the presence of Belanger and two other members of the *Conseil*. This humiliation must have made it clear to Liouville that he was no longer wanted at the school, but he remained. During the next semester, 1837–1838, however, the *Conseil* continued to watch the mechanics course, and Belanger undertook on his own a comparison between a copy of Liouville's lecture notes edited by one of his students from 1836–1837 and a copy of Coriolis' notes from 1830–1831. When he presented the *Conseil* with his reflections, some members suggested that the mechanics course begin with a separate treatment of analytic geometry and even differential and integral calculus. However, Liouville answered that he preferred to include analytic geometry in the course in order to better catch the attention of the students; moreover, he had found ways of avoiding the infinitesimal calculus.

This remark seems to indicate that Liouville may not have been as theoretically biased as Pothier's account suggests, and that the problems may have been exaggerated by Belanger. At any rate, when the *Conseil* again impressed on Liouville that he should examine the students regularly and that he should transfer interesting, but not strictly necessary, subjects to the end of the course, he decided he had enough. On April 7, 1838 the *Conseil des études* received his resignation, and at the same meeting Belanger resigned his job as *Directeur des études* in order to replace Liouville as a professor of mechanics. Belanger taught at a lower level of abstraction and was therefore able to get the rational mechanics across to his students. To this story of failure, one should add that in addition to the mechanics course Liouville also taught an optional course on differential calculus on Sundays at 8 o'clock in the morning! This seems to have been a success, for

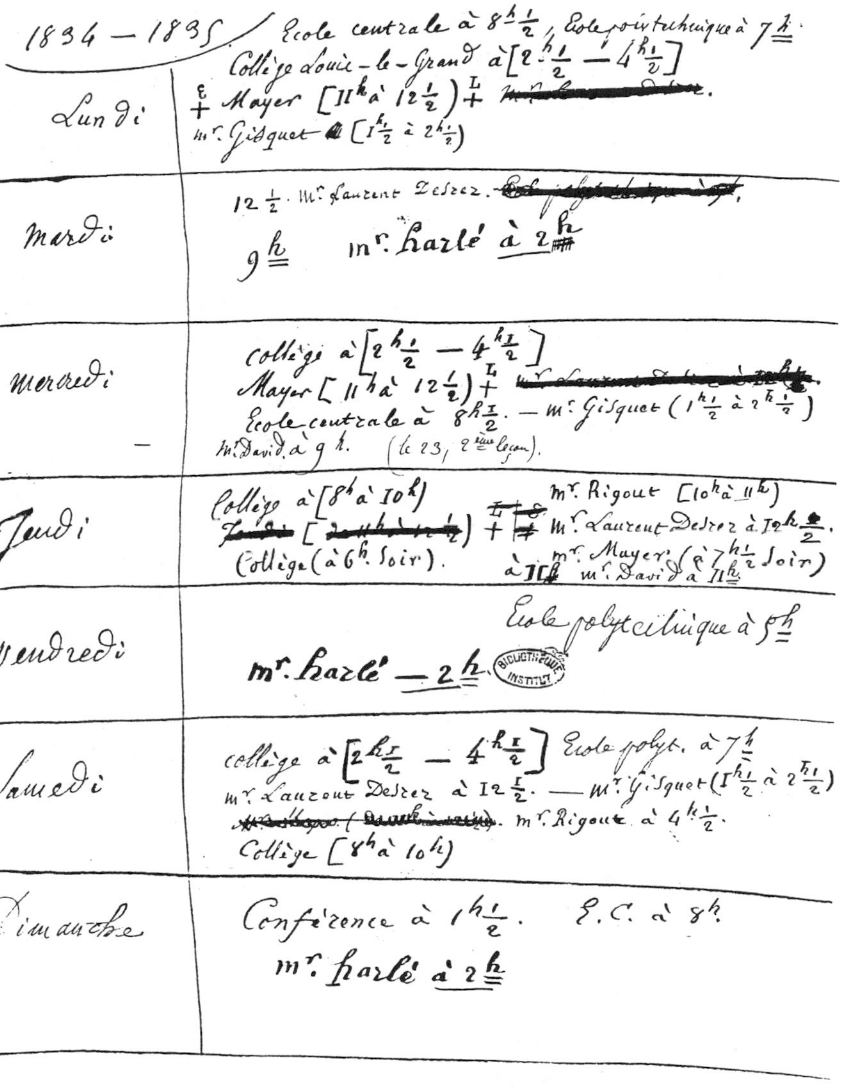

1834 — 1835	École centrale à 8h½, École polytechnique à 7h.
Lundi	Collège Louis-le-Grand à [2h½ — 4h½] + Mayer [11h à 12½) + Mr. Gisquet [1h½ à 2h½)
Mardi	12½. Mr. Laurent Desrez. 9h Mr. harlé à 2h
Mercredi	Collège à [2h½ — 4h½] Mayer [11h à 12½) + École centrale à 8h½. — Mr. Gisquet (1h½ à 2h½) Mr. David à 9h. (le 23, 2e leçon).
Jeudi	Collège à [8h à 10h) Mr. Rigout [10h à 11h) Mr. Laurent Desrez à 12h½. Collège (à 6h. soir). Mr. Mayer (à 7h½ soir) Mr. David à 11h.
Vendredi	École polytechnique à 5h. Mr. harlé — 2h
Samedi	Collège à [2h½ — 4h½] École polyt. à 7h. Mr. Laurent Desrez à 12½. — Mr. Gisquet (1h½ à 2h½) Mr. Rigout à 4h½. Collège [8h à 10h)
Dimanche	Conférence à 1h½. E.C. à 8h. Mr. harlé à 2h

PLATE 4b. Liouville's schedule from 1834–1835. It shows that the young Liouville taught at least 34 hours per week. (Bibliotèque de l'Instut de France, Ms 3615 (4)p 93r)

in December 1834, the third-year students thanked the *Conseil des études* for arranging it while recommending that it be moved to the first year of study [Ms 3615 (5), p. 93r, Weiss 1982].

It is not so surprising for a modern reader to see that Liouville had to take on many different jobs in private institutions, at a Collège, at the *École Polytechnique*, and the *École Centrale* in the beginning of his career, but it is hard to believe that he worked at all of them at the same time. From his schedules of 1833 and 1834–1835, which are still preserved (Plate 4b), one can see that he taught at least 34 hours a week—40 hours being probably closer to the truth.

Was Liouville greedy? No, he just joined the widespread game of accumulating jobs, the so-called *cumul*. This practice was responsible for much of the bad climate in the academic world, for it resulted in a never-ending battle of bitter competition for jobs. For some scientists, *cumul* had become a lifestyle or a game, but for others it was a way to make up for low academic salaries. At the highest educational institutions, a professor would make around 6,000 francs. Two or three such jobs were the norm for a senior scientist. Assistants and teachers on lower levels made much less: for example, Liouville made 157 francs a month as a *répétiteur* at the *École Polytechnique*, which of course did not make ends meet. As late as November 1833, his father, therefore, still sent him 400 francs, and his brother 60 francs. The brother Félix had settled down as a lawyer in Paris after having finished his more lucrative studies in 1829. He specialized in civil law and after a few years acquired a great reputation as a legal advisor and orator.

After Joseph Liouville was appointed to the *École Centrale*, he did not need the support of his family. He and his wife could live comfortably from the extra monthly 100 francs from this school, the 100 francs from Mr. Mayer, and the approximately 100 francs from each of the other private institutions or students [Ms 3615 (4), p. 92r, receipts of November 1833 to August 1834]. On top of this full teaching schedule, Liouville found time to pursue his research.

Scientific Societies; the Société Philomatique

Although teachers in the private schools, the university institutions, and the professional schools obtained their jobs on the basis of their scientific work, these jobs did not involve any research. The social setting for research consisted of more or less formalized learned clubs or societies. Colladon's recollections show that around 1830 there were several groups of scientists who met regularly and that Liouville belonged to at least one of them comprising Sturm, Colladon, Coriolis, Fresnel, and Elie de Beaumont (later professor of geology at the Collège de France).

In addition, on August 4, 1832 (later lists of members have August 25, 1832), Liouville was elected a member of the more formal *Société Philo-*

matique, which was also nicknamed the antechamber of the *Académie* [cf. Berthelot 1888, Mandelbaum 1980]. It had been formed during the years 1787–1788 around a group of young scientists and had acquired a certain prestige in 1793 when a large number of *Académiciens* joined it after the suppression of the *Académie des Sciences*. They remained *associés libres* when in 1796 the *Académie* was reorganized as the first class of the *Institut de France*. Thus, when Liouville became a member in 1832, the list of *associés libres* included celebrities such as Lacroix, Poisson, Ampère, Arago, Puissant, Binet, de Prony, Biot, Gay Lussac, Hachette, Girard, and Dulong. However, at no time did the older generation take an active part in the meetings of the society. It remained primarily a meeting place for young scientists, and a good one, because from 1791 the *Société Philomatique* published a *bulletin* containing papers by its members. When Liouville entered, the *Société Philomatique* counted among its ordinary members: Francoeur, Savary, Coriolis, Duhamel, Sturm, and Savart, and on the same day as Liouville, Lamé was also elected.

Despite this impressive list of members, the *Société Philomatique* had lost most of its importance in the 1830s. The publication of its *Bulletin* had been interrupted in 1826, and except for two volumes covering 1832 and 1833 it was not resumed until 1836 when it came out as a rather uninteresting collection of *Extraits des procès verbaux des séances*, which had only a limited circulation. The *archives* of the *Société* at the *Sorbonne* contain virtually nothing pertaining to Liouville's early years as a member, and the published *procès verbaux* suggests that Liouville was never active in this society. It even seems doubtful that he came to the meetings regularly although he continued as a member. In 1849 [Soc. Phil. P.V.] Liouville was listed among the *membres honoraires* (the former *associés libres*), just like all other members elected before 1837 (except two who had resigned).

As a forum for scientific discussions and collaboration, these private societies had great importance, but the administrative power was concentrated at the *Académie*; therefore, Liouville continued to submit his works for final judgement to this learned society.

The Creation of New Fields

During the early 1830s, Liouville had a very productive period when he created two new theories and contributed to several other fields. In the previous chapter, I pointed out the surprising fact at Liouville only published two papers during his first year as a full-time mathematician. The reason was that he was working hard on a great new project: fractional calculus. He defined and studied differential operators of arbitrary complex order and showed how to apply them to a wide range of mathematical and physical problems. It was probably one of these physical problems that inspired him to create his new theory, namely, the central problem of Laplacian physics, how to determine the microscopic elementary interactions. Following many Laplacian physicists, in his theory of heat Liouville had inferred the microscopic interactions by analogy with macroscopic phenomena. However, in his work on electricity, he followed Ampère's more positivistic rule, that the microscopic interactions must be deduced solely from their integrated consequences, which can be investigated by experiments. Liouville saw more clearly than any of his contemporaries that the functions representing the microscopic interactions must therefore be found as solutions of integral equations. He also discovered that these integral equations could be transformed into differential equations of fractional order and thereby realized the importance of differential operators of arbitrary order.

His notebooks reveal that by early 1832 he had composed his first two memoirs on this subject, and at the end of the year, he had made all his major discoveries in the field, including methods for solving fractional differential equations. Half a year later, he almost stopped his research in this area. The publication of the results dragged out until 1837, but that does not change the fact that already at the age of 23, and in less than two years, Liouville had created the first comprehensive theory of fractional calculus. This impressive achievement was appreciated later in the 1830s by a number of British mathematicians and has gained current interest by recent developments in the field. However, the young, ambitious French mathematician wanted immediate approval by the Parisian establishment, and he did not succeed with the fractional calculus. He sent most of these memoirs to the *Académie des Sciences*, but the appointed committees never composed any reports. This may be a result of their uneasiness with Liouville's highly untraditional, and not always rigorous, methods.

He had more success with another great creation, which he took up in the fall of 1832 after having worked out his fractional calculus. This was the theory of integration in finite terms, the main goals of which were to decide whether given algebraic functions have integrals that can be expressed in finite (or elementary) terms, that is by means of algebraic functions, logarithms, and exponentials, and to find the finite expressions when they exist. Like the fractional calculus, this theory concerned the very heart of analysis, but unlike the former and most of Liouville's earlier research, it had a purely mathematical motivation. He had finally made up his mind in favor of mathematics rather than physics.

His first papers on this subject from the winter of 1832–1833 [1833b,c], which may have been inspired by an allusion in Laplace's *Théorie analytique des probabilités* [1812], completely solved the question of integration of algebraic functions in algebraic terms. They were reviewed before the *Académie* by Poisson, and in contrast to his earlier destructive attitude to Liouville's physical theory of heat, he was now generally positive toward Liouville's ideas. He found that "his research in a difficult and important area seems to us to be worthy of the attention of the geometers" and concluded:

> We think that the two memoirs of this young geometer, who is al-
> ready known from others works, deserve the approval of the *Académie*
> and we suggest that the *Académie* decide to publish it in the *recueil
> des Savants étrangers*.
>
> [Poisson 1833b, p. 213]

Thus, for the first time, Liouville managed to get a paper printed in
an official publication of the *Académie*. Poisson, however, also pointed out
shortcomings in Liouville's papers, and suggested that he extend his investigations to the general problem of integration in finite terms. With
his remarks Poisson appears as Liouville's second mentor next to Ampère.
Later in life Liouville recalled that these papers "also brought me the first
testimony of esteem from Mr. Poisson" [Liouville, 1839k, p. 795], but for
several years their relations were somewhat unstable. Thus, in a letter from
around 1837, Liouville informed Catalan: "Together we will visit Mr. Poisson with whom I am at present on the best of terms" [Jongmans, 1981,
p. 291].

In 1834, [1834c] and 1835 [1835a], Liouville developed his theory of integration in finite terms along the lines suggested by Poisson. These two
papers have become his most celebrated papers in the area because they
contain the first proof that the integrals of certain algebraic and simple
transcendental functions cannot be expressed in finite terms, and because
they contain the fundamental theorem in this branch of mathematics, the
first of his theorems, which has later been named "Liouville's theorem."

Unfortunately for Liouville, the *Académie* reports on these papers were
not composed by Poisson, but by Libri, who was unable to do full justice
to them. He praised them in general terms:

> In this work which concerns the most difficult questions of analysis,
> as well as in his earlier works, the author has shown evidence of great
> skills and sagacity.... To sum up, your commissioners think that Mr.
> Liouville's learned research merits all the praise of the *Académie* and
> that one should encourage the author to pursue it and to complete
> as soon as possible the work which he is preparing on this matter,
> and of which this memoir can only be considered as a preliminary.
>
> [Libri 1834, pp. 482,483]

Like Poisson, Libri pointed to weaknesses and suggested extensions of
the theory, but for the most part, his remarks resulted from his lack of
understanding of Liouville's ideas, in particular the underlying classification
of elementary functions. Still, they made Liouville undertake a thorough
analysis of this classification, which he published in 1837 [1837d].

Liouville planned to write a book on his new theory. That is the "work" to
which Libri alluded in the above conclusion of his report. However, the plan
never materialized, possibly because Liouville could not carry through his

projected chapters on integration of differential equations in finite terms. He began research in that direction in the spring of 1833, but got stuck, and not until 1839–1840 did he present his partial, yet impressive results to the *Académie*, [1839e], [1840i], [1841a], at a time when his career was secure.

During the crucial early 1830s, Liouville obtained the highest recognition for a work that later sunk into oblivion. At the request of Poisson [Neuenschwander, 1984a, II, 1], he had, during his early investigations of heat, tried his hand at stationary heat conduction in a two-dimensional homogeneous plate, which is unequally polished on the edges. At first [1830, §1], Liouville observed that the problem offered "almost insuperable difficulties," but it remained a challenge to which he often returned during the following years [Ms 3615 (3,4)]. Finally, in 1834, he succeeded in solving the problem in a number of special cases. This was his first paper on potential theory. Employing Fourier's techniques, he devised ways of solving the integral equations he had found for the Fourier coefficients. He presented the paper to the *Académie* on March 17, 1834, and 10 months later, Poisson read a report in which he expressed his unconditional praise:

> The memoir which the *Académie* has given us to examine will certainly add to the favorable opinion which the geometers have formed about the talent of the author and his study and knowledge of analysis. Moreover, it is desirable to call their attention to the difficult and interesting question which Mr. Liouville has treated...
> [Poisson 1835b, p. 644]

Liouville's paper was published in 1836 [1836b; see also 1836i]. In historical perspective, this paper proved less important than the papers on fractional calculus and integration in finite terms. Still, it is natural that at the time it got the best review of all Liouville's papers, for it dealt with a traditional subject in which there was a common agreement about what was important, difficult, correct, and ingenious, and Liouville's investigation qualified for all that. The two other fields were so novel that it was harder for the establishment to judge their value—great scientists are often first recognized for a difficult contribution to a traditional area.

The Creation of Liouville's *Journal*

Recognition by the *Académie des Sciences* was important for the young mathematician, but clearly he also wanted to see his results published for a wider audience. The *Académie* published two journals, one for members and one for non-members, or its *étrangers*, as they were called. When recommended by the examining committee, the papers submitted to the *Académie* were printed in the latter but with varying delays. Thus, Liouville's papers on integration in algebraic terms presented to the *Académie* during the winter of 1832–1833 did not appear in the *Mémoires des Sa-*

vants Étrangers until 1838, together with the paper on potential theory from 1834. No wonder Liouville tried faster ways of publication.

However, publication of mathematical papers had grown more difficult since Liouville's earlier works had appeared. Gergonne had been appointed *recteur de l'Académie de Montpellier* and found himself so overloaded that he was compelled to disrupt the publication of his *Annales des Mathématiques Pures et Appliquées* in the middle of the 1831–1832 volume, which succeeded the one containing Liouville's long paper. That was the end of the world's first journal devoted entirely to mathematics after 21 years of existence. Also, Férussac's *Bulletin* stopped appearing in 1831, the year Liouville had published his short electromagnetic note in it.

Instead, Liouville turned to the *Journal de l'École Polytechnique*, the official journal of this school. Its editors accepted his major papers on fractional calculus and integration in finite terms, including those that later appeared in the *Mémoirs des Savants Étrangers* of the *Académie*. The *Journal de l'École Polytechnique* was widely circulated, but apparently Liouville felt that his results would be better known if they were published in the only remaining specialized mathematical journal: *Journal für die reine und angewandte Mathematik*, published in Berlin by A. L. Crelle (Quetelet's *Correspondance mathématique et physique*, published in Bruxelles from 1825 to 1839, was of much less importance).

As the name indicates, Crelle had created this journal as a German counterpart to Gergonne's *Annales* in 1826. It had from the start obtained recognition for publishing papers by the best mathematicians, such as Abel, Jacobi, Dirichlet, and Steiner, but only a few French mathematicians (mostly outsiders such as Poncelet and Sophie Germain) had used this outlet for their productivity. Frenchmen were of the justifiable opinion that their capital was the center of science and thus found it natural for the whole world to be in scientific contact with Paris and unnatural for themselves to publish in peripheral, i.e., foreign journals.

However, in 1833, Liouville began to send his papers to Crelle, thereby showing a sense of internationalism and an understanding of the growing mathematical powers of Germany. Crelle agreed to publish abridgements of Liouville's two papers from the *Journal de l'École Polytechnique* and new results pertaining to his two new theories. Moreover, he published Poisson's review of Liouville's memoir on two-dimensional heat conduction followed by a summary by Liouville of the mathematical techniques in the memoir.

These publications brought Liouville into his first contact with the mathematical world outside France, and it also made him aware of the insufficient publication facilities in the world's scientific center. To be sure, the level of information about scientific activity increased in 1833 with the creation of *l'Institut*, a journal devoted to the events in various academies that contained summaries of papers presented to these learned assemblies. Two years later, the *Académie des Sciences* began to publish its own *Comptes Rendus des Séances*, which appeared weekly with news about the latest meeting. This publication was mainly the creation of the *Secrétaire Perpétuel* Arago. Still, these publications did not diminish the need for a French mathematical journal, for they covered all the sciences and only accepted short papers.

Therefore, Liouville decided to create a new French mathematical journal, often described as his single most important creation. He had already faced the problems of journal editing in connection with his own papers. For example, it had been necessary to add an erratum to his first paper in Crelle's *Journal*, and therefore he asked Crelle's permission to read the proofs of the second article, for, as he wrote to Crelle in December of 1833 in the accompanying letter:

> as far as I can see, this will hardly retard the printing of the journal a few days; that is a very small inconvenience compared with the resulting great advantages. I do not fear printing errors, for they can be corrected in Berlin just as well as in Paris, but rather the mistakes which have escaped my attention while copying my manuscript.... The request which I have the honor of making you is in the interest of the journal as well as in my own interest.
>
> [Ms 3640 (1841), Neuenschwander 1984b, II, 1]

The problems of delay were also known to him, for having had no reply to the above letter for two months, he wrote to Crelle:

> Therefore, I have absolutely no idea whether the manuscript...has arrived at the correct address or if I should consider it to be lost. This uncertainty is the more unpleasant for me as it is very important that my memoir be printed promptly. Therefore, you will do me a great favor by writing me a few lines to inform me if you have received my memoir *yes* or *no* .... If you could speed up the publication, it would be excellent.
>
> [Ms 3640 (1841), Neuenschwander 1984b, II, 2]

Thus, the young Liouville had already formed such clear ideas about certain aspects of journal editing that he dared to make suggestions to the senior editor Crelle. Still, he respected Crelle's "excellent journal" so highly [Ms 3640 (1841), Neuenschwander 1984b, II, 1] that he decided to design his own journal along similar lines and to give it the same name *Journal de Mathématiques Pures et Appliquées*, which was a slight variation of the name of Gergonne's *Annales*. Ironically, Gergonne, who had so severely criticized Liouville's paper in his *Annales*, now gave his consent for Liouville to declare the new journal a continuation of his *Annales*.

This appears from the *avertissement* in the first issue of Liouville's journal, in which the editor explained his editorial policy. As the title indicated, the *Journal* was intended for papers in all fields of pure and applied mathematics, including their histories. By way of example, Liouville held out prospect of publishing some letters by Leibniz and Huygens that Libri had found. Liouville intended to accept both original papers on advanced subjects and new approaches to more elementary mathematics. The *Journal* was intended mostly to contain papers sent by their authors to Liouville, and he would only rarely reprint papers or summaries of papers already published elsewhere, as had been done in Ferrusac's *Bulletin* and *l'Institut*. In fact, over the years, Liouville published several translations of papers in foreign journals. When refereeing a paper, Liouville took the mature and progressive position, to judge its merits rather than to censure its shortcomings. Liouville also made it clear that he would not allow his *Journal* to degenerate into a forum for the everlasting and often slanderous quarrels in the competitive Parisian academic circles:

> During the last few years, a peculiar spirit of denigration has seized some critics and we have seen them heap abuse on one after the other of the men who in various fields of science have honored France with the greatest dignity. Here, our profession of faith will be frank and positive: this sharp and peremptory style, which is so much in fashion at present, will never be mine, for it dishonors both the character and the talent of those who adopt it.
>
> [Liouville 1836j, p. 3]

If a controversy should nevertheless arise in the *Journal*, Liouville promised to "restrict myself to the role of a spectator and faithfully transmit to the public the documents of the case." Liouville's fears were needless, for except for occasional claims of priority, the only person who used the *Journal* for polemical purposes was the editor himself, in which cases his role as spectator can be debated. The absence of quarrels in the *Journal* was probably not so much a result of Liouville's editorial abilities as of the appearance of the *Comptes Rendus*. From 1835, this journal published accurate accounts of the discussions in the *Académie*, providing the scientists with a faster outlet for their aggression. Liouville ended his *avertissement*:

> Now, it is up to the geometers, in particular, the French geometers, to make this enterprise prosper. The most distinguished among them have already promised us some articles, and doubtless they will keep their promise. We dare say that it is in the interest of their reputation: the failure of a useful journal which they have refused to support would be honorable neither for them nor for France.
>
> [Liouville 1836j, p. 4]

Indeed, the first volume of Liouville's journal contained papers by a large number of the most famous French mathematicians, such as Coriolis, Lamé, Sturm, Chasles, Ampère, Lebesgue, and Liouville himself, and also by the two famous German mathematicians, Jacobi and Plücker. With this display of great names, it was manifest from the start that Liouville's journal completely outmatched the scientific level of its predecessor, Gergonne's *Annales*, which had only contained occasional pearls among a majority of elementary or mediocre contributions.

As his publisher Liouville chose Bachelier, whose son and partner had been his fellow student at the *École Polytechnique* [Fourcy 1828, Marièlle 1855]. The economic aspects of the publication are unclear [cf. Neuenschwander 1984 b]. Liouville stressed in his *avertissement* "that this is the question of a truly scientific enterprise and not a commercial speculation." Indeed, it is uncertain whether the journal made any profit for Liouville at all. By way of comparison, it may be mentioned that Crelle sustained a personal loss on his journal, although it made a profit for the publisher [Neuenschwander 1984b, p. 11]. It is possible that the government supported Liouville's journal from the start either by direct subvention or by subscribing to a fixed number of copies. Actually, later French governments did support certain journals [Neuenschwander 1984b, p. 12], and the following quote from a letter Liouville wrote to the Minister of Education, J. Simon, indicates that the *Journal de Mathématiques* was also favored by the state. "[I am flattered] that after Mr. José Etchegaray praises the *Journal de mathématiques*, whose publication I have directed for almost 40 years (with the support of the Ministry of Public Education, for from the regrettable Mr. de Salvandy until you, the Ministry has constantly supported this thankless and difficult task)..." [Neuenschwander 1984b, II, 25; the letter is dated December 12, 1872, and the words in parentheses are crossed out in Liouville's sketch].

Risky as the creation of the *Journal de Mathématiques* may have been economically, it was at least as daring scientifically. It should be kept in mind that the editor who took on the duty of judging the works of the best mathematicians of the world was only a 26-years-old *répétiteur*, who still had his own work evaluated with a view to possible promotions. Liouville must have had great confidence in himself.

The fact that Liouville could persuade so many excellent mathematicians to contribute to the first volume shows that by 1835 he had many contacts in academic circles in Paris and some abroad, and that these people trusted his gifts. The only conspicuous omissions from the list of contributors to the first volume were Poisson and Cauchy, but their absence was not the result of hostility toward Liouville. Indeed, Poisson published eight of his last papers in Liouville's journal, beginning in the second volume. This volume also contained two papers by Cauchy, who later contributed five more papers to the journal. For unknown reasons however, Cauchy preferred to publish in the *Comptes Rendus*, and when it did not have space enough for the results of his incredible productivity, he issued his own "journals," containing only papers by himself.

While Liouville had had many scientific contacts in 1835, he became world famous in 1836. His journal became a great success with a reputation on the level of its German counterpart, and the recognition it conferred on its editor would clearly help him in his career, which had not made much progress since he had been appointed a professor at the *École Centrale*.

Defeats at the Académie and at the École Polytechnique

As far as his career was concerned, Liouville had two goals: to become a professor at one of the prestigious schools and to be elected a member of the *Académie des Sciences*. However, the opportunities were few. In fact, the French academic structure was rather stable, so new positions were rarely created at the higher educational institutions, and except for a slight modification, the number of seats in the *Académie* had been fixed since 1803. Moreover, the members of the *Académie* and the professors were appointed for a lifetime, so there were only openings in the system when one of the old guard died. At the *Académie*, the opportunities became very rare as a result of the division into sections. Thus, the geometry section, which was the natural one for Liouville to join, consisted of only six members, as did all the other sections. In addition to the geometry section, the division for the *sciences mathématiques* consisted of the following sections: *mécanique*, *astronomie*, *physique générale*, and *géographie et navigation*. The second division for the so-called *sciences physiques* was divided into sections for *chimie, minéralogie, botanique, economie rurale et art vétérinaire, anatomie et zoologie*, and *médicine et chirurgie* [Institut de France 1979, Aucoc 1889]. Each division was led by a *Secrétaire perpétuel*. It was therefore a great opportunity when Legendre died in 1833, leaving a vacant seat in the geometry section. Liouville was a natural candidate, and his name was put on the list together with that of Libri, Sturm, and Duhamel, who were all his seniors.

The Count Guglielmo Brutus Icilius Timoleon Libri-Carucci della Sommaja was born in 1803 in Florence. After taking part in an unsuccessful liberal conspiracy, he fled to France in 1830. At this time, a number of his papers on algebra, some of which had appeared in Crelle's *Journal*, gained him a name in mathematical circles, and with his courteous manners, he soon found supporters in the *Académie*, first of all, the powerful *Secrétaire perpétuel* from the mathematics division, Arago. Because of him, Libri was elected as a *correspondant* at the end of 1832, but after having obtained French citizenship in February the following year, he was eligible as an ordinary member [Billy 1959].

Arago was also the patron of the other foreign candidate, Sturm. After having written the paper on compression of fluids together with Colladon, Sturm turned to mathematics in which he proved the celebrated theorem (now named after him) on the number of real roots of a polynomial [1829a, 1835]. Thanks to another of his patrons, Ampère, he became the main editor of the mathematical parts of Ferussac's *Bulletin* in 1829, and the same year he gave a summary of Liouville's first published paper in this journal. Around this time, the two developed a devoted friendship to which Liouville's letter quoted earlier bears witness. This same letter shows that

PLATE 5. Gugliemo Libri-Carucci (1803–1869). At first he supported Liouville, but from 1837 he became Liouville's arch-enemy. (Archives de l'Académie des Sciences de Paris)

Sturm did not promote his own career systematically, and so, being a foreigner and a protestant, he had a hard time finding a post in Paris. After the revolution in 1830, however, Arago secured him a professorship at the *Collège Rollin* [Speziali 1964].

The oldest of the candidates, Jean-Marie-Constant Duhamel, was born in 1797, but he was a late starter. He presented his first memoir to the *Académie* in 1828, the same year as the 12-years-younger Liouville, and his election as Arago's *répétiteur* in the geodesy course at the *École Polytechnique* also coincided with Liouville's election. Duhamel's early publications mostly dealt with the mathematical theory of heat. This was also true of his most important paper containing the celebrated Duhamel's principle. This paper which had been presented to the *Académie* in 1830 appeared at the time of the election [1833].

When the *Académie* was going to choose between the four candidates on March 18, 1833, Poisson had only just read the first report on a mathematical work of Liouville, so Liouville was probably not too surprised that he only obtained one vote. Libri definitely won the election with 37 votes. Duhamel came in second with 16 votes, whereas Sturm found no supporters. If subsequent history is taken into account, the academicians had almost reversed the order of the talents of the candidates, but from their point of view, it was a natural ordering. Only Sturm had good reasons to be disappointed.

Liouville's next chance to get at least a temporary job on a higher level came two years later at the *École Polytechnique*. An accident prevented the examiner at the entrance exams from carrying out his job, and so a committee was formed to fix an order of priority of the possible substitutes. It consisted of the two professors of analysis and mechanics, Mathieu and Navier, and the *Directeur d'Études*, Dulong. They proposed the two *répétiteurs*, Liouville and Duhamel on the same *ligne* (level), followed by Courtial, a *répétiteur* in descriptive geometry. On July 13, the *Conseil d'Instruction* of the school had to decide the matter. It is likely that Poisson supported Liouville, for only six months before, he had highly praised Liouville's work on two-dimensional heat flow, and less than a month before Liouville had, in the *Académie*, proved the value of his fractional calculus by applying it successfully to a problem posed by Poisson [Liouville 1835d]. Still, Duhamel was elected.

The defeat only lasted five days, for on July 15, Liouville was elected substitute for another of the entrance examiners, Reynaud, who had fallen ill. This time the same committee proposed Liouville alone as the first candidate, but changed their minds for the second place, where they now put Auguste Comte, who made part of his living as a *répétiteur* of analysis and mechanics while developing his philosophy of positivism.

The elections at the *Académie* and at the *École Polytechnique* made it evident that Duhamel was Liouville's prime opponent, so it was a great event for Liouville when he defeated him in an election for the first time. That happened the following year on July 8, 1836, when Reynaud was ill again, and the choice as his substitute once again was between Liouville, Duhamel, and Comte. The change of sympathy in the *Conseil d'Instruction* was probably mainly a result of the renown Liouville had gained as a founder of his *Journal*; that he had become *Docteur ès-sciences Mathématique* at the university on January 29, 1836, may also have played a role. His thesis, a collection of two papers published elsewhere as well [Liouville 1834d and 1836f], was signed on December 30, 1835, by the dean of the Paris science faculty, the chemist Thénard [Bibl. Inst. Fr. 4°, N.S. Br 137 (JJ)].

The same summer, Liouville also examined the second-year students in analysis and mechanics, replacing the professor, Navier, who had fallen ill. It turned out to be a serious illness that resulted in Navier's death a few months later. Thereby one of the strongly desired permanent professorships in analysis and mechanics at the *École Polytechnique* became vacant. This was a crucial moment in the competition between Duhamel and Liouville. As could be expected, they both applied for the job together with Comte, and on October 7, a committee consisting of Dulong, Mathieu, Savary, Leroy, and Lamé put them in the following order of priority:

1. Liouville,
2. Duhamel, and
3. Comte.

Their highly qualified fellow *répétiteur*, Coriolis, had refrained from applying as a result of his weak health. After a discussion of their qualifications, Liouville was elected the candidate of the *École Polytechnique* a week later by a majority of the votes in its *Conseil d'Instruction*. This corroborated the decision taken three months earlier.

However, the battle was not yet won, for when professors were to be appointed, both the school and the *Académie des Sciences* had to choose a candidate, and His Majesty's government then made the final choice. On October 27, the complete geometry section of the *Académie*, consisting of Puissant, Libri, Lacroix, Biot, Poinsot, and Ampère, presented an undecided ordering of the candidates, putting Duhamel and Liouville in the first rank (*ex aequo*) and Comte in the second rank. This left the decision to the entire *Académie*, the majority of whose members had only vague ideas about mathematics. The vote was a thriller. At the first round, Duhamel got 20 votes, Liouville 19, and Comte 2. Since none of the candidates had obtained a majority of the votes, the *Académie* proceeded to a second vote between the two winners. Here, Duhamel obtained 23 votes, and Liouville only 20. Thus, Duhamel was declared the candidate of the *Académie*.

Now, the decision had to be made by the minister of war, under whom the *École Polytechnique* belonged. Apparently, he found it hard to decide, for on November 19, when the courses were about to start, the *Conseil d'Instruction* still did not know the name of their new professor, and therefore asked Comte to take over the teaching temporarily. Six days later, they received the minister's decision. Duhamel was officially appointed.

Magnanimity toward Sturm

This must have been a great disappointment for Liouville, but he had no time to despair, for a new race had begun: the race for the *Académie*. Two seats had become vacant, one in the mechanics section by Navier's death, and another in the geometry section by the death of Liouville's former teacher and supporter, Ampère.

Ampère had not seriously worked with mathematical physics since finishing his electrodynamics in 1827. To be sure, he had done Liouville the favor of submitting a paper to the first volume of his *Journal*, but it was an old paper, written originally for his course at the *Collège de France*, the year Liouville had taken it. It was the last paper he published in his lifetime, for he died on June 11, 1836, having perhaps just had the chance to see his paper appear in the June issue of Liouville's *Journal*.

Liouville decided that he would not apply for the mechanics seat. This is a bit surprising, for the boundaries between the different sections were not very sharp, and after all, he was also a *répétiteur* of mechanics. However, he had only published one paper that could qualify as a mechanical paper, so he gave up in advance. He spent much more energy on a campaign for the geometry seat. He submitted a series of results to the *Académie*, one concerning celestial mechanics and the rest in a completely new and revolutionary field, later called Sturm-Liouville theory.

Sturm had opened this field in 1833 with two papers, of which he presented one to the *Académie*. They dealt with the qualitative behavior of the solutions to the eigenvalue problem found by separating variables in the equation describing the heat flow in a non-homogeneous bar. At the time, only summaries of the papers appeared in the journal *l'Institut*, but in 1836, Liouville persuaded Sturm to publish them in full in his new journal. Liouville realized that Sturm's work provided him with a solid basis for the study of the convergence of the corresponding Fourier series, a problem he had earlier attacked with questionable methods in his paper in Gergonne's *Annales*.

In November 1835, Liouville had presented a paper in which he attempted to prove that the Fourier series have the expected sums. Immediately after Ampère's death, he submitted an alternative proof for one of Sturm's theorems, and on November 14, 1836, he discussed a generalization to a third-order differential equation, which Poisson had earlier suggested to him as a hard nut to crack. Finally, at the eleventh hour, Liouville devised an ingenious convergence proof for the Fourier series of the second-order Sturm-Liouville problems. There was no time to write a proper memoir, but after a few days, he could send a letter containing a sketch of the proof to the *Secrétaire Perpétuel*, Arago, who read it to the *Académie* on November 21. The following week, Liouville seized the chance to make Arago read an elaboration on the proof, just before the geometry section announced their ordered list of candidates.

Liouville later published this convergence proof and another one in his journal [1837c and i], and he generalized Sturm's theory to a certain class of higher order equations [1838h]. A later quote will show how highly he valued his achievements in Sturm-Liouville theory:

> ... the general theory of linear [differential] equations where I have undertaken, not to integrate these equations, but to study in the differential equations themselves the properties of their integrals. Mr. Sturm was the first to consider the equations of the second order from this point of view and I have later extended... the majority of his theorems.... If I am allowed to express my own opinion of the relative value of my Memoirs, I shall not hesitate to put the above said works in the first rank.
>
> [Liouville 1839k, pp. 795-796]

History has proved him right; his contribution to Sturm-Liouville theory was the most ingenious, and later also the most influential, work he had done up to then. From late 1835, Liouville and Sturm's research in this area was carried out in close collaboration. They discussed each others work before its publication, and in print they always praised each other and even covered up each other's mistakes (see Chapter X). In May 1837 they presented a long joint paper on Sturm-Liouville theory to the *Académie* of which unfortunately only a fraction was published [Liouville and Sturm 1837a].

Shortly before the competition in November 1836 at the *Académie*, the two friends published their first joint paper, in which they gave a method for finding the number of complex roots of a polynomial inside a given contour [Liouville and Sturm 1836]. In this way, Liouville supported Sturm in a controversy with Cauchy, who had claimed priority for another generalization of Sturm's theorem on the number of real roots. This was an awkward situation, for as anybody could guess, both Sturm and Liouville were interested in the vacant seat at the *Académie*. How did Liouville solve that problem? Many French scientists in similar situations had competed bitterly with their (former) collaborators and friends. At first, it appeared as though Liouville had chosen this usual behavior. He obtained permission to address the *Académie* in person on November 14, and he began his talk on third-order Sturm-Liouville theory with the traditional historical survey aimed at making his colleagues aware of his new ideas:

> In order to explain the subject of this memoir, I first have to summarize some details of the history and recall the remarkable improvements made in recent years in the theory of partial differential equations. However, I shall abbreviate this digression as much as possible. I know quite well how difficult it is to give a clear verbal exposition of complicated analytical studies. The *Académie* is aware that I am not in the habit of abusing its precious moments for, having during the last six years written more than twenty memoirs on various subjects, I have nonetheless only taken the floor once.
>
> [Liouville 1837a, CR III, 1836, p. 572]

One can imagine the academicians listening with one ear to the subsequent account only to wake up when Liouville came to the point where he should convince them about the importance of his own work, for he did that by highly praising his opponent:

> ...Mr. Sturm.... This geometer has made a profound study of the various functions V .... Sturm's research on this subject has appeared in my journal. I am proud that I have been the first to do justice to these two beautiful memoirs, which the impartial future will rank on the same level as Lagrange's most beautiful memoirs.
>
> [Liouville 1837a]

This was an unusual message to give the *Académie* only three weeks before the election. In his letters to Arago, Liouville continued in the same vein to praise Sturm, while presenting his own results on the Fourier series:

> The difficulty of proving the convergence has been mentioned by Mr. Sturm himself in his last memoir.... I must say that after I had

> communicated my work to Mr. Sturm he found a few days later a
> second proof of the convergence of the series, which is as simple as
> mine and based on his own principles.
>
> [Liouville 1836h, p. 655]

At the meeting where Arago read this message to the *Académie*, the
geometry section presented their list of applicants in order of priority: (1)
Sturm, (2) Liouville, (3) Duhamel, (4) Lamé, (5) Boucharlat. There is little
doubt that Liouville's propaganda had been instrumental in obtaining the
top seat for Sturm. Liouville's own reference to an "impartial future" suggests that Sturm was not well liked by all his fellow scientists, and the zero
votes he had obtained at the election in 1833 points in the same direction.
Therefore, Liouville and Duhamel still had a chance of winning the final
election. However, on the very day of the election, December 5, both Liouville and Duhamel shocked the *Académie* by announcing that they stood
down in favor of Sturm. As a result of that, Sturm was elected with an
overwhelming majority of 46 votes from a total of 52.

Liouville's behavior clearly shows that from the start he had decided to
help his senior friend Sturm to obtain the seat. Why then did he apply, and
why did he so feverishly pursue his research and present his results until the
last moment before the election? Of course, it gave him the opportunity
officially to support Sturm, but the real reason was probably Duhamel.
In the election on October 27, Liouville had painfully discovered that the
Académie valued Duhamel higher than himself. He therefore had to prove
his talents to the *Académie* at the same time as he indicated Sturm's higher
genius. His plan succeeded, and for the first time the *Académie*, or at least
its geometry section, ranked him higher than Duhamel. This could be of
importance in a future election. However, in the geometry section, there was
not another opening for a long time. To be sure, some of its members were
not youngsters, but they enjoyed good health. As a matter of fact, another
seven years passed without any replacements. This throws Liouville and
Duhamel's unselfish act into perspective. Duhamel and Lamé both applied
for Navier's seat in mechanics, but both lost to Coriolis.

Success at the Collège de France
and at the École Polytechnique

The following year brought an unexpected success in Liouville's career,
not at the *Académie* nor at the *École Polytechnique*, but at the *Collège de
France*. Here, as in the other institutions for higher education, the professors were appointed for life, but they often exercised their right to have a
substitute to do the teaching. Thus, the influential physicist and astronomer
Biot, who occupied the chair of *Physique générale et mathématique* for the
extremely long period from 1801 to 1862, and who had been substituted by

Cauchy twice in the 1820s and by Levy and Libri in the courses of 1832–1833 and 1833–1834 respectively, on March 19, 1837, asked the *Assemblée des Professeurs* of the *Collège* "for permission to let Mr. Liouville substitute for him during the second semester" [P.-V. Coll Fr.]. The *Assemblée* gave their permission, and Liouville began his teaching at the most highly esteemed institution in France. However, there were two drawbacks to this job: It only lasted as long as Biot wanted and as long as the *Assemblée* accepted. Moreover, it only paid about one third of the professor's salary, the remaining two thirds being paid to the professor for the onerous task of carrying the title [Coll Fr. Pers. file, letter, Nov. 19, 1879]. After having resigned from the *Ecole Centrale* in 1838, he was left without any permanent position in the public institutions.

Therefore, it was convenient for Liouville that his professor at the *École Polytechnique*, Mathieu, was, in the summer of 1838, promoted to an examiner's job left by de Prony. The *Procès Verbaux* from the meetings during this period in the *Conseil d'Instruction* are lost, so the candidate of the school for the vacant chair is unknown, but with the earlier elections in mind, we can be fairly certain that it was Liouville. Moreover, the *Compte Rendus of the Académie* tells us that on September 17 Liouville was elected unanimously as the candidate of the *Académie*. He was formally nominated as Professor of Analysis and Mechanics on October 23, 1838, just in time to start teaching the second-year students. As his *répétiteur* and successor, he and the *Conseil* chose Sturm, who thereby got his first job at the school.

Thus, Liouville had reached one of his goals in a Parisian academic career. Half a year earlier the French State had bestowed its first honor on him by naming him *Chevalier de la légion d'honneur* on April 29, 1838 [Neuenschwander 1984a, Annexe, p. 119]. Still, he did not lose sight of the second goal: membership to the *Académie des Sciences*. During October 1837, he submitted a solution to a problem treated earlier by Duhamel; Liouville used the methods of Sturm-Liouville theory, his aim being obviously to show the force of these methods compared to those of his competitor.

A Vacancy in the Astronomy Section of the Académie

Though not a member, Liouville kept abreast of events in the *Académie*. For example, on July 9, 1838, possibly as a part of his campaign for the professorship, he presented an analytic solution of a problem on attractions of spheroids, which Poisson had just solved synthetically the previous week. Poisson had declared that "it will be very difficult, to say the least, to derive these results with analytic means," but Liouville [1838g] showed how to do it along lines similar to those Poisson had used in an earlier paper. Apropos to this paper on wave propagation [Poisson 1831], two weeks later Liouville presented the *Académie* with a simplification of one of the calculations in it. This method was in turn generalized by Poisson in a paper that Liouville printed in his *Journal* in December of the same year [Poisson 1838], and the following month [Ms 3616 (3), p. 9v–10r], Liouville further generalized and simplified it [1839a]. Liouville's courses at the *Collège de France* during 1839–1840 touched on the theory of sound, but otherwise it is remarkable that in spite of his interest in this aspect of his mentor's work, he did not make further contributions to the otherwise rapidly expanding theory of waves in elastic media, his only further work being another comment on Poisson's work [1856b, p. 148].

With his constant interest in the transactions of the *Académie*, Liouville was immediately alert when the astronomer Lefrançois-Lalande died on April 8th 1839. To be sure Lalande was a member of the astronomy section, but impatient from waiting for an opening in the small mathematics section, Liouville decided to run for the vacant seat. As early as the following week, he presented the first in a series of papers dealing with astronomy to the *Académie*.

Liouville was prepared for the challenge, for he had earlier done sporadic work in celestial mechanics. The first remarks on the subject in his notebooks date back to June and July 1834 [Ms 3615 (4), p. 52v–53r,59v]. Commenting on one of these notes, dealing with the tautochrone, Liouville later explained to Bertrand in an undated letter: "My note...is a part of a sort of commentary on the first two volumes of the Mécanique céleste [i.e., Laplace's masterpiece], which I composed around 1834 and 1835 for my own use" [Neuenschwander 1984a, III, 3]. Among the other subjects discussed in the notes were the series expansions of planetary theory and attraction of a spherical shell. These studies enabled him to respond when Jacobi, on the following October 20, sent the *Académie* a letter announcing that there exist rotating fluid ellipsoids in equilibrium having three different axes. This was a very surprising result, for everyone since Maclaurin had believed that if a mass of fluid, such as a fluid earth, rotated around an axis, its only equilibrium figure was an ellipsoid of revolution, i.e., one having its two equatorial axes equal. Indeed, Lagrange had "proved" this result previously [1811–1815, vol. 1, §199–204].

In his letter to the *Académie des Sciences*, Jacobi announced that he would "take up the *Mécanique céleste* in the pitiable state in which Laplace left it" [Liouville 1834d, pp. 290–291, footnote]. Liouville regarded this remark as a challenge to French mathematicians [Liouville 1834d, pp. 290], and the following week he proved to the *Académie* that Jacobi's theorem was but an easy consequence of formulas in Laplace's *Mécanique Céleste*. This paper was his first publication in celestial mechanics [1834d], but he did not think highly of the result:

> Although I must admit that this theorem does not seem to me
> to present the great importance that the illustrious geometer of
> Königsberg seems to attach to it, I thought that I ought to look

PLATE 6. Carl Gustav Jacob Jacobi (1804–1851). In 1834, Liouville did not think highly of Jacobi's work on fluid ellipsoids, but later Liouville became a great admirer of Jacobi's work and counted Jacobi among his friends. (Math Dept., University of Copenhagen.)

for a demonstration of it (∗).

In a note, he added:

> (∗) This phrase has been blamed by several persons. Nevertheless, I ought to preserve it in the printed paper because it expressed the feeling I had when I found it so easy to prove a theorem which had been addressed, so to speak, as a challenge to the French geometers.
> [Liouville 1834d, pp. 290–291]

Later, however, he came to admire Jacobi's works as a whole, and this theorem in particular, and the hard feelings this comment might have caused were soon forgotten. In fact, we have already seen that Jacobi was one of the two foreigners who contributed to the first volume of Liouville's Journal. Liouville's answer to Jacobi indicates that although Liouville was internationally minded, he was also patriotic in the sense that he did not like a foreigner to attack French work.

The following year Liouville continued with a more traditional problem in celestial mechanics. He had noticed that the planes of the orbits of Jupiter, Saturn, and Uranus cut each other almost along a line, and he asked the question whether that would continue to be the case. His long calculations of April 7, 1835, answered the question in the negative [Ms 3615 (4), pp. 68v–74r]. In spite of the significance of the problem, Liouville did not publish the result immediately, but after the death of Lalande in 1839, he dug up his old notes and was thereby able to produce a paper for the *Académie* within a week.

During the first months of 1836 Liouville made a serious study of planetary theory, the three-body problem, the motion of the minor planets, Vesta and Ceres, and similar problems (cf. Chapter XVI, §12). Out of these calculations grew two publications. In the first [1836d], Liouville proved that certain double integrals, introduced by Poisson, could in certain cases be replaced by simple integrals. In the second [1836g], he showed how to use elliptic integrals to estimate the remainder of the series occurring in perturbation theory. Poisson refereed the first of these positively in the *Académie* and pointed out "that it will in principle be useful for the calculation of the inequalities with long periods" [Poisson 1836].

During the following two years, Liouville did only scattered research in perturbation theory, but in January 1838, he was led to a very famous discovery. He had observed that when applying the method of variation of constants to solve certain nth-order differential equations in the time variable, one often had to use a specific determinant. Liouville showed that given certain conditions this determinant is a constant independent of time [Ms 3615 (5), pp. 28r–29r]. Since this observation was of importance in perturbation theory, he sent it, on January 17, 1838, in the form of a note, to the *Bureau des Longitudes*, the highest French authority in astronomical matters. Later the same year, he published the note in his *Journal* [1838e].

In the published note, Liouville also applied his observation to a system of n simultaneous first-order differential equations satisfying certain conditions. Later, it was observed that these conditions are satisfied by Hamilton's equations, and that the determinant in this case can be interpreted as a volume ratio in phase space. Therefore, Liouville's theorem implies that when a bounded domain in phase space evolves according to Hamilton's equations its volume is conserved. This theorem is now known as Liouville's theorem. It has acquired great renown, because it turned out to have substantial implications in statistical mechanics as developed during the 1870s by Boltzmann and Gibbs. Liouville probably never heard of this application, but he may well have understood its applicability to rational mechanics (cf. Chapter XVI, §23).

Distracted by other matters, Liouville pursued his research on the figure of fluid planets with only little energy. In 1837, he showed how a gap he had discovered in Laplace's treatment could be filled using techniques inspired by Sturm-Liouville theory. His interest was revived in 1838 by a paper by the astronomer Ivory on the shape of the Jacobi ellipsoid. Liouville saw that Ivory's conclusions contradicted the results of an investigation that Poisson had encouraged him to undertake several years earlier. Ivory's paper came at a suitable moment, for it gave Liouville the opportunity to correct Ivory's mistakes and publish his own results in the April volume of his *Journal*, that is, the very month Lalande died.

As already mentioned, Liouville presented his work on Jupiter, Saturn, and Uranus to the *Académie* on April 15, one week after Lalande's death. At the next meeting, he presented a purely mathematical paper on definite integrals and Γ-functions and elaborated on a paper that Dirichlet had published in the *Comptes Rendus* a few weeks earlier. Two weeks later, Liouville reminded the *Académie* about his two earlier papers on perturbation theory, [1836d] and [1836g]. He emphasized their importance by setting them into a historical perspective (the usual trick) and alluded to a generalization of his method by elliptic integrals.

Opposition from Libri

Three important papers during four meetings was a good start for Liouville's campaign for the vacant astronomy seat. To be sure, his natural opponent, Pontecoulant, also presented noteworthy astronomical works, but Liouville's greater general fame gave him better chances. However, on the fifth meeting after Lalande's death, his campaign suddenly met with opposition, not from another potential applicant, but from Libri. In order to understand this opposition, we must analyze the relationship between Libri and Liouville as it had developed after Libri had joined the *Académie* in 1833 ahead of Duhamel, Liouville, and Sturm.

At first, Libri had patronized Liouville as Ampère and Poisson had earlier. In connection with the uniqueness of the expression of explicit algebraic functions, Liouville had already referred to a similar theorem by Libri on algebraic numbers [Liouville 1833b, p. 142]. This theorem inspired Liouville to go on with expressions containing only square roots of rational numbers: $\alpha + \sqrt{a} + \sqrt{b} + \sqrt{c} + \cdots + \sqrt{d}, a, b, c, \ldots, d \in Q$. In May 1834, he wrote in his notebook, "The research on the square roots, which I have recorded in two letters to Mr. Libri, deserves to be elaborated further" [Ms 3615 (4)]. Libri did not present Liouville's letters to the *Académie*, but he did help his *protégé* by spreading his ideas in other ways. I have already mentioned Libri's rather favorable, albeit incompetent review of one of Liouville's papers, and he also saw to it that it was published. In a letter to Crelle of February 9, 1834, Liouville wrote:

> Having kindly undertaken to forward my small packet to you, Mr. Libri has given it to one of your friends in common who returned to Berlin.
>
> [Neuenschwander 1984b, p. 59, Ms 3640 1841]

Liouville also obtained support from Libri when he founded his *Journal*. Indeed, Libri's projected publication of historical papers were the only particular contributions mentioned explicitly in the *Avertissement*:

There exist a great number of unpublished letters by Huygens and Leibniz on various scientific questions. We have eagerly accepted Mr. Libri's benevolent offer to communicate them to us, and we shall publish them in our Journal. In this way, we shall, so to speak, place it under the patronage of these geniuses of the 17th century who have so forcefully contributed to the progress of mathematics and in whose works one still finds, even today, the germ of more than one new and profound theory.

[Liouville 1836j]

However, the only paper Libri ever published in Liouville's *Journal* was a purely mathematical paper in Volume 1 [Libri 1836]. The historical papers never appeared because of bad feelings between Libri and Liouville. The first sign of the changed relations between the two is a small paper with the innocent title *Sur une lettre de d'Alembert à Lagrange* written by Liouville in the July 1837 issue of his *Journal*. It was not a positive act by Libri that made Liouville study the correspondence between the two great mathematicians. On the contrary, the paper served only one purpose: to show that Libri did not limit himself to finding "the germ of more than one new and profound theory," but sometimes found whole theories that he published under his own name.

To be more precise, in 1829, Libri had published a method differing from the well-known Lagrangean method of reducing the order of a linear differential equation if a number of solutions to the corresponding homogeneous equation were known. As one of his claims to fame, this theorem had been included in his supporter Lacroix's famous textbook *Traité élémentaire du Calcul différentiel et du Calcul intégral* (5th edition) [Lacroix 1837], with the following introduction: "Mr. Libri has summarized the theory of linear differential equations in a very elegant and very profound way." Moreover Libri had repeated the procedure in his paper in Liouville's *Journal*. In the small paper mentioned above, Liouville accused Libri of having plagiarized the method, but he phrased it with French elegance:

I think it is my duty to announce that the method in question does not belong to Mr. Libri but to a French geometer [national pride again], namely, to d'Alembert who has given it in a letter to Lagrange printed in Volume III of the *Miscellanea Taurinensia*, page 381 [d'Alembert 1766]. I do not understand how this passage has escaped Mr. Libri who has for a long time taken an interest in the history of mathematics.

[Liouville 1837h p. 246]

With this beautifully phrased insinuation, Liouville opened a life-long antagonism. Perhaps personal matters, unknown to us, started the fight, but it is also entirely possible that reading through the *Miscellanea Taurinensia* Liouville discovered Libri's method and suddenly saw through the shallowness of his mathematical production, including the paper in his own journal. Indeed, it is now generally agreed that there is at most one good idea in Libri's mathematics: the attempts to transfer theorems from the solution of algebraic equations (Galois theory) to the field of differential equations, but Libri never got far with his idea. [Demidov 1983].

Liouville struck again in the survey article opening the third volume of his journal, *Sur les deux derniers cahiers du Journal de M. Crelle*. In his *avertissement* to the first volume, he had announced that he would only include such abstracts in exceptional cases, and in fact, this isolated paper of that kind was only a pretext for citing Dirichlet's critical remarks about Libri concerning the series

$$s = \sum_{i=1}^{p-1} \cos \frac{2i^2 \pi}{p} \qquad \text{and} \qquad t = \sum_{i=0}^{q-1} \sin \frac{2i^2 \pi}{q} ,$$

where p is a prime of the form $4m + 1$, and q another prime, $4m + 3$. It is easy to see that $s = \pm\sqrt{p}$ and $t = \pm\sqrt{q}$, but it is hard to show that the upper sign is the correct one. Gauss had given the first proof in 1811. Libri published another in 1832, and in the paper reviewed by Liouville, Dirichlet [1837c] provided yet a new proof. In the review, Liouville quoted the following translation of Dirichlet's remarks concerning the ambiguity in the sign:

> Gauss's method has until now been the only way to overcome the difficulty concerning the ambiguous sign. Despite its ingenuity, the method given by Mr. Libri does not seem suitable to solve this problem because it reduces the sums to the solution of a quadratic equation. In order to remove the ambiguity created in this way, the learned author has recourse to the expression transformed into a product without indicating in any way how to obtain this transformation. However, this passage from the sum to the product is itself the only problem, for once it has been carried out, it dispenses of all further analysis, since the product is among those that Euler has determined a long time ago by the simplest considerations.
>
> [Dirichlet 1837c, quoted in Liouville 1838a, p. 3]

It seems that Liouville had become obsessed with the plan of discrediting all Libri's mathematics. Still, when he made his third stroke with his *Observations sur un mémoire de M. Libri relatif à la théorie de la chaleur*, he pretended that he had only by accident found a new mistake. He explained that during his current research on "the general laws of the motion of heat" (for which there is no evidence in his notebooks), he had consulted the published memoirs on the subject:

> especially the one published by Libri in Florence in 1829 and in Berlin in 1831 in Mr. Crelle's *Journal* [Libri 1827–1831]. This double [added in a note: "read triple, for a first edition of the same memoir has appeared in Pisa in 1827"] publication guaranteed in my eyes the care with which the author has checked his analysis... and so I had to examine it with attention.... However, imagine my surprise when I

discovered that the formulas published by Mr. Libri are incorrect and that the general principle on which they are based is inadmissible.

[Liouville 1838f, pp. 350–351]

Liouville sent his extensive *Observations* to the *Académie* whose *Secrétaire perpétuel* for the mathematical sciences and Libri's former patron, Arago, analyzed them verbally at the meeting on February 19, 1838. The *Comptes Rendus* only published five lines indicating Liouville's criticism, followed by a note by Libri reminding the reader that Fourier had once approved his methods and results. For the rest, Libri refrained from defending himself before the *Académie* until he had read Liouville's criticism. However, the defense did not come, nor did a report from the evaluating committee consisting of Biot, Poisson, Poinsot, and Sturm.

Therefore, when Liouville published his *Observations* in the July issue of his journal, he could add:

P.S. Mr. Libri has not replied to the objections contained in the preceding Note, and so the dispute between us has been settled by this very fact.

Moreover, he added that someone had called his attention to Kelland's recent theory of heat [1837] in which the author also criticized Libri's work.

After Liouville's attack on Libri's theory of heat, there was a cease-fire of more than a year, during which Liouville closely followed the steps of his opponent, who was now favored by the king. Thus, on April 12, 1838, Liouville noted in his notebook:

I have been told that the amiable Libri has been appointed librarian of the *Bibliothèque royale* and Treasurer!!!

[Ms 3616 (2), p. 63v]

The exclamation marks proved to be prophetic, for it was the job as a librarian that created the circumstances that finally led to Libri's disgrace. But, let us not anticipate events. In 1838, Libri was not defeated, as his missing answer might have made Liouville believe. He was just waiting for the right moment to strike back. This moment came more than a year later, when Liouville applied for Lalande's seat in the *Académie*, and that leads us back to the fifth meeting after Lalande's death.

In the *Comptes Rendus* of that meeting, Libri inserted a *Mémoire sur la théorie générale des équations différentielles linéaires à deux variables* [Libri 1839a], in which he claimed to have shown that all properties of algebraic equations can be transferred to linear ordinary differential equations. The only result he mentioned explicitly was the following: If a homogeneous linear differential equation of order $m + n$ is given and if we know another homogeneous linear differential equation of order m, all of whose solutions are also solutions of the first equation, then it is possible to find a third homogeneous linear differential equation of order n such that the solutions of the two latter span the solutions of the first.

This theorem generalized his (or d'Alembert's) earlier theorem, so it provided a good excuse for reviving the old fight. Claiming that he had not known d'Alembert's paper, he pointed out that Liouville should be the first to acknowledge that it is impossible to know the complete history of the subjects one works with; after all, Liouville had admitted that he did not know Abel's work when he wrote his first papers on integration in finite terms. Here, Libri started his counterattack, for Abel, he claimed, had left nothing for Liouville to discover. Sure enough there were many open questions, but Liouville had not shed light on them since "unfortunately the geometers have approved of neither these [Liouville's] methods nor these classifications." Libri claimed that the *Académie* committee in polished phrases had rejected Liouville's entire analysis. The reader can leaf back to be convinced that Libri's "rejection" in the report to which he referred was indeed so polished that it clearly looks like a general approval, although it also voiced Libri's superficial criticism of the classification.

For these reasons, Libri would not classify Liouville's works under the heading "general theory of differential equations," which Libri himself worked with, nor would he classify Sturm's works under this heading, because "they relate in particular to the numerical solution of these equations." Libri finally reminded Liouville about his *avertissement* in the first volume of his journal, and in particular, the following quote:

> A celebrated author says that all this criticism is the lot of four or five small, unfortunate authors who have never by themselves been able to excite the curiosity of the public. They always wait for some work to succeed and then attack it, not because of jealousy, for why would they be jealous, but because they hope that someone takes the trouble to answer them and save them from the oblivion where their own works have left them all their life.
>
> [Liouville 1836j p 3]

Liouville's repeated attacks showed, Libri wrote, that Liouville was one of these unfortunate authors whom he had himself condemned. This refined insult begged for a reply. Sturm was the first to answer at the following meeting of the *Académie* on May 20, not because he wanted to defend himself against the offense Libri had directed against him, but for the following reason:

> Unfortunately, Mr. Libri has reopened a discussion which seemed to have been brought to a conclusion. In the Memoir which he read during the last meeting of the *Académie* he asserted that he found Mr. Liouville's observation without any foundation and he in turn attacked some works by this geometer. Mr. Libri has the advantage over Mr. Liouville of being a member of the *Académie*, and he has chosen to accuse Mr. Liouville of making errors at a moment where the latter has presented himself as a candidate for the Astronomy section. It would be unfortunate if the silence of the Commission, who had been asked to decide the controversy, was interpreted in a

way unfavorable to Mr. Liouville.

[Sturm 1839]

Sturm also explained the silence of the commission:

> By publishing his Note after some time in his Journal [Liouville 1838b] Mr. Liouville relieved us of the obligation of writing a report which could not be favorable to Mr. Libri.

[Sturm 1839]

After Sturm had delivered his Note, Libri rose, and even the dry factual report of the *Compte Rendu* hints of the dramatic scenes that took place:

> While Mr. Libri continued to devote himself to an examination of Mr. Sturm's note, the latter abruptly left the discussion despite repeated entreaties and Mr. Libri's fruitless efforts to keep him back.

[Libri 1839b]

In Sturm's absence, Libri explained that he did not know that the geometer Liouville was a candidate for an astronomy seat. That is hard to believe, although his explanation contained the point that if he had wanted to interfere in Liouville's career, he could have done so already when Liouville was elected professor at the *École Polytechnique*, an election which had taken place after Liouville's criticisms.

After Libri, Arago rose to give "a very detailed analysis" of a six-page-long *Observations sur le Mémoire de M. Libri* [Liouville 1839k], which Liouville had sent to him. In this note, Liouville started head-on with a criticism of Libri's general theory of differential equations, observing that "for the most part, the results do not seem to me to be either very new or very difficult to prove." In order to support the latter statement, he had enclosed a one-page proof of the theorem that Libri had stated without proof in his memoir [Liouville 1839j]. This proof was later quoted by Moigno [1840–1844, vol. 2, pp. 579–583]. Further, Liouville defended both his works on integration in finite terms and the Sturm-Liouville theory, and having reminded the *Académie* about the flaws he and others had found in Libri's papers, he ended his *Observations* by denying the stubbornness that Libri had attributed to him:

> In the very few articles which I have devoted to criticism, I have always used a polite but clear geometrical language. I could not and would not allow myself to do anything else. How can Mr. Libri confuse my expressed contempt for an ignorant and coarse polemics with the commitment to abstain from any scientific discussion.

[Liouville 1839k, p. 798]

During Arago's summary, Libri had in great haste written down his own proof of his theorem on a small piece of paper and had handed it over to Arago. It proved to be almost identical with Liouville's. Afterward, "Mr. Libri rose and replied in the greatest detail to all the assertions contained in this note." His only new point was that the report on Liouville's theory of integration in finite terms, which he maintained to have been very negative, was also signed by Lacroix and Poisson, so in fact Liouville was attacked by all of them. With this in mind, Libri found Liouville's behavior peculiar:

> What does Liouville then do? Without ever speaking about the report, without repudiating the objections contained in it, he continues for five years to publish works based on the same principles as though these principles had been approved.
>
> [Libri 1839c, p. 799]

Here, Libri has "forgotten" that Liouville in fact defended his classification in 1837 [1837d] and that Libri himself had ended his review by encouraging Liouville to pursue his researches.

At this point in Libri's answer, Arago interrupted to remind him that members of committees often signed the reports without reading them. Although this fact was doubtless known to everyone, it was easy for Libri to parry this remark by acting shocked that a member of the *Académie* could dream of taking his duties so lightly.

In his printed reply, Libri also attributed to Arago a remark similar to Sturm's about the unfortunate time Libri had chosen for his counterattack, but in a footnote, Arago, who saw the *Comptes Rendus* through the press, called three of his fellow academicians to witness that he had not brought up this point.

It must have been comforting for Liouville to see that not only Sturm, but also the much more powerful *Secrétaire perpétuel*, Arago, so actively backed him. This support came just at the right time, for at the same meeting, it was decided with 35 votes against 6 to proceed with the replacement of Lalande. At the following meeting, therefore, the astronomy section presented a list of candidates, with the following order of priority:

1. M. Liouville,
2. M.G. de Pontécoulant, and
3. M. Francoeur [CR May 27, 1839a].

The following discussion of the qualifications of the candidates took place behind closed doors, but we may guess that it was animated. Another week had to pass before the election could take place. On June 3, then, the 51 present academicians cast their votes as follows:

> Mr. Liouville obtains 29 votes
> Mr. de Pontécoulant obtains 16 votes
> Mr. Francoeur obtains 2 votes

> There are two blank ballots. Having obtained the absolute majority of the votes, Mr. Liouville is declared elected. His nomination will be submitted to be approved by the King.
>
> [CR June 3, 1839]

The following week, the *Comptes Rendus* announced the official nomination:

> The Minister of public education addresses a copy of the royal order confirming the election of Mr. Liouville to the seat in the astronomy section which has been left vacant by the death of Mr. Lefrançois-Lalande. Mr. Liouville, who is present, is asked to take his seat among the members.
>
> [CR June 10, 1839]

Thus ended the dramatic events that led to Liouville's election to the *Académie* as a representative of astronomy, an area with which his name is no longer connected. In practice, he became at least as attached to the geometry section. The election also put an end to the longstanding competition between Liouville and Duhamel. It may be considered as a symbolic settlement that Liouville shortly afterward published the first paper of Duhamel in his *Journal* [Duhamel 1839a].

Liouville and Dirichlet against Libri

However, the election did not put an end to the animosities between Libri and Liouville. On the contrary, the two academicians now met every week in the *Académie*, where they could both freely vent their spleen for each other and even have the pleasure of seeing their insults published in the *Comptes Rendus*. The first time they had the chance to quarrel was in February of the following year. A result by Dirichlet provided the scientific context. It was stated in a letter to Liouville in which Dirichlet suggested that they begin a scientific correspondence.

Despite his French sounding name, Gustav Peter Lejeune Dirichlet was a German, born in Düren in 1805. He had pursued his mathematical studies in Paris from 1822 to 1826, where he had been attached to the circle around Fourier. For all his great talents and his powerful mentor, he had not been able to find a suitable job in Paris, so in 1826 he accepted an offer from Breslau University. Thus, he left Paris at a time when the four-years-younger Liouville was only a student at the *École Polytechnique*, so it is no wonder that they did not come to know each other then. As is witnessed by a much later letter from Liouville to Dirichlet (of February 10, 1853), a devoted friendship between the two began to develop during Dirichlet's stay in Paris in 1839:

> However, Mr. Holmgren can tell you that although I do not write I speak, that your name is continually spoken by me, that your works are always cited in my courses as the true classical works, and finally that I preserve all the friendship for you which arose when I had the pleasure of seeing you in 1839.
>
> [Tannery 1910, p. 22]

Perhaps, they made each others acquaintance at a dinner arranged by Cauchy. In any case, they were both invited together with Catalan, through whom Cauchy sent Dirichlet the following invitation, dated July 25, 1839:

> Dear sir and fellow member,
>
> I would like to ask you to honor the promise you have made me the other day by coming to dine with us at Sceaux next Tuesday. We will dine at six o'clock and Mr. Liouville will also come. I hope there will be no obstacles and I shall receive the confirmation with a great pleasure.
>
> Tout à vous,
>
> Baron Augustin Cauchy, rue de Voltaire, n° 49

By 1839, Dirichlet had been professor in Berlin for ten years and had gained a great reputation as one of Germany's best mathematicians, on the level with Jacobi and second only to the titanic Gauss. Liouville was very flattered when Dirichlet, after his return to Berlin, saw to it that Liouville was elected corresponding member of the Berlin Academy in 1839 [c.f. Biermann 1960] and suggested that they should begin a scientific correspondence:

> My dear Mr. Dirichlet,
>
> First allow me to express the great pleasure I felt when I received the diploma from the Berlin Academy. I am proud of being twice your colleague [Dirichlet had been elected corréspondant of the Paris Académie after Libri had become an ordinary member in 1833]. Please present my sincere and warm greetings to the Academy. I must add that your letter [unfortunately lost] pleased me at least as much as the diploma; there are no geometers whom I have respected and appreciated as much as you even before I met you personally, and now that I know you I have certainly not changed my opinion. You can imagine the eagerness with which I accept your idea of a mathematical correspondence. To be sure, my eagerness may look a bit like egoism because it is I who have everything to gain by this correspondence.
>
> [Tannery 1910, pp. 1–2]

The correspondence had a rather dramatic opening. In his first letter, Dirichlet had included the following interesting algebraic result and its proof:

> If the equation
>
> $$s^n + as^{n-1} + \cdots + gs + h = 0, \qquad (1)$$
>
> with integer coefficients has no rational divisors and if at least one of its roots, $\alpha, \beta, \ldots, \omega$, is real, then I say that the indeterminate equation
>
> $$F(x, y, \ldots, z) = \varphi(\alpha)\varphi(\beta) \ldots \varphi(\omega) = 1 \qquad (2)$$

where I have introduced the abbreviation

$$\varphi(\alpha) = x + \alpha y + \cdots + \alpha^{n-1} z,$$

has always an infinity of integer solutions.

[Dirichlet 1840, p. 286]

This and the rest of the mathematical contents of the letter were received with enthusiasm by the Parisian mathematicians; in Liouville's answer, we read:

> I communicated the mathematical details in your letter to various persons, and they were all delighted and told me that I had to present an excerpt to be printed in the *compte rendu*. I resisted and answered that I did not have your authorization. However, they would not listen to my reasons and told me that your letter was so well written that you would not have to change a word if you wanted to publish it, etc., etc. In short, they insisted so much that I had to give in and so an excerpt of your letter was included in the *compte rendu* of the meeting of February 17, and later in the February issue of my journal, which appeared only yesterday.
>
> [Tannery 1910, pp. 2–3]

This publication came for Libri as a welcome occasion to take revenge against both of his critics. At the following meeting, he submitted a note criticizing Dirichlet. Apropos to the theorem cited above, Libri wrote:

> The conditions expressed in the statement of the theorem concerning the reality of at least one of the roots, α, β, ..., ω, and the impossibility of decomposing the equation (1) in rational factors (conditions which are repeated at the end of Dirichlet's letter) may make one think that the learned geometer has believed that his theorem is not satisfied except when these conditions are fulfilled. However, there exist cases in which they are not necessary. That is evident for example when $h = \pm 1$, no matter which form the roots of the equation (1) have, and in spite of the rational factors which this equation may have. In fact if you put

$$x = 1, \quad y = g, \ldots, z = a,$$

you will always have

$$\varphi(\alpha)\varphi(\beta) \ldots \varphi(\omega) = (-\alpha^n)(-\beta^n) \ldots (-\omega^n) = \pm 1,$$

and subsequently

$$[\varphi(\alpha)\varphi(\beta) \ldots \varphi(\omega)]^2 = 1.$$

It is well known that this equation can be written in the form

$$F(x_1, y_1, \ldots, z_1) = \varphi(\alpha)\varphi(\beta) \ldots \varphi(\omega) = 1,$$

when you give the new unknown quantities

$$x_1, y_1, \ldots, z_1,$$

suitable values, and it is easy to deduce from that an infinity of solutions for the equation (2), by way of the equation

$$[F(x_1, y_1 \ldots z_1)]^p = [\varphi(\alpha)\varphi(\beta), \ldots, \varphi(\omega)]^p = 1,$$

where p is an arbitrary positive integer. Therefore, when the quantity h in the equation (1) is equal to one the theorem stated by Mr. Dirichlet is verified even if the conditions he has given are not fulfilled.

[Libri 1840a p. 311–312]

Thus, Libri tried to discredit Dirichlet's theorem, which obviously gives sufficient conditions, by showing that the conditions are not necessary! Moreover, Libri stated:

> that I think it is possible to deduce from a well-known theorem by Euler from the year 1770 the property of quadratic forms which contain an infinity of prime numbers [cf. Dirichlet 1840, p. 285]. I shall conclude this Note by replying to an ancient remark by Mr. Dirichlet [1837] on a passage in one of my Memoirs [Libri 1832], a remark which has often been reproduced in France. Mr. Dirichlet said à propos a certain transformation which I have employed "this passage from the sum to the product is itself the only problem." I must admit that I have never completely understood Mr. Dirichlet's observation.

After a shallow defense of his earlier proof, the "peace-loving" Libri had not yet quite finished:

> You may wonder why I have waited so long before making this small correction. However, in general, discussions of this kind seem to me to be of too little importance to deserve to be treated in a separate piece of writing, and I find it more useful to attach it to some scientific point. That is why I ask for permission to profit from the actual situation to remark that I solved before Abel (cf. [Libri 1833]) the equations related to the division of the Lemniscate. I would have been very flattered if Mr. Jacobi ... had done me the honor of citing me next to the illustrious geometer of Christiania.

[Libri 1840a, p. 314]

Libri communicated this note to the *Compte Rendu* without reading it in the *Académie*. Therefore, Liouville had to wait until the next meeting on March 2, 1840, to defend his correspondent. He came straight to the point:

> Why does Mr. Libri say that one might think that Mr. Dirichlet regarded the conditions in question as necessary? Absolutely nothing justifies such a hypothesis.

[Liouville 1840j, p. 343]

About his reaction to the Euler reference, Liouville wrote to Dirichlet:

> He did not even say which theorem by Euler he was thinking of. I
> have laughed a lot at this sentence, but I was careful to declare that
> I had nothing to say to such a prudent and vague assertion.
>
> [Tannery 1910, p. 5]

On the other hand, he went into great detail explaining Dirichlet's remark, about the sign in Gauss's formula, which Libri had declared he had
never understood. "I think," he wrote to Dirichlet, "that in my answer I
have made it understandable, if not to him, at least to the public" [Tan-
nery 1910, p. 5]. Finally, Liouville pointed out that since Jacobi had not
cited Libri's work on the division of the lemniscate "one may perhaps believe that the illustrious geometer of Koenigsberg did not find his claim
of priority well founded." [Liouville 1840j]. Now, Libri was clearly on the
defensive. He answered immediately [Libri 1840b], but except for repeating
what he had already said, he only answered one of Liouville's questions. He
"showed" that Dirichlet had thought that his condition was necessary by
pointing out that the condition was repeated at the end of the letter: "You
see this condition repeated twice; why should Mr. Dirichlet have repeated
it in this way if he did not believe it was necessary?" After this nonsense,
Liouville gave up the discussion declaring that he maintained the exactness
of his observations.

When Liouville sat down to write to Dirichlet a few days later (March
5, 1840), he discovered that he could make his reply to Libri even more
spicy. Recall that Libri's disproof of the converse of Dirichlet's theorem was
based on a general theorem stating that if only $h = \pm 1$ in (1), then (2) has
always infinitely many integer solutions independent of whether Dirichlet's
assumptions are fulfilled or not. Now, Liouville could easily disprove this
claim; for example, if (1) has the simple form $s^2 + 1 = 0$, Libri's general
"theorem" would imply that $(x + iy)(x - iy) = x^2 + y^2 = 1$ has an infinity
of integer solutions, which is obviously absurd. In his comment to Dirichlet,
on this point, Liouville did not mince his words:

> What madness of Mr. Libri to present objections to a very concise
> letter, objections which would have been ridiculous even if it had
> been a question of a detailed Memoir. Above all he should not have
> stated a false theorem. Wouldn't you say that by wanting to put
> dots over the i's of your letter the silly critic has deleted them? $\cdots$
> Nevertheless you see what a stupid role one plays when one wants
> to rule by quackery.
>
> [Tannery 1910, pp. 4–5]

At the following meeting in the *Académie*, Liouville triumphantly pointed
out Libri's mistake and analyzed why Libri's proof of his "theorem" was
wrong. Libri immediately answered that he never meant to say that "his
theorem" was always valid, but only that there were examples for which
it held true. Liouville's victory must have been obvious even to the nonmathematical members of the *Académie*. A few months later (May 26,

1840), Liouville summed up the declining popularity of his opponent by writing to Dirichlet that Libri was "a man who, at least in the *Académie*, is beginning to be despised almost as much as he deserves." However, the scandalous events three years later at the *Collège de France* were to show that Libri had not lost his support in all circles.

Election to the Bureau des Longitudes

Before these events, Liouville enjoyed the triumph of being elected a member of the *Bureau des Longitudes*. Like the *École Polytechnique*, the *Bureau des Longitudes* was a product of the great revolution. It had been founded on 8. messidor an III (June 25, 1795) as the supreme astronomical society in charge of the observatory and the *École Militaire* [see Bigourdan 1928–1933]. The members met once a week—during Liouville's membership on Wednesdays from 3 or 3:30 to around 5 p.m.—to discuss administrative and scientific subjects related to astronomy, for example, which instruments to procure and which observations to carry out, and to inform each other about the latest advances in astronomy and geodesy. All the activity of the *Bureau des Longitudes* was in a sense directed toward the improvement of the astronomical tables of whose publication the *Bureau* was in charge. This principal *raison d'être* for the *Bureau des Longitudes* was not for its theoretical value, but for its enormous practical value, in particular for navigation. The tables were published three years in advance in the *Connaissance des Temps*, a publication that also contained a few scientific papers. In addition, the *Bureau* occasionally published astronomical papers in its *Annuaire*.

Membership in the *Bureau des Longitudes* carried a prestige similar to membership in the *Académie des Sciences*, and it had the additional advantage of being a well paid position, worth almost as much as a professorship in one of the advanced schools. There were ten full members (*membres titulaires*), namely, four astronomers, two geometers, two navigators, one geographer, and one artist. In addition, there were some eight co-opted members. From the start, the greatest scientists had been members, such as Laplace, Lagrange, J. de Lalande, and Delambre. When Liouville joined the *Bureau*, the leading member was Arago. He had been appointed adjoint member in 1807, and in 1822 he became *membre titulaire*. His great influence in the *Bureau* was accentuated when he was appointed director of the Paris Observatory in Paris in 1830. During the 1830s and 1840s, he was usually elected *Secrétaire* of the *Bureau des Longitudes* (an annual job), which placed in his hands the task of writing the *Procès Verbaux* of the meetings. Compared with his job as *Secrétaire perpétuel* of the *Académie des Sciences*, this job was less powerful, because the *Procès Verbaux* was not published. The originals are still preserved at the *Bureau des Longitudes* except for the volume covering the period 1853–1876 (a copy covers the period 1868–1876). They constitute the main source material concerning the work of this learned institution.

They reveal that during the late 1830s, the most active member after Arago was Biot (elected 1825), with whom Arago had strained relations. Together with Lefrançois-Lalande and A. Bouvard, they constituted the astronomy section of the *Bureau*. The mathematics section consisted of Poisson and Prony. On July 29, 1839, Prony died, leaving an opening that was clearly desirable for the new member of the astronomy section of the *Académie*, Liouville. However, the main actor in the following farce was Cauchy.

True to his conservative, catholic ideas, Cauchy had refused to take the oath of loyalty, toward the Citizen King, Louis-Philippe, after the 1830 revolution. Since this oath was required from the professors of the higher schools, Cauchy lost his job at the *École Polytechnique* and went into voluntary exile first in Italy and later in Prague, where he served as a tutor for the duc de Bordeaux, the grandson of Charles X and the legitimate heir to the French throne. After eight years abroad, Cauchy decided to return to Paris. Here, only his seat in the *Académie* had remained open to him, so he considered the vacant position at the *Bureau des Longitudes* as a welcome opportunity to resume his position in French scientific life. According to Cauchy's biographer, Valson, Cauchy's candidacy was strongly supported by Arago:

> One even recalls that Mr. Arago received the other candidates rather poorly. To all they did he invariably answered, "Mr. Cauchy is on the list of candidates," and if you insisted, he contended himself with repeating with a humorous gesture "but, I just told you, Mr. Cauchy is on the list."
>
> [Valson 1868, pp. 97–98]

With Arago's support of Liouville in the *Académie* in mind, it seems most unlikely that he now thought poorly of him; he may simply have valued Cauchy more highly, and with a good deal of justice. This was of the utmost importance since the members of the *Bureau des Longitudes* had the right to elect the new members themselves. At the meeting of October 29, 1839, a committee was formed with the purpose of presenting the candidates. The following Wednesday, the majority consisting of Arago and Biot suggested a list containing the names Cauchy, Liouville, and Sturm, whereas the third member of the committee, Poisson, wanted to honor Lacroix, then more than 70 years old, with membership. The majority won, so in the *Procès Verbaux* of the following meeting (November 13) one reads:

> Mr. Cauchy gets the majority of the votes and is nominated member of the *Bureau*. The nomination is submitted to be approved by the King.[1]

The last statement, however, contained the origin of great problems, for the king insisted that Cauchy should take the oath of loyalty, and Cauchy still refused. Despite the efforts of the ministers and the members of the

Bureau des Longitudes, neither the king nor Cauchy could be made to dissent from their attitudes, so, although elected a member, Cauchy was not allowed to take his seat in the *Bureau*.

Thus, when Poisson died a few months later (April 25, 1840), the *Bureau* had no active members of the geometry section, and so the nominating committee consisted of three members of the astronomy section, namely Biot, Arago, and Arago's brother-in-law and Liouville's former superior at the *École Polytechnique*, Claude-Louis Mathieu, who had succeeded Lefrançois-Lalande. The committee was elected on June 24, but it lasted until November 11, until it presented its report. Its unusual discussions of principle reflect the difficulties they encountered:

> November 11, 1840 — The committee in charge of presenting a list of candidates for the vacant seat in the Geometry Section of the *Bureau* reads its report. It is concerned with the nature of the work which the *Bureau* can expect from its members of the Geometry Section. The analytical derivations related to the calculation of the planetary inequalities are indispensable. Adopting this point of view, shouldn't the *Bureau* nominate a geometer who can actually look after these often difficult calculations? Or should we on the contrary regard the seats of the *Bureau* as retreat after earlier well done work? The latter opinion is held by the public. But, when the number of observers increases and produce more exact and numerous foundations for the theory, it is doubtless desirable that the theories themselves can go in front and maintain the reputation of the *Bureau* in all of Europe.
> [P. V. Bur. Long. in Bigourdan 1930, pp. A11–12]

Since being second on the list one year earlier, Liouville had not advanced his own astronomical research much further. Still, he had secured his reputation as an astronomically interested mathematician in the *Académie*, where he had reported on two astronomical works, the one being a paper on refraction by Ritter, the other being Le Verrier's paper on the stability of the solar system, which he praised highly. Moreover, Liouville had been engaged in a discussion of Pontecoulant's *Théorie analytique du Système du Monde*. Finally, when Cauchy, on August 3, 1840, raised doubts about the rigor of certain procedures in Laplace and Lagrange's theory of perturbation of the planets, Liouville immediately (August 17) defended the old masters. He claimed that the research he had undertaken in his active astronomical period some four years earlier had convinced him that:

> Even though the methods of Lagrange and Laplace are not rigorous, they can be made rigorous by means of very small modifications.
> [Liouville 1840n, p. 253]

Where Liouville had earlier criticized a passage in Laplace's *Mécanique Céleste* [Liouville 1837g], he now defended him. He announced that the details would appear in a memoir, but the only thing one hears of it later, is an *Addition* which he presented to the *Académie* on November 2, 1840. The *Comptes Rendus* did not even contain a summary of this addition [CR 10, p. 678].

Liouville had clearly shown himself capable of supervising the calculation of the tables that the *Bureau des Longitudes* was in charge of, and the commission presented him as its first candidate under the assumption that this should be regarded the prime duty of the new mathematics member. They continued their report as follows:

> But, after having exhibited these considerations with impartiality, the Committee judged it best to make a double presentation and leave it to the *Bureau* to appreciate the motives that apply to each of them. Thus on the one hand, it presents
>
> Mr. Savary and Mr. Liouville in the first rank; Mr. Leverrier in the second rank.
>
> It adds that Mr. Savary has declared that he will this time abstain from any competition and withdraw his candidacy while acknowledging his deep gratitude for the honorable distinction one wanted to grant him.
>
> On the other hand, the Committee presents Mr. Lacroix and Mr. Poinsot.
>
> [Bigourdan 1930, p. A 12]

The report immediately shows that the committee gravitated towards the young and active mathematician as opposed to an honorary member. In the subsequent discussion, Biot made clear that this was his point of view: "Mr. Biot will not allow the seats of the *Bureau* to be regarded as retreats" [Bigourdan 1930, p. A 12].

The astronomer Bouvard supported this opinion by declaring that "it is necessary to nominate a geometer who is capable of leading, if necessary, the students in their theoretical calculations." Also Arago "returns to the necessity of finding a geometer with special interests in astronomical calculations." However, Arago muddled this strong support of Liouville's candidacy by explaining that he "regards it as his duty to make known to the *Bureau* that... Mr. Poinsot has told him that he had a large number of completely prepared works on the *système du monde.*"

At the following meeting, on November 18, 1840, Arago further informed the *Bureau* that for the above reasons Poinsot thought that he would qualify for a place on both of the lists of candidates. Since we can no longer hear Arago's tone of voice, it is hard to decide whether he gave this information purely as a duty, as he says, or whether he actively supported Poinsot. If the latter is the case, he did not succeed, for at the subsequent election Liouville obtained six votes and Poinsot only one. Two weeks later, Liouville's nomination was confirmed. The *Procès Verbaux* explains:

> The president deposits the Royal order according to which the nomination of Mr. Liouville is confirmed. Mr. Liouville can take his seat after the formality of the oath required by the law (these terms are explicitly expressed in the letter sent by the Minister of Public Education).
>
> [P.V. Bur. Long. 9, Dec. 1840][2]

As this quote shows, the Cauchy case was far from dead. The government was careful to make explicit that the oath was required by law, and the secretary of the *Bureau* was careful to state that this was the opinion of the minister and not of the *Bureau*. Seemingly, he thereby wanted to support Cauchy, who claimed that the oath was not statutory. As Cauchy, Liouville was opposed to the reign of Louis-Philippe, although for very different reasons, and only a few months earlier, he had taken direct action against the government. Still, he choose not to let politics interfere with his scientific career and took the oath. Thereby he followed all the other scientists opposed to the government except for Cauchy.

To honor the deceased Poisson, in whose chair he was elected, Liouville wrote a memoir and devoted a chapter to Poisson's *Mécanique*. However, it was not published until 1856 [Liouville 1856p].

Thus, in his thirty-first year Liouville reached the summit of his career, being a member of the *Académie des Sciences* and *the Bureau des Longitudes* and a professor at the *École Polytechnique*.

III

Professor, Academician, and Editor (1840–1848)

Setting the Stage

Having reached the highest positions in French academia in 1840, Liouville had the opportunity to bury himself in his research without the urgent need to publish feverishly. Such may have been his own dream about his life as an academician and a member of the *Bureau des Longitudes*, but as it often happens, he soon lost his illusions. Teaching and administrative work for the *Académie*, the *Bureau*, and his *Journal* took up an increasing amount of time. In 1840, he poured out his troubles to Dirichlet:

> Overwhelmed by annoying occupations I have no time for steady research.
>
> [Tannery 1910, p. 12]

Then, further during the next summer:

> All these annoying occupations of the year these eternal and tedious courses, the lectures of which hardly one out of six offers any interest, have this year almost put me off mathematics.
>
> [Neuenschwander 1984a, II, 1]

Nevertheless, Liouville enjoyed instructing the students, and in particular guiding and promoting the young mathematical talents; most of the following generation in France were his students. Through this instruction and through his work in the *Académie* and the *Bureau* as well as his journal editing, he was virtually bombarded with new ideas from all branches of mathematics. An open-minded scholar, Liouville often found shortcuts, clarifications, and generalizations of the theories he was presented with, and thus his production consisted to an increasing degree of additions to the theories of his contemporaries. Thus, where his earlier research had often resulted in developed theories presented in long articles, it now primarily became separate results published in shorter papers or notes. He continued to develop large-scale theories during the 1840s, but he did not always find enough connected time to publish them. Among these and his shorter works of this period, one finds some of his most ingenious and celebrated contributions to mathematics. Liouville was, during the 1840s, at the peak of his scientific creativity.

As a whole, Liouville's professional life during this period looks very flickering. Before 1840, Liouville had pursued some clear paths to secure his own career; during the following 20 years, the promotion and development of other mathematicians' ideas became the central issue. In describing this period, I shall first analyze Liouville's activities at the various teaching institutions and learned societies and then turn to his collaboration with a selection of his students and colleagues. Presenting Liouville's work in this way will naturally emphasize some connections at the expense of others, which will merely be indicated. Still, I hope that the reader will get a picture of a densely woven net of ties connecting all of Liouville's activities during the 1840s.

Although teaching often inspired Liouville, the summer holidays provided the most intensive period for connected research, and he often structured his ideas and wrote his papers during these periods. His letter of 1841 to Dirichlet (quoted above) continues:

> However, with the arrival of the vacation the love for science has returned, and I have set myself to work. I have begun five or six memoirs simultaneously.
>
> [Neuenschwander 1984a, II, 1]

Liouville spent the summer holidays in Toul, where he had grown up. However, even so far from Paris, his work could be interrupted, as he had explained to Le Verrier in a letter of September 18, 1833:

> Everybody has been ill in our poor house; my father and one of my aunts have had the flu, my wife has had an attack of smallpox and still has the red spots, although she does not seem to have to fear to remain marked. This may make you understand the delay of my letter.
>
> [Neuenschwander 1984a, III, 1]

At this moment, the young couple apparently lived in the house of Liouville's father; later, they moved to 8 rue Salvateur, close to Toul's gigantic Gothic cathedral. But, even there, many things prevented him from studying. Among the unpleasant ones was Liouville's failing health. As we have seen, he was already of a weak constitution as an *élève ingénieur*, and it seems to have grown worse. In 1840, he wrote to Pelouze:

> For our part we are well in the sense that we are not ill; but my wife has suffered a great deal these days; as for myself my bile and my blood work each on its own. Again this morning I vomited a lot; it is odd, that this vomiting has no influence on my stomach.
>
> [Neuenschwander 1984a, I, 1]

These indispositions were so frequent that Liouville emphasized to Dirichlet (August 4, 1841) when he was well:

> Goodbye my dear friend. Our excellent Pelouze will deliver this letter to you and will give you all the details you may want regarding my health, my position (today quite good) and our friends.
>
> [Neuenschwander 1984a, II, 1]

Moreover, Liouville had also to take care of journal editing when in Toul, but he still enjoyed the relatively relaxed months there. Fortunately, the summer vacations were long, lasting from the end of July to around November 10th. During most of the remaining year, the Liouvilles lived in an apartment in the Latin Quarter in Paris. In 1833, they lived in number 1 of the former rue de l'Est [Bull Soc Philo 1833] (now the southmost part of Boulevard Saint Michel) that is on the corner of the former rue d'Enfer to which he alluded in the letter to Colladon. From this place close to the observatory, the Liouvilles later moved to the more central place, 3 rue Sorbonne [cf. Connais des Temps pour 1849, publ. 1846]. Here they lived until 1853 [letter to Dirichlet, Neuenschwander 1984a, II, 3], a few blocks from the *Collège de France* and only a few minutes walk from the *École Polytechnique*.

The École Polytechnique

ADMINISTRATIVE DUTIES

As a military institution, the *École Polytechnique* was placed under the ministry of war. The daily management of the school was put in the hands of two military persons, the so-called *commandants*. The *Directeur des Études*, on the other hand, was a scientist. He shared the planning of the teaching with the professors who were all members of the *Conseil d'Instruction*. The more general questions, such as the evaluation of the examinations of the year and of the achievements of the professors, were discussed in another council with the ambitious name *Conseille de Perfectionnement*. Each year the *Conseil d'Instruction* elected three professors to be members of the *Conseil de Perfectionnement*, which in addition consisted of the two *commandants* and the *directeur*, five examiners at the final examination, one examiner at the entrance examination, three members of the *Institut de France*, and ten representatives of the military and civil public services, who were to employ the graduated students [Fourcy 1828, Marielle 1855].

Liouville took an active part in the administration of the school, in particular during his first years as a professor. He never became a leading figure in the *Conseil d'Instruction*, but his colleagues demonstrated their faith in his administrative skills by electing him member of the *Conseil de Perfectionnement* during the years 1840–1841, 1841–1842, 1842–1843, 1843–1844, 1845–1846, 1847–1848, and 1849–1850 [Éc. Pol. Cons. Instr.]. Moreover, Liouville was a member of a great number of other committees. In 1840, 1841, and 1842, he was a member of committees that nominated candidates for vacant professorships. The first of these committees suggested Sturm as a candidate for the other professorship of analysis, and the second committee recommended another of Liouville's good friends, the geometer Chasles, for the chair of geodesy. Both choices were confirmed first by the *Conseil d'Instruction* and then by the minister. Thus, after two years of subordination as Liouville's *répétiteur*, Sturm finally obtained a suitable post. For the next ten years, the two friends taught the analysis and mechanics classes alternately, Liouville being in charge of the classes beginning in the odd-numbered years.

In a number of committees, Liouville worked to maintain the high scientific level of the school and its strictly scientific elitist and competitive structure. For example, in 1840, he took part in designing a new system to classify the students, and the following year he was a member of a committee that turned down "the proposal to introduce in the admission jury, and with the right to vote, the examiner in charge of classifying the literary essays of the candidates" [Éc. Pol. Cons. Instr., April 23, 1841]. The committee felt that it would be enough to listen to the advice of the examiner. (Remember that literary composition had not been Liouville's favorite when he studied at the school.) In 1842, the minister of education, who was in charge of the civil education but not of the *École Polytechnique*, tried more drastic means to promote a broader knowledge of the humanities among the students. He suggested that a *Diplome de Bachelier-ès-lettres* be required from the students who were admitted to the school. A similar proposal had already been turned down twice before, and it suffered the same fate in 1842 in a committee consisting of Liouville, Dubois, and Regnauld.

During parts of his first years as a professor, Liouville was a member of the *Conseil d'Administration*, but in October of 1840, he was replaced by his good friend, the chemist Théophile-Jules Pelouze, in order that Liouville could devote himself to the *Commission du Journal et de la Biblioteque*, which he had entered in August of that year. With the experience Liouville had from his own journal, it was natural that the *Conseil d'Instruction* entrusted the *Journal de l'École Polytechnique* to him. The *Comptes Rendus* of the *Conseil d'Instruction* does not often mention the library and the journal, but apparently Liouville was in charge of them from 1840 to 1847 or 1848. In 1849, he was not reelected to this committee.

The only permanent committee at the school that Liouville never joined was the *Conseil de Discipline*, so he certainly gave his share in the administration of the school.

Liouville's "Cours d'Analyse et de Mécanique"

Liouville gave his lectures on analysis and mechanics on Thursdays and Saturdays from 8:30 to 10. The curriculum that Liouville reportedly "followed exactly" [*Commission mixte*, report 1851] had changed little since he studied at the school. In the mid-1840s, the programme described that the course opened with differential calculus of functions of one or more variables and its application to geometry, going as far as expressing the line element ds of a surface in space. Then, followed integral calculus and its geometric applications. Much of the elementary statics that Liouville had learned in his first year 1825–1826, was later assumed known upon entrance, so in the 1840s, this part of the course was rudimentary. Instead, Liouville taught much of the dynamics previously taught in the second year, such as the energy principle, or the *principe des forces vives* as it was called, the conservation of angular momentum, or the *principe des aires*, and their applications to planetary motion.

The second year began with a study of ordinary and partial differential equations, including reduction of the order when an integral is known and integration of the wave equation. It was followed by an introduction to calculus of variations, calculus of finite differences and the study of the lines of curvature of a surface. The mathematics course ended with "elements of calculus of probability and social arithmetic," which had earlier been taught in another course. The second year mechanics course was primarily concerned with a study of rigid bodies, based, in the tradition of d'Alembert and Lagrange, on the static principle of virtual velocities. The course ended with a short treatment of hydrostatics and the derivation of the equations of hydrodynamics.

Did Liouville really follow this programme exactly, as the *Commission Mixte* reported, or, like Cauchy, did he introduce his own new ideas? In order to answer this question, we must consult lecture notes taken by his students, for Liouville never published his *Cours d'Analyse*. The *École Polytechnique* still preserves three lithographed copies of the analysis notes from his first-year course for the years 1843–1844, 1845–1846, and 1847–1848, one undated copy of the second-year lecture notes of analysis and mechanics, and the notes on *Calcul Intégral* from the two years 1847–1849. [Ec. Pol. lecture notes, Liouville]. Moreover, Liouville's own plans of the courses from 1838–1840 are preserved in his notebooks [Ms 3616 (4)].

These notes reveal that although Liouville followed the programme much closer than Cauchy had done, he still made some interesting digressions, such as a treatment of the gamma and beta functions, a historical remark about Roberval's kinetic method for finding tangents, and an extended treatment of surfaces in space, for example, developable surfaces. The next sections will display other examples of such digressions. It is characteristic that they are all of a theoretical mathematical nature rather of a kind that would teach the engineering students to apply the theory to practical problems. We have seen that a much more moderate version of this theoretical style of teaching cost Liouville his job at the *École Centrale*. At the *École Polytechnique*, it was tolerated; indeed, it was the rule rather than the exception. In the following section, I shall analyze the publications by Liouville that were influenced by his teaching at the *École Polytechnique*, but first his attitude toward rigor in analysis will be discussed.

RIGOR

Lagrange and Cauchy, in particular, had made high standards of rigor the hall mark of the *Cours d'analyse*. What was Liouville's view on this issue? In his early writings on fractional calculus, he expressed himself as a protagonist of Cauchian rigor with remarks such as "divergent series which most often lead to erroneous results, ought to be completely banished from analysis." Still, as shown in Chapter VIII, §41–49, he often failed to live up to this ideal. He used infinitesimals in a highly un-Cauchian way, confused convergence and absolute convergence, and only rarely checked the domain of convergence or indeed the exact assumptions under which his theorems were valid, dismissing this problem with the remark:

> we have neglected certain details and even here and there passed silently over the exceptions to which our rules are subject. It is sufficient that the reader is warned once and for all.
>
> [Liouville 1832b, p. 88]

In his later papers, Liouville showed a deeper understanding of the question of rigor. Two cases deserve to be mentioned: One is the convergence proof for the "Fourier series" of general Sturm-Liouville problems and, in particular, his apprehension of the interplay between assumptions, proofs, and results, manifested in this connection (Chapter X). The other is an understanding of the problem of interchanging two limit processes. This problem had been overlooked by Cauchy, for example, in his *Cours d'Analyse*, Chapter 6, §1, to which Liouville referred as follows in his lecture notes from a course at the *Collège de France* (1858–1860).

> Inexact theorem of M. Cauchy (pointed out by Abel) that if $\sum u_n$ is a convergent series and u_n continuous function of x, $\sum u_n$ is also a continuous function.
>
> [Ms 3625 (1), lectures 5 and 6]

Abel [1826d] had pointed out that discontinuous Fourier series give counter examples to Cauchy's theorem. Later, an analysis of the weak point in Cauchy's reasoning led Seidel and Stokes to the concept of infinitely slow convergence and brought Cauchy close to a formulation of uniform convergence, which in its classical form is due to Weierstrass [cf. Grattan-Guinness 1970]. Abel merely formulated and proved a special case where Cauchy's theorem is correct. Abel's theorem states that if $\sum_{i=0}^{\infty} a_i$ converges toward A, then $\sum_{i=1}^{\infty} a_i \rho^i$ is convergent for $0 < \rho \leq 1$ and tends to A for ρ tending to 1. Liouville often thought about this theorem (there are several proofs of it in his notebooks), and in 1863 he recalled:

> One day while talking with my excellent and much lamented friend Lejeune Dirichlet, I told him that I found it rather difficult to explain (and even to understand) the proof which Abel has given of this important theorem. Dirichlet offhand and before my very eyes began to write, with the only purpose of coming to my help, the following note which has been a great help to me and which the readers will

now be grateful to see published. Its method of proof can be carried over to a great number of applications and has often been useful to me in my lectures at the *Collège de France*.

With these words, Liouville published Dirichlet's note in his journal [Dirichlet 1862].

In his lectures at the *Collège de France*, Liouville further called attention to

> Other errors pointed out by Dirichlet. Especially if $v_n = u_n (1 + e_n)$ and if $\sum u_n$ converges, $\sum v_n$ also converges; counter example $u_n = \frac{(-1)^{n+1}}{\sqrt{n}}$, $e_n = \frac{(-1)^{n+1}}{\sqrt{n}}$ - various forms of the series treated by Dirichlet.
>
> [Ms 3625 (1)]

These problems of convergence and absolute convergence had been discussed by Dirichlet in 1829 and 1837 [1829,1837b]. In his lectures at the *École Polytechnique*, Liouville limited his considerations to a completely rigorous proof of a less interesting version of Abel's theorem:

Abel's theorem

> Let $u_1, u_2, \ldots, u_n$ be the positive and negative terms of a convergent series; let moreover $\alpha_1, \alpha_2, \alpha_3, \ldots, \alpha_{n-1}, \alpha_n$ be numbers or rather coefficients all of the same sign (say positive) and such that each of them is larger than its successor, at least from a certain index which we shall take as the origin; I say that the series
>
> $$\alpha_1 u_1, \ \alpha_2 u_2, \ \alpha_3 u_3, \ldots, \alpha_n u_n$$
>
> is convergent.
>
> [Éc. Pol. lecture notes, Liouville 1845–1846]

In general, Liouville's lectures at the *École Polytechnique* show a noticeable influence from Cauchy's famous *Cours d'Analyse*. Thus, he imitated Cauchy's treatment of the convergence of infinite series, and he solved the simple functional equations which Cauchy had made the basis of the study of transcendental functions, although neither of these subjects were mentioned in the programme at all. Introducing the basic concepts of analysis, such as continuity, the derivative, etc., Liouville also used Cauchy's approach. As we saw above, in rare cases, he even went further than Cauchy, pointing out foundational problems. Another example is his mention of Dirichlet's function:

> Moreover among the discontinuous functions we shall never deal with those that are constantly discontinuous, such as the function of x which is zero when x is rational and equal to one when x is irrational.
>
> [Ec. Pol. lecture notes, Liouville 1847–1848, p. 2]

It is characteristic that while pointing to this pathological function Liouville in the same breath excluded such functions, for, as a whole, Liouville did not insist so much on rigor as Cauchy had done. He passed more quickly over the fundamental notions in order to devote more time to the technically more complicated matters; for instance he did not discuss the general existence theorems for integrals (defining integration as the inverse operation of $\frac{d}{dx}$) and solutions of differential equations, but went in great detail with the various methods of integration. Thus, although he taught his own method of successive approximations (cf. Chapter X, §32) and proved its convergence, he did not present it as an existence proof but as a method of solution. This reflects a general feature of Liouville's attitude toward mathematics. While accepting the problems of rigor pointed out by others, such as Abel and, in particular, Dirichlet, he did not raise fundamentally new questions of rigor himself nor did he contribute essentially to the arsenal of concepts and methods to circumvent the problems. His genuine interests did not lie in the foundations of mathematics but in the advancement of its upper layers. He was a problem solver who did not let methodological questions get in his way.

We may conclude that while teaching according to the programme at the *École Polytechnique* Liouville emphasized its theoretical side and added some of Cauchy's points of view and some subjects of his own particular liking. A similar approach can be found in Sturm's *Cours d'Analyse* and his *Cours de Méchanique* edited and published by E. Prouhet after Sturm's death [Sturm 1857–1859, 1861]. In fact, their great similarity to Liouville's lecture notes suggests that the two friends and colleagues discussed the content of their lectures with each other. Their courses differed strikingly from that of Liouville's predecessor, Navier, who had emphasized applications and tuned down theory as can be seen from Navier's lecture notes, which were published posthumously in 1840 by Liouville and Catalan.

Notes in Navier's Résumé

The *Résumé des leçons d'analyse donné à l'École polytechnique* [Navier 1840] was a printed version of the lithographed notes that had been distributed among Navier's students. The only changes made in the printed version had been indicated by Navier himself. Liouville's role in the edition was therefore limited: "The proofs have been read partly by Mr. Liouville and partly by Mr. Catalan" [Navier 1840, avertissement]. However, at the end of the book, Liouville appended six notes, which I shall now analyze in order to illustrate how Liouville's teaching at the *École Polytechnique* influenced his publications.

The first note deals with Maclaurin series. As a *répétiteur*, Liouville had observed that in order to employ Lagrange's remainder,

$$R_n = \frac{h^{n+1}}{n!} f^{n+1}(x + \theta h),$$

where $\theta \in [0, 1]$, in Taylor's formula

$$f(x + h) = f(x) + \frac{h}{1} f'(x) + \cdots + \frac{h^n}{n!} f^n(x) + R_n,$$

one has to assume that f has a derivative of order $n+1$. He thought that "it is somewhat inconvenient to introduce this derivative in the calculations." Instead, he suggested the remainder

$$R_n = \frac{h^n}{n!}[f^n(x + \theta y) - f^n(x)],$$

which can be used as long as f^n is continuous. He published this observation in 1837 [1837k], and as a professor, he later taught it to his students at the École Polytechnique. However, in the note in Navier's *Résumé*, Liouville preferred a remainder due to Cauchy.

The second note also deals with a theorem "due to Mr. Cauchy I think," namely, "On the fractions which appear in the form $\frac{\infty}{\infty}$." Cauchy's proof, as repeated by Liouville, proceeded by reducing the problem to l'Hospital's well-known rule. However, Bertrand, in a paper published in Liouville's *Journal* [1841], pointed out that this reduction did not cover all cases, and he presented a new proof. The following year, Liouville, in turn, published yet another proof [1842b].

In the third note, Liouville expounded the most fundamental properties of the Γ and beta functions, and the fourth note contained a new version of Stirling's formula. Liouville's presentation was similar to Lacroix's in his classical textbook *Traité élémentaire du calcul différentiel et du calcul intégral* [1837], but he completed Lacroix by providing an upper limit for the error committed in this approximation of $\log(\Gamma(x))$. More specifically, he proved that

$$\log(x!) = \log(\sqrt{2\pi}) + \left(x + \tfrac{1}{2}\right) \log x - x + \frac{\mu}{12x},$$

where $\mu \in [0, 1]$. Liouville had already published this result in 1839 [1839d], and the note in Navier's book is a verbatim quote of this paper.

In the fifth note, Liouville proved the fundamental theorem of algebra. His first contribution to this theorem appeared in 1836 in a joint paper with Sturm in which they gave an elementary proof of a method due to Cauchy for finding the number of roots of a polynomial inside a given contour in the complex plane (cf. Chapter XIII, §3). Following Cauchy, Sturm and Liouville deduced the easy corollary that inside a large enough contour there will be as many roots as the degree of the polynomial indicates; that is the fundamental theorem of algebra.

Liouville's interest in this theorem was aroused again when he read Lefébure de Fourcy's elementary *Leçons de Géométrie analytique* in which a proof due to Mourey [*Vraie théorie des quantités négatives et des quantités prétendues imaginaires, 1828*] was cited. He transcribed this obscurely phrased proof into ordinary mathematical language [Liouville 1839 h] and pointed out an incompleteness in the proof, namely, that a certain curve was not shown to be connected. Instead, he gave a new proof inspired by Gauss's third proof [Gauss 1816], which he had just read a year earlier [Ms 3616 (3), pp. 4v–5r]. The central idea in the proof is that one can interchange the order of integration in a certain double integral. The note in Navier's book was a verbatim quote of this alternative proof. (See also [Liouville 1840r,s].)

One month after his original criticism of Mourey, Liouville realized that it was possible to "supply easily the details which Mr. Mourey has left out, no doubt in order to abbreviate" [1840s] (typical French shamming). Having completed "the beautiful proof of Mr. Mourey" Liouville used the occasion to remind his readers of Gauss's first proof of the fundamental theorem in his doctoral dissertation [1799].

Liouville's contribution to the fundamental theorem of algebra was highly esteemed in France. Thus, in a review of Serret's *Cours d'Algèbre* [1849] in the *Nouvelle Annales de Mathématique*, the reviewer, possibly the editor Terquem, remarked after having rebuked Serret for not having proved this theorem:

> Euler and d'Alembert have understood that it needed a proof and have presented some attempts. Mr. Gauss has succeeded; later, Mr. Cauchy has developed Mr. Gauss's procedure, and Mr. Liouville and Mr. Sturm have clarified this development.
> [Nouvelle Ann. Math. 13 (1854), pp. 350–351]

Liouville returned to the fundamental theorem of algebra in 1844, where he derived it from his famous principle on doubly periodic functions (cf. Chapter XIII, §18). However, this deduction, which was by far his most interesting contribution to the fundamental theorem of algebra, remained unpublished.

The last note in Liouville's Navier edition was devoted to "the integration of a class of differential equations." This note had also been published earlier, namely, in 1838 [1838b]. Then Liouville had revealed that he had been inspired by Binet's and Jacobi's work on rational mechanics and the calculus of variations. During the years following the publication of Navier's *Résumé*, Liouville "in the interest of the students" added to the usual "treatises of integral calculus" two other classes of differential equations in which the order can easily be reduced [1841e, 1842a]. He taught the first of these methods to his students [Liouville, Éc. Pol. lecture notes, 1848–1849, p. 70].

The *avertissement* in Navier's *Résumé* reveals that Liouville had intended a more extensive supplement:

The notes at the end of the volume are written by Mr. Liouville. The author had hoped to add more notes and to make them more extensive, but since his other occupations do not permit him to devote himself to this work at present, he will return to it later and collect it in a supplement to Navier's work.

[Navier 1840, *avertissement*]

This projected supplement never appeared. But, Liouville continued during the early 1840s to publish notes that were motivated by his teaching at the *École Polytechnique* and his reading of elementary texts.

OTHER RELATED WORKS; TRANSCENDENTAL NUMBERS

The most interesting papers that the teaching at the *École Polytechnique* inspired Liouville to write were a series concerning the number e. Although it was not stated in the programme there was a tradition of proving that $(1 + \frac{1}{m})^m$ converges to e when m tends to infinity. Liouville followed that tradition, but by February of 1838 he discovered that the usual proof found in Cauchy's *Cours d'Analyse* [1821, Chapter VI, p. 167] was not rigorous because - as we would now say - the order of two limiting processes had been incorrectly interchanged. Liouville overcame this problem in a paper published in 1840 [1840d] and included the result in his lectures. This confirms what I said in connection with Abel's counterexample, namely, that Liouville realized the problem of interchangeability more clearly than most of his contemporaries; still, it is noteable that this note was also inspired by Dirichlet, to whom Liouville attributed the essence of the proof.

Even more important were two papers published earlier in 1840 that dealt with the transcendence of e (cf. Chapter XII). In the first of these papers, Liouville summarized his result as follows:

One proves in the elements [this Euclidean term was generally used to refer to the curriculum of the introductory university and polytechnical courses] that the number e, the base of the Neperian logarithm, does not have a rational value. It seems to me that one ought to add that the same method also shows that e cannot be a root of a quadratic equation with rational coefficients.

[Liouville 1840b]

In addition to this result, he showed that e^2 cannot be a root of a rational quadratic equation either [Liouville 1840c]. There is little doubt that Liouville's aim was to prove that e is transcendental, i.e., it is not a root of any polynomial with rational coefficients. The problem of transcendental numbers has a long and interesting history. Implicitly, the problem had its root in antiquity, namely, in the problem of the quadrature of the circle. It had been conjectured by the ancients that it was impossible to construct a square with the same area as a given circle using compass and ruler only. During the 17th century, the problem was transformed into a discussion of the algebraic nature of the number π. Before any progress was made in this direction, Euler had introduced e, the base of the natural logarithm and proved that e and e^2 are irrational [Euler 1737]. Thirty years later, Lambert showed that π is also irrational and confirmed Euler's result concerning e. Both Euler and later Legendre conjectured that π is transcendental, but until 1844 nobody had shown the existence of a single transcendental number (cf. Chapter XII).

During that year, however, Liouville resumed his work on the transcendence of e. His new interest in this problem was triggered by reading a passage in the correspondence between Christian Goldbach and Daniel Bernoulli, which had been published the previous year by P. M. Fuss. This passage dealt with certain infinite series, whose sum the correspondents, without proof, claimed to be transcendental. Inspired by this and following the idea in Lambert's proof of the irrationality of e and π and using results by Lagrange, Liouville soon succeeded in finding a class of continued fractions that converge so quickly that they represent a transcendental number. He presented the result to the *Bureau des Longitudes* on March 27, 1844. In its *Procès Verbaux*, one reads under that date:

> Mr. Liouville talked to the *bureau* about the research he has done on a class of quantities which are neither rational nor reducible to algebraic rationals.[1]

Almost two months later, on May 13, he also announced his discovery in a speech given at the *Académie des Sciences* and printed in *Comptes Rendus*. His original proof relied on inequalities for which he referred to Lagrange, but a week later, in a note in the *Comptes Rendus*, he gave a self-contained proof. In his talk, Liouville mentioned a simpler example of transcendental numbers, namely

$$\frac{1}{l} + \frac{1}{l^{2!}} + \frac{1}{l^{3!}} + \cdots + \frac{1}{l^{n!}} + \cdots,$$

where l is a natural number greater than one. This is most easily proved by an approach avoiding continued fractions, which Liouville published in his journal seven years later [1851e]. The basic theorem of this paper states that if a series of rational numbers converges toward an irrational limit faster than any power of the inverse of its denominator then this limit is a transcendental number. This beautiful and important theorem is now often called after Liouville (cf., e.g., [Maillet 1906] and [Schneider 1957]). Liouville's notebooks also reveal that in 1844 his goal was to prove the transcendence of e. Thus, he must have been happy to see that it was one of his protégés (Hermite) who finally succeeded in providing the proof in 1873. Liouville also lived long enough to hear about Lindemann's proof of the transcendence of π in 1882.

Liouville's investigations of transcendental numbers were not only a step on the way to proving the transcendence of e and π. His methods were a first substantial step on the way to understanding transcendental numbers, and any modern book on this subject begins with a chapter on Liouville's numbers. Later mathematicians have developed his ideas much further (cf. [Schneider 1957]). I shall conclude this section by mentioning a few more works by Liouville related to the curriculum at the *École Polytechnique*. In Cauchy's *Cours d'Analyse* [1821], Note II, Théorème 17, it is proved that the arithmetic mean of n positive quantities is always larger than or equal to their geometric mean. Cauchy first proved it for $n = 2, 4, 8, 16, \ldots$ and reduced the general case to one of these by a trick. Liouville remarked in 1839 [1839g] that Cauchy's theorem could be proved by an elegant mathematical induction. It should be kept in mind that although the principle of mathematical induction had been formulated generally by Pascal it had not at Liouville's time obtained its central, and still less its axiomatic, place in mathematics. Finally, in 1842 [1842c], Liouville warned his students about a problem one can encounter in the determination of extreme values of a function that is not everywhere defined.

It is remarkable that all of Liouville's works related to his teaching at the *École Polytechnique* were concentrated around his first years as a professor (a new proof of the decomposition of rational functions into simple fractions [1846l] being an exception). This indicates that Liouville's interest in the problems of elementary teaching soon declined. This picture is only slightly confused when we count four papers on such topics published from 1844 to 1849 in Liouville's *Journal*, by a certain M. Besge, three being on the integration of simple differential equations and one on the evaluation of a definite integral. In 1902 [1902], Brocard presumed that Besge was a pseudonym for Liouville, and recently this has been confirmed by Neuenschwander, who cites a letter from Darboux to Houël to the effect that:

> Mr. Besge writing to Mr. Liouville is the same as Mr. Liouville writing to Mr. Besge.
>
> [Neuenschwander 1984a p. 60]

INFLUENCE

Both as a *répétiteur* and as a professor at the *École Polytechnique*, Liouville guided a number of the young people who later were to form the next generation of the French scientific community, people such as Le Verrier, Catalan, and Bertrand. By chance, during the 1830s and 1840s, a majority of the future mathematicians were admitted to the school in the even-numbered years, which means that they were taught analysis by the other professor, i.e., by Sturm. Among these students were Ossian Bonnet, Hermite, Camille Jordan, P. M. H. Laurent and J. A. Serret. We shall see that Liouville also exerted a strong influence on several of them.

To conclude, one must admit that Liouville's teaching at the École Polytechnique was of limited importance. For all the improvements, his lectures were rather traditional, and except for the works on transcendental numbers, only rather elementary articles were inspired by them. He got his first protégés at this school, but the inspiring effect of his teaching, both on the young generation of mathematicians and on his own work, was much greater at the *Collège de France*.

Collège de France

INSPIRING COURSES

Liouville's teaching at the *Collège de France* was analyzed by Bruno Belhoste and me in 1984. Therefore, only the essentials shall be discussed here.

From 1837 to 1843, Liouville acted as a substitute for Biot in the chair of applied mathematics, and he usually chose the subjects of his lectures accordingly. Thus, during the first three years, he lectured on "planetary perturbations, astronomical refraction, heat, electricity, differential equations applied to physics" [Neuenschwander 1984 a, III, 3]. In particular, during the first semester of 1837–1838, he discussed "the general methods of integration which are of use to the physico-mathematical problems." During these lectures, he developed the generalization of Sturm-Liouville theory, which was published in 1838 [1838h]. The second semester of 1839–1840 was devoted to "thermo-mechanical phenomena, theory of waves and potential theory," mainly according to Poisson, and was followed by a two-semester course on "celestial mechanics," most probably given in order to show that he was worthy of the recent appointments to the astronomy section of the *Académie* and to the *Bureau des Longitudes*. It is possible that celestial mechanics was also on the programme during 1841–1843, but we only know that the first semester of 1841–1842 dealt with an "analysis of memoirs by Laplace, Fourier, and Poisson on mathematical physics."

The first semester of 1839–1840 is an exception to the applied character of Liouville's lectures during this period. It dealt with definite integrals. In his choice of this subject, one can see the mark of the newly formed friendship with Dirichlet, who had lectured in Berlin on this subject starting in 1835 [Biermann 1959, pp. 33–39]. Dirichlet's lectures on definite integrals, which he repeated with irregular intervals, have become famous through the book G. Arendt edited in 1904 based on his lecture notes from Dirichlet's 1854 course. Liouville's lectures, on the other hand, have remained unpublished, as have the rest of his lectures.

However, Bruno Belhoste has found the lecture notes of the 1839–1840 course taken by one of Liouville's students, the mathematician and physicist A. Barré de Saint-Venant [Liouville/Saint-Venant 1839–1840]. They illustrate Liouville's free and impulsive style, with him frequently flying off onto tangents. During the first ten lectures, he often made small digressions, for example, to Cauchy's theorem on convergence of Maclaurin series and Dirichlet's proof of the convergence of Fourier series. The rest

380   PARIS. — Le Collège de France. — LL.

PLATE 7. Collège de France. Postcard from around 1920. Liouville taught here first as a substitute for Biot (1837–1843) and later as a Professor of Mathematics (1851–1882). (Royal Library, Copenhagen)

of the course consisted almost exclusively of three long digressions, one on number theory inspired by Dirichlet, another on his own theory of integration in finite terms, and the last on his work on fractional calculus. On the whole it was a very rich course on many diverse subjects with only a peripheral connection to the original theme. The notes clearly reveal an inspired lecturer. That he was able to convey his enthusiasm to the audience is witnessed by the astronomer Hervé Faye, who took Liouville's courses during this period. At Liouville's funeral, he declared:

> Monsieur Liouville was one of the most brilliant professors one has ever heard. His lectures impressed me so strongly in my youth that today I still have a vivid recollection of the startling clarity with which he was endowed.

At the same occasion, Laurent spoke in the same vein:

> Liouville was an eminent professor. Unfortunately I was only able to follow his lectures at a time when he was already ill and tired and did not have the ardor of his youth. His speech was no longer warm and vibrant but he had a great talent for putting the important points of a question into relief. He did not confine himself to a simple exposition of facts; he taught you to search and he never neglected

to show the way which had been followed to find the truths which
he explained with remarkable lucidity.

[Laurent 1895, p. 131]

These testimonies must be confronted with the evidence of his apparently
traditional teaching at the *École Polytechnique* and his unsuccessful lectures
at the *École Centrale* and the *Collège Louis le Grand*. The apparently
conflicting views seem to reflect that Liouville was a better lecturer on
high levels than on low levels and, perhaps, that his teaching was mainly
aimed at the best of his students.

Unfortunately, the teachers at the *Collège de France* were not required
to keep a list of their students, so there are only a few mathematicians of
whom we know for certain that they took Liouville's courses: in addition
to Faye, Laurent, and Saint-Venant, one can mention the foreigners Bjerknes, Mittag-Leffler, and Holmgren. Probably, his audience at the *Collège
de France* also contained many of the other young mathematicians whom
he guided, for example, Bertrand, Hermite, Le Verrier, Bonnet, Catalan,
Mannheim, and Hirst, but this is guesswork.

A SCANDALOUS ELECTION; LIOUVILLE, CAUCHY, AND LIBRI

The course at the *Collège de France* was clearly Liouville's favorite teaching,
but as a *suppléant*, his future at this institution was insecure. Therefore,
when Lacroix died on May 26, 1843, leaving the chair of pure mathematics
open, Liouville immediately applied for it, and so did his ardent enemy
Libri, who had served as Lacroix' substitute at the *Collège de France* since
1836. Moreover, certain unnamed members of the *Académie des Sciences*
urged Cauchy to apply for the job, and a week later he added his name
to the list hoping "again to be able, as a professor of mathematics, to be
useful to the friends of science and to my country" [Ac. Sci. Doss. Libri].
Recall that Cauchy had not been entrusted any teaching duties at the state
institutions after his return from exile.

During the first years of Libri's stay in France, Cauchy had supported
him, and from his exile, Cauchy had often communicated his works to the
Académie through Libri. However, a discussion in 1842 in the *Académie* on
discontinuous functions [CR 15 (1842), August 29, pp. 410–411] indicates
that by then their relations had become less cordial. Liouville and Cauchy,
on the other hand, enjoyed a relationship of mutual respect. Bruno Belhoste
has established that Liouville was Cauchy's most regular collaborator at
the *Académie* [Belhoste 1982], and as we saw earlier, they were on visit-
ing terms. Still, Cauchy's awkward manners disagreed with Liouville, as
with many other mathematicians. One incident was a result of Liouville's
1839–1840 course at the *Collège de France*. Liouville had proved some of
Dirichlet's theorems on quadratic residues and had offered some generalizations. However, as he told Dirichlet, he did not think much of them:

Such a generalization is so self-evident that you have the priority in

advance. One has to be stupid if one does not discover it after having read just a few lines by you devoted to this subject.
[Liouville to Dirichlet, spring 1840, in Tannery 1910, pp. 10,11]

Liouville told Cauchy about his lectures, and as was his habit, Cauchy generalized the formulas even further. He presented his results to the *Académie* on March 16 and May 11 1840 [Cauchy 1840a,b], giving several references to Liouville, but hardly mentioning Dirichlet. Embarrassed by this unbalanced credit, Liouville complained to Cauchy. As a continuation of the letter quoted above, he wrote to Dirichlet:

Imagine, therefore, my surprise when I saw that Mr. Cauchy had cited me in the two notes, which he had taken it into his head to publish after your paper, in particular the one of May. I can hardly see anything but an amplification of a good schoolboy, drawn up on the text of the master; in the second, in particular, the plagiarism is evident. I have frankly had it out with him regarding the foolish role he has made me play by citing me without letting me know, and on the glaring injustice which he has committed toward you. I should say that in the conversation he seemed to me to be more fair than on paper. He even promised me to delete my name and to take your rights better into account in a second edition which will appear, I think, in the *Mémoires de l'Académie*. We will see what he will do. If he does not restore your rights, I advise you to take it up yourself when you publish some new work.
[Neuenschwander 1984a, p. 69]

Liouville's criticism of Cauchy was justified. Cauchy had the habit of publishing his ideas as fast as possible, often forgetting to credit the person who had inspired him. This habit proved disastrous to many young mathematicians who entrusted him with their ideas. Since he rushed into print before he had thought the subject through, he usually published series of papers containing improvements and additions to the first crude paper, thereby creating a continuous fast-running stream of papers. Although Liouville was not a Gauss who polished his papers until perfection, his style of publication was more selective than Cauchy's, of which he spoke ironically, at least later in his life. Thus, in 1866, he wrote:

Mr. Cauchy was not the kind of person who did not publish what he had discovered or even sometimes rediscovered; he would rather publish it ten times than forget to publish it once.
[Liouville to Fleury, January 16, 1866; Neuenschwander 1984a, III, 12]

In a similar vein, Mittag-Leffler recalled from his Paris visit in 1872:

However, regarding his [Liouville's] evaluation of contemporary mathematicians, one must not forget that political antipathies and different ways of working - about Cauchy he used to say among other things that he suffered from "diarrhé mathématique" — to a certain degree obscured his opinion of his great contemporary and compatriot.
[Mittag-Leffler 1925, p. 29]

PLATE 8. Augustin Louis Cauchy (1789–1857). While respecting Cauchy for his great mathematical results, Liouville still took exception to his conservative political views and his way of publication—Liouville characterized it as "Mathematical diarrhoea" (Lithograph by Belliard after painting by Roller. Royal Library, Copenhagen)

Still, his critical attitude toward Cauchy's ways of publishing did not prevent Liouville from recognizing Cauchy's scientific genius and treating him with respect. This is most clearly demonstrated by Liouville's official application for the chair at the *Collège de France*, sent on June 9, 1843, to the administrator of the *Collège*, J.-A. Letronne:

> I am writing to ask you to present me as a candidate for the chair of mathematics at the *Collège de France*, which has become vacant by the death of our venerable colleague Mr. Lacroix. The members of the *Conseil* can decide if the memoirs which I have published as a geometer and the students which I have trained as a professor will entitle me to obtain their votes. I am at least sure that my exactitude and my zeal has made me worthy of them. Nevertheless I must declare that if the *Conseil*, by disregarding all obstacles foreign to the

science, chooses Mr. Cauchy, far from being hurt by this nomination, I shall be the first to applaud it. To do justice to the superior merits of a colleague is in my eyes a holy duty which it would be pleasant for me to do at any time and over which mean considerations of personal interest would never prevail in my heart.

[Coll. Fr., Dos. Pers. Liouville]

Thus, as when Sturm was elected to the *Académie*, and even more explicitly this time, Liouville again did the unusual: to recommend his competitor for a position. It demonstrates his high esteem and respect for Cauchy. Cauchy felt touched by this gesture and returned it in a note intended for the *Comptes Rendus* of June 26, 1843, in which he said about himself:

What encouraged Mr. Cauchy to offer himself as a candidate was that he saw that the motives which could make him take the decision were understood by all his colleagues. In particular, Cauchy was, in this case, deeply touched by the noble and loyal conduct of Mr. Liouville. He was all the more touched because this disinterested conduct gave Mr. Cauchy an unquestionable testimony of esteem and affection. For, certainly, the qualifications and the works of Mr. Liouville speak so highly in his favor that he could very well simply present himself as a candidate without dealing at all with any other candidacy than his own.

[Smith Collection of Columbia Library, (16), unpublished proof sheets with corrections in Cauchy's hand; private communication by I. Grattan-Guinness]

Probably, Liouville had a second, less generous motive for supporting Cauchy, namely, to hinder the election of Libri. However, he did not succeed. On June 18, Libri was elected the candidate of the *Collège de France* in a thrilling election. In the first round, Libri obtained 12 votes against Liouville's 9 and Cauchy's 3; the second round brought Liouville in front with 12 votes, one more than Libri. Because of one vote in favor of Cauchy, however, Liouville had not won the majority. So, there was a third round, in which Libri won with 13 votes against Liouville's 10.

When Liouville heard about the election, he reacted immediately by resigning from his position at the *Collège*. The following letter, addressed to Letronne, was sent the day after the election:

I am profoundly humiliated as a person and as a geometer by the events that took place yesterday at the *Collège de France*. From this moment it is impossible for me to lecture at this institution. Receive, therefore, herewith my resignation from the post as a substitute.

[Coll. Fr., Dos. Pers. Liouville]

As a supporter of Libri, Letronne felt hurt by the step taken by Liouville and immediately answered that this was none of Liouville's business, since he was nothing but a substitute:

I can understand that you are not satisfied with the events taking place last Sunday at the *Collège de France*, but I do not understand so well that you are "humiliated" and even "profoundly humiliated as a person and as a geometer." Apparently you forget that it is a question of one of your fellow members in the *Institut*, of your senior in age and in the *Académie*, of a member in the geometry section, of the successor to Legendre, of the person which Lacroix himself chose to substitute for him. The remark you venture, and which your bad temper has suggested to you, shows that one can acquire a great mathematical knowledge without acquiring good manners to the same extent. At any rate, in the position you occupy at the *Collège de France*, you do not have a resignation to hand in; I have none to receive. You depend solely on the professor who has chosen you. You can work it out with him. It has nothing to do either with the assembly or with the administrator.

[Coll. Fr., Dos. Pers. Liouville]

In spite of their defeat at the *Collège de France*, Liouville and Cauchy still had a chance to be elected candidates for the *Académie des Sciences*, in which case the minister had to decide. Indeed, Libri had probably won at the *Collège* because the majority of its teachers were nonscientists who could not evaluate the mathematics of the candidates. What they did know, however, and what Libri during his campaigning reminded them, was that he enjoyed the protection of the King and that he was an outspoken opponent of the unpopular Jesuits, who supported Cauchy. In the *Académie*, with its majority of mathematically minded members, such political reasons played a lesser role. Already in 1840, Liouville had told Dirichlet that Libri "in the *Académie*, at least, is beginning to be despised almost as much as he deserves." [Tannery 1910, p. 10].

The following year, Libri added to his unpopularity by a series of disputes with some of the most influential members of the *Académie*. It started on March 22, 1841, when Biot presented a general method by which he solved a particular differential equation that was important in optics. Afterward Libri rose and claimed that the solution was undoubtedly a consequence of his own general theory of differential equations. Biot disagreed and sarcastically remarked:

Mr. Libri has a very easy way to change my opinion, namely, to arrive at the same results, or at equally simple results by solving, with his own methods, the differential equation which I am eager to communicate to him.... And if he does that, he would do a real service to the geometers who have perhaps until now hesitated to use his procedures fearing that they would make the solution very complicated or at least more complicated than the work involved in the direct elimination.

[Biot 1841]

However, instead of producing the desired solution, Libri, at the same meeting, launched an attack on Chasles, who had collected a catalogue of historical records of shooting stars from which he had detected that they came in showers at particular times of the year. Libri correctly criticized

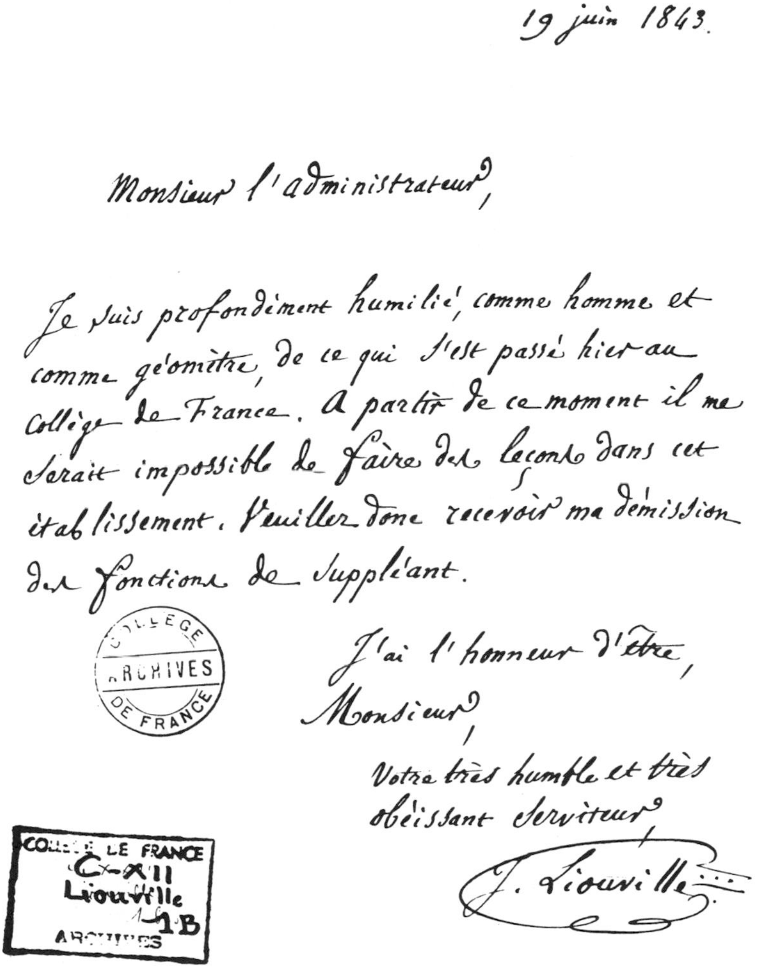

19 juin 1843.

Monsieur l'administrateur,

Je suis profondément humilié, comme homme et comme géomètre, de ce qui s'est passé hier au Collège de France. A partir de ce moment il me serait impossible de faire des leçons dans cet établissement. Veuiller donc recevoir ma démission des fonctions de suppléant.

J'ai l'honneur d'être, Monsieur,

Votre très humble et très obéissant serviteur,

J. Liouville.

PLATE 9. Liouvilles' letter of resignation from the Collège de France. (Archives du Collège de France)

Chasles for not having corrected the dates from the Julian to the Gregorian calendar, but then cut a ridiculous figure by changing the dates the wrong way. Moreover, as Chasles pointed out, the change has no influence on the conclusions that Libri had tried to falsify (we now know that Chasles was right). Libri also indicated that it was absurd when Chasles had the idea that Saturn's rings consisted of a large number of circulating satellites, but he dropped this point when Arago informed him that his own compatriot Cassini had made this conjecture independently of and long before Chasles.

Halfway through the one-and-a-half-months-long dispute, Libri was imprudent enough to criticize the work of the *Bureau des Longitudes* and to intimate that the editor of the *Comptes Rendus* did not treat all members of the *Académie* equally. Hurt by these insinuations, Arago entered the debate on Chasles's side, and shortly afterward Chasles launched a counterattack, criticizing Libri on the grounds that in his *Histoire des Science mathématique en Italie* [Libri 1838–1841] he had unjustly deprived Viète of some of his glory giving it instead to the Italian Fibonacci. Faced with this theme of national pride, Libri gave up the discussion.

However, the criticism of his book continued. The following year, Arago inserted a paper in the *Annuaire du Bureau des Longitudes* in which he claimed that Galilei did not have priority over Fabricius as the discoverer of the sunspots, as Libri meant to have established in his book. For once, modern opinion is on Libri's side in this historical question, but that does not change the fact that the repeated quarrels undermined whatever popularity Libri may have enjoyed at the *Académie*.

Thus, Liouville and Cauchy had very good chances to beat Libri as the candidate of the *Académie* for the professorship. Nevertheless, they were both so upset by the election at the *Collège de France* that they withdrew their candidacy for the job. Therefore, the geometry section of the *Académie*, which had been charged with the evaluation of the candidates, had no choice to make. Still, on June 26 when they presented the only candidate, all the members of the *Académie* could read between the lines that the majority of the geometry section did not support the candidate:

> The only person who has officially offered himself as a candidate for this chair is Mr. Libri, member of the *Académie*. Consequently, we confine ourselves to announcing his candidacy.
>
> [CR 16 (1843), p. 1457]

Sturm and Lamé opposed Libri more explicitly in a letter addressed to the president of the *Académie* and read at the meeting. The letter was not printed in the *Comptes Rendus*, but is preserved in Libri's file at the *Académie*. It reads:

> The two oldest members of the geometry section have thought that any section which was called to designate the candidates of a vacant chair ought to confine itself to making known the names of those that offer themselves when they belong to the *Académie*. Mr. Lamé and I had thought that we had to give in on this point although the

principle seemed questionable to us. However, we have reserved for ourselves the right to explain to the *Académie* our individual opinions on the inaptitude of the only candidate who has offered himself to the section. Mr. Libri has demonstrated in his public teaching his inability as a professor. One can even fear that his reputation as a geometer and as an academician has suffered. The audience is missing at his courses even though the professor abridged them. These facts are common knowledge.

We most truly regret the voluntary or enforced resignation of the two eminent geometers, Mr. Cauchy and Mr. Liouville, who at first offered themselves for this chair to which they both have an incontestable right. After them many other geometers known for their success in teaching can still lay claim to this chair and therefore it is only left for us to declare our formal opposition to Mr. Libri's candidacy. Our protest has no other aim than the interests of the mathematical sciences and those who cultivate them. Finally we think that we ought to remind the *Académie* that, according to the very wording of its report, the section does not present Mr. Libri for the election of the *Académie*, it only announces that Mr. Libri presents himself.

Mr. Lamé, who is absent, has authorized me in a letter deposited on my desk to explain his opinions, which agree entirely with mine, to the *Académie*. C. Sturm

[Arch. Ac. Sci., Dos. Libri]

A week later, on July 3rd, the *Académie* elected its candidate. Not surprisingly, Libri was elected, but the election was a miserable act. Libri only obtained 13 out of 45 votes; three voted for Cauchy, one for Liouville, and 28 returned a blank ballot paper. Still, this was sufficient for Libri. On July 5, 1843 he was officially nominated professor of mathematics at the *Collège de France*.

Thereby Liouville had lost an important position in French education. This is probably the reason why on August 8, 1843, he asked to be reintegrated into the *Corps d'Ingénieurs* as *Ingénieur des Ponts et Chaussées Honoraire*. However, his request was refused because he had left the school while still *élève*, that is, before he had properly received his title as *Ingénieur*. [Arch. Nationale ANF[14] 2271[1]].

Thus, Liouville had to continue to feed his family on his salary from the *École Polytechnique* and the *Bureau des Longitudes*. It was also an intellectual loss, for now he could no longer use the *Collège de France* as an outlet for his scientific thoughts. To make up for that, he began to arrange private courses. I shall return to two of these later.

The large number of blank ballots at the election mentioned above reflects Liouville's influence in the *Académie des Sciences*, to which we shall now turn.

Académie des Sciences

THE ACTIVE EXAMINER

"Liouville whose opinion was law then"—this phrase, referring to the 1840s, is taken from an obituary of J.A. Serret written by Ossian Bonnet [1885]. Since it was written several years after Liouville's death, there is little doubt that it contains a great deal of truth. More specifically, Bonnet was referring to Liouville's great influence in the *Académie*. As in the lecture room, Liouville displayed his oratorical gifts in front of this audience. Fay recalled:

> Later when I had the luck to hear him talk to the *Institut*, wasn't I surprised to see the effect it produced on our fellow members, who were filled with wonder that by following in his footsteps they had been able to penetrate for a moment into the most difficult subjects of higher analysis. Nobody else has ever had this effect in the same degree, perhaps with the exception of Arago. Certainly, Mr. Liouville was in the prime of life a powerful scientific orator.
>
> [Faye, 1882]

When in Paris, Liouville attended the meetings of the *Académie des Sciences* on Monday afternoons, and when in Toul, he followed its activity through the *Comptes Rendus* and sometimes addressed it through letters to the *Secrétaire Perpétuel*, Arago.

During the first 10 to 15 years of his membership, he was among the most active members. To be sure, his administrative efforts were much less than those of Arago, for example, and his scientific contributions to the meetings were inferior to those of Cauchy, who literally flooded the *Comptes Rendus* with papers. Still, a quick glance at the *Comptes Rendus* reveals that Liouville contributed more to this journal than the average 20 pages per year available to each member. A more detailed investigation shows that his influence was much greater than this crude counting of pages indicates.

We have seen that Liouville was elected to the *Académie* on the strength of a great number of works submitted to this society. After his election, Liouville continued to present some of his results to the *Académie*, but now he often confined himself to presenting short abstracts if the complete memoir was too long to go into the *Comptes Rendus*. Most of the works he presented to the *Académie* were later published in his own *Journal*. Thus the *Académie* served him as a forum where he could quickly announce his ideas.

A more integrated part of the work of the *Académie* was the evaluation of the works sent by nonmembers to the *Académie*. Liouville was the member of an almost innumerable number of judging commissions. Although he only edited a fraction of the reports of these committees, they made up a substantial proportion of Liouville's contributions to the *Comptes Rendus* during his first years as a member. For example, he composed favorable reviews of Ritter's memoir on refraction (November 21, 1839), Le Verrier's solution of the stability problem for the planetary system (March 30, 1840), Steiner's geometrical research (1841), Binet's rational mechanics (March 21, 1842), Bertrand's work on equipotential surfaces and Hermite's great contribution to Abelian functions (August 14, 1843), J.A. Serret's early work on elliptic curves (July 30, 1845) and Delaunay's famous new method in dealing with the motion of the moon (January 4, 1847).

On occasion, the *Académie* also entrusted Liouville with the editing of reports on subjects far outside his own broad field of interest. Thus, on January 15, 1844, he gave favorable mention to a work by his fellow student at the *École Polytechnique* and the *École des Ponts et Chaussées*, Ernest Lamarle on the ultimate strength of guns, and earlier, on September 6, 1841, he judged a note in which Passot criticized a "dynamometric break" invented by Poinsot. The latter is one of the rare instances where Liouville wrote a negative review. The circumstances were also unusual. Originally, Arago had taken it upon himself to edit the report, but after Passot had issued a lampoon against him, Liouville took over the job. Thus, one may guess that Arago's views may be discernible in the report in spite of the assurance in a footnote in the *Comptes Rendus* to the effect that Arago had nothing to do with it. The report tore Passot's arguments to rags.

On another occasion, when Liouville was asked to judge a memoir by Fonvielle on the metric system, which apparently Liouville did not think highly of, he simply stated that it was not of a nature that required a report.

In addition to these official reports, Liouville from time to time on his own initiative told the *Académie* of the works of young talents. For example, in 1845 he brought out reports on the analytical and algebraical works of Lamarle, the differential geometric works by M. Roberts from Dublin, and other works by Serret. He even communicated some of Liebig's results in chemistry to the *Académie* (1841).

Liouville often contributed to the general discussion of the works of others with additions, suggestions, or criticisms. Thus, he presented alternative proofs to the above-mentioned works of Serret (December 8, 1845), and earlier the same year he commented on communications by Cauchy on complex function theory. The year before he had also discussed series expansions with Cauchy (December 9, 1844) and had urged Chasles to generalize some of his geometrical results.

On rare occasions, he also pointed out mistakes in the work presented to the *Académie*. The first instance occurred in 1842 (August 29), when Liouville, in a letter written while in Toul to Arago [Liouville 1842f], pointed out that the proof of the invariability of the major axes of the planetary orbits, which Maurice [1842a] had printed in the *Comptes Rendus*, was not exact. This led to some discussion in which Wantzel [1842] took part (cf. [Maurice 1842b and c]).

PLATE 10. The Palais de l'Institut in 1860. Here the Académie des Sciences met and still meets. Liouville's notebooks are preserved in its library (Archives de l'Académie des Sciences de Paris)

FERMAT'S LAST THEOREM

A more wide-ranging intervention of Liouville's had to do with Fermat's last theorem, which states that $x^n + y^n = z^n$ has no solution in natural numbers x, y, z if n is a natural number greater than or equal to 3. Liouville's role in the investigations of this theorem has been discussed in detail in Edwards' interesting papers [1975, 1977] on which the following account is based.

The cases $n = 3$, 4, and 5 had been disposed of by Fermat, Euler, Legendre, and Dirichlet, and in 1839 Lamé presented to the *Académie* a proof in the case of $n = 7$. Liouville, who had entered the *Académie* a few months earlier, was appointed member of the judging committee, as was Cauchy. The latter wrote a favorable review, and Liouville agreed to print the complicated proof in his *Journal* [Lamé 1840].

Eight years later Lamé believed he had solved the general case using a method for which he credited Liouville:

> For a long time, I looked for a type of proof which would be applicable to all the cases and in some way independent of the size of the exponent when one day, a few months ago, I spoke with Mr. Liouville. He seemed to me to be convinced that the negative property announced by Fermat must depend on certain complex factors,

which had recently been studied by geometers working with the theory of numbers. This was a new way which I had never explored; I followed it and I have arrived at the method of proof which I shall now explain, and which appears to me to justify Mr. Liouville's prediction.

[Lamé 1847, p. 310]

Thus spoke Lamé on March 1, 1847, to the *Académie*. The idea that Liouville had suggested to Lamé was to factor the expressions in question in cyclotomic numbers, that is, complex numbers of the form $\sum_{i=0}^{n-1} a_i r^i$, where $a_i \in N$ and r is an nth root of unity, i.e., $r^n = 1$. However, Liouville was not enthusiastic about Lamé's ideas regarding the general case. After Lamé had sketched the proof to the *Académie*, Liouville rose and declined the credit Lamé had given him. He had two reasons for doing so. First, he pointed out that the idea of using such complex numbers was not new and could be traced back to works of Euler, Lagrange, Gauss, and, most recently, Cauchy and Jacobi. Second he cast doubts on Lamé's proof:

A few attempts make me believe that first one must try to establish for the new complex numbers a theorem analogous to the elementary proposition concerning the ordinary integers, namely, that a product can be decomposed into prime numbers in only one way. There is a need for this theorem, and yet I do not see that our fellow member has entered into the details of the subject as far as the matter requires. Isn't there a gap to be filled here?

[Liouville 1847c]

Indeed, this gap cannot be filled. What enabled Liouville instantaneously to put his finger on the weak point of Lamé's proof? As he himself admitted, "I have not even taken a serious interest in the equation $x^n - y^n = z^n$." He had only published one note on it [1840h], pointing out that if $u^n + v^n = w^n$ is impossible in natural numbers, then $z^{2n} - y^{2n} = 2x^n$ is impossible as well. As argued by Edwards [1975], Liouville had probably been alerted to the problems of unique factorization by a paper of Jacobi in which it was pointed out that "if λ is a divisor of $p-1$, the prime p can be represented in several ways as a product of two complex numbers made up of λ'th roots of unity." Liouville was familiar with this paper for he had asked Hervé Faye to translate it for his *Journal* in 1843 [Jacobi 1839b]. In his objection to Lamé's proof, Liouville specifically referred to this paper: "particularly in a paper by Mr. Jacobi one can find useful information."

Liouville's doubts proved to be well founded. A few months later he received a letter from Kummer in Breslau informing him that a year earlier Kummer had published the proof that the unique factorization in prime numbers "does not hold in general" for cyclotomic integers. Kummer had introduced his so-called ideal numbers to save the theorem, and he was full of hope that they would soon lead him to a proof of Fermat's theorem.

Liouville read Kummer's letter to the *Académie* (CR 24, 899–900) and published an extract in his *Journal*. In a footnote, he very tactfully referred to works of his compatriots Cauchy and Lamé and ended:

> Here we have not examined to what extent the authors which we have cited agree or differ, nor their respective priority to this or that discovery. Time will show the value of their works and put everything in its proper place.

In the following two issues of his *Journal*, he then published Lamé's "proof" and a paper Kummer had sent to him.

As a result of the stir about Fermat's last theorem, the *Académie des Sciences* decided to pose this problem for its next great mathematics prize. On July 9, 1849, the following report was read by the committee appointed the previous year to decide on the subject:

> Since the recent works of several geometers have called attention to Fermat's last theorem and made considerable advances even in the general case, the *Académie* proposes to remove the last difficulties which remain in this area. Therefore it poses the following problem for the *grand prix de Mathématique* to be awarded in 1850:
>
> "Find for an arbitrary integer exponent n the solutions in integer and unequal numbers of the equation $x^n + y^n = z^n$, or prove that there are no such solutions."
>
> The prize consists of a gold medal with a value of three thousand Francs. The Memoirs must be handed in to the office of the *Académie* before October 1, 1850.... The names of the authors should be contained in a sealed note which will be opened if the work is awarded the prize.

The committee was composed of Cauchy, Sturm, Arago, Poinsot, and Liouville. It seems likely that it was Liouville who suggested the subject, for it was he who read the report.

Since no proofs were received, the prize was not awarded in 1850, but in 1857, the committee decided to award the prize to Kummer for his many results in this area. Liouville was the driving force behind this decision, but it was Cauchy who wrote the report "although he basically opposed the proposal, against which he had at first only presented some very bad arguments." This quote comes from a letter from Liouville to Dirichlet that has been published by Edwards [1975, pp. 231–232]. Liouville wrote the letter in great haste, because after the committee had decided on the award, Cauchy confronted its other members with "a serious objection which I admit would have made me [Liouville] retreat to a better informed person if it had been announced more clearly earlier and not when we were already committed to the *Académie*" (Liouville to Dirichlet in [Edwards 1975]). Liouville informed Dirichlet of this objection and of another weakness he had himself found in the paper, for which Kummer was awarded the prize, and he asked if Dirichlet or Kummer could help him out of this embarrassing situation.

Only five days later [Tannery 1910, pp. 30–33] Dirichlet relieved Liouville by telling him that Cauchy's objection had already been solved by Dedekind, one of Dirichlet's students in Göttingen. Moreover, he referred

Liouville to a new paper by Kummer. In this paper, Kummer met the criticism of Liouville, but Dirichlet did not stress this in his letter. Whether or not Liouville discovered that his objection had been met, the committee stuck to its decision, and Kummer was awarded the medal.

INTERNATIONAL CONTACTS

I have summarized Edwards' story about Fermat's last theorem in order to emphasize the great role Liouville's foreign contacts played on French mathematics at a time when Germany began to overtake France as the leading mathematical power. Most French mathematicians thought of Paris as the eternal center around which all mathematical activity turned, and they did not bother to keep abreast of the developments in the rest of the world. It is symptomatic that even Cauchy was so ignorant of the work of the Germans that even after Liouville's opposition to Lamé's proof he was convinced of its prospects and tried to get part of the credit by pointing to his earlier papers on the subject [CR 24, p. 316]. In fact, Edwards convincingly argues that Cauchy never took the trouble to read Kummer's works carefully.

The astronomers in Paris, for example Arago and Le Verrier, had many foreign contacts, but among mathematicians, Liouville seems to have been the one who was best informed about the development of his science outside of Paris. Much of his contact with German mathematicians went through Dirichlet, and when Kummer addressed him in the above-mentioned letter, he carefully started out by emphasizing that it was his friend Dirichlet who had urged him to write. We shall later see that Liouville had many mathematical friends in other countries as well.

PRIZE COMPETITIONS

The *Académie des Sciences* made the most of Liouville's broad knowledge of the international trends in mathematics by frequently appointing him member of committees in charge of finding subjects for prize competitions. From 1844, Liouville served on the majority of the committees dealing with the mathematical prizes and also on many of those taking care of the astronomical ones, both those suggesting themes and those judging the answers. This soon became Liouville's primary administrative job at the *Académie*. During his first years at the *Académie*, the only mathematical prize was the *Grand Prix de Mathématique*, which Kummer received in 1857. In principle, a prize problem was posed every second year, but when the *Académie* did not receive any satisfactory solutions before the expiry of the time limit, they often decided to repeat the question, as happened with the Fermat prize.

A second mathematical prize, the *Prix Bordin*, was founded in 1835, but was not awarded until 1856. Like the more prestigious *Grand Prix de Mathématique* it was awarded with irregular intervals to a piece of work on a specified subject. The *Prix Poncelet* started in 1869, on the other hand, was awarded regularly every year to "the most important mathematical work in the preceding ten years." Likewise, the astronomical *Prix Lalande* had no specific subject but was awarded almost each year to "an exceptional observation or memoir." Liouville served on the commissions of all these prizes. Moreover, he was often appointed to the commissions judging the annual competitions at the *École des Ponts et Chaussées* and to those suggesting new members at the *Académie* and at the *Bureau des Longitudes*.

The Bureau des Longitudes

Members of the *Bureau des Longitudes* were paid almost a professor's salary. Nevertheless, Liouville seems to have been less active there than in the *Académie*. Thus, although Bonnard, in arguing for Liouville's election, had put forward that he would be able to supervise the calculators working on the astronomy tables, I have found no trace that Liouville ever undertook this job. In fact, the *Procès Verbaux* from the meetings indicates that Liouville played a rather passive role in the duties of the *Bureau*. During the 1840s, Arago, Biot, and to some extent Mathieu were the leading persons taking care of the practical astronomical and surveying activities, which constituted the central task of the *Bureau*. Still, Liouville did serve on a few commissions; for example, in 1846 he was appointed, together with Arago and Laugier, to referee a paper by Petit for the *Connaissance des Temps*. Moreover, he was elected president three times during his lifetime, for the years 1843, 1847, and 1872.

Cauchy's membership?

During Liouville's first period as a president, it fell upon him to clarify the delicate question of Cauchy's membership. The preceding year, one of the members of the *Bureau* had tried to call the minister's attention to the problem. In the *Procès Verbaux* of January 5, 1842, it is reported thus:

> Mr. Beautemps Beaupré analyzes the conversation he has had with the Minister of Education on the situation at the *Bureau* after the nomination of Mr. Cauchy was not accepted. Mr. Beautemps Beaupré remarks to Mr. Villemain, that the question ought to be settled by the government. The *Bureau* can take no initiative in the matter.[2]

However, the initiative of the government failed to appear, and so, on November 19, 1843, Liouville sent an official letter to the minister reminding him of the unsettled situation and continued:

> It is not up to him [the president] to inquire the reasons that have detained the government from confirming this nomination, but it is his duty to assure you, Mr. le Ministre, that the actual situation

cannot be prolonged without causing real harm to the astronomical
sciences.

[P. V. Bur. Long., Nov. 19, 1843][3]

Liouville's intentions with this letter were probably to try to support
Cauchy in his reintegration into the scientific institutions after his exile.
The date of the letter supports this interpretation. Five months earlier
he had magnanimously recommended Cauchy for the professorship at the
Collège de France only to see both Cauchy and himself humiliated by the
election of Libri. The intervening months Liouville spent in Toul where he
apparently thought out this new plan of supporting his great colleague.
Liouville was usually back in Paris around November 10–15, so he seems
to have carried out the plan as soon as possible. However, if I am right in
supposing that Liouville tried to persuade the minister finally to sanction
Cauchy's election he failed miserably. Ten days later the minister wrote
back in order to:

ask the *Bureau des Longitudes* to proceed to a new designation (of a
geometer), because the first designation (that of Mr. Cauchy) could
not be ratified and therefore remained ineffective, after the candidate
had refused to fulfill an obligation prescribed by law.

[P. V. Bur. Long., Nov. 29, 1843][4]

The *Bureau* immediately obeyed the order by appointing a commission
consisting of Liouville, Arago, and Mathieu, who at the following meeting
presented their candidate to the vacant seat. They chose Poinsot, who had
been passed over when Liouville had been appointed a member three years
earlier. Obviously upset, Cauchy wrote an open letter to the President
of the *Bureau des Longitudes* in which he rejected the arguments of the
Minister and asked for permission to present his arguments orally to the
Bureau. Probably annoyed with Cauchy's public action, the *Bureau* rejected
the application [Belhoste, 1985 p. 171], and at the following meeting, on
December 13 Poinsot was elected unanimously.

Thus, one of Liouville's few contributions to the administration of the
Bureau put an end to the delicate vacancy, although probably in a way
different from the one he had hoped for.

PRESENTATION OF NEW IDEAS

Liouville seems to have regarded the *Bureau des Longitudes* as a second
Académie, rather than an organizing body for practical undertakings, that
is, he considered it as a forum in which he could promote his own scientific
ideas and those of his colleagues, friends, and students. Other members also
shared scientific news with their colleagues at the *Bureau*, but nobody else
exploited this activity more consistently than Liouville. Thus, he was the
only member who had notes containing new results in celestial mechanics
and related fields pasted or copied into the *Procès Verbaux* of the meetings.
I have found more than a dozen such notes. It is difficult to explain why

Liouville wanted to insert his notes in the unpublished *Procès Verbaux*, which have been sitting virtually untouched on a shelf in the *Bureau des Longitudes* since they were written. The only reason I can think of is that in this way he could secure his priority. Indeed, his most famous note in the *Procès Verbaux* served this purpose. This *Note sur l'intégration des équations différentielles de la dynamique* [1855d] states a theorem, now called "Liouville's theorem," according to which Hamilton's equations can be integrated by quadrature in an interesting case. I shall return to it in Chapter V.

The *Procès Verbaux* of the years before 1848 do not contain such pearls, but still Liouville presented interesting ideas. Most of them dealt with subjects connected to astronomy or geodesy, for example, the stability of a fictitious moon introduced by Laplace (September 29 and October 6, 1841), cf. [1842g], the generalizations of a formula due to Biot on refraction (December 22, 1841) [1842e], a proof of a theorem in rational mechanics indicated by Jacobi (April 27, 1842), a generalization of the tautochronous problem solved by fractional calculus (February 7, 1844), geodesics on ellipsoids (December 4, 1844; January 28, 1846), adjoint differential operators (February 12, 1845), potential theory (April 30, June 11, and June 25, 1845), and the libration of the moon, considered a pendulum with variable length (June 3 and 10, 1846). A limited number of his communications dealt with subjects outside the scope of the *Bureau*. I have already mentioned that he chose to disclose his important discovery of transcendental numbers to the *Bureau des Longitudes*, although this forum was probably not the most receptive to this purely mathematical result. Moreover, on May 26, 1847:

> Mr. Liouville talks about the discovery which has been made in Germany of some of Pascal's mathematical manuscripts which Leibniz had copied and which contain, it is said, the treatise on Conics.
>
> [P. V. Bur. Long., May 26, 1847][5]

The communication of this important addition to our knowledge of the work of the young Pascal is one of the rare instances where Liouville showed a genuine interest in the history of his science.

Liouville often advertised the results of his students, colleagues, and friends. I shall return to the more important cases, so let me just in passing mention the names: Le Verrier, Delaunay, Sylvester (on roots of equations, (October 20 and 27, 1841)), Jacobi, Chasles, Ossian Bonnet, Gauss, Green, W. Thomson, Serret, Hermite, and Michael Roberts. The merits of several of these mathematicians were mentioned repeatedly. In particular, Liouville often emphasized Jacobi's work, revealing thereby his admiration for the great mathematician who inspired him to much of his own work in celestial and rational mechanics.

The *Bureau des Longitudes* provided the institutional framework for most of Liouville's research in these fields, and he published several memoirs in its publication, the *Connaissance des Temps*. His works on astronomy and mechanics can be divided into four groups. The first represents a brief continuation of the activity in celestial mechanics that had been forced on Liouville by the prospects of seats in the *Académie* and the *Bureau*. The second was a highly interesting but mostly unpublished work on the figure of a rotating mass of homogeneous fluid.

In 1834, Liouville had proved - as we have already seen - Jacobi's theorem stating that a three-axial ellipsoid can form the equilibrium surface of a homogeneous rotating mass of fluid. Five years later he briefly returned to the subject [1839b] in order to correct formulas published by Ivory [1838] on the shape of the Jacobi ellipsoid, and in the fall of 1842, he undertook a large-scale investigation of the equilibrium ellipsoids. In the spring of 1843, he published a small paper on the shape of the equilibrium ellipsoids expressed as a function of the angular momentum, but it did not reveal the main objective of his research: the study of their stability. He established that as long as the angular momentum is so small that the Maclaurin ellipsoid is the only equilibrium shape then it is stable. For higher momenta, the Maclaurin ellipsoid becomes unstable, and the Jacobi ellipsoid is stable. Liouville's unpublished results were soon forgotten because these surprising results were only indicated in two brief addresses to the *Académie* [1842h, 1843f], and only the beginning of the hydrodynamical considerations were published ten years later [1852f]. The results were rediscovered 40 years later by Poincaré [1885] and Liapounoff [1884] independently, of whom only the latter was acquainted with Liouville's published notes. In Chapter XI, I shall analyze Liouville's unpublished theory and show how it led to Liouville's interest in Lamé functions, which were of the greatest importance in many of his later works in potential theory [1845d; 1846e,f]. Liouville's interest in three-axial ellipsoids also aroused his interest in the geometry of these surfaces, about which he published many papers beginning in 1843 with a geometric interpretation of the effective gravitational force on the surface. These geometric papers, with their obvious relevance for geodesy, make up the third group of papers related to Liouville's membership in the *Bureau des Longitudes*.

The fourth group was a series of papers on rational mechanics beginning in 1846 [1846h,i; 1847i; 1855c,d,h]. It was a continuation of his works in celestial mechanics from 1834 to 1835 and 1839 to 1840 in the sense that the subject was point mechanics. However, instead of his earlier Laplacian orientation, Liouville now took his point of departure in Lagrange's mechanics, and in particular in Hamilton and Jacobi's elegant reformulation. The first two, and the longest, works Liouville wrote in this area established a broad class of cases where Lagrange's equations - or the Hamilton-Jacobi equations - for one [1846i] or a finite number [1847i] of point masses can be integrated. These cases are still known as Liouville's integrable systems. [Campbell 1971, p. 87], (cf. Chapter XVI).

Journal Editor

One of Liouville's greatest services to mathematics was the creation of his *Journal* in 1836 and his editorship during its first 39 years. I have discussed its creation earlier. Its editorial policy and contributors have been compared to that of three other nineteenth-century journals by E. Neuenschwander [1984b]. From the letters published in Neuenschwander's paper, it appears that Liouville occasionally asked his friends, such as Bertrand and Chasles, to referee papers for the *Journal*, but generally he did the work himself. In order to overcome this work, he sometimes accepted papers by his friends and by well-reputed mathematicians on trust, whereas he was more critical toward unknown beginners. This is not to say that he only published works by established mathematicians; on the contrary, he deliberately used this vehicle for propagating the ideas of his protégés, such as Hermite, Le Verrier, Mac Cullagh, de Saint-Venant, J.A. Serret, W. Thomson, Chebyshev, and Delaunay. The international spirit of the *Journal* is visible in this list and is underlined by the fact that the *Journal* counted among its authors people of the calibre of Cayley, Dirichlet, Eisenstein, Gauss, Jacobi, Kronecker, Kummer, and Plücker. The *Journal* consisted of original papers sent to Liouville by the authors and papers from other journals that Liouville had some of his friends translate into French. In one sense, Liouville's *Journal* was less international than Crelle's: it only published papers in French.

Relatively few letters or other written material pertaining to Liouville's work as an editor have been preserved, but it must have required a large part of his time to run the edition single-handed. To judge from his letters, the monthly issues usually appeared on time, and although Liouville often prepared them several months in advance, he seems to have succeeded in keeping the backlog down to a minimum. He often published shorter papers within a month, and when longer papers had to wait for more than a few months, he made excuses to the author. This regularity and speed, which modern authors may envy their 19th-century colleagues, was a result of three factors: a laborious and skilled printer, a swift and regular postal service (at least comparable to present-day standards), and an editor that must have given this work top priority. Even during his summer vacations, manuscripts and proofs circulated between Toul and Paris.

The work for the *Journal* gave Liouville an exceptional overview over the state of affairs in all branches of mathematics, and this knowledge in turn enabled him to select the most important foreign papers for translation in his *Journal*.

Liouville did not just sit and wait for papers to be sent to the *Journal*. He often encouraged his colleagues and students to write papers, and he was often inspired to pursue ideas in the papers sent to the *Journal*. For example, in 1839, after Liouville had told Jacques Binet about Rodrigues' work [1838] on the number of ways one can divide a polygon, Binet wrote a paper on the same subject [1839]. It inspired Catalan [1839] and Lamé [1838] to similar studies, and a few years later Liouville [1843c] contributed to the discussion with a critical mention of a paper by Nicolas Fuss. In turn Binet commented on this in a note [1843]. It is also probable that Liouville's application of Sturm-Liouville theory to yield a convergence test [1840g] was inspired by Duhamel's paper on convergence tests published in the 1839 volume of Liouville's *Journal*. It was followed up in 1841 when Lebesgue translated a paper from Crelle's Journal by the Zürich mathematician Raabe [1834]. Its general results comprised those of Duhamel.

The 1841 volume of Liouville's *Journal* also contained notes by Bertrand and Blanchet and allusions to Duhamel's ideas on the "values" of functions $\frac{f(x)}{g(x)}$ at points where both f and g become infinite. They triggered off Liouville's own solution the following year [1842b].

The above examples suggest that Liouville purposely tried to collect papers on the same subject for his *Journal*. Another example is provided by three papers on the propagation of light in crystalline media in the 1842 volume. The first of these was written by the Irish mathematician James Mac-Cullagh [1842]. It was followed by a claim of priority by Cauchy [1842], in fact the first paper he wrote directly for Liouville's journal. Finally, Liouville put the discussion into relief by publishing a translation of a paper by F.E. Neumann [1835]. Four years later Chasles [1846d] defended his priority against Mac-Cullagh.

Occasionally, Liouville appended notes to the papers in his *Journal*, as he did in 1845 in a note on Serret's elliptical curves [1845f] and in 1847 in an important *Note sur deux lettres de M. Thomson...* [1847f]. In 1847, he further drafted a *Note relative à un mémoire de M. Lebesgue* [Ms 3629 (9), pp. 49v–54r] in which he elaborated on his own sketch of a proof of the theorem of quadratic reciprocity [Liouville 1847d]. He felt "forced" to this elaboration "by the benevolence with which M. Lebesgue mentions my proof in the paper mentioned above" [Ms 3629 (9), p. 50r]. However, possibly because he planned to publish the full proof in a memoir on number theory, Liouville in the end did not include the note, and the reference to Liouville was suppressed in Lebesgue's published paper [Lebesgue 1847]. Liouville's following up of the papers in his *Journal* is similar to the behavior of Cauchy, who often wrote papers inspired by his students and colleagues. However, where Cauchy often forgot to promote the original paper and even forgot to quote it, and thereby got a bad reputation among young mathematicians, Liouville always carefully referred to the important ideas that had inspired him. The beginning of his *Note au sujet de l'article précédent* [1847f] is a good example of his tact:

> Mr. Thomson's letter has suggested to me some remarks which I think I ought to present here because, as far as I can see, they show even more clearly the great importance of the work of which the young geometer from Glasgow has given us a brief excerpt.
>
> [1847f]

As far as I know, Liouville's development of ideas sent to his *Journal* never brought him any enemies.

Instead of prolonging the list of instances where Liouville influenced the content of his *Journal* or was himself inspired by its contents, I shall now illustrate how Liouville used this and all the other channels of influence to promote some of his protégés and how their ideas in turn inspired Liouville.

Guiding Young Talents

I have already quoted Faye and Laurent to the effect that Liouville was a brilliant lecturer, at least on the higher levels. His influence on the interested students went far beyond the classroom, as Laurent's following recollection shows:

> Everybody who knew him knew how benevolent his welcoming was and how interesting his conversation. You rarely spoke with him without profiting in some way. He liked to encourage young people and to help them with his advice. When he finished his lectures at the *Collège de France,* he loved to let some of the students follow him home, and it often happened that he forgot lunchtime during the long and interesting conversations he had with them.
>
> [Laurent 1885, p. 131]

Later in life Liouville explained to Mittag-Leffler why he always helped and guided young and talented mathematicians. At the Scandinavian congress of mathematicians in 1925, Mittag-Leffler said:

> In particular, I recall my first meeting with Liouville. He welcomed me more than kindly with the following remark: "I consider it my duty to cordially welcome all young mathematicians with a prospective future." "Can you imagine, I have seen Abel without knowing him and I have examined Galois".... Liouville always counted it among the greatest griefs which ever overcame him that he met Abel but missed the opportunity to recognize in him a great mathematician, in his opinion the greatest who ever crossed his path.

I shall now describe some examples of Liouville's great efforts to guide young scientists.

Le Verrier (Catalan and Delaunay)

From irresolution to authority

In his first class as a *répétiteur* (1831), Liouville instructed a scientist whose career he was to influence strongly. Urbain Jean Joseph Le Verrier, who is still celebrated for his discovery of the planet Neptune, was only two years younger than Liouville. A friendship developed between the two, and when graduating from the *École Polytechnique*, Le Verrier turned to Liouville for advice on the direction of his future career. Le Verrier wavered between entering the *École des Ponts et Chaussées* or becoming *élève ingénieur* of the state tobacco company. He wrote to Liouville in Toul to ask his opinion. I have already quoted passages of Liouville's answer from September 18, 1833, to illustrate his life in Toul and his generally positive attitude toward his old school, the *École des Ponts et Chaussées*. Here, I shall quote another passage from his only letter preserved from this time, to illustrate that Liouville had not yet developed his later ability to guide his students with a firm hand.

> ... you have reckoned without your host if you think that I would be capable of putting an end to your irresolution. In order to give good and decisive advice in these matters, it would be necessary to know the different services to which the *École Polytechnique* leads, and nobody knows that. In your situation, I was formerly as undecided as you, wavering between the *Ponts et chaussées* and the *Génis*. To be sure, today I do not hesitate to prefer the *Ponts*. As for the *Tabacs*, which did not exist in my time, I do not see any serious objection to it. It is an immense advantage for your future goal to remain in Paris as long as possible. As for the fear you express of getting less scientific qualifications in the *Tabacs* than in the *Ponts*, this fear is unfounded, for your scientific qualifications are those of a former student at the *École Polytechnique*.
>
> [Neuenschwander 1984a, III, 1]

I doubt whether he would have argued in this way had he known that in 1843 his application for entry in the engineering corps was turned down because he had not formally received a title from the *Ponts et Chaussées*. After a few more arguments for and against, he summed up his opinion with self-deprecation.

> After this illuminating advice, you do not know if I definitely prefer the *Ponts* or the *Tabacs*, and I do not know either. All that I can repeat to you is that if I were to start over again I would do what I have done, I would enter the *Ponts et Chaussées*. But, that is simply due to the fact that I know it personally, whereas I know the *Tabacs* only by hearsay.

The irresolute side of Liouville's character revealed in this letter soon gave way to a great authority, witnessed by his guidance of Catalan. This rebellious student had been accepted in Liouville's class of 1833 at the *École Polytechnique* and expelled from the school in 1834. Reintegrated in 1835, he graduated and taught for the next two years at the *collège* of Châlons-sur-Marne. During this period, Liouville helped him search for a job in Paris. The following passage is quoted from an undated letter from Liouville:

My dear Catalan,

Excuse me. I am very lazy when it comes to writing. Laziness is the only cause for my silence toward you. Besides, I do not think that your affairs are in such a bad state as you think. We will discuss them soon when you visit Paris during the Easter holidays. Together we will visit Mr. Poisson, with whom I am at present on the best of terms. I regard this trip to Paris as almost indispensable; you need a very serious reason to stop you from making it. Allow me to answer your different questions very briefly: 1° I care little about all those alterations in the theory of incommensurable quantities; and the theory of Legendre seems quite correct to me. 2° We shall talk about your theorem on numbers. Perhaps you have made a mistake when transcribing the statement which I do not understand. 3° If I am not mistaken, you wrote to me that they have refused to let you pass your examination as *Bachelier ès lettres*. Let me assure you that this is not possible. By allowing you to instruct in the public institutions, they have implicitly granted you the right to register for the required examination (without any certificate). Send your request to the Minister through Mr. Poisson and it will be accorded. After that you can and must be successively *Bachelier ès Sciences*, *licencié* and *aggrégé* at the end of this year. If you do not carry out this plan, which is very practicable, you lack energy and you will regret your weakness forever.

You can be sure my dear Catalan that if you really want to succeed you will succeed. But it is necessary to will in order to succeed, and I am always afraid of seeing you weaken.

Goodbye my friend, come at Easter. You must make this trip, then we will talk; a conversation of a quarter of an hour is better than a hundred letters.

All the best,
J. Liouville

P.S. please give my regards to Mrs. Catalan. It would be sad for you to leave her in a month. However, if her condition does not make you worry, I repeat, you must come.

[Jongmans 1981, no. 30]

Liouville was now familiar with the rules governing academia, and he not only advised but pushed his protégés up the ladder; he succeeded, and in 1838, Catalan was nominated *répétiteur* of descriptive geometry at the *École Polytechnique*. Catalan, however, was a difficult person to promote, for he remained his entire life an uncompromising leftist, to a degree that threatened his scientific career. This point of opinion is supported by the view expressed by the chemist Henri Sainte Claire Deville in a letter to Catalan dated 1865, when at long last Catalan had obtained a professor's chair in Liège after having held only subordinate positions in Paris:

Your best friends will also reproach you slightly for having subordinated your scientific and material interests to your political ideas. I do not doubt that otherwise you would have been professor at the *Sorbonne.*

[Jongmans 1986]

There was no such problems with the career of Le Verrier. Although Liouville in his letter had gravitated toward the *Ponts et Chaussée*, Le Verrier had decided to enter the school of tobacco from which he resigned two years later in order to pursue a scientific career. Initially, he worked in both astronomy and chemistry, but having been elected (1837) *répétiteur* of geodesy at the *École Polytechnique*, he devoted himself completely to celestial mechanics, in particular the general problem of the stability of the solar system. At this decisive moment of his career, Liouville supported him strongly by reviewing his great memoir very favorably to the *Académie* on March 30, 1840. He threw Le Verrier's results in relief by pointing out that although Laplace had theoretically proved that there exists a planetary system that is stable, neither he nor Lagrange had proved that the relationships among the axes, eccentricities, and masses of our system ensures stability. This was Liouville's second report to the *Académie*. Probably Liouville had also been responsible for having Le Verrier's own summary of the paper printed in the *Comptes Rendus* already when it was sent to the *Académie* on September 16, 1839 [Le Verrier 1839]. Thus, by 1840 Liouville had learned how to promote the interests of his students.

QUARREL WITH PONTÉCOULANT

He soon experienced that even such unselfish work could lead to trouble. The troublemaker, Compte Gustave de Pontécoulant (1795–1874), was a *polytecnicien* and colonel, who had obtained a name as an astronomer mainly because of his *Théorie analytique du Système du Monde* [1829–1843] of which the third volume had appeared in 1834 . However, it was pointed out by Poisson that some of the calculations of the secular variations of the elliptic elements of the planets were flawed. Pontécoulant acknowledged the mistakes and on June 1, 1840, he sent a corrected version of page 25 of the third volume to the *Académie*. By then Pontécoulant's relationship with the professional scientists of the *Académie* was already strained, after he had been defeated by Liouville in the election for a successor to Lalande. In a letter accompanying the corrected page and printed in part in the *Comptes Rendus*, Pontécoulant [1840] further succeeded in treading on the toes of at least two members and two nonmembers.

First, he complained that he had for almost two months (from May 4) unsuccessfully tried to get permission to read a reply to Poisson's criticism in the *Académie*. I remark in passing that this means that Pontécoulant had carefully postponed his answer until just after Poisson's death on April 25, 1840. Pontécoulant's complaint was directed toward Poncelet, who after the death of Poisson had taken his place as president of the *Académie* in 1840. Immediately after Pontécoulant's letter had been read, Poncelet rose.

He rejected the complaint and criticized Pontécoulant for having published his insinuating letter in the journals before sending it to the *Académie* [CR 10, p. 874].

Secondly, Pontécoulant put the blame for the flaw on one of the *élèves astronomes* at the observatory, Eugène Bouvard. He was the nephew of the astronomer Alexis Bouvard (1767–1843), who had become famous for his Uranus table. Though calculated with great skill and industry this table proved to be in poor accordance with observation, a fact that Bouvard in his correspondence with Airy had conjectured might be due to an unknown planet. It was Alexis Bouvard who had recommended his nephew to Pontécoulant.

During the meeting of June 1, 1840, Poncelet, the president, received Eugène Bouvard's reply [CR 10, pp. 874–876]. Bouvard pointed out that he had done most of the calculations in 1833, as a very young man, before he was appointed *élève* at the Observatory. Therefore, he had only calculated - accurately - what Pontécoulant had asked of him, without being able to judge the relevance of the calculations:

> Let us remark that the most careful calculator must arrive at absurd results if the geometer directing him gives him false or badly combined formulas.
>
> [CR 10, pp. 875]

As the one responsible for the formulas, Pontécoulant had earlier taken the entire honor for the results. In the *Système du Monde* he had written: "Mr. Eugène Bouvard has been so kind to help me with this tedious work, and here are the results I have arrived at." Having recalled this phrase, Bouvard continued his answer:

> Thus, while Mr. Pontécoulant thought that the results were good he attributed all the merit to himself. For him, I was only a helper who kindly gave him a hand. But, as soon as the calculations are attacked it is no longer he who has made them, it is an astronomer at the Observatory.
>
> [CR 10, p. 875]

Finally, Bouvard took exception to the following bragging remark in Pontécoulant's letter:

> Having resumed the entire calculation in question, on Mr. Poisson's request, I arrived in a few days and without the least difficulty to results which were in perfect accordance with those given by Lagrange.
>
> [Pontécoulant 1840]

Bouvard pointed out that even the best calculator would need several months to completely repeat the calculations, so that if Pontécoulant had indeed found the correction in two days, this proved that it was not the calculations but the methods that were wrong.

Having claimed to have reproved the results of Lagrange, Pontécoulant finally launched his last offense:

Therefore, I am convinced that the author of the Memoir which was presented to the *Académie* on September 16th and which accused the formulas of Lagrange of being *completely inaccurate* is himself *completely* mistaken. Lagrange's formulas are irreproachable not only for their simplicity but also for their correctness.

[Pontécoulant 1840]

The author being accused of error was Le Verrier. The reason why Pontécoulant tried to discredit the young astronomer was probably not the disrespectful remarks about the great master, which did in fact occur in Le Verrier's summary mentioned above, but rather the continuation; Le Verrier had written:

Mr. de Pontécoulant has resumed this work in the third volume of the *Système analytique du monde*; but the numbers which he has recorded are all violated by the gravest errors and sometimes exceed the correct values 20,000 and even 40,000 times. Moreover, by taking the time to be zero in his formulas, one finds values for the eccentricities and for the positions of the perihelia which do not have the slightest resemblance with those determined by observation.

[Le Verrier 1839, p. 271]

Thus, Pontécoulant's remark was a counterattack. Not being a member of the *Académie*, Le Verrier could not answer for himself, but he had a good advocate in Liouville. In the following issue of the *Comptes Rendus*, he lambasted Pontécoulant in a four-page–long sarcastic note. Since he knew that Pontécoulant would find his remarks harsh, he explained his motivation for publishing them:

The *Académie*, which last Monday heard Mr. de Pontécoulant complain in a very sour way of the report from the Committee in charge of judging the work of Mr. Le Verrier, who in particular heard him direct the most bitter and unjustified criticism toward the person who wrote the report, mustn't this *Académie* think that Mr. de Pontécoulant has been sharply censured in the report which he complained about? Well, Mr. de Pontécoulant is not even mentioned in this report. By a sense of decency, which one will appreciate, I have abstained from any allusion to Chapter XI of the *Théorie analytique*. Now, one can judge if Mr. de Pontécoulant has been grateful for this reserve. One may wonder if it is worth while to be considerate to such an adversary in the future.

[Liouville 1840m]

Thus, Liouville's note was not only meant as a defense of his student, but also as a reaction to some derogatory remarks (suppressed in the *Comptes Rendus*) about Liouville himself. Considering that both Pontécoulant and Liouville at this time had applied for a vacant seat at the *Bureau des Longitudes*, there is no doubt that Liouville welcomed this chance to put Pontécoulant in his place. Still, from the start Liouville underlined the unimportance of Pontécoulant's utterings by affecting an air of indifference:

Nevertheless, I do not attach great value to answering in detail the peculiar and sharp assertions contained in Mr. Pontécoulant's Note...

Do I need to make a great effort to prove that one does not at all undermine the immortal reputation of Lagrange either by stressing the previous observations or by saying that he has not obtained, and indeed could not obtain, satisfactory numerical results when he based them on incorrect observational data, for example a mass of Venus which differs very much from the value which the astronomers assume today? Isn't the considerable influence of Venus on the variations of the obliquity of the ecliptic well known? Doesn't the criticism which Mr. de Pontécoulant directs toward me thus fall to pieces before the eyes of the attentive reader? Does his pompous tone hide for the geometers the contradictions he has propagated?

[Liouville 1840m]

Liouville continued with a simple mathematical analysis of the new version of the notorious place in Pontécoulant's book in order to show that it was still flawed. Moreover, he listed a few other mistakes in Pontécoulant's work. According to Liouville, some of Pontécoulant's gravest errors were due to a rejection of Laplace's correct proofs and a lack of knowledge or understanding of the more recent work of Sturm, and in particular Cauchy:

Should a memoir by such a geometer [Cauchy] which carries the title *Sur l'équation à l'aide de laquelle on détermine les inégalites séculaires du mouvement des planètes*, remain unknown to a person who boasts that he has for a long time devoted his evenings to celestial mechanics?

[Liouville 1840m]

Liouville ended his attacks on Pontécoulant with the words:

However, this is enough, I have attained my object. The *Académie* can now see whether I in any way provoked the violent attack which Mr. de Pontécoulant has directed against me. It would be useless to correct all the analytical mistakes in the *Théorie analytique du Système du Monde*. How is it possible to agree with an author who does not understand (even when you point it out to him) that a definite integral $V = \int \pi dm$ (vol I., p. 186) depends neither explicitly nor implicitly on the variables x, y, z, with respect to which you integrate, and can only be a function of the limits of these variables and the parameters contained under the integral sign?

[Liouville 1840m]

With his harsh remarks, Liouville put an end to this discussion but not to Pontécoulant's astronomical production. In 1843, he published the fourth and last volume of the *Théorie Analytique du Système du Monde*, and after a break of 16 years, he again took part in astronomical discussions in the 1860s. With hindsight, his *Système du Monde* is a good popularization of Laplace's work, but Liouville was right in pointing out that his talents were not comparable to those of Le Verrier.

THE NAME NEPTUNE

During the early 1840s Liouville continued to support Le Verrier by publishing his works in the *Journal de Mathématique pures et appliquées*. The first of Le Verrier's papers [1840a] dealt with the intersection of the orbit planes of Jupiter, Saturn, and Uranus, precisely the problem Liouville had discussed in 1839 [1839f]. The second work [1840b] was discussed above. Moreover, in 1841, Liouville saw to it that the *Bureau des Longitudes* published two other papers by Le Verrier. At the same occasion:

> Mr. Liouville gave a verbal account of a memoir which Mr. Delaunay proposes to present to the *Académie* at the next meeting; it deals with the motion of the earth around its center of gravity.
>
> [P. V. Bur. Long., April 7, 1841][6]

Delaunay, who was later to compete with Le Verrier for the position as the leading French astronomer, had attended Sturm's class at the *École Polytechnique*. In 1838, he began to contribute mathematical papers to Liouville's *Journal*, but soon his interests turned to celestial mechanics. His works in this area were also promoted by Liouville. In addition to the above-mentioned talk, Liouville told the *Bureau* about Delaunay's *Calcul de deux inégalités d'Uranus qui sont de l'ordre du carré de forces perturbatrices* (March 2, 1842), and on March 13, 1844, about his *Mémoire sur la théorie des Marés*. The first of these was a continuation of the work of Alexis Bouvard (mentioned above). Urged by Arago, Le Verrier took up this problem as well and carried it to a solution on June 1, 1846, by showing that the irregularities in the motion of Uranus could indeed be accounted for by the assumption of a new planet, such as Bouvard had conjectured. LeVerrier's ingenious and accurate calculations determined the position of the new planet with such a precision that the Berlin astronomer Galle found it one day after hearing about Le Verrier's calculations, and only 52' from the predicted position.

On September 30, 1846, the news of Galle's observation was broken at the *Bureau des Longitudes*. Arago could also say that the preceding night the observation had been confirmed with three different instruments at the Paris Observatory. The following week the *Bureau* had a welcome chance to acknowledge Le Verrier's outstanding discovery. A seat as adjoint member of the astronomy section had become vacant after Largeteau and the section pointed to Le Verrier and Faye as possible successors. Although Liouville also had contacts with Faye, who translated for his *Journal*, he had no doubts about for whom to vote. From his summer residence in Toul, he wrote to Arago:

Toul, October 1846

Dear Sir and fellow member,

> Enclosed please find my vote for the election next Wednesday at the *Bureau des long.* It is sealed, as you have requested, but I

have written Le Verrier's name on it. After the magnificent discovery which we owe to him, no hesitation necessary. Moreover, I hope that Faye, who has also published a work which I think is very beautiful [probably Faye 1846] will soon receive his reward in another place. Please convey my sincere congratulations to the <u>inventor</u> of the new planet...

[Draft Ms 3618 (5), p. 31v]

At the following meeting, on October 14, Le Verrier was unanimously elected adjoint member of the *Bureau des Longitudes* and thereby obtained the right to participate in the meetings. A few months later this led to friction with his mentor. Thus, the *Procès Verbaux* of November 18, 1846, explains:

Mr. Liouville speaks about a memoir which Mr. Jacoby [sic] has published in German on the secular inequalities. He wanted the *Bureau* to arrange to have it translated. Afterward, the entire memoir or only an excerpt could be published in the *Connaissance des Temps*.

[P. V. Bur. Long., Nov. 18, 1846][7]

This proposal was adopted at the following meeting:

Mr. Faye will be asked by the *Bureau* to translate Mr. Jacoby's memoir on the secular inequalities.

[P. V. Bur. Long., Nov. 25, 1846][8]

But, Le Verrier made a reservation:

Mr. Leverrier points out the great interest of the work of the German geometer while remarking that it must be entirely worked over again because of the action of the new planet.

[P. V. Bur. Long. Nov. 25, 1846][9]

Liouville was probably irritated to see his great idol criticized by his young protégé. Still, he remained silent, although Le Verrier's remark was somewhat off the mark, considering that Jacobi's paper was purely theoretical and did not give numerical values.

The following year Le Verrier's membership developed into a burden for the *Bureau*. The reason was a dispute with Arago about the name of the new planet. I shall give an account of this episode because the little-known *Procès Verbaux* from the meetings in the *Bureau des Longitudes* nicely sets the stage for the later hostility between Le Verrier and Liouville. Moreover, Liouville was involved in the trouble in his capacity as President of the *Bureau* for 1847.

Ironically, the dispute started with Arago's strong support of Le Verrier the preceding year. At the first meeting of the *Bureau des Longitudes*, after Galle's confirmation of the existence of the new planet, the question of its name was discussed:

The members discuss the proposals which have already been made regarding the name of Mr. Leverrier's planet, and in particular the name Neptune.

[P. V. Bur. Long., Sept. 30, 1846][10]

As pointed out by Morando [1987], Le Verrier hoped that the new planet would be named after him. In fact, he tried to pave the way for such a decision when he summarized his discovery in a paper entitled *Recherches sur les mouvements de la planète Herschel (dite Uranus)*, which appeared in the *Connaissance des Temps* for 1849 (printed in 1846). Here, he explained that he considered it his duty to make the name Uranus disappear completely by calling this planet Herschel after its discoverer. Still, he knew that it would be too embarrassing if he explicitly suggested that the new planet be called Le Verrier, and so he gave this task to Arago. At the following meeting of the *Académie*, Arago gave a eulogy of the achievements of his friend and concluded by making the suggestion that Le Verrier had hoped for:

> Mr. Arago announces to the *Académie* that having received a very flattering job, the right to name the new planet, he has decided to call it by the name of the person who has skillfully discovered it, and hence to call it Le Verrier.

Would it be justified, he continued, that celestial objects were named after Halley, Encke, and many others if:

> the name of the person who by an admirable method without precedent has demonstrated the existence of a new planet, who has pointed out its position and dimensions, was not inscribed in the firmament!!! No, No! This would shock reason and the most common principles of justice. In conclusion Arago said: I commit myself never to call the new planet by any other name than Le Verrier. In this way, I think I will give an unimpeachable token of my love for science and follow the inspiration of a legitimate national sentiment.
>
> [CR 23, p. 662]

One week later Arago's nationalism burst into flames when Le Verrier's priority was contested from various sides. Arago announced a defense of Le Verrier [CR 23, p. 716], which he gave the following week (October 19) after Le Verrier had been elected member of the *Académie*. The most serious claim came from England, where John Herschel, Airy, and Challis pointed out that John Couch Adams (1819–1892) from Cambridge University had the preceding year predicted, on the basis of calculations, the orbit of a new planet very close to the one predicted by Le Verrier. In particular, on October 15, Challis had addressed a letter to the paper *Athenaeum*, containing the following passage - as translated in the *Comptes Rendus*:

> The part Mr. Adams has taken in the theoretical research of the new planet might justify that he suggest a name for it. With his consent, I mention *Oceanus* as a name which could probably be adopted by the astronomers. I have been authorized to affirm that the research of Mr. Adams will soon be published in detail.
>
> [CR 23, p. 751]

Arago flatly rejected the claims:

> Mr. Challis exaggerates the merits of the clandestine work of Mr.
> Adams so far that he to a certain degree attributes to this young
> geometer from Cambridge the right to name the new star. This pre-
> tension is not welcomed. The public owes nothing to a person who
> has not helped it or taught it anything. What do you think? Mr. Le
> Verrier has confided his research to the entire learned world; everyone
> has been able to see the new planet dawn through the first formulas
> of our learned compatriot... and quickly appear in all its radiance.
> And today somebody suggests that he should share this fame... with
> a young man who has not communicated anything to the public and
> whose more or less incomplete calculations, with the exception of
> two, are totally unknown in the European Observatories. No, No!
> the friends of the sciences cannot permit that such a glaring injus-
> tice to be committed. The journals and the letters I have received
> from several English scientists show that in that country also the
> respectable rights of our fellow countryman have zealous defenders.
> To sum up.... Mr. Adams does not have the right to appear in
> the history of the discovery of the planet Le Verrier either with a
> detailed citation or even with the faintest allusion. In the eyes of all
> impartial men, this discovery will remain one of the most magnificent
> triumphs of theoretical astronomy, one of the glories of the *Académie*
> and one of the most beautiful distinctions of our country....
>
> [CR 23, p. 754]

Arago did not have it his way. In later histories of astronomy, Adams'
early predictions are duly acknowledged.

The name of the new planet had become a matter of national pride and
one in which Arago had engaged himself so personally that he became very
upset when he found evidence that Le Verrier himself in his correspondence
with foreign astronomers opposed him behind his back, pretending that
Arago's suggestion came as a surprise to him and that he opted for the name
Neptune. The *Procès Verbaux* of the *Bureau des Longitudes* of February 24,
1847, describes the incident:

> Mr. Arago shows Mr. Le Verrier, who at that moment is sitting
> close to him, but only as a private and individual communication,
> a printed letter from Mr. Encke to Mr. Schumacher, in which the
> Berlin astronomer includes the following passage quoted verbatim
> from a letter Le Verrier had written to him on October 6, 1846:
> "I have asked my illustrious friend Mr. Arago to choose a name
> for the planet with care. I was a bit embarrassed by the decision
> which he took in the midst of the *Académie*."
> Mr. Le Verrier, who apparently preferred an official explanation
> to a private discussion, reads the above passage aloud and declares
> that he could never or should never have written anything of this
> sort. I would be to blame, he added, if the quoted passage was by
> me; but I have preserved the draft of my letter. I shall look for it
> and you will see that my words have not been quoted faithfully.

Mr. Le Verrier leaves the meeting.

[P. V. Bur. Long., Feb. 24, 1847][11]

Indeed, Encke advocated the name Neptune.

The *Procés Verbaux* probably gives a too peaceful picture of the incident. In fact, several members of the *Bureau* left the meeting so that the president, Liouville, had to close it. Le Verrier did not only leave the meeting, he also sent in his resignation from the *Bureau* to the Minister of Education [P.V. Bur. Long., March 10, 1847]. During the following weeks, Arago informed the Minister about the "unfortunate discussion" and the resignation [P.V. Bur. Long., March 3, 10, 1847], and he continued to search for evidence of Le Verrier's treachery. He made Humboldt supply him with a copy of Le Verrier's letter to Encke and informed the *Bureau* that it contained precisely the quote he had shown to Le Verrier [P.V. Bur. Long., March 10, 1847].

At the meeting on March 17, Arago read two letters from Struve and Airy, which had been printed in the *Paris Observer*:

> The directors of the observatories of Pulkova and Greenwich have made up their minds in favor of the name Neptune after having received a letter, dated October 1, 1846, in which Mr. Le Verrier told them that the *Bureau des Longitudes* had chosen this name.
>
> It is observed by reading the *Procès Verbaux* and by the recollection of all the members present that the *Bureau* has never taken a decision on this matter.
>
> [P. V. Bur. Long., March 17, 1847][12]

When reading these accounts, one must keep in mind that it was Arago, in his capacity of secretary, who wrote the *Procès Verbaux*, so one may suspect that they are not completely disinterested. Indeed, the *Procès Verbaux* of September 30, 1846, is completely crossed out, and the much shorter statement about a discussion of the name, and in particular the name Neptune, was written between the lines. As far as I can see, the original text did not state that the name Neptune had been chosen, but it is possible that it tended in that direction, and that Arago therefore changed it afterward. As president of the *Bureau des Longitudes*, Liouville had to report the incident to the Minister and to inquire what the *Bureau* should do after Le Verrier's resignation, a highly extraordinary situation in a society where members would stay until they died. His letter of March 10 was also signed by the secretary—Arago. Apparently, the Minister did not want to declare Le Verrier's seat vacant, so on July 23:

> it is considered how to ask the Minister's opinion regarding Mr. Le Verrier's resignation. It is decided to put a committee in charge of the question.
>
> [P. V. Bur. Long., undated but probably July 28, 1847][13]

The committee met with the Minister, but they were unable to clarify the matter, so on August 18, the president, Liouville, had to write to the

Minister to ask him whether Le Verrier ought to be mentioned in the list of members in the new volume of the *Connaissance des Temps*. The Minister answered by sending a letter he had received from Le Verrier. It was read at the following meeting:

> Mr. Le Verrier, in his letter, announced to the Minister that he will resume his position as adjoint member of the *Bureau des Longitudes*.
>
> [P. V. Bur. Long., August 25, 1847][14]

In the meantime, Arago had definitely lost the name battle on July 28, but he did not admit it in the *Procès Verbaux* until the day Le Verrier reentered the *Bureau*:

> At the meeting of July 28 this year, the *Bureau* decided to give the planet, whose existence was pointed out by Mr. Leverrier, the name Neptune, which prevails today among the astronomers. It was decided that a note to this effect should be published in the *Connaissance des temps* for 1850.
>
> Mr. Leverrier, through Mr. Beautemps Beaupré, demands that the following words be added to this note: Mr. Leverrier was not present at the meeting.
>
> [P. V. Bur. Long., August 25, 1847][15]

In fact, despite his reentry as an adjoint member, Le Verrier stayed away from the meetings for several years. Still, the annoyance this created among many members of the *Bureau*, including probably Liouville, was only a sign of things to come.

Hermite, Bertrand, and Serret

TWO REPORTS

We shall now turn to Liouville's relationship with three young mathematical talents: Joseph Alfred Serret (1819–1885), Charles Hermite (1822–1901), the leading French mathematician between Cauchy and Poincaré, and Joseph Bertrand (1822–1900), who after a meteoric career succeeded Arago as the administrative leader of the exact sciences in Paris. On August 14, 1843, Liouville advanced the careers of the latter two by presenting enlightening and commendatory reviews of their works submitted to the *Académie des Sciences*.

Bertrand's *Developpements sur quelques points de la théorie des surfaces isothermes orthogonales* dealt with equipotential surfaces and continued the work on orthogonal curvilinear coordinate systems initiated by Lamé (who was also on the two judging committees). The subject was also related to Liouville's own geometrical research originating in the rotating masses of fluids, and probably reinforced Liouville's interests in these matters. Having praised the work under review and Bertrand's earlier works on the theory of electricity, Liouville concluded:

To the merit of having solved with sagacity the questions he has been dealing with, he has joined the merit of chosing the questions very well. This is the sign of an excellent mind.

[Liouville 1843m]

In the subsequent review of Hermite's first famous work on Abelian functions, Liouville was even more laudatory:

It is with a great pleasure that we present today the results of the examination to which we have been devoted. In fact, a few words suffice to show the great importance of the work of our young fellow countryman.

[Liouville 1843d]

Hermite's celebrated result in the paper says that the equations describing the division of an Abelian integral can be solved by radicals. In order to explain the problem to his fellow members, Liouville reminded them that $\cos mx$ can be expressed as a rational expression in $\cos x$. Therefore, it is elementary to find $\cos mx$ from $\cos x$, whereas to find $\cos x$ from $\cos mx$ requires the solution of algebraic equations. The similar problem of finding $\varphi(x)$ from $\varphi(mx)$, when φ is an elliptic function, is called the problem of division. Abel had shown that it, too, leads to algebraic equations that could be solved by radicals. Later, Jacobi simplified Abel's ideas and further showed that the problem of division of Abelian functions led to a system of algebraic equations. These were the equations that Hermite proved to be solvable by radicals.

In his review, Liouville found several occasions to emphasize the genius of Jacobi, both in order to pay tribute to a mathematician he admired and as a way to put Hermite's results into relief. In this way, he happened to tread once more on Libri's sore toes, starting thereby the last big quarrel between the two archenemies. I shall briefly summarize this incident before returning to the three young talents.

LAST CLASH WITH LIBRI

After Liouville had refereed Hermite's work, Libri rose and expressed his exception to the following phrase in the report: "Abel was the first to develop the general theory of the division of the elliptic functions." Repeating what he had already said in earlier quarrels with Liouville, Libri claimed that he had, before Abel, proved Gauss's assertion about the division of the lemniscate, that is, the simple elliptic integral $\int \frac{dx}{\sqrt{1-x^4}}$. This time Libri even extended the claim to all elliptic functions, maintaining that his method was universal.

Although it had been only one month since Liouville had been humiliated by the election of Libri to the *Collège de France*, I do not think that Liouville consciously aimed at a confrontation. If that had been his goal,

PLATE 11. Charles Hermite (1822–1901). Liouville strongly supported the young Hermite and was in turn inspired by Hermite's ideas on elliptic functions. (Math. Dept. University of Copenhagen)

he would probably have mentioned Gauss's remarks on the lemniscate and Abel's first proof of that. He could not have foreseen that Libri would be so presumptuous as to claim credit for the general solution. But, when he did, Liouville probably did not mind the chance this offered him to get the better of his enemy.

Liouville answered [1843g] that even if Libri had solved the division of the lemniscate his phrase about Abel still held true, and he promised to prove at the following meeting that even the lemniscatic problem had not found its proof in Libri's papers. He stuck to this despite Libri's subsequent attempts to make Liouville amplify his assertions immediately.

The next Monday Liouville opened the meeting by reading an eight-page-long analysis, or rather slating, of Libri's work. In his paper of 1825 (published in 1832 [1832]) Libri had claimed that the division of the lemniscate could be solved by a theorem on the solution by radicals of polynomial equations, whose roots can be found from one root by repeated application of a rational function. Liouville pointed out that (1) the theorem does not solve the division of the lemniscate, (2) Libri's theorem contains an unnecessary restriction, (3) Libri did not give a proof of this theorem until after Abel had published his simple proof, which had according to Liouville almost been given by Poisson, and (4) even after having seen Abel's paper, Libri garbled the proof when it was published in 1830, basing it on a wrong lemma, namely, that all powers of a root of the equation can be expressed linearly by the other roots. Liouville concluded:

> Moreover, in this Memoir of 1830, the false and the true alternate rapidly among vague phrases. From the first pages, one can see the

PLATE 12. Joseph Louis François Bertrand (1822–1900). The young Bertrand made a brilliant mathematical career, supported by Liouville. He later became secretary of the Académie des Sciences and wrote biographies of many former members but unfortunately never one of Liouville. (Archives de l'Académie des Sciences de Paris)

> author applying to irrational functions some operations which only make sense for rational functions. But this must be enough, or too much, perhaps. I abuse the precious moments of the *Académie* by continuing this discussion.
>
> [Liouville 1843h, p. 334]

Having expressed his regret that Liouville's contribution had degenerated to a simple conversation instead of a serious and elevated discussion, Libri postponed his answer for two weeks because Liouville informed him that he would be absent at the next meeting.

Although meant differently, Libri's description of Liouville's talk gives a good idea of Liouville's great oratorical gifts (cf. [Laurent 1895]).

> I must admit that if I had not been slightly sensible to the external circumstances, the form of this speech would have impressed me strongly. Everybody recalls his peremptory tone, his commanding gestures, and his often repeated words like *false demonstration* and *grave errors*, which he directed toward me in the form of an indictment.
>
> [CR 17, p. 432]

According to Libri, Liouville's oral and untechnical argument had perhaps convinced the nonmathematical members of the *Académie*, although the question could not be "clarified completely without analytical signs and notation." Nevertheless, Libri's own 14-page answer did not contain one single formula. Instead, he put forward two good counterarguments: (1) It was wrong of Liouville to state that everybody had rejected Libri's proof. Lacroix and Crelle had accepted it. (2) It was not true that Libri's theorem had been known earlier. Indeed, here Liouville had gone too far. Moreover, Libri confronted Liouville with two sins from his youth. It was known that Liouville had earlier claimed that a theorem was uninteresting just because its proof was easy. He referred to the remark on Jacobi's ellipsoid, which I think Liouville would rather have forgotten. Also, Libri pointed out that even if his proofs did not cover all cases, they need not be rejected, especially by Liouville, who had written in 1832 [1832a, p. 8]: "All the formulas which I use in this Memoir are subject to restrictions... which I do not feel I need to make explicit. The way in which we prove them indicates sufficiently clearly when and how they are correct."

Libri admitted that he had left out certain simple arguments because he had not intended to write for students, but for his fellow academicians. When he taught the theory of equations at the *Collège de France* and some of his students had raised the same questions as Liouville concerning this paper, Libri had given them "some useful indications" and "after a little reflection everyone had understood it." After this elegant insult, Libri continued with a rather vague defense of his theorems, maintaining that they were all correct. For example, he argued that Liouville was wrong when he believed that one could leave out the restriction in his theorem and still conclude that the equations in question were solvable by radicals. For in that case, "the radicals are all rational. Thus, there will not be any radicals." [CR 17, p. 441]. Libri must have known that here he was completely misusing the terminology "solvable by radicals," but it may have convinced some of the nonmathematical members.

As for the lemma, Libri asked for the counterexample that Liouville had promised and concluded:

> Finally, my adversary has shown in his criticism that he has not applied himself sufficiently to the theory of equations, a theory one must know in depth if one wants to discuss such delicate questions. If he had done that, as one would have expected from him, he would have saved himself the trouble of proclaiming that there were grave errors in places where he should have met only elementary difficulties.
>
> [CR, 17 p. 445]

Naturally, this provoked Liouville to answer back immediately. He carefully went over a concrete counterexample to Libri's lemma, and concerning the superfluous restriction in the theorem, he asked Poinsot to state his opinion. Not surprisingly, Poinsot affirmed "in the most positive way all that I [i.e., Liouville] have said on this matter," after which Liouville left that subject because "my feeble speech cannot add to the authority of his [Poinsot's]." Liouville finally used the opportunity to announce that he had discovered that Galois' papers contained a profound treatment of the question of solution by radicals and that he intended to publish an annotated edition of these papers. This was the only constructive remark in the whole discussion.

After two more weeks, Libri answered back on September 18 with a two-fold remark on Liouville's counterexample. First, he remarked:

> First, it should be observed that even if the lemma in question was not always true it does not follow that it must be rejected.
>
> [Liouville 1843i]

Second, he argued that Liouville's counterexample was invalid because he had assumed that the coefficients in a certain equation must be rational, which was not true. Finally, he amplified his criticism of Liouville's fractional calculus by referring to Peacock's rejection of it (Chapter VIII(48)). It is amazing that Libri had not discovered this weakness in Liouville's production earlier. Now, Liouville could easily reject the criticism by referring to other British mathematicians (e.g., Greatheed [1839]), who had defended Liouville's work against Peacock. Otherwise, Liouville repeated himself in his answer, and so did Libri in his answer to the answer.

Apparently, Liouville had become very upset during this long quarrel, for Libri remarked:

> When one of the two opponents in a scientific discussion is carried away by his passion and instead of expressing his arguments calmly pronounces utterances full of bitterness, he shows in the eyes of the less clearsighted that he gets angry only because he lacks good arguments. In this respect, I ask Mr. Liouville to receive my sincere thanks for the form he has adopted in this discussion.
>
> [CR 17, pp. 551-552]

Despite his state of agitation, Liouville was wise enough not to respond to Libri's last remarks, and so this discussion ended. But, Libri continued his fight against the establishment in the *Académie*. Parallel with his discussion with Liouville he had been engaged in a debate with Arago (August 7 - September 4, 1843) on a historical subject concerning Galileo's manuscripts, and three weeks after his quarrel with Liouville had come to an end, he launched a new attack against Arago (October 9). Triggered by a priority dispute between two biologists, Libri returned to Arago's criticism of his Galileo research in the *Annuaire du Bureau des Longitudes*. In this new debate, Liouville intervened on Arago's side [CR 17, p. 769]. On the same day, Liouville presented the last of two papers that were offshoots of his and Libri's debate.

The first [Liouville 1843k], presented to the *Académie* on October 2, 1843, dealt with the division of the complete lemniscate by complex Gaussian integers. In this way, he filled in a hole that Abel had left behind at his death (cf. Chapter XIII, §4–8). Proceeding according to the ideas of Abel, Liouville established that the roots of the equation in question can be expressed rationally by one of them, say x, such that if $\theta_i(x)$ and $\theta_j(x)$ are any two other roots then

$$\theta_i[\theta_j(x)] = \theta_j[\theta_i(x)].$$

According to a theorem of Abel, this implies that the equation is solvable by radicals.

In the second note [Liouville 1843l], presented the following week, Liouville remarked that the above "Abelian" criterion for solvability could be exchanged with others. In particular, he stressed that the central idea was "groups of roots... rather than... isolated roots." Despite the vagueness of the hint, it indicates that Liouville had started to read and understand Galois.

HERMITE AND DOUBLY PERIODIC FUNCTIONS

The review of Hermite's paper on Abelian functions, the quarrel with Libri, and the subsequent two notes were the first public signs that Liouville had entered into two new fields of research: elliptic functions and Galois theory. In the discussions described above, the two areas were closely connected through the problem of division of elliptic functions by radicals, but otherwise, Liouville treated them separately. I shall follow Liouville, starting with the elliptic functions. Liouville's work in this direction closely interacted with that of Hermite, so let me resume the story of their scientific relationship.

Like Liouville, Hermite was born in Lorraine. At the *Collège Louis le Grand* in Paris, he learned his first mathematics from Galois' previous teacher, Mr. Richard, whom Hermite described as "an excellent man of superior merits." When Hermite was finally admitted to the *École Polytechnique* in 1842 after having failed the entrance examinations the previous year, he received a mathematical education that satisfied him less:

> Mr. Sturm's lectures were of less use to me than those of my dear master Mr. Richard, but I was acquainted with Mr. Liouville to whom I am indebted for having given my parents the advice, which they followed, of allowing me to leave the *École* and follow my propensity towards mathematics. That was in 1843, and since then I have been entirely devoted to the study of elliptic functions and the *Disquisitiones arithmeticae* by Gauss.
>
> [Hermite 1984, letter 80]

Although Liouville did not teach Hermite's class, he rapidly detected the extraordinary gifts of the young student, and he immediately introduced him to an even more competent judge. In January of 1843, only two months after his start at the school, Hermite wrote a letter to Jacobi telling him about his result on the division of Abelian functions. It began:

M. Liouville urged me to write to you in order to submit this work.

[Hermite 1846]

Hermite presented his discovery to the *Académie*, and as I have mentioned, Liouville reviewed it very favorably. Later the same year Liouville inserted the review in his *Journal*, followed by Jacobi's answer to Hermite, which began:

> I thank you sincerely for the beautiful and important work you have communicated to me regarding the division of Abelian functions. By the discovery of this division, a new field of research has been opened to you, and the new discoveries will give a new boom to the analytic art.
>
> [Jacobi 1843, p. 505]

On June 17 of the following year, Liouville received another letter from Hermite on Abelian functions which was published in the *Comptes Rendus*. Until then, it was Hermite who had profited most from the relation with Liouville, but then Liouville began to get his reward. Hermite's work had opened a world to Liouville that he previously had only known superficially. In connection with the notes following the discussion with Libri, Liouville had begun a detailed study of Abel's work on elliptic functions, a study he continued the following year [Ms 3617 (5)]. At first, they had led him to new investigations on integration of differential equations in finite form, but he had concluded the notes with the remark "All this has made little progress since our first investigations." Indeed, there are no published results from this research. Soon after, however, a small remark by Hermite set Liouville off on his own fruitful research on elliptic functions. (For a more detailed discussion of this, see Chapter XIII.)

Abel and Jacobi had revolutionized the theory of elliptic functions by allowing complex variables and by focusing on the inverse functions to the elliptic integrals. These ideas were used by Hermite and Liouville in their 1843 works. Abel and Jacobi had shown that the inverse functions have two rationally independent periods, of which at most one can be real. In August 1844, Hermite suggested to Liouville that one may use trigonometric series in order to prove that there is only one real period. It was an easy task for Liouville to carry through this demonstration, but at first he thought that this approach was farfetched. After a few months, however, he realized that a slight variation of this proof established what he called *our principium*:

> A well-determined doubly periodic function which never becomes infinite reduces to a simple constant.
>
> [Ms 3617 (5), p. 113v]

This is what we now call Liouville's theorem, in the case of a doubly periodic function. Liouville honored it with the name "our principle," because he made it the basis of a simple theory of doubly periodic functions. Thus, he soon discovered that a doubly periodic function cannot have just one simple pole in a period parallelogram, and when it has more poles, it can be expressed rather simply by inverses of elliptic integrals. In this way, Liouville turned the entire theory of elliptic functions upside down. Where everybody before him had started with the integrals and their inverses, Liouville, within a few months - probably weeks, created a general theory of doubly periodic functions based on his principle, from which he then drew conclusions about the elliptic functions. This is the idea we still apply.

Although it is one of Liouville's most important contributions to analysis, it had to take a detour before it was handed down to us, for Liouville published only a brief outline of the theory in an appendix to a note on geodesics. Liouville communicated it verbally to the *Académie des Sciences* on December 9, and had it printed in the *Comptes Rendus* [1844g]. On $\frac{3}{4}$ of a page, he mentioned his principle and concisely explained that it gave rise to a theory in which "the integrals that have given rise to the elliptic functions and even their moduli disappear in a way, leaving visible only the periods and the points for which the functions become zero or infinite."

He also mentioned a few results he had derived from his principle, among them a result that Jacobi had stated without proof. At this point, he expressed his great admiration for Jacobi, and even mentioned that "Hermite (who has devoted himself to these questions in his profound research) has before me found a proof, about which he has kindly informed me. Besides, the method which he has used is very different from mine" [Liouville 1844g, p. 1263].

It is characteristic of Liouville during this period of his life to so graciously give credit to his colleagues and protéges, even at a moment when he announced one of his own great discoveries.

Cauchy, present at the meeting, immediately saw that Liouville's principium was an easy consequence of his own earlier theorems on complex integrals, and at the following meeting, he presented two proofs based on his calculus of residues. Cauchy's proofs immediately established Liouville's theorem in its full generality: "If a function $f(z)$ of a real or imaginary variable z always remains continuous and consequently always finite it simply reduces to a constant" [Cauchy 1844b, p. 1342].

Thus, in fact Cauchy was the first to announce what we call Liouville's theorem. Despite Cauchy's claims of priority, Liouville nevertheless deserves credit for it, because not only had he discovered its great importance in the theory of elliptic functions, he also deduced the general theorem from his "principle" by a simple composition with Jacobi's elliptic function sinam. This unpublished proof seems to predate that of Cauchy (cf. Chapter XIII, §17–21). Moreover, Liouville discovered the now well-known elegant proof of the fundamental theorem of algebra based on Liouville's theorem: if $f(x)$ is a polynomial without complex roots, $\frac{1}{f(x)}$ is bounded in C and therefore constant. Thus any nonconstant polynomial has a complex root (Since Liouville deduced this result before he derived the general version of Liouville's theorem, his proof was slightly different (cf. Chapter XIII, §18)).

It is remarkable that Liouville did not use Cauchy's theory of complex integration in the development of his theory. Still, after Cauchy's intervention, he began to discuss elements of Cauchy's theory openly at the *Académie*. On the same day as Liouville presented his remarks on doubly periodic functions to the *Académie*, he pointed out that Cauchy had announced two versions of his famous theorem stating that the Maclaurin series expansion of a complex [meromorphic] function is convergent if the modulus of the variable is smaller than that for which the function has a singularity. The problem was how "singularity" should be understood. In the original version of the theorem published in 1831 in Turin, Cauchy only required that "the function $f(x)$ stops being finite and continuous." He repeated this in a letter to Coriolis, published in the *Comptes Rendus* of 1837. However, when he rephrased the theorem in the first volume of his *Exercices d'Analyse* of 1840 and 1841, he required that both f and f' cease to be bounded and continuous [Cauchy 1831, 1837, 1840d]. Liouville pointed out the discrepancy and said that he preferred the first version.

Liouville's remark is not contained in the *Comptes Rendus*, so it may just have been a private communication to Cauchy. Perhaps, motivated by Liouville's remarks on doubly periodic functions, the two academicians had a chat about complex functions after the meeting. Our source for this discussion is Cauchy's answer, printed in the *Comptes Rendus* of the next meeting, and also containing his generalization of Liouville's theorem:

> Thus, it follows, in accordance with the judicious observation which Mr. Liouville communicated to me last time, that among the two statements of the theorem given in my memoir from 1831 and in my *Exercices d'Analyse* it seems convenient to chose the first.
>
> [Cauchy 1844b]

Kline [1972, p. 639] writes that it was Liouville and Sturm who convinced Cauchy that he should add the assumption on the derivative of f. I have not been able to confirm this statement, which would imply that Liouville had changed his opinion completely. (Botazzini [1986, p. 179] treats later views on Cauchy's theorem.)

On March 31 the following year, Liouville commented on a paper Cauchy had presented on the approximation of integrals:

> Mr. Liouville verbally presented some remarks regarding Cauchy's communication and also regarding the report which the learned academician read during the meeting on March 17th when Mr. Liouville was not present. Liouville's observations exclusively concerned some of the assertions contained in this report and not the conclusions themselves, with which he gladly agrees.
>
> [CR 20, p. 927, March 31, 1845]

Cauchy's report dealt with Le Verrier's memoir on the motion of Pallas. It had been very positive and had been followed by three long notes in that issue and three more in the next issue of the *Comptes Rendus* in which

Cauchy used his own methods to confirm Le Verrier's results. Liouville in his remark probably expressed misgivings about this behavior, which had completely outshone Le Verrier's original contribution. But, these exchanges of opinion did not change the mutual respect between Liouville and Cauchy. Thus, in 1846 when Liouville's friend Ernest Lamarle published a paper criticizing Cauchy's formulation of the theorem discussed above, Liouville let Cauchy defend himself in a paper published a few months later. In this paper, Cauchy once more referred to Liouville's decisive remark:

> However, a remark by Mr. Liouville called my attention to this subject, and urged me to examine it again in a *Mémoire sur quelques propositions fondamentales du calcul des résidus et sur la théorie des intégrales singulièrs.*
>
> [Cauchy 1846a, p. 327]

The interaction between Liouville and Cauchy's research on complex functions remained rather weak, whereas the recurrent mention of Hermite's name in Liouville's notes on doubly periodic functions suggests that the latter two developed their theories in close contact, though in somewhat different directions. Liouville continued to support Hermite, for example, by giving an account of Hermite's theory of equations to the *Bureau des Longitudes* on June 16, 1845. Moreover, in February of that year, Liouville, on behalf of a commission, announced that the subject of the *Académie's Grand Prix des Mathématiques* of 1846 would be "Abelian functions." It seems fairly certain that it was Liouville who had suggested this theme in order to give Hermite a chance to win this prestigious prize. However, after having postponed the deadline in 1846, the *Académie* finally awarded the prize to Jacobi's student Rosenhain (1816–1887) in 1851, for his breakthrough in the problem of inversion of Abelian integrals. His ideas, independently discovered by Göpel, were later extended by Hermite.

The most direct evidence of a collaboration between Liouville and Hermite can be found in a letter from Liouville to Jacobi of 1846:

> There [in Toul] I am the neighbor of Mr. Hermite, who usually lives in Nancy. We shall read your works again, and perhaps this will soon give us a scientific occasion to write to you.
>
> [Liouville to Jacobi, June 1, 1846, in Jahnke 1903]

Partly as the result of Liouville's propaganda, Hermite was finally appointed admission examiner and *répétiteur* at the *École Polytechnique* in 1848.

Although Liouville did not publish his theory of doubly periodic functions, he lectured on it. In 1847, he gave a private lecture to two foreign students, Borchardt and Joachimsthal. Recall that at this time Liouville had no position at the *Collège de France*, and therefore he lacked the forum of public lectures on the subject. The two young Germans seem to have had close relationships with Liouville, and before they left Paris, Liouville wrote from Toul to Joachimsthal to encourage him to continue his own and Liouville's earlier research on geodesics in the light of Jacobi's "great Memoir." He ended the letter:

> Goodbye my dear Mr. Joachimsthal. Do not forget, together with Mr. Borchardt, to say a few words about my *principium* to Mr. Jacobi. If he does not laugh too much at my pretensions regarding elliptic functions, I shall seriously put myself to work. Please remember me to my good friends in Berlin, in particular to Mr. Jacobi and Mr. Dirichlet.
>
> [Neuenschwander 1984a, III, 5]

One can imagine that Liouville was anxious to hear Jacobi's opinion, for not only was Jacobi the great master of elliptic functions, he was also among Liouville's greatest idols in other fields, such as rational mechanics. In 1846, Liouville showed his great admiration for Jacobi by strongly supporting his candidacy for the position of *associé étranger* at the *Académie*. In the letter, from which I have quoted above, he informed Jacobi that on that very day he had been elected with 46 out of 47 votes.

> Allow me to congratulate myself for having done my humble best for this happy result as a member of the commission nominated to prepare a list of candidates. Our list of associates begins with the names of Newton and Leibniz, the names of Gauss and of Jacobi figures with dignity next to them. You alone can console us for the great loss we have suffered by the death of Mr. Bessel.

When he returned to Berlin, Borchardt edited the notes he had taken during the lectures and sent one copy to Liouville, one to Dirichlet, and one to Jacobi [Liouville 1880]. On September 28 the following year (1848), Borchardt finally informed Liouville about Jacobi's reaction:

> Mr. Jacobi writes to me that he has read your theory with great plea-sure, that he has drawn much information from it, that he strongly wishes to see it published soon, and that he plans to write to you then.
>
> [Borchardt to Liouville in Liouville 1855i, p. 208]

However, by 1848 Liouville was busy with political issues, so he did not find time to put himself to work as seriously as he had promised Joachim-sthal. The publication Jacobi was waiting for never appeared. Not until 1880, two years before Liouville died, did Borchardt publish his notes in Crelle's journal, which he now edited. By then, however, Liouville's ideas had become known through the circulation of Borchardt's notes and through a course that Liouville gave in 1851. I shall return to this later.

J.A. SERRET; "ELLIPTIC CURVES"

A year after having discovered his principle, Liouville was introduced to another aspect of the theory of elliptic functions through his protégé Joseph

Alfred Serret (1819–1885). In Ossian Bonnet's obituary of Serret, Liouville's encouragement of the young Serret was described as follows:

> All these works...caught the interest of the geometers. Liouville whose opinion was law at the time, welcomed them with praise, inserted them in his *Journal de Mathématique* and made them the object of several reports to the *Académie* always concluding with a publication in the *Recueil des Savants étrangers.*
>
> [Bonnet 1885]

Serret graduated in 1840 as a mathematics student of Duhamel, wrote his first paper in Liouville's *Journal* in 1842, and thereafter became a regular contributor to this journal. Among his first works is one on fractional calculus [Serret 1844]. Otherwise, it is remarkable that hardly any of the young mathematicians Liouville guided chose to continue his work. In general, his students came to Liouville with their ideas, he listened to them, became interested, studied the matters himself, discussed their ideas, guided them toward fruitful questions and perhaps started investigations of his own in the area. Also, J. A. Serret soon went his own way and composed a memoir on representation of elliptic and ultraelliptic integrals. This was the first work of Serret's that Liouville refereed with praise to the *Académie*. It took place on July 30, 1845, a few weeks after he had told the members of the *Bureau des Longitudes* about its contents. Elliptic integrals derive their name from the fact that one of them expresses the arc length of an ellipse; another gives, as we have seen, the arc of a lemniscate. In his paper, J.A. Serret addressed the general problem of determining algebraic curves, the arclength of which is represented by an elliptic or hyperelliptic integral. Liouville concluded his report:

> Mr. Serret has shown us an infinity of algebraic curves whose arcs also have the property of having differentials equal to those of an elliptic function of the first kind. From there a new field is opened for geometric speculations.
>
> [Liouville 1845e, p. 283]

Liouville immediately entered this new field. In his report, he remarked that he had assured himself that a certain parameter, which Serret had assumed to be an integer, could also be fractional without altering the conclusions. This observation greatly enlarged the number of elliptic integrals to which Serret's argument applied. When Liouville printed Serret's great paper later the same year in his *Journal* [Serret 1845a], he added a proof of his claim in a note [Liouville 1845f].

Later the same year, Serret [1845b] also generalized and simplified his results in another note in Liouville's *Journal*, and on December 8 Liouville pushed these ideas still further at the *Académie* [1845g]. Liouville returned to the subject of elliptic curves in the 1850s in a series of unpublished notes (cf. Chapter XVII, §27–29).

Serret and Liouville reached such friendly terms that in the summer of 1847 Serret visited the Liouvilles in Toul. "Mr. Serret came to stay a week

PLATE 13. Joseph Alfred Serret (1819–1885). Serret's first works were praised by Liouville, but during the 1850s the two former friends had a falling out. (Archives de l'Académie des Sciences de Paris)

with us. Afterward, we have sunk back into our monotonous existence" [letter from Liouville to Laugier, Toul, August 27, 1847; Ms 3623 (9), p. 57v].

By then, Liouville had probably introduced Serret to one of his other future areas of research: Galois theory.

GALOIS THEORY

I have previously mentioned (p. 85) that in one of his contributions to the discussion with Libri in 1843 Liouville announced the discovery of Galois' ingenious results:

> At the end of a discussion containing so many algebraic equations, I want to excite the *Académie* by announcing that in the papers of *Evariste Galois* I have found a solution which is both correct and profound of the following beautiful problem: given an irreducible equation of prime degree, decide if it is solvable with the aid of radicals.
>
> [Liouville 1843i]

It is among the well-known stories in the history of mathematics how the young revolutionary Evariste Galois (1811–1832) failed to get his novel ideas across to the mathematical community of his day; how he failed to enter the *École Polytechnique*, how he was expelled from the *École Normale*, and how Cauchy did not finish a report he was supposed to write on the first paper Galois sent to the *Académie*; how his second paper was lost after Fourier's death; and how Poisson judged the third paper to be incomprehensible. Taton [1971] has shown that Cauchy did sense the importance of Galois' work, but when Galois died, only 20 years old after a duel in 1832, nobody really understood his ideas. In a letter written the night before the fatal duel to his friend Auguste Chevalier, Galois sketched the main results he had obtained in the theory of equations and on elliptic functions and asked that Chevalier show them to Gauss and Jacobi. Chevalier saw to it that Galois' letter was published, but there was no reaction from Gauss, Jacobi, or anyone else.

Ten years later, some of Galois' friends persuaded Liouville to study Galois' *Nachlass*. This is attested by a series of notes from 1842 [Ms 3617 (2), (3)] in which Liouville applied Galois' ideas, in particular his term *groupe*, in an unsuccessful new attack on his old problem of integration of differential equations in finite terms. Chevalier had prepared an annotated copy of Galois' *Mémoire sur les conditions de résolubilité des équations par radicaux*, which Poisson had found incomprehensible, and having been convinced about its importance, Liouville had this paper typeset for the December 1843 issue of his *Journal*. However, at the last minute, he decided to replace it with papers by Serret and others. Whether Liouville had become uncertain about some detail in the paper or whether he had already decided to prepare a more comprehensive edition of Galois' works is unclear, but when the *Mémoire* finally appeared in the 1846 volume of the *Journal*, Liouville added a fragment of another unpublished paper, the last letter to Chevalier, and all of Galois' previously published works.

While announcing the importance and correctness of Galois' methods in the *Académie* in 1843, Liouville also admitted that "Galois' *Mémoire* is perhaps written in a too concise way," so he promised "to complete it by a commentary that will leave no doubt concerning the reality of the beautiful discovery of our ingenious and unfortunate fellow countryman" [Liouville 1843i]. This intention was repeated in Liouville's preface to Galois' works in 1846, but his commentaries never appeared. What was the reason for the three years' delay of the publication, and why did Liouville never publish his promised notes?

The most plausible reason for the delay is that Liouville found it harder and more time-consuming than expected to penetrate the ideas of Galois, in particular into the deep connection between the structure of groups and the solvability of equations. In his *éloge* of Galois, Bertrand later recalled:

> When Liouville published the Memoir, which Poisson had found obscure, fifteen years after the death of Galois, he announced a commentary which he has never given. I have heard him declare that the proof was very easy to understand. When he saw that I made a gesture of astonishment he added, "It is sufficient to devote a month or two to it, without thinking of anything else." These words explain

and justify the embarrassment to which Poisson dutifully admitted and which was no doubt met by Fourier and Cauchy as well.

[Bertrand 1902, p. 342]

Liouville took to the traditional method used by teachers who want to understand a new subject:

> In order to understand completely the work which he was planning to annotate, Liouville invited some friends to follow a series of lectures on Galois' work. Serret was present at these lectures.
>
> [Bertrand 1902, p. 342]

We do not know when Liouville gave his private lectures on Galois theory, but the winter of 1843–1844 is a plausible date. It may have been around this time that he wrote a series of notes on Galois' *Mémoire sur les conditions...* [Galois 1831–1846], which are now bound together with Galois' papers. They show that Liouville had understood Galois' ideas well enough to fill the biggest holes in his arguments, in particular, in the central theorem on the reduction of the Galois group when a root of an auxiliary equation is adjoined to the "known quantities."

The existence of these notes, of which the last version is almost ready for press, accentuates the question of why were they not published in 1846? One possibility is that Liouville came to realize that there was more to Galois theory than his notes revealed, and therefore postponed his commentary until he could write a separate and more comprehensive work. Such a plan is attested by Serret in the first edition of his *Cours d'Algèbre Supérieure* [1849]:

> Galois has given the necessary and sufficient condition that an irreducible equation of prime degree is solvable by radicals. My friend, Mr. Liouville, has told me that he intends to publish one day some developments of this remarkable work. It is only through these developments, of which Liouville has kindly communicated some to me, that I have come to understand certain points of Galois' *Mémoire*.
>
> [Serret 1849, 26. leçon]

Although this plan of publishing a work on Galois theory never materialized, Liouville's activity in Galois theory had wideranging influence on the subsequent development of algebra. Thus, it is probable that Cauchy's renewed interest in permutations was directly or indirectly inspired by Liouville. In any case, Cauchy, who had studied permutations in his youth (cf. [Belhoste 1985, p. 65]), suddenly returned to this subject in 1845 when he was appointed reviewer of Bertrand's *Sur le nombre de valeurs que peut prendre une fonction quand on permute les lettres qu'elle renferme*. Bertrand's interest in permutations, in turn, probably stemmed from his mentor, Liouville, either through the private course or through private discussions, which are attested by the remarks quoted above and by a letter of 1847 [Neuenschwander 1984 a, III, 2].

This influence of Liouville's course may be conjectural, but there is no doubt about the importance of his publication of Galois' work. From this edition, one of the most fruitful branches of algebra was developed by Italians, in particular Betti; Frenchmen, such as Serret and Jordan; and Germans, for example, Dedekind, Weber, and later Artin (cf. [Kierman 1971/1972]).

For a fuller analysis of Liouville's role in this development, the reader is referred to Chapter XIV and to Chapter V, where I shall discuss the friction it created between Liouville and Serret.

Foreign Visitors

STEINER, THE DUBLIN SCHOOL, GEOMETRY

During the 1840s, Liouville introduced many foreigners to the French scientific world. One of the earliest, Jacob Steiner, can hardly be grouped among the young protégés. He was a professor of geometry in Berlin, and 13 years Liouville's senior. When he came to Paris in 1840, he was introduced to Liouville through letters from Dirichlet, but Liouville replied:

> When you sent me your letter, I was in Toul. I am still there and consequently, I have not met Mr. Steiner. I am pleased to hear that he will spend the winter in Paris. We will make the most of this occasion and ask him to communicate the treasures of his geometry to us. He will learn French and I will try to set to study a little German.
>
> [Tannery 1910; for dating, see Neuenschwander 1984a, p. 69]

Half a year later, Dirichlet inquired about Steiner:

> What does Mr. Steiner do in Paris? Does he confine himself to savor the delights of your capital? Try to push him to work a bit. He had very beautiful projects when he left us, so if he returned without having produced any of his *treasures*, one would accuse you and your friends for letting him be completely absorbed in your charming company. In particular, I call your attention to a manuscript which he has brought with him and whose subject is a summary of his most important research. If Mr. Steiner would resign himself to not saying all that he knows on this subject, which is almost inevitable for such a deep author, I am sure that it will not take him a long time to put this work into a publishable state and that it would be a splendid success. Although I am but a simple amateur in Geometry, I still know enough to be sure that the competent judges will be very interested to see a work which from a common source derives theorems which have not been connected until now.
>
> [Tannery 1910, p. 15]

This almost fatherly plea helped. On March 15, 1841 the, academicians listened to an extract from Steiner's paper, and on May 29, Liouville [1841g] read a clear and commendatory review of it to the same assembly. Later the same year, Liouville published it in Wertheim's French translation [Steiner 1841]. Steiner's great paper dealt with maximum and minimum properties of geometrical figures. He first established the isoperimetric property of the circle with a beautifully simple proof, which, however, did not establish the existence of a figure of maximal area (cf. [Monna 1975]). From this, he then derived many other theorems on extreme values. A single note in Ms 3617 (3) bears witness to Liouville's genuine interest in Steiner's works, which the following year resulted in two short notes on geometric maximum and minimum problems [Liouville 1842c,d]. In his report, Liouville particularly emphasized the ease with which Steiner could transfer theorems from plane geometry to spherical geometry. Five years later, Liouville carried this idea further by generalizing a max-min theorem from the plane and spherical cases, where they had been shown by Chasles, to the surface of a three-axial ellipsoid [Liouville 1847e,h].

Indeed, if Liouville's geometric interests had been awakened in 1841 by Steiner's work, it was kept alive and grew because of the contact with Chasles, who became a very good friend of Liouville's. Liouville's geometrical works are analyzed systematically in Chapter XVII, so here I shall just place them in the general history of geometry in the mid-nineteenth century.

During this period, the geometers were divided into two camps: the analytical camp, which, following Descartes, Leibniz, Euler, and others, used algebraic methods (and differential calculus) to obtain geometric results, and the synthetic camp, which, following Monge and Dupin, advocated for a purely Euclidean descriptive, or projective, approach. The establishment generally belonged to the first camp, whereas Poncelet, Steiner, and Chasles were the main members of the latter.

Liouville revealed his membership in the first camp in his first and most important paper on algebraic geometry, presented to the *Académie* just after Steiner's visit [CR 13, August 23, 1841, 1841d]. Here, he proved with algebraic means a theorem that Chasles had proven geometrically and whose algebraic proof he had declared would be difficult to find. Also, in most of Liouville's papers from 1842–1843 on the geometry of the surface of an ellipsoid, he followed Jacobi's analytical approach [Liouville 1843b, 1844c, 1847e].

Despite his great analytical talents, however, Liouville was not dogmatic, and he fully appreciated the synthetic methods employed by Steiner and Chasles. In his review of Steiner's memoir, Liouville commented on some of the theorems that Steiner had derived by a succession of simple deductions:

> These propositions... would certainly have caused the most skilled geometer difficulty, especially if they would have tried to tackle them with analytical methods.
>
> [Liouville 1841g, p. 934]

In order to counterbalance this acceptance of synthetic methods, Liouville quoted Steiner's admission of the analytic alternative:

> I think that the analytic method and the geometric method do not exclude or repel each other; on the contrary, they are both indispensable to overcome the great difficulties of the matter.
>
> [Steiner quoted in Liouville 1841g, p. 935]

In accordance with this wise remark, Liouville later published several synthetic arguments (cf. Chapter XVII, §2–7), of which the most noteworthy dealt with the interplay between the geometry of ellipsoidal surfaces and Abelian integrals. After Chasles had presented a geometric investigation of elliptic integrals to the *Académie* on December 9, 1844, Liouville [1844g] suggested that he might extend the results to Abelian integrals. In particular, he suggested that Chasles might try to find a geometric derivation of a simple first integral that Liouville had found analytically for the Abelian equations for the geodesics of a triaxial ellipsoid. (It was at the same occasion that Liouville announced his new approach to doubly periodic functions.)

Chasles soon found the desired proof, and so did Liouville [1846b,c]. Moreover, Liouville's remarks on geodesics gave rise to a series of papers by three Dublin mathematicians, James Mac-Cullagh and the brothers Michael and William Roberts. Liouville had met the former in Paris, and he corresponded with the brothers, giving them the following advice:

> Please give your brother my compliments, you must enjoy working with him. Imitate the brothers Bernoulli, but not their disputes.
>
> [Liouville to M. Roberts, March 13, 1847, Neuenschwander 1984b, p. 60]

Liouville started to promote the works of the Roberts brothers in 1845, when Michael sent him an application of Liouville's integral to the geodesics on ellipsoids. Liouville first presented Michael's paper to the *Bureau des Longitudes* on December 24, 1845, and five days later to the *Académie*. In 1847, Liouville presented new results by Michael Roberts to the *Bureau des Longitudes*, and during these years, he published six of his papers in the *Journal de Mathématiques*. William Roberts had no less than 26 of his papers printed by Liouville in the period from 1843 to 1857.

Thus, Liouville had close contacts with the flourishing center of mathematics in Dublin. He also knew the works (at least those in mechanics) of the greatest of the Dublin mathematicians, William Rowan Hamilton, and admired him so much that around 1845 he arranged for Hamilton to get a free copy of his *Journal* [Neuenschwander 1984b, p. 61]. However, Liouville's closest British friend and most talented foreign protége was William Thomson.

WILLIAM THOMSON — LORD KELVIN

During the 1840s, Liouville was fortunate to get as a student the young William Thomson (1824–1907), who was later to become one of the world's leading physicists, and ennobled under the name Baron Kelvin of Largs. His life has been recorded in a very charming and highly informative two-volume work by Silvanus P. Thompson. His relation to Liouville is very well described in Thompson's book, so in a sense it is superfluous here. Still, since this is one of the best documented examples of Liouville's work as a supervisor of the next generation, I shall extract the information on Liouville from Thompson's book and combine it with the other evidence we have about their collaboration.

A long time before Thomson met Liouville he had developed an admiration for him. Like Liouville, Thomson had learned the beauty of mathematics through the reading of the great French masters from the time of the revolution, in particular Fourier. It is therefore not surprising that in his early works he soon found "himself to have been particularly anticipated by Liouville in his papers" [Thompson 1910, p. 29] (probably one of the papers on Sturm-Liouville theory). A few years later, he also wrote a small "Note on the Law of Gravity at the Surface of a Revolving Homogeneous Fluid" [1844], where he referred to Liouville's paper [1843b]. In addition to his early appreciation for Liouville's scientific works, Thomson also regarded Liouville's journal editing as an ideal. Thus, in discussions in 1844 with the editor of the *Cambridge Mathematical Journal*, R.E. Ellis, Thomson proposed to enlarge this journal "so as to make it something of the nature of Liouville's if possible" [Thompson 1910, p. 79].

Thomson's ideas about Liouville, as about the rest of the world, were very heavily influenced by his father James Thomson, a mathematician at the University of Glasgow. James Thomson had the greatest ambitions for his son and ruled his academic career with a firm hand. Having managed to have William matriculated as a student at Cambridge University, he continued to urge him on in his letters, explaining, for example, the importance of the right contacts:

> Think, therefore, of every fair, honorable, and practicable way of preparing for the contest: and when you hear of the bad health of others, use every precaution in your power regarding your own. Try to become known to persons of name and influence - Liouville, for instance.
>
> [Thompson 1910, p. 63]

William's admiration for Liouville's *Journal* was also influenced by his father. Thus, when William used a mathematics prize of £5 to buy an illustrated Shakespeare, his father asked him why he had not bought Liouville's *Journal* instead [Thompson 1910, p. 46]. This, by the way, was only one of a great number of letters, in which the typical Scottish father rebuked his son for his carelessness with money. "Do not spend a sixpence unnecessar-

PLATE 14. William Thomson—Lord Kelvin (1824–1907). This photograph was taken in 1846 one year after Thomson had visited Paris where Liouville had become his teacher and friend. Liouville's famous theorem on conformal mappings grew out of Thomson's work on electrical images. (Reproduced from Andrew Gray's "Lord Kelvin: An Account of His Scientific Life and Work," London and New York, Dent, 1908. Courtesy of George Weidenfels and Nicolson Ltd.)

ily" was his advice [father to son, April 2, 1845, Thompson 1910, p. 129].

After his graduation from the university in the winter of 1844–1845, William Thomson set out for Paris. The evening before his departure something happened that was important for his meeting with Liouville. Thomson had earlier found a reference to an "Essay on the Application of Mathematical Analysis to the Theories of Electricity and Magnetism," which had been privately published by a certain Mr. Green [1828]. Thomson had tried to find a copy of the essay, but he had not succeeded until that evening, when by accident he learned that his tutor, Mr. Hopkins, had three copies of it. Hopkins gave Thomson one of them and asked him to deliver another copy to mathematicians in Paris.

When Thomson and a friend of his, Hugh Blackburn, arrived in Paris on January 31, 1845, they took a room in Rue M. le Prince, in the Latin quarter. The following day he paid a visit to Liouville, whose apartment was only a few blocks away, and gave him a copy of Green's essay. "I called on Liouville on Friday, Jan. 31. and he received me very freely..." Thomson wrote in his first report to his father [Thompson 1910, p. 115]. In his next letter, he told more about his first encounter with Liouville:

> I called on Liouville, shortly after I arrived here, with a paper which Cayley had given me to carry to him. We very soon became acquainted, and he began directly to work with pen and paper at various subjects of conversation, and when I went away he invited me to return again "pour causer de toutes ces choses." I called again, bringing with me another paper which I had received from Cayley. I found Chasles there, to whom Liouville presented me, and we had a long conversation on mathematical subjects. I had lent Liouville a memoir of Green's, of which I had received two copies from Hopkins before I left Cambridge, in which I found that he [Green] in 1828 had given almost all the general theorems in attraction which have since occupied Chasles, Gauss, etc., and I also lent him my copy of the *Camb. Math. Jour.*, as far as it had appeared, which he wished to look over. Liouville showed them to Chasles, who seemed very much struck with both. Liouville asked me to come and dine with him on Thursday along with Chasles. [Pelouze was also present].
>
> My French is exceedingly bad in conversation, but I hope it will improve by practice. I am writing at present a short paper for Liouville, which will also serve for a French exercise.
>
> [Thompson 1910, p. 117–118]

In order to throw Liouville's friendly reception of Thomson into relief, it is instructive to read what he wrote to his father on February 23, when he had finally summoned up his courage to pay a visit to the great Cauchy:

> I called on Cauchy on Friday (I had previously called, but was told by the porter that he is only visible on Friday between $3\frac{1}{2}$ and five) and presented Prof. Forbes' introduction. When he read the letter he commenced asking me questions about how much mathematics I had read, the first being whether I knew the Diffl. Calc. When he found that I knew enough to be probably able to understand him, he commenced telling me what he was working at, and various things which he had done, some of which were very interesting. He gave me two copies of the paper, which I enclose, one of which he wished me to send you, in which he gives an account of a memoir he had presented to the Academy containing a method as complete as that of Sturm for separating the roots of equations. From the explanations he gave me, and from what is contained in the paper, I think his method must be complete, though it is much more complicated than Sturm's. He says that he has since applied it to the imaginary as well as real roots, but I have not seen any of the memoirs of which he spoke. I am to go to his soirées if I choose, given by his "belmère" on Tuesday evenings."
>
> [Thompson, 1910 pp. 120–121]

The contrast is striking. Thomson could visit Liouville freely whereas he had to wait for Cauchy's $1\frac{1}{2}$-hour weekly audience. Liouville immediately began to work with pen and paper; the mistrustful Cauchy began with an examination. Liouville was immediately fascinated by the subjects that interested Thomson, in particular Green's essay; Cauchy told him about his own work, some of which was interesting, in particular about his priority over Sturm. Liouville invited Thomson to contribute to his *Journal*, Cauchy did not seem to have cared for Thomson's work. Both Liouville and Cauchy invited Thomson to return, but Liouville's invitation was personal and rather informal, Cauchy's was to the soirées of his mother-in-law. It is no surprise that it was Liouville and not Cauchy who became the mentor of the next generation of mathematicians.

I must do justice to Cauchy with the following quote from Thomson's letter to his father a month later:

> I see Cauchy very often, and when I call on him he always has a great deal to say, and tells me what he is working at, and all the fine things he is discovering. He has either one or two memoirs for the meeting of the Institute every week.
>
> [Thompson 1910, p. 125]

In addition, Cauchy took an interest in the young man's soul, trying to convert him to Roman Catholicism [Thompson 1910, p. 119].

Like Thomson, Liouville immediately sensed the importance of Green's essay and began to spread the news of its existence to his friends and officially informed the *Bureau des Longitudes* about it. In its *Procès Verbaux* from February 12, 1845 we read:

> Mr. Liouville gave an account of a beautiful theory of electricity and magnetism published by a certain Mr. Green. The great merits of this theory does not seem to have been recognized during the lifetime of its author[16].

He also talked about the essay with Sturm, who got so enthusiastic about it that he ran off to see Thomson immediately. Thomson had already tried to contact Sturm at his home, but he had not been in. Thomson wrote to his father:

> As he [Sturm] had heard from Liouville that I had a copy of Green's memoir, he called the same day here when I was out, and left a card, "C. Sturm, Membre de l'Institut," and said he would call again next morning. However, about ten o'clock the same evening Blackburn and I were astonished by his entrance. He soon began talking on mathematical subjects, and did not lose much time in asking about Green's memoir, which he looked over with great avidity. When I pointed out to him one thing which he had himself about a year ago in Liouville's Journal, he exclaimed, "Ah! mon Dieu, oui."
>
> [Thompson 1910, p. 121]

With his enthusiastic propaganda for Green's work, Liouville thus played a role in saving Green's ingenious work from oblivion. The main person responsible, however, was Thomson, who later (1853) had the whole essay reprinted with an introduction in Crelle's Journal. One may wonder why Liouville did not print a translation of it himself. Maybe he considered it too long for his *Journal.*

After Liouville read Green's essay, he took up his own research on potential theory, which he had developed in connection with the rotating masses of fluid in 1842. On May 12 and June 2 [1845b,c], he communicated two notes on this subject to the *Académie des Sciences* and on June 11 a third one to the *Bureau des Longitudes.* The first two were printed in the *Comptes Rendus,* reprinted in the *Journal des Mathématique* [1845d], and the following year greatly expanded in two published letters to P.H. Blanchet, who had asked Liouville for the lacking details [Liouville 1846e,f]. These papers contained many new relations between Lamé functions, including those of the second kind introduced independently by Liouville and Heine, and their applications in potential theory. Essentially, the results stemmed from his research on fluid masses in 1842 and 1843, but occasionally Liouville now referred to Green's work. Moreover, Liouville referred in rather vague terms to a generalization that he "did not hesitate to... regard... as being of the utmost importance in analysis" [1845d]. This theory never appeared, but I have been able to reconstruct it from his notebooks (cf. Chapter XV). It is indeed highly remarkable that in these notes Liouville anticipated many central ideas and results concerning spectral theory of integral operators by about half a century. In particular, it should be emphasized that he used the Rayleigh-Ritz method of finding eigenvalues three decades before Rayleigh and Ritz, and with an elegance reminiscent of Hilbert.

At their first meeting, Liouville had invited Thomson to write a small paper for the *Journal.* When it was finished on February 22, Thomson "took it to Liouville, but as he was not well I could not see him" [Thompson 1910, p. 122]. On March 16, Liouville was still ill:

> I received aunt's letter a few days ago, containing your [his father's] message about Jerrard's work. I have not been able to see Liouville yet to tell him, as he has been unwell, but I communicated it to Sturm yesterday, who received it with great avidity.
>
> [Thompson 1910, p. 124]

Thomson spent an increasing part of his time in Paris working as an assistant in the laboratory of the young, later famous, experimental physicist Regnault at the *Collège de France.* It was his father's idea that he should demonstrate that he was "not merely an expert x-plus-y man" [Thompson 1910, p. 130]. The reason was that the health of the professor of natural philosophy in Glasgow was wavering and father Thomson planned for his son to succeed him. As usual, William Thomson did what he was told, but it was the theoretical aspects of the experiments that interested him the most, and even in the letters to his father, he did not hide that he found the great collection of books at the *Collège* at least as fascinating as the experiments. In his spare time, he studied Poisson's memoirs on electricity [Thompson 1910, p. 126], so he was highly motivated when Liouville suggested a subject for research in this area to him. He wrote to his father on Sunday March 30, 1845:

> I sometimes go to the laboratory as early as 8 in the morning, and seldom get away before 5, and sometimes not till 6. I generally breakfast before I go, and so have the time free when Regnault and M. Izarn, his assistant, go to breakfast. On Thursday I made use of that time by going to see Liouville (at 12 P.M., for the French breakfast very late), whom I found recovered, as he said, from his illness. He did not let me away till $4\frac{1}{2}$, as he had a great many things to speak about, besides, working a good deal, and reading the *Nova Comm. Petr.* through. He asked me to write a short paper for the Institute, explaining the phenomena of ordinary electricity observed by Faraday, and supposed to be objections fatal to the mathematical theory. I told Liouville what I had always thought on the subject of those objections (*i.e.* that they are simple verifications), and as he takes a great interest in the subject, he asked me to write a paper on it, and said he would get it translated if I choose to write it in English.
> . . .
> Arago, it seems, has recently heard of Faraday's objections, and the uncertainty thus thrown on the theory prevented, as Liouville told me, its being made the subject for the mathematical prize of the Institute this year, and instead of it Abelian functions have been proposed. However, as Poisson before he died wished Liouville to do anything he could for it, I think it will very likely be proposed again. Liouville said he would lend me Chasles's *Précis historique de la géométrie* if I would go back for it again. I went yesterday, and he worked through a memoir of Jacobi's with me. - Your affectionate son,
>
> [Thompson 1910, p. 128–129]

It is characteristic that Liouville had the idea of designing a prize question for Thomson, but despite Poisson's wishes, it never materialized. As we saw earlier, the alternative prize problem was probably primarily aimed at Hermite, but it is likely that by reading Jacobi with Thomson, Liouville attempted to interest him in this subject as well. However, the works of Abel and Jacobi were too theoretical for Thomson's liking [Thompson 1910, p. 124]. On the other hand, he was enthusiastic about Liouville's idea of comparing Faraday's experiments with traditional French electrodynamics. Indeed, this project became the origin of some of Thomson's most interesting works and ideas. Among the ideas that grew out of this research project was the concept of analogies, which became central in Thomson's later way of thinking. If two physical situations are described by the same mathematical equation, they are analogous in the sense that one can carry over conclusions from one situation to the other by reinterpreting the quantities entering the equation, even though one does not solve the equation.

For example, electrostatics and stationary heat conduction are analogous because they are both described by Laplace's equation. Ole Knudsen discussed Thomson's use of analogies [1985]. I think it should be emphasized that Thomson owed a great deal to his French teachers and colleagues, in particular to Chasles, who in 1837 investigated the analogy mentioned above in a chapter entitled "Analogies between the properties of the attraction of an ellipsoidal layer and the laws of heat in motion in a body in temperature equilibrium" [Chasles 1837b, p. 297].

It is interesting to see that during Thomson's visit Liouville used a typical argument by analogy to establish the maximum principle for a harmonic function. He remarked that a harmonic function can be thought of as the temperature distribution of a stationary conduction of heat and that such a distribution cannot have a maximum in a point of the interior, because then the surroundings would cool off that point, and the temperature distribution would not be stationary [Liouville 1845d] (cf. Chapter XV(19)). There is little doubt that Thomson and Liouville discussed such analogies during Thomson's stay in Paris. It is less clear, however, who passed ideas to whom. Still, when Liouville shortly afterward congratulated Thomson for some new results on moments of inertia, he explained the virtues of analogies in a way that looks like a teacher explaining some important point to his student rather than a person recalling something he had heard from the colleague whom he addressed:

> These theorems give us a new example of how mechanical and physical theories of a very different nature can in a way meet and unite on a common geometrical ground. An example is Fresnel's wave surface, which you introduce in the theory where Mr. Lamé's isothermal surfaces already appear. Such connections are not only curious, they are of the greatest importance for science. They can simultaneously be used to extend it and to simplify it.
> [Liouville to Thomson, July 25, 1846, Thompson, 1910 p. 181]

It is amazing that Liouville, according to Thomson, had retained a great interest in electrodynamics 15 years after he had contributed to Ampère's theory. He probably felt that the new phenomena had to be compatible with some variation of Ampère's ideas, and he may even have had the idea that Faraday's experiment could be used in a complete determination of Ampère's force law, which he had shown could not be determined by traditional experiments. Whatever his motives were, the choice of this subject for Thomson illustrates his great abilities as an advisor of young talents.

Less than a month later, Thomson finished the comparison of the English and French approach to electrodynamics:

> My dear Father — I got my paper on Electricity finished yesterday. I wrote it myself in French and got M. Izarn at the laboratory to look it over. I left it at Liouville's house, but I have not seen him, and so do not know what he will do with it.
> [Thomson to father, April 26, 1845; Thompson 1910, p. 131]

In passing, it may be remarked that although Thomson had improved his written French, his pronunciation still left something to be desired. This is beautifully illustrated by a small incident that he described to his father:

> I went to every book-shop I could think of, asking for the *Puissance motrice du feu*, by Carnot. "Caino? Je ne connais pas cet auteur." With much difficulty I managed to explain that it was "r" not "i" I meant. "Ah! Ca-rrr-not! Oui, voici son ouvrage," producing a volume on some social question by Hippolyte Carnot; but the *Puissance motrice du feu* was quite unknown.
>
> [Thompson 1910, p. 133]

Later in life, Sadi Carnot's important book inspired Thomson to make important contributions to thermodynamics, but he did not draw Liouville's interest in this direction. When Liouville saw Thomson's paper on electricity, he decided to publish it in his *Journal*, instead of presenting it to the *Académie* as he had intended. Perhaps to make up for not propagandizing for Thomson at this learned society, he presented some of Thomson's ideas to the *Bureau des Longitudes* a few days later:

> Mr. Liouville speaks of a theorem concerning the attraction from an electric layer in equilibrium on an exterior point situated close to the surface of this layer. This theorem, which Mr. Poisson in his memoirs on electricity attributes to Laplace, can already be found in Coulomb's memoirs (*Mem. de l'accad. des sciences pour 1788*). This remark has been communicated to Mr. Liouville by an English [sic] geometer, Mr. William Thomson.
>   Mr. Liouville announces that this geometer has undertaken an extensive research on the theory of electricity.
>
> [P. V. Bur. Long., April 30, 1845][17]

After Thomson returned to Cambridge, he continued to correspond with Liouville, who printed the mathematical contents of three of the letters in his *Journal*. Moreover, in 1847, Liouville published a very important paper of Thomson's containing the first statement and proof of Dirichlet's principle according to which one can find harmonic functions with given conditions on a boundary surface. The name of this principle stems from Riemann, who had learned it from the lectures of his teacher Dirichlet. In 1870, Weierstrass pointed out that Dirichlet's proof of the principle (and Thomson's as well) was insufficient. In fact, Liouville had realized this in the 1850s and that may have convinced him that there was a hole in his own generalization mentioned above (cf. Chapter XV, §45–48).

By 1847, Thomson had little reason to publish in Liouville's *Journal*, for on his return to Great Britain, he had taken over the edition of the *Cambridge Mathematical Journal*, whose name he changed to the *Cambridge and Dublin Mathematical Journal*. The first volume of the new series contained a short paper by Liouville [1846o], followed by an extract of a letter from Thomson to Liouville. The roles of contributor and commenting editor had temporarily been changed.

During the summer of 1846, Liouville was back in his role as the authority. The physics professor in Glasgow had finally died at a very convenient moment for Thomson (and his father), and so Thomson asked all the influential people he knew to write letters of support for his candidacy. On July 1, 1846, he wrote in his diary:

> Some results of the co-ordinates explained (Oct 27th), including the actual solution by means of them of the problem for two spheres in contact, I sent to Liouville in a letter last week - I was writing him, at any rate, for a testimonial.
>
> [Thompson 1910, p. 159]

I shall return to the mathematical contents of this letter of June 26. One month later, Liouville sent a testimonial to his friend Armitage in Cambridge. Having highly praised the works and talents of Thomson as being a worthy "successor" of the "illustrious George Green," he ended the letter:

> From the above, you can see how much I have it at heart to be able to help Mr. Thomson in the decisive situation he is in. I passionately love the sciences and those who cultivate it with distinction, in whichever country they may be. Therefore, please help me if you know any of the judges and if you think that my humble opinion may carry any weight in their eyes. Tell them all that I think; tell them that I consider Mr. William Thomson destined to figure in a high rank in the midst of this pleiade of *savants* of which England is justifiably so proud. Do what you like with this letter and be sure that our young friend will day by day come up to my praise by new successes.
>
> [Thompson 1910, p. 180]

The following day, July 26, 1846, he sent a letter directly to Thomson, the end of which shows with even more clarity how much Liouville cared for the well-being of his young friend:

> Why am I not in a position where I can help you in this important situation? Why can't I speak to the judges and tell them what I think of you? I have at least tried to take a part through one of my friends. But, is that the same as being helpful to you? I do not know. Whatever will happen, be careful not to lose heart. By the nature of your talent, you seem to me to be called to make up for the loss which the sciences have had by the death of your fellow countryman, the illustrious Green, whose great merit you have been one of the first to appreciate and who I shall attempt at every occasion to do him the justice which he did not receive during his lifetime. You are luckier than him; believe me, your future is happy.
>
> [Thompson 1910, p. 182]

Liouville need not have worried. Both of the letters were presented separately to the faculty, together with a printed pamphlet of 28 pages of recommendations from other British scientists and from Regnault. Thomson got the job.

In a way, Thomson paid back Liouville's generous guidance and help by inspiring him to one of his most beautiful results. During his stay in Paris, Thomson had explained his "principle of electrical images" to Liouville. By this principle, one can transform an electrostatic problem into another that is perhaps easier to solve. The transformation takes any point A in space into another A' on the line AO, where O is a fixed point, such that $OA' = \frac{R^2}{AO}$. Thus, it transforms the interior of a sphere with center O and radius R on the space outside it and vice versa. Among geometers, this transformation was known as inversion in a sphere; this is still its most widespread name, but in analytical and physical contexts it is sometimes called the Kelvin transform, after its first user in these areas.

After his return to Cambridge, Thomson committed his ideas on paper in a letter to Liouville, who immediately published them in the September 1845 issue of his *Journal*. Thomson followed up these ideas the following year when he wrote to Liouville to ask him for a testimonial. The diary entry continues:

> I have resolved, if possible, to send him [Liouville] some more in acknowledging his answer, as I wish to get some of the results published at once, seeing that I have made many abortive attempts to commence my *treatise* on Electricity.
>
> [Thompson 1910, p. 159]

In fact, Thomson did send "some more" on September 16, 1846, but Liouville did not publish the letters "at once." Not until the following summer when Thomson passed through Paris, trying in vain to visit Liouville, could Liouville inform him in a letter from Toul about the projected publication of the letters:

> My dear Mr. Thomson,
>
> I deeply regret that I was not in Paris when you came to visit us. I would have enjoyed speaking with you and showing you the proofs of the letter, or rather the two letters concerning the mathematical theory of electricity which you sent to me last year, and a note which I have added, but which could not be inserted in the same issue of the *Journal de mathématique*. Your letters complete the June issue, and the above-mentioned note will be at the beginning of the July issue. Perhaps all this will be too late; perhaps you have already written a more extensive publication on the same subject in the *Cambridge and Dublin mathematical Journal*, issues of which I have not seen for a long time. However, I shall always be pleased to have used this occasion to compliment you publicly for your beautiful research.

Indeed, extracts of Thomson's letters [Thomson 1847a] and the accompanying note [Liouville 1847f] appeared as planned, the latter ending with the promised compliments:

> I hope the reader will pardon me for these further developments, which I thought I could publish next to Mr. Thomson's very interesting letters, without getting into the way of his research. My object

has been attained, I repeat, if they could help emphasize the great importance of the work of this young geometer and if Mr. Thomson himself would regard it as a new token of my friendship with him and of the esteem I have for his talent.

[Liouville 1847f, p. 290, July 1847]

Liouville's note, which was in fact longer than Thomson's letters, contained several interesting results. Liouville showed how Thomson's transformation, which he called *transformation par rayons vecteurs réciproques*, could be applied in synthetic (projective) geometry. With his examples, Liouville, apparently without knowing it, connected Thomson's ideas to a growing tradition in geometry (cf. Chapter XVII(23)). Moreover, he pointed out more explicitly than Thomson had done the simple way in which the important differential equations of physics transforms. This explains the great applicability of inversions in analysis and mathematical physics.

Last but not least, Liouville pointed out that the Kelvin transform exhausted a rather general class of transformations: consider a transformation $T : \mathbf{R}^3 \to \mathbf{R}^3$ and assume that T transforms the distance between any two points $\overline{x}, \overline{x}' \in \mathbf{R}^3$ so that

$$|T(\overline{x}) - T(\overline{x}')| = p^2(\overline{x})p^2(\overline{x}')|\overline{x} - \overline{x}'|, \tag{$*$}$$

where p is a (sufficiently nice) function $\mathbf{R}^3 \to \mathbf{R}$. Then, Liouville proved that T can be composed of displacements and inversions in spheres. This is, of course, a strong theorem, but three years later, Liouville discovered that it could be made much stronger.

In the note to Thomson's letters, Liouville pointed out that a transformation satisfying $(*)$ is conformal, that is, angle preserving, and in a one-page note [1850a], he then stated that *all* conformal mappings can likewise be produced as compositions of displacements and inversions in spheres. This theorem is often called Liouville's theorem. It is a rather surprising fact that there are so few conformal mappings in space, considering that all holomorphic functions are conformal in the plane (and the only ones). The latter famous theorem implicitly used by Gauss had been mentioned by Liouville in an earlier note [1843a]. Liouville discovered his theorem while he was preparing the fifth edition of Monge's *Application de l'analyse à la géométrie* [Liouville 1850g], and he inserted the proof as a note in this treatise. I shall return to some of Liouville's other notes in this edition later.

After 1847, Liouville and Thomson went in different directions. After some time, Liouville stopped publishing on applied mathematics, and Thomson, in his chair of natural philosophy, turned more and more towards physics and even engineering. Still, he continued to be loyal toward mathematics. The following story was told to Thomson's biographer by his assistant:

> The father of a new student when bringing him to the University,
> after calling to see the Professor, drew his assistant on one side and
> besought him to tell him what his son must do that he might stand
> well with the Professor. "You want your son to stand well with the
> Professor?" asked MacFarlane. "Yes." "Weel, then, he must just have
> a guid bellyful o' mathematics"!
>
> [Thompson 1910, p. 420]

Thomson also continued to have the greatest respect for his former mentor.
Thompson told the following anecdote:

> Once when lecturing he [Thomson] used the word "mathematician,"
> and then interrupting himself asked his class: "Do you know what a
> mathematician is?" Stepping to the blackboard he wrote upon it:
>
> $$\int_{-\infty}^{\infty} e^{-x^2}\,dx = \sqrt{\pi}.$$
>
> Then, putting his finger on what he had written, he turned to his
> class and said: "A mathematician is one to whom *that* is as obvious as
> that twice two makes four is to you. Liouville was a mathematician."
> Then he resumed his lecture.
>
> [Thompson 1910, p. 1139]

Further, when the Institut de France celebrated its centenary, Thomson
came to Paris to give the speech on behalf of the Royal Society. Recalling
his stay in Paris, he said:

> I am eternally grateful to Regnault and Liouville for the kindness
> they showed me and for the methods they taught me in experimental
> physics and mathematics during the year 1845.
>
> [Thompson 1910, p. 949]

A Coherent Mathematical Universe

At a first glance, Liouville's works in the 1840s, covering as they do almost
all parts of pure and applied mathematics, give a very flimsy impression.
It has been one of the aims of this chapter to show the bombardment of
impulses that gave rise to these works. However, I shall now end with a
second glance at his works from this period in order to show that in fact
they are bound together in a great web by a large number of connections.

In the 1830s, Liouville's works had fallen into three or four nicely separated categories. But, it is characteristic for mathematics that the more one
learns of it the more connections one sees. This fact, which is perhaps the
most important factor in the accumulation of mathematical knowledge, is
amply demonstrated in Liouville's case. While his teaching, editing, reading and his protégés and colleagues introduced Liouville to a multitude of
new fields, they also wove these different parts together, so that toward the

end of the 1840s, when Liouville was at the peak of his mathematical abilities, mathematics appeared to him as one body of interconnected ideas. Let me end this chapter by enumerating a number of these connections. We can begin with the work on a rotating mass of fluid. It had roots in celestial mechanics and linked up with potential theory (Thomson's work on electrostatics), which in turn had a connection through integral equations to ideas of Sturm-Liouville theory. The rotating fluids were directly linked to geometry, through the study of the geometry on the ellipsoidal surfaces and indirectly through Thomson's work on inversion in spheres. The use of ellipsoidal coordinates used in the work on rotating fluids was of importance in rational mechanics, which in turn was linked to his geometrical research through the interpretation of geodesics as traces of freely moving particles on the surface. The analytic methods linked geometry to algebra, which was tied to the theory of doubly periodic functions, both through the proof of the fundamental theorem and via Galois theory and the division of elliptic functions. Ellipsoidal coordinates (Chasles) and Serret's elliptic curves provided a direct link between geometry and doubly periodic functions. Finally, doubly periodic functions, or rather elliptic integrals, were of importance in celestial mechanics and the inspiration for questions on integration in finite form.

To be sure, Liouville cultivated only limited areas at different places of this mathematical landscape, but there is no doubt that he had a great view over it.

IV

The Second Republic (1848–1852)

Banquets Réformistes (1840)

Of all the political changes that took place during Liouville's lifetime, the bourgeois revolution of 1848 was the one that had the strongest impact on him and his career. In fact, Liouville had been involved in paving the way for this revolution, whose initial ideals he completely shared.

An avowed republican, Liouville had taken political action in the campaign for universal suffrage in 1840. The previous year, a popular movement had demanded an extension of the franchise, which was limited to rather wealthy men, but the new, more liberal government formed during March 1840 under Thiers turned down the idea. In the Chamber of Deputies, it was only the small radical group of about a dozen men who agreed when on May 16, 1840, one of the leading radicals stood up and proposed universal suffrage and "a new organization of work." This courageous radical was Liouville's good friend and colleague at the *Académie* and at the *Bureau des Longitudes*, François Arago [Jardin and Tudesq 1973, p. 154]. Having been elected a member of the Chamber in 1831 as a deputy from his home region Perpignan in the eastern Pyrenees, Arago became the leader of the republican party, or radical faction as it was called during the July Monarchy, where it was not openly allowed to threaten the constitution.

After Arago's speech in the Chamber of Deputies, the opposition arranged a number of *banquets réformistes*. This was the most popular, legal, way to express opposition to the government. During the banquets, toasts were proposed on various political issues, in 1840 often on the electoral reform. The first banquet in Paris on July 1 was followed on Bastille Day, July 14, by many banquets in the larger cities of France—except for Paris, where the manifestation was banned by the government. These were exciting and promising days for the republicans, as can be seen in a letter to Liouville from his friend Catalan, dated September 1. He wrote from Arcueil

> under the shadows illuminated by Bertholet, Dulong, Arago, Biot, Laplace etc. so that although my letter is badly written, it might still have a scientific flavor.

In addition to the scientific matters, he reported:

I was content with the list published last Sunday in the *National*. When they see the names of the men who form the Committee, our opponents may end up understanding that the electoral reform is a serious matter. However, what can one expect from people who have learned nothing from the last fifty years of our history?

Yesterday there were rumors in Arcueil and in the surroundings of the super-banquet in Châtillon. We counted the oxen, the calves, sheep, pâtés etc. I have not received news yet, but I prefer to believe that everything went well.

[Jongmans 1981, p. 293]

(Originally founded by Thiers, the newspaper the *National* was now the propagator of the radical or reformist ideas.)

The unrest continued all through the fall. In October, Arago arranged *banquets réformistes* in and around his *département*, [Jardin and Tudesq 1973, II, p. 57]. This may have been instrumental in persuading Liouville to take action as one of the leading radicals in Toul. In a letter to his like-minded friend and colleague in the *Académie*, the chemist Pelouze, he described his feverish activity; the letter is dated "Toul 30 7bre 1840," 7bre meaning "september."

My dear Pelouze

Next Sunday, October 4th, we shall have a *banquet réformiste*. I have been asked to be the chairman, and I have accepted. On the 8th, in the morning, I shall set out for Paris where I will arrive on the 9th at noon if the roads are good. It is a little less than ten days ago that my father gave me clear permission to get involved in politics and I have lost no time. But by going so fast I naturally meet all the inconveniences characteristic of an improvisation. Thus, I cannot count on more than 100 persons for our *banquet*, but after all, that is already something for the town of Toul; it is more than they had gathered at the historical *banquets* in 1830 and 1831, since when there have not been any. We are also short of copies of the radical petition. They had promised to send us some in Nancy, but we have not received any. Perhaps underneath they are vexed by a small local vanity since our reformist committee is the first, and until now the only, committee organized in this *département*. In this extreme situation, I shall fall back on you, my dear Pelouze. Please rush to the *National* and ask for copies of the radical petition. Two or three will suffice for the moment, but twenty won't hurt. Quick, send them by mail under open cover; we are burning with the desire to sign. Besides, isn't it convenient that the guests at a *banquet réformiste* have signed a petition before they arrive?

All of this has made me neglect mathematics a bit.

[Neuenschwander 1984a, I, 1]

Liouville's fears of the limitation of the banquet proved unfounded. After the *banquet*, almost 1000 inhabitants of the *arrondissement* signed the reformist petition, which Liouville had asked Pelouze to get at the *National* [Liouville 1848d].

The political situation was soon aggravated by economical problems, strikes among the workers, and Thiers' explosive anti-English and anti-Turkish policy, which confronted France with an alliance between all the other great European powers. The threat of war was so imminent that Thiers decided to fortify Paris. In this connection, Liouville wrote to Pelouze in the letter quoted above:

> Having become the director of the fortifications of Paris, General Vaillant has no doubt left us. So much the better for him, but too bad for us. It is an irreparable loss. May heaven direct our poor School [*The École Polytechnique*]
>
> [Neuenschwander 1984a, II, 1]

The danger of war also led to patriotic and revolutionary manifestations, as, for example, when England displayed its power by bombarding Beirut. Catalan later recalled how he and another of Liouville's protégés, Le Verrier took part:

> This odd fellow named Leverrier came to fetch me on October 5th or 6th to join in the riots at the opera on the occasion of the bombardment of Beirut. We played the *Marseillaise*...
>
> [Catalan's diary 1858–1862, Jongmans 1986]

The sarcastic tone is a sign of the subsequently diverging political opinions of Liouville's two students: Catalan remained a radical, whereas Le Verrier was attracted to the policy of Louis-Napoléon.

The unrest in 1840 did lead to a political change, but a disappointing one for all the republicans. The liberal monarchist Thiers was replaced by the conservative monarchist Guizot, who soon ended the internal and external unrest.

Political Opposition (1840–1848)

Although Liouville remained in opposition to the July Monarchy, he did not let his political opinions stand in the way of his scientific career. In this respect, his conduct differed from that of the more radical Catalan and from the ultraconservative legitimist (i.e., Bourbon supporter) Cauchy, who refused to take the oath of loyalty to the King that was required of the professors. Liouville took the oath like his many republican colleagues, including Arago. In fact, there was a disproportionately large number of republican scientists, and according to Jardin and Tudesq [1973, I, p. 224], the *Académie des Sciences* was the only one of the five academies in the *Institut de France* that was critical of the regime.

Liouville's relations with some of his colleagues was no doubt influenced by the political views he shared with his friends Pelouze, Arago, and Catalan. His animosity toward Libri also had a political element, if we can trust a letter Libri wrote to Betti in 1851:

> [works on equations] some of which, several years before, were praised by Liouville in one of his papers before he turned my bitter enemy for political reasons etc.
>
> [Bolletino Unione Matem. Italiana 1953, p. 320]

Indeed Libri, who had come to France as a political refugee from Italy, had soon sided himself with the Citizen King, the "revolutionary" alternative to the Bourbons. So, there were certainly grounds for political disagreements between Libri and Liouville. One may wonder that Liouville's collaboration with Cauchy in the *Académie* went so relatively well. But, in addition to the reciprocal respect between the two scientists, republicans and legitimists were in a sense united politically in this period, around the common goal of overthrowing the ruling king.

In 1844, Liouville's political convictions seem to have caused him trouble at the *École Polytechnique*. Professor Jongmans found a letter from the Belgian statistician Quetelet to Catalan dated December 27, 1844, that alludes to this episode:

> It is with regret that I hear what you tell me about the *École Polytechnique* and what you tell me concerning Mr. Liouville. Their reactions are really unfortunate in every respect. ····. Please give Mr. Liouville my compliments in this situation and tell him how much I have been distressed to hear the news that concerns him
>
> [Correspondence of Catalan, Univ. Liège]

It is known that shortly beforehand the *École Polytechnique* had, for political reasons, rejected Catalan as Francoeur's substitute in the algebra chair, preferring instead the young Bertrand (cf. [Jongmans 1986]). But, what problems had Liouville encountered, and who's reactions were disagreeable? Being unable to answer these questions with precision, I shall conjecture that the incident had to do with sending home the students in 1844 (cf. [Pinet 1887, pp. 233–236]). During the July Monarchy, the students at the *École Polytechnique* had often flouted discipline, and during the spring of 1844, there were antigovernment demonstrations. So, when the students left the school in August refusing to be examined by Duhamel, who by direct intervention of the government and behind the back of the *Académie des Sciences* had accumulated the two incompatible jobs as examiner and director of studies, the Minister acted severely. The students were sent home and the examinations were postponed until January 1845, after a new decree had removed the influence of the *Académie* on appointments at the school.

> The circumstances leading to the disbandment served as a pretext for the *Constitutionnel* and the *Revue de Paris* to publish "jesuitic and calumnious" articles in which the discipline of the school, the

opinions of the students, the administration and even the teaching staff were the objects of violent attacks.

[Pinet 1887, pp. 234–235]

Pinet mentions that Arago, who sided with the students, was under attack, and the political overtones make it probable that Liouville, too, fell victim to the scurrilous conservative campaign. This is probably the problem alluded to in Quetelet's letter.

The 1848 Revolution

The elections to the Chamber of Deputies in 1846 were a victory for the conservative Guizot. Indeed, for the first time during the July Monarchy, there was a stable majority in the Chamber behind the ministry. However, after its economic success of its first six years in power, the government was now shaken by sudden agricultural and industrial crises. A growing social and political unrest was followed in the summer of 1847 by a new wave of banquets. As in 1840, the banquets were arranged in order to support an electoral reform, in this case one the Chamber had turned down on March 27. The first banquet, on July 9, 1847, in Château-Rouge in Paris, was arranged by the royalist opposition to Guizot, and toasts were also proposed for the King. Soon after, however, the movement was radicalized and on February 22, 1848, a crowd gathered in Paris, violating the government's prohibition of a reformist banquet. The following day, barricades were erected. Louis-Philippe tried to stop the popular uprising by naming Thiers head of a liberal government, but it was too late. On February 24, the King fled from Paris, and on the same evening, a republican provisional government was formed at the *Hôtel de Ville* (cf. [Agulhon 1973]).

Its composition was a delicate question, for there were various factions in the republican movement. The people behind the paper the *National*, for example, Arago, were moderate liberals usually representing the bourgeoisie. At the other end of the spectrum were the various socialist factions rooted in the lower classes. Their mouthpiece was the paper *la Réforme*, founded in 1843.

In the provisional government, the first group was in a majority, represented by the former deputies Dupont de l'Eure, Arago, Crémieux, Marie, Garnier-Pagès, and the editor of the *National*, Marrast. Flocon, the editor of *la Réforme*, Louis Blanc, and Albert were the only true representatives of the latter group. As a link between the groups were the former deputies Lamartine and Ledru-Rollin, who, in particular the latter, shared many ideas with the socialists. For the first weeks after the revolution, the entire nation was intoxicated with a feeling of *fraternité*. The old ties had been broken, and the provisional government from the outset made it clear that it would with all its might avoid a régime of terror like the one that had followed the great revolution of 1789. People from all levels of the society embraced each other in the streets, and everybody believed that justice and hope were alive. Most of them were disappointed sooner or later.

Among the first to be disappointed was Cauchy. Now, when the oath of loyalty to the Citizen King no longer stood in the way, he saw his chance of taking a more active part in official academic life. His first goal was the *Bureau des Longitudes*, which had elected him a member in 1839, thanks to the support of Arago, among others. However, as we have seen Cauchy had never functioned as a member, because the King had not sanctioned the membership. This barrier having been removed, Cauchy believed he could finally take his seat in the *Bureau*. The *Procès Verbaux* of March 1, the first meeting of the *Bureau* after the revolution, begins:

> Before the meeting formally opened, Mr. Cauchy entered the meeting room, without having told anybody about it in advance, and signed the attendance sheet.
>
> Since Mr. Cauchy is not a member of the *Bureau des Longitudes*, the President urged him to withdraw, and with the unanimous consent of the *Bureau*, he was asked to strike his name off the attendance sheet.
>
> [P. V. Bur. Long., March 1, 1848][1]

Thus, with his usual lack of tact and sense of occasion, Cauchy forced the *Bureau des Longitudes* to a very quick confirmation of its election in 1843 of Poinsot instead of Cauchy.

Candidate for the Constituting Assembly

A few days after this incident the provisional government issued a decree announcing the election to an *Assemblée constituante*. This was the first French election with universal suffrage and was scheduled for April 9, but was later delayed to April 23. Among the candidates was Liouville. Apparently [Neuenschwander 1984a, note 77], Liouville had been encouraged to take this step by the Minister of the Marine, Arago, who on March 19 recommended Liouville's candidacy to de Ludre, *commissaire* of the provisional government in the *département de la Meurthe*, where Toul was situated:

> I recommend the candidacy of Mr. Liouville, member of the *Institut*. Mr. Liouville is one of my best friends. He is a very eminent man, a patriot, an experienced republican. God grant that the National Assembly will contain many members of this caliber. I tell you in all sincerity that I do not know anybody who would be more worthy than Mr. Liouville to compete for the establishment of the Republic.
>
> [Neuenschwander 1984a, I,2]

After this recommendation from one of the leaders of the Republic Liouville was naturally nominated a candidate. There were no formalized

PLATE 15. The Provisional Government. The physicist and astronomer François Arago (1786–1853) supported Liouville in his scientific and political careers. (Bibliothèque Nationale, Paris)

parties, so each candidate had to present his own political programme. Liouville had two sheets printed, one in Paris by the printer of his *Journal* [Bibl. Nat.] and one in Toul [Acad. Sci. Dos. Pers. Liouville], both carrying the text:

To the electors
of the Département de la Meurthe

Dear Citizens

I shall ask you to include me among your Representatives in the National Assembly which is going to constitute the French Republic on the eternal basis of freedom, equality and brotherhood.

If you want to compose this Assembly of inveterate republicans you can take it from me that I am today what I was in 1840 when I presided at the *banquet réformiste* in Toul, after which almost a thousand signatures were collected for our radical petition in one single *Arrondissement* in spite of threats from the authorities. I am as frank in my speech as I am firm in my convictions.

I have never concealed any of my thoughts, I have nothing to change in any of my declarations. Naturally, I agree to the principles of the Republic; these principles have always been mine. But essentially moderate by character and by reason, I must admit that I do not have that feverish fervor, that quick-tempered zeal which is flaunted by men who even yesterday we saw as ardent servants of the fallen government. My devotion to the Republic is more profound but calmer. It will more easily resist difficult trials.

May I join the standards of my friends, the great citizens Arago and Dupont (de l'Eure) and may their names vouch for me.

Being an ancient student at the *École Polytechnique*, and having for the last seventeen years been a professor at this national institution, I am honored to share the sentiments of my comrades of all ages. Devoted to their country, fraternally united with the people whose esteem and affection they have won by their services rather than by flattery, supporters both of order and of liberty, they have in a sense presided over the birth of the Republic.

May the spirit that moved them, may the spirit which moved the people of Paris during the miraculous days of February reign forever among us.

Receive, dear citizens, the assurance of my deep devotion.

J. Liouville (of the *Institut*)

[Liouville 1848d]

Although undated, the reference to the Constituting Assembly shows that this programme stems from the election of 1848. Its main aim was to place Liouville in the republican spectrum. By pointing to the 1840 banquet, he emphasized that he belonged to the rather limited group of so-called *républicains de la veille*, who had fought the long battle for the republic under the July Monarchy. In this way, he kept aloof from the large

AUX ÉLECTEURS
DU DÉPARTEMENT DE LA MEURTHE.

CHERS CONCITOYENS,

Je viens vous demander de m'admettre au nombre de vos
Représentants à l'Assemblée nationale qui doit constituer la
République française sur la base éternelle de la liberté, de
l'égalité et de la fraternité.

Si vous désirez composer cette Assemblée de républicains
d'ancienne date, je vous offre cette garantie. Je suis aujourd'hui
ce que j'étais en 1840, quand je présidais à Toul ce banquet
réformiste à la suite duquel près de mille signatures vinrent,
dans un seul arrondissement et en dépit des menaces de
l'autorité, couvrir nos pétitions radicales. Aussi franc dans
mon langage que ferme dans mes convictions, je n'ai jamais
dissimulé aucune de mes pensées; je n'ai rien à changer à
aucune de mes paroles. Je me trouve tout naturellement à la
hauteur des principes de la République : ces principes ont
toujours été les miens. Mais essentiellement modéré par carac-
tère et par raison, je n'ai pas, je l'avoue, ces élans de ferveur

PLATE 16. Liouville's election poster 1848. (Bibliothèque Nationale, Paris)

number of enthusiastic people who had abruptly turned republicans during the last days of February 1848, and who, for a great part, only considered the Republic as a stepping-stone to some other system, such as a liberal monarchy, a return of the conservative Bourbons, or the Empire. This group was called the *républicains du lendemain*, a term to which Liouville alluded when he pointed out that they had served the July Monarchy "yesterday."

In his programme Liouville further stressed the moderate character of his republican beliefs by placing himself in the group around the *National*. This was underscored when he set up as his ideals the two oldest members of the provisional government, its rather powerless president Dupont de l'Eure and Arago. The final reference to the services he had done the people through his ties to the *École Polytechnique*, as opposed to the flattery of other candidates, may be another dissociation from the theoretical socialists, like Louis Blanc. Even by his signature, Liouville associated himself with Arago, who had signed the proclamation of the provisional government issued on February 25 "Arago (de l'Institut)," being thereby the only member of the government to indicate title or job, except for "Albert, worker."

Member of the Constituting Assembly

At the elections on April 23, the moderate republicans won a great victory. With about 500 of the 800 *représentants du peuple* in the National Constituting Assembly, they completely crushed the socialists who obtained less than a hundred seats. The rest of the seats were occupied by imperialists and monarchists of one persuasion or another. It was also a personal victory for Liouville. He became number 2 out of the 11 *représentants du peuple* elected in the *départément de la Meurthe*, despite the fact that his home base was Toul, and not the "capital" of the *départément*, Nancy.

On May 4, the Constituting Assembly met for the first time to proclaim the Second Republic officially, and a few days later they elected an Executive Commission to take over the executive power after the provisional government and until a president could be elected. As a clear demonstration of the moderate predominance of the assembly, they elected the five members of the commission in the following order: Arago, Garnier-Pagès, Marie, Lamartine, and Ledru-Rollin. The last named was only elected because Lamartine, the leader of the provisional government, had made Ledru-Rollin's election a condition for his own participation.

In the assembly, Liouville was seated among the moderate democratic majority and voted for the discontinuation of the rule of the Orleans for the abolishment of the death penalty, against a progressive system of taxation, and with a minority against the incompatibility of functions [Cougny and Robert 1890]. The latter law limited the number of posts a public servant could hold to two. At this time, the law had no consequence for Liouville for he only held his seat in the *Bureau des Longitudes* and his chair at the *École Polytechnique*. In addition to the daily work of the Assembly, Liouville also worked in the committee of finances.

The comfortable majority for the moderates would have made the work in the Assembly easy had it not been for unrest among the workers of Paris. Dissatisfied with the social reforms of the government, they invaded the Assembly on May 15 with socialist slogans. This threat to the authority of the Assembly and the force with which it was met by the National Guard started a polarization of the population, and the state of *fraternité* was definitively broken the following month. On June 21, the Executive Commission decided to close its only socialist experiment, the *Atteliers Nationaux*, created by the provisional government in order to secure work, and thus food, for the unemployed masses. The following day Arago tried to explain this step to a crowd on Place du Panthéon, but he was met with the shout of a worker: "Ah, Mr. Arago, you have never been hungry" [Agulhon 1973, pp. 68–69]. Feeling justly that the bourgeoisie had taken over their revolution, the workers of Paris opened a revolt. It was brutally suppressed by the Minister of War, General Cavaignac, who afterward singlehandedly replaced the five members of the Executive Commission.

During the uprisings, the socialist *représentants du peuple* who supported the rebels were imprisoned. Furthermore, when the loyalty of Louis Blanc and Caussidière toward the republic was unjustly questioned, they fled to London for fear of a commission which ,on August 26, the majority of the Assembly had authorized to investigate the matter. On this occasion, a minority of 252 *représentants du peuple*, including Liouville, voted against the proceedings against the two. This shows that Liouville, contrary to the majority of the nonsocialist republicans, remained faithful to the democratic ideals of the revolution.

The work on the new constitution continued, but the reactions to the uprisings and the complementary elections of June 4 indicated that the French people might elect a president who was in principle against the republican ideals. Cavaignac and his supporters tried to avoid that situation with the so-called *amendement Grévy*, but Lamartine convinced the majority of the Assembly, including Liouville, to maintain the public election of the president. Both Lamartine and Liouville probably regretted this decision soon after the Assembly, with 739 votes, including Liouville's, against 30, passed the entire constitution on November 4. For, on December 10, the people elected as the president of the Republic Louis-Napoléon Bonaparte, a nephew of the great emperor. He obtained 5.4 million votes against 1.4 million for Cavaignac, 400,000 for Ledru-Rollin, and less than 8,000 for Lamartine. This was a great victory for the Bonapartists, the legitimists (in favor of the Bourbons), the Orleanists, and the Church which had agreed to join forces behind Louis-Napoléon, and it was a shattering defeat for the moderate republicans. In Liouville's *Département de la Meurthe*, there was also an absolute majority for Louis-Napoléon.

In an attempt to gain time to change public opinion, the republicans in the Constituting Assembly tried to extend its own period of function, but after pressure from the president, the election to the Legislative Assembly, its replacement according to the constitution, was set for May 13, 1849. Just before that date, the Constituting Assembly for the last time demonstrated its democratic ideals by protesting that Louis-Napoléon had ordered a so-called negotiating expedition to attack Rome in order to end the revolutionary rule there and reinstall the Pope as political leader. Liouville voted for the protest.

The Bitter Defeat (1849)

Liouville accepted renomination to the Assembly. On April 17, 1849, he wrote to President Serre of the *Union électorale des villes et des campagnes pour le département de la Meurthe*:

> I hasten to answer the letter that you have honored me by writing on behalf of the *Union électorale des villes et des campagnes*. Yes, I accept a candidacy in the coming elections and I adhere readily to your programme. The principles that it contains are those that were always my guide when I voted in the *assemblée constituante*. I will naturally continue to follow them in the *assemblée législative*, if our compatriots will honor me with their votes a second time.
>
> [Neuenschwander 1984a, I,3]

However, the polarization of the French people did not diminish after the presidential election, and so the election of the Legislative Assembly was a disaster for the moderate republicans of the "National" type. They obtained less than 100 seats out of 700 and even such prominent people as Dupont de l'Eure and Lamartine were not reelected. The great winner was the conservative *Partie de l'ordre*, which supported Louis-Napoléon and got around 500 seats, leaving approximately 200 for the socialist *montagnards*. In the Meurthe, the *Partie de l'ordre* also won absolute majority, and Liouville found no favor in the eyes of his electors.

The political situation had become absurd. The second republic was in the hands of an imperialist Prince President, and an Assembly with a large majority of monarchists. Its fate was sealed. Supported by the *Partie de l'ordre* and the Catholic church, Louis-Napoléon increasingly pursued conservative, repressive political aims, and by a coup d'état on December 2, 1850, he prolonged his presidential period to 10 years and suppressed the Assembly. Formally, the Second Republic continued for one more year, until Louis-Napoléon was elected Emperor, under the name Napoléon III, by an overwhelming majority of the French people.

The elections in 1849 ended Liouville's political career. He was disappointed to see the political ideals for which he had fought for 10 years defeated. Still, it was a comfort for him that the attacks he had personally suffered in his *département* could be explained as a result of a general shift of public opinion. Soon after the election (on June 15, 1849), he wrote from Paris to the principal of the *Collège* in Toul, Mr. Grebus:

> But, don't run away with the idea that this or that change of opinion produced by calumnies thrown against me (during my absence) from the right and from the left can prevent me from living as I used to and staying the course. It is not at all humiliating not to be re-elected when Dupont de l'Eure, Lamartine and many others who the entire France honor are not re-elected either. In the midst of the storms which the extreme parties (to the right as well as to the left, to the left as well as to the right) like to raise, I have always been able to conserve this strong moderation to which one sooner or later pays tribute. I live in peace with my conscience, and I can go with my head high.
>
> [Neuenschwander 1984a, I,4]

Although Liouville claimed that he could face his electors with his head high, he did not do so. Having arrived at Toul on July 23, he wrote to his brother Felix:

> Here I am settled in my study. I have not met and shall not meet anyone in Toul. At least, I shall make no visits. Here I shall only work and bring my backlog of tasks up to date.
>
> As for the meeting in Nancy, I carefully missed it. Liouville and all my friends give you a thousand compliments. I have sent Lombard packing. I am too old to stick to that type of friends...
>
> [Ms 3618 (10), p. 14r]

Liouville also stayed away from social events in Paris. Toward the end of May 1849, he sent an excuse to a banquet at the *École Polytechnique* [Ms 3618 (9)], although he was touched by the invitation:

> I thank you for your greetings from the *École Polytechnique*. Nothing can make me more happy than a testimony of friendship from my former students.

Thus, he wrote to one of his former students, Souchère, in Marseille on May 30, 1849 [Ms 3618 (8)]. In May 1851, Liouville similarly excused himself from a celebration at the *École des Ponts et Chaussées*, losing thereby "a precious occasion to meet my former comrades again," as he wrote to Mr. Cavenne on May 20, 1851 [Ms 3633 (1)].

The following year a new misery befell Liouville. His father, with whom he and his family had a close relationship, died:

> Dear and excellent friend [he wrote Dirichlet], I drained the bitter cup of grief the day I lost my dear father, whom you have seen, whom you have even visited in Toul, as far as I remember. This father, raised above all price, who made up half of my life and half of the life of my brother, died last April 25.
>
> [Tannery 1910, p. 22]

This grief reinforced Liouville's decision to shun society. This can be seen from a letter of September 22, 1852, to his old friend Vogin, whom he had known since his student days at the *École Polytechnique* and who had also been a member of the Constituting Assembly.

> Alas! for you, for me, for our friends the situation is very sad. I have had the grief of losing my good father. He died on April 26 [sic!] and the effect of the political shocks was perhaps not without influence on him. It is for me an immense and irreparable emptiness and yet another reason for me to sink into solitude. Thus, I meet no-one in Toul. In a few days, I shall no doubt go to Nancy to shake hands with Laflize and Charon. However, I shall not meet our excellent Viox who in his exile in Belgium is condemned to much distress: his old father is alone in Lunéville, his wife and his son, Camille, have gone to Termonde to meet him, and poor Camille is reportedly very ill. Thus, our friend has been hit at once in every part of his heart. May God at least preserve his child.
>
> [Neuenschwander 1984a, I,5]

Laflize, Charon, and Viox were friends of Liouville and Vogin's from the Constituting Assembly. The intellectuals were generally against Louis-Napoléon, and Viox was only one of more than 1000 intellectuals who had been arrested or prosecuted or had left the country during the coup d'état in December 1851. Among mathematicians alone, 47 had been arrested or prosecuted [Agulhon 1973, p. 236].

The political defeat changed Liouville's personality. In his earlier letters, he was often depressed because of illness, and he could vent his anger toward his enemies, such as Libri, but he had always fought for what he believed was right. After the election in 1849, he resigned and became bitter, even toward his old friends. When he sat down at his desk, he did not only work, as he had written to his brother, he also pondered his ill fate. During this period, his mathematical notes were interrupted with quotes from poets and philosophers, in particular from Voltaire, from whom he took his first quote:

> tolerance is as necessary in politics as in religion.
>
> [Ms 3618 (10), p. 5v]

It is significant that the enlightened ideas of Voltaire were considered a threat by the ultra-Catholic, often intolerant, part of the *Partie de l'ordre*.

Liouville's bitterness toward the society that had rejected him is also revealed in a quote he copied from Plato's seventh letter to Dion:

> I am convinced that all the states are rather badly governed. There are hardly any good institutions or good administrations. One lives, so to speak, from day to day according to fortune rather than according to sagacity.
>
> [Ms 3618 (10), p. 65v]

Reduced Mathematical Activity

As Liouville wrote to his brother, his mathematical productivity had suffered from his political work. During the mid 1840s, he published around ten papers and notes each year, but in 1848 he only published two and in 1849 only four, none of which were very important. An additional reason for his small scientific output during these years was that some of his available time was spent on preparing a fifth edition of Monge's *Application de l'analyse à la géometrie*. Considering that Monge was known as the great revolutionary creator of the *École Polytechnique*, Liouville may have had political reasons for planning this edition in 1848 [Ms 3618 (9), p 62r].

At the start, Liouville had the idea of adding half a dozen notes to Monge's text, containing some of his own results in differential geometry and some results of his contemporaries. However, while working on the notes, Liouville changed their content, and so the six notes published in 1850 differed strikingly from those he had sketched in 1848. For example, Liouville had originally planned to include a discussion of Thomson's transformation by reciprocal radius vectors, but it was apparently not until late 1849 or early 1850 that he discovered the theorem called after him on conformal mappings (cf. Chapter III). Its proof makes up the most original note in the Monge edition.

In the other notes, Liouville made substantial contributions to the theory of surfaces, following the approach Gauss had used in his *Disquisitiones generales circa superficies curvas* [1828]. In 1847 [1847g], Liouville published a new proof of Gauss's celebrated *Theorema egregium*, which states that the curvature of a surface can be determined from the line element $ds^2 = L dx^2 + 2M dx\, dy + N dy^2$ and is therefore conserved under bending of the surface. Liouville's new proof had awakened an interest in the theory of surfaces in France and had in particular inspired Bertrand, Puisseux, and Diguet to present other proofs [Reich 1973, p. 291]. In 1850, Liouville spread the knowledge of Gauss's work further by including his proof of the *theorema egregium* and other important developments of Gauss's ideas in the six notes to the Monge edition and by appending Gauss's *Disquisitiones*. I shall analyze the mathematical content of Liouville's notes in Chapter XVII.

During the years 1850–1853, Liouville was also active in the *Bureau des Longitudes*. He discussed dispersion (January 23, 1850), improvements of the calculation of perturbations of comets (June 13, 1850), the speed of sound (December 17, 1851), error analysis as presented by Bienaymé (January 28, 1852), Euler's work on the propagation of sound (May 26, 1852), etc. He also gave a "synthetic demonstration of the motion of a simple pendulum with regard to the influence of the rotation of the earth around its axis." This exposition, given on February 12, 1851, was evoked by Léon Foucault's celebrated experiment performed a few weeks earlier. Liouville had originally presented his observation to the *Académie* in connection with a communication by Binet [1851] in which Binet had announced that he had deduced the effect on the pendulum analytically from the equations of motion; the calculations involved were rather complicated, whereas Liouville's synthetic argument was extremely simple:

> At the occasion of Binet's Memoir, Mr. Liouville explained verbally and in detail a synthetic method which also seemed rigorous to him. This method is based on the successive examination of the behavior of 1° a pendulum oscillating at the pole; 2° a pendulum oscillating at the equator, either in the equatorial plane or in the plane of the meridian, or finally in an arbitrary vertical plane. One can pass to the general case of a pendulum oscillating at any latitude by the considerations that Mr. Binet talked about.... "The idea is very simple" said Liouville "it must have occurred to everybody after the communication of Mr. Foucault, which has made everything easy; but the developments which I have added constitute, I think, a mathematical demonstration which is sufficient in itself and provides everything that the calculation can give."
>
> [Liouville 1851b]

The final remark was probably not a sign of Liouville's modesty. Rather, it was meant to belittle the analytical calculations of Binet. In fact, Binet was an ultra-Catholic and thus a political adversary of Liouville. Biot may have had similar political motives when he rose after Liouville had explained his methods in the *Bureau des Longitudes*: "This report from Mr. Liouville gives rise to some considerations from Mr. Biot" [P. V. Bur. Long., February 12, 1851].

Judging from the sparse evidence we have, the most interesting work presented by Liouville to the *Bureau des Longitudes* during the Second Republic concerned atmospherical refraction. On July 25, 1850, he discussed a generalization of his formula from 1842 [1842e] to the case of an atmosphere that is not in thermal equilibrium, and on December 4 of the same year, he extended his theory even further (cf. Chapter XVI, §16). However, he never published these results.

Apparently Liouville was not satisfied with the salary he received from Louis-Napoléon's government for his membership in the *Bureau* and so:

> Mr. Liouville insists that, in the letter which is to be sent to the Minister, the Treasurer of the *Bureau* brings out the difference which exists between the prescribed salaries of the *Bureau* and the sum actually received.
>
> [P. V. Bur. Long., March 30, 1850][2]

In addition to this somewhat egoistic concern for the administration of the *Bureau*, Liouville took an active part in the publication of the *Connaissance des Temps* [P. V. Bur. Long., May 15, 1850] (however, a publication he had promised for this journal in 1851 did not reach completion), and he was a member of the delegation from the *Bureau* that visited the president of the Republic on the occasion of the new year 1851 [P. V. Bur. Long., December 18, 1850].

These activities at the *Bureau*, however, should not conceal the fact that the last years of the Republic were generally a miserable and rather unproductive period in Liouville's life. His republication of earlier ideas on transcendental numbers [1851e], rotating masses of fluid [1851f, 1852f], and doubly periodic functions [1851c] betrays his want of new ideas. Still, the revolution had a lasting triumph to offer Liouville, namely, a professorship at the *Collège de France*.

The Second Election at the Collège de France; Retirement from the École Polytechnique

Libri had taught at the *Collège de France* during his two first years as a professor, but from 1845–1846 he had been allowed to let B. J. Amiot replace him. He was busy in his job as the chief inspector of the French libraries. This post gave him an occasion to augment his collection of rare books and manuscripts. He played a very bold game donating large collections to the libraries with one hand while emptying other collections with the other. Finally, however, his thefts were discovered and a secret report was sent to Guizot. However, Guizot protected Libri, so the scandal was not publicly revealed until 1848.

During the revolution, Libri fled to England — and brought all his stolen goods with him — in order to avoid prosecution. Officially, he remained a professor at the *Collège de France*, but when he wrote to the Assembly of Professors to ask permission to let Amiot replace him once more, they rejected the proposal and decided to leave Libri's name out of the progamme of courses for the spring of 1849. Moreover, they asked the Minister of Education for permission to hire a special replacement for Libri's course. He agreed, and the Assembly unanimously elected Hermite, who used this occasion to lecture on his results on the theory of numbers and elliptic functions.

On June 23, 1850, Libri was sentenced in absentia to 10 years in prison, and by request of the Minister, the Assembly of Professors at the *Collège* decided to grant Libri five months' respite. If he had not returned by then to serve his sentence, he would be considered retired from the *Collège*. However, on October 30, Arago discovered that Libri had even "forgotten" to return some manuscripts that he had borrowed in 1836 from the *Bureau des Longitudes*, and the President of the Republic, without awaiting the expiry of the five month respite, pronounced that the Chair of Mathematics at the *Collège de France* was vacant.

The presidential decree was read in the *Assembly* on November 15, and at the same meeting, Barthélemy Saint-Hilaire read the following letter of application from Liouville:

My dear fellow member,

> Please declare to the Professors at the *Collège de France* that I present myself as a candidate for the vacant Chair of mathematics.
>
> [Coll. Fr. Dos. Pers. Liouville]

At the same occasion, the astronomy professor J. Binet announced the candidacy of his friend Cauchy. Thus, the two candidates from 1843 met again, but this time without Libri.

Several other things had changed since 1843. First, the republic had removed the hated oath of loyalty, so that Cauchy could finally with good conscience serve in the public educational system. Indeed, in March 1849, he obtained the professorship of mathematical astronomy at the Sorbonne. But, the relationship between Cauchy and Liouville had also changed since 1843. Liouville's fame had grown, Cauchy had become older, and the revolution had accentuated the political antagonism between them. At the end of 1849, the relation between the two had deteriorated. After Cauchy, on November 19, 1849, read a *Mémoire sur les intégrales continues et les intégrales discontinues des équations différentielles aux dérivé partielles* [Cauchy 1849] to the *Académie*, "Mr. Liouville presents some critical observations which give rise to a discussion between the two academicians." [CR 29 (1849), p. 557].

It seems probable that this "discussion" was more unfriendly than their earlier exchanges of views, for Liouville followed it up in the Bureau des Longitudes on December 5th:

> On the occasion of a mémoire recently read to the *Académie des sciences* by Mr. Cauchy, Mr. Poinsot and Mr. Liouville presented some considerations on the nature of functions as well as on the nature of differentials and integrals which are related to them.
>
> [P. V. Bur. Long., December 5, 1849][3]

The following month, Liouville apparently tried to belittle Cauchy's work on the dispersion of waves by referring to Fresnel's classical ideas:

> The matter at issue is a mémoire in which Mr. Cauchy gives an analytical expression of the dispersion. Mr. Liouville mentions that the physical condition for the dispersion, according to Fresnel, is that the radius of the sphere of activity of the molecules of the ether must be comparable with the wave length.
>
> [P. V. Bur. Long., December 5, 1849][4]

Despite these disagreements, Liouville seems at first to have been prepared to leave the chair at the *Collège de France* to Cauchy, whom he had openly supported in 1843. Only at the last minute did he decide to apply. This appears in a note by Cauchy from January 13, 1851:

> Changes made in a programme urged a candidate, who had not presented himself this time, to put his name on the list shortly before the election.
>
> [Belhoste 1985, p. 205]

The candidate in question was obviously Liouville, but to which programme was Cauchy referring? Certainly not the one at the *Collège de
France* for the chair continued to be in the subject *Mathématique*, and
there was no official programme at all. No, the programme referred to was
the one of the analysis and mechanics course at the *École Polytechnique*,
which was indeed in the process of being changed.

As a reward for its support of the *Partie de l'ordre* and of Napoléon, the
Church obtained the right to create its own higher educational institutions,
a right they had fought for ever since they lost influence over the educational
system during the great revolution. Cauchy had been very involved in this
fight for "liberty of education." It was guaranteed in the *Loi Falloux* of
March 15, 1850. This law also regulated the university system and the
other public institutions of higher education.

The new political winds were also felt at the *École Polytechnique*. Prompted by LeVerrier, a *Commission mixte* had been set up to reform the entrance requirements and the programmes of study. In the government decree that set up the committee, it was emphasized that "the aim of the
School is to produce engineers rather than students who seem destined for
a job as a professor," prescribing therefore "to restrict the study in the
mathematical courses by eliminating masses of abstract theories, which are
bound to find no applications in any of the public services, and to introduce
instead some practical questions" [Arch. Ec. Pol. III, 1 carton no. 2, Proc.
Verb. Comm. Mixte, July 18, 1850].

Le Verrier was nominated spokesman of the *Commission Mixte*, which
was otherwise composed mainly of civil and military engineers. During 25
sessions, from July to November 1850, it inquired into all aspects of the curriculum and in spite of the objections from the director of studies, Duhamel,
everything was called into question. On November 4, a ministerial decree
conferred the power of the *Conseil de Perfectionnement* to the *Commission
Mixte* [P. V. Ec. Pol. Cons. Perf.], thereby reducing the self-determination
of the school. Two days later, Le Verrier's report of the work of the *Commission Mixte* was published, including the new programme for the year
1850–1851. Considerable modifications were introduced in the curriculum,
as well as in the methods of teaching. The mechanics course was reduced
to a course of applied mechanics. It was detached from the analysis course
and joined with the machine course. The curriculum of analysis was simplified by systematically replacing theories with applications. As for teaching
methods, the professors were required to arrange written examinations every fourth lecture and to supervise more closely the work of the *répétiteurs*,
whose work load was considerably increased.

It is no wonder that the reforms made Liouville want to find a new job
and therefore to announce his candidacy for the chair at the *Collège de
France* nine days after Le Verrier's report had appeared.

Having discussed the qualifications of the candidates on November 25,
the Assembly of professors at the *Collège de France* proceeded to the nom-

ination of the candidate of the *Collège* one week later. The *Procès Verbaux* of the meeting recounts:

> There are twenty-two members present, the majority is twelve. Mr. Cauchy obtains eleven votes, Mr. Liouville ten. There is no result.
>
> One proceeds to a second ballot. There are twenty-three voters. Mr. Liouville has twelve votes, Mr. Cauchy eleven. Having obtained the majority of the votes, Mr. Liouville will be presented as the candidate of the *Collège de France*.
>
> [P. V. Coll. Fr. November 25, 1850]

Five days later, the Administrator of the *Collège de France*, Barthélemy Saint-Hilaire, informed the minister of education that Liouville had been nominated the candidate of the *Collège*, explaining at the same time why Cauchy was not nominated [Coll. Fr. Dos. Pers. Liouville]. Apparently, Saint-Hilaire had foreseen the inevitable disagreements. At the next meeting, a member of the *Assemblé* objected to the *Procès Verbaux* of the preceding meeting, pointing out that there had in fact been two blank ballots during the first vote. However, a greater question concerning the blank ballots had already been raised in a letter to the Minister by three of Cauchy's friends at the *Collège de France*, the professor of Turkish language (Desgranges), the professor of Hebrew (Quatremère), and the astronomy professor (Binet) [Coll. Fr. Doc. Pers. Liouville]. The question was: should the nomination be based on a majority of the members present or of completed ballots? In the letter, they explained that after the first vote on November 25, which raised this problem, no one remembered the rules or any precedents. Saint-Hilaire therefore suggested a new vote, and no one objected at the meeting. However, the three signers of the letter had found a similar case from 1837 when Letronne had been nominated with the majority of the completed ballots, although he had not obtained the majority of the assembled professors. Having referred to similar decisions in other assemblies, they concluded that Cauchy had in fact been nominated as the candidate of the *Collège de France* during the first vote.

They repeated the arguments in the *Assemblée des Professeurs* of the *Collège* on December 8 but were answered by Saint-Hilaire, who maintained that article 24 of the regulation decided the question by stating that the nominations were settled by a majority of the members present rather than of the completed ballots [P. V. Coll. Fr., Dec. 8, 1850].

On the motion by Laboulay, the discussion was postponed indefinitely, but on December 21 the Minister asked the Administrator to convoke the *Assemblée* immediately to settle the matter [Coll. Fr. Dos. Pers. Liouville]. Enclosed in the letter he sent a copy of the complaint that the supporters of Cauchy had not told the *Assemblée* about. Three days later, Saint-Hilaire answered that the election of Liouville remained in force. Finally, the minister ended the exchange of declarations on December 30. He cut the Gordian knot by referring to article 24 in the regulations, which stated that the voting must continue until one of the candidates obtained the majority of "the titular members in office." Since there were 25 professors in office at that time, the majority was 13, and so the *Collège* had not elected a candidate in any of the two previous votings. Therefore, the Minister once more asked Saint-Hilaire to convoke an extraordinary assembly of the professors to take the necessary vote.

Still, when the professors met on January 5, 1851

> a letter from Mr. Cauchy is read. Mr. Cauchy requests that the 1st ballot in the election on November 25, 1850, be considered the only valid one and declares that he will withdraw his candidacy if the *assemblée* decides to have another ballot.
>
> [P. V. Coll. Fr. January 5, 1851]

Cauchy had earlier sent the same request to the Minister. However, the meeting proceeded as though no one had read either the letter from Cauchy or the letter from the Minister:

> A member demanded that before the ballot the *assemblée* decides what is the legal status of the blank ballots.
>
> The *assemblée* decides unanimously that the blank ballots shall not be taken into account in the formation of the majority.
>
> The members proceed to a new vote to elect a candidate for the chair of mathematics. The number of voters is 23, and the majority is 12. Mr. Liouville obtains sixteen votes and Mr. Cauchy seven. Mr. Liouville is proclaimed the candidate of the *Collège* for the chair of mathematics.
>
> [P. V. Coll. Fr. January 5, 1851]

So, in the end, Liouville was elected with a great majority. But, the election was a farce. The very decision to take a new vote removed Cauchy as a candidate, so the vote was in fact superfluous. Moreover, by deciding that blank ballots should not be counted, the Assembly not only overruled the regulation that the minister had referred to, but also adopted an interpretation according to which Cauchy had been elected in the first ballot. Finally, by counting the majority before seeing the number of blank ballots, the *Assemblée* clearly overruled the decision it had just taken. It is amazing that the professors of law, who were present (e.g., Jean Reynaud), could swallow this procedure.

The change in opinion of many of the professors in favor of Liouville may have been influenced by the fuss made by Cauchy and his three friends, and was probably also inspired by the nomination of Liouville as the candidate of the *Académie des Sciences*. At the election in the *Académie* on December 30, 1850, Liouville had obtained 29 votes against Cauchy's 16.

Disappointed and humiliated by the defeat, Cauchy read a note at the following meeting of the *Académie des Sciences* relating how he failed to be appointed to the *Bureau des Longitudes* in 1839 and to the *Collège de France* in 1843 and now in 1850–1851. Cauchy concluded:

... a question which at first appeared entirely scientific, had a solution which was strongly influenced in 1843 by the political requirements and in 1850 and 1851 by a combination of a change of programme and a blank ballot.

[Belhoste 1985, p. 206]

Cauchy's complaint had no effect. On January 18, 1851, Louis-Napoléon formally nominated Liouville professor of mathematics, thereby giving him the opportunity to lecture again on the highest level after an interruption of $7\frac{1}{2}$ years. The nomination also gave Liouville the financial ability to send his resignation to the *École Polytechnique* a few days later. In fact, a law of 1848, which he had himself voted against in the Constituting Assembly, allowed him to hold only two posts at the same time, and being a member of the *Bureau des Longitudes*, he had to choose between the *Collège de France* and the *École Polytechnique*. He ended his teaching at the *École* on February 1, after 20 years of service to this school [Ec. Pol. Dos. Pers. Liouville].

Liouville was only one out of many who left the *École Polytechnique* as a consequence of aggravated conditions for its teachers. Catalan, still only a *répétiteur* of descriptive geometry, resigned on November 28 in protest at the new regulations that demanded that he follow the lectures of Professor La Gournerie, who was a much less experienced mathematician, without raising Catalan's salary. In his letter of resignation, he also complained that the law of June 5, 1850, "has given the *Commission mixte* the power to upset the education of mathematics or rather to destroy it" [Letter: Catalan to *Directeur des. Études Univ. Liège*]. However, the Minister of War refused to accept his resignation, but fired him instead on December 7, because he had refused to do his job, i.e., attend the lectures, (cf. [Jongmans 1981]; Prof. Jongmans has kindly sent me the relevant letters).

Duhamel was next. In January, he resigned from his job as director of study after the *Commission mixte* appointed two of its own members to supervise the work of the teachers. These two, Colonel Morin and Mary, an engineer, did a thorough job analyzing the teaching of all the professors in the minutest details. For example, on January 11, 1851, in their analysis of the *situation des cours* in the *Conseille des Perfectionnement*, they criticized Liouville for his formulation of a particular exercise. Liouville had asked his students to calculate the moment of inertia of an ellipsoid with the three axes $K = 15$, $K' = 14$, and $K'' = 10$ and the density 15, around an axis making the angles l and c with the first two axes. The two supervisors criticized Liouville for not having mentioned the units of length and density and for not having specified the angles. Except for this stupid remark, however, they found no fault with Liouville's teaching, concluding that "Liouville follows his programme exactly."

Many other professors on the other hand were severely blamed for not following their programme, in particular, Liouville's friends, Chasles and Sturm. Chasles immediately resigned, and Sturm was soon forced to leave the school as a broken man, after severe persecution by his superiors. Clearly, Liouville got the job at the *Collège de France* just in time.

Lectures at the Collège de France

Liouville started his lectures at the *Collège* on February 27, 1851. As he had done when he taught as a substitute for Biot in the chair of applied mathematics, he choose to lecture on the subject that presently occupied him the most, namely, "entire and homogeneous differential forms of several variables, such as $M\,dx + N\,dy$, $L\,dx^2 + 2M\,dx\,dy + N\,dy^2$, etc., which often turn up in pure analysis, geometry, and mechanics" [Ms 3640 (1845–1851)]. Elaborating the ideas from his earlier works in differential geometry, in particular the notes in Monge's *Applications*, he was led to a new derivation and a new form of Gauss's *Theorema Egregium*, expressing the curvature of the surface in terms of the geodesic curvature of certain curves on it. At the end of the course, Liouville informed the *Académie* about this new result [Liouville 1851d]. In passing, it should be mentioned that Liouville introduced the term *geodesic curvature* and gave an intrinsic definition for it (cf. Chapter XVII, §42, 43).

In the introduction to the first lecture, Liouville contemplated the ideal plan for his course:

> In an completely regular and methodical study, one ought to discuss first the linear forms, then pass on to the quadratic forms, and then to the entire or fractional forms of an arbitrary degree. However, the small number of lectures which we have at our disposal this semester compel us to pass immediately to the quadratic forms, which offer a great number of very interesting applications.
>
> [Ms 3640 (1846–1851)]

What Liouville had in mind was a systematic study of differential forms "treated in themselves and in all their generality" [Liouville 1852e], but the 12 lectures that were left of the first semester of 1850–1851 only gave him time to expound "the first rudiments."

Liouville published a brief sketch of the basic ideas in this course [1852e], but he never got round to publishing an extensive account of his grand programme for differential forms. This is regrettable, for he seems to have anticipated the idea of a general, differentiable manifold, which Riemann developed a few years later (cf. Chapter XVII, §45–48). In particular, Liouville explained how the action integral of a point mass on a surface under influence of an external force could be treated just as an arc length, $\int ds$. He thereby equipped the surface with a Riemannian metric different from the one it inherits from the space it is embedded in. This pointed to an entirely new concept in differential geometry, and led Liouville to new insights in rational mechanics, which he later included in a course at the *Collège de France* (cf. Chapter XVI, §45–48).

Liouville continued to teach geometry during the first semester of 1851–1852 in a course that he titled *Elliptic coordinates and the analytic transformations that relate to them* [Ms 3619 (6)]. By means of these old favorites of his, he found solutions to three minimum questions concerning tangents to an ellipse. The questions were inspired by Steiner's work. Chasles had already [1843] solved one of these problems by purely geometrical means, and after the course, J. A. Serret's brother Paul Serret, who had been in the audience, found a geometric solution for the two others [P. Serret, 1852]. Liouville postponed the publication of his own proof until 1856 [1856c].

In between these two geometric courses, Liouville lectured on doubly periodic functions. It is most probable that he had originally planned a second geometric course as a continuation of the first very brief one, but a discussion with Cauchy changed his mind. On March 31, 1851, Cauchy reviewed a *mémoire* by Hermite *relatif aux fonctions à double période* to the *Académie* [Caucy 1851a]. Cauchy highly praised Hermite's paper, in particular, its main theorem, which stated that "one can reduce to a rational function of elliptic functions every doubly periodic function that always remains continuous, together with its derivative, as long as it does not become infinite." Considering that this had also been one of the central results in Liouville's theory, it is not surprising that Liouville rose after Cauchy's report:

> Without being opposed to the conclusions of the report, Mr. Liouville thinks that he should recall that he also, a long time ago, found a general theory of doubly periodic functions which he incidentally made known to the *Académie* in 1844 at the occasion of a mémoire of Mr. Chasles· · ·.
>
> [Cauchy 1851a, p. 450]

Liouville continued with a summary of his methods and results. In order to substantiate his claims, he hurried home to fetch a copy of Burkhart's notes from his 1847 lectures, which he then deposited on the desk of the président before the end of the meeting. Its table of contents was printed in the *Comptes Rendus* next to a summary of his talk.

Moreover, Liouville's talk made Cauchy add a footnote to the printed version of his review stating that:

> Already in 1844 Mr. Liouville obtained, and announced in the presence of Mr. Hermite, the above reduction [of doubly periodic functions to elliptic functions], using a method which is quite different from the one used by Mr. Hermite.
>
> [Cauchy 1851a, p. 450]

The greatest difference between the two methods is Hermite's extensive use of Cauchy's theory of complex integration. Cauchy also added a two-page note repeating with more vigor than in 1844 his claim of priority of Liouville's theorem in its general form. Indeed, he indicated a new proof of this theorem and added in a footnote:

> Mr. Hermite, who I told about this proof based on the calculus of
> residues, told me that he had already discovered it and given it in a
> lecture at the *Collège de France.*
>
> [Cauchy 1851b]

But, Cauchy went further. He claimed that the basic method used in
Liouville's work stemmed from his *Exercices Mathématiques.* One senses a
hostility between the two mathematicians, who had been opponents at the
Collège de France only a few months earlier. Still, compared with Liouville's
quarrels with Libri and Pontécoulant, these polite and sober exchanges of
views were moderated by their mutual scientific respect.

Liouville's relationship with Hermite also seems to have changed. In 1843,
Liouville defended Hermite against Libri. In 1851, on the other hand, Liouville emphasized his own results at the expense of Hermite, who, judging
from the footnote, was now more closely attached to Cauchy. Indeed, during
a severe attack of smallpox, Hermite had adopted Cauchy's religious and
political beliefs [Hermite 1984, Freudenthal 1972], so his relationship with
Liouville must have been strained during the revolution and the Second
Republic.

In order to underscore his priority, Liouville choose "the theory of elliptic
functions or rather of doubly periodic functions which can be reduced to
elliptic functions" [Ms 3640 (1846–1851)] as the subject of his lectures at
the *Collège de France* during the second semester of 1851–1852. According to Bertrand [1867, p. 3], he lectured "for a select audience" including
Charles-Auguste Briot (1817–1882) and Jean-Claude Bouquet (1819–1885),
who later included Liouville's ideas in their famous treatise, *Théorie des
fonctions elliptiques* [1859]. Liouville still refrained from publishing anything.

The Disappointing Outcome of the Second Republic

Apparently Liouville enjoyed being able once more to lecture on the subjects that occupied him, but his professorship at the *Collège de France*
seems to have been the only reward he ever got from his enthusiastic in-
volvement in the establishment of the Second Republic. He experienced how
the reign of Louis-Napoléon soon came to resemble that of Louis-Philippe.
From Liouville's point of view, the educational policy was even worse because of the influence of the Church, as guaranteed in the *Loi Falloux,* and
because of Le Verrier's influence, which led to a reduction in mathematics
education at the *École Polytechnique.* After the *coup d'état,* Liouville had
to take an oath of loyalty once more, first to a president and later to an
emperor, who had himself broken the oath to the constitution for which
Liouville had worked and voted.

The reintroduction of the oath of loyalty again put Cauchy in an awkward situation. For, although he preferred Louis-Napoléon to the liberal or socialist republicans and was in fact in favor of the *Loi Falloux*, he was still true to the oath he had sworn to the Bourbons. Therefore, he chose to leave the *Faculté des Sciences*. However, after a few years, Napoléon III exempted Cauchy as well as Arago from the oath, thereby showing a greater flexibility than his predecessors.

Still, this does not hide the fact that absolutism and conservatism had won over the democracy and liberalism for which Liouville had fought.

V

The Last Flash of Genius (1852–1862)

Imperial Politics; Its Influence on Liouville's Family

The beginning of the Second Empire was a period of progress and glory
for France [Plessis 1979]. Industry and agriculture flourished. The railroad
network was greatly extended, and in 1852 the line between Paris and
Strasbourg through Toul was completed, thereby facilitating Liouville's
travel between his two homes. However, the economic expansion also had
annoying effects on Liouville's life, caused by Napoléon III's decision to
transform his medieval capital into the grandiose city we can still admire. In
addition to imposing buildings like the *Opéra*, the *Bibliothèque Nationale*,
and the *Gare du Nord*, Baron Haussmann planned the broad boulevards,
such as *Boulevard St. Michel*, which cut through the densely populated
part of the Latin Quarter, where Liouville lived. His house on the *rue de
Sorbonne* was demolished in 1853, and the family moved to 13 *rue de Condé*
in the vicinity of the *Place de l'Odéon*, that is, closer to the *Institut* but
further from the *Collège de France*.

Externally, France also triumphed. First, together with England, it won
the Crimean War in 1856, and only three years later, Napoléon III returned
victoriously to Paris after a war he had fought in Italy against Austria. Liouville followed the movements of the French armies with particular interest, for, following family tradition, two of his brothers-in-law had military
careers. They served with the colors during the almost one-year-long siege
of Sebastopol [Ms 3621 (5)] and won glory and promotion during the attack
on the town on June 7, 1855.

Because of the invention of the electric telegraph, it was for the first
time in history possible for the newspapers to publish daily reports from a
distant front. No wonder Liouville was anxious when he read of the decisive
but very bloody assault on the Malakoff Tower on September 8, 1855. One
week later, he told Chasles:

> We have been informed by the Ministry of War that *Commandant*
> Balland [one of Liouville's brothers-in-law] does not figure on the
> list of major State Officers killed or wounded at Sebastopol. But,
> the telegraphic information until...does not report...
> [Ms 3621 (2) the rest of the draft is missing]

The other brother-in-law also "had the luck of escaping the double dangers of disease and bombs; his decoration cost him neither arms nor legs.

Therefore, we must thank God." (Liouville to Dirichlet, November 5, 1855) [Neuenschwander 1984a, II 8].

For a republican like Liouville, the success of the Second Empire must have been viewed with mixed feelings. He must have been constantly aware of the oppression of scientists and others, and he did not take to the toadyism in the surroundings of the Imperial Court. Thus, in his notebooks, he wrote "Vanitas Vanitarum" below a clipping from the *Moniteur* of November 13, 1865, containing a report from the Minister of Education to Napoléon III in which he glorified the French participation in the World Exhibitions and in particular the great contribution of the Emperor. In fact, this large scale industrial fair was held in Paris in 1855 and in 1867.

Apart from the enforced clashes with Imperial town planning and war making, Liouville did not get involved in politics, but concentrated his energy on science. From almost total unproductivity during the first years of the Second Empire, Liouville's scientific output and creativity reached a peak during 1856 and the first half of 1857. In 1857, it decreased again as a result of his appointment as a professor at the *Faculté des Sciences*. This appointment was the last triumph in his professional career.

During the early 1850s, Liouville continued to have a great influence in the *Académie*. He still helped young new talents, such as Chebyshev and Holmgren, on the way, but more often than before he was on collision courses with his colleagues, and many of his earlier protégés and students turned against him. He seems to have enjoyed the role of "master," guiding inexperienced but talented youth, but apparently he found it difficult to accept when his students began to act as his equals. In fact, during the 1850s, Liouville saw some of his protégés surpass him, Hermite scientifically, for example, and Le Verrier politically, at a time when his own "teachers," whom he had never excelled, passed away (Arago in 1853 and Cauchy in 1857). Consequently, his own time as the master and head of French mathematics only lasted for a few years at the end of the 1850s. From this period stems the following eulogizing report written by Chasles, Lamé, and Lefébure de Fourcy:

> This learned Academician has for a long time been known for his numerous and important works on various parts of pure... mathematics. For a long time, his voice has been heard with great authority from public posts. With a constant energy he alone runs the *Journal des mathématiques*, which he founded and which is the only publication of its kind in France. His name is often mentioned by us in our courses, either in connection with the fields of research he has created or in connection with the improvements he has introduced in the works of the masters who have preceded him. Finally, as a result of his numerous works and his vast erudition, he is today the head of the French geometers and he is not inferior to any of the European geometers, of great repute.
>
> [Neuenschwander 1984a, p. 129]

PLATE 17. Rue de Condé. From 1853 Liouville lived in number 13 which is the house in the middle of the picture. (Photographed by Jesper Lützen)

This chapter deals with the peak of Liouville's institutional and public influence and fame, but as we shall see, the period was not a glorious one for Liouville himself. Increasing health problems and depression sometimes made life unendurable for him, and by the end of the period, his creative scientific talents had diminished.

Ernest Liouville, Statistics, Bienaymé (1852–1853)

During the early 1850s, Liouville enjoyed the triumph of seeing his son embark on a promising scientific career. Toward the end of the 1840s, Ernest had been sent to the *Collège* in Toul, where Liouville had studied 25 years earlier. Ernest seems to have done well in school, although one can find an accidental testimony of bad behavior in a letter that Liouville wrote to the head of the *Collège*, Mr. Grébus:

> I thank you for the information you have been so kind to send me regarding my son Ernest. Madame Liouville has written him a letter in the form of a sermon, which I hope will suffice. I did not want to write myself for that would have considerably aggravated the reprimand and that seems unnecessary to me because his marks have been fine until now.
>
> [Ms 3618 (9), p. 80v]

After *Collège*, Ernest joined the *Faculté des Lettres* and sat for the examination of *Baccalaureat-es-lettres* during the spring of 1851 [Ms 3633 (1)]. The following year he embarked on a career as an astronomer. Under the date July 28, 1852, one reads in the *Procès Verbaux* of the *Bureau des Longitudes*:

> Mr. Arago proposes Mr. Ernest Liouville as a trainee candidate with the title of *Élève astronome de l'observatoire de Paris*.
>
> This proposition is submitted to be approved by the *Bureau*.
>
> There are 8 voters, the vote gives 8 ballots carrying a "Yes."
>
> [P. V. Bur. Long., July 28, 1852][1]

Thereby Ernest was included among the assistants who the aging Arago gathered around himself at the Observatory. They had their lodgings in the Observatory building and helped Arago with the observations and calculations, but they also had time to do their own research. Thus, during the autumn, Ernest informed the *Académie* about observations made at the Observatory on November 18 and 20 of an asteroid discovered a few days earlier by Hermann Goldschmidt (1802–1866) [CR 35, p. 757].

Early the following year, Ernest presented a work on the influence of personal errors on astronomical data to the *Bureau des Longitudes*. His choice of subject may have simply been due to its popularity at the time, but it may also reflect the interest his father had acquired in statistics the previous year while preparing a review of a work by Jules Bienaymé (1796–1878) on "the probability of errors according to the method of least squares."

This work was among Bienaymé's most important contributions to statistics, and although it was written in his 56th year, it was the first paper he presented to the *Académie des Sciences*. He had read papers to the *Société Philomatique* during the 1830s, and according to Heyde and Seneta [1977, p. 32], Liouville had several times emphasized his results in this forum. This seems to indicate amicable relations between the two. A job as a civil servant had delayed Bienaymé's scientific career, but after 1848 he resigned in order to pursue the science of probability and statistics. The very favorable review that Liouville read to the *Académie* on January 19, 1852, on behalf of Chasles, Lamé, and himself, was probably instrumental in securing for Bienaymé a seat in the *Académie* the same year.

The method of least squares had earlier caused bitter priority disputes between Gauss and Legendre. Liouville therefore selected his words with care in the historical introduction to the report:

This celebrated method [of least squares], which is very frequently used today, was first found by Legendre... in 1805, and since it has been the subject of research of a great number of geometers among whom we cite Laplace and Mr. Gauss.

[Liouville 1852a, p. 90]

As a whole, the methods of mathematical statistics were very weakly founded at this time, and as soon as Bienaymé had entered the *Académie*, a second round of disagreements about the method of least squares flared up between him and Cauchy. No wonder that Liouville expressed his support of the practical utility of the calculus of probability with some reservations:

However, the calculus of probability, to which the impressive names of Pascal, Fermat, Huyghens, Jacques Bernoulli, Laplace, Fourier, Poisson etc. are attached is not an abstract pure speculation. Restricted by its proper limitations, it is useful in practice. To be sure, it has often been abused, but that is no reason for proscribing its use.

[Liouville 1852a]

Bienaymé was a Laplacian in the sense that most of his papers were generalizations, corrections, and new interpretations of Laplace's works. In 1852 [1852], he extended Laplace's work on the method of least squares to situations in which there is more than one source of uncertainty. The *Académie* review contained no explicit evaluation of this idea, but when Liouville told the *Bureau des Longitudes* about Bienaymé's work a week later, he strongly supported Bienaymé's conclusions:

Mr. Liouville explains some details in the work of Mr. Bienaymé regarding the calculus of probability and in particular an application he has made of the method of least squares to the calculus of probability of the error of the mass of Jupiter as determined by Mr. Laplace and Mr. Bouvard. Mr. Bienaymé has shown that the formula given by Laplace in this connection was ~~wrong flawed~~ because it gave the probability of the mass as though it was the only unknown of the problem.

[P. V. Bur. Long., January 28, 1852][2]

In the above quote, the word "wrong" [faute] was first replaced by "flawed" [défautive], but that was also crossed out, leaving an empty space. This suppression of Liouville's own words suggests that even 25 years after Laplace's death it still caused offense to criticize the master of *Mécanique Céleste*.

Liouville's announcement of Bienaymé's ideas seems to reveal a certain knowledge of, and interest in, the theory of probability. Indeed, at the *École Polytechnique*, Liouville had taught this subject, and as we saw earlier it inspired him to write one small note. Bienaymé's paper temporarily aroused his interest in the more complicated applications of this theory, and so on August 11, 1852

> Mr. Liouville spoke to the *Bureau* [*des Longitudes*] about a problem
> concerning the most probable position of a point determined by the
> intersection of three straight lines drawn through three points given
> in position and having directions which are also known.
>
> [P. V. Bur. Long., August 11, 1852][3]

Liouville became so interested in the calculus of probability that he chose
this as the subject of his lectures at the *Collège de France* during the
first semester of 1854–1855 and on two later occasions (1862–1863, second
semester, and 1872–1873, second semester). The first of these lectures inspired him to publish a small note [1856j] in this field, but a long memoir,
which he announced there, never appeared.

As pointed out, these interests may have inspired his son to the work on
the compensation of personal errors, which he described to the *Bureau* on
February 2, 1853 (cf. also [Bigourdan 1932, pp. A55-A56]). Liouville was
apparently proud of his son's progress, and nine days later he wrote to
Dirichlet:

> Therefore come. In France, at the Observatory, you will see a tall,
> beautiful young man called Ernest Liouville.
>
> [Tannery 1910, p. 22]

However, one year later Ernest's career was abruptly cut short by the
dramatic changes at the Observatory.

Liouville Opposing Le Verrier (1852–1854)

On October 2, 1853, Arago died, and French science lost its great leader.
Many scientific, as well as administrative, posts had to be filled, first of
all the directorship at the Observatory. On January 31, 1854, the Minister
appointed Senator Le Verrier to this post, without consulting either the
Bureau des Longitudes or the *Académie* [Bertrand 1889, pp. 157–191]. From
a scientific point of view, this was a natural choice, but it turned out to be
disastrous for the Observatory in the long run. It was also an unpleasant
choice from Liouville's point of view. Le Verrier's quarrels with Arago,
his conservative political, catholic, and Napoleonic views, and his work in
the *Commission Mixte* at the *École Polytechnique* turned Liouville from
his former student, and from 1852, when Le Verrier, strongly encouraged
by the members of the *Bureau* [P. V. Bur. Long., October 6, 1851], again
occupied his seat as an adjoint, Liouville began to criticize his work openly.
For example, one reads in the *Procès Verbaux* of September 11, 1852:

> On the occasion of the last opposition of Neptune, Mr. Liouville in-
> sists that one should check if the observations of this star continue to
> be in accordance with a major axis of approximately 30 and differing
> therefore by 1/6 from the one Leverrier adopted hypothetically as a
> basis of his calculations.
>
> [P. V. Bur. Long., September 11, 1852][4]

PLATE 18. Urban Jean Joseph Le Verrier (1811–1877). Liouville strongly supported his first prominent student Le Verrier at the beginning of his career. After Le Verrier was appointed director of the Observatory in 1853 the two became bitter enemies. (Archives de l'Académie des Sciences de Paris)

It is now known that Le Verrier had indeed assumed a value that was too large for the major axis of Neptune, but by adjusting the value of the eccentricity, he had come out with a rather accurate approximation of the orbit around 1846. The erroneous major axis therefore does not diminish the great importance of Le Verrier's discovery, and Liouville just seems to have used the last opposition as a pretext for calling Le Verrier's calculations into question. A few months later, he sharpened the tone considerably:

> On the occasion of a new asteroid discovered by Mr. Hind, one recalls Mr. Olbers' well-known hypothesis that these small bodies are the result of an explosion of a planet which in former times circulated in the region between Mars and Jupiter. At this occasion, Mr. Biot and Mr. Babinet cite a theorem due to Mr. Leverrier, according to which no star in this region with a very small mass can conserve an orbit with a small eccentricity and only slightly inclined to the ecliptic under the perturbing influence of Jupiter and Saturn.
>
> Mr. Liouville observes that Mr. Leverrier's analysis, mentioned above, rests on an inexact basis and on an approximation which is

badly suited for the case under consideration and the one to which
Mr. Leverrier believed he could extend it. This can easily be seen
without calculations from the result itself and using a very simple
argument. "But," Liouville added, "one can treat the question by
way of rigorous formulas, for I have made sure that the simultaneous
differential equations related to the calculation of the mutual inclination of the orbits of three mutually perturbing planets, for which the
eccentricities are very small and the inclinations are arbitrary, can
always be integrated whatever the masses of the three planets are
[probably Liouville 1839f]. The integration becomes the more simple
when one of the masses is imperceptible. This simple case has been
the point of departure for my research; its scope has gradually been
extended, and I have made it the subject of a memoir which the
current discussion has made me present to the *Bureau* a little earlier
than I had otherwise planned to do."

[P. V. Bur. Long., December 22, 1852][5]

In fact, Liouville's memoir was not published.

A few weeks after this episode, Liouville joined forces with Arago and
Chasles in an argument against Le Verrier and Faye. This quarrel, con-
cerning the new curriculum at the *École Polytechnique*, took place at the
Académie (cf. [CR 35 (December 27, 1852), p. 930, and 36 (1853), p. 3]).
A letter from February 1853 to Dirichlet shows how much it embittered
Liouville to see his former students turn against him:

> ...our *Académie* where the [les] Leverrier and the [les] Faye have
> replaced the [le] Libri, ... I continue to perform my duties being
> hardly afraid of these ill-mannered people of wicked hearts, my for-
> mer students, alas! who I helped all too much when they were cold
> and weak. Since then, they have warmed up as La Fontaine's snake.
> But, you know that I believe in God, so do not fear for me.
> [Tannery 1910, pp. 21–22 complemented by Neuenschwander]
> [1984a, note 66]

After Le Verrier was appointed Director of the Observatory, the battle
between him and an increasing number of French scientists continued in
various institutions. The events in the *Bureau des Longitudes* have to be
extracted from Bigourdan [1828–1933], since the *Procès Verbaux* covering
the period from Arago's death until 1868 are lost (one may wonder if that
is an accident or if someone has deliberately removed this testimony of a
dark period in the history of the *Bureau*).

While Arago dominated both the Observatory and the *Bureau*, these
two institutions had worked harmoniously together. However, Le Verrier
immediately created tensions by trying to transfer as much power and as
many resources as possible from the *Bureau*, where he was only an associate
member to the Director of the Observatory. He succeeded in transferring the
calculators of the *Bureau* to the Observatory, which from now on became

responsible for the publication of the new observations. However, at the meeting in the *Bureau* on February 8, 1854:

> Mr. Liouville observes that according to the decree the reduction and publication of old observations is within the competence of the *Bureau des Longitudes.*
>
> [Bigourdan 1932, p. A92]

Liouville continued by pointing out that the *Bureau* needed calculators who could take care of these reductions and work out the tables for the *Connaissance des Temps* and the *Annuaire*, of which the *Bureau* remained in charge. Clearly, he was trying to oppose Le Verrier's growing power, but he did not succeed. On the contrary, Le Verrier took revenge three days later by informing Liouville's son that he was no longer wanted as an *élève astronome*. This was only part of a large plan that completely emptied the Observatory of all the helpers that Arago had gathered around him during his last years. How this, more than anything else, infuriated Liouville can be seen in his letter to Dirichlet from April 17, 1854:

> My dear friend
>
> You will no doubt excuse the fact that I have taken so long to write to you. You are probably aware of the painful emotions I have had to face during the six months which have passed since your nice visit. The death of our excellent secretary Mr. Arago, the expulsion of his family and his friends from their formerly peaceful stay at the Observatory, the degradation of the *Bureau des Longitudes*, the suicide of poor Mauvais, whose life was made unbearable by too much trouble, this is but a small part of what we have had to suffer. What is even worse is the hideous spectacle of hatred and lying which Le Verrier, Biot and other *fellow* members of the same kind have offered us. More than once I have planned to try and draw you a faithful picture of their folly and of their rage, but my heart revolts from thinking of them and the pen falls out of my hand.
>
> Only two words about my son - Saturday, February 11, he received the following letter from Leverrier (Leverrier did not even know him by sight) "Sir, the discharge of my duties absolutely requires that the rooms you occupy at the Observatory at my disposal Monday morning. Sincerely etc." This letter was short but clear, and although the interval between Saturday evening and Monday morning is short, my son (who, by the way, was just as eager to leave as Leverrier was eager to throw him out) moved his home during this time to the first apartment he could find, namely, 199 *rue de Vaugirard*, very far from us but close by the workshop of Mr. Brunner. When leaving the Observatory, he felt he ought to thank the *Bureau des Longitudes* which had appointed him *élève* fifteen months earlier. Below, I have copied his letter to the President which shows that he has worked well: "Mr. President, the circumstances that force me to leave the Observatory, together with Mr. Arago's family and with my

dear and excellent masters Mr. Mathieu, Laugier and Mauvais do not prevent me from preserving a deeply felt gratitude for the kindness with which the *Bureau des Longitudes* appointed me one of its *élèves*. I have tried to justify this great favor with my work. I have inscribed almost seven thousand observations in the meridian registers, and I can assure the *Bureau* that I have not tried to obtain the number at the expense of the difficulty. Moreover, I have been rewarded for my work by the benevolence of the illustrious man whom France mourns for and whose honorable behavior and scientific dignity I shall always follow. I am sincerely etc."

[Neuenschwander 1984a, II, 5]

After Dirichlet expressed his sympathy to Liouville and his son [Tannery 1910, pp. 24–26], Liouville returned to his son's fate in the following letter:

...let me correct an error which I committed by writing that Ernest was at the Observatory for 15 months when Leverrier came and dismissed everybody. He was there for 18 months, and during 15 months he was allowed to inscribe his observations in the registers; that makes 15 months of service as a real astronomer after a 3-month training period. Alas! this poor boy really had the sacred fire!

[Neuenschwander 1984a, II, 6]

During the next couple of years, Liouville published three of his son's papers on astronomy in his *Journal* [E. Liouville 1854a,b; 1855], and in 1856 Ernest presented a *Note sur deux étoiles variables* to the *Académie* [CR 42, p. 546] based on his observations in 1853. He had planned to continue the observations, but as he remarked "I do not have the means for this work." In fact, after the unpleasant experience in 1854, Ernest gave up his scientific career and turned to his uncle's profession, law. Many years later he combined this profession with his earlier interest in the theory of probability and wrote a paper *Sur la statistique judiciaire* for his father's *Journal* [E. Liouville 1873].

It is not surprising that until the end of his life Le Verrier replaced Libri as Liouville's major adversary in Parisian scientific circles. Like Libri, Le Verrier soon became unpopular among French scientists because of his dictatorial behavior, and like Libri he was supported by the politicians in power, but unlike Libri he was no charlatan and was widely recognized abroad as the greatest French expert in theoretical astronomy. For this reason, Le Verrier was a much harder adversary than Libri, and it is much more difficult to evaluate in a just way who was right, Le Verrier or Liouville. However, I feel it is fair to say that while Liouville's opposition to Le Verrier is very understandable from a human point of view, he was often scientifically skating on thin ice.

This is certainly the case when Liouville tried to bar Le Verrier from becoming a titular member of the *Bureau des Longitudes*. This problem arose in July of 1854 when the Minister of Education invited the *Académie des Sciences* to propose two candidates for a vacant seat at the *Bureau*. At first, the President suggested that the decision should be made by the geography and navigation section and the astronomy section, but Liouville demanded that the geometry section also be heard, for "among the members of the *Académie* who also belong to the *Bureau des Longitudes* there are two, Mr. Biot and Mr. Poinsot, who do not belong to either of the two sections but to the geometry section" [CR 39 (1854), p. 39]. One may infer that his real reason was to involve his friends from the geometry section in a fight against Le Verrier.

Whether due to Liouville or not, the three *Académie* sections pigeonholed the matter, and the seat was left vacant. By the same token, several seats, which became vacant at the *Bureau* during the next couple of years, were left open. Bigourdan [1933] interprets this as a way to hinder Le Verrier from becoming an ordinary member. This trick worked until 1861. In order not to anticipate events, I shall postpone a discussion of Liouville's role in the dramatic conflict in 1861 until I have analyzed other aspects of his career during the 1850's.

Friendship with Dirichlet

Liouville's correspondence with Dirichlet is one of the best sources for Liouville's life during the 1850s. From 1841 to 1851, no letters between the two are preserved, but that seems to be due to the incomplete transmission of the correspondence rather than to an actual pause in their relation. At any rate, the letters from the 1850s testify to a close friendship between the two mathematicians and more generally to the cordial relations between a group of French scientists and friends of Dirichlet, namely, Chasles, Sturm, Lamé, Poncelet, Pelouze, and Duhamel and a less well-defined group of German scientists, including von Humboldt, Steiner, Borchardt, and others, who Liouville called "our friends."

The friendship between Dirichlet and Liouville was reinforced during Dirichlet's visits to Liouville's home. The first visit during the 1850s was enthusiastically suggested by Liouville on February 10, 1853: "Madame Liouville claims that you have previously promised her to come to France with Madame Dirichlet. Do come, do come" [Tannery 1910, p. 21]. However, Dirichlet replied that he feared that the "present political affairs" - probably the reintroduction of the Empire - would make it hard for him to see France again, and so he suggested that the two families meet elsewhere during the summer holiday [Tannery 1910, p. 24]. During the spring, the political winds were tempered, and the two families decided to meet in Toul after all. Liouville welcomed this idea with great joy:

> You are not only a great geometer, you are also an excellent man, and that I do not appreciate less. A thousand thanks for the nice visit you promise.... No doubt you do not come alone; we count on

your family as well. Madame Liouville will be very happy finally to see Madame Dirichlet. Do not fear that you will disturb us; we have room for all of you in our house, and if needed we have my brother's house at our disposal, but that won't be necessary. Do come very soon.

[Liouville to Dirichlet, August 21, 1853, Neuenschwander 1984a, II, 4]

The visit, which took place during the fall of 1853 (cf. the letter of April 17, 1854, quoted above), was soon eclipsed by the problems with Le Verrier. Still, in the letter in which he told about his son's hardships, Liouville also wrote about a triumph that he and his friends had won on Dirichlet's behalf:

Again today, we have to fight the Biots and the Le Verriers regarding the position as foreign associate which has become vacant after Mr. de Buch's death [the *Académie* had 9 such positions]. However, this time we are allowed to defend ourselves; and in the Committee [appointed to propose the candidates] headed by Mr. Combes and composed of Mr. Biot, Elie de Beaumont, Thenard, Chevreul, Flourens and me, Mr. Biot was the only one against you, whereas the others put you in the first rank on the list presented to the *Académie*. - We are going to vote; my heart contracts. The result of the vote: 51 voters; you obtained 41 votes, Mr. Airy (Mr. Biot's candidate) 6 votes, etc. Thus, my dear Dirichlet, victory! Now you are our associate.

After all, the geometric sentiment is not extinct in us. Last time you saw that we did justice to Steiner.

[April 17, 1854, Neuenschwander 1984, II, 5]

Steiner had been elected correspondent of the *Académie* one week earlier [CR 38]. When Dirichlet heard of his own election, he thanked Liouville and also asked him to thank the members of the committee, who "without knowing me personally, and for the most part are foreign to the mathematical sciences, have treated me so favorably on your authority, which is justifiably very powerful but which your friendship with me could make a bit suspect," [Tannery 1910, pp. 25–26].

The following year, Liouville again invited Dirichlet to come:

P.S. Come and visit us, and if the traveling geometer Borchardt comes in this direction, tell him to think of us and of the town of Toul (rue Salvateur), where we will receive him with open arms.

[August 8, 1855, Neuenschwander 1984a, II, 7]

It took long for Dirichlet to answer, and first Liouville hoped that "instead of writing you would come and visit us; your presence is better than a letter" [Liouville to Dirichlet, November 5, 1855. Neuenschwander 1984a, II, 8]. However, Dirichlet did not come until the following March. At this time, Liouville was in Paris, and Dirichlet used the opportunity to attend a meeting of the *Académie* in the capacity of foreign member. Liouville had

PLATE 19. Gustav Peter Lejeune Dirichlet (1805–1859). Was Liouville's intimate friend from 1830 to Dirichlet's death. Dirichlet's works on the theory of numbers inspired Liouville to his last long series of papers. (Math. Dept. University of Copenhagen)

again invited Mrs. Dirichlet to join her husband [Neuenschwander 1984a, II, 9], but she could not come because she was not well. Upon returning to Göttingen, where Dirichlet had been appointed Gauss's successor in 1855, he found his wife "very ailing... the fatigue of moving [to Berlin] and a chill that she caught afterwards have had a strong influence on her nerves, which you know, Madame, are far from being in a normal state" [Tannery 1910, pp. 26–28]. The *Madame* is Liouville's wife to whom Dirichlet sent his warmest gratitude when he returned from Paris. This letter reveals that Dirichlet was now a close friend of the entire Liouville family. Mrs. Liouville bought a dress for Mrs. Dirichlet, which "always fills the Göttingen society with admiration" [Dirichlet to Mrs. Liouville December 12, 1856, Tannery 1910, p. 29], and when Dirichlet asked her to take care of his niece during a stay in Paris, he expressed his gratitude with a mathematical metaphor:

> Yet, I do not say that it would increase the feeling of gratitude which the long lasting kindness of the Liouville family has given me, for being married to Geometry, you know as well as I that the infinite quantities are not susceptible to an increase."
> [Dirichlet to Mrs. Liouville, May 21, 1857, Tannery 1910, p. 34]

Dirichlet repaid what Liouville had done for him by arranging for him to obtain the Gauss medal and be elected foreign member of the Academy of Göttingen. Of course, officially Liouville's thanks went to the *secretaire perpétuel* of the Göttingen Academy, Haussmann [Tannery 1910, p. 29], "but I owe you in particular eternal gratitude," he wrote to Dirichlet [Tannery 1910, p. 28].

After 1856, Liouville and Dirichlet do not seem to have met again despite many invitations from both sides. In July 1857, Liouville tried to lure Dirichlet to "come soon to Toul" so that they could go over a French translation of Dirichlet's obituary of Jacobi together [Neuenschwander 1984a, II, 11]. The following year, Dirichlet invited the Liouvilles to Göttingen, but they could not come because they had to supervise the wine harvest in the wineyard (8000 m^3 in 1859) that Liouville had inherited from his father [Neuenschwander 1984a, note 50]. Instead, Liouville suggested that the Dirichlet's came for a "grape cure" in Toul, but they had already planned a cure in Bavaria. Still, since this place was only five or six hours from Toul, Dirichlet would have agreed to come had it not been for a principle of number theory:

> However, I think that I must gain the necessary moral force to stay away in order not to violate once more a law which ought to be sacred, in particular to us *viri arithmetici*, as Jacobi called us, namely the law of reciprocity.
>
> [Dirichlet to Liouville August 27, 1858, Tannery 1910, p. 35]

In fact, Dirichlet traveled to Switzerland instead [Kummer 1860, p. 339]. There he suffered a heart attack from which he recovered slowly. When Liouville heard of it, he was shocked and wrote back urging Dirichlet to send him a "nice and long letter full of mathematics" [November 21, 1858, Neuenschwander 1984a, II, 16]. When he received no letter, Liouville wrote again the following month:

"I would like to think that you are now completely cured and I wish that the year 1859 will do you good. ... Console us, dear friend, with a few lines." [Neuenschwander 1984a, II, 17]. Dirichlet soon wrote a few lines, but they brought no consolation: his wife had died. Liouville answered:

> Dear friend
>
> What terrible news. I have no consolation to offer you! Such an excellent wife, such a fine mind, and such a good heart, all that disappeared in one day! - What shall I say to you dear friend, except that we all suffer from the loss that grieves you - that you need courage in this life; that you also need it for your children, even when you no longer have any for yourself.
>
> Madame Liouville adds a few words on the following page....
>
> [Liouville to Dirichlet January 7, 1859, Neuenschwander 1984a, II, 18]

The trembling handwriting of the draft in Liouville's notebook shows how shaken he was, but he would certainly have been worse had he known that this was to be his last letter to his friend (at least the last preserved letter). On May 5, 1859, Dirichlet died from another heart attack. However, before we return to this distressing event, we shall see what the correspondence between Liouville and Dirichlet, and other sources, tells us about the life and work of the former during the 1850s.

Mathematical Production (1852–1857)

As mentioned earlier, Liouville's scientific output almost stopped during the revolution. It gradually grew during the following years to five publications in 1852 and then declined again in 1853 to only one paper in his own name and two notes under the pseudonym Besge. Liouville explained this low productivity to Dirichlet on May 10, 1853:

> I have had to change my address, the house on rue de Sorbonne where I lived is going to be demolished. Here I am at 13 rue de Condé, very badly settled, with my books in disorder, very bad for the work. Add to all this, that during the move I have married off my eldest daughter, Céline, and you will understand why I have had to neglect the x's and y's.
>
> [Neuenschwander 1984a, II, 3]

Despite Dirichlet's visit, the holidays of 1853 were unproductive, and after the following two semesters where the controversy with Le Verrier had added to the ordinary fatigue from teaching, Liouville wrote to Dirichlet:

> Now, I am finally free and perhaps tomorrow after the meeting of the *Académie*, or the day after at the latest, I believe that I shall leave for the nice town of Toul, where you definitely have to visit us. There I can finally get to work and send you some scientific letters, after having wasted almost one year on the worries and the dirty tricks we have to defend ourselves against every day. ...
>
> Ah! my dear friend, with what pleasure shall I walk in my garden far from all these troubles. I have considerably increased my country library. To be sure, I do not have all I need, but at least I have things to read and I can work comfortably.
>
> [July 9, 1854, Neuenschwander 1984a, II, 6]

Liouville's scientific output during 1854 became almost as meager as the previous year, but he probably used the time in Toul to prepare some of the material that made the first half of 1855 his most productive period since 1847. Although he was "continually ill or at least very ailing" [Neuenschwander 1984a, II, 7] during this period, he published three papers on definite integrals and two papers on rational mechanics, plus a number of other works.

The papers on definite integrals were probably by-products from Liouville's course in 1852 at the *Collège de France* on "definite integrals and in particular the Γ-functions of Legendre." At the end of the course, Liouville had told the *Académie* about some new results he had proved in the lectures [Liouville 1852d], and that induced Cauchy to call attention to his own work from 1843 on the same subject. At the end of his talk to the *Académie*, Liouville claimed that he had obtained other new results that had to wait "for the publication of my lectures." However, these lectures shared the fate of Liouville's other lectures and were never published. Still, the papers of 1855 [1855b,f,g] probably contain some of the announced results.

The following year Liouville continued this trend by publishing a small paper on the Γ-function [1856d] (read to the *Académie* the very day Dirichlet was present), followed by three papers on the application of the Γ-function to various definite integrals [1856e,n,o]. According to Liouville, the first of these notes was but an extract from an extensive *Mémoire sur la réduction de classes très-étendues d'intégrales multiples*; although there are many, possibly important, notes on this subject in Liouville's notebooks from these years, I have not found any trace of such a complete memoir. On the other hand, the extract was republished in the first volume of *Schlömilch's Zeitschrift*. As with Liouville's paper in Thomson's *Journal*, this may be interpreted as an approving gesture from the established editor toward his newly fledged colleague.

However, later the same year, Schlömilch repaid this encouragement by claiming that in 1848 he had proved the results published by Liouville in 1856 [1856n,o]. Cayley sent a similar claim of priority to Liouville, who loyally quoted the two letters in his next paper [1857a], while pointing out that he had in fact discovered his own proof in 1845, when he examined a paper by William Thomson [1845a]. Indeed the 1856 article [1856o] had the title *Démonstration nouvelle d'une formule de M. Thomson*. Thomson may very well have been one of the major inspirations for Liouville's work on definite integrals.

The publication of Liouville's two 1855 papers on rational mechanics [1855d,h], and in a way [1855e] also, were prompted by a work by the young differential geometer and mechanician, Edmond Bour, which Liouville refereed very favorably for the *Académie* [Liouville 1855c]. Still, the main result was drawn from his lectures of 1853 at the *Collège de France* on *La formation et l'intégration des équations différentielles - mécanique rationnelle*. The theorem, which had already been presented to the *Bureau des Longitudes* in 1853, states that if half of the integrals of Hamilton's equations are found and if they are independent and in involution, i.e., have vanishing Poisson brackets with each other, then the system can be integrated by quadrature. This theorem, with some global geometric superstructure foreign to Liouville, is now called Liouville's theorem [Arnold 1978, p. 271]. It is the last theorem that was named after Liouville.

In 1856, Liouville published two more excerpts from his 1853 lectures [1856g,k], plus possibly [1856f], of which [1856k] is a very important analysis of the principle of least action using ideas from differential geometry. He simply transformed the integral of least action as though the trajectories of a point, influenced by forces, were geodesics in a different "Riemanian metric" and found in this way a shortcut to the Hamilton-Jacobi equations. This idea of including the forces into the Riemanian metric characterizing the space was carried further by Lipschitz, Levi-Civita, and Ricci and finally led Einstein to the general theory of relativity (cf. Chapter XVI, §45–48).

The published notes alone show that Liouville's course on rational mechanics went further than the lectures given by Bertrand the previous semester, and that they must have been important in diffusing German ideas, in particular, those of Jacobi, in France. In the published notes, Liouville claimed that he had developed his ideas further during his lectures, and that he had also included other new results. Therefore, it is a pity that I have not been able to find his lecture notes. On the other hand, his notebooks from 1856 and 1857 contain several highly remarkable notes on rational mechanics, some of which are analyzed in Chapter XVI(49–54). They emphasize that during the mid-1850s Liouville reached a level of inventiveness in this area equaling the level he had reached in other areas during the two preceding decades. It was probably his last flash of genius.

The occupation with mechanics gave Liouville the opportunity to publish some *Développements sur un chapitre de la "Mécanique" de Poisson* [1856p], which he had originally written in honor of Poisson shortly after Poisson's death in 1840. In 1856, Liouville further published a comparison of two of Poisson's works on the propagation of sound [Liouville 1856b]. This note probably had its origin in his course during the first semester of 1855–1856 on *L'intégration des équations aux différences partielles qui se présentent dans les questions de physique mathématique*. The notes taken by G. Lespiault [Liouville-Lespiault 1855–1856] reveal that the lectures were based on Poisson's methods, but according to Jim Cross, to whom I have shown the notes, Liouville also introduced Stokes' equations on the theory of waves, which was unusual at this time in France. In 1856, Liouville also began his first profound research in the field that was to become his favorite for the rest of his life: the theory of numbers, and in particular the theory of quadratic forms. In February of 1856, Liouville alluded to the "inventor" of number theory when he called himself "a Fermat" [Neuenschwander 1984a, II, 9]. He did so in a letter to Dirichlet, who was at that time the greatest expert on analytic number theory. When Dirichlet visited Paris the following month, he stimulated Liouville's interest in the subject, and on May 18, Liouville wrote to Dirichlet:

> I have worked arduously during the last weeks. In particular I have read with care, and I dare say in depth, your beautiful Memoirs on the number of quadratic binary forms with a given determinant. I have no doubt been well predisposed, for I find this matter very simple. On occasion I have also had to study Gauss, and thanks to

> you I now know the composition of forms very well. You will see
> that little by little I shall become bold enough to send you one day
> some ideas on the theory of numbers! *Audaces fortuna juvat.* Still,
> I shall devote the forthcoming holidays to a very different subject.
> Therefore, pray to the good God to let me live if you want to see
> me follow in your footsteps on the arduous path of transcendental
> arithmetic.
>
> [Neuenschwander 1984a, II, 10]

The very different subject was probably either definite integrals or rational mechanics, and indeed he did work on these subjects during the fall
of 1856. However, his notebooks [Ms 3622 $(9_1, 9_2)$, 3623 (1)] reveal that
he could not refrain from working on his new favorite subject, analytical
number theory, as well. He penetrated so far into the subject that he chose
to lecture on it during the two following semesters at the *Collège de France*
[Belhoste and Lützen 1984].

> Do you believe [he wrote to Dirichlet] that I have had the audacity to
> chose as the subject for my course the applications of the infinitesimal
> analysis to the theory of numbers. And in the audience I found (but I
> only knew it when it was too late) one of your students, Mr. Bjerknes
> from Christiania.
>
> [July 17, 1857, Neuenschwander 1984a, II, 11]

During the summer of 1856, the results of Liouville's new interest began to appear. In July, he presented one note on quadratic forms to the
Académie [1856h], and during his stay in Toul, two more appeared [1856l,m].
The following year Liouville's number theory output exploded. He published no less than 13 papers in this area, of which the most important,
Sur quelques fonctions numeriques [1857f], was printed as a serial in four
parts. The note [1857o] "on the occasion of a memoir of Mr. Bouniakowsky"
points to an inspiration other than Dirichlet's. Indeed, Liouville wrote to
Dirichlet in December 1857:

> I also have some theorems of the same type as those of Bouniakowski
> on prime numbers. The method which this geometer has devised
> seems profound to me, and I think it has a wider scope than he has
> discovered himself.
>
> [Neuenschwander 1984a, III, 13]

Later, Liouville often returned to Bouniakowsky's methods in his notebooks, and he continued to refer to his Russian colleague in his published
work [e.g., 1858b,c].

According to Liouville [1861u], all the basic ideas entering his subsequent papers on number theory were discovered during the years 1856 and
1857. His papers appearing in the first issues of 1858 in his *Journal* were
probably written during the 1857 holiday. These include *Démonstration
d'un théorème sur les nombres premiers de la forme* $8\mu + 3$ [1858b], which
Dirichlet commented on in the most favorable way in a letter of August 15,
1858:

It is with the greatest interest that I have read your ingenious memoir on the primes $8r + 3$ in the March issue. The theorem you prove and the theorem of Mr. Bouniakowsky (whose memoir I have not seen yet) seem to me the more remarkable because they are of an entirely new type. When I express myself in this way, I believe that I appreciate the importance of these investigations better than Mr. Bouniakowsky himself, who seems to speak (in the *Bulletin* etc.) badly of an analogy to the law of reciprocity.

[Copy in Ms 3625 (5)]

There is no doubt that Liouville was very happy to hear this positive judgment of his first works in his new field of interest, in particular because it came from the master of analytical number theory. Two weeks later Dirichlet was again full of praise after he had occasion to read the first part of the important *Sur quelques formules générales qui peuvent être utiles dans la théorie des nombres* [1858c], which appeared in 12 parts over the next 3 years:

Concerning arithmetic, I must express the great pleasure I felt when I read the beautiful theorem which you have published in the April issue.

[August 27, 1858, Tannery 1910, p. 35]

Still, it was no great problem for Dirichlet to penetrate to the core of Liouville's ideas; he continued his letter:

I did not know of it until yesterday evening in the library, because my own copy has not yet arrived. Since this first paper does not contain the proof of this theorem, which is so full of consequences, I shall risk communicating to you the proof I found on my evening stroll just after having read the paper.

[Tannery 1910, p. 35,36]

The subsequent proof was based on principles that Dirichlet had used in a paper in Liouville's *Journal* [Dirichlet 1856]. In his answer, Liouville admitted that Dirichlet had indeed guessed his method, and that Dirichlet's paper had been his point of departure. "But later I shall give some formulas that require other principles" [Liouville to Dirichlet, October 20, 1858. Neuenschwander 1984a, II, 15].

We may conclude that the years 1856 and 1857 became two of the most productive years in Liouville's career. Not only did he publish notable works on rational mechanics, definite integrals, and number theory, he also laid the foundation for all his subsequent works in the latter field. He had finally recovered from the depressions caused by the political events in 1849, and he was as full of scientific enthusiasm as he had been before the election in 1848. His decision to begin a new series of his *Journal* after 20 volumes in the first series, may reflect this scientific rebirth. However, the state of passionate productivity was abruptly cut short toward the end of 1857 when Liouville succeeded Sturm in the chair of rational mechanics at the *Faculté des Sciences*.

Sturm's Death

Having been persecuted during the reorganization of the *École Polytech-nique*, Sturm suffered a nervous breakdown in 1851 and resigned his chair at this school and obtained a leave of absence from the *Faculté des Sciences*. Two years later he resumed the teaching at the *Faculté* [Speziali 1964], but the mental illness continued, and on December 15, 1855, the 52-year-old "Sturm died insane" as Catalan wrote in his diary [Jongmans 1986]. This was a great loss for Liouville. At the tomb, Liouville, as a representative of the *Académie des Sciences*, gave a moving speech, the end of which bears witness to the intimate friendship that had bound them together:

> Oh, dear friend, it is not you we should pity. Escaped from the an-guish of this earthly life, your immortal and pure spirit lives in peace in God's bosom and your name will live as long as science.
>   Good-bye, Sturm, Good-bye.
>
> [Liouville 1855j]

Two months later, when writing to Dirichlet, Liouville was still grieving: "Come and you will be welcome. Together we shall mourn our poor Sturm" [Neuenschwander 1984a, II, 9]. During the following semester, Liouville honored his late friend by devoting his lectures at the *Collège de France* to "the analysis of the principal memoirs of Mr. Sturm." Moreover, probably through Sturm's bereaved wife, with whom the Liouvilles kept in touch, Liouville got access to Sturm's *Nachlass*. In his talk at the funeral, Liouville had pointed out that in addition to his published works, Sturm had many great ideas:

> May the very precious manuscripts that some of us have caught a glimpse of be found intact in the hands of his family! By publishing them, we would not spoil the masterpieces we have admired so much.
>
> [Liouville 1855j]

While searching Sturm's papers, Liouville in fact found a "very old Note" [CR 42 (1856), p. 1087], which he published in the *Comptes Rendus* of May 26, 1856, under the title *Note sur les fonctions élliptiques* [CR 42 (1856), pp. 988–990]. However, as one of Sturm's friends and students, Despeyrous, pointed out to Liouville, the note was really a quote from a paper that he, Despeyrous, had sent to Sturm for comments and that had been published afterward by the Academy of Dijon. At one of the following meetings of the *Académie des Sciences*, Liouville admitted his mistake [cf. Liouville 1856g, p. 1088], and he never published any of Sturm's other "precious manuscripts."

PLATE 20. Michel Chasles (1793–1880) was a good friend of Liouville and inspired many of his works on geometry. (Royal Library, Copenhagen)

Chasles's Substitute?

However distressing Sturm's death was to Liouville, it still opened the possibility that he could succeed him at the *Faculté des Sciences*. Liouville strongly desired to teach at the *Faculté*, and when the advertisement of the post was postponed, he planned on obtaining a footing by becoming Chasles's substitute in the chair of geometry. He got the idea when Chasles, during a short visit in Toul, told Liouville that he would prefer to use all his time to complete a second volume of his *Traité de géométrie supérieure*. On October 23, 1856, Liouville wrote to Chasles to suggest that he could relieve him of his teaching duties by acting as his substitute for the following semesters. Liouville promised to follow exactly Chasles's line of procedure "and to be, as much as anyone can be, another you..." [Neuenschwander 1984a, I, 6]. As his guide, Liouville would use the first volume of Chasles's *Traité*, and if time permitted, he would finish the course by applying the methods to potential theory, following again Chasles work (cf. Chapter XV, §13, 14). Liouville even seems to have attended Chasles second course as a professor at the *Faculté des Sciences*, at the time when he began to take an interest in geometry. At any rate, Liouville's notebook [Ms 3618 (6), pp. 18v–39r] contains a long series of notes in his own hand under the heading *Cours de M. Chasles 1847–1848*.

As we have seen, the jobs as substitutes were the traditional training grounds for promising young talents, so it is not surprising that Chasles found that this post would be below the dignity of Liouville. The next day he wrote to Liouville:

> My dear friend,
>   The title of substitute is certainly not a job which suits the position you occupy in the sciences.... Therefore, much as I desire to see you at the *Faculté*, I cannot allow myself to propose this method.... I shall be pleased to carry out this project if they postpone the nomination to the chair which naturally ought to go to you with all your qualifications, or if we encounter difficulties, which we may foresee and fear. I think it would be best to keep silent until we know what they will do about the vacant chair. However, the *Faculté* will perhaps be summoned at the end of the month or during the first days of November in order to fix the days when the different courses start and to settle the advertisement.
>
> [Neuenschwander 1984a, I, 8]

However, the advertisement announcing the courses was issued before any meeting had been held, and on October 28 Chasles informed Liouville that to his regret Puiseux had temporarily been put in charge of the mechanics course:

> Thus, apparently they do not want to nominate anybody at this moment. I regret it greatly. For I have met Mr. Delaunay who talked to me about it, and we were full of hope, because apparently the Minister will not permit any of the methods that they have used earlier to influence either the *Faculté* or the *Conseil académique*, and he will accept the rights and initiatives of the faculties.
>
> [Neuenschwander 1984a, I, 8]

As for the plan of having Liouville as his substitute, he now definitely declined the idea because the geometry course usually had a small audience compared to the mechanics course, and so if Liouville took it, he would compare poorly to Puiseux.

Therefore, Liouville had to wait half a year until the Minister of Education asked the *Faculté des Sciences* and the *Conseil académique* to suggest a successor to Sturm [Neuenschwander 1984a, pp. 127–129]. As Chasles had feared, this led to a great battle, not with the ministry, but with Serret.

Competition with J. A. Serret

As late as February 1853, Liouville and Serret had collaborated in the spirit of friendship that they had established in the 1840s. Liouville had suggested that Serret could extend his earlier investigations on surfaces with plane lines of curvature to surfaces with circular lines of curvature. More specifically, he had indicated that Thomson's transformation by reciprocal radius vectors would do the trick. This was in tune with the ideas Liouville worked on in his notebooks around this time (cf. Chapter XVII(26–28)). When Serret had completed the investigations, he communicated them to Liouville, who inserted them in the *Comptes Rendus* [Serret 1853].

The following year, however, the spirit of collaboration ended. Recall that in the first edition of his *Cours d'Algèbre supérieure* [1849] Serret had refrained from including Galois theory because Liouville intended to publish a treatise on this subject. Serret had also declared that it was only because of Liouville that he had understood any of it. When Serret published the second edition of his book in 1854, he left out this reference to Liouville, and included a note (number 13) on Galois theory. He was careful to point out that the contents of the note did not go beyond what Hermite had written to him in a letter. He thereby implicitly made Hermite the expert on Galois theory instead of Liouville. In fact, Bertrand later recalled that behind the screen Serret had withdrawn even more from his former mentor:

> When Liouville's project [of writing a treatise] seemed to be abandoned, Serret wrote on Galois theory; I think I recall that he devoted sixty-one pages to it; they were printed, and I read the proofs.
>
> As I was astonished not to find Liouville's name, he answered me: "I attended his lectures, that is true, but I understood nothing." However, after some reflection he saw that this declaration was difficult to make, and at Liouville's imperial request, he decided to suppress the sixty-one pages, and in order to fit the typography (the following pages having been printed), he had to write the same number of lines on a very different subject.
>
> [Bertrand 1902, p. 342]

Although Serret followed Liouville's request, there can be little doubt that he was upset to see his former teacher try to monopolize this field without even working in it. Finally, when the third edition of Serret's book appeared in 1866, it contained a whole chapter on *les équations résoluble algébriquement*. By then it was ten years since the disagreement between Serret and Liouville, caused by their competition at the *Faculté des Sciences* and to which we shall now return, had burst into flames.

The *Faculté* had appointed a committee consisting of Chasles, Lamé, and Lefébure de Fourcy to select two candidates for the chair of mechanics among the three applicants: Liouville, Puiseux, and Serret. On June 26, 1857, they presented their list: Liouville in the first rank and Puiseux in the second rank. They justified their choice of Liouville in a eulogizing report which I have quoted on p. 176. The election of Liouville might have been nonproblematic had it not been for the fact that a successor to Cauchy as a professor of astronomy was to be appointed at the same time. For this chair, a committee consisting of Le Verrier, Chasles, and Lefébure de Fourcy suggested Puiseux as the first candidate and Serret as the second candidate. Serret's behavior on this occasion irritated Liouville a great deal. He wrote to Dirichlet:

> Why do I sadden you by telling that the ungrateful J. A. Serret was my fierce and disloyal competitor under the auspices of Leverrier and

> Dumas (my former teacher who I pardon). He had first declared that
> he would only compete with Puiseux (for the Astronomy chair) and
> leave the field free for me in Mechanics. And he ended up doing the
> contrary. But, let these dirty tricks lie.
>
> [December 30, 1857, Neuenschwander 1984a, II, 13]

From a more impartial point of view, however, it seems as though Liouville's supporters, who were in the majority on both committees, played Serret a rather dirty trick. For in order to ensure Liouville's election, they choose Puiseux as the second candidate for the mechanics chair, although he was also the first candidate for the astronomy chair. In fact, at the meeting in the *Faculté*, several members protested that Serret had been passed over in this way and pointed out that the *Académie des Sciences* had preferred him to Puiseux. The Committee stuck to its choice of candidates, but Serret's supporters voted for him anyway. He obtained 3 votes against Liouville's 10 votes in the election of the first candidate for the mechanics chair. In the election of the second candidate for this chair, Serret almost beat the official candidate, but with 7 votes against 6, Puiseux was elected. In the election of the candidates for the astronomy seat, the members of the *Faculté* likewise followed the recommendations of the committee.

The *Conseil académique* also put Liouville first on their list of candidates for the mechanics chair. Liouville attributed this success to a large degree to Poinsot, who, despite his moderate republican views, was a senator and occupied an influential post in the *Conseil de l'Instruction Publique*. In a letter of July 20, 1857, Liouville thanked Poinsot for his help while excusing himself for not having written earlier:

> I have been so long in thanking you for all that you have done for me,
> because I thought that by waiting a few days I could enclose with
> my letter a Memoir which I have been working on for a long time
> and which I plan to dedicate to you as a testimony of my gratitude. I
> propose to give a considerable extension of Lagrange's and Poisson's
> theorems concerning the quantities $[a, b]$ and (a, b) the consideration
> of which have greatly simplified the calculus of perturbations and
> from which Jacobi has drawn so many useful consequences from the
> point of view of integral calculus. I think I have discovered much
> more than they have done in this area. Mathematically speaking,
> my memoir is accomplished, but the editing, which I want to make
> worthy of you, turns out to be more difficult than I imagined.
>
> [Neuenschwander 1984a, I, 9]

This memoir, which Liouville also mentioned in a letter to Bertrand from July 15 [Ms 3622 (1)], never reached a publishable state. However, I have found more than 100 pages in Liouville's notebooks that are without a doubt the unfinished drafts of this paper. These notes begin with the considerable extension of the Poisson and Lagrange brackets mentioned above and gradually develop into a study of how noncanonical systems of differential equations can be changed into canonical ones by a transformation of the variables. This approach to differential equations is an extension of the use of canonical transformations, which Jacobi had loosely sketched in one of his earlier papers. Since these drafts contain the most far-reaching proposals Liouville ever made in rational mechanics, I have summarized their content in some detail in Chapter XVI, §52–59.

Although he had been chosen as the first candidate for the chair of mechanics by both the *Faculté* and the *Conseil académique*, Liouville could not be sure that he would get it. As he wrote to Poinsot, "now it is up to the Minister to decide, and I hope your kind words have made up his mind." A week later Poinsot answered Liouville:

> My dear and very excellent fellow member,
>   It is up to me to thank you for the precious honor which you will do me; and if I soon hear of your success in the matter that interests you it is again up to me to show you my gratitude, for it will certainly make me more happy than you. - I would have preferred to answer your nice and friendly letter earlier, but I am still a bit feeble. I can hardly hold the pen...
> [Poinsot to Liouville July 27, 1857, Ms 3622 (1)]

The above is a copy made by Liouville and headed by the words "Can one be more amiable?"

Although the Minister postponed the nomination for several months, Liouville seems to have been rather certain that he would not pass him over for political reasons. On October 16, he wrote to Pelouze:

> The token of affection I have lately received from my friends would completely console me if I needed consolation; but, in fact, I am quite tranquil, my health is rather stable, and I have worked a lot during the last two months. If I succeed it will be without a fight, and I think that only the intriguers are to be pitied.
> [Ms 3623 (2)]

Professor of Mechanics at the Faculté des Sciences

One day after the above letter was written, Liouville was appointed professor of mechanics at the *Faculté des Sciences*. He read about his success in the *Moniteur* of October 20, and the following day he sat down, overwhelmed with joy and gratitude, to write to his friends and supporters. First, he wrote to Dirichlet "to inform you about this nice piece of news" [Neuenschwander 1984a, II, 13], and then to one of Pelouze's collaborators Frémy: "This is our first victory" [Ms 3623 (2)]. To Jobert de Lamballe, he wrote: "I do not know of all you have done for me, because I have not been told any details, but I know that you have been very active and very good" [Ms 3623 (2)]. Liouville's future superior, the Rector of the *Académie de Paris*, R. J. B. C. Cayx, got a more formal letter:

> I shall hasten to send my thanks to the Minister and to prepare my leave. Before the end of the week, I shall be at your disposal, Mr. *Recteur*, and receive your instructions.
>
> I know which part you have taken in this matter and what I owe to your kind conduct.
>
> [Neuenschwander 1984a, I, 10]

A few days after having written these letters, Liouville rushed off to the capital to thank his supporters orally and to receive instructions for his teaching.

His duties consisted of lecturing at the *Sorbonne* every Wednesday and Friday at 10 o'clock and, what turned out to be the most exhausting burden, to examine all the young people who earned one of the degrees conferred by the *Université*. The audience at his courses consisted of students from various colleges and from the *École Normale Supérieure*. This school, which had been set up during the great revolution in order to educate teachers, had been overshadowed by the *École Polytechnique* during the first half of the 19th century. However, after Le Verrier's "reform" of the latter, the *École Normale* gradually appeared as an alternative possibility for students with serious scientific interests, and from the 1860s several of the greatest French mathematicians were educated at this school. Thus, Liouville once again had the possibility of forming the next generation of scientists, but he did not really succeed.

His course, which was spread over two semesters, was regulated by a curriculum with the basic ingredients of statics, dynamics, and hydrodynamics. The surveys preserved in Liouville's notebooks [Ms 3626 (2), 3629 (14)] show that his lectures were rather similar to the mechanics part of the *Cours d'analyse et de mécanique*, which he had taught six years earlier at the *École Polytechnique*, though perhaps on a slightly higher level. By way of example, I have quoted his plan of the course for 1864–1865 in Appendix III. Liouville stuck to this general scheme, although there were annual fluctuations in content and emphasis. For example, in 1861–1862 and 1868–1869, he discussed potential theory in more depth [Ms 3626 (2)], and in 1862–1863, 1867–1868 and 1868–1869 he expanded more on celestial mechanics [Ms 3626 (2), 3629 (14)]. Occasionally, when the students from the *École Normale* were unable to attend, Liouville gave a lecture "outside of the framework," for example, at the first lecture of the second semester of 1866–1867, he entertained the university students with Clairaut's law of attraction, $A/r^2 + B/r^3$. Over the years, the content of hydrostatics and hydrodynamics gradually decreased. In 1868–1869, Liouville only devoted $3\frac{1}{2}$ hours to this subject, and the following year he was allowed to leave out this part altogether, when the second semester was shortened, probably because of political unrest.

Liouville's approach continued to be a highly mathematical mixture of Lagrangian variational methods and Laplacian potential theory, and it is striking that he never tried to present the more recent ideas of Hamilton, Jacobi, or himself to his students. It is equally noticeable that his own research in mechanics stopped entirely at the very moment he got a chair in this subject. This complete division between research and teaching made his course at the *Sorbonne* of little value to the young mathematical talents.

Courses at the *Collège de France*

At the *Collège de France*, however, Liouville continued through the 1850s to give inspiring and important courses, which were closely linked with his own research interests. I have already discussed the courses on differential geometry (1850–1851 (1) and 1851–1852 (1)), on definite integrals (1851–1852 (2)), on rational mechanics (1852–1853), on the calculus of probability (1854–1855 (1), 1862–1863 (2)), and on doubly periodic functions (1850–1851 (2)). Recall that Briot and Bouquet were among the audience of this latter course. Four years later they used what they had learned in their first note on doubly periodic functions [Briot and Bouquet 1855]. At this occasion, Liouville, who apparently wanted to call attention to his own priority, reprinted his *Comptes Rendus* notes [1844g, 1851c] in his *Journal*, together with a few additional remarks [1855i], but still refrained from publishing an extensive account of his method. Therefore, Briot and Bouquet were the first to write a textbook on elliptic functions based on Liouville's principles. In their preface to *Théorie des fonctions doublement périodiques et en particulier des fonctions elliptiques* [1859], they acknowledged their debt to Liouville:

> We owe much to Mr. Liouville. Some years ago this eminent geometer chose "elliptic functions" as the subject of his course at the *Collège de France*; his learned lectures were the point of departure of our own research.
>
> [Briot and Bouquet 1859, p. XXIV]

Liouville, who still considered the approach to be his property, was not satisfied with this preface, and repeated his course on doubly periodic functions at the *Collège de France* during the second semester of 1858–1859. In his old age, Liouville explicitly accused his former students of theft. In his notebooks, he wrote:

> Mr. Briot and Mr. Bouquet, vile thieves but highly dignified Jesuits. Elected as thieves by the *Académie*!!!!
>
> [Ms 3640 (1877–1879)]

Thus, as in the case of Galois theory, Liouville tried to monopolize this field of research and tried to prevent his former students from developing it, despite the fact that he had abandoned his plans for publishing his own results. The contrast with his encouraging behavior during the 1830s and 1840s is striking.

Others of Liouville's lectures at the *Collège de France* were devoted to the work of mathematicians whom he admired, such as Sturm (1855–1856 (2), Dirichlet (1859–1860 (1+2), and Euler (1860–1861 (1+2)). From the mid-1850s, Liouville's interest in number theory also made itself felt at the *Collège de France*. During the second semester of 1854–1855 he lectured on the *Calcul approché des formules qui contiennent de grands nombre* and both of the semesters of 1856–1857 were devoted to the *Applications de l'analyse infinitesimale à la théorie des nombres*. The following semester the lecture was simply entitled *Elements de la théorie des nombres*, and in 1858–1859 (1), the title shrunk to *Théorie des nombres*, a title used by Liouville 11 times during the subsequent 20 years. Moreover, the theory of numbers constituted a major part of Liouville's other courses at the *Collège de France* from 1860 onward.

Chebyshev

> *Fac. des Sciences*: The words of the Minister Mr. Roeland when he received me: Mr. L. One should not be foolishly kind, but one ought to be paternal.
>
> [Ms 3625 (10)]

Liouville probably jotted down this remark after returning from the meeting, because it was in harmony with his way of treating his students. Although he was unable to keep his friendship with some of his former students, he continued to encourage and help new talents, including visitors from abroad, such as Chebyshev, Hj. Holmgren, and Bjerknes.

Pafnuti Chebyshev, the founder of the famous St. Petersburg school of mathematics, was appointed to the university of that town in 1852. On that occasion, he wrote an autobiographical report, which throws light on his relationship with Liouville during his stay in Paris earlier that year:

> When I arrived in Paris I went to see the celebrated geometer Liouville, member of the *Académie des Sciences de Paris* and editor of a journal of mathematics with which I have collaborated since 1842. With the kind help of this geometer, I had occasion to get to know the *savants* whose support was of great importance for the success of the journey...
>
> I spent the afternoons sometimes at the *Conservatoire des arts et métiers* [studying mechanical devices], sometimes in factories... and my evenings were partly devoted to discussions with Mr. Cauchy, Liouville, Bienaymé, Hermite, Serret, Lebesgue and other savants, and sometimes with theoretical studies related to my studies of various mechanical systems or to questions of analysis to which the conversation with the savants turned my attention. Thus, Mr. Liouville and Mr. Hermite suggested that I should develop the principles which are at the basis of my dissertation, presented in 1847 to the University

of St. Petersburg *pro venia legendi.* In this work, I have examined the cases where the differential to be integrated includes the square root of a rational function - the simplest and most frequent in the applications. But, it would be interesting from several points of view to extend these ideas to radicals of arbitrary degree.

[Chebyshev 1852, pp. VIII, IX]

It is only natural that Liouville encouraged Chebyshev to elaborate these ideas, for they constituted a direct continuation of Liouville's work on integration in finite terms. In fact, the half-a-dozen papers that Chebyshev published in this field over the next ten years contained the only important additions to this theory until the end of the 19th century (cf. Chapter IX, §52 ff). Moreover, Liouville gave Chebyshev private lectures on his theory of doubly periodic functions:

On August 28th Liouville departed for Toul after having put me in charge of publishing two of my papers in his *Journal.* I spent the evenings editing them while continuing to occupy myself with applied mechanics.

[Chebyshev 1852, p XV]

These two papers, concerning analytical number theory, appeared the same year in Liouville's *Journal,* and a revised French version of Chebyshev's thesis appeared in the following volume.

It is not surprising that Chebyshev chose integration in finite terms as the subject of his thesis, for his fellow countryman, Ostrogradsky, had already contributed to this field (cf. Chapter IX). However, it is remarkable that Chebyshev's approach was much more similar to that of Liouville than to that of Ostrogradsky. This indicates that Chebyshev had other contacts with Liouville in the early 1840s than just being a contributor to the *Journal de Mathématiques.* In fact, Butzer and Jongmans [1989] have suggested that Chebyshev visited Paris in 1842. Whether he met Liouville at this occasion, or whether the two just began a scientific correspondence at this time, would also explain Chebyshev's lifelong interest in the figures of equilibrium of rotating masses of fluid. Indeed, 1842 was the very year Liouville did his important research in this area. Thus, Chebyshev may very well be the missing link between Liouville and Liapounoff, a student of Chebyshev's, who in 1884 continued Liouville's work, independently of Poincaré (cf. Chapter XI, §12, 13).

Chebyshev and Liouville remained friends for the rest of their lives, and there can be little doubt that Chebyshev was the driving force behind Liouville's election as a foreign member of the St. Petersburg Academy. Liouville referred to this membership in the first of his two surviving letters to Chebyshev:

Mr. Chebyshev of the Academy of Sciences of St. Petersburg. Correspondent of the *Institut de France.* March 19, 1864

PLATE 21. Pafnuti Lvovich Chebyshev (1821–1894). During a stay in Paris in 1852 (perhaps earlier) Chebyshev made friends with Liouville whose interests in integration in finite terms and figures of rotating masses of fluid he shared (Math. Dept. University of Copenhagen)

My dear Fellow Member,
    I wish to introduce you to my nephew, Henri Liouville, who wants to complete his medical education by visiting hospitals in St. Petersburg. He accompanies Doctor Lefort, who has been sent on a scientific mission by our hospital administration. Allow me to profit by this occasion to send my best wishes to our fellow members of the Academy. I can never thank them sufficiently for Euler's arithmetical works, which they have sent me as a present. Naturally, I think most often of you. When do you return to visit us in Toul, if only to eat nuts?

Sincerely...
    [Copy in 3630 (14)] (A Russian translation of the original can be found in [Chebyshev 1951, p. 438])

Chebyshev did visit Toul at least once more, namely, in 1873. Afterward, he went to Paris, from where he wrote to Liouville:

I regard it as my pleasant duty to send you a little Memoir *Sur les quadratures*, which I had occasion to tell you about in Toul.

It is not necessary to tell you how much I would be pleased if
Sur les quadratures could appear in your respected *Journal* which is
so highly estimated by all geometers.

I take the opportunity, Monsieur, to ask you to give my respects
to Mrs. Liouville....

[Chebyshev to Liouville, Paris October 1873, Chebyshev 1851]

Scandinavian Students

In the period 1850–1853, the Swedish mathematician Hj. Holmgren also
visited Paris. In his diary, which Jan Erik Roos has studied, one learns, for
example, that on April 24, 1851, he "met Liouville in the street. He talked
about his lectures on elliptic functions, which were to begin on the 28th.
That made me curious to attend them" [personal communication from J. E.
Roos]. Holmgren soon published a paper in Liouville's *Journal* [Holmgren
1851], and the two developed a close friendship, attested by a letter that
Holmgren's student Mittag-Leffler wrote to his teacher during his stay in
Paris in 1874:

I have been to see Liouville with the work of the Professor [i.e. Hj.
Holmgren].

Liouville was deeply moved and had tears in his eyes when I told
him, that I came on behalf of the Professor. He talked about the
happy years when the Professor was in contact with him in Paris,
about his old friendship with the Professor, and how this had not
diminished during 20 years of separation. "Greet him and say," he
asked me, "que je l'aime toujours."

He had received the three memoirs earlier and even read them
with the greatest interest. He had been able to follow the two memoirs in Swedish quite well, in spite of the language. He talked about
them with the greatest satisfaction and admiration, and said that
the Professor, in his *Index Calculus* "l'avait beaucoup dépassé", and
that the study of this and the last memoir had given him much to
think about....

[Mittag-Leffler to Hj. Holmgren, Göttingen April 25, 1874,
communicated to me by J. E. Roos, Uppsala University Library,
Holmgren Archive]

Holmgren, on his side, was glad to hear about Liouville's reaction to his
works:

That Liouville has seen them makes me very happy. I have not sent
them to any person abroad other than him.

[Holmgren probably to Mittag-Leffler. Communication by J. E. Roos]

The works in question dealt with fractional calculus and introduced its
modern definition for the first time. It is interesting to note that, with the
exception of one paper by Serret [1844], it lasted almost 20 years before

Liouville found students who were receptive to his early ideas, Holmgren to fractional calculus and Chebyshev to integration in finite terms. It may indicate that Liouville, in the 1850s, tried harder to get his own ideas across to his students, rather than listening to their interest.

The young Norwegian mathematician, C. A. Bjerknes, also experienced Liouville's readiness to guide foreign talents:

> I tried only to a minor degree to acquaint myself with the mathe-
> matical professors in Paris. Professor Liouville, to whom I had been
> recommended by Dirichlet in Göttingen, gave me a great attention,
> and I had many opportunities to talk to this eminent mathematician,
> mostly after the end of his lectures. In this way I received many useful
> explanations and very valuable guidance.
>
> [Bjerknes 1858]

As Liouville wrote to Dirichlet, he became a bit nervous when he discovered that Bjerknes, who had just come from Göttingen, would attend his first course on number theory, and indeed, Bjerknes was not impressed:

> At the *Collège de France*, where the lectures did not begin until the
> beginning of December, I heard lectures by Professor Liouville and
> by Bertrand.
>
> The former lectured two hours a week about the use of infinites-
> imal calculus in the theory of numbers. By the way this new theory
> is due to Professor Dirichlet, and I had already heard its own inven-
> tor lecture on it in Göttingen. Therefore the lectures on this theory,
> which is in fact of a particular interest, could not be much more than
> a repetition of what I had earlier heard Dirichlet himself develop in
> a very instructive way.
>
> [Bjerknes 1858]

According to a biography, written by his son, Bjerknes did not find among the Parisian mathematicians anyone who "possessed Dirichlet's brilliant art of lecturing or penetrated so deeply as Riemann" [V. Bjerknes 1925]. He was surprised to see how few people attended the lectures of Liouville and his colleagues:

> Incidentally, the lectures that I attended had a very small audience.
> While the elementary lectures about e.g. integral calculus and ana-
> lytic geometry etc. were attended by rather many students, it turned
> out that the higher purely mathematical subjects, even in Paris, only
> attracted a small audience. Thus Liouville's and Cauchy's lectures
> were only attended by 3, 4 or 5 persons, and the lectures of Bertrand
> had an audience of 6–8. It must, however, be remarked that the great
> difficulty that is always connected with presenting, in an oral lec-
> ture, such abstract themes, as several of those that are connected
> with higher analysis, influences this to a high degree. It is therefore
> always more or less true that the number of auditors of a series of
> mathematical lectures decreases as the semester goes on.
>
> [Bjerknes 1858]

The mathematicians in Paris had a different explanation of the declining interest in their subjects. Thus, in Mittag-Leffler's letter quoted previously one reads:

> Universally and bitterly is heard in Paris among older mathematicians the lamentation of the lack of young scientific talents. One ascribed this mainly to the detailed regulations which the Empire has introduced in all branches of education and which have been especially suited for eradicating all greater aptitudes. By the way, at present the problem is not in the least better, but if possible, rather worse. Every course—Hermite's at the Sorbonne, however, is an exception—is determined in detail in advance, and as little freedom as possible is left for the lecturer. At the *École Polytechnique* one has introduced a host of subjects, entirely foreign to the original plan of the school, and at the same time decreased the study of the mathematico-physical [subjects]. Hermite's *Cours d'Analyse...* is not as he has given it, but such as he wanted to make it.
> [Personal communication from J. E. Roos]

Actually, this letter was written in 1874, but its content also applied to the 1850s. The curriculum at the *École Polytechnique* had been regulated in 1851, and indeed in the 1850s and 1860s where Germany and Italy produced large numbers of mathematicians of the first rank, there was a remarkable ebb in the new stars in Paris. Jordan and Darboux were the only outstanding new mathematicians between Hermite and the new great period starting around 1880 with Appell, Picard, and not least, Poincaré. To be sure, lesser talents, such as Bour, Mathieu, Laguerre, and Halphen arose in the 1850s and 1860s, but in a smaller number than both before and afterward.

The lamentable state of French mathematics, and in particular science, may in part be due to the centralized regulation of the courses, but the peculiar appointment policy certainly added to the problem. As we have seen, the positions, except at the *Bureau des Longitudes*, were pure teaching posts. Research was done during the spare time, but since the salaries were low and *cumul* therefore spread, there was very little time left for research. No doubt the *cumul* was in part also responsible for the relatively bad lectures, compared to German standards.

Great Teaching Load—No Research

Liouville's appointment as a professor at the *Faculté des Sciences* gives a clear-cut example of the dangers of the *cumul.* After the first year in this chair, he poured out his troubles to Dirichlet:

> I already know from Mr. Weber of all the obligations imposed on you, and this year I have myself felt all the difference there is between

a course at the *Collège de France* and a university post. After 64
long lectures at the *Sorbonne* (and on several occasions during these
lectures that did not stop until the middle of July), I have been
absorbed by the examinations, an agreeable relaxation which starts
at 8 o'clock in the morning and lasts until 2, 3, 4 o'clock in the
afternoon, sometimes even longer. You won't believe it, but it goes
on until the end of August. However, being unable to continue, I
have asked Delaunay and Puiseux to replace me, and I have finally
come to rest a bit in Toul.

[August 22, 1858, Neuenschwander 1984a, II, 4]

He almost regretted that he had applied for the chair:

My position at the Sorbonne has burdened me with very tedious
work and has made me much more dependent. It can happen that
I may be suddenly called to Paris, and you can imagine that after
having so strongly desired a nomination, which my friends barely
obtained for me, it is of vital concern to me to justify what they
have done for me by scrupulously performing all my duties.

Besides I can tell you that my teaching this year has been a great
success. I hope I will do even better next year, thanks to the good
advice of some intelligent professors, who have kindly followed my
course, de Faurie in particular.

[continuation of the letter above]

Thus, he also exhausted himself by striving for perfection. This was hard,
because the rigid plan did not leave him time to penetrate as far into the
matter as he wished. This continued to put pressure on Liouville. In his
private notes from the lectures of 1861–1862 [Ms 3626 (2)], he started out
with a brief introduction of the basic mechanical concepts, distinguishing
between statics and dynamics of rigid bodies by the fact that in the earlier
branch only the forces and their points of application matter, whereas the
mass distribution in the body is of importance in the latter. A few pages
further on he began to elaborate this idea, but broke off with the remark:

I know that I ought to write all this clearly once and for all
    But the time! But! But!
    But the position! But! But!
    Ah! the freedom -
    And yet I must admit that I am rather free at the *Sorbonne*, at
the *Institut*, at the *Collège de France*. But, what to do, and what
shall one say about the *Bureau des Longitudes*?

[Ms 3626 (2), p. 3v]

Here, Liouville seems to be speaking about his political freedom, i.e.,
the independence from the conservative Le Verrier party rather than the
liberty to teach.

Having once more tried to write down his introduction of the basic notions of mechanics, he gave up with the words:

Why don't I have a stenographer? I yawn while writing this - let us go to bed.

[Ms 3626 (2), p. 4v]

After having lectured for less than two months at the *Sorbonne*, Liouville foresaw that the new job would disturb his work seriously. On December 30, 1857, he wrote to Dirichlet:

The happy event, which has sheltered me from financial straits, entails on the other hand a considerable loss of time. At the moment when I was informed about the nomination, on which I had almost stopped counting on, I was in an inspired process of working, and I was simultaneously dealing with integral calculus and the theory of numbers. I had vigorously begun to study the latter part of science with a new ardor. But the forced and sudden interruption of my writings has disturbed everything. God knows when I can conveniently continue this task. For the next two years, I cannot do any steady and thorough research. After that I will be rather free again (I hope). In the meantime I shall try to publish some short notes that can keep me spellbound.

[Neuenschwander 1984a, II, 13]

Indeed, Liouville's productivity, which had gradually increased over the last three years, fell drastically, but he did keep alert by publishing half a dozen short notes of a few pages, in both 1858 and in 1859. Almost all of them dealt with number theoretic results that he had found in 1856 and 1857, and none of them contained proofs. Apparently, Liouville felt an urge to have his results published, but could not find time to write down the proofs. Thus, the sudden increase of Liouville's teaching load may be the origin of a highly singular habit Liouville maintained for the rest of his life, namely, of publishing his results piecemeal and without proofs. This behavior reduced the value of his works and later lead to the disapproval of his colleagues.

Bad Health

In addition to the great teaching load, Liouville's failing health was also responsible for his reduced scientific creativity after 1857. We have seen that Liouville was of weak constitution as a student at the *École des Ponts et Chaussées* and that he had had frequent attacks of vomiting as early as 1840. In 1847, we hear of his first strong attack of gout, which he tried to cure by an early move to Toul from where he wrote to Joachimsthal on June 30:

Neither the journey nor the stay in the country have improved my health yet. Three days ago the gout even came back in my right foot, but the attack is rather light, and if the pain does not grow worse, I may avoid staying in bed almost all the time. As a result of all this trouble my work does not progress very well. It is sad to have only a few days free each year and then spend them suffering and drinking herb tea.

[Neuenschwander 1984a, III, 5]

Liouville described his attacks in more detail in a letter presumably from the same year:

My dear friend,
    The night before I received your letter I a new attack of this damned gout, almost as in January, that is, with vomiting, unbearable and permanent pains in the back etc. However my breast did not hurt this time, and I was again on my legs faster, thanks to the heat no doubt. But if there is no longer any illness there is continual uneasiness, and I tremble at the thought of the approaching winter.

[Letter from Liouville, probably to Laugier Ms 3618 (9), pp. 47v–48r]

This letter indicates that in the late 1840s Liouville had such strong attacks of gout that even in the periods when the pain eased off the fear of new pain affected him psychologically and hindered his work. This impression is confirmed in the following letter to Thomson from July 29, 1847:

...do not acquire the habit, as I have done during the last 6 months, of suffering from the gout. It is a sad companion, who gives nothing but sad ideas. Newton was in good health when he wrote the *Principia.*

[Neuenschwander 1984a, III, 6]

In August, however, he could assure Laugier that "I am in fact a bit less feeble than in Paris" [Ms 3623 (9)]. During the politically active years of the Republic, we do not hear of troubles with Liouville's health, and except for a few complaints (e.g., in 1850 [Ms 3619 (2) p. 15]), he seems to have enjoyed reasonable health until the winter and spring of 1855, when he was "incessantly ill or at least very ailing" [Neuenschwander 1984a, II, 7]. Again, in January 1857, Liouville felt "a bit ailing" [Edwards 1975, Appendix I], and at the beginning of the summer he complained to Dirichlet:

This year I have been very ailing (almost without interruption), my daughter Louise has been very ill for a long time, a thousand troubles of every sort have assailed us. When your kind niece came to visit us, she found me in a state in which I was hardly able to speak. I coughed all the time, and I had to leave Paris abruptly for a change of air.

[July 17, 1857, Neuenschwander 1984a, II, 11]

The descriptions of this affliction sound less serious than the gout, and as he wrote to Pelouze, he felt better during the second half of the year. Indeed, we have seen that the summer and fall of 1857 were a productive period in Liouville's life. The following summer, however, Liouville was so exhausted by his new job at the *Faculté* that he could not finish his examinations, and for the first time his bad health seriously inhibited his work throughout his stay in Toul. On October 20th 1858 he wrote to Dirichlet:

> In four or five days I shall leave Toul where no one has come to visit me and where I have been ill almost all the time. I have worked very little this year and I have hardly regained a spark of the beautiful fire that inflamed me last year at the moment when I had to leave abruptly for Paris. And now I will lapse into examinations and courses.
>
> [Neuenschwander 1984a, II, 15]

When Liouville wrote this letter, he was so weak that the following day he had to repeat a mathematical argument, contained in it, in order to make sure that it was understandable:

> I suffered so much yesterday when I wrote to you that I strongly fear that I have explained the elements of the formula I informed you about very badly.
>
> [Tannery 1910, p. 38]

Thus, illness deprived Liouville of the little time his new job left for research. When teaching and bad health went together, as they did after Liouville returned to Paris in 1858, it became almost unbearable:

> As for me, I have returned very ailing from my vacation. I can only repeat what I told you some weeks ago; I drag myself to my courses and, truly, to live like this is not to live.
>
> [Liouville to Dirichlet, December 30, 1858; Neuenschwander]
> [1984a, III, 17]

One-and-a-half years later, Liouville could not even drag himself to his lectures. He canceled them and "I even had to give up attending the meetings of the *Académie*" [Ms 3625 (14)]. The following winter he barely kept his courses going.

> I have been and I am still very ailing. I hardly leave my bed except when I have to give my lectures, which in this state of health are naturally more tiring than usual. Nevertheless I should not complain when I look about me; when I see the [?] of our poor friend Viller, when I think of the cruel illness that has attacked our cousin Gerard and his family...
>
> [Liouville to a certain Ferdinant, January 10, 1861, Ms 3626 (3)]

In June of the same year, Liouville was so ill that his son had to take care of the most pressing correspondence related to his *Journal* (cf. letter from Ernest to Cremona in [Neuenschwander 1984b, pp. 67–68]). This was not the last time that, during periods of illness, Ernest had to act as his father's secretary (cf. letters of 1865: [Neuenschwander 1984b, p. 71] and [Ms 3637 (1) p. 26v]).

It is characteristic that as Liouville's own health deteriorated he became increasingly concerned with the health of his friends, and the death of many of them had a great effect on him. To be sure, he had reached an age when failing health and death visited his circles, but he seems to have developed an affinity to these tragedies. Thus, from 1859, he began to enter death-days in his notebooks, often in long lists on the covers. A typical entry looks like: "Death of Madame Liouville, wife of Liouville the pharmacist, former Mayor of Commercy – Funeral on the 2nd 9^{ber} [November] 1860 as for Madame Lefort in Paris" [Ms 3625 (19)]. In connection with this death Liouville wrote to his relative Odile Chenneval "Our correspondence is nothing but an obituary" [Ms 3625 (19)].

Liouville was terribly struck when some of his friends and colleagues died, which many of them did during the 1850s. The first was Jacobi. When Liouville got the news, he immediately wrote to their common friend, Dirichlet:

> A newspaper which caught my eye announces the death of our ex-cellent Jacobi. What sad news, is it really true? What emptiness around you. And what a loss for Europe, for the whole world. Oh! I would wish that you would reassure me. These newspapers often de-ceive us. Could they have lied this time? — Goodbye. I do not have courage to add anything; the thought of this unexpected misfortune has completely upset me.
>
> [February 28, 1851, Neuenschwander 1984a, II, 2]

We have already seen that Liouville was greatly distressed when he lost his father the following year, and that the death of Arago in 1853 affected the scientific career of Liouville and in particular that of his son. I have also mentioned Liouville's grief after Sturm's death in 1855. The same year Gauss died. Although Liouville was among the French mathematicians who best understood the greatness of Gauss's work, he, like most other mathe-maticians, had no personal relationship with the "prince of mathematics," so Gauss's death was not a personal loss to him. Still, Liouville had met with one of Gauss's sons in 1836. In a letter to Schumacher, dated January 3, 1837, Gauss wrote:

> My son came back to town a month ago. He had been received very well in Paris by Poisson, Libri, Liouville and several others.
>
> [Gauss and Schumacher 1861]

In 1856, C. J. Gauss sent Liouville some rare copies of various memoirs of his late father, and Liouville thanked him in a letter that concluded:

> During your journey to Paris (perhaps you recall it) I had the chance to express my deep admiration and veneration that I felt for the great geometer whom we have unfortunately lost too early. This feeling has only increased with each new work he has given us.
>
> However, it is up to Mr. Dirichlet to praise Gauss and to settle his place among the *savants* of the first order. As for me I will sit

down to study and follow at least, thanks to the means you have sent me, the footprints of the giant, with whom I do not venture to compare myself.

[Liouville to C. J. Gauss, September 8, 1856; Neuenschwander]

[1984a, III, 9]

This letter bears witness to Liouville's admiration for Gauss and also to his high esteem of Dirichlet.

Liouville's Final Opinion of Cauchy

In July 1857, Liouville informed Dirichlet that:

> ... the death of Dufrenoy, my former teacher at the *École des Ponts et Chaussées*, the death of Mr. Cauchy and finally the death of Mr. Thenard have greatly affected me singularly.
>
> [Neuenschwander 1984a, II, 11]

It is a bit surprising that Cauchy's death greatly affected Liouville, for their relationship, which had suffered from the competition at the *Collège de France* had remained rather cool during Cauchy's last years. On the occasion of the republican Bertrand's election to the *Académie* in May 1856, Liouville told Dirichlet that "Mr. Cauchy covered himself with ridicule."

One month later the tension between the two mathematicians was increased when the *Académie* asked Cauchy, Liouville, and Bertrand to write a report on a memoir by Briot and Bouquet. This paper, which dealt with the use of complex and in particular, doubly periodic functions in the investigation of integration in finite terms, would clearly have given Liouville the opportunity to call attention to his own early works on integration in finite terms and to repeat his claims of priority concerning doubly periodic functions (Briot and Bouquet [1855] having provided the last opportunity). However, to Liouville's apparent annoyance, Cauchy did not even consult Liouville before he wrote the review and had it typed for the *Comptes Rendus*. When Liouville received the proofs of the review in Toul, he first wanted to withdraw from the judging committee, and so he drafted the following letter to Cauchy:

> Dear Sir and Fellow Member,
>
> I think that it is best to say things as they are, and that it is most natural to suppress my name when it is a question of a report in which I have no share, concerning a Memoir which I have neither read nor seen and which has been presented to the *Académie* without the committee in charge of examining it having been summoned once. However, I can understand that *in a case of urgency* a committee goes ahead without worrying about the absence of one of its members when the others agree.
>
> [July 9, 1856, Neuenschwander 1984a, III, 7]

Liouville planned to send this letter to Cauchy by way of the printer in order to have his name suppressed before it was too late. In the accompanying letter to the printer he wrote:

My dear Monsieur Bailleul,
    While asking you to give the included note to Mr. Cauchy I can without indiscretion, and I even must, ask you to read it before sending it to his address. In fact, it concerns the *Compte rendu* of this week. You may act as if I had written you something similar [?] make the correction immediately, except perhaps [?] to send new proofs to Mr. Cauchy.

[Ms 3623 (1), p. 63r]

Having cooled off by drafting these two notes in his notebook, Liouville reconsidered the matter, crossed out the drafts, and wrote a new and more moderate letter to Cauchy, in which he accepted having his name on the report while still repudiating the procedure:

Dear Sir and Fellow member,
    Leave my name at the head of the report which you have kindly written on the Memoir of Mr. Briot and Bouquet. Although this Memoir has never been communicated to me I am easily persuaded that it deserves all your praise. Yet, as a general principle, I think that all the members of a Committee ought to read the Memoir which they have been asked to judge; and I add that the conclusions ought to be deliberated in common and after an official summoning. The work done at such a meeting belongs to everybody, even to the absent persons who have been officially ordered to state their opinion.
    However, if a motive of urgency has forced you to use a faster procedure, I can understand you and I accept your reasons in this particular case, which is not so important.

[Neuenschwander 1984a, III, 8]

Indeed, Liouville's name figures next to Cauchy's and Bertrand's in the printed review [CR 43 (1856), pp. 26–29].

Over the years, Liouville's initial high esteem for Cauchy seems to have been somewhat overshadowed by an irritation at Cauchy's ridiculous behavior in academic matters and his legitimist agitation. As we have seen, Liouville later in his life often made fun of the "mathematical diarrhea" of his late colleague. Still, as he wrote to Dirichlet, he was strongly affected when he heard of Cauchy's death. This is corroborated by a letter written from Toul on July 25, 1857, probably to Briot or Hermite:

My dear Fellow Member,
    Please thank our illustrious Dean for having thought of me and tell him that he can count on my zeal. Like you, I hope that this homage to the memory of the great Geometer we have lost will bring some consolation to the profound grief of his family.

[Bibl. Inst. France Ms 4896, item 90]

I do not know how the *Faculté* paid their respect or how Liouville supported this action, but the draft of the above letter shows that the geometer in question was Cauchy, for it contains the following deleted phrase "and in the middle of the grief which the death of Mr. Cauchy has caused me..." [Ms 3622 (1)].

Liouville Commemorating Dirichlet

However much Liouville mourned the death of Jacobi, Sturm, and Cauchy, there is little doubt that the death of Dirichlet took even more out of him. As we have seen, Liouville had not only considered him one of his closest friends, he had also admired him as the only mathematician who could succeed Gauss. I have found no trace of Liouville's immediate reaction to Dirichlet's death, but it may not be a coincidence that 1859 was the year when he began to fill his notebooks with notes about deaths of friends and colleagues.

As a more outward reaction, Liouville decided to commemorate his late friend by devoting the courses of the following year at the *Collège de France* to his works. He sketched the first lecture as follows:

> First lecture (Thursday 8. X^{ber} [December] 1859)
> We shall present *l'analyse des principaux Mémoires de M. Lejeune Dirichlet*. However, here is a double difficulty: the variety of the works of this illustrious geometer, the insufficient training of some of the audience. Let us try to evade the latter inconvenience by dealing first with some simple papers published in particular in the first issues of Crelle's Journal. Let us not worry too much about the mixture of material at first. Later, we shall come back to that and envisage Dirichlet full face and consider him in his totality - Today we shall talk about the very elementary paper which Dirichlet included in Crelle's *Journal* (Vol. 3 p. 390)...
>
> [Ms 3625, p. 1r]

The note concerning the following lecture reveals that Liouville also included biographical information about his late friend:

> Second lecture (Saturday 10. X^{ber} 1859)
> In the last lecture we indicated Dirichlet's visit to France close to General Faye [?] etc. - Fourier's influence on him as well as on Sturm, etc. ...
>
> [Ms 3625 (1), p. 2r]

In connection with his subsequent summary of Dirichlet's proof of the convergence of Fourier series, Liouville, on several occasions, referred to "Cauchy's error" concerning the continuity of the sum of a convergent series of continuous functions. The rest of Liouville's notes show that during

these lectures of 1859–1860 Liouville paid worthy homage to Dirichlet by
explaining a great variety of his works on series expansions, definite integrals, potential theory, and analytical number theory.

In addition, Dirichlet's works continued to be a basic ingredient in Liouville's subsequent lectures at the *Collège de France*, as they had been before
Dirichlet's death (cf. Chapter III). For example, during the first lecture of
his next course on *Calcul différentiel et calcul intégral*, Liouville said:

> Therefore, I only treat certain questions regarding the differential
> and integral calculus, which I connect in particular with Dirichlet
> (whose works in the area I did not study sufficiently in the semester
> which I had devoted to his good memory).
>
> [Ms 3625 (1), January 9, 1862]

Indeed, a large part of the 1862 course was devoted to Dirichlet's work in
potential theory. In particular, I shall emphasize the seventeenth lecture:

> I spoke about the attraction from an ellipsoid inversely proportional
> to the 4th and in general the power $2n$ of the distance, n integer
> > 1. Dirichlet wrote to me that in these cases the integrations can
> be carried out.
>
> [Ms 3625 (1)]

In fact, the letter in question, dated August 15, 1858, is preserved in a
copy in Liouville's notebooks. I have already quoted the beginning of it.
The part concerning potential theory reads:

> In the course on integral calculus, which I gave this fall, I have made
> a remark concerning the attraction of ellipsoids, which I think I ought
> to communicate to you because it would perhaps interest our friend
> Mr. Chasles, who is so skilled at treating these questions in a purely
> geometrical way. The remark is the following: the attraction of a
> full ellipsoid can be expressed without an integration sign when the
> elementary attraction is inversely proportional to the fourth power
> [of the distance]. In fact, it follows from my formulas (the easiest way
> is to go back to the expression of the potential, which I have published
> in the Mem. of Berlin, and which is not included in the excerpt
> included in your Journal) that for an exterior point the attraction
> is (as it is for Poisson's layer in the ordinary case) normal to the
> homofocal ellipsoid passing through the attracted point and has the
> value $\frac{Mq}{\alpha'\beta'\gamma'\delta^2}$, where M is the mass, q the perpendicular dropped
> from the center onto the plane tangent to the homofocal ellipsoid
> [with the semiaxes] $(\alpha', \beta', \gamma')$ passing through the attracted point
> and δ^2 the quantity with which $\alpha'^2, \beta'^2, \gamma'^2$ surpass the squares of the
> semiaxes of the given ellipsoid. I do not speak of an interior point
> because this case can be deduced immediately from the above, if
> one can at all speak about the attraction of an interior point when
> the attraction is proportional to a negative power greater than the
> third. However, one can have a result free from all indeterminacy by
> considering a layer of a finite thickness.
>
> [Ms 3625 (5), pp. 9v–10v]

This important letter, which has not been published before, was followed 12 days later by a letter in which Dirichlet indicated how to extend this result to forces proportional to $\frac{1}{r^{2m}}$, m being an integer greater than one (cf. [Tannery 1910, p. 37]). From Liouville's reply [Neuenschwander 1984a, II, 15], we know that Liouville talked to Chasles about Dirichlet's ellipsoids, and as we saw above, he used his course at the *Collège de France* to difuse the knowledge of this theorem, which was probably one of the last discoveries Dirichlet made. In addition, Liouville published Dirichlet's elegant proof of a theorem by Abel.

"I myself, who only like my hole"

As it appears from Liouville's letter of August 22nd, 1858 Liouville never repaid Dirichlet's visits. To be sure, the wine harvest was his excuse that year, but generally Liouville was a stay-at-home. As he wrote the same year to his brother Félix, who was on a holiday in Italy for his health:

> What a beautiful journey you are on, as long as the medicine does not interfere, and how Sophie [Félix' daughter], who is young and strong, must enjoy all that she sees. I myself, who only like my hole, cannot think without enthusiasm of the pleasure I would enjoy by taking a walk in this beautiful country full of great places.
>
> But, if God has denied me this pleasure forever, he has at least granted me in the study of mathematics, a great consolation within my reach.
>
> [December 16, 1858, Neuenschwander 1984a, I, 11]

The only travels I have found traces of are those related to his training periods at the *École des Ponts et Chaussées* and the journeys he made as external examiner at the *École Polytechnique* during the late 1830s. The lists of marks from the years 1837–1839 are preserved in his notebooks and carry the names of the following French towns: Montpellier, St. Cyr, Nimes, Forêst, Mailler, Marseille, Lyon, Dijon, etc. One entry even shows that Liouville was outside France, namely, in Bruxelles. Apart from these journeys, Liouville's life seems to have been confined to Paris and the region around Toul, including Commercy and Nancy.

Considering how open Liouville was toward foreign scientists and how exceptionally well informed he was about their works, it is striking that the adventures of other countries did not attract him. He did know at least the basics of some foreign languages. In school, he had learned classical languages, and as a scientist, he clearly had to master Latin. Peiffer claims [1983, p. 215] that he also mastered classical Greek, but one has to be careful not to be mislead by the many notes in his notebooks which appear as though they are written in Greek. In fact, they are written in French with Greek letters. For example, in April 1861, when he examined his students at the *Faculté des Sciences*, he wrote his comments about each student in this way, e.g.:

$$\iota\nu\upsilon\tau\iota\lambda\epsilon\mu\epsilon\nu\tau \quad \lambda o\upsilon\gamma$$

or $\tau\rho\epsilon\sigma \quad \lambda o\upsilon\gamma \quad \epsilon\tau \quad \mu\alpha\lambda$

or $\tau\rho\epsilon\sigma \quad \varphi\alpha\iota\beta\lambda\epsilon$

[Ms 3627 (12)]

Moreover, he was able to read mathematics in English, Italian, Swedish, and other languages, and he even admitted that German could be written "with elegance and precision" [1838a, p. 1], at least when Dirichlet was the author. Still, as we saw above, he complained when Dirichlet's papers were written in German, because then they had to be translated into French in order to be published in his *Journal*, in which he only allowed French papers. He seems to have always asked some of his colleagues or students to translate the German papers for his *Journal*, so, although he had tried to "start to learn a little German" when Steiner visited Paris in 1840 [Tannery 1910, p. 17], he probably never mastered Dirichlet's native language, and the new major language of mathematics.

The Quarrels with Le Verrier Continued

While Liouville lost several of his friends and colleagues, while his gout worsened and his scientific creativity decreased, his trouble with the conservative catholic part of the scientific community continued. Let us now return to the controversy between Le Verrier and Liouville where we left it in 1854, when the academic establishment had played a trick on Le Verrier by refusing to decide whether to appoint him an ordinary member of the *Bureau des Longitudes*. As revenge, Le Verrier, in his capacity as Director of the Observatory, annoyed the members of the *Bureau* in many ways. He only allowed them access to the Observatory when the *Bureau* had its meetings, which meant that the *Bureau* could not use the library as it had previously. Moreover, Le Verrier often "forgot" to heat the meeting room on days where the *Bureau* met [Bigourdan 1932–1933].

The main subject of the quarrel between Le Verrier and the *Bureau* was the two annual publications, the *Connaissance des Temps* and the *Annuaire du Bureau des Longitudes*. In 1856, Le Verrier claimed in the *Comptes Rendus* that "The *Connaissance des Temps* has not been a scientific journal for a long time" [CR 50, p. 349]. Indeed, this publication was no longer up to the standards of the foreign astronomical tables; one reason was the lack of appropriations for calculators, a problem that Liouville raised when Le Verrier drained the *Bureau* of its resources in 1854. After Le Verrier repeated his open criticism of the two publications to the

Académie on February 6, 1860, Liouville and Mathieu protested. At the following meeting, they pointed out that Le Verrier had chosen a bad time for his criticism, because the Minister had just raised the funding for this undertaking, thereby heralding improvements. Moreover, they found it unbecoming that Le Verrier had addressed himself to the *Académie*, i.e., to the public, rather than to the *Bureau*, of which he was an associate member (cf. [CR 50 (1860), pp. 348–3590]).

Apparently, Liouville also used the opportunity to criticize Le Verrier's scientific work. His arguments are not quoted in the *Comptes Rendus*, but they were met in the strongest terms by Le Verrier:

> Mr. Liouville, for his part, has in particular tried to make a useful diversion by blaming Mr. Le Verrier for some mistakes which he believed he had discovered in his printed works. None of these mistakes, which incidentally are insignificant, have the least reality. It is Liouville who, in his precipitation, has committed grave errors. We shall prove this clearly at the next meeting. However, one cannot expect us to demean ourselves by reproaching Liouville with some errata which he has been able to add to his works according to our instructions. We ask the *Académie* for permission to ignore these people [Mathieu and Liouville] and to concentrate exclusively on a calm examination of the question of analysis.
>
> [CR 50 (1860), p. 351, February 13]

Le Verrier's detailed rejection of Liouville's argument never occurred, but the quarrel continued at the following meeting. Now, Delaunay actively took Liouville's and Mathieu's side.

Delaunay, Liouville's other astronomical protégé, had been elected a member of the *Académie* in 1855. The following short letter from Liouville to C. P. M. Combes, dated March 10, 1855, shows that Liouville had been vigorously campaigning for Delaunay:

> My dear Fellow member,
>   Delaunay tells me that next Monday you have to set out on a journey. I beseech you, do not leave before the vote. You cannot imagine what efforts are afoot against us. I do think that we will carry it off, but we have no votes to loose. I repeat, come and vote next Monday; come if you can before 3 o'clock to the library, where you know that matters are sometimes decided. This is not a cry of distress but a cry of prudence.
>
> [Neuenschwander 1984a, III, 14]

This campaign, which seems to have been directed against the Le Verrier camp, was won on Monday, March 12, 1855, when Delaunay was elected with 33 votes against 24 [CR 40 (1855), p. 567].

In 1860, the week after having defended Liouville and Mathieu, Delaunay summarized his own developing theory on the motion of the moon and refuted the doubts that Le Verrier had earlier cast on its correctness. Before

PLATE 22. Charles-Eugène Delaunay (1816–1872). Liouville sided with Delaunay in his fight with Le Verrier and was instrumental when Delaunay replaced Le Verrier as the Director of the Paris Observatory in 1870. (Royal Library, Copenhagen)

reading his note, "Delaunay expressed his regret at seeing that Mr. Le Verrier is not present at the meeting." However, he only had to wait one week to read the long answer of his adversary in the *Comptes Rendus.* Le Verrier used one of the old tricks in these academic discussions when he deeply regretted that he had been dragged into this quarrel against his will. He felt that instead of attacking him Delaunay "should rather use his time to correct his own mistakes" [CR 50 (1860) pp. 454–455]. Now, Delaunay was out of town (cf. [CR 50 (1860), p. 458]), so he had to wait a week before he could present his very long answer [CR 50 (1860), pp. 510–520], to which Le Verrier immediately replied [pp. 520–530].

It is peculiar to see Le Verrier arguing against Delaunay despite the fact that Delaunay's arguments resemble the arguments that had led Le Verrier to the discovery of Neptune. Since Liouville had reported favorably on Delaunay's treatment of the theory of the moon in 1847, Delaunay had refined it to a point where he could detect discrepancies between the observations and his calculations. In particular, the secular acceleration was off by about 6 seconds. Delaunay had concluded that hitherto unknown factors must be responsible for this discrepancy. Le Verrier, on the other hand, attributed it to Delaunay's flawed theory. Time showed Delaunay right. In 1865, he suggested that the problem was caused by a deceleration of the rotation of the earth due to the friction caused by the tides. This is still considered to be the correct explanation and is Delaunay's most celebrated discovery.

In the discussion of 1860, Liouville had the final word. In the *Comptes Rendus*, after the printed version of Le Verrier's answer to Delaunay, one can read:

> At a certain moment, Mr. Liouville interrupted Mr. Le Verrier and said "Monsieur you pretend by *grace* to relinquish the critical examination of Mr. Delaunay's old works, but all that you have written on this subject (added in a footnote: In a publication we have all received today) you have already said at an earlier time to the *Académie*, and the *Académie*, after a thorough discussion, answered you then, by admitting Delaunay to its midst by a great majority."
>
> [CR 50 (1860), p. 531]

After more than a year of cease-fire, the battle opened again in July 1861, when the Minister asked first the *Bureau des Longitudes* and then the *Académie des Sciences* to present a list of candidates for *three* vacant positions at the *Bureau*. Obviously, it was a hard problem for Liouville and his allies to avoid the candidacy of the most famous French astronomer. Bigourdan has described [1833, pp. A11–A17] how they succeeded:

> In the *Bureau des Longitudes*, a committee was formed consisting of Liouville, Deloffre and Mathieu. The question was soon raised whether the adjoints (read Le Verrier) had the right to vote, and when it became known that the Minister was of the opinion that they had, Liouville demanded to have a written confirmation. The President of the *Bureau*, Maréchal Vaillant procured this answer.
>
> [Bigourdan 1933, pp. A3–4]

On July 10, the committee presented its list of candidates. It took great pains to explain that it had only considered those candidates "who have themselves submitted a demand to the *Bureau*" [Bigourdan 1833, A12, quote from P. V. Bur. Long.]. In this connection,

> Mr. Liouville asks permission to read a letter that will be addressed to the Minister of Public Education at the same time as the presentations, if a situation arises, which he considers probable. He believes that the reading of the letter might with advantage precede the vote of the *Bureau*. The permission is not granted.
>
> [Bigourdan 1933, A12, quote from P. V. Bur. Long.]

There can be no doubt that this letter concerned Le Verrier, for having probably, and for good reasons, considered himself the most obvious candidate, he had not dreamed of bringing up his own name. Therefore, the committee had found an excuse for leaving him off of their list. Instead, they had chosen Laugier, Delaunay, and Peytier as first candidates and Puiseux, Faye, and Begat as second candidates. One can well imagine that these people had been urged to present their names to the *Bureau*. After this list had been accepted through a ballot, "Mr. Liouville returns to the letter which he wanted to read; the letter is not read. The presentations [the list of candidates] will immediately be communicated to the Minister" [Bigourdan 1933, p. A12, quote from P. V. Bur. Long.].

Of course, this decision stirred up emotions. The President of the *Bureau* is said to have exclaimed, "Thus, it has been decided that you will reject Mr. Le Verrier!" [Bigourdan 1933, p. A12].

The *Académie des Sciences*, which had also been asked to present two candidates for each of the three positions, decided to let the sections of astronomy, navigation, and geometry propose a list. Thereby they followed Liouville's proposal from 1854. On July 29, the entire *Académie* was to state its position on the list. As usual in such cases, the discussion took place in a secret sitting from which the *Comptes Rendus* did not report anything. Still, we get a firsthand impression of the events, and in particular of Liouville's part in them, in a letter written by one of the applicants, Delaunay, to the wife of another applicant, Mrs. Laugier, dated August 1, 1861:

> When the public was asked to withdraw, because of the secret sitting, I left the meeting room in order to wait in the Library. There I received from time to time news of what went on from Frémy and Longet. Mr. Valenciennes must have told you everything in detail; Liouville's *admirable* improvisation (I underline the word admirable which I have heard being used by all those who have talked to me about it), the unbelievable self-confidence of Serret who first protested against what the Commission had done by not presenting Le Verrier for the astronomy seat (but, Liouville at once said, you forget that in the first ballot Mr. Laugier obtained the unanimity of the votes, and consequently yours also). A bit disconcerted by this crushing answer, Serret soon recovered and asked that the *Académie* add the name Le Verrier to each of the three lists of candidates. Having got some opposition, he restricted his proposal to one list, namely the one concerning Poinsot's seat. They voted and of 37 ballot papers there were 16 *yes* and 20 *no* and one blank.
>
> Mr. Flourens asked that this incident should not be inscribed in the *procès verbal*; they did not authorize that. Serret, on the other hand, wanted that it be printed in the *Compte Rendu*; this was refused as well.
>
> On Monday, the great battle, which I hope we will win.
>
> [Bigourdan 1933, pp. A14-A15]

During the secret sitting, Biot read a letter in which he too protested that the name of Le Verrier was missing [quoted in Bigourdan 1933, pp. A13–14]. Hermite, who was not present had authorized Biot to declare that he was of the same opinion.

However, Liouville's camp was the strongest. From the many votes during the following meeting on August 5, 1861, Le Verrier got only 15 votes or less, so that the final list of candidates ended up being identical to the one sent in by the *Bureau des Longitudes*.

This was a great victory for Liouville. It probably gave him satisfaction for the humiliations he felt he had endured from his earlier students. He had shown that his oratorical gifts and his power in the *Académie* were still as strong as in his youth. However, there is one great difference. In most of his earlier interventions, Liouville's objective had been to advance an idea or a student. This time it was not so much a question of having Delaunay elected, for example, as it was a battle against Le Verrier. Negative objectives had taken hold of Liouville.

The whole maneuver proved to be completely in vain. As Liouville had feared from the start, the Minister refused to bypass his fellow partisan, the Senator Le Verrier, so on March 26, 1862, in addition to nominating the new members, he also raised the associate members to the rank of full members. This political intervention finally installed Le Verrier in the seat to which his scientific merit had for a long time entitled him. But, the confrontations were not over. He continued to annoy the *Bureau* with his despotism, and nobody could work with him at the Observatory. Combined with his conservatism, this led to his defeat in 1870, to which I shall return in the next chapter.

Declining Influence in the Académie; Bour

The outcome of the elections in 1861 indicates that Liouville was still a powerful person in the *Académie*. This is particularly true as far as the prizes are concerned. For example, when the *Académie* tried to bring a little order to this otherwise rather chaotic domain of its activities by listing all the prizes proposed for the years 1858–1861 [CR 46, 1858, pp. 299–306], Liouville was the only academician who figured in all the committees related to the prizes of the exact sciences, except for one in experimental physics. He was the head of the committee on the prizes in number theory, the theory of the tides, and on thermoelectricity, and he was an ordinary member of the commission of the prizes regarding capillarity and elasticity. This is representative of his very wide-ranging work in the prize commissions.

However, in 1862, the decline of Liouville's powers was heralded by his first great defeat in an election. After the death of one of Liouville's traditional opponents, Biot on February 3, there was a feverish competition between Ossian Bonnet and Bour, who both had a series of papers printed in the *Comptes Rendus* in order to get the vacant seat at the *Académie*. On April 7, Chasles, on behalf of the geometry section, presented the following list of candidates: (1) Ossian Bonnet, (2) Bour, (3) Blanchet and Puiseux, (4) Briot and Bouquet [CR 54 (1862), p. 770]. The list was discussed at a secret sitting of the *Académie*, which lasted more than two hours. Here, Liouville strongly supported Bour and succeeded in convincing the entire geometry section (except for Lamé) that their "first candidate ought to be ranked not only after Mr. Bour but also after Mr. Blanchet and Mr.

Puiseux." Thus, Liouville wrote to his fellow member of the *Académie*, Montagne, who had only been able to attend the beginning of the meeting. Liouville clearly wanted to follow up his victory by manipulating other members of the *Académie* to vote for his candidate. Still, he ended his letter with the words:

> I naively write you what I think (after a profound study) of course, without pretending to dictate your opinion. I love freedom so highly that I would not curtail anybody else's freedom. However, I cannot refuse to express my sincere opinion to a Fellow member who asks it from me and in a matter about which I am perfectly competent.
>
> [April 13, 1862, Neuenschwander 1984a, III, 15]

However, Liouville did not convince enough members of the *Académie*. On the following Monday (April 14, 1862), Bonnet was elected with 29 votes against Bour and Blanchet, who each got 14 votes.

Even after this defeat, Liouville continued to support Bour strongly. On May 5, he told the *Académie* that he had generalized some of Bour's results on ordinary and partial differential equations, but he would withhold the results until Bour had published the complete memoir, "for in the too few pages he has inserted in the *Comptes Rendus*, each word is an idea." According to Liouville, "Mr. Bour has established his rank next to the masters" [Liouville 1862a]. Liouville never got around to publishing his own results, which are sketched in his notebooks [Ms 3628 (6) and (17)].

With this personal evaluation of Bour, Liouville behaved almost as high-handed as Cauchy had done when writing the report on Briot and Bouquet's work, for Liouville was actually a member of a committee that was going to report on Bour's work to the *Académie*. Bertrand and Serret were the other two members of the committee, and at least the former was more indulgent toward Liouville than Liouville had been toward Cauchy. After Liouville's speech, Bertrand rose to assure the *Académie* that:

> although Mr. Liouville has on his own signed the communication he has just given, and which is in reality a report on Mr. Bour's work, one should not conclude that there is any difference in opinion between him and his two fellow members concerning the importance of the discoveries about which he has spoken.
>
> [CR 54 (1862), p. 942]

After having failed to get Bour elected, Liouville seems to have played a lesser role in academic policy.

Official Honors

At the point where Liouville's scientific and administrative powers began to fail, he had the pleasure of seeing Imperial France recognize his services by nominating him Officer of the *Légion d'Honneur*, of which he had been *Chevalier* since 1838. He was nominated on August 13, 1861, and immediately thanked the Minister of State, Mr. Walewsky, for "this unexpected favor, and I am flattered to know that it has been granted on your initiative" [Toul August 17, 1861, Ms 3625 (18)]. When Liouville returned from Toul he took care of the arrangements of the formal promotion ceremony and sent an autobiography [Ms 3628 (2)] and other documents for which the *Grand Chancelier* had asked. Further, he asked the *Grand Chancelier* if the reception in connection with his nomination could take place under the auspices of his friend Pelouze, Director of the Mint and Commander of the *Légion d'Honneur* [Ms 3628 (2)]. Liouville's wish was granted, and the reception took place on December 10, 1861, followed by a dinner at Pelouze's.

At this time, Liouville seems to have been a well-known public figure, both in Paris and Toul. The Liouville family continued to occupy various high public and civil posts in Toul and the surrounding towns. That they were rather distinguished members of the *haute bourgeoisie* is indicated by a plan of the seating arrangements for a lunch with the *Préfet*, Mr. Lambert, on October 26, 1856, which Liouville sketched in his notebook [Ms 3622 (8), p. 13r]. Here, he and his brother Félix sat next to the *Préfet*. Félix's sons, Henry and Albert, were also present together with nine important persons of the *Département*.

Félix was also a well-known figure in Paris. In 1858, he was a candidate for the republican opposition of the capital, but he was not elected. Still, when he went to Italy later the same year in order to recover from an attack of cancer of the stomach [Ms 3626 (9)], Joseph wrote to him:

> Please remark that I can not take one step in Paris without being accosted and questioned, even by strangers, "how is your brother? Where is he? What is he doing? When does he return? etc. etc.."
>
> [December 16, 1858, Neuenschwander 1984a, I, 11]

If this is true, it indicates that Joseph Liouville was also a rather well-known figure. Indeed, Félix himself called his brother "the glory of our family" [Ms 3626 (9)]. This phrase stems from the testament Félix wrote in August 1859, and which was opened on April 7th the following year, when he finally succumbed to the cancer. His death was a great loss to the family, and when Liouville was nominated Officer of the *Légion d'Honneur*, he wrote to his two nephews, Albert and Henry:

> Why is it that my poor brother is no more? And how could one let him die without according him these honors which he deserved more than I!
>
> [Ms 3625 (8)]

After the death of his brother, Liouville became the head of his branch of the family.

VI

Old Age (1862–1882)

Mathematical Work

With the death of Dirichlet, Liouville lost his main correspondent, inspiration audience, and judge of his number-theoretical works. He tried to compensate for this loss by discussing his results with Hermite, whose great mathematical talents he still admired despite their different political and religious views. Thus, in 1856, he wrote to Dirichlet how glad he was that Hermite had been elected to the *Académie* [Neuenschwander 1984a, II, 10] and as late as 1874, he declared to Mittag-Leffler [letter to Hj. Holmgren April 25, J. E. Roos, private communication] that he considered Hermite to be "the only [French mathematician] who can with real success compete with the Germans."

A couple of letters printed in the *Comptes Rendus* of August 5, 1861, bear witness to the arithmetical discussions between Liouville and Hermite. Hermite's letter begins:

> Since our last conversation on the questions of arithmetic, which are the subject of your research and for which you have given me a new example of the great fruitfulness of the methods whose principles you keep to yourself, I think I have succeeded, to a certain degree, in satisfying a wish you have expressed to me several times concerning Mr. Kronecker's beautiful theorems on quadratic forms. However, these theorems, which by their nature seem to belong to the sphere of your studies of numerical functions, have remained isolated and belong to a very different universe of ideas to which the theory of the complex multiplication of elliptic functions seems to provide the only access.
>
> [Hermite 1861, pp. 214–215]

This letter reveals that Liouville strove for an elementary proof of Kronecker's results (cf. [Liouville 1860d] and [i]), that is proofs that circumvented elliptic functions and used Liouville's number theoret functions instead. It also shows that although he had often discussed "questions of arithmetic" with Hermite, he had not revealed the principles behind his methods. Having provided a proof of Kronecker's theorems, which he believed to belong to Liouville's universe of ideas, Hermite concluded the letter:

> I hope, my dear fellow member, that you do not forget that you have also promised me an arithmetical letter which somewhat withdraws the veil with which you have covered yourself until now.
>
> [Hermite 1861, p. 228]

Liouville answered with a letter printed in the *Comptes Rendus* next to Hermite's letter, in which he wrote that Hermite's proof was not the one he had in mind, for it still built on functions defined for all real values, whereas Liouville's simple functions only need be defined for integer values of the variable. However, instead of revealing his principles, he added a few more theorems, remarking about their origin:

> I found these formulas in 1857. Since then I have, so to speak, added nothing because I have been overwhelmed with other occupations and continually disturbed in this work that requires my full attention. The dozen papers which I have published do not contain half of what I knew four years ago; and I still leave aside the particular applications that present themselves in great numbers, but whose real value can only be gained through the choice one makes and through the order one establishes between them.
>
> [Liouville 1861μ, p. 230]

These applications pertained to quadratic forms. Liouville had begun publishing on quadratic forms in 1856. In 1860, he inserted more than a dozen notes on this question in his *Journal*, and in 1861, he ran amuck, publishing more than 30 notes of one or two pages each, all with the same structure: a theorem stating that numbers of a particular form, $a + b\mu$ (a, b are specified numbers, μ a variable), can be written in a given number of ways, by way of a particular quadratic form, for example, $Ax^2 + By^2 + Cz^2 + Dt^2$ (A, B, C, D are specified numbers). The theorems were not proved, but merely illustrated with a particular value of μ. Thus, not only did Liouville keep the proofs of his theorems about number-theoretical functions to himself, he also hid how they could be applied to quadratic forms.

For two subsequent years, he continued to pour out more than 30 notes a year along these lines, and he published nothing else. Even the fictitious correspondence with his pseudonym Besge dealt with similar questions. During 1864 and 1865, the stream was slowed down to around 20 notes a year, and included more general formulas on number-theoretical functions [1864s] and an elementary remark about Rolle's theorem [1864e]. However, not until 1866 did the stream fade out (11 notes), and in 1867 it stopped.

How can we explain Liouville's abnormal policy of publication in the 1860s? First of all, why did he devote his publications exclusively to number theory? The reason can perhaps be found in the events of 1857. Inspired by Dirichlet, he had a breakthrough in his investigations of number-theoretical functions, and he seems to have planned a line of continued research and a wide field of applications, when suddenly his appointment to the *Faculté des Sciences* deprived him of his research time. As we saw in Chapter V, he had planned to stay in limbo for a couple of years and then resume his work in number theory. Writing to Hermite on November 7, 1861, Liouville stuck to this plan:

> Now, I am a grand-father for the second time. My grand-son was yesterday followed by a grand-daughter. This warns me that I am

old. Therefore, I must finally hasten to edit what I have done, first of all that which concerns the integers. I have succeeded in the subject which occupies both of us, and I hope I will be able to talk to you about it soon. At the moment, I am at the examinations.

[Neuenschwander 1984a, I, 12]

Thus, Liouville felt that his time was running out and that he had to publish his ideas fast before it was too late. That explains the feverish rate of publications, but it does not explain why he continued to publish rather uninteresting applications of his general ideas instead of revealing the core of these ideas and his methods of proof. Twenty years later, Hermite was in no doubt about Liouville's motives. Having assured Stieltjes that Liouville had not published his principles even after he had been urged to do so, Hermite continued:

However, Mr. Liouville, to whom I have expressed this wish, did not want to satisfy it, no doubt in order to keep for himself the entire harvest of all the consequences of his original discovery.

[Hermite to Stieltjes, November 5, 1883, Hermite-Stieltjes 1905, I, pp. 46–47]

If Liouville hid his methods, like renaissance mathematicians, in order to impress his colleagues with his results, he did not entirely succeed. This can be seen in a letter from Hermite to Catalan, probably written shortly after 1865, when Catalan had moved to Liège:

For a long time, I have shared your sentiments of regret concerning Liouville's last arithmetical publications. The secret behind his numerous theorems has not been long in becoming known (P. Pépin has proved them). He would have gained much by showing his principles and his methods at once instead of keeping them to himself; his meager and monotonous verifications make one smile a little.

[Jongmans 1981, p. 294]

So, instead of creating a marvel, Liouville's notes made his former protéges worry about him and laugh behind his back.

Lack of time may also explain Liouville's behavior. He often lamented that it took him a long time to turn his crude notes into publishable memoirs, and there is little doubt that the steady stream of notes caused him much less effort than a memoir presenting his basic ideas and providing the missing proofs. This puts his number-theoretical ideas in the same category as his ideas on doubly periodic functions, Galois theory, the theory of rotating fluid masses, potential theory, and many other results. But, his unwillingness to explain his ideas orally to Hermite puts his number-theoretical principles in a category of their own. He may have had the idea of publishing them before discussing them with Hermite, and then never found the time or energy for this enterprise. But, such secrecy was far away from his earlier complete openness about his unpublished research. I can not find a definite explanation of Liouville's singular publishing practice in the 1860s, but it is hard to escape the impression that he had become more secretive about his ideas, and there is no doubt that this contributed to a depreciation of his mathematical work.

Still, there were light spots in the monotony of these publications, in particular during 1866 [1866a,k] when Liouville succeeded in proving a theorem on the number of ways a number can be represented as the sum of 10 squares. After Eisenstein had given up finding this result, Liouville tried for several years before he could tell the *Académie* about his success in June 1865 and April 1866. These papers were among his last scientific communications to this learned assembly.

Liouville's number-theoretical works attracted little interest. Stieltjes referred to them in his correspondence with Hermite [Hermite, Stieltjes 1905, I, pp. 44–47, 54–59, and 60], but otherwise his ideas have been dormant until recently, when several Soviet number-theorists, such as Kogan and Lomadze, have taken them up again (see *Mathematical Reviews* after 1960).

After a break of two years, Liouville again published a few papers in 1869, one in 1870, one in 1873, and finally two in 1874, plus a few under his pseudonym. Now, it was clearly old age that prevented him not only from composing papers but also from obtaining new ideas. As Mittag-Leffler wrote to Holmgren in 1874 concerning Holmgren's papers on fractional calculus (cf. [Chapter VIII]):

> ... the study of this... had given him [Liouville] much to think about which he would have liked to inform me about if his force of thought during the last years had not weakened so much that he could not any longer work on difficult matters. Liouville is indeed not that old, but during the last couple of years, his strength has been broken by long illness, and I think he is right when he says that he is no longer in as full possession of his unusual intelligence as he was before during his active years.
>
> [April 25, 1874, private communication by J. E. Roos]

Liouville did not quite give up all his publication plans, however. For example, in 1873, he planned a series of papers and even drafted the introduction to them in his notebook [Ms 3636 (6)]:

Thursday April 10, 1873. Paris

> Miscellaneous [Variétes]; By M. J. Liouville. I will set out to edit successively and to publish under this title several articles from the notes on various subjects that I have jotted down [?] embarrassment of the siege of Paris and at Rambouillet, where I was lucky to find a [?] during the horrors of the Commune.
>
> [Ms 3636 (6)]

These *Variétes* never appeared. Neither did the many number-theoretical results that Liouville promised to publish at the end of his last note read to the *Bureau des Longitudes* on January 21, 1874[1].

Apparently, Liouville also planned to publish some results on celestial mechanics as late as 1876. At least when the *Bureau des Longitudes* discussed the next volume of the *Connaissance des Temps*, they counted on the "inclusion of a memoir that Mr. Liouville has promised to edit during this year" [P. V. Bur. Long., February 9, 1876]. However, this plan was not carried out either, so except for Borchardt's notes on doubly periodic functions [Liouville 1880], Liouville's publications ended for good in 1874.

Still, he continued to work on mathematics in his notebooks. During the 1860s, he filled more of these than ever before, and from the 1870s more, than 70 notebooks are preserved. They seem to contain every calculation Liouville made, in contrast to his early notebooks, which only contained the more promising ideas. Liouville had obviously stopped using scrap paper in addition to his notebooks.

The notes are mostly concerned with number theory, but also contain calculations on definite integrals, rational mechanics, and other fields that had occupied him earlier, and a quick glance indicates that they contain little news of any scientific interest.

I shall only mention one new subject: non-Euclidean geometry. In Ms 3635 (3), one finds some notes on a curve equidistant from a given line. They were probably made in connection with a paper that his friend, Ernest Lamarle, had sent to him. However, as is revealed by the reply, this was among the few subjects that had never caught Liouville's interest; the letter is addressed to Ernest Lamarle in Calais and dated March 28, 1870:

> I am ailing and I have never before had any taste (to know it enough) for the [?] of the theory of parallels which has given rise to so many alleged proofs which have immediately been recognized to be false by the authors themselves. Therefore, I have asked one of my friends, Mr. Lionnet, who is perfectly competent in these matters, to examine your proof.
>
> [Ms 3635 (10)]

The fact that the "proof" was not published in Liouville's *Journal* indicates that it was a new fallacious proof of the parallel postulate, in which case Liouville was wise to let an expert have a look at. However, it is worth noting that Liouville did not reject the proof a priori, for this shows that he had apparently not digested the recent progress in this field. Since 1860–1863, Gauss's conviction of the possibility of a non-Euclidean geometry had been known through the publication of his correspondence with Schumacher, and in 1866 and 1867 the works of Lobachewsky and Bolyai were rescued from oblivion through the translation into French by Houël of their most important works. Houël had earlier helped Liouville translate other works for his *Journal*, and in 1869 he further translated Beltrami's important paper from the previous year, in which the author identified the non-Euclidean plane with a surface of constant negative curvature. This description gradually convinced most of the mathematical community of the logical consistency of non-Euclidean geometry.

However, this change of view did not happen overnight, and as late as December 1869, Bertrand claimed in an *Académie* report that a memoir by a certain Carton contained a correct proof of the parallel postulate. However, this was too much for Liouville. In a letter to Houël, Darboux described how:

> Liouville rose and opposed the conclusions of the Report, only for the singular reason that the *Académie* must not deal with Euclid's postulate.
>
> [Gispert 1987]

That Liouville's opposition was not founded on a knowledge of the newest ideas shows that by 1869 he was no longer as good at spotting really important new trends in mathematics as he had been earlier.

Lecturer and Promoter

Loria has suggested that Liouville might have explained the purpose and method of his long series of papers on quadratic forms in his courses at the *Collège de France*. Indeed, in the first semester of 1866–1867 [Ms 3633 (12)], he did present his theory of numerical functions, and on several occasions he discussed quadratic forms (for example 1855–1856, first semester [Ms 3631 (10)], 1867–1868, first semester [Ms 3634 (2)], 1869–1870, first semester [3634 (9)], and 1872–1873, first semester [3636 (8)]). However, it is hard to see from his notes how much of his ideas he revealed. The major part of his lectures, entitled *Théorie des nombres*, were devoted to the works of Gauss and Dirichlet, while the ideas of Kummer were rarely mentioned [Ms 3636 (8), 1872–1873, first semester], despite the fact that Liouville had been an advocate for Kummer's works in France.

On occasion Liouville even presented his audience with number-theoretical results that he did not publish. For example, a proof of Waring's theorem for biquadratic numbers, which he used in his lectures, was published in 1859 by V. A. Lebesgue in his *Exercices d'Analyse numerique* [Lebesgue 1859].

After 1860–1861, most of Liouville's courses at the *Collège de France* bore the title *Théorie des nombres*, as above (11 semesters), or *Diverses questions d'analyse* (15 semesters), with occasional variations of this title, *Calcul différentiel et intégral* (3 semesters), *Diverses questions qui dépendent du calcul différentiel et du calcul intégral* (3 semesters), and *Séries et intégrales définies* (2 semesters). Moreover, as mentioned on earlier, two semesters during this period were devoted to the calculus of probability.

All these titles are rather imprecise. Indeed, Liouville insisted on the right of the professors at the *Collège* to chose their subjects as they pleased. When, in 1873, the Minister of Education asked the teachers to present detailed programmes before the start of the courses, Liouville remarked "that the words "detailed programmes" should not be taken rigorously" [P. V. Coll. Fr., March 2, 1873].

Hence, the use of the same title for many years does not mean that Liouville always repeated the same subjects. Still, it cannot be denied that there exist analogies between many of these courses. Those on *séries et intégrales définies*, for example, resembled that of 1839–1840 (cf. Chapter XVII, §45) but were not as extensive. Liouville often treated potential theory in the lectures on *diverses questions d'analyse* as well as definite integrals, infinite series, and continued fractions (in particular in connection with transcendental numbers), differential equations, sometimes with applications, elliptic functions, and differential geometry (in a simpler way than in 1850–1851, cf. (Chapter XVII, §45).

The level of the courses varied considerably. For example, in 1861–1862 in his course on *le calcul différentiel et le calcul intégral*, he taught rather hard problems of potential theory in the first semester [Ms 3625 (1)], whereas in the second semester, he introduced the elementary transcendental functions (cos, sin, exp, log) in the complex domain [Ms 3615 (1)], a subject that was taught to the first-year students at the *École Polytechnique*. In general, the courses entitled *Calcul différentiel et intégral* were more elementary than those on *diverses questions d'analyse*, but as Liouville pointed out in 1862 [Ms 3625 (1)], they were not as elementary as those of his former teacher Lefebure de Fourcy, nor did they have a historical or philosophical stamp as those of Duhamel at the Sorbonne.

Despite the choice of imprecise titles, Liouville often went outside the subject. For example, in the last four lectures of the second semester of 1872–1873 on the calculus of probability, he studied fractional calculus, the wave equation, and the Γ-function [Ms 3636 (8)]. But, most digressions were concerned with the theory of numbers, which constituted between a third and a half of the content of the lectures on *diverses questions d'analyse*. Indeed, during the 1860s and 1870s, the theory of numbers saturated most of Liouville's courses to a degree that having begun the course *Sur diverses questions d'analyse* of the first semester of 1870–1871 with six lectures on number theory, he opened the seventh lecture with the words: "Here begins a digression on differential calculus, series and integral calculus!" [Ms 3635 (7)].

Contrary to the systematic courses on doubly periodic functions, most of Liouville's courses during the 1860s and 1870s were loosely organized or even chaotic, in the style of his 1839–1840 course (cf. (Chapter III)). As an illustration I shall outline the content of a typical course, *Sur diverses questions d'analyse*, namely, that of the first semester of 1864–1865 [Ms 3631 (9)].

Liouville first explained how Euler had developed the product $(1-x)(1-x^2)\ldots(1-x^n)$, and then in the third lecture, he proved an extension of Rolle's theorem to complex roots, which he had introduced in an earlier course and which he had just published in his *Journal* [1864e]. The three subsequent lectures were devoted to algebraic and transcendental numbers; he began by studying the roots of algebraic equations by Gauss's method and continued with his own famous proof of the existence of transcendental numbers (cf. (Chapter XII)) and with his research on the number e. From the seventh to the eleventh lecture, he considered elliptic and doubly periodic functions, using Jacobi's approach. In the twelfth lecture, he deduced Brounker's continued fraction and Leibniz's series for π, and he presented during the two subsequent lectures polygonal numbers and Gauss's proof of the fundamental theorem of algebra. He continued his course by showing how Leibniz, Bernoulli, Herman, Euler, and others had "reduced quadratures to rectification" and by summarizing Euler's study of synchrones. The eighteenth lecture was devoted to perfect numbers, and the two subsequent ones to the solutions given by Euler, Lagrange, and Jacobi of the elliptic equation:

$$\frac{d\varphi}{\sqrt{1-k^2\sin^2\varphi}} + \frac{d\psi}{\sqrt{1-k^2\sin^2\psi}} = 0.$$

The twentieth lecture dealt with the quadrature of the parabola following one of Archimedes' methods. In the twenty-first lecture, he analyzed two memoirs by Euler, one on infinite series and the other on a nonspecified problem of inverse tangents. After yet another lecture on a paper by Euler, this time on plane differential geometry, Liouville began a series of lectures (from the twenty-third to the thirtieth) on the theory of numbers, in particular, on quadratic and biquadratic residues. Finally, he returned in the thirty-first lecture to the problem of inverse tangents studied in the twenty-first lecture and ended his course with a lecture dealing partly with four formulas of Euler's from integral calculus and partly with a formula from the theory of numbers.

This disorder worsened during the 1860s and the 1870s to a point where he sometimes, in the same lecture, treated three unrelated subjects. The scientific level also decreased. In the letter from 1874, Mittag-Leffler continued:

> He [Liouville] lectured this semester at the *Collège de France* on "definite integrals and summations of series," but he never passed beyond the treatment of some very elementary questions. Several times he started more difficult investigations, but then always broke off for one reason or another.

Thus, Liouville's lectures had less and less merit, but he still continued to try to be of help to young talents and to most foreigners. I have already mentioned his great kindness toward Mittag-Leffler in 1874, and the note "M. Woepecke (22 rue Bréda)" in [Ms 3628 (10)] indicates that he also helped Woepke, whose papers on mathematics and the history of Arabic mathematics were published in Liouville's *Journal* from 1859 to 1865. The St. Petersburg geographer, N. de Khanikof, also followed one of Liouville's courses at the *Collège de France* (namely, that of 1866–1867 [cf. 3633 (3)]), and the two became close friends [Neuenschwander 1984a, note 93]. For example, Liouville wrote to him on July 27, 1869:

> We always hope that you will visit our hermitage (8 rue du Salvateur, the number is of little importance, but the name of the street is useful in order to distinguish us from my nephews). We will prepare your lodgings in advance; it is up to you to chose the day and the hour.
>
> [Ms 3634 (11)]

Moreover, distinguished professors such as Cayley and Sylvester, followed the beginning of Liouville's course of 1861–1862 (first semester) [Ms 3625 (1)]. Liouville also made new friends among his younger colleagues, for example, with Emile Mathieu (1835–1890), whom he gladly accepted as his guest in Toul in 1868 [Ms 3634 (7), letter Liouville to Matheiu, September 16, 1868], and probably with the Austrian-born astronomer Loewy, whose candidacy for the *Bureau des Longitudes* he strongly supported in 1872 (cf. [CR 75 (1872) p. 251]) by trying to influence the members of the *Académie des Sciences* with "some words in favor of Loevy" [Ms 3636 (15)]. A great many young scientists asked Liouville to write letters of recommendation and introductions; such letters abound in his notebooks from the 1860s and 1870s. However, Liouville did not support everybody uncritically. For example, in 1868, he refused to help Niewenglowski to a position as a *répétiteur* (probably at the *École Polytechnique*), although he answered his father that he loved the young man [Ms 3629 (14)]. The same year he also showed that he could still be sarcastic when confronted with pompous incompetent work: a certain Mr. Allégret had a note on elliptic functions printed in the *Comptes Rendus* without having it refereed, and at the following meeting Liouville pointed out that the "remarkable discovery" announced in this paper was a simple consequence of formulas already found by Euler and was an intuitive fundamental property according to the modern theories of doubly periodic functions. He concluded:

> I make these remarks... mainly because Mr. Allégret, in pompous terms, announces a sequel to his work. I think it is best to tell him that he should first consider studying the works of his predecessors. It would not be too much if he devoted some months or even some years to this study.
>
> [Liouville 1868b]

Yet, with hindsight, we can see that as Liouville grew old he did not always support the right persons. Thus, the year before his death he supported Mannheim as a candidate for a vacant position at the *Académie des Sciences* against two obvious candidates, Camille Jordan and Darboux (for the maneuvers behind the screen, see [Hermite 1984]). Jordan was elected.

From 1862, Liouville was in position of a new means of promoting young talents in addition to his *Journal*, the numerous prize committees, etc., namely, the *Journal des Savants* for that he had been chosen editor at the suggestion of J. P. Flourens [Neuenschwander 1984a, note 75]. However, after only three years in charge, he quit the job with the following letter to the Secretary of the *Bureau du Journal des Savants* dated November 21, 1865:

> I am obliged to recognize that I have assumed a burden which is too heavy for me, when I asked to become one of the Editors of the *Journal des Savants*. To be sure, my health has improved much; and perhaps I could finally turn into a convenient collaborator, if (as I thought at one moment) I would stop the publication of the *Journal de Mathématique* in its thirtieth year. I am perfectly in my right to do so; for when I founded this periodical in 1836, I reserved myself the right to do what I wanted in this free enterprise. However, I must admit that I do not have the courage to sacrifice my child. Moreover, there I am sure that my work bears fruit. It is different with the *Journal des Savants*: by leaving it, I feel a very great regret at losing some excellent colleagues, but I know that one can replace me without problems and with advantage for the public. Therefore, I ask you to accept my resignation.
>
> [Neuenschwander 1984a, II, 24]

Liouville nourished his child, the *Journal de Mathématique Pures et Appliquées*, for another ten years. It continued to include papers by the old guard, of whom de Saint-Venant stood out with 15 papers in the second series. In fact, Liouville wrote to him in 1871: "we will be pleased to receive all that you can give us for the *Journal*, of which you are today the most regular contributor" [Neuenschwander 1984a, III, 17].

Toward the end of the second series, the young generation, for example, Jordan and Darboux, also began to publish some of their interesting work in Liouville's *Journal*. But, it was no longer monopolizing French mathematics. I have already mentioned the creation in 1842 of the *Nouvelle Annales de Mathématique, journal des candidats aux Écoles Polytechnique et Normale*, or Terquem's *Journal*, as it was often called after one of its founders. In 1864, the *École Normale* began issuing *Annales scientifiques de l'École Normale supérieure*, and, in 1870 Darboux and Houël revived Ferussac's former *Bulletin* under the name *Bulletin des sciences mathématiques et astronomiques* (after 1876, only *Bulletin des sciences mathématiques*). Further, the *Société mathématique de France* began issuing its own *Bulletin* in 1872. In addition, around 1870 came a large number of new foreign journals (cf. [Neuenschwander 1984b, pp. 7–8]), which also tempted French mathematicians, so it is no wonder that Liouville's *Journal* lost its leading position toward the end of Liouville's time as an editor.

In 1875, when Liouville finally left his "child" in the care of Resal, whose mechanical works he had followed since 1857 [CR, June 1, 1857], it almost died. During 1875–1878, it continued to contain papers by the famous French mathematicians, but during the early 1880s, they left the *Journal* and it became a forum for mediocre authors. In 1883, Hermite informed Mittag-Leffler that "I have been told that Mr. Resal's *Journal* approaches its end, and one attributes this to the success of the *Acta*... because all the interesting works go to Sweden" [Hermite 1984, p. 226].

Indeed, thanks to Hermite's propaganda, the *Acta Mathematica*, founded that year, attracted many first-rank papers from France, in particular those by Poincaré, but the new editor of Liouville's *Journal* was probably also responsible for the hard times of the *Journal de Mathématique*. Liouville certainly made a great mistake by electing a not very prominent member of the *Académie* as his successor. At the last minute, Liouville's *Journal* was saved by Jordan, who succeeded Resal as its editor in 1885. He immediately attracted great authors again, thereby reassuring the reputation of the *Journal*, which it has kept to this day.

Public Life

Liouville continued to be a respected member of society during the 1860s. For example, he was acting as a juryman in 1864, which forced him to cancel some of his classes at the *Faculté des Sciences* [3629 (14)], but he never returned to active political work in spite of repeated requests. That this was not due to lack of political interests, but to a preference for his work as a teacher, can be seen from an answer he wrote on March 12, 1863, to the lawyer Granpierre from Bar-le-Duc, who had officially asked him to run for the Legislative Assembly:

> I have received your nice letter and I am really sorry that I have to answer in the negative. You are probably not astonished that I have used some days to ask my friends advice. Not that I had not already made up my mind; I made myself clear on this subject a long time ago on the occasion of other openings, but then it was not a question of Lorraine, and I therefore wanted to think and consult again.
>
> The law is such that in order to enter the Chamber, I must entirely abandon not Science and the Institute, but my active career, leave the *Bureau des Longitudes* forever, give up my courses forever. None of my friends would advise me to do that; on the contrary they said to me: Will you really be useful in the Chamber? And in any case, don't you serve the Public better in your present position? Going into detail they added other conclusive motives. I had to give in to this evidence, but be assured, Monsieur and dear fellow countryman, that it is not without regret that I resign on this occasion.
>
> [Neuenschwander 1984a, I, 13]

Liouville also rejected other less burdensome official tasks, such as a post in the commission in charge of the Intellectual Exposition, in which the Minister wanted to involve him in 1866 (cf. [Ms 3631 (10), p. 41]). Moreover, when Grandpierre, in 1868 tried once more to make Liouville run for the Legislative Assembly, Liouville again refused:

Paris, May 30, 1868

Monsieur,

The motives I have given you in the course of time for abstaining from any electorial candidacy still subsist. In a sense, they are even more valid than ever; for next year I complete my sixtieth year, so for want of a few months I will lose my soul, that is, my pension of retirement which neither I nor my family want to be missing. Moreover, what good can a deputy do when he is at the same time a novice and old? And I must add in such bad health! It is necessary to say the truth to one's self.

Therefore, I would with great pleasure see the votes go to my excellent and honorable nephew, Mr. Picard: but I doubt that he can run as a candidate at Bar-le-Duc without lessening his chances of reelection in Paris.

[Neuenschwander 1984a, I, 14]

The above-mentioned Ernest Picard had married Sophie Liouville, a daughter of Joseph Liouville's brother, Félix, in 1860. From the letter that Liouville sent to Sophie when she informed him about the subject of her choice, it appears that Ernest Picard was a student of Félix Liouville and that Joseph Liouville had known him for some time:

Toul, August 17, 1860

My kind Sophie,

I am happy to receive the news you announce to me; I feel much esteem and friendship for Monsieur Ernest Picard, and I think that you have made the best choice. No doubt nothing will ever console you for the loss of your excellent father, but at least you will find his honorable character and his talent and even his affection in the student which he has formed and which he loved.

Your aunt [?] and Louise and Marie share our sentiment and join me in congratulating you.

I am entirely at your disposal at the time you choose.

Please give my respectful greetings to Madame Deschamps, my friendship to Céline, to Albert, to Henri, to Mr. Picard and receive, my beloved niece, the expressions of all our wishes for your happiness. Your devoted uncle....

[Ms 3636 (4)]

By then, Picard had, for three years, been one among the handful of republicans in the Legislative Assembly, of a total of 267 deputies [Pradalié 1979 p. 30].

In 1868, Ernest Picard joined forces with Grandpierre in order to persuade Liouville to run for the election, because he felt that France needed intelligent and patriotic representatives [Neuenschwander 1984, note 92]. As we have seen, Liouville stuck to his decision, but he was no doubt happy to see that at the election in May 1869 the opposition made great progress for the first time during the Second Empire. The left-wing opposition, republican or socialist, together with the right-wing, obtained 3,333,000 votes, but the government maintained the majority with 4,438,000 votes, mainly obtained in the rural areas.

Since 1867, an increasing liberal movement had forced Napoléon III to accept political reforms. For example, the legislative corps gained the right to demand that the government answer questions pertaining to its work, and the censorship of the press had been eased. Under the pressure of the election of 1869, Napoléon appointed a new, liberal government in January 1870, headed by Emile Ollivier, a sincere republican from 1848. Together with the emperor, he designed liberal reforms that were accepted with an overwhelming majority in a referendum in May.

These liberal winds were felt at the *Bureau des Longitudes* and at the Observatory. On April 3, 1868, a ministerial decree considerably reduced the power of the Director of the Observatory, Le Verrier (cf. [Ms 3635 (12)]). A council of nine members was appointed to supervise the work of the Director, and in case of disagreements between the Director and the council, the Minister was to make the decision. This new rule had been designed by a commission, not comprising Liouville, appointed by the *Académie* after a new quarrel between Le Verrier and his confrères, among whom he had now lost all support (cf. [CR 65 (1867)] and [CR 66 (1868)]).

However, the council was not able to calm the unrest, and at the end of January 1870, all the highest ranking employees at the Observatory collectively resigned their positions as a protest against Le Verrier. On January 29, the Emperor demanded an inquiry into the situation, and the new Minister of public education, Legris, asked Liouville to become a member of the commission of inquiry:

> I have received your letter of February 4 concerning the observatory.
>
> I am at your service,

Liouville answered on February 5 [Ms 3635 (10)].

On the same day, Napoléon signed a decree by which Le Verrier's appointment was revoked. In addition to the unrest at the Observatory, Napoléon justified his decision with the fact that despite the inquiry demanded by the Emperor, Le Verrier had directly demanded the Senate, of which he was a member, to ask the government about their view on the situation at the Observatory, an action that Napoléon found against "all the hierachical rules and discipline." Liouville was so happy about the fall of his ardent enemy that he glued the decree into his notebook [Ms 3635 (13)].

On February 8, the Observatory commission met for the first time. A subcommittee consisting of Liouville, Serret, and Briot was formed to plan the future organization of the Observatory. They met regularly, sometimes in Liouville's apartment, and two days after the second meeting of the entire commission on February 22, the subcommittee decided to demand the Minister appoint a new director [Ms 3634 (9)]. On March 2, the minister followed their advice, nominating Delaunay as the successor to Le Verrier.

Under the rule of Delaunay and the new regulations designed by the Observatory commission, the Observatory prospered for two years, and then Delaunay suddenly died. To the obvious dismay of Liouville, who served as the president of the *Bureau des Longitudes* in 1872, Le Verrier was reinstalled as Director. Just before Delaunay's death "a discussion [in the *Bureau*] took place between Mr. Liouville and Mr. Le Verrier on the problem that the equipment of the *Bureau* be brought in harmony with the competence of its members" [P. V. Bur. Long., July 10, 1872][2].

Although there does not seem to be any public record of further quarrels between Liouville and Le Verrier, the animosity between them continued. Thus, in connection with the election of Loewy to the *Académie* Liouville wrote in his notebook:

> Le Verrier was odious and ridiculous. What will he achieve? However, I judge Le Verrier to be devoted to a final impenitence and a very prompt fall; he will kill a few more poor people, but he will himself inevitably succumb to his sins and one cannot be sorry for him. Such a stupid tyrant cannot last.
>
> [Ms 3636 (16I), April 4, 1874]

Le Verrier lasted four more years. When he died, Liouville celebrated by gluing the announcement of his death and funeral in the *Figaro* of September 27, 1877, into his notebook [Ms 3637 (10)].

The Franco-Prussian War and the Commune

By following the Le Verrier story, we left the political situation in May 1870 when the strong public support of the liberal line of the Imperial government seemed to herald a new, stable period. However, the following year became one of the most traumatic in French history, leading to the installment of the Third Republic in a way that affected every Frenchman, in particular, the citizens of Paris, including Liouville.

Having taught Denmark and Austria a lesson, Bismarck used the succession to the Spanish throne as a pretext to provoke France to declare war on Germany on July 19, 1870. Napoléon III and most other European countries expected that the French army would rather quickly march to Berlin. However, Bismarck knew better. By uniting the small German states into a strong militaristic unity, he had created a formidable war machine that soon threatened the French border with Germany and Belgium, driving before them the Imperial army in which Liouville's family was represented by "Colonel Balland *chef d'état, major de la division de grenadiers de la garde impérial du Rhin*" [Ms 3636 (10)].

From his summer residence in Toul, Liouville carefully followed the movements of the armies and kept track of technical details, such as the names of the French generals and the number of divisions commanded by each. As a curiosity, it can be mentioned that this list [Ms 3634 (4)] includes General Bourbaki, who later become a celebrated figure, not for his efforts in the war but through the work of the fictitious mathematician named after him.

At the beginning of August, the war entered a dangerous phase. On the 4th, the French army lost a battle at Wessembourg, followed by defeats at Forbach and Wörth on the 6th and at Gravelotte on the 18th. Seeing the German army advance from this half circle around Toul, Liouville decided to interrupt his holidays and flee to Paris. This move seems to have happened headfirst. At least a letter written on July 30, in which he authorized Ernest to fetch his salary [Ms 3634 (4)], indicates that at that moment he had no plans of leaving Toul; on the other hand, he had arrived in Paris before the German army.

His nephew, Henri, went the other way. Having received his doctorate of medicine, he arrived at Toul just the day before the siege of that city. During the German bombardment, he was cited by the commandant in the army daily report for his brave behavior. After the fall of Toul, he joined the Loire army, where he served as chief of the ambulance service. He was well prepared for this job, for on his way to St. Petersburg in 1864 (cf. Chapter V) he had studied ambulance service in Denmark during its war with Germany.

On September 1, Napoléon III surrendered to King Wilhelm I at Sedan, but the republican government, which was proclaimed in Paris three days later and headed by Gambetta, Jules Favre, and the veteran Thiers, decided to continue the war effort. Ernest Liouville was called to serve the republic in Toulon, in which connection his father recommended him to the "Amiral Chopert, Préfet maritime à Toulon"

> Dear and ancient comrade,
>   I hope you will excuse me for being so free as to recommend to your benevolence my son, Ernest Liouville, who has been appointed to serve in Toulon as Prosecutor of the Republic. You will remember that his first name is that of our excellent friend Lamarle.
>
> [Ms 3634 (9)]

If this letter, dated September 15, 1870, was ever delivered, it was among the last messages to leave Paris before the Germans began a siege the following day, which isolated the town for the next 134 days. Shortage of food was soon felt and according to contemporary Parisian cartoons, rats became a great delicacy. Scientists became engaged in efforts to alleviate the hunger, and at the *Académie des Sciences* nutrition and conservation of food were regularly discussed [CR 71,72], as was the construction of corn mills [CR 72 p 17]. Even Liouville entered the lengthy discussion about the value of gelatin as nourishment for human beings:

> Mr. Liouville recalls that, at a time when the question of the properties of gelatin as food was also discussed very keenly, Mr. Arago had occasion to visit the hospital in Metz and to question the patients if the addition of gelatin to their ordinary allowance appeared as

unfortunate to them as one had pretended, and he learned that not only did they accept this addition without reluctance, they would even be very angry if it was suppressed.

[CR 71 (1870), pp. 759–760]

Except for this change of emphasis of research, the academic community showed its resistance to the external pressure by keeping academic life running as undismayed as possible. As it was pointed out in the annual report at the first meeting of the *Académie* in 1871, "the issues [of the *Comptes Rendus*] have appeared every week with their usual punctuality." This report was delivered by Liouville, who had been the president of the *Académie* during the difficult year of 1870, having been elected vice-president in 1869 by a large majority of 36 votes against Bertrand's 4 [CR 68 (1869), p. 13].

In his capacity as president of the *Académie*, Liouville issued a strong protest on December 26, 1870, when news penetrated the German lines that the academician, P. Thenard, had been taken prisoner of war:

After the lecture of the verbal proceedings, M. le Président rose and spoke as follows:

"From the accounts in the journals, the *Académie* has learned about the recent arrest of our excellent fellow member Mr. P. Thenard who has been sent to Bremen by the order of the Prussian generals. If Mr. Thenard has been taken sword in hand, defending his country, we must respect him even more and bow to the sort of arms that betrayed his courage, but if the only motivation for this measure is Mr. Thenard's well-known fortune and his qualifications as a distinguished *savant* and Member of the *Académie des Sciences*, then I shall not hesitate to say that such an arrest is simply an infamy that every one of us must remember to his last hour and for which the perpetrator will be punished one day or the other by divine justice."

The *Académie* declares that they completely agree with the words of the President and decides that they be inserted in the *Compte Rendu* of the meeting.

[CR 71 (1870), p. 911]

All Liouville's admiration for the nation who had produced the greatest mathematicians of his time had now given way for a deeply felt hatred against the besieger. This feeling was certainly reinforced when he heard that two days later his relative, Edmond Liouville, had been "killed on the plateau d'Avron [?] by the Prussians" [Ms 3634 (1)]. By this time, the starvation of the masses in Paris had become critical, and on top of that Bismarck decided to advance the capitulation by bombing the town, in particular, the Odéon quarter, where Liouville lived came under heavy fire. Liouville, who had begun his courses at the *Faculté des Science* [Ms 3629 (14)] and at the *Collège de France* [Ms 3635 (7)], had to cancel them from January 6, and seek refuge with his family in a less threatened part of the town. In his lecture notes from the *Faculté des Sciences*, we read under "10th lecture, Wednesday, January 18th":

There was no lecture on Wednesday, January 11, nor on the 13th because of the bombardment; from Monday 9th we have sought refuge at the *Palais royal* at Barbey's after having the previous night heard the blast of the shells that did damage to Mr. de Vertillac's (21 rue de l'Odéon) and across the road (rue de Condé) at Mr. Petit's. Today, the 17th, the courses are suspended until new orders.

[Ms 3629 (14)]

The day this was written Bismarck proclaimed Wilhelm Emperor at Versailles, and only ten days later, Liouville's "nephew," Ernest Picard, joined Jules Favre in Versailles to arrange the capitulation of Paris. After the election on February 8, 1871, Picard became Minister of the Interior in Thiers' cabinet, which on February 26 was forced to sign a humiliating peace treaty ceding Alsace and the northeastern parts of Lorraine to Germany and accepting German occupation in certain French towns until the indemnities were paid. Fortunately for Liouville, Toul remained French territory.

Liouville's foreign friends had for good reasons feared for his life, and on February 11, Khanikof, who was in London, send him the following letter:

Dear Monsieur,

In great haste I will use a reliable opportunity to send you this letter in order to tell you how much I and all your English friends have worried for your well-being and for the well-being of your kind family during the horrible siege and bombardment and all the numerous calamities which have beaten down on Paris with a severity and a fierceness almost without precedent in history.

My worry for you has been the greater because, according to the letters from the correspondents of the English newspapers, the bombardment has in particular ravaged the unfortunate Odéon quarter, but I hope that the houses in the rue de Condé having prime numbers have been spared and that I will again have the honor of presenting my homage to Madame Liouville in her hospitable salon in number 13.

Mrs. Cayley, Sir W. Thompson [sic!], Hirst and Sylvester have often asked me for news about you, but the last I have had was that from our friend Tchébychef; it was good and I hope it has not changed since then.

[Neuenschwander 1984a, I, 15]

After the German army had humiliated the Parisian population completely with a great victory parade through the town on March 1, they left, and Liouville started on his second semester at the *Faculté des Sciences*. However, on March 27, the Rector canceled all courses because of the proclamation a few days earlier of the Paris Commune by the radicals, whom Thiers had unsuccessfully tried to disarm after the German troops had left. Two months of horror, worse than the German siege, began. The mob plundered the town and set the Observatory on fire, destroying among other things one of its meridian circles (cf. Villarceau's and Delaunay's accounts in [CR 72 (1871), p. 611 and p. 662]).

Liouville preferred to flee "from the horrors of the Commune" [Ms 3636 (6)]. He found rescue at Rambouillet, southwest of Paris, where he could quietly work on various mathematical subjects, while Thiers's armies laid siege to the capital. On May 21, the government troops entered Paris, beginning a week-long bloody battle with the communards, which ended with the execution of the last rebels. Liouville returned and resumed his teaching at the *Faculté*, and on June 12 an ultrashort second semester of only four lectures began at the *Collège de France*.

During the Commune, Ernest Picard had been criticized by both the monarchist and the republican press, and in May he resigned as a Minister. Later in the year, he was sent as ambassador to Brussels. Upon returning in 1873, he resumed his seat in the left center of the National Assembly, and in 1875, he became a life senator. By the time he died in 1877 (May 13), the Liouville family already had a new representative in the Assembly, namely, the doctor, Henri, who had been elected in February 1876 as a republican deputy of the arrondissement of Commercy, close to Toul. Reelected in 1877, 1881, and 1885, he remained a member of the republican left-wing until his death in 1887. He was succeeded by Raymond Poincaré, who later became President of the Republic. In this capacity, he recaptured Alsace-Lorraine during the first world war, thereby giving France revenge for the humiliation of 1871, which had imprinted a strong hostility toward Germans in the French population.

Liouville shared this sentiment, but he soon got a chance to prove that "the most ardent patriotism does not dispense with personal recognition" [Neuenschwander 1984a, I, 17].

Foreign Member of the Berlin Academy

In 1875, the Berlin Academy lost two of its foreign associates, and all the members of the geometry section agreed that the vacant seats be offered to Liouville and Chasles. However, before officially proposing this to the Academy "we think... we ought to make sure that such a nomination will be agreeable to you and that you will ratify it, accepting therewith the new tie that we want to establish between our Academies." The political situation taken into account, it was not a matter of course that Liouville would accept the nomination, so Kronecker asked him about it in a letter dated December 16, 1875, from which the above quote stems [Neuenschwander 1984a, I, 16]. With the letter he included some small memoirs "of which one may have caught your attention by the close links that exist between the arithmetical results which I develop there and those that you have formerly been led to by your celebrated research concerning the theory of numbers" [Neuenschwander 1984a, I, 16].

The two friends accepted the offer, Liouville with the following letter dated December 22, 1876:

My dear Monsieur Kronecker

Yes, I would be flattered to obtain through my feeble works the votes of your geometers. I have obtained my first encouragements from the Academies of Berlin and Torino, that is from Lejeune Dirichlet, from Jacobi, from Alexander Humboldt and from Plana, who honored me, when they lived, with their friendship and whose memoirs are always dear to me. I will feel a similar gratitude for every new favor your illustrious Academy will show me. The most ardent patriotism does not dispense with personal recognition.

[Neuenschwander 1984a, I, 17]

Recall that Liouville had already been elected corresponding member of the Berlin Academy in 1839. Upon receiving this acceptance, Weierstrass wrote an official "Wahlvorschlag," which was subsequently signed by himself, Borchardt, Kronecker, and Kirchhoff. Although it has been published by Biermann [1960], I shall quote it in full, because it throws light on the choice of Liouville and Chasles, and in particular because it offers the most competent contemporary appraisal of Liouville's work:

Berlin, January 16, 1876

At present, France has two analysts of the first rank, Liouville and Hermite. The prominent scientific position they occupy in their own country and the general recognition their contributions have found among their foreign colleagues would in our opinion completely justify that the Academy simultaneously include as foreign members both these men, who have for a long time been corresponding members — Liouville from 1839, Hermite from 1859. Yet, we limit ourselves to make petition for the first of them, partly because he is much older than Hermite and partly for the reason that we thought that in order to complete the list of foreign members, which since Riemann's death has not contained any mathematicians. We ought to suggest an important and illustrious representative of the Geometry in addition to an important analyst.

Liouville's scientific works are so numerous and so extensive that we must renounce recording them all and discussing them in detail. However, the subsequent remarks will suffice to give an idea of the indefatigable and successful activities of this man, even to those Herrn Colleagues who are not experts.

Liouville, who was born in the year 1809, began his career as a writer when he was 23 years old. Already his first works, which appeared in 1832–1837 in the *Journal de l'école polytechnique*, dealing primarily with problems of mechanics and mathematical physics in addition to a series of analytical questions, found so much recognition that at the age of 24 he obtained a professorship at the polytechnical school and soon thereafter at the *Collège de France*. In 1839, soon after he had been elected correspondent to our Academy on the recommendation of Dirichlet, the doors of the French Institute opened for the hardly thirty-year-old man, who immediately afterwards joined the members of the *Bureau des Longitudes*.

From 1837 Liouville published almost exclusively in the mathematical Journal which he had founded and which has only recently passed into other hands. This Journal included almost all parts of mathematics: the algebra, the theory of functions, the application of analysis to geometry, mechanics, physics, and number theory. Hardly any important question has moved the mathematical world during his long period of activity in which he has not been engaged with sagacity and not infrequently with great success. This manysidedness of the work was in Liouville's case not officiousness but a result of his extensive erudition in all parts of our science. From the long series of papers that he published in the first twenty volumes of his Journal, we shall in particular emphasize the note on the posthumous works of his fellow countryman, Galois, who died young, his research on the theory of elimination, on differentials with fractional index, on the general principles, found in connection with his friend Sturm, concerning the possibility of developing arbitrary functions in series of integrals of certain linear differential equations, which are of interest in problems in mathematical physics, and finally a number of smaller, but highly interesting and very elegant geometrical works dealing with the theory of surfaces and space curves.

In particular we must mention a very important work by Liouville, which has had a curious fate, namely his theory of elliptic functions in which all the properties of these transcendentals are derived from their double periodicity. This theory of which the essentials are completely developed [by Liouville] is distinguished by the originality of the basic ideas. For reasons we do not know, Liouville has not published any of his research on this matter but he has given detailed talks on it at the *Collège de France*, so that two of his students, the authors of a well-known textbook, could earn the gratitude of having passed on to the mathematical literature the essential content of this important work — a merit which by the way would not have become smaller if the authors of the mentioned book, instead of just mentioning in the preface that they owe much to their teacher Liouville, had according to the truth recognized that the entire basis of their book was the work of Liouville. That this is in fact the case can be seen in an edition of the Liouvillian theory written by one of our colleagues almost thirty years ago from lectures which Liouville gave to him and to another German mathematician, and which according to Liouville's explicit wishes have been communicated to Jacobi, Dirichlet and several other colleagues.

During the last 18 years, Liouville has almost exclusively published number-theoretical studies. These have not obtained such an unanimous approval as his earlier works. The reason for that is no doubt partly the piecemeal publication, which greatly impedes the understanding and the coherence of this research and partly also the fact that, next to the essential, he has also given much less essential material. However, these works by Liouville also offer to the attentive reader a mass of important results and fine methods.

[Weierstrass 1876 in Biermann 1960, pp. 48–49]

Borchardt's publication four years later of the notes from Liouville's lectures on doubly periodic functions may be seen as a way of supporting Weierstrass's (too) harsh remarks about the work of Briot and Bouquet.

After this masterly report, it is not surprising that the Berlin Academy elected Liouville foreign member, together with Chasles, at its meeting on March 15, 1876.

In return, the Paris *Académie* elected Borchardt correspondent on April 3, despite a strong anti-Prussian campaign against his candidacy in all the republican newspapers. Jongmans has shed light over this fishy episode in his interesting paper [1986], indicating that the election of Borchardt was part of a reciprocal arrangement between the Paris and Berlin academies. Chasles was the most active proponent of Borchardt, whereas Le Verrier supported Catalan, repairing thereby their ancient friendship that had been broken for many years. The friends of Catalan at first thought that Liouville was on his side [Jongmans 1986, p. 11], but discovered that he sided with Chasles. The newly elected deputy, Henri Liouville, tried to arouse the anti-German feelings in his uncle, and at Catalan's request, Ernest Picard tried in the same vein, but without success; he answered Catalan on April 6:

> Mr. Liouville makes his decision for reasons on which we have little influence, and I have often heard him speak of you in very affectionate terms; thus, there is nothing to win in this respect, everything possible having been obtained.
>
> [Jongmans 1986, p. 597]

During his lifetime, Liouville became a member of a great number of other academies, such as the Royal Society and the academies of Bologna, Brussells, Copenhagen, Geneva, Göttingen, Lille, Madrid, Napoli, Philadelphia, St. Petersburg, Stockholm, Toulouse, Turin, and Vienna. In 1872, he was nominated Extraordinary Commander of the Spanish Order of Charles III [Neuenschwander 1984a, II, 24], and on August 4, 1875, the French Republic raised him to the rank of *Commandeur de la Légion d'Honneur*, an honor limited to 1000 persons. However, in spite of all these distinctions, his last years became miserable.

The Last Courses

The dramatically abbreviated course of 1870–1871 was the last one that Liouville gave in its full extent at the *Faculté des Sciences*. The following year, his failing health forced him to ask Bouquet to replace him from May 1872 [Ms 3635 (13)]. During the summer of 1872, he wrote to Bouquet that he had decided to teach again the following year [Ms 3635 (13)], and he sat down to plan the lectures: "One will see; for I must give the second part of my course in 1872–1873" [Ms 3636 (10)].

However, his health did not allow him to resume his course, and from 1872–1873 to his death he was supplemented first by Darboux and later by Tisserand. Taton [1973] writes that Darboux supplemented him from 1874, but the note "Darboux is finally nominated my substitute for 1872–1873 at the Sorbonne" on the cover of Ms 3636 (16) shows that Darboux started earlier. When Darboux's salary of 1500 francs had been deducted, Liouville still made 6500 francs annually as a professor, a handsome amount indeed for a job that involved no work. As a comparison, he received 10,000 francs from the *Collège de France* and 5000 francs from the *Bureau des Longitudes*. Moreover, he sold wine from his vineyard in Toul (this brought him more than 5000 francs in 1859 [Ms 3624 (5), p. 1]), and his notebooks bear witness to certain financial transactions that may also have yielded an additional income. Thus, there is no reason to believe that Liouville had to fight financial troubles during his final years.

At the *Collège de France*, he continued his teaching with some interruptions until his death. But, from 1870, advancing illness forced him to cancel numerous lectures. Certain semesters he only taught about half the lectures, but, according to Laboulaye [1882], his urge to instruct and talk with young people kept him going. Laurent also recalled:

> Until his last moment, he continued to teach his course at the *Collège de France*. I assiduously followed his lectures, and seeing him very tired, I asked him if he had not considered being replaced. "In that case," he then answered, "I should be very ill."

In the same vein, Henri Lebesgue [1922] wrote:

> One of our attendants still recalls Liouville in his last years as a professor. He was an old man, whom it was necessary to guide and who by a new miracle at every lecture regained all his quick-wittedness and youthful appearance when he was at the blackboard.

Still, during the first semester of 1879–1880, his health had deteriorated so much that he had Ossian Bonnet replace him. The following two semesters Liouville resumed his duties but on May 8, 1881, he wrote to the Administrator of the *Collège* to apply for leave because "my health no longer permits me to teach my lectures of the second semester" [Coll. Fr. Dos. Pers. Liouville]. At this moment, he was so ill that he had somebody else write the letter. It was signed by Liouville in a trembling hand. Liouville obtained this leave without a substitute, although the Minister asked the Administrator of the *Collège* if it would not be better if a substitute took over the course [Coll. Fr. Dos. Pers. Liouville]. This allowed Liouville to cash his entire salary. In December 1881, Liouville was still in bad health, and the Minister accorded him another semester's leave [Coll. Fr. Dos. Pers. Liouville]. He resumed his teaching during the second semester of 1881–1882, but only gave five lectures [P. V. Coll. Fr., July 30, 1882]. This course, on *le calcul intégral et en particulier les intégrales définies*, became his last.

We have seen that in the middle 1850s Liouville, as well as his colleagues, had only a handful of students for their lectures on higher mathematics.

This number even decreased. For example, during the second semester of 1876–1877, he remarked under the thirteenth lecture:

> Lecture, very badly taught by an ill Professor in the presence only of Amiot.
>
> [Ms 3637 (10)]

Other lectures were also attended by Roger Liouville and Réalis, and at the end of the course Liouville noted:

> I owe a debt of gratitude to my friends Amiot and Réalis, who have supported me until the end. Thanks to them I have been able to complete my two semesters.

The following year, Réalis was often the only who listened to Liouville, and at the eighth lecture, he promised to be there the next two times in order that the lecture could continue. In an entry of October 29 from his diary of 1876, to which we shall return shortly, Liouville wrote:

> I have been told that Lallemand and Roger Liouville will follow my course. Fine.
>
> [Ms 3637 (1), p. 35v]

Roger Liouville (1856–1930) was, according to the *Grande Encyclopédie*, a cousin of Henri Liouville and thus a relative of Joseph Liouville. He later became a professor at the *École Polytechnique* and a distinguished mathematician. His works on integration of differential equations, in particular in finite form, and on geodesics have some similarities with those of his more famous relative (cf. the biography written by Paul Lévy [1931]). If Roger Liouville attended Joseph Liouville's lectures, he was among the few of the boom in French mathematicians of the last half of the 1870s who did. Liouville did not attract the new, young generation because he had no new ideas to offer, and although he livened up in front of the blackboard, there is no doubt that his teaching had deteriorated. For this reason, he had lost most of his students at the *Faculté des Sciences* before leaving the course to Darboux. Thus, in a letter to Houël dated February 13, Darboux wrote:

> And your course? Do you have zealous students? My students or rather those that Liouville has bequeathed to me are few in number. Alas! There was nobody attending his course anymore; everybody took measures to learn (or not learn) the mechanics without him.
>
> [Guispert 1987]

This testimony seems to contradict that of Laurent, who recalled that even late in life Liouville could still "set the important points of a question into relief" and still taught his students how to search for the "truths which he explained with remarkable lucidity" [Laurent 1895, p. 131].

If we are to trust Laurent at all, he is probably writing about an earlier period, or he had in mind the private discussions that Liouville mastered much longer according to the following phrase in a letter of 1874 from Mittag-Leffler to Hjalmar Holmgren:

> During personal talks, he [Liouville] was highly interesting, and it was impossible to discover any weakness of old age. However, he will never discuss mathematics except in great generality.
>
> [Mittag-Leffler letter]

Domestic Life

Thus, when he was well, Liouville maintained his conversational powers, and they were put to use not only when he discussed mathematics with his colleagues but also when he discussed politics with Henri Liouville and Emil Picard and philosophy with his son-in-law, Célestin de Blignières (1823–1905). The latter was a polytechnician and a student of Auguste Comte who had created his own version of the positivistic philosophy (cf. [Blignières 1857]). In a footnote in [Galois 1906, p. 226], Tannery describes the relationship between Liouville and Blignières:

> During the nine years (1874–1882), a very active intercourse took place between Liouville and Mr. de Blignières. Of this intercourse, which both enjoyed very much, Mr. de Blignières kept a singularly lively and present recollection until he died.

To judge from his notebooks, Liouville continued to have a wide circle of acquaintances, both in Paris — mostly scientists –and in Toul — lawyers, military persons, civil servants, etc., well into the 1870s. However, the circle shrunk. We have seen that in the 1850s Liouville began to fill his notebooks with dates of deaths; as he grew older, his friends passed away in increasing numbers, and Liouville even began to note anniversaries of the deaths of his best friends. He certainly made new friends, in particular through family ties, but they could not replace his lost friends. The following letter to his very good friend and a relative of his wife, Odile Chenneval, from Commercy conveys this feeling; it is undated, but was written during 1872 when Faye was the president of the *Académie des Sciences*:

> My dear Odile,
>
> A long time ago, when I offered you the *Comptes rendus*, it was by way of Mr. Arago [?]. Today, when Mr. Arago is dead and I am no longer anything, it is much more difficult for me to be useful to you. Yet I have used my influence with the president, Mr. Faye, and with the *secrétaire perpétuel*, Mr. Elie de Beaumont, and I have heard that the administrative committee has decided to grant you the *Comptes rendus* from now on.
>
> [Ms 3636 (10)]

Sometime between 1871 and 1875 [Neuenschwander 1984a, I, 17] Liouville moved to 6 rue de Savoi, close to his former apartment in rue de Condé but nearer to the Seine (did the bombs hit his former apartment?), but he seems to have felt increasingly tied to his house in Toul where he continued to spend the summers. There he had a private library, which according to the following note from 1866 must have been rather large:

Works sent to Toul in April 1866:
Mems. of St Petersburgh and London - Guizot, 7 vols, Conn. des
Temps 6 vols - [?] - Lalande astronomie 3 vols, Journ. de Mat. 1865
- Seneca 7 vols - Various novels and scientific memoirs - Societé des
Sciences du Luxembourgh - Sauri 5 vols - [?]

[Ms 3632 (16)]

However, as the following note indicates, Liouville did not always have
the opportunity to study in solitude while at Toul:

Left Paris on Thursday evening (July 5, 1877); we came to Toul
on Friday the 6th in the morning, Louise and I with our servant
Charles and a new maid Marie. The same Friday around 2 o'clock
Ernest (Councillor at the Court of Nancy) hastened to come and see
us. He has been called to Nancy the following Monday, then he came
to Toul to spend his vacation. We also had Madame Céline Barbey
and her daughter Thérèse here.

[Ms 3637 (10)]

Louise and Céline were two of his daughters. Although these months
in the bosom of the family provided Liouville some consolation for the
hardships during his last years, they were far from being happy. This was
revealed in a diary he kept during the summer and autumn of 1876 [Ms
3737 (1), pp. 17r–37r]. Despite the monotony of these notes, I shall give a
rather full account of them because they give the best idea of Liouville's
family life as a whole and his painful old age in particular.

The Stay in Toul, Summer and Autumn 1876

We leave Paris headed for Toul on Monday June 19, 1876... arrive
at Toul on Tuesday 20.

Thus begin the notes that continued until November 12 with a daily
entry. On days of no particular interest, the notes are brief, as exemplified
in the following sequence:

29° Monday, July 17. I get up very ill; but in general the day is not
as bad as I feared. Nevertheless, I am suffering greatly when I go to
bed and I cannot sleep [?] the night.
30° Tuesday, July 18. 1876. Still very ill.
31° Wednesday, July 19. 1876. 32° Thursday 20, Ill.
33° Friday, July 21, 1876. Eternal sufferings.
34° Saturday, July 22, 1876. Ah! death, what a relief.
35° Sunday, July 23, 1876. My suffering increases every day.
36° Monday, July 24, 1876. Still more and more ill. Besides, stormy
weather provoking nervous pains.
37° Tuesday, July 25, 1876. Bitter cold this morning.
38° Wednesday, July 26. Nice weather, suffering. I find in Cousin,
on Kant, this quote which Kant took from the Bible "Our life lasts

70 years, at most 80, and the best is but fatigue and work." As a whole a tolerable day.

39° Thursday, July 27, 1876. I have had a very bad night. Yet, when I get up, I find the weather very fine. Deceitful omen, severe pains.

40° Friday, July 28, 1876. I get out of my bed completely exhausted and I find very fine weather. In all, a tolerable day although very tiring.

[Ms 3637, (1) p. 20r,v]

This is depressing reading and there is no doubt that Liouville was extremely depressed by these eternal pains. He had hoped that after the attacks of gout during the winter his health would improve, but on September 7, he admitted "I do not expect anything from this stay in Toul as far as my health is concerned." Indeed, his health, about which we find daily bulletins, goes from "very ill," "I suffer too much," "horrible day," "cruel pains," and "I have never suffered more" to days where he simply "suffers as usual" intercepted by occasional "tolerable" days with less suffering, such as:

Friday, July 14, 1876. When I got up I felt ill but much less than yesterday. And that continued until the evening. God is good

He usually slept extremely badly and therefore had a bad start to his days. His suffering completely overshadowed everything else in Liouville's life. He hardly wrote a letter without alluding to his illness. For example, congratulating Odile Chenneval on the occasion of his daughter's engagement to Mr. Brion, he stressed that his best wishes for their "happiness in this hard and difficult time" would probably be heard:

for God loves the youth. He favors it and reserves his hardships and his rheumatoid arthritis for the old ones like us, who are only good for suffering and dying. However, one summons up one's courage and tries to suffer and to live as long as one can; from time to time one still has a little pleasure. I would have had some if I had been able to go and see you at Commercy, but the gout holds me back.

[July 2, 1876, Ms 3637 (1), p. 18r]

Also, when the secretary of the *Faculté des Sciences* asked him to send the curriculum for his course, Liouville explained that the answer was delayed because he had had a violent attack of gout; he continued his letter of July 7, 1876:

Thus I cannot come to the meeting at the *faculté* summoned for tomorrow at $4\frac{1}{4}$. You can imagine that I dare not contemplate teaching my course. You know that this was my ambition, and I have spent all the spare time that I have had here preparing its elements. As before, I would have announced on the poster of the 1st semester that I would treat "the composition of forces and general facts concerning equilibrium and motion." However, this ancient, and at the same time sudden, illness leaves me no choice. I therefore ask you, Monsieur the Dean, to keep my substitute, Mr. Darboux (if he agrees), or else give me another...

[Ms 3637 (1), pp. 18v,19r]

During the first weeks in Toul, the illness confined Liouville to his house, but on Sunday August 6, he reported: "He [Michon] will come tomorrow at 7:30 to make me go for a walk." The following day was dramatic:

50° Monday, August 7, 1876. I wait for Michon, and although I am very tired after a very bad night, we will go where he wants. In fact, we leave at 7.30 and go for a [?] walk until 9 o'clock, passing by the station. Rather nice weather and very hot. I do not expect anything good to come out of this walk. And it turned out that later in the day Louise fell from a very high ladder in the dining room. She became quite confused. Ah!! what a stay we have had in Toul this year.

51° Tuesday, August 8. For sure my walk yesterday did me no good. However, one must be particularly worried for Louise; her condition has grown worse since yesterday, as one would expect. She really needs a week before she [?] Alas, Alas - This evening, a bit better.

[Ms 3637 (1), p. 22r]

Later during his stay he often went for a walk of a length dictated by his health. Otherwise, his day was mostly devoted to reading his newspaper, the *Journal des Débats*, and corresponding with Paris acquaintances mostly about the salary he received from the *Faculté*, the *Collège*, and the *Académie*. Moreover, he continued to keep track of the deaths of everybody, from Mr. Claude, *député* de Meurthe et Moselle to "Madame Bastien, the baker." However, his sufferings did not always allow him to indulge into these morbid interests because they inhibited him from attending the funerals. Likewise, he could not attend the commemorative service for the resistance against the Germans:

Saturday, September 23. Sad night! Sad awakening! succumb from fatigue. And the weather is fine. At 11 o'clock a funeral service is celebrated at the Cathedral for the peace of the souls of the victims of the siege of Toul. I received an invitation from the mayor, Edouard Deligny, but I could not go to this ceremonial. Alas! are we condemned to become Prussians?

[Ms 3637 (1), p. 28]

During July and August, Liouville only received a few guests such as Henri Liouville, his daughter, Louise Ruau, and "the old teacher of Joseph Ruau who really made this poor lady suffer." During the first days of September, however, the family flooded 8 rue Salvateur:

Friday, September 1: Still bad weather, cruel suffering — Ernest came yesterday in the middle of the day. Sophie, Picard &c also arrived in Toul. So, here is the family [?], and Thérèse will come today from their [?] My God, give me a little peace and tranquility. I need it

badly. However, [?] who will come with her daughter. I am really
content, but will I have the strength to endure even these pleasures?
We will see tomorrow morning.

Saturday, September 2. Day of suffering which, however, could
have ended worse.

Sunday, September 3. Picard, Sophie and the two children, André
and Paul, are in Toul with Mad. Barbey, Lucien and Thérèse, with
Lucien, his chief, Mr. Napoli; when we add Henry, deputy of Commercy, and complementing with senator Picard, then Madame Liouville and me and Ernest, we are 12 at the table.

The supper was very good, but the political opinions diverse. I
agree more with the deputy than with the senator. As for my health,
I think it has hardly improved, but at least it has not grown worse
today. Besides, it is the first day of the market; I stayed at home.

[Ms 3637 (1), pp. 25r,v]

September 5 was blessed with "a visit by our very kind Henry, who
invited us for Thursday at 6 o'clock. Accepted with pleasure." However,
when the day of the dinner came, Liouville was less enthusiastic:

It is raining cats and dogs and tonight we must dine with Henry. I
have never found the stay in Toul more annoying than this year.

He returned wet and tired from the dinner, but the following day he felt
better and for the first time he contemplated work:

Today even more rain than yesterday. Let me at least try to work.
— I have not been able to work, but at least when I go to bed I am
well recovered by this period of rain.

[Ms 3637 (1), p. 26v]

On September 11, there were again 12 family members and friends for
dinner at Liouville's, and later during the stay, Liouville's hospitable home
opened its doors for other members of the family and for Emile Mathieu:

Friday, October 20, 1876. Still cruel suffering and no work. During
the day over long visits by Mr. Peaucellier and in particular by Emile
Mathieu.

[Ms 3637 (1), p. 33v]

Liouville discussed a theorem of mechanics with Peaucellier, one of the
few remarks concerning science in these notes. The entries for July contain
only a few formulas concerning definite integrals. However, as we have
seen, from September 8 Liouville planned to work and on September 17 he
succeeded for the first time.

Finally, this evening I go to bed very tired after having worked a
little.

[Ms 3637 (1), p. 27v]

Primarily, he prepared his courses at the *Collège de France*: "I can prepare my course a little, at least as far as the theory of numbers is concerned," he wrote on October 6 [Ms 3637 (1), p. 32r]. Still, on most days, he could not work:

Sunday, October 8, 1876. Still ill.
  Monday, October 9, 1876. No work possible.
  Tuesday, October 10, 1876. No work, still ill.
  :
  :
  Sunday, October 15, 1876. Still a little ill, eternal sufferings, work zero.
  Monday, October 16, 1876. When I get up after a very painful night, I try to work. However it is in vain.
[Ms 3637 (1) pp 32v–33v]

Having longed for departure since October 19, Liouville left Toul with his family on the 25th, arriving in Paris "about 11 o'clock and very tired. When we arrived at least we found the kind Lucien; but the rest is bad." [Ms 3637 (1), p. 35r]. Back in Paris he met with his colleagues at the *Académie* and at the *Bureau des Longitudes*:

Monday, October 30, 1876...At the Institute I met Mr. Darboux who way very polite and I shook his hand in a friendly manner.
  Tuesday, November 8, 1876. I attend a long meeting of the *Bureau des longitudes*, where by the way we are almost all present. It is very nasty weather, and I am completely wet when I reach home.
[Ms 3637 (1) p. 36v]

Longing for Death

As one might expect, this miserable life made Liouville long for death. I have already quoted the diary entry from July 22, 1876: "Ah! death, what a relief." The entry of July 29 ends less dramatically: "Alas. Alas! And deliver us from evil. Amen," but on October 19 he had enough, "Ah my God, deliver me from this life that cannot offer me anything but the unbearable." The following year, after the death of Ernest Picard, Liouville wrote:

It is on April 2, 1876, that my brother-in-law General Balland, died, then on the 22nd of the same month, his mother, Madame la Comtesse de Verteillac. Ah, my God, which visitations have you sent in quick succession to our pour family. Fortunately, for me it is soon my turn. Life has become impossible for me in the middle of all these Jesuits set against me.
[Ms 3639 (15), Neuenschwander 1984a, note 37]

Liouville generally felt that he was the victim of fate and of his surroundings:

Monday, August 21, 1876... After a very bad night, I remain an old ill man as a consequence of all the insults and all the persecutions, but I live.

⋮

Thursday, August 24, 1876... God is good, but there are many wicked people.

[Ms 3637 (1), p. 23r,v]

His self-pity and bitterness grew when he was alone, as in the following entry after his return to Paris:

Thursday October 28, 1876. I did not sleep tonight and when I get up I am too tired to go to Versailles to have dinner with Marie. They tried to make me do that and that finishes me. Therefore, I am alone for dinner. No work tonight [?]. When one crushes me from one side, it is only fair that one hastens to crush me from the other side. This is what my life is composed of. Pillaged, insulted or rather insulted, then pillaged. Ah! death is gentle when it relieves us from such a life! There is no more justice on Earth.

[Ms 3637 (1), p. 35v]

His bitterness was often more concretely directed toward particular persons. His notes from 1877 abound with attacks of the kind:

Briot and Bouquet, unworthy thieves! but Jesuits, and from then on model Jesuits!!!! Ah! What a time are we living in.

[Ms 3640 (1877–1879), note on Γ-fct., p. 5v]

Mr. Briot and Bouquet, vile thieves but very dignified Jesuits. Elected as *thieves* by the *Académie*!!!!!

[Ms 3640 (1877–1879), note on Γ-fct., p. 3v]

It is not surprising to see Liouville attack Briot and Bouquet, who had published his theory of doubly periodic functions and who had also been accused of stealing by Weierstrass. More surprising are assaults such as the following:

Briot, Bouquet unworthy thieves! *Gauthier Villars* and *Bertrand* depressing bastards, Chasles unworthy animal, who I have saved from death and who has killed me!...

Gauthier the vile and the Jesuit, abnormal thief, Bouquet thief, arch Jesuit, Briot thief and Jesuit &c...

Ah! which boors have Picard and Jules Simon handed over to me! The innocent Picard, the Jesuit Simon...; Bertrand, the Jesuit! Helper of the Jesuit Jules Simon...

Ah, Republic of sacrificing and of Jesuits, it is you who is but a name! Those scoundrels have no doubt been my assassins, but they are not the only ones.

[Ms 3640 (1877–1879), messy notes]

Here, Liouville is aiming at several of his former friends. Chasles had been among his closest friends during most of his life, and we have seen that as late as 1876 the two collaborated in getting Borchardt elected member of the *Académie*. However, later in 1879, Liouville also accused Chasles of vanity [Ms 3638 (5), November 29, 1879]. Bertrand, too, used to be one of Liouville's allies and Gauthier-Villars had a fine relationship with Liouville when they collaborated on the publication of the *Journal de Mathématique pures et appliquées*. Thus, on March 22, 1866, Liouville wrote to his republican friend from 1848, the philosopher and political writer Barthélemy Saint-Hilaire:

> My dear Fellow Member,
>
> Allow me (on the occasion of the choice of a librarian that you must make soon) to warmly recommend Monsieur Gauthier-Villars whose house I know very well, for he replaced Mr. Bachelier more than two years ago. Moreover, Monsieur Gauthier-Villars is a former student at the *École polytechnique* who graduated from the school with a very fine rank. He has been in the telegraph service for thirteen years. I think that you will be pleased with him.
>
> [Ms 3632 (16)]

(As far as I know Barthélemy Saint-Hilaire never published anything with Gauthier-Villars.)

With the often repeated pun "Gauthier the Vile," Liouville, during his later years, seems to have turned his back to Gauthier-Villars as he seems to have done on several of his former friends. Yet, other evidence seems to suggest that during the early 1870s the common opposition against the Prussians and against Le Verrier had reconciled Liouville with some of his former adversaries, such as Serret and Faye. Therefore, it is possible that one should not take the critical notes above too seriously. Indeed some of the accusations are really farfetched. Thus, calling the liberal Bertrand a Jesuit supporter is just as much off the mark as accusing the new republic of being run by the Jesuits. If the notes do not reveal real animosities, they show how desperate Liouville had become because of his miserable life.

In 1869, he willed what he owned to his wife [Ms 3633 (3)], but this will was never executed, for as a last blow of fate, Liouville lost his devoted Louise on June 19, 1880, and, he also lost his son Ernest. Concerning these losses and Liouville's last years, Faye recalled in his obituary:

> Some time ago Liouville underwent a bodily collapse, pulled down by the infirmities of age and in particular by the very cruel bereavements of his family. He lost, in a nearly inconceivable catastrophe, a charming and excellent wife, who was his support and guide, and a son, dead at a young age, who was adviser at the Appeal Court of Nancy, whose kind and delicate spirit some privileged friends knew. However, his great intelligence remained intact. He worked until the end; as late as last Tuesday he attended the meeting of the *Bureau*

des Longitudes whose works he followed with the greatest interest. However, he seemed to long for release. This hour suddenly came to him the day before yesterday. He has left us, leaving among us a great emptiness as a traveler who goes ahead of us to where we hope to join him again.

[Faye 1882, p. 469]

In a similar vein, Laboulaye reported:

For some years his health greatly worsened. Gout weakened him, sorrow overcame him. Mr. Liouville had the misfortune of people who live long: he outlived those who had been his support and consolation during his old age. The unexpected loss of his wife and his son was the last blow. From that moment he gradually languished, in spite of the care of a numerous family; he was but a shadow of himself; this year he could not even complete his course.

[Laboulaye 1882, p. 469]

Posthumous Reputation

Liouville died on September 8, 1882, in his home on 6 rue de Savoie at the age of 73. The invitation to his funeral on the 10th was signed by his three daughters and their husbands and children. At the tomb, Faye gave a speech on behalf of the *Académie des Sciences*, the *Faculté des Sciences*, and the *Bureau des Longitudes*, and Laboulaye spoke on behalf of the *Collège de France*. Furthermore, a representative of the *École Polytechnique* honored its former professor [CR, 95, p. 467].

Having several times had occasion to quote from these speeches, I shall here only include the more personal appreciations by Faye, who was the one who had known him longest:

The Astronomy section of the Institute, the *Faculté des Sciences* of Paris, and the *Bureau des Longitudes* have asked me to express their sorrow at this tomb. By losing Mr. Liouville, these three learned institutions have lost one of their oldest and one of their most illustrious members; and I, if you will allow me to add a personal feeling to the expression of this regret, I lose a venerated master, whom I came to know when I was a young man sitting on the seats of the *École*, and who for almost fifty years, has not ceased giving me his encouragement and his support.

Mr. Liouville belonged to a distinguished family of which I have the honor of knowing several members in Lorraine. In their country, the Liouvilles have always been considered as people of honor and of spirit. Paris also knows of eminent men of this family who have come here to make their name known and loved. However, our learned fellow member had added a European fame, for abroad, as in France, he was considered one of the foremost geometers of our era.

[Faye 1882]

Liouville was buried on the *Cimitière du Montparnasse* in *division 13, ligne 5 nord, numéro 20 ouest*, next to his wife, his wife's brother, Joseph Balland, and their mother the Countesse de Vertillac, who had been married to Joseph Liouville's uncle. One can still visit the tomb, but only when the sun sets is it possible to read the text under the moss:

> Joseph Liouville. Membre de l'Institut
> Professeur au Collège de France
> à la Faculté des Sciences
> Membre du Bureau des Longitudes
> Commandeur de la Légion d'Honneur e.t.c.
> Decedé le 8 séptembre 1882 a l'age de 73 ans.

The day after the funeral, when the *Académie des Sciences* met, the president, É. Blanchard, gave a brief speech in honor of Liouville: "one of the oldest members, who was also one of the most brilliant," and concluded: "The learned world is deeply affected by the regrettable loss that it has suffered. Expressing the general sentiment of the *Académie* I close the meeting" [CR 95, p. 469]. This gesture was not a matter of course. For example, the meeting had not been closed when Cauchy died.

After Liouville's death, however, the grief of the scientific community was accentuated at the next meeting of the *Bureau des Longitudes* when its president also decided to "close the meeting in the memory of the death of our colleague Mr. Liouville, a distinguished and honorable man, loved by all his colleagues" [P. V. Bur. Long., September 13, 1882].

After his death, his hometown, Toul, honored him by renaming the rue Salvateur, where he had lived for almost half of his life, rue Liouville. Paris has not given him a similar honor.

Apart from the two obituaries mentioned above, only one biographical sketch of Liouville was written by a person who knew him, namely, Laurent's sketch in the centenary volume from the *École Polytechnique* [Laurent 1895]. It is regrettable that Bertrand did not write down his recollections from his long acquaintance with Liouville. As a secretary of the *Académie*, he was so active writing obituaries of the academicians that they have been published in two separate books [Bertrand 1889 and 1902]. In his obituary of Cordier, Bertrand wrote:

> The most illustrious [members of the *Académie*] have for a long time awaited, some still await, a homage worthy of our regret and of their fame. Dulong, Cauchy, Chevreul, Cordier, Clapeyron, Sturm, Liouville, A de Jussieu, Becquerel, J. B. Dumas, Claude Bernard have not been honored yet.
>
> [Bertrand 1902, p. 59]

In 1888, Chrystal wrote a superficial and rather incorrect obituary, and in 1902–1907 a few biographical details about Liouville were revealed in the *Intermédiaire des mathématiciens* [Brocard 1902–7] as an answer to W. Ahrens' question, no. 2285 [V 9]:

PLATE 23. Liouville's grave at the Montparnasse Cemetery. (Photographed by Jesper Lützen)

> Are there any biographical Notes (academical obituaries, etc.) on Joseph Liouville, except the speeches given at the funeral by Faye and Laboulaye?
>
> [Interm. des Math. 1902, p. 36]

The first major biography of Liouville, however, was written in 1936 by Gino Loria [Loria 1936], and it was not surpassed until Taton's biography in the *Dictionary of Scientific Biography*. But no biography, no street name, no plaque on a house wall is a better monument to this great mathematician than the title page of the *Journal de Mathématiques*, which has been decorated with his name on every issue from its start until this day.

Part II

Mathematical Work

VII

Juvenile Work

Electrodynamics

1. Apart from a brief encounter with synthetic geometry, Liouville did his
first research in electrodynamics. He only published two short notes [1829]
and [1831] on this subject and neither were particularly original or influential. Still, it is interesting to have a closer look at their genesis because it
sheds light on the formation of Liouville as a scientist and illuminates the
origin of his theory of differentiation of arbitrary order.

As we saw in Chapter I, Liouville became interested in electrodynamics
when he followed Ampère's course at the *Collège de France*. He took notes
[AC AL Ms 1], and after Ampère had corrected them, he made a fair
copy [AC AL Ms 2]. On the basis of these notes, he prepared a memoir
that also contained some of his own results [AC AL Ms 3], and finally he
wrote a completely original memoir [AC L Ms 2], which he presented to the
Académie on June 30, 1828. After a somewhat critical report by Maurice,
only a few pages of this memoir appeared [1829] as Liouville's first published
paper. Both this and the second published note on electrodynamics [1831]
dealt with Ampère's force law between conducting elements. In order to
appreciate these notes and Liouville's work in electrodynamics in general,
we need to consider briefly the state of affairs in this discipline.

AMPÈRE'S ELECTRODYNAMICS

2. When Liouville attended Ampère's lectures, electromagnetism was only
six years old. The field had been opened in 1820 by H. C. Ørsted's discovery
of the interaction between a magnet and an electrical current. Before that
the mathematical theory of electricity and magnetism had been limited to
electro- and magnetostatics, where Coulomb, around 1790, had discovered
the elementary laws of force.

According to Coulomb, these forces were due to elementary central forces,
similar to gravitation, acting between two electric and two magnetic fluids
of which the first could move freely in conducting media whereas the latter
could only get displaced giving rise to magnetized molecules. In the Laplacian conception of the world, all physical phenomena could similarly be
reduced to central forces between molecules (cf. [Fox 1974]), so in his and
Poisson's mathematical theories of electrostatics and the latter's theory of
magnetostatics [Poisson 1824], the electric and magnetic fluids continued

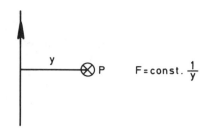

$F = const. \dfrac{1}{y}$

FIGURE 1

to be fundamental. However, it was difficult to adapt the model to Ørsted's experiment.

When Ørsted's experiment became known in Paris, Jean-Baptiste Biot (1774–1862) and Felix Savart (1791–1841) immediately examined the effect quantitatively. Observing the oscillations of a magnetic needle, they found that the force from an infinite rectilinear conductor on a magnetic pole was inversely proportional to the distance between them (Fig. 1).

They informed the *Académie* about this law on October 30, 1820, and on December 18 of the same year, they confirmed Laplace's conjecture that the force on the pole could be considered as being composed of forces from infinitesimal elements of the conductor. They "showed" that this elementary force is proportional to $\frac{\sin \omega}{r^2} ds$, where (Fig. 2) ω is the angle between the conducting element ds and the line connecting it with the magnetic pole, and r is the length of this line. The direction of this elementary force is orthogonal to ds and to the connecting line. This is Biot's and Savart's law.

Biot and Savart deduced their law from measurements of the force between a V-shaped conductor and a magnetic pole as in Figure 3. At first, they found that the force was proportional to the angle i and claimed that this implied their law. However, Ampère's collaborator, Savary, pointed out in 1823 that if $\frac{\sin \omega}{r^2} ds$ is integrated along the wire (Fig. 3), one finds that

FIGURE 2

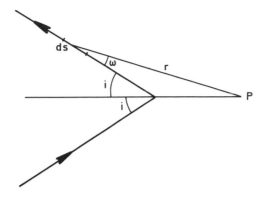

FIGURE 3

the force is proportional to $\tan \frac{i}{2}$ and not to i. Thus, Biot's and Savart's original deduction of their law was fallacious. More accurate measurements later allowed them to confirm that the force was in fact proportional to $\tan \frac{i}{2}$.

With Biot's and Savart's law, the electromagnetic force had also been reduced to an elementary force, not between molecules but between an element of a conductor and a magnetic pole. However, the new elementary force was not a central force, and unlike earlier elementary forces it acted between different "fluids."

3.   The gradual build up of Ampère's new electrodynamics has recently been described in Christine Blondel's excellent thesis [1982], which together with [Knudsen 1980] has been of great help in the composition of this section. In order to overcome the problems raised by Biot and Savart's work, Ampère suggested that the new phenomena should be explained by central forces between two conducting elements. He demonstrated the existence of such an interaction between two currents soon after hearing about Ørsted's experiment and claimed that the interaction between a conductor and a magnet was due to interactions between the conductor and rotating currents inside the magnet.

In 1826, Ampère presented the results of six years of work in his book *Théorie des phénomènes électrodynamiques, uniquement déduit de l'expérience*. Here, he first deduced the law of attraction between two conducting elements. He implicitly assumed that it was a central force, and he found that its numerical value was

$$F = \frac{ii'\, ds\, ds'}{r^2} \left(\sin \theta' \sin \theta \cos \omega - \tfrac{1}{2} \cos \theta \cos \theta' \right) \tag{1}$$

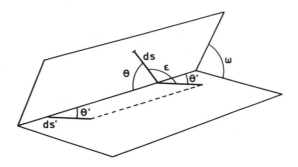

when the two conducting elements ds and ds' carry the currents i and i', θ and θ' are the angles these elements make with the line connecting them, and ω is the angle between the planes containing one of the elements and the connecting line (Fig. 4).

If ε is the angle between the two elements, this can also be written

$$F = \frac{ii'\, ds\, ds'}{r^2} \left(\cos \varepsilon - \tfrac{3}{2} \cos \theta \cos \theta'\right). \tag{2}$$

Having found the elementary force law, he showed by integration that it could account for the laws established experimentally by Coulomb and Biot and Savart for the interaction between two magnets and between a magnet and a conductor, respectively, under the assumption that magnetism is due to rotating electricity.

4. Ampère's elementary force does not agree with modern electrodynamics. In the modern description, the element ds creates a magnetic field equal to $k\, i\frac{d\mathbf{s}\times\mathbf{r}}{\mathbf{r}^3}$, and this impresses the force

$$\overline{F} = kii'\frac{d\mathbf{s} \times (d\mathbf{s} \times \mathbf{r})}{r^3} \tag{3}$$

on ds' (Fig. 5). The force is not directed along $\mathbf{r}$, and its numerical value is not equal to (1). Moreover, it is affected by the reversal of ds and ds',

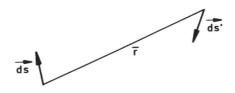

so Newton's third law of action and reaction is not satisfied. It is therefore very different from the elementary force Ampère was looking for.

In fact, the concept of an elementary force acting at a distance does not really make sense in the modern field theory due to Faraday and Maxwell. Clearly, only forces between electric circuits have a physical reality, and indeed, one will find that the action of a closed circuit on an element ds' is the same whether it is found by integrating Ampère's expression (1) or the modern expression (3) around the circuit.

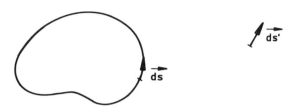

That explains why Ampère's law was corroborated by experiments. However, it also makes clear that Ampère's force law (1) was not deduced solely by experiment or experience as he claimed. It was highly influenced by his a priori conception of an elementary force as a central force that satisfies Newton's third law.

5.   Liouville's contributions to electromagnetism were all closely connected to Ampère's deduction of the force law. Therefore, I shall summarize the deduction as given in Ampère's course of 1826–1827 at the *Collège de France*. This proof, found in Liouville's notes, also sheds new light on Ampère's theory, for two of the four crucial experiments that constitute the basis of the deduction are different from those published in his book in November 1826, i.e., simultaneously with the start of his course.

The first experimental fact used by Ampère in the course is:

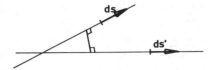

First experimental fact

> 1° Two rectilinear currents exert the one upon the other an action
> which is attractive if both of them point towards or away from their
> common perpendicular. On the contrary they repel each other if one
> of them points towards and the other points away from it.
>
> [AC AL Ms 2]

Moreover, if the current changes its direction in one of the conductors,
the numerical size of the force does not change. From these experimental
facts, Ampère concluded that "an infinitely small portion of electric current
exerts no action on another infinitely small portion of a current running in
a plane perpendicular to and through the midpoint of the first."

The second experimental fact is the following:[1]

Second experimental fact

> 2° The action of a sinuous current is precisely the same as the action
> of a rectilinear current between its end points.
>
> [AC AL Ms2]

Here, a "sinuous current" is only supposed to deviate infinitely little
from the straight line. Ampère obviously did not want to imply that a
closed circuit had no effect.

6.   In order to determine the fundamental force law, Ampère first considered two parallel elements of conductors ds, ds', which are perpendicular
to the line connecting their midpoints.

He argued that the force between them must be proportional to their lengths ds and ds' and to the currents i, i' in the two elements. Moreover, it must be a function of their distance, and Ampère assumed that it is proportional to $\frac{1}{r^n}$. Thus, the force between the two elements is proportional to

$$F = \frac{ii'\,ds\,ds'}{r^n}.$$

Ampère adjusted the unit of current such that the proportionality factor is 1 (he did not use the unit "ampere").

Similarly, Ampère argued that the force between two collinear elements must be

$$F = k\frac{ii'\,ds\,ds'}{r^n},$$

where k is a yet undetermined factor.[2]

To find the force between two elements $ds = MN$ and $ds' = M'N'$ in the same plane (Fig. 6), he drew the line joining their midpoints and dropped the perpendiculars MQ, NP, $M'Q'$, and $N'P'$ onto this line. According to the second fact, the action between ds and ds' is equal to the action between the broken lines $MQPN$ and $M'Q'P'N'$ if they carry the same currents as ds and ds'. But MQ and PN can be united to one element mn perpendicular to QP' through Q, and similarly with $M'Q'$ and $P'N'$. In this way, the interaction is split into four. The action between mn and $Q'P'$ is zero because $Q'P'$ lies in a plane that cuts orthogonally through the midpoint of mn. Similarly, $m'n'$ has no interaction with QP. If the angles made by the conducting elements and their connecting line are θ and θ', respectively, the length of mn and $m'n'$ are $ds\sin\theta$ and $ds'\sin\theta'$, respectively, and since they are orthogonal to the line joining their midpoints, the force between them is

$$\frac{ii'\,ds\,ds'}{r^n}\sin\theta\sin\theta'.$$

Similarly, the force between QP and $Q'P'$ must be

$$k\frac{ii'\,ds\,ds'}{r^n}\cos\theta\cos\theta'.$$

FIGURE 6

The total force between ds and ds' is therefore

$$\frac{ii' \, ds \, ds'}{r^n} (\sin \theta \sin \theta' + k \cos \theta \cos \theta')$$

or

$$\frac{ii' \, ds \, ds'}{r^n} (\cos(\theta' - \theta) + (k - 1) \cos \theta \cos \theta').$$

By a similar decomposition, he concluded that two arbitrarily placed elements attract each other with a force equal to

$$\frac{ii' \, ds \, ds'}{r^n} (\cos \varepsilon + (k - 1) \cos \theta \cos \theta'), \tag{4}$$

where ε is the angle between the two elements (Fig. 4).

7.   The problem is then to determine the constants n and k. To this end, Ampère used two experimental facts:

> 3° A closed circular circuit [Fig. 7] can never produce a continued motion in one direction by acting on a movable conductor of an arbitrary form, but closed, or starting at a point on the axis of the circle and ending at another point on this axis, if the movable conductor can only move by turning around this line. Conversely, the closed conductor cannot move the movable circle around its axis.
>
> [AC AL Ms 2]

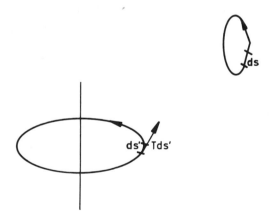

Figure 7

By integration of the general force law, Ampère found that the total tangential force on the circular conductor caused by the arbitrary closed conductor is equal to

$$\int T \, ds \, ds' = ii' \frac{2k + n - 1}{n + 1} \int ds' \int \frac{\cos \varepsilon \cos \theta' \, ds}{r^n},$$

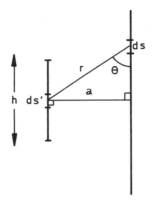

FIGURE 8

where one must integrate around the circle and the other closed contour. According to the third fact, this force must be zero, and since Ampère could find a circuit for which the last integral is not zero, this can only happen if

$$2k + n - 1 = 0.$$

Finally, Ampère determined n, and thereby k, by a fourth experimental fact:

> 4° ...one can easily prove that $n = 2$ by letting a finite current of length h act on a infinite [indefinite] parallel conductor at a distance a...[Fig. 8]. The experiment [or experience] shows that the action depends only on the ratio $\frac{h}{a}$ and remains invariable when one varies h without changing their ratio.

In this case, $\varepsilon = 0$ and $\theta = \theta'$ in (4), so it is easy to find that the total force between the two conductors is

$$-\frac{ii'h}{a^{n-2}} \int_0^\pi \sin^{n-2}\theta (1 + (k-1)\cos^2\theta)\, d\theta,$$

which depends on $\frac{h}{a}$ only, if $n - 1 = 1$, i.e., if $n = 2$. Since $2k + n + 1 = 0$, Ampère concluded that $k = -\frac{1}{2}$. Thereby he established the law of attraction (2) between two conducting elements.

8.   The beginning of the proof and the first two crucial experiments were the same in the lectures as in Ampère's book, but the determination of n and k is simpler in the lectures. In the book, Ampère had not only used the fact that the whole circle, in fact 3°, could not be rotated by an arbitrary closed circuit. He had also used the fact that each element ds' of this circle

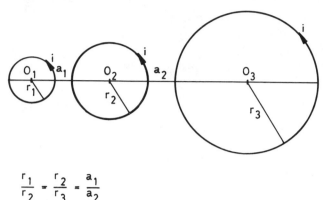

$$\frac{r_1}{r_2} = \frac{r_2}{r_3} = \frac{a_1}{a_2}$$

FIGURE 9

could not be brought to rotate about its axis. Ampère admitted in a note that the instrument designed to demonstrate this fact was neither precise nor convincing. In the lectures, however, he convincingly demonstrated fact 3° in connection with Faraday's experiment with a conductor rotating continuously around a magnet.

Moreover, in the book, Ampère determined n from the fact that if the two extreme of three homologous circular conductors (Fig. 9) are fixed then the middle conductor is in equilibrium.

The book contained a drawing of an apparatus aimed at making this equilibrium situation evident, but in a note, Ampère admitted that he had not yet built the instrument, and it is most unlikely that he ever did. It is clear from the comments to this experiment that in 1826 the deduction of the force law had become more of a mathematical game than a search for the secrets of nature. As such, the replacement of this experiment with the fact about the finite and infinite parallel conductors in the lectures was an improvement, in that the deduction of $n = 2$ became easier. From an experimental point of view, this new approach may also have been an improvement, but there is no indication in the notes that this experiment was carried out either.

LIOUVILLE'S CONTRIBUTIONS

9.   In the above derivation of the force law, Ampère used the fact that the force between two conducting elements is directed along the line connecting their midpoints. In the introduction to the lecture notes, Ampère argued loosely for this fact by remarking that otherwise the couple of conducting elements would begin to rotate, and thus one would get a movement "with-

out the action of an exterior force." Ampère must have known that this argument was invalid, not so much because it neglects the energy loss in the battery, for that had not been well understood at the time, but because a similar argument would render Faraday's experiment impossible.

In the concluding remarks of the lecture notes, one finds a new and more convincing derivation of the central action of the electric forces. This is the raw sketch of the argument which Liouville later published as his own. Since Ampère never contradicted its attribution to Liouville, I think that the proof in the lecture notes is due to the young student. That also explains why it does not occur in the introduction, but slightly out of context in the conclusion, where Ampère discussed whether his force between conductors, or Biot's force between a conductor and a magnet, is the true elementary force. Since this argument is probably Liouville's first contribution to science, I have reproduced this part of the lecture notes in Appendix I.

10. The main idea in the argument is, following Ampère's proof of the force law, to decompose the conducting elements in three orthogonal directions and to use symmetry arguments for each pair of components. The symmetry arguments are based on Ørsted's observation that the action of a rectilinear conductor is symmetric with respect to rotations around the conductor. They easily take care of the interaction between (1) two elements lying on the same line and (2) two parallel elements with a common perpendicular bisector. Liouville then claims that "as we have seen" the interactions between any two elements can be reduced to these two cases. Here, he has used Ampère's result that there is zero action between two elements, of which one lies in the perpendicular bisecting plane of the other (cf. §5), which in turn had been derived from Ampère's first experimental "fact." However, in this fact, Ampère had already talked about attractive and repulsive forces and had thus assumed what was to be proved, namely, that the elementary forces were central.

In the memoir that he presented to the *Académie* [AC L Ms 2], Liouville gave a more detailed version of the above deduction and argued that he had not gone in a vicious circle. In fact, he so persistently rebutted the accusations of a vicious circle that it seems to suggest an awareness of the weakness I have pointed out above. He did admit that someone had objected to Ampère's first fact because it could not be observed directly. Although Liouville did not consider this a serious objection, he still sketched an alternative deduction of the centrality of the elementary force that only built on symmetry arguments. He showed that if both conducting elements are orthogonal to the line connecting their midpoints, the force must be zero, but he admitted that he had found no symmetry argument to ensure that if one element lies on the normal to the midpoint of the other (Fig. 10) then the force perpendicular to II' is zero. Thus, the second proof was incomplete.

FIGURE 10

In their review of Liouville's memoir, Maurice and Ampère wrote that they considered the original proof completely rigorous, and since Liouville did not consider the objection to it serious either, he ought to have left out the second incomplete proof. Therefore, the published version [Liouville 1829] only contained the first proof without any remarks about possible objections. However, Liouville retained the remark about having avoided "going in a vicious circle."

11. I shall now pass to the unpublished novelties in Liouville's memoir [AC L Ms 2]. The first relates to the dependence of r. Already in [AC AL Ms 3], Liouville observed that it was easier to deduce that the elementary force was proportional to $\frac{1}{r^2}$ before determining the dependence of the angles, but he still followed Ampère in assuming a priori that the force is proportional to $\frac{1}{r^n}$ for some n. In his own memoir [AC L Ms 2], however, he only assumed that the force is of the form

$$F = \frac{ii'\, ds\, ds'\, \varphi(\varepsilon, \theta, \theta')}{f(r)},$$

where f is an "arbitrary" function of r. In order to determine f from fact 4° (§7), Liouville developed $\frac{1}{f(r)}$ in a series:

$$\frac{1}{f(r)} = \sum A_j r^{\alpha_j}.$$

By integration, he easily deduced that the force between the finite and the infinite conductors of fact 4° is equal to

$$-\sum A_j\, ii'h\, a^{\alpha_j+1} \int_0^\pi \frac{\varphi(0, \theta, \theta)\, d\theta}{\sin^{\alpha_j+1}\theta}.$$

Since this expression must depend only on $\frac{h}{a}$, he concluded that one of the exponents, α_j, must be -2 and all the rest must be zero. Therefore, $f(r) = r^2$.

Later, Liouville discovered [1832a, pp. 26–27] that this conclusion is only valid if all the integrals $\int_0^\pi \frac{\varphi(0,\theta,\theta)}{\sin^{\alpha_j+1}\theta}$ are different from zero. As long as φ is not known, one cannot say anything to this effect, and when Ampère's expression for φ is introduced, the integral is actually zero for a certain value of $\alpha_j \neq -2$ (cf. Chapter VIII, §8)). This error was not discovered by the examining committee, but still they did not consider the generalization of Ampère's proof as a great improvement:

> We think, for more than one reason, which it is unnecessary to explain, that this greater generality given to the well known proof can not be considered as an important improvement of the science. Admittedly the procedure of the calculations employed is without doubt conducted with discernment but in themselves they present nothing new.
>
> [Maurice 1828, p. 119]

However, this proof was probably the starting point for something new, namely, Liouville's theory of differentiation of arbitrary order (cf. Chapter VIII, §8 (10)).

12.   The second unpublished novelty in Liouville's memoir was an alternative derivation of Biot and Savart's law. In the lectures, Ampère had rather scornfully pointed out that Biot, his main opponent but also his fellow professor at the *Collège de France*, had originally given an erroneous derivation of the law. Ampère "forgot" to say that Biot and Savart had later improved the derivation. Still, one of his central arguments for the identity of a magnet and a coil consisted of showing that a coil satisfied Biot and Savart's law.

Liouville obviously felt uneasy about basing the new model of magnetism on a formula that had been derived in a fallacious way, for having recorded Savary's counterargument in the lecture notes, he left three pages empty, clearly intended for the true derivation. He did not supply the derivation in the notes, but in his paper, written in Ampère's name [AC AL Ms 3], he gave a new proof of the law, despite the fact that by then he knew about Biot and Savart's improved derivation. He introduced his new proof with the words:

> However, we still do not know how Biot, by using the calculus to analyze a false experiment, could be led to a correct result, and one that contradicts the result that ought to have been deduced from his experiment. Be that as it may, the procedure he finally used is susceptible to a high degree of precision and serves as a confirmation of our theory.
>
> One can verify it even more completely by trying to determine, in every detail, the law with which a magnetic pole influences a conducting wire. This is precisely what has been done by Mr. Liouville by generalizing the ideas that led him to determining the direction of the interaction between two voltaic elements, and by imitating, on the other hand, the way in which I have determined the constants n and k, which enter in the formula, by the consideration of equilibrium situations.
>
> [AC AL Ms 3]

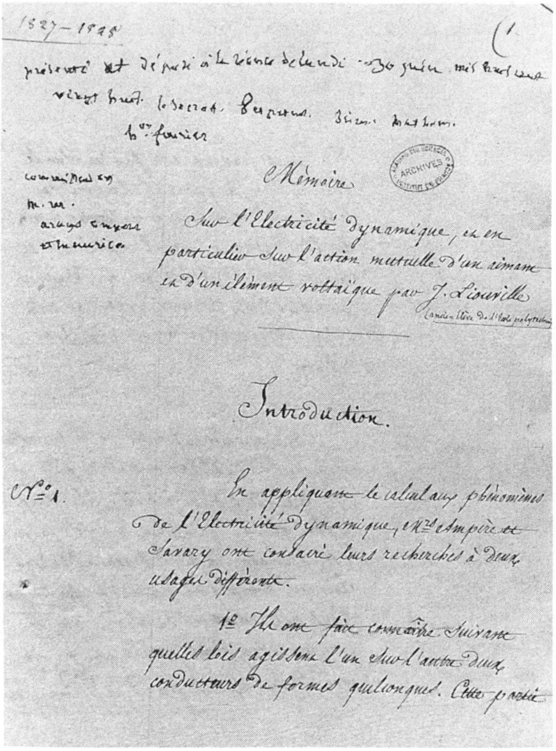

Mémoire

Sur l'Electricité dynamique, et en particulier sur l'action mutuelle d'un aimant et d'un élément voltaïque par J. Liouville

Introduction

N° 1. En appliquant le calcul aux phénomènes de l'Electricité dynamique, M.rs Ampère et Savary ont consacré leurs recherches à deux usages différents.

1° Ils ont fait connaître suivant quelles lois agissent l'un sur l'autre deux conducteurs de formes quelconques. Cette partie

PLATE 24. Liouville's first Memoir. The trembling hand at the top of the first page belongs to the Secrétaire Perpetuel Fourier. It testifies that the Memoir was presented to the Académie des Sciences on June 30, 1828. On the page with the figures an alignment of small circles represents a coil or equivalently (according to Ampère) a magnet. (Archives des l'Académie des Sciences de Paris. File of meeting June 30, 1828.)

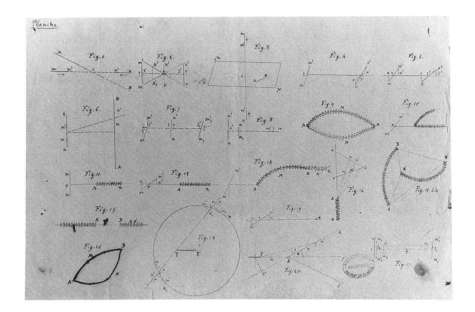

PLATE 24. *continued.*

13.   In his memoir [AC L Ms 2], Liouville elaborated the argument. By symmetry arguments, he first proved what Biot had taken for granted, that the force "is perpendicular to the plane section which has the element as its basis and the pole as its vertex" and that its point of application is the midpoint of the element. To this end, he used the facts:

1° The force of a magnet only depends on the position of its poles, but is independent of the shape of the magnet.

2° The direction of the force is reversed when the direction of the current is reversed.

3° A sinuous conductor can replace a straight conductor with the same end points.

Liouville then argued that the force must be of the form

$$F(\omega)\,(Ar^n + Br^m + \cdots)\,ds$$

(cf. Fig. 2) and found $F(\omega) = \sin\omega$ and $(Ar^n + Br^m + \cdots) = k\frac{1}{r^2}$ from two other experimental facts, namely,

1° Faraday's experiment, and 2° "a circle cannot turn around its axis under the action of a magnet having its two poles on this axis." This new derivation was not received favorably by the examining committee, who considered Biot's improved derivation sufficiently rigorous:

> But, here as above, Mr. Liouville seems to us to use overlong arguments to derive the determination of the force exerted by a magnet on an element of a conducting wire from other experiments, and we do not think that his method here is basically preferable to the known methods nor that it can substitute for them in education with any advantage.
>
> [Maurice 1828]

14.   The last novelty in Liouville's memoir was a new deduction of the fact that coils act like magnets. Liouville based his argument on the experimental fact that a closed coil has no effect on a conductor. This was heavily criticized by Maurice in his report, for, as he pointed out, the action of a coil can be determined from the elementary force between conductors alone, and therefore, it is a logical imperfection to introduce a new superfluous experimental fact into the theory.

However, in Liouville's defense, it must be mentioned that the action of a coil is more easily determined when the new experimental fact is taken into account. Thus, it follows from this fact that the action of a coil only depends on its end points and therefore the action of a semi-infinite coil AB on a conducting element DE is equal to the action of the rectilinear coil AC lying along the line from the midpoint of the conductor through the end point of the coil (Fig. 11). The integration involved in the determination of the latter force is easier than that involved in the former. Moreover, Liouville showed that his new experimental fact yielded a new derivation of Ampère's law of forces.

15.   In *Note sur l'électro-dynamique* [1831], Liouville pointed out that since conductors are necessarily closed, it is in principle impossible to determine with certainty the elementary forces between conducting elements:

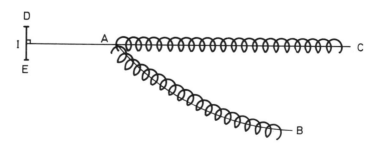

FIGURE 11

In fact, if one puts two conductors face to face in practice the movable conductor may be arbitrary, but the conductor acting upon it is always closed or susceptible of being considered closed....

Since the acting conductor is closed, we see that if there are elementary forces between two elements whose integral disappears when one of the conductors is followed our experiments can never show these forces. The practical results will be the same as if they were zero.

[Liouville 1831, pp. 29–30]

As an example of a force that has zero effect when integrated along a closed circuit, Liouville mentioned the Coulomb expression of the force between two magnetic molecules (dipoles). In the comparison (§4) between Ampère's and the "modern" expression of the elementary force, I mentioned that the difference between these two expressions is another example of such a force.

A modern scientist would immediately draw the conclusion that the deduction of Ampère's force law including Liouville's proof of its central action must therefore be invalid. However, Liouville was careful not to hurt Ampère:

My remarks do not tend to attack Mr. Ampère's beautiful theory. On the contrary, they throw new light on its validity by characterizing its nature better.

[Liouville 1831, p. 30]

Therefore, he expressed himself more vaguely:

The mathematical theory of electro-dynamical phenomena is based on general facts considered as being given by observation, and from which one deduces the direction and the intensity of the force with which two Voltaic elements act on each other. The formula expressing this force was given a long time ago by Mr. Ampère; as for the direction I have later proved that it follows the straight line connecting the midpoints of the two infinitely small conductors. All these consequences have the certainty of the general facts from which they are derived; and since they are very important, the way in which the experience indicates them must be strictly examined.

[Liouville 1831, p. 29]

After the examination, Liouville still believed that Ampère's formula was the most probable expression for the elementary force, and he never explicitly mentioned the possibility of a noncentral action:

However, it is best to observe that a systematic theory, where the properties of conducting wires are explained by primitive forces from atom to atom or by vibrations of a fluid, need not be rejected if the formula deduced for this system differs from Mr. Ampère's formula. It is only necessary and sufficient that the two coincide when the moving conductor is arbitrary and the one acting on it is a continuous circuit.

[Liouville 1831, pp. 30–31]

Later continental mathematical physicists worked according to this concluding remark, probably without knowing about Liouville's clear statement of it.

Liouville also composed another short manuscript on closed circuits, *Action d'un courant fermé sur un élément de courant* [AC AL Ms 5], dealing with Ampère's *directrice*, which corresponds to the modern notion of the magnetic field due to the circuit. The manuscript reveals that Liouville had a good grasp of this part of Ampère's theory as well, but it contains nothing new.

16.   In conclusion, Liouville's papers on electrodynamics and in particular his memoir [AC L Ms 2] were rather long-winded and often rather confused in their structure. Still, they bear witness to an able mathematician with a strong analytical ability and with several bright ideas, two of which his contemporaries found worth publishing. The first of them contained a circular argument, and in the second paper, Liouville almost admitted that the result of the first was uncertain.

In retrospect, the importance of these works was to point out the importance of integral equations to Liouville. In fact, the central idea in Ampère's deduction of the force law is to find this infinitesimal elementary force from its integrated consequences, i.e., by solving an integral equation. However, in Ampère's deduction, the mathematical technique of solving integral equations was completely avoided, and in his criticism of Biot in his lectures, he not only did not point out the incorrect deduction of his law but even seems to have suggested that such a deduction was impossible:

> It is very strange that this expression [Biot's and Savart's law] has been found by analyzing, using the calculus, the results of an experiment such as the one we have just described [cf. §2], since that presupposes that it is possible from a definite integral to deduce its differential by way of the calculus.
>
> [AC AL Ms 2]

Liouville repeated this mocking statement in both [AC AL Ms 3] and [AC L Ms 2]. However, later he came to realize that the solution of integral equations described so accurately here was indeed the fundamental mathematical process in the determination of the elementary laws of nature. More specifically, he soon realized that his determination of $f(r)$ in the force law (§11) was in fact a solution of an integral equation. I shall argue in the next chapter that it was a study of integral equations connected with the determination of the elementary forces of nature, and the determination of $f(r)$ in (23) in particular, that led Liouville to his first important original theory: that of differentiation of arbitrary order.

Theory of Heat

17.   Like most of his contemporaries, the young Liouville also occupied himself with the theory of heat, a slightly earlier triumph of mathematical physics in the 19th century. Following Fourier, and in particular the Laplacian Poisson, Liouville presented three memoirs on this subject to the *Académie*: *Théorie analytique de la chaleur* (June 29, 1829), *Recherches sur la théorie physico-mathematique de la chaleur* (Prize essay February 15, 1830, and again August 16, 1830), and *Mémoire sur les questions primordiales de la théorie de la chaleur* (November 2, 1830). The two latter are still extant [AC L Ms 4 and 5]. In Chapter I, I described their presentation to and reception by the *Académie*; here, we shall have a closer look at their mathematical and physical contents.

The three memoirs deal with two distinct problems. The first concerns the solution of the heat equation for a homogeneous unequally polished bar. This is dealt with at the beginning of the first two memoirs, and the treatment is perfected in a paper [Liouville 1830] printed in Gergonne's *Annales*. The second is a typical problem of Laplacian physics (cf. Chapter I), namely, to determine a molecular model of heat conduction. Liouville discussed it from two points of view at the end of the first two memoirs and extended both approaches in the last memoir mentioned above.

18.   In the paper published in Gergonne's *Annales*, Liouville solved the heat equation for a one-dimensional bar:

$$g(x)\frac{\partial u}{\partial t} = \frac{\partial(k(x)\frac{\partial u}{\partial x})}{\partial x} - l(x)u, \qquad (5)$$

given first-order linear boundary conditions at its end points. Since his method is a precursor of his later work on Sturm-Liouville theory, I shall postpone a discussion of it to Chapter X. Here, I shall just mention that it introduced the method of successive approximations as a means of analyzing solutions of differential equations.

In the third paragraph of the memoir [AC L Ms 4], Liouville further applied his methods to heat conduction in a homogeneous sphere when the external temperature and the conductivity of the surface are functions of the latitude ω only. He limited his investigations to a thin shell close to the surface and assumed that the temperature varied linearly with the radius in this region. Thus, the problem was reduced to a one-dimensional problem like the previous region. This part of Liouville's memoir was rejected by Gergonne, as well as the last paragraph dealing with the physical, molecular theory of heat conduction. The rest of this section will be devoted to the latter subject.

Laplace, Fourier and Poisson on the heat equation

19.   The question of how heat is conducted had been raised by Laplace in connection with Fourier's derivation of the heat equation. In the basic paper submitted to the *Académie des Sciences* in 1807, Fourier had based his derivation on the idea that a molecule, or an infinitesimal thin slice of a bar, exchanges heat only with the neighboring molecules or slices with which it is in contact. If we neglect the heat radiation from the surface, Fourier's argument ran as follows [Grattan-Guinness 1972].

Consider a metallic insulated bar AB of cross section ω, conductivity K, and specific heat[3] per unit volume C (Fig. 12) and assume that the temperature $u(x)$ is constant on each plane orthogonal to the x axis AB. Divide the bar by such planes in slices of infinitesimal thickness dx and consider three consecutive slices with the abscissae x, x', x'' and the temperature u, u', u''. Fourier then used Newton's law, which states that the heat flow between two bodies is proportional to the difference in their temperature. Therefore, the slice x' receives in a time interval dt from the slice x a quantity of heat proportional to $(u - u')dt$ or to $-du\,dt$. However, in order to avoid problems with the dimensions, Fourier argued that the heat flow is proportional to $-\frac{du}{dx}$:

> The quantity of heat which in an instant $\delta t[dt]$ flows through an arbitrary section of the first slice is equal to the product of the conductivity K, the surface S, the quotient $-\frac{dz}{dx}[-\frac{du}{dx}]$, and the length of the instant.
>
> [Fourier 1807]

Thus, the heat flow from the first to the second slice is equal to

$$-KS\frac{du}{dx}\,dt.$$

Similarly, the flow from the third slice into the second is

$$KS\frac{du'}{dx}\,dt\,,$$

so that the total heat flow into the middle slice is

$$KS\left(\frac{du'}{dx} - \frac{du}{dx}\right)\,dt \quad\text{or}\quad KS\frac{d^2u}{dx}\,dt.$$

In order to obtain the increase in temperature du of the slice, one must divide this expression by $CS\,dx$ so that

$$du = \frac{KS\frac{d^2u}{dx}\,dt}{CS\,dx}\,,$$

or

$$\frac{du}{dt} = \frac{K}{C}\frac{d^2u}{dx^2}.$$

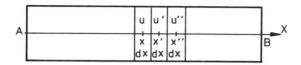

FIGURE 12

Similarly, Fourier found that when the heat is conducted in all three dimensions of the body, it is governed by the equation

$$\frac{du}{dt} = \frac{K}{C} \left(\frac{d^2u}{dx^2} + \frac{d^2u}{dy^2} + \frac{d^2u}{dz^2} \right).$$ (6)

20.   Laplace was on the committee that reviewed Fourier's paper and turned it down, mainly because of Lagrange's opposition to Fourier's use of Fourier series, the generality of which Lagrange had denied in his youth. Laplace was generally sympathetic to the ideas of Fourier (cf. [Grattan-Guinness 1972]), but he was not convinced of Fourier's use of $\frac{du}{dx}$ instead of du in the above derivation. In fact, in a paper of 1809, one of the first to announce some of Fourier's results publicly, he argued that if Fourier's model of heat conduction was adopted, the heat conduction from the first to the second slice would be an infinitesimal of the same order as du ($K\,du$) and so would the heat conduction from the second to the third slice. Thus, their difference would be an infinitesimal of the second order (Kd^2u) "whose accumulation in a finite time cannot raise the temperature of the bar by a finite quantity." [Laplace 1809, p. 332]

Laplace suggested solving this problem by assuming that heat inside a conducting medium is not only communicated at the point of contact between its elements but is radiated some distance from the molecules, just as it is between two macroscopic bodies. In order to derive the heat equation from this hypothesis, Laplace considered three slices sitting at the points with abscissae $x - s$, x, and $x + s$, with the temperatures u', u, and u_1 (Fig. 13).

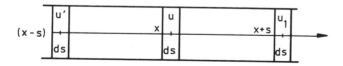

FIGURE 13

According to Newton's law, the heat flux from $x - s$ and $x + s$ to x is proportional to $(u' - u)$ and $(u - u_1)$, respectively, so that the total heat flux from these two slices to x is proportional to $(u' - 2u + u_1)$. This must be multiplied by "ds and by the function that expresses the law of the heating action relative to the distance, a law that we denote $\varphi(s)$." Thus, the heat flux from the two slices is $K_1(u' - 2u + u_1)\varphi(s) ds$, and therefore the total heat flux to x is obtained by integrating this from 0 to ∞:

$$K_1 \int_0^\infty \varphi(s)(u' - 2u + u_1) ds.$$

Laplace now set $u' - 2u + u_1 \sim s^2 \frac{d^2u}{dx^2}$, so that the increase of temperature in the time interval dt is

$$du = K_2 \frac{d^2u}{dx^2} \int_0^\infty s^2 \varphi(s) ds\, dt\,,$$

or

$$\frac{du}{dt} = K_2 \frac{d^2u}{dx^2} \int_0^\infty s^2 \varphi(s) ds.$$

This is the one-dimensional heat equation.

21.   In his papers on heat conduction from 1823, Poisson took the same point of view as Laplace. He copied Laplace's derivation of the one-dimensional heat equation [Poisson, 1823a pp. 8–22] and transformed it into a derivation of the three-dimensional heat equation. I shall give a summary of Poisson's derivation because it was used in an essential way by Liouville.

Consider (Fig. 14) two infinitesimal volumes $d\nu$ and $d\nu'$ inside the material situated at the points (x, y, z) and $(x + x', y + y', z + z')$ and having the temperatures u and u'.

> The heating action of the element $d\nu'$ on the element $d\nu$ during the instant dt produces in $d\nu$ an increase in the quantity of heat proportional to the duration of this instant, to each of the two volumes $d\nu$ and $d\nu'$, and to the excess of the temperature of $d\nu'$ over the temperature of $d\nu$ and which has, moreover, a factor which is a function of the distance expressing the interior law of radiation in the material of which the body is composed. Thus, if we represent this factor by P, this quantity of heat is expressed by
>
> $$P(u' - u) d\nu\, d\nu'\, dt\,.$$
>
> [Poisson 1823a, §46]

The total flux of heat, $\mu\, d\nu\, dt$, into $d\nu$ during the time interval dt is therefore

$$\mu\, d\nu\, dt = d\nu\, dt \int P(u' - u) d\nu'.$$

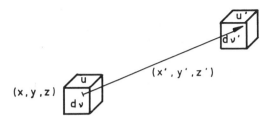

FIGURE 14

Poisson then developed $(u' - u)$ in a Taylor series:

$$(u' - u) = \frac{du}{dx}x' + \frac{du}{dy}y' + \frac{du}{dz}z' + \frac{d^2u}{dx^2}\frac{x'^2}{2} + \frac{d^2u}{dy^2}\frac{y'^2}{2} + \frac{d^2u}{dz^2}\frac{z'^2}{2}$$
$$+ \frac{d^2u}{dx\,dy}x'y' + \frac{d^2u}{dx\,dz}x'z' + \frac{d^2u}{dy\,dz}y'z' + \cdots$$

and changed the expression to polar coordinates:

$$z' = s\cos\theta, \quad y' = s\sin\theta\sin\psi, \quad x' = s\sin\theta\cos\psi,$$
$$dv' = ys^2\sin\theta\,ds\,d\theta\,d\psi.$$

Now,

$$\int_0^{2\pi}\int_0^{\pi} x'^n y'^{n'} z'^{n''} \sin\theta\,d\theta\,d\psi$$

is zero whenever one of the natural numbers n, n', or n'' is odd and

$$\iint x'^2 \sin\theta\,d\theta\,d\psi = \iint y'^2 \sin\theta\,d\theta\,d\psi = \iint z'^2 \sin\theta\,d\theta\,d\psi = \frac{4s^2}{3}.$$

Therefore, if we follow Poisson and assume that P is spherically symmetric (isotropic material) and neglect terms of fourth and higher order in the Taylor expansion of $(u' - u)$, we get

$$\mu\,dv\,dt = dv\,dt \int P(u' - u)\,dv' = k\left(\frac{d^2u}{dx^2} + \frac{d^2u}{dy^2} + \frac{d^2u}{dz^2}\right),$$

where

$$k = \frac{2\pi}{3}\int_0^{\infty} P\,s^4\,ds. \tag{7}$$

But, since the increase in temperature of dv in the time interval dt is $\frac{du}{dt}\,dt$, its increase of heat is $c\frac{du}{dt}\,dv\,dt$, where c is the specific heat per unit volume. Thus,

$$c\frac{du}{dt}\,dv\,dt = \mu\,dv\,dt = k\left(\frac{d^2u}{dx^2} + \frac{d^2u}{dy^2} + \frac{d^2u}{dz^2}\right),$$

or

$$\frac{du}{dt} = \frac{k}{c}\left(\frac{d^2u}{dx^2} + \frac{d^2u}{dy^2} + \frac{d^2u}{dz^2}\right). \tag{8}$$

Poisson pointed out that in his and Laplace's derivation of the heat equation the precise form of φ and P does not matter[4] as long as they fall off rapidly with s, i.e., when the heat radiation is only short ranged compared to the variations of u. This is necessary in order to neglect higher order terms in the Taylor expansion of $(u' - u)$ in the whole region where the integrand is not negligible.

> The form of this function is not known at all; it is possible that it depends on the material of the bar or that it does not differ from one substance to another except by the size of a coefficient independent of s.
>
> [Poisson 1823a, p. 10]

> However, it must be decided by experiments [experience] if the sphere of activity of this radiation is as limited as we suppose. If, on the contrary, its range is just a little bit longer, the form of the differential equation will change.
>
> [Poisson 1823a, p. 6]

LIOUVILLE'S CONTRIBUTION

22.   It was this problem of the communication of heat between the molecules of a conducting medium to which Liouville devoted parts of his first two memoirs and the entire third memoir on the theory of heat. He based his theory on a model that had already been indicated by Laplace:

> It seems natural to admit this internal radiation of heat in the interior of dense bodies; however, the interior radiating heat is totally intercepted by the molecules very close to the molecule that heats them and whose heating action therefore only extends a very short distance.
>
> [Laplace 1809, p. 333]

According to this model, elaborated by Liouville, each molecule radiates heat equally in all directions. In empty space, where the heat flux through all spheres centered at the molecule is the same, the heat flux per unit area must fall off as $\frac{1}{r^2}$. However, Liouville assumed that if the heat hit another molecule it would be totally absorbed, so the heat flux would be zero in the cone behind the molecule (Fig. 15).

Based on these assumptions, he devised two slightly different derivations of the heat flux, or rather the mean flux, between two molecules of the solid as a function of r. In the first, he considered what happened in a narrow cone with the emitting molecule as vertex, and in the second, he considered the whole heat flux through spheres centered at the molecule. He used the

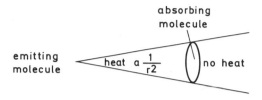

absorbing
molecule

emitting
molecule

heat $a\,\dfrac{1}{r2}$

no heat

FIGURE 15

first method in the first paper and the second method in the second paper and presented them both in the third. I shall briefly summarize the second method, since Liouville considered it the more elegant.

Let α be the mean molecular distance and μ^2 the mean cross section of a molecule and assume that $\frac{\mu^2}{\alpha^2} \ll 1$. Liouville considered the matter around the emitting molecule M as composed of concentric spheres with mutual distance α. On a sphere of radius r are sitting $\frac{4\pi r^2}{\alpha^2}$ molecules, which take up a total area $\frac{4\pi r^2}{\alpha^2}\mu^2$. If θ is the heat flux reaching this sphere, it will absorb

$$\theta\frac{(4\pi r^2/\alpha^2)\mu^2}{4\pi r^2} = \theta\frac{\mu^2}{\alpha^2}.$$

Thus,

$$\Delta\theta = -\theta\frac{\mu^2}{\alpha^2},$$

which, by integration, yields

$$\theta = A\left(1 - \frac{\mu^2}{\alpha^2}\right)^{r/\alpha},$$

where A is the total heat flux from the molecule per unit of time. Thus, on an average, a molecule at distance r will receive per unit of time the amount of heat

$$i = \frac{\Delta\theta}{4\pi r^2/\alpha^2} = \frac{A\mu^2\left(1 - \frac{\mu^2}{\alpha^2}\right)^{r/\alpha}}{4\pi r^2} \sim \frac{A\mu^2}{4\pi}\frac{e^{-\mu^2 r/\alpha^3}}{r^2}.$$

Therefore, the heat exchange in the period dt between two molecules M, M' with the total heat emission A, A' per unit of time and placed at a distance r apart is

$$(i' - i)\,dt = (A' - A)\frac{\mu^2}{4\pi}\frac{e^{-\mu^2 r/\alpha^3}}{r^2}\,dt. \qquad (9)$$

Thus, the shade effect contributes an exponentially decreasing factor to the heat exchange between the molecules of the conducting medium, and according to Liouville, this makes it fall off fast enough to make Poisson and Laplace's arguments acceptable. However, he made no numerical estimates.

23.   Liouville continued to determine the variation of the net heat exchange between two molecules as a function of their temperatures. Laplace and Poisson had accepted Newton's law, which states that the heat exchange between two bodies is proportional to their temperature difference. However, in [1820], Dulong and Petit discovered that when the heat is radiated between two bodies, this law is only correct for moderate temperature differences. For larger differences, they found that the heat exchange was proportional to $a^{u'} - a^u$, where u and u' are the temperatures of the two bodies and a is a constant ~ 1.0077. Their experiment was highly praised by contemporaries as an ideal of exactness, and in the temperature range from $0°$ C to $300°$ C, it agrees very well with Stefan-Boltzmann's law, which states that the heat radiation of a body is proportional to T^4, where T is its absolute temperature.[5]

Liouville applied Dulong and Petit's law to molecular radiation and obtained for $(A' - A)$ in (9):

$$A' - A = kB(a^{u'} - a^u) \sim kBa^u \log a \, (u' - u),$$

where B is the specific heat of the molecule. I do not know the reasons Liouville had for believing that $A' - A$ was proportional to B (considerations of dimensions?), but taking this for granted, we can conclude with Liouville that the net exchange of heat between the two molecules per unit time is

$$(i' - i) = \frac{kB \log a \, \mu^2 a^u}{4\pi r^2} e^{-\mu^2 r/\alpha^3}(u' - u). \tag{10}$$

Thereby Liouville solved the first of the two problems that he asked himself in the introduction to the last memoir [AC L Ms 5]:

> Here, I propose to answer the two following questions: 1° According to which function of the distance and the temperature does the quantity of heat vary when it is transmitted in a solid between two very closely spaced molecules? 2° According to which law does the interior conductivity vary when the temperature changes?
>
> [AC L Ms 5, no. 1]

24.   Liouville solved the last problem by comparing Fourier's and Poisson's expressions for the proportionality factor in heat equations (6) and (8). From these two equations, he concluded that Poisson's constant k is equal to the conductivity K, i.e.,

$$K = \frac{2\pi}{3} \int_0^\infty P(r) r^4 \, dr. \tag{11}$$

Recall that Poisson had defined $P(r)$ in such a way that the net heat exchange between the elements dv and dv' is $P(r)(u - u') \, dv \, dv'$. Since there are $\frac{d\nu}{\alpha^3}$ and $\frac{d\nu'}{\alpha^3}$ molecules in these elements, Liouville's formula (10) equivalently gives the heat exchange:

$$\frac{kB\log a\,\mu^2 a^u}{4\pi r^2}e^{-\mu^2 r/\alpha^3}(u'-u)\frac{d\nu}{\alpha^3}\frac{d\nu'}{\alpha^3}.$$

Comparing these two formulas, Liouville could express Poisson's function P in terms of the molecular quantities of the substance:

$$\begin{aligned}
P(r) &= \frac{kB\log a\,\mu^2 a^u}{4\pi r^2}e^{-\mu^2 r/\alpha^3}\frac{1}{\alpha^6}\\
&= \frac{kC\log a\,\mu^2 a^u}{4\pi r^2}e^{-\mu^2 r/\alpha^3}\frac{1}{\alpha^3}
\end{aligned}$$

because $C = \frac{B}{\alpha^3}$. Thus, Liouville concluded that

$$\begin{aligned}
K &= \frac{2\pi}{3}\int_0^\infty P(r)r^4\,dr = \frac{2\pi}{3}\frac{kC\log a\,\mu^2 a^u}{4\pi\alpha^3}\int_0^\infty e^{-\mu^2 r/\alpha^3}r^2\,dr\\
&= \frac{kC\log a\,a^u}{3}\alpha^2\left(\frac{\alpha^2}{\mu^2}\right)^2.
\end{aligned}$$

Now, α^3 is proportional to $\frac{1}{D}$, where D is the density, which is a function of u, so that

$$\frac{K}{C} = n\frac{a^u}{D^2}, \tag{12}$$

where n is a constant of the material.

In this way, Liouville found the conductivity as a function of the temperature. The derivation given here is the one presented in [AC L Ms 5] whereas the one in [AC L Ms 4] is slightly different.

Poisson had already conjectured that K might vary with the temperature:

> It should also be observed that for high temperatures, the specific heat of solids varies with their temperatures, and perhaps this also holds for their conductivity. The coefficient $a^2[= \frac{k}{c}$ in (8)] is therefore also a function of u whose nature has not yet been determined experimentally.
>
> [Poisson 1823a, §47]

This was the problem Liouville tried to solve by equation (12). He suggested verifying his theory by testing (12) experimentally, but does not seem to have made any experiments himself. If he had made only a few measurements, he would have seen that his law was entirely wrong.

In fact, for temperatures that are not too low, and certainly for the temperatures that could be reached around 1830, the specific heat C of a solid is almost constant. Dulong and Petit discovered that the molar specific heat is almost the same constant for all simple solids. Around room temperature, the thermal conductivity of metals decreases slightly with the temperature. For example, for iron it decreases from 0.86 at 250 K to 0.74 at 350 K. This contradicts formula (12), according to which the conductivity should increase with the temperature.

In passing, Liouville used his formula for $P(r)$ to discuss the validity of replacing sums, ranging over the molecules of a material, with integrals. This problem had occupied Poisson for some time (cf. [Arnold 1981]). In [1823a], he had replaced $\Sigma_{\text{molecules}} P(r)r^4$ by $\int_0^\infty P(r)r^4 \frac{dr}{\alpha}$, but later he expressed great doubts about the legitimacy of such a procedure. By simple calculations, Liouville showed that with the value of P he had found earlier the two expressions give the same result. However, since he did not discuss whether in general one can make such a substitution in the molecular domain, his remark was of little impact.

25.   Until now, I have bypassed the question: what was heat considered to be in the early 19th century? Although Liouville claimed that his results were independent of the nature of heat, the question played a role in his papers, and in particular in the reception of them, so I shall briefly discuss it here.

There were at this time two competing theories of heat: a particle theory and a wave theory. According to the old particle theory, heat was due to weightless particles, *calorique*, which could be absorbed and emitted from the molecules of the material. The *calorique* particles acted like a chemical element and were introduced by Lavoisier as the element carrying heat, when he overthrew the phlogiston theory. It was natural for chemists to maintain this point of view. Around 1830, however, many physicists turned to a wave theory of heat due in part to the experiments made by Rumford, Davy, and others showing that heat could be created by friction. If heat was a substance, these experiments would contradict the conservation of matter. Instead, the experiments suggested that heat was due to vibrations of the molecules, and according to the wave theory, heat was transferred by waves created by the molecules in an ether surrounding them.

The discussions of the nature of heat ran parallel with the discussions of the nature of light. In that discipline, Young and Fresnel's undulatory theories were gradually superseding Newton's corpuscular theory during the 1820s, and that greatly promoted the wave theory of heat. For more information on the wave theory of heat, I shall refer to Brush's interesting paper [1970].

The conflict between the two theories of heat was already apparent in the introduction to Lavoisier and Laplace's joint article, *Mémoire sur la chaleur*, from 1780. However, they pointed out that their results were independent of the nature of heat. Liouville had the same view of his theory and claimed that it had an "almost mathematical certainty... destined to survive changing theoretical opinions regarding the nature and essence of heat" [AC M Ms 5, no. 11]. Still, Liouville admitted that he preferred "the system of waves" to "the theory of emission." He probably inherited this preference from his teacher, Ampère, who was in turn influenced by his friend Fresnel. Ampère later published the first real paper on the wave theory of heat [1835].

26.   However, Liouville's claim that his results were valid under both theories of the nature of heat was rejected by Poisson. In a four-page note (unsigned but obviously in Poisson's awful hand) entitled *Observations sur le mémoire de M. Liouville*, attached to Ampère's report of [AC L Ms 5], he pointed out that Liouville's reasoning was in fact invalid under both hypotheses about the nature of heat. According to Poisson, it was not true in either theory that the heat flux falls off as $\frac{1}{r^2}$ until it hits the first molecule in a given direction, and then is totally intercepted behind it, i.e., Figure 15 is incorrect; for, in Poisson's version of the emission theory, "the heat is absorbed by each molecule, in accordance with the action that it exerts on the caloric particles that pass at a certain distance from this molecule." The attraction or repulsion depends on the quantity of caloric already absorbed in the molecule. If the forces are attractive, some caloric particles will be deflected into the cone behind the molecule, and if they are repulsive, some will be reflected from the molecule. Thus, according to Poisson, Liouville's mistake is to assume that the action between the caloric particles and the molecules is inelastic impact, where in fact "the forces act at a distance."

In the wave theory, "each molecule would modify the wave of heat in a way analogous to what happens in the diffraction of light." Thereby the $1/r^2$ law would be destroyed in front of the molecule and "there would be some heat in the part of the cone that is situated behind the molecule; and it would be very difficult to know what would happen."[6]

In his *Observations*, Poisson further pointed out that under the assumptions made by Liouville his derivation of the radiation as a function of distance is equal to "the one concerning the absorption of light in the atmosphere, which was solved a long time ago." Finally, he argued that the use of Dulong and Petit's law of radiation was inadmissible in the molecular domain. This law had been established only for the radiation between two macroscopic surfaces and "the law it follows as a function of the temperature has no known relation with the law of communication of heat from one molecule to another."

27.   In spite of this devastating criticism, Poisson took over Liouville's ideas in the first two chapters of his book, *Théorie mathématique de la chaleur* [1835]. Poisson subscribed to the emission theory and derived the dependence of the radiation as a function of r, taking into account the action at a distance. This complicated the argument considerably, but the result was similar to Liouville's law. He derived the dependence of the temperature from Dulong and Petit's law just as Liouville had done and seems to have completely overcome his scruples about its validity in the molecular domain. In fact, having stated Dulong and Petit's law, he continued:

> An analogy indicates, and I have in fact assumed, that the same
> holds true in the interior of bodies.
>
> [Poisson 1835, p. 6]

On page 13, he even explained that analogies of the kind he had criticized
in Liouville's paper constituted the true foundation of the theory of heat:

> The mathematical theory of heat is founded on this general hypoth-
> esis of a molecular radiation, considered, no matter what causes it,
> as a deduction from experiment [experience] and analogy.
>
> [Poisson 1835, p. 13]

Thus, in spite of basing his arguments explicitly on the emission theory,
Poisson took over most of Liouville's ideas. It is not surprising that he
changed his mind, since Liouville's ideas fitted so well into his own theories
and had been devised as supplemental to them. However, it is perhaps not
quite according to scholarly honesty that he never so much as mentioned
Liouville's name in this connection.

28.   Poisson's originally critical attitude toward Liouville's ideas was in
sharp contrast to the official report read by Ampère to the *Académie des
Sciences* on April 4, 1831. Ampère summarized the memoir and concluded
by remarking that Liouville had only used known theorems from physics:

> ... but he presents a new and important application of these laws,
> and we think it deserves to be approved by the *Académie* and to be
> inserted in the *recueil des savants étrangers*.
>
> [AC A Ms 1]

The *Procès-Verbaux* briefly reported:

> Mr. Ampère and Mr. Navier reported on the Memoir of Mr. Liouville.
> After Mr. Poisson's observations, the decision of the *Académie* is
> delayed.
>
> [P. V. Acad. Sci. April 4, 1831]

Thus, the disagreement between Poisson and the official committee as
to the merits of Liouville's theory was discussed at the meeting, and the
Académie abstained from making any official conclusion. The results were
that Ampère's report was not included in the *Procès-Verbaux* and that
Liouville's memoir was not published, as Ampère had recommended. [7]

29.   The memoirs on the conduction of heat provide a good basis for a
discussion of the philosophy of nature Liouville was brought up with. Two
basically different ideals presented themselves during the 1820s. Fourier
had advanced a phenomenological, positivistic point of view according to
which the physicists should only describe observable phenomena mathematically, without making hypotheses about their possible microscopic causes.

Most physicists, on the other hand, derived their theories from various microscopic models of the observable quantities. Among these, there was a division between the orthodox Laplacians, as Poisson and Biot, and the modernists, as Ampère, Fresnel, Arago, Dulong, and Petit. The Laplacians maintained that nature ought to be explained by interactions between particles of ordinary matter or of imponderable fluids, such as the electric, the magnetic, and the caloric fluid, and they believed in the corpuscular nature of heat. The modernists, on the other hand, were in favor of the wave theory of light and heat, and they rejected the imponderable fluids in favor of Ampère's electrodynamics (cf. [Fox 1974]).

Generally, the young Liouville sided with the last group, but in his papers, one can find traces of all these points of view. In the concluding section of [AC L Ms 5], he admitted that Fourier's theory of heat had the advantage of being independent of physical assumptions, but he preferred Poisson's procedure because it makes it possible to know "the nature of the bodies more intimately." This remark only concerned the constitution of matter, for with regard to the nature of heat, he followed Fourier and claimed that his results were independent of the various models. Still, he personally preferred the vibrational model to Poisson's particle model.

Liouville also acted as a follower of Fourier when he pointed out that Ampère's electrodynamical force law could not be derived from experiment, as Ampère had claimed. However, in his other electrodynamic papers, Liouville followed Ampère and analyzed the microscopic interactions carefully. Thus, Liouville's philosophy of physics seems to have been shaped mainly by Ampère and to a less extent by Poisson. His preference for Poisson's theory of heat suggests that Fourier had only small part in his education (cf. Chapter I).

30.   Liouville's period as a "microscopic" physicist was limited to his student years. The only papers in which he tried to analyze the microscopic structure of nature are those on electrodynamics and heat discussed in this and the preceding sections.[8] In 1830, he turned his back to physics, and during the next five years, he devoted all his papers to pure mathematics, although these continued to be inspired by physical phenomena. When he returned to mathematical physics and astronomy in 1836, he only derived mathematical consequences from the fundamental laws of nature and abstained from a physical analysis of the laws.

Differential Equations

31.   Before 1831, Liouville also wrote four purely mathematical memoirs on differential equations. The first three papers entitled *Premier (Second, Troisième) Mémoire sur le calcul aux différences partielles* [AC L Ms 3a,b,c] were all presented on December 1, 1828, to the *Académie des Sciences*, in

whose archive they are still preserved. The last paper, which deals with the Riccati equation, was presented to the *Académie* on November 2, 1830, but only a published extract [Liouville 1833a] remains.[9]

The memoirs on partial differential equations were composed in the interval between Liouville's two electrodynamical memoirs. As mentioned in Chapter I, they contain overly complicated solutions to rather artificial problems, so it is no wonder that they were never reviewed by the *Académie*. Still, there are two reasons why I think it is worth while to summarize their contents. First, they illustrate that the young Liouville had not yet learned to direct his analytical talents if no physical problem suggested a fruitful way; as we shall see in Chapter IX, he acquired this purely mathematical skill very early. Second, the 1828 memoirs contain the vague roots of some of the problems and methods that became important in Liouville's later works.

The central tool in all the three 1828 manuscripts is a theorem shown by Parseval [1806]. It states that if

$$f(\omega) = \sum_{n=0}^{\infty} \omega^n \psi_n$$

and

$$F(\omega) = \sum_{n=0}^{\infty} \omega^{-n} \alpha_n \,,$$

then

$$\sum_{n=0}^{\infty} \psi_n \alpha_n = \frac{1}{2\pi} \int_0^{\pi} \left[f(e^{iz})F(e^{iz}) + f(e^{-iz})F(e^{-iz}) \right] \, dz.$$

The proof that Liouville repeated in the third of the memoirs [AC L Ms 3c] consists of computing the right-hand side and remarking that only the terms containing ω^0 survive. (Its historical connection to Cauchy's integral formula would be worth studying.)

32.   Liouville applied the formula to the solution of linear partial differential equations. In all the papers, he began by considering first-order equations with two independent variables x and y. Then, according to Liouville, the solution can usually be expressed in the form

$$\varphi(x,y) = \sum_b c_b e^{bx} F(y,b), \tag{13}$$

where b can vary over an arbitrary set, and c_b are arbitrary constants. In fact, Liouville implicitly assumed that the coefficients of the equation are only functions of y, for otherwise separation of variables does not yield the factors e^{bx}. His goal was to find another expression of the solution that did not contain a sum. He remarked that the continuous version of (13)

$$\varphi(x,y) = \int \theta(b)e^{ux} F(y,b) \, db \,, \tag{14}$$

was a possible "finite" expression, but he rejected it because he believed it was difficult to determine the arbitrary function $\theta(b)$, which makes φ satisfy a given "initial value" condition $\varphi(x, 0) = f(x)$. Indeed, the inversion of the Laplace transform was not well understood then.

Instead, he found another transformation of (13). In the first memoir [AC L Ms 3a], he assumed that $F(y, b)$ could be expanded in a series:

$$F(y, b) = \sum_{n=0}^{\infty} \alpha_n(y) \frac{1}{b^n}. \tag{15}$$

Writing $\frac{c_b}{b}$ instead of c_b in (13), this yields the expression

$$\varphi(x, y) = \sum_{b} \sum_{n=0}^{\infty} \frac{c_b}{b^{n+1}} \alpha_n(y). \tag{16}$$

Liouville then introduced the function

$$\psi(x) = \sum_{b} c_b e^{bx}, \tag{17}$$

which is an arbitrary function determined by the arbitrary constants c_b. Since

$$\int^n \psi(x) \, dx^n = \sum_{b} \frac{c_b}{b^n} e^{bx},$$

(16) can be written

$$\varphi(x, y) = \sum_{n=0}^{\infty} \alpha_n(y) \int^{n+1} \psi(x) \, dx^{n+1}. \tag{18}$$

On the other hand,

$$\int e^{bt} \frac{(x - t)^n}{n!} \, dt = \frac{e^{bx}}{b^{n+1}}, \tag{19}$$

so that (16) can also be written

$$\varphi(x, y) = \int \psi(t) \sum_{n=0}^{\infty} \alpha_n(y) \frac{(x - t)^n}{n!} \, dt. \tag{20}$$

Parseval's theorem applied to $F(y, b)$ (cf. (15)) and to

$$e^{\omega(x - t)} = \sum_{n=0}^{\infty} \omega^n \frac{(x - t)^n}{n!}$$

yields

$$\sum_{n=0}^{\infty} \alpha_n(y) \frac{(x - t)^n}{n!} = \frac{1}{2\pi} \int_0^{\pi} \left[e^{(x-t)e^{iz}} F(y, e^{iz}) + e^{(x-t)e^{-iz}} F(y, e^{-iz}) \right] dz$$

so that

$$\varphi(x,y) = \frac{1}{2\pi} \int \psi(t)\,dt \int_0^\pi \left[e^{(x-t)e^{iz}} F(y, e^{iz}) + e^{(x-t)e^{-iz}} F(y, e^{-iz}) \right] dz.$$

(21)

This is the expression of the solution $\varphi(x,y)$ Liouville was after. It contains the arbitrary function $\psi(x)$. When the development, (15), of F only contains a finite number of terms, one can determine $\psi(x)$ from $\varphi(x,0)$ by solving (18), which is then a linear differential equation with constant coefficients. However, Liouville was not able to determine ψ from $\varphi(x,0)$ if (15) was an infinite series.

33.   Liouville continued to show how his method could be generalized to more variables and to higher order equations. In the second and third memoirs [AC L Ms 3b,c], Liouville further adapted the procedure to the situation where F, instead of an expansion of (15), has a power series expansion around a point b_0:

$$F(y, b_0 + \beta) = \sum_{n=0}^{\infty} \alpha_n(y)\beta^n.$$

In the third memoir [AC L Ms 3c, no. 7], Liouville applied his method to the equation

$$\frac{d\varphi}{dy} = \frac{d^3\varphi}{dx^3},$$

"which has been suggested to us by Mr. Poisson, because it eludes several of the known procedures." In this case Liouville wrote the solution in the form

$$\varphi(x,y) = \sum_b c_b e^{bx} F(y,b)$$

(22)

and found

$$F(y,b) = e^{b^3 y}.$$

(23)

His formulas yielded the solution

$$\varphi(x,y) = \frac{1}{\pi} \int_0^\pi dz\, e^{x\cos z} e^{y\cos 3z}$$

(24)

$$\times \left(\cos(x\sin z + y\sin 3z + z) \int e^{-x\cos z} \cos(x\sin z)\psi(x)\,dx \right.$$

$$+ \left. \sin(x\sin z + y\sin 3z + z) \int e^{-x\cos z} \sin(x\sin z)\psi(x)\,dx \right),$$

where $\psi(x) = \varphi(x,0)$.

This is probably not a very helpful formula, especially considering that Liouville had to describe how to fix the arbitrary constants in the integrals (cf. Chapter X, §50), for his later approach to this equation).

34.   We shall now investigate which of the more or less implicit ideas in these manuscripts anticipate Liouville's later important works.

First of all, the investigation of the expansion of functions in series of eigenfunctions, such as (13), were at the heart of all his work on Sturm-Liouville theory. However, before his ideas could bear fruit, his vague ideas concerning (13) had to be shaped by physical applications, in particular in the theory of heat (cf. Chapter X). From these applications, he learned that in order to discuss solutions of partial differential equations, he should make explicit the equations under consideration. In [AC L Ms 3a-c], the whole discussion was extremely vague because Liouville never wrote down any partial differential equation. He discussed solutions to equations that were never mentioned. Applications also taught Liouville to appreciate boundary value problems, for which the spectrum, i.e., the possible values of b in (13), is better determined.

Liouville's goal in his three papers on partial differential equations seems to have been to express the solutions in closed form or rather by quadrature, i.e., using integrals but no infinite series. Since he could not in general determine ψ, he did not really obtain this goal. However, a few years later, when he detached the question from the spectral theory, it led to one of his most celebrated theories (cf. Chapter IX).

The development, (17), of the arbitrary function $\psi(x)$ in exponential series represented a central idea in another of Liouville's later theories: fractional calculus. Furthermore, formula (19) confronted him with a problem he later hit upon in his fractional calculus, where such convolution integrals became of great importance: the integral must be taken over the interval $[-\infty, x]$ if b is positive and over $[\infty, x]$ if b is negative. Thus, (20) does not make sense except when the development of φ or ψ contains only negative or only positive values of b, or at least complex values of b that have all positive or all negative real parts.

However, in his early memoirs, Liouville did not reflect too deeply on these problems, and it did not disturb him that in the final formula, (21), $F(y, b)$ was integrated over purely imaginary values of b, where the earlier formulas made no sense. This shows the formalist attitude with which the young Liouville approached pure mathematics. Later in life he avoided such formalistic tendencies in his analytical research by being more aware of rigor and primarily by staying closer to physics.

35.   His *Mémoire sur l'équation de Riccati* from 1830 was much clearer than the 1828 papers. It dealt with the Riccati equation:

$$\frac{d\nu}{dx} = a\nu^2 + bx^m ,\tag{25}$$

so called after Count Riccati of Venice (1676–1754), who derived it as a means to solve certain second-order equations. By the opposite transformation, to the one used by Riccati, Liouville derived the equivalent equation:

$$\frac{d^2u}{dx^2} = x^m \, u. \tag{26}$$

This is a special case of the equation

$$\frac{d^2u}{dx^2} = f(x)u, \tag{27}$$

which governs the stationary temperature distribution in a homogeneous, unequally polished bar. Liouville discussed this equation in his earlier memoir on heat [Liouville 1830] (cf. Chapter X, §30–34). His first approach to this equation consisted of approximating $f(x)$ by a piecewise linear function, solving the equation in this case, and letting the number of sides in the polygon tend to infinity.

Thus, Liouville had to solve the equation for $f(x) = ax + b$, a problem he could easily reduce to the case $f(x) = x$, i.e., to (26) for $m = 1$. His solution of the Riccati equation in this case can be found in his published paper on heat [Liouville 1830, sections v-x]. The more general solution of (26) that he gave in *Mémoire sur l'équation de Riccati* is a rather simple generalization of his earlier solution.

The point of departure of Liouville's investigations was the series expansion of the solution. It is easily seen[10] that the general solution to (26) is the linear combination $AX + BY$ of the two infinite series:

$$X = 1 + \frac{x^{m+2}}{(m+1)(m+2)} + \frac{x^{2(m+2)}}{(m+1)(m+2)(2m+3)(2m+4)}$$

$$+ \frac{x^{3(m+2)}}{(m+1)(m+2)(2m+3)(2m+4)(3m+5)(3m+6)} + \cdots,$$

and

$$Y = x + \frac{x^{(m+2)+1}}{(m+2)(m+3)} + \frac{x^{2(m+2)+1}}{(m+2)(m+3)(2m+4)(2m+5)}$$

$$+ \frac{x^{3(m+2)+1}}{(m+2)(m+3)(2m+4)(2m+5)(3m+6)(3m+7)} + \cdots. \tag{28}$$

Liouville's aim in the memoir is similar to that of the memoirs on partial differential equations; but in 1830, he expressed it explicitly:

> However, in this memoir I propose to express each of the two series (which I call X and Y) in finite form by means of multiple quadratures.
>
> [Liouville 1833a, p. 2]

Liouville first considered the series X when m is a natural number. He wrote the general term in the form

$$\frac{P_n x^{np}}{n!},\qquad(29)$$

where $p = m + 2$, and P_n is recursively determined by $P_0 = 1$, and

$$P_{n+1} = P_n(pn + 1)(pn + 2)\cdots(pn + m).$$

In order to simplify the determination of P_n, Liouville wrote it as the product $P_n = P_n' P_n'' \cdots P_n^{(m)}$, where $P_n^{(i)}$ is recursively determined by $P_0^{(i)} = 1$, and

$$P_{n+1}^{(i)} = (pn + i)P_n^{(i)}.\qquad(30)$$

Liouville then searched for a function $\varphi_i(\alpha)$, such that

$$P_n^{(i)} = \int \varphi_i(\alpha)\alpha^n\,d\alpha \quad .^{11}\qquad(31)$$

By inserting (31) into the equation (30), Liouville found

$$P_n^{(i)} = C^{(i)} \int_0^\infty e^{-\frac{\alpha}{p}} \alpha^{n-1+\frac{q}{p}}\,d\alpha,$$

such that the product $P_n = P_n' P_n'' \cdots P_n^{(m)}$ can be expressed as

$$P_n = A \int_0^\infty d\alpha \int_0^\infty d\beta \cdots \int_0^\infty d\omega\, M(\alpha, \beta, \ldots, \omega)(\alpha\beta\cdots\omega)^n,\qquad(32)$$

where

$$M(\alpha, \beta, \ldots, \omega) = e^{-\frac{1}{p}(\alpha+\beta+\cdots+\omega)}\alpha^{-1+\frac{1}{p}}\beta^{-1+\frac{2}{p}}\cdots\omega^{-1+\frac{m}{p}}.$$

Liouville then inserted the value of (32) for P_n into the expression for X, and interchanging the order of integration and summation,[12] he found

$$X = \int_0^\infty d\alpha \int_0^\infty d\beta \cdots \int_0^\infty d\omega\, \lambda(x, z)\, M(\alpha, \beta, \ldots, \omega),\qquad(33)$$

where $z = \sqrt[p]{\alpha\beta\cdots\omega}$ and

$$\lambda(x, z) = \sum_{n=0}^\infty \frac{x^{np} z^{np}}{(np)!}.$$

He observed that λ is a solution of the differential equation

$$\frac{d^p \lambda}{dx^p} = z^p \lambda,$$

which has the complete integral

$$\sum_{i=0}^{p-1} C_i \, e^{\mu^i z x} \, ,$$

where $1, \mu, \mu^2, \ldots, \mu^{p-1}$ are the pth roots of unity. Since

$$\lambda(0, z) = 1 \quad \text{and} \quad \frac{d^i \lambda}{dx^i}(0, z) = 0 \quad \text{for} \quad 1 = 1, 2, \ldots, p-1 \, ,$$

all the C_i's must be equal to $\frac{1}{p}$, so that

$$\lambda = \frac{1}{p} \sum_{i=0}^{p-1} e^{\mu^i z x} . \tag{34}$$

When inserted into (33), this gives the finite expression for X, which Liouville was after.

Using a similar argument, he determined Y for positive integer values of m. In order to find expressions for X and Y for positive rational values of m, he only had to adjust the proofs slightly. By an ingenious little trick [Liouville 1833a, p. 15], he further reduced the expressions for m negative and rational to the case where m is positive, except when m belongs to the intervals $(-3, -2) \cup (-2, -1)$. In this way, Liouville solved the Riccati equation in "finite form" for all rational values of m except those in the above intervals.

36.   The central tool in Liouville's transformation of the series solution was the use of what we now call the Mellin transform (31), to integrate the finite difference equation, (30). This idea was due to Laplace [1812, ed., p. 111], to whom Liouville explicitly referred. In later work, Liouville often used the similar Laplace transform to solve differential equations (cf. Chapter VIII, §20) and [Deakin 1981]).

At the end of the published extract of the memoir, Liouville briefly told what had been the content of the rest of the memoir:

> In the memoir on the Riccati equation, which I presented to the Institute at the end of 1830, and of which the present is but an extract, I have in fact shown that one can, by convenient transformations of equation [25], integrate several of the equalities for which the general procedure that I have explained here does not work. In this indirect way, I have even rediscovered the condition of integrability of equation [25] which was discovered at the beginning of the integral calculus, namely that m must be of the form
>
> $$m = -\frac{4i}{2i \pm 1} \, ,$$
>
> where i is an integer.

But since these studies do not contain anything really new and do not lead to simple results, I do not think that I need to mention them here.

[Liouville 1833a, pp. 18–19]

The "condition of integrability" referred to by Liouville had been found by Nicolaus Bernoulli in his correspondence with Goldbach in 1721 (cf. [Fuss 1843]). He had shown that for the above values of m it is possible to separate the variables in Riccati's equation and therefore to determine the solutions by quadrature. Daniel Bernoulli published the solution in the Acta Eruditorum of 1725. For more information on the Riccati equation, the reader is referred to [Cantor, 1898, pp. 456–461].

In 1840, Liouville returned to the problem of solving the Riccati equation in finite form [Liouville 1841a]. By then, however, he had adapted a more limited definition of a "finite expression" or rather a "solution found by quadrature." Thus, he excluded integral operators of the kind in (33), and allowed only indefinite integrals, $\int_0^x f(x)\,dx$, in the expressions. He then proved that the Riccati equation can only be integrated by quadrature for the values of m found by the Bernoullis (cf. Chapter IX, §51). When Liouville, in the quote above, wrote about "the condition of integrability," there is little doubt that he only thought of a sufficient condition and not of a necessary condition. Nevertheless, the 1830 memoir on the Riccati equation bears witness to a continued interest in solutions in closed form and anticipates his later theory in general and the paper [1841a] on the Riccati equation in particular.

VIII

Differentiation of Arbitrary Order

Introduction

1. Out of Liouville's early work on electromagnetism grew an interest in the theory known today under the name fractional calculus. It is a theory that generalizes the meaning of the differential and integral operators $\frac{d^\mu}{dx^\mu}$ from the well-known cases where μ is an integer ($\frac{d^{-1}}{dx^{-1}}$ meaning integration) to the cases when μ is a real rational or irrational or even complex number. Liouville's own term, *calcul des différentielles à indices quelconques* (differential calculus of arbitrary index (or order)), is therefore more appropriate than the misleading "fractional calculus." This was the first field in which Liouville published extensively and to which his name is still attached, in that the modern definition of differentiation of arbitrary order is called the Riemann-Liouville definition. He published 9 papers on the subject during the period from 1832 to 1837 and one late comer in 1855.

The three first papers were printed in succession in the twenty-first *cahier* of the *Journal de l'École Polytechnique* of 1832 and opened with a definition of differentiation of arbitrary order. Liouville took the rule $\frac{d^n}{dx^n}e^{mx} = m^n e^{mx}$ as his starting point:

> I suppose that y represents a function of x, and I develop this function in an exponential series, such as
>
> $$A_1 e^{m_1 x} + A_2 e^{m_2 x} + A_3 e^{m_3 x} + \cdots$$
>
> or more briefly $\sum A_m e^{mx}$. This being done, I call the differential or rather the derivative of y of order μ that function which one deduces from y by multiplying each term $A_m e^{mx}$ of the series by the power μ of the corresponding exponent, i.e., by m^μ, and I denote this derivative by $\frac{d^\mu y}{dx^\mu}$. Thus, I have
>
> $$\frac{d^\mu y}{dx^\mu} = \sum A_m e^{mx} m^\mu. \tag{1}$$
>
> [Liouville 1832a, p. 3]

Liouville often used the symbol $\int^\mu dx^\mu$ instead of $\frac{d^{-\mu}}{dx^{-\mu}}$, in particular when μ is positive. He devoted his first paper to illustrating the applicability of his new calculus to geometry and mechanics, postponing a thorough examination of the foundations of the theory to the second paper. He chose this order of presentation not only because this was probably the historical order of his own research but also as a means of catching the attention of his readers, before entering on the more tricky problems of the foundations; or as Liouville expressed it himself:

> I hope that the solutions of the geometrical and mechanical problems of the first paper suffice to explain to the geometers that this new calculus may one day be of great use, and urge them to pay some attention to it.
>
> [Liouville 1832a, p. 2]

In the second paper [1832b] of a further 91 pages, Liouville investigated the possibility of expanding arbitrary functions in series of exponentials, a problem basic to his definition. Further, he proved the rules for differentiating sums and products, and he determined the fractional derivative of the most common rational and transcendental functions. A particular problem in the theory was the so-called complementary functions. Liouville argued that just as the n-fold integral of a function contains an arbitrary polynomial of degree $n - 1$ so every derivative of arbitrary order (except for integer orders) must contain a complementary function, and this function must be a polynomial with arbitrary coefficients and of arbitrary order.

The last paper in the twenty-first *cahier* of the *Journal de l'École Polytechnique* [Liouville 1832c] was devoted to a solution of the hypergeometric equation

$$(mx^2 + nx + p)\frac{d^2y}{dx^2} + (qx + r)\frac{dy}{dx} + sy$$

by way of the fractional calculus.

2.   In the two next papers published in the eleventh and twelfth volume of Crelle's *Journal*, both of 1834, Liouville elaborated two ideas from his first papers. In the first [1834a], he gave an alternative proof of his theorem on complementary functions, and the second [1834b] was devoted to the fundamental formula

$$\int_0^\infty \varphi(x + \alpha)\alpha^{\mu-1}\,d\alpha = (-1)^\mu\Gamma(\mu)\int^\mu \varphi(x)\,dx^\mu, \qquad (2)$$

which had been at the basis of the applications given in [Liouville 1832a]. Liouville provided a new proof of the formula, applied it to a new problem, and indicated several general ideas on the new calculus. Because of its reasonable size and its clarity, I recommend this paper to the reader who wishes to get a first-hand impression of Liouville's theory.

The following year, Liouville published a third paper in Crelle's *Journal* [1835b] in which he analyzed how well his own definition of fractional differentiation was in accordance with a definition given earlier by Fourier. Fourier [1822 p. 561] had defined $\frac{d^i f}{dx^i}$ by the formula

$$\frac{d^i f(x)}{dx^i} = \frac{1}{2\pi}\int_{-\infty}^\infty d\alpha\,f(\alpha)\int_{-\infty}^\infty dp\,p^i\cos\left(px - pa + \frac{i\pi}{2}\right), \qquad (3)$$

which is the result obtained for integers i by formally differentiating under the integral sign in Fourier's formula. However, because of a peculiarity in the fractional derivative of trigonometric functions, pointed out in [Liouville 1832b], this definition does not agree with Liouville's definition, but in the paper [1835b], Liouville provided two related formulas to replace it.

In [1832b] Liouville had not included rules for differentiating composite functions, or as he phrased it, for changing variables in arbitrary derivatives. He never solved this problem in general, but in [1835c] he published the rules for expressing $\frac{d^\mu}{d(\sqrt[n]{x})^\mu}$ in terms of $\frac{d^\nu}{dx^\nu}$. He showed how these rules could help solve certain differential equations of fractional order. A more general theory of solution of fractional differential equations had been foreshadowed by Liouville in [1832c, p. 185] and in [1834a, p. 14], but a paper on this subject was not published until 1837 [Liouville, 1837m]. In this paper, Liouville provided methods for solving linear fractional differential equations with constant or rational coefficients.

I have passed over a small note [Liouville 1835d] in which Liouville applied his new calculus to solve an integral equation that Poisson had encountered in the theory of heat. With these papers, Liouville ended his contributions to the theory of differentiation of arbitrary order. He only returned to the subject once more in print 20 years later. In a small note [1855a], he defended himself against Tortolini [1855], who believed he had found a mistake in Liouville's formula (2). Liouville pointed out that Tortolini had not taken into account the conditions that Liouville had explicitly formulated, and he provided an alternative formula to be used under the assumptions implicitly made by Tortolini.

3.   The above chronology of Liouville's papers gives a distorted picture of his process of discovery. In his first paper, Liouville pointed out that he had applied the fractional calculus for a long time. We do not know for how long a time, for Liouville's early notes are not preserved. However, the very first of his systematically arranged notebooks [Ms 3615 (3)] reveals that by early 1832 he had composed his two first papers on fractional calculus, and by the end of that year he had discovered all the main ideas that entered into his later papers, including the rules for changing variables and the solution of linear fractional differential equations with rational coefficients. Toward the summer of 1833, the notes on this subject became infrequent, and among the notes of 1834, one mainly finds revisions of earlier ideas. After that year, Liouville only returned to the subject in one note of 1836 [Ms 3615 (5), pp. 17r–18r] three lines in 1838 [Ms 3616 (3), p. 5v] and 40 pages in 1843–1844 [Ms 3617 (4), pp. 70v–91r], but these investigations did not lead to any publications. The delay of Liouville's papers was due in part to difficulties with presentation and probably also to an unfulfilled wish to improve the initial methods.

4.   Liouville was the founder of the theory of differentiation of arbitrary order in the sense that he was the first to attempt to compose a connected

theory and to show its applicability to a great variety of mathematical and physical problems. However, he was not without predecessors. Great mathematicians such as Leibniz, Euler, Laplace, Lacroix, and Fourier, had reflected on the meaning of the symbol $\frac{d^\mu}{dx^\mu}$ when μ is not an integer, but none of them had made much more than passing mention of the problem. Liouville referred to all of these predecessors in his papers [1832a, pp. 2 and 69], [1832b, p. 71], [1834a, pp. 16–19], and [1835b]. Thus, Gino Loria [1936, p. 149] was mistaken when he wrote that Liouville probably did not know his predecessors. Still, he could not learn much from them except the mere existence of the problem of defining such fractional derivatives.

Niels Henrik Abel had also made use of the fractional calculus in his solution of the tautochrone problem, but as I shall argue in the following sections (§10–12), it is most improbable that Liouville knew this.

5. I shall not review the prehistory or the subsequent development in much detail for these have already been described in three excellent papers by Bertram Ross [1974], [1975], and [1977]. The first is a very valuable chronological and annotated bibliography, and in the last two, Ross gives surveys of the history of the fractional calculus from Leibniz to our day. Though I shall not presuppose any knowledge of either the fractional calculus or its history, I recommend the reader consult Ross's papers, in particular the last one, in which the structure of the modern theory is outlined.

Ross's papers do not render this chapter superfluous, for although Liouville played a key role in the history, the scope of Ross's papers has allowed him to discuss only briefly Liouville's definition of the fractional derivative and his introduction of complementary functions. The rest of Liouville's theory is passed over in silence. Since it was the attempt to create a large-scale connected theory that distinguished Liouville's work relative to his predecessors, rather than any single definition or theorem, a broad overview of the entire theory should not be omitted from this biography.

The existence of Ross's papers has made the composition of this chapter much easier, but the subject matter is still difficult to treat, for, because of the concept of complementary functions and several other peculiarities, Liouville's theory is rather vague and imprecise. Except for the applications, probably none of Liouville's theorems are entirely true in the modern theory. Therefore, although I shall tie some of Liouville's concepts to their modern counterparts, I shall not attempt to supply a complete critical analysis of Liouville's papers from a modern point of view. Instead, I shall try to explain the inner coherence that characterizes Liouville's system despite its obvious weaknesses. The following analysis follows the order of Liouville's publications, starting with the applications, then analyzing the foundations, and ending with the solution of differential equations of arbitrary order.

The papers on fractional calculus are among the few where Liouville reflected on the rigor of his methods. Therefore, I have added a section in which I analyze the young Liouville's attitude toward the new movement of rigor created mainly by Cauchy. Liouville's mature view on rigor is discussed in Chapter III.

Applications, the Source of Interest

6.   For Liouville, the applications of the fractional calculus were not just a pedagogical means to arouse the interest of other mathematicians; applicability was the *raison d'être* for the theory. This attitude is very different from that of his predecessors, whose sole aim seems to have been a generalization of the usual terminology.

The applicability of the fractional calculus is based on a few formulas that transcribe integrals of arbitrary order into definite convolution integrals. The fundamental formula is

$$\int^\mu \varphi(x)\, dx^\mu = \frac{1}{(-1)^\mu \Gamma(\mu)} \int_0^\infty \varphi(x + \alpha)\alpha^{\mu-1}\, d\alpha, \tag{2}$$

where Γ is Euler's gamma function defined for $\mathrm{Re}\, x > 0$ by

$$\Gamma(x) = \int_0^\infty e^{-\theta}\theta^{x-1}\, d\theta. \tag{3a}$$

Liouville provided two proofs of this theorem. In [1832a pp. 8–9], he first remarked that for $\varphi = e^{mx}$ with $\mathrm{Re}\, m < 0$ the formula is an easy consequence of the definition $\int^\mu e^{mx} = \frac{1}{m^\mu}e^{mx}$. If φ is of the form $\Sigma A_m e^{mx}$, the formula is verified by summation under the integral sign. This proof is, according to Liouville, "the simplest and the most direct of all" [Liouville 1834b, pp. 273–274]. Still, he gave a second proof in [1834b, pp. 274–275], which is based on an expression of the fractional derivatives as limits of "difference quotients" (cf.§27). In both proofs, Liouville explicitly assumed that the development of φ in a series of exponentials, $\varphi(x) = \Sigma A_m e^{mx}$, contains only exponents m with negative real parts:

This condition is evident because without it the integral

$$\int_0^\infty \varphi(x + \alpha)\alpha^{\mu-1}\, d\alpha$$

would be infinite. It amounts to saying that the function $\varphi(x)$ becomes zero when you set $x = \infty$.

[Liouville 1832a, p. 8]

When Tortolini [1855] found a formula that was at variance with (2), Liouville immediately pointed out [1855a] that Tortolini had implicitly assumed that the exponential expansion of φ has only exponents m with

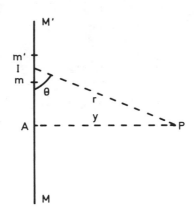

FIGURE 1

positive real parts. Under this assumption, Liouville established the alternative formula

$$\int^{\mu} \varphi(x)\, dx^{\mu} = \frac{1}{\Gamma(\mu)} \int_0^{\infty} \varphi(x-\alpha)\alpha^{\mu-1}\, d\alpha , \tag{4}$$

which is in accordance with Tortolini's result. As pointed out by Liouville, this formula had been derived 11 years earlier by Serret [1844].

In the applications, Liouville always worked with functions that tended to zero at plus infinity, and therefore he used formula (2). From this, Liouville deduced a series of similar formulas [1832a, formulas A-G] and [1834b, formulas A-E]. For example, by substituting x^2 and α^2 for x and α in (2) he found [1832a, p. 10]

$$\int_0^{\infty} \varphi(x^2+\alpha^2)\alpha^{2\mu-1}\, d\alpha = \frac{(-1)^{\mu}\Gamma(\mu)}{2} \int^{\mu} \varphi(x^2)\, d(x^2)^{\mu}. \tag{5}$$

By similar substitutions, Liouville discovered that the complete elliptic integrals of the first and second kind (cf. Chapter IX, §5) could be expressed as simple fractional integrals of their moduli [1834b, p. 280].

7.   I shall illustrate Liouville's application of formula (2) or (5) with the simplest physical example from his first paper [1832a, pp. 15–20]. There he derived Biot and Savart's law of attraction between an element mm' of a conductor and a magnetic pole P (cf. Chapter VII, §2) when the attraction $f(y)$ between P and an infinite conductor MM' is known for any value of the distance y between the pole P and the conductor (Fig. 1)

Liouville took for granted that the attraction between P and mm' is of the form

$$\phi(r)\sin\theta\, ds, \tag{6}$$

where ds is the length of mm', and θ is the angle made by MM' and the line, of length r, connecting p and the midpoint of the small element mm'. The problem is to find ϕ. Integration of (6) along the entire wire MM' gives the total force $f(y)$:

$$f(y) = \int_{-\infty}^{\infty} \frac{y\phi(\sqrt{s^2+y^2})}{\sqrt{s^2+y^2}} \, ds = 2 \int_{0}^{\infty} \frac{y\phi(\sqrt{s^2+y^2})}{\sqrt{s^2+y^2}} \, ds ,$$

or

$$\frac{f(y)}{2y} = \int_{0}^{\infty} \frac{\phi(\sqrt{s^2+y^2})}{\sqrt{s^2+y^2}} \, ds. \tag{7}$$

This is an integral equation which Liouville had to solve for ϕ. He accomplished this by transcribing the right-hand side to a fractional integral, by way of (5) with $F(r^2) = \phi(r)/r$ and $\mu = \frac{1}{2}$. In this way, he was led to the simple fractional differential equation:

$$\frac{\sqrt{-1}\,\Gamma(1/2)}{2} \int^{1/2} F(y^2) \, d(y^2)^{1/2} = \frac{f(y)}{2y}.$$

By substituting $y^2 = z$, Liouville obtained

$$\int^{1/2} F(z) \, dz^{1/2} = -\frac{\sqrt{-1}}{\sqrt{\pi}} \frac{f(\sqrt{z})}{\sqrt{z}} \tag{8}$$

since $\Gamma(\frac{1}{2}) = \sqrt{\pi}$. Then, he applied the operator $d^{1/2}/dx^{1/2}$ to both sides of (8), obtaining

$$F(z) = -\frac{\sqrt{-1}}{\sqrt{\pi}} \frac{d^{1/2}(f(\sqrt{z})/\sqrt{z})}{dz^{1/2}}.$$

Changing z into r^2 and reintroducing $\phi(r) = rF(r^2)$, he was thus led to the solution

$$\phi(r) = -\frac{\sqrt{-1}}{\sqrt{\pi}} r \frac{d^{1/2}(f(r)/r)}{d(r^2)^{1/2}}. \tag{9}$$

In their very first experiment, Biot and Savart demonstrated that $f(r) = k/r$ (cf. Chapter VII, §2). In that case, the solution becomes $\phi(r) = k/2r^2$ because $d^{1/2}(x^{-1})/dx^{1/2} = \sqrt{-1}\sqrt{\pi}(2x)^{-3/2}$ (cf. §26). This is Biot and Savart's law (cf. Chapter VII, §2).

8.   Thus, Liouville applied fractional calculus to solve a problem that he had already investigated with more primitive mathematical methods and with different experimental evidence in his earlier unpublished memoir [ACL Ms 2] (cf. Chapter VII, §11). The connection between Liouville's fractional calculus and his early work on electromagnetism is even more striking in the next example of [1832a] in which Liouville derived Ampère's

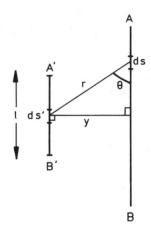

force law between two elements of conductors from precisely the same experimental "fact" as in [ACL Ms 2]. As in the derivation of Biot and Savart's law, he only discussed how the force depends on the distance between the two elements. He considered two elements of conductors ds, ds' that conducted currents i, i', making an angle ε with each other and angles θ, θ' with the line connecting their midpoints and let this connecting line have the length r. Then, he took it for granted that the attraction between them is expressed by the formula (cf. Chapter VII, §4)

$$ii'ds\,ds'\varphi(r)(\cos\varepsilon + (k-1)\cos\theta\cos\theta') \qquad (10)$$

and is directed along the line that connects them [Liouville 1829]. The problem Liouville needed to solve was to determine φ.

To this end, he used the experiment he had learned from Ampère's lectures (cf. Chapter VII(7)). Consider two parallel conductors, (Fig. 2) one, AB, being infinitely long, and the other, $A'B'$, of length ℓ, both conducting currents $i = i' = 1$. Then, according to Ampère, the interaction between the two conductors only depends on $\frac{\ell}{y}$, where y is the distance between them. On the other hand, the interaction is obviously proportional to ℓ, so it must be equal to $\alpha\frac{\ell}{y}$, where α is a constant.

The interaction between AB and $A'B'$ can also be found by integrating the infinitesimal interactions (10) (with $i = i' = 1$, $\varepsilon = 0$, $\theta = \theta'$). After some transformations of this double integral, Liouville found

$$\int_0^\infty \left(1 + (k-1)\frac{s^2}{s^2+y^2}\right)\frac{\varphi(\sqrt{s^2+y^2})\,ds}{\sqrt{s^2+y^2}} = \frac{\alpha}{2y^2}, \qquad (10a)$$

which is an integral equation for the unknown function φ. Again, by (5) this integral equation can be transformed into a fractional differential equation:

$$\int^{\frac{1}{2}} F(z)\, d(z)^{\frac{1}{2}} + \frac{1-k}{2} \int^{\frac{3}{2}} \frac{F(z)}{z}\, dz^{\frac{3}{2}} = \frac{\alpha}{\sqrt{-1}\,\Gamma\left(\frac{1}{2}\right) z}, \tag{11}$$

where $\frac{\varphi(y)}{y} = F(z)$ and $y^2 = z$. Liouville took the derivative of order $\frac{1}{2}$ of both sides and obtained an ordinary differential equation for $\int \frac{F(z)\, dz}{z} = P$:

$$z\frac{dP}{dz} + \left(\frac{1-k}{2}\right) P = \frac{\alpha}{2z\sqrt{z}}.$$

It has the complete integral

$$P = \frac{C}{(1-k)/z^2} - \frac{\alpha}{k+2}\frac{1}{z\sqrt{z}},$$

where C is an arbitrary constant. Reintroducing the old variables, he obtained

$$\varphi(r) = -\frac{(1-k)C}{2r^{-k}} + \frac{3\alpha}{2(k+2)}\frac{1}{r^2}. \tag{12}$$

The last term is the usual expression for the force $\varphi(r)$, but Liouville pointed out that one cannot exclude the first term on the basis of the experiment used here. He explained that Ampère had not discovered this problem because he had a priori assumed $\varphi(r)$ to be of the form Ar^m. On the other hand, he did not reveal that he had himself overlooked the problem in his unpublished work [ACL Ms 2, no. 6]. Implicitly, however, he analyzed in [1832a, pp. 26–27] the mistake made in the early memoir, for he remarked that the term $\frac{A}{r^{-k}}$ in φ contributes to the total force between AB and $A'B'$ an amount

$$-2A\ell y^{k+1} \int_0^{\pi/2} (1 + (k-1)\cos^2\theta)\sin^{-(k+1)}\theta\, d\theta,$$

which is zero because the integral is zero. This is precisely the observation Liouville missed in [ACL Ms 2] (cf. Chapter VII, §11).

9. In his first paper [1832a], Liouville also showed how fractional calculus made possible the determination of the law of gravitation if one of the following interactions is known experimentally:

   a. the attraction between two parallel lines, one finite, the other infinite;

   b. the attraction between a parallelepiped and a mass element;

   c. the attraction between two parallelepipeds;

   d. the attraction between a circle and a mass element on its circumference (for varying radii); and

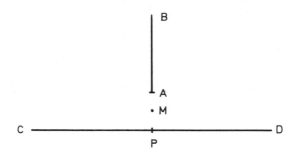

FIGURE 3

e. letting CD be an infinite line and AB a semi-infinite orthogonal line and letting a mass point M be situated halfway between the end A of AB and the intersection P between its prolongation and CD (Fig. 3), it is assumed that the attraction of M toward CD is twice its attraction toward AB for all values of the distance AP.

10.   As all these examples show, Liouville emphasized the use of the fractional calculus in the following general complex of problems: determine the microscopic interactions between molecules, conducting elements, etc., from experimental evidence concerning macroscopic bodies. This was a type of problem he had studied in connection with his earlier work on Biot and Savart's law and Ampère's law, and it seems most probable that these physical investigations constitute the origin of Liouville's calculus of arbitrary order. A likely reconstruction of his way to fractional calculus between 1828 and 1832 goes as follows:

In his unpublished memoir of 1828 [AC L Ms 2], he showed how to do away with the a priori assumption that the elementary forces are proportional to $r^{-\alpha}$ for some α. Instead, he assumed the r dependence to be of the form $1/\Sigma\, A_i\, r^{\alpha_i}$. It is most likely that under the impact of Cauchy's critical analysis of infinite series Liouville became dissatisfied with this approach as well and tried to find the r dependence $\varphi(r)$ directly. This must have led him to integral equations such as (7) and (10a). With formulas such as (VII, 19) in mind, he discovered that the convolution integrals entering the integral equations could be transcribed into derivatives of fractional order. His earlier work on partial differential equations and in particular the development of arbitrary functions in exponential series (VII, 17) and the formula (VII, 19) further suggested that the fractional derivatives could be defined by way of such series.

Thus, the fractional calculus grew naturally out of Liouville's early unpublished memoirs, and there is no reason to believe that he was inspired by Abel's Norwegian paper [1823] on the tautochrone, as Ross [1975, p. 2] suggests. Yet, it is possible that it was a knowledge of Abel's German paper on the tautochrone [1826b] which made Liouville show [1832a, pp. 49–54] that this problem can also be solved by fractional calculus. However, this is also uncertain, for he did not refer to Abel's paper until in [1834b, p. 283], and the first time it is mentioned in Liouville's notebooks is in a note from March 1833 [Ms 3615 (3), p. 91v].

But, even a knowledge of this paper would not have told Liouville anything about fractional calculus because Abel used only classical differential calculus in it. The fractional calculus is only to be found in the original 1823 Norwegian version to which Liouville never referred. Indeed, it is highly improbable that he would have known of this paper, which appeared in an obscure Norwegian popular science magazine.

11.   The generalized tautochrone problem was stated as follows [Liouville 1832a p. 49]:

> Determine the curve AMB [Fig. 4] so that the time it takes for a heavy body, sliding on this curve, to go from M to A, is a given function $f(h)$ of the vertical height $MP = h$, that separates M and A.

Let $\varphi(x)$ be the unknown arc length AM expressed as a function of the height x. By conservation of energy, Liouville found that the time of descent from height h can be expressed as

$$\frac{1}{\sqrt{2g}} \int_0^h \frac{\varphi'(x)\,dx}{\sqrt{h-x}} = f(h). \tag{13}$$

By a formula similar to (2), the left-hand side is reduced to a fractional integral so that (13) is equivalent to

$$\sqrt{-1}\,\sqrt{\pi}z \int^{\frac{1}{2}} \frac{dz^{1/2}}{z\sqrt{z}} \varphi'\left(\frac{1}{z}\right) = \sqrt{2g}\sqrt{z} f\left(\frac{1}{z}\right). \tag{14}$$

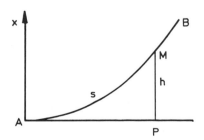

FIGURE 4

Liouville differentiated (14) $\frac{1}{2}$ a time and obtained φ first as a fractional derivative and later [1834b, p. 284] as an ordinary integral:

$$\varphi(x) = \frac{\sqrt{2g}}{\pi} \int_0^x \frac{f(\theta)\, d\theta}{\sqrt{(x - \theta)}}. \tag{15}$$

This is identical to Abel's formula. For the ordinary tautochrone, $f(h)$ is a constant, and in this case Liouville found that formula (15) led to the well-known cycloid.

12.   According to Ross [1977, p. 78], Abel solved this problem similarly in [1823]. Ross writes:

> In this problem, the time of slide is a known constant k such that
>
> $$k = \int_0^x (x - t)^{-\frac{1}{2}} f(t)\, dt. \tag{15a}$$
>
> Abel wrote the right side of [15a] as $\sqrt{\pi} \dfrac{d^{-\frac{1}{2}}}{dx^{\frac{1}{2}}} f(x)$. Then, he operated on both sides with $\dfrac{d^{\frac{1}{2}}}{dx^{\frac{1}{2}}}$ to obtain
>
> $$\frac{d^{\frac{1}{2}}}{dx^{\frac{1}{2}}} k = \sqrt{\pi} f(x).$$
>
> [Ross 1977, p. 78]

I admit that in [Lützen 1982b, p. 377] I followed Ross and claimed that Abel solved the tautochrone problem as Liouville later did. However, if I had read Abel's paper more thoroughly, I would have discovered that this is not true. In fact, Abel never solved the tautochrone problem by fractional calculus but merely showed how the solution, found by other means, could be written as a fractional derivative. In order to correct this error, I shall briefly summarize what Abel did in [1823, pp. 55–63]. He showed that the problem led to the integral equation

$$\psi(a) = \int_0^a \frac{ds}{(a - x)^n},$$

where s is the arc length of the desired curve, and ψ is a known function (in the ordinary case $n = \frac{1}{2}$). Abel then developed ψ and s in infinite power series and found without mention of fractional calculus that

$$s = \frac{1}{\Gamma(n)\Gamma(1 - n)} x^n \int_0^1 \frac{\psi(xt)\, dt}{(1 - t)^{1-n}}. \tag{15b}$$

Having determined s in the special case $n = \frac{1}{2}$ and in particular for $\psi = $ const, he remarked:

One can also express s in another way, which I will mention for its peculiarity, namely:

$$s = \frac{1}{\Gamma(1-n)} \int^n \psi x \, dx^n = \frac{1}{\Gamma(1-n)} \frac{d^{-n}\psi(x)}{dx^{-n}}. \qquad (15c)$$

[Abel 1823, p. 61]

Abel's proof of the peculiar expression ran as follows. Let $\psi(x) = \sum \alpha_m x^m$. Then,

$$\frac{d^k \psi(x)}{dx^k} = \sum \alpha_m \frac{\Gamma(m+1)}{\Gamma(m-k+1)} x^{m-k}.$$

If the well-known expression

$$\frac{\Gamma(m+1)}{\Gamma(m-k+1)} = \frac{1}{\Gamma(-k)} \int_0^1 \frac{t^m \, dt}{(1-t)^{1+k}}$$

is inserted in the formula for $\frac{d^k \psi(x)}{dx^k}$, one obtains

$$\frac{d^k \psi(x)}{dx^k} = \frac{1}{x^k \Gamma(-k)} \int_0^1 \frac{\psi(xt) \, dt}{(1-t)^{1+k}},$$

or, if $k = -n$,

$$\frac{d^{-n} \psi(x)}{dx^{-n}} = \frac{x^n}{\Gamma(n)} \int_0^1 \frac{\psi(xt) \, dt}{(1-t)^{1-n}}.$$

Abel then inserted this value into the solution (15b) and obtained the desired formula (15c). Thus, Abel simply showed that the solution (15b) could be written in this peculiar way, but he did not solve the problem by way of fractional calculus as Ross and I stated. This deprives Abel of the honor of being the first to apply fractional calculus, bestowed on him by Ross, and conveys it unambiguously onto Liouville.

13.   In his paper [1834b, pp. 281–283], Liouville also solved the tautochrone problem in a resisting medium when the resistance is proportional to the square of the velocity. His notebook [Ms 3615 (3), pp. 6r–8r] reveals that Liouville solved this problem in January 1832 and that he immediately thereafter [Ms 3615 (3), pp. 8r–9r] tried his hand at the generalized tautochrone in a resisting media, i.e., where the time of descent is an arbitrary function of the height. The latter problem, however, led to a fractional differential equation that he could not solve. Liouville often returned to variations of the tautochrone problem in his notes. For example, the following month [Ms 3615 (3), pp. 13v–14v] he considered a family of curves (Fig. 5) and asked for a curve OMN such that the time of descent from a variable point M to O in a nonresisting medium is a given function $G(\alpha)$ of the index of the curve $f(x, \alpha)$ on which M lies. Again, he found a fractional differential equation, but was unable to solve it.

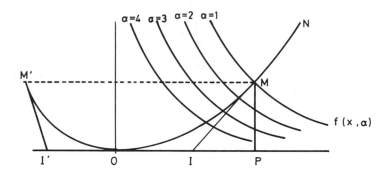

FIGURE 5

Later that month [Ms 3615 (3), pp. 23v–24v], he asked that the time of descent be $m \cdot \text{arc}\, OM + n \cdot$ height MP, and after a few fruitless attempts [Ms 3615 (3), pp. 24v–25r, 25v–26r], he succeeded in solving this problem during March 1832 [Ms 3615 (3), pp. 26v–28r]. Liouville's last notes on the fractional calculus, from 1843 or 1844, also deal with generalized tautochrone problems. He easily found the curve OM such that the time of descent is proportional to $MI + PM$ (Fig. 5), where I is the intersection of the x-axis and the tangent at M [Ms 3617 (4), p. 74v]. With more difficulty he also indicated a solution to the problem where the time of descent is proportional to the area MIP (Fig. 5).

Finally, he found the equations of two arcs, OM and OM' (Fig. 5), such that (1) the time it takes for a point to slide from M to a point M' at equal altitude is independent of M and (2) the excess of the time of descent over the time of ascent is proportional to the difference $MI - MI'$ between the two extreme tangents. Only the last of these results was passed to Liouville's colleagues. It was presented to the *Bureau des Longitudes* on February 28, 1844, and a note in Liouville's hand was glued into the *Procès Verbaux* of the meeting.[1]

14. In his first paper, Liouville also applied his new methods to solve three geometrical problems. They are all rather artificial and not at all as interesting as the physical ones. I shall limit the discussion of these examples to the enunciation of the first problem:

> It is a matter of tracing a curve AMB [Fig. 6] which has the following property:
>
> You draw an arbitrary ordinate MP and describe a parabola PQR having its vertex at P, its major axis on Ox, and its parameter $= 2OP = 2x$. Then, you construct a third curve PNV whose ordinate for each abscissa is equal to the product of the corresponding ordinates of the first two, so that $NS = BS \times QS$.

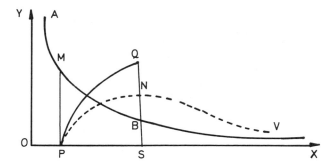

FIGURE 6

This being set out, one demands that the infinite area $xPNV$ be constant and equal to a^2 for all the positions of the ordinate MP.

[Liouville 1832a, p. 11]

15. Much more interesting were Liouville's remarks about the applicability of the fractional calculus in the solution of initial and boundary value problems for partial differential equations [1832a, pp. 57–64]. He illustrated this idea with a geometrical problem which led to the equation

$$y\frac{\partial^2 z}{\partial x \partial y} = \frac{\partial z}{\partial y} - \frac{3}{2}\frac{\partial z}{\partial x}. \tag{16}$$

The solution of this equation is of the form

$$z(x,y) = \int_0^\infty e^{-\alpha}\sqrt{\alpha}\varphi(x+\alpha y)\,d\alpha. \tag{17}$$

Liouville wanted to determine the arbitrary function φ such that $z(x,y)$ is a given function $f(x)$ for $y = 1$. Then, φ must be a solution to the integral equation:

$$\int_0^\infty e^{-\alpha}\sqrt{\alpha}\varphi(x+\alpha)\,d\alpha = f(x). \tag{18}$$

By formula (2), the left-hand side can be transformed into

$$-e^x\sqrt{-1}\frac{\sqrt{\pi}}{2}\int^{3/2} e^{-x}\varphi(x)\,dx^{3/2}, \tag{19}$$

so that $\frac{d^{3/2}}{dx^{3/2}}$ applied to (18) yields the desired value for φ:

$$\varphi(x) = -\frac{2e^x}{\sqrt{-1}\sqrt{\pi}}\frac{d^{3/2}(e^{-x}f(x))}{dx^{3/2}}. \tag{20}$$

Liouville also indicated how this method could be used to solve physical problems, such as propagation of sound and oscillations of chains. He even claimed that his methods were simpler than Fourier's [Liouville 1832a, p. 60].

16.   Therefore, it is not surprising that he also tried to use the fractional calculus to solve problems in the theory of heat, which occupied him during the same period. In January 1832, he wrote a note with the headline: "Application of the new calculus to the theory of heat" [Ms 3615 (3), pp. 4r–5r]. It gives an unsuccessful attempt to solve his favorite problem in the theory of heat at that moment (cf. Chapter XV, §12a), namely, "the problem proposed by Mr. Poisson, which consists of taking into account the variation of the radiating power at various points of the surface of the hot body." In particular, Liouville studied the steady-state temperature distribution $u(x, y)$ in the half-plane $y > 0$. He remarked that it must satisfy Laplace's equation:

$$\frac{\partial^2 u}{\partial x^2} + \frac{\partial^2 u}{\partial y^2} = 0.$$

He made an error and wrote u in the form

$$u = \int_{-\infty}^{\infty} \frac{x f(\alpha)}{x^2 + (y - \alpha)^2} \, dx , \tag{21}$$

where f is an arbitrary function. Liouville apparently did not discover that this expression for u does not satisfy Laplace's equation, and continued to investigate the usual boundary condition:

$$\frac{\partial u}{\partial y} + F(x)u = F_1(x) \qquad \text{for } y = 0. \tag{22}$$

Inserting expression (21) for u, this condition gives the integral equation for f:

$$\int_{-\infty}^{\infty} \frac{f'(\alpha) \, d\alpha}{x^2 + \alpha^2} + F(x) \int_{-\infty}^{\infty} \frac{f(\alpha) \, d\alpha}{x^2 - \alpha^2} = F_1(x).$$

At this point, Liouville remarked in the margin:

Everything is reduced, I think, to integrating:

$$\frac{d\left(x \frac{d^{1/2}y}{dx^{1/2}}\right)}{dx} + F(x)\frac{d^{1/2}(y\sqrt{x})}{d\left(\frac{1}{x}\right)^{1/2}} = f(x) \quad [F_1(x) \text{ in my terminology}].$$

In order to do that, it is necessary to give $\frac{d^{1/2}(y\sqrt{x})}{d(\frac{1}{x})^{1/2}}$ another form.

[Ms 3615 (3), p. 4r]

Either he later found the mistake in the beginning of the note or he could not solve the last problem. At any rate, he never continued this approach.

In later manuscripts and published papers [1836b] and [1838h] on heat conduction in thin plates, Liouville used only Fourier series and Fourier integrals. Only once did he publicly use fractional calculus in the theory of heat, namely, in the solution of a problem mentioned at the end of a long note on heat conduction in infinite plates:

> It only remains to extend this analysis to the case of a sphere of finite radius; this is in fact the question proposed by Mr. Poisson.
> [Ms 3615 (4), p. 15r]

Poisson had shown in [1823a, p. 299] that in this case the boundary value equation is of the form

$$F(r) = \tfrac{1}{2}\sqrt{\pi}r^{n+1} \int_0^{\pi} \psi(r\cos\omega)\sin^{2n+1}\omega \, d\omega. \tag{23}$$

Further, he claimed that it was easy to find ψ for "the simplest values of n, such as $n = 0$, $n = 1$ &$_c$. However, without entering into this determination, one can obtain another value of N [which determines the temperature distribution] that does not include the function ψ..." [Poisson 1823a, p. 299]. Thus, Poisson evaded the solution of (23), but Liouville showed in a note [1835d] that by way of his fractional calculus it could be solved along the same lines as the problems in [1832a].[2]

In a note from May 1836 [Ms 3615 (5), pp. 17r–18r], Liouville remarked that a formula due to Jacobi similarly allowed him to solve integral equations of the form

$$F(r) = \int_0^{\frac{\pi}{2}} \psi(r\cos\omega)\cos i\,x\,dx.$$

He repeated this remark in [1841c, p. 73].

17. I shall now conclude this section by summarizing the general lines of Liouville's application of the fractional calculus. His new calculus applied to such problems that led to integral equations, as he believed that most physical problems did:

> The solution of most physico-mathematical problems basically depends on a question similar to those we have dealt with, namely, the determination of an arbitrary function placed under the integral sign.... Thus, the properties of differentials of arbitrary order are linked with the most tricky and most useful mathematical theories.
> [Liouville 1832a, p. 15]

Liouville's programme for solving these integral equations consisted of transforming the definite integrals into fractional integrals, by formulas like (2) and (5), and then solving the resulting differential equation of fractional order. The solution is then usually expressed as a fractional derivative, but it can be transformed into a definite integral if that is preferable.

As I pointed out [Lützen 1982b], the integral equations encountered in this theory are of the first kind. The reader is referred to this paper for an appraisal of the place of fractional calculus in the history of integral equations. Here, I shall only point out that Liouville's method sketched above presented the first programme for the solution of integral equations.

Foundations

18.   Today the derivatives of arbitrary order are usually defined by way of the so-called Riemann-Liouville definition suggested by Holmgren [1863–1864]:

$$_cD_x^{-\nu}f(x) = \frac{1}{\Gamma(\nu)} \int_c^x (x-t)^{\nu-1} f(t)\, dt. \quad ^3 \tag{24}$$

Thus, there is a whole family of definitions, one for each value of the constant c. This definition is closely related to Liouville's fundamental formula

$$\int^\mu \varphi(x)\, dx^\mu = \frac{1}{(-1)^\mu \Gamma(\mu)} \int_0^\infty \varphi(x+\alpha)\alpha^{\mu-1}\, d\alpha. \tag{2}$$

Indeed, if we substitute $(x+\alpha) = t$ in the right-hand side of (2), we obtain

$$\int^\mu \varphi(x)\, dx^\mu = \frac{1}{(-1)^\mu \Gamma(\mu)} \int_x^\infty \varphi(t)(t-x)^{\mu-1}\, dt,$$

or equivalently,

$$\int^\mu \varphi(x)\, dx^\mu = \frac{1}{\Gamma(\mu)} \int_\infty^x \varphi(t)(x-t)^{\mu-1}\, dt, \tag{26}$$

which is Equation (24) for $c = \infty$. Thus, when the functions involved contain only exponents m with negative real parts in their exponential series $\sum A_m e^{mx}$, i.e., when (2) is valid, Liouville's fractional derivatives correspond to the modern $_\infty D_x^{-\nu}$. On the other hand, when the exponents in the exponential series have all positive real parts, formula (4) holds and that is equivalent to

$$\int^\mu \varphi(x)\, dx^\mu = \frac{1}{\Gamma(\mu)} \int_{-\infty}^x \varphi(t)(x-t)^{\mu-1}\, dt, \tag{27}$$

so in this case Liouville's derivatives correspond to $_{-\infty}D_x^{-\nu}$. Ross [1977, p. 85] has only mentioned the case $c = -\infty$, which is the case applied the least by Liouville. It is helpful to have in mind that Liouville's theory corresponds to two different values of c, for that elucidates some difficulties encountered by Liouville in the discussion of fractional derivatives of trigonometric functions, which do not fall off quickly at either plus nor minus infinity. Also, Liouville's peculiar ideas on differentiation of polynomials can be understood better when we keep in mind that in modern fractional calculus x^n $(n > 0)$ is usually differentiated by using $_0D_x^{-\nu}$ and not by either of the two operators that correspond to Liouville's theory.

19.   After this digression into modern theory, I shall return to the nineteenth century and try to explain Liouville's own ideas on the foundations of the theory. He justly criticized his predecessors, both for having neglected applications and also for having left the concepts in obscurity:

> But, the geometers whom I have quoted only deal with this matter in passing and have not gone deeper into the theory. The little they have said, which is no doubt very valuable, is far from being free from a certain obscurity. As far as I can see, one does not find an explicit and convenient definition of differentials of arbitrary order, and one sees still less what purpose these differentials may serve.
>
> [Liouville 1832a, p. 71]

In his second paper [1832b], he set out to remove these obscurities. Although his applications were all based on formula (2), he never seems to have considered this as a suitable definition. Instead, he preferred the definition in §1. Since this definition applies to exponential series, it became of vital importance for Liouville to show that all functions can be expanded in such a series. In the first paper [1832a, pp. 4,5], Liouville argued that in order to expand $F(x)$ in an exponential series one simply has to develop $F(\log z)$ in a Maclaurin series:

$$F(\log z) = \sum A_m z^m ,$$

and then reintroduce the variable $x = \log z$ or $e^x = z$:

$$F(x) = \sum A_m e^{mx}.$$

Poisson [1817] and [1823a, p. 370] used this argument to show that a solution to a partial differential equation could be expanded in an exponential series, and Liouville simply extracted it from its context (with due reference). The proof is clearly invalid because $F(\log z)$ need not have a Maclaurin expansion and also because $\log z$ has a singularity at zero.

In [1832b, pp. 74–76], Liouville gave this argument a twist and based it on the idea that "the simplest algebraic function x is in fact nothing but an exponential function with infinitely small exponents, because, for example, x is the limit of the ratio

$$\frac{e^{\beta x} - e^{-\beta x}}{2\beta} ,$$

when β becomes zero, after having been decreased indefinitely." [Liouville 1832b, p. 74]. He proved this theorem by expanding $e^{\beta x}$ and $e^{-\beta x}$ in power series. That yields

$$\frac{e^{\beta x} - e^{-\beta x}}{2\beta} = x + \frac{\beta^2 x^3}{3!} + \cdots ,$$

which reduces to x for $\beta = 0$. Therefore, by replacing x by $\frac{e^{\beta x} - e^{-\beta x}}{2\beta}$ in the expression for a function $F(x)$, it becomes a function of $e^x : f(e^x)$, and its power series expansion is transformed into an exponential series[4].

20.   Having thus acquired "exact and precise notions" concerning exponential expansions (the modern reader may not quite agree), Liouville showed how the ordinary functions can be expanded. In the first paper, he only needed to expand functions of the form $\frac{1}{x^n}$ for $n > 0$:

$$\frac{1}{x^n} = \frac{1}{\Gamma(n)} \int_0^\infty e^{-\alpha x} \, \alpha^{n-1} \, d\alpha. \tag{28}$$

He also considered such an integral to be a sum. In this way, the development into exponential series corresponds to taking the Laplace transform (cf. [Deakin 1981, pp. 369–371]). In [1832b], Liouville further found more peculiar expansions of $\frac{1}{x^n}$ for $n < 0$, $n \notin \mathbb{Z}$, of $\log x$ and of the trigonometric functions, whose expansions are provided by Euler's formulas, for example:

$$\cos x = \frac{e^{x\sqrt{-1}} + e^{-x\sqrt{-1}}}{2}. \tag{29}$$

As remarked above, Liouville expanded x in exponential series by using infinitely small exponents:

$$x = \frac{e^{\beta x} - e^{-\beta x}}{2\beta} \quad (\beta \text{ infinitely small}). \tag{30}$$

He pointed out that this expansion is not unique. Expansion in a power series, as above, shows that

$$x = \frac{e^{\beta x} + 2e^{4\beta x} - 3e^{-\beta x}}{12\beta} \quad (\beta \text{ infinitely small}) \tag{31}$$

will also do the job. It is easy to see that there is an infinity of ways of expanding x in exponential series with infinitely small exponents. Multiplying n of these expressions leads to an expansion of x^n. Thus, Liouville proved that:

> Every algebraic and entire function such as $c_0 + c_1 x + c_2 x^2 + \cdots + c_n x^n$ can be transformed into a series of exponentials $\sum A_m e^{mx}$, where all the exponents m are infinitely small.
>
> [Liouville 1832b, p. 89]

With an argument, which is incomplete, even by his own standards, he also believed he had established the converse:

> and conversely, if in exponential function $\sum A_m e^{mx}$ none of the exponents have a finite value, this quantity is equivalent to an entire algebraic function.
>
> [Liouville 1832b, p. 89][5]

21. Having thus expanded all the simple functions in exponential series, it should in principle be easy to find their derivatives of arbitrary order. However, Liouville encountered several difficulties. Except for the exponential function, the easiest functions to differentiate were $\frac{1}{x^n}$ for $n > 0$. Using the definition and formula (28), Liouville [1832a, p. 7] showed that for $n + \mu > 0$:

$$\frac{d^\mu \frac{1}{x^n}}{dx^\mu} = \frac{(-1)^\mu \Gamma(n+\mu)}{\Gamma(n)} \frac{1}{x^{n+\mu}} . \tag{32}$$

In [1832b, pp. 143–162], Liouville generalized this formula to other values of n and μ by way of an argument involving complementary functions. Liouville described these considerations as "very tricky" [1834a, p. 3], but his readers would probably have called them complicated, unelegant, and confused. However, in [1834a], Liouville conversely wanted to determine the complementary function from the extension of (32), and therefore he provided a generalization, that was independent of the complementary functions. It turned out to be much simpler than the original generalization of (32).

The last method of generalization was simpler because Liouville based it on the extension of the Γ-function to values of x with negative noninteger real parts. At the negative values of x, the integral (3a) diverges, but by successive use of the functional equation,

$$\Gamma(1 + x) = x\Gamma(x) ,$$

Legendre [1811–1817, vol. 2, pp. 60–63] extended the Γ-function — or the Eulerian integral of the second kind, as he called it — to the negative half-axis, except $x = -1, -2, -3, \ldots$ Liouville remarked that this extension also works for complex values of x. Later, Liouville learned of Gauss's definition of the Γ-function [Gauss 1813, §20], and in [Liouville 1855f, p. 157], he acknowledged that it provided "the best basis for the theory of the Γ-function."

Having extended the Γ-function, Liouville proved that formula (32) is valid for $n + \mu \notin Z_- \cup \{0\}$ and $n \notin Z_- \cup \{0\}$. The first step in this proof is to show that (32) is also applicable if $-1 < n + \mu < 0$ and $n > 0$. In this case, $n + \mu + 1 > 0$, so that (32) yields

$$\frac{d^{\mu+1} \frac{1}{x^n}}{dx^{\mu+1}} = \frac{(-1)^{\mu+1} \Gamma(n+\mu+1)}{\Gamma(n) x^{n+\mu+1}} .$$

Integrating this formula once, Liouville found

$$\frac{d^\mu \frac{1}{x^n}}{dx^\mu} = \frac{(-1)^\mu \Gamma(n+\mu+1)}{(n+\mu)\Gamma(n) x^{n+m}} ,$$

which yields (32) because $\frac{\Gamma(n+\mu+1)}{(n+\mu)} = \Gamma(n + \mu)$. In this way, Liouville recursively established (32) for real variables, and he remarked that his proof was equally applicable to complex values.

22.   However, formula (32) does not apply when $n \in Z_- \cup \{0\}$, i.e., it leaves undetermined the derivatives of a polynomial. In fact, because of the ambiguity of the exponential expansion of the functions x^n, $n = 0, 1, \ldots$ mentioned in §20, the derivatives of arbitrary order of polynomials are undetermined. Liouville argued [1832b, p. 144] that if

$$\psi = A + Bx + Cx^2 + \cdots + Kx^n$$

then "among all the exponential developments of ψ, there is one for which $\frac{d^\mu \psi}{dx^\mu} = 0$, while more generally $\frac{d^\mu \psi}{dx^\mu}$ is an algebraic entire function." [Liouville 1832b, p. 144].

Thus, the derivative of any polynomial is any other polynomial. In particular, the fractional derivative of zero can be any polynomial. This particular important theorem was proved as follows [Liouville 1832b, pp. 95–101] for $\mu = \frac{1}{2}$. Liouville first showed that $\frac{d^{\frac{1}{2}}}{dx^{\frac{1}{2}}}(0)$ can be any constant. To this end, he remarked that

$$\psi = \sqrt{\beta}\frac{e^{\beta x} - e^{-\beta x}}{2\beta} \qquad (\beta \text{ infinitely small})$$

is equal to zero (cf. (30)). Therefore,

$$\frac{d^{1/2}0}{dx^{1/2}} = \frac{d^{1/2}\psi}{dx^{1/2}} = \frac{e^{\beta x} - e^{-\beta x}\sqrt{-1}}{2} = \frac{1 - \sqrt{-1}}{2}.$$

On the other hand, when c is an arbitrary complex number, $c\psi$ is also zero, so that

$$\frac{d^{1/2}0}{dx^{1/2}} = \frac{d^{1/2}(c\psi)}{dx^{1/2}} = c\frac{1 - \sqrt{-1}}{2},$$

which is an arbitrary complex constant. He similarly showed that $\frac{d^{1/2}0}{dx^{1/2}}$ can be any first-degree polynomial $c_0 + c_1 x$, and it is clear that this proof can be generalized to any polynomial $c_0 + c_1 x + c_2 x^2 + \cdots + c_n x^n$.

Now, since $f(x) = f(x) + 0$, Liouville concluded that

$$\frac{d^\mu f(x)}{dx^\mu} = \frac{d^\mu f(x)}{dx^\mu} + \frac{d^\mu 0}{dx^\mu} = \frac{d^\mu f(x)}{dx^\mu} + c_0 + c_1 x + c_2 x^2 + \cdots + c_n x^n.$$

Therefore, whenever a fractional derivative is performed, one must add a *complementary function*, which is a polynomial of arbitrary degree and with arbitrary coefficients.

Liouville also argued [1832b, pp. 101–106] that the complementary function cannot be more general than a polynomial. For assume that $\sum A_\beta e^{\beta x} \beta^\mu$ is a complementary function so that

$$\frac{d^\mu \sum A_m e^{mx}}{dx^\mu} = \sum A_m e^{mx} m^\mu + \sum A_\beta e^{\beta x} \beta^\mu, \qquad (33)$$

then by integration μ times,

$$\sum A_m e^{mx} = \sum A_m e^{mx} + \sum A_\beta e^{\beta x}, \tag{34}$$

so that $\sum A_\beta e^{\beta x} = 0$. But, this implies that all the exponents β are infinitely small, otherwise the [numerically] largest exponent would make the sum tend to plus or minus infinity for x tending to $\pm\infty$. Thus, $\sum A_\beta e^{\beta x} \beta^\mu$ contains only infinitely small exponents and is therefore a polynomial, according to the theorem in §20. This argument, however, is wrong, for in the deduction of (34) Liouville has forgotten the complementary function which could a priori be any function.

From the ordinary calculus, however, it is known that an n-fold integral ($n \in \mathbf{N}$) only allows a complementary polynomial of degree $n-1$ and a derivative allows no complementary function at all. Liouville was also able to explain these restrictions in the complementary functions. For returning to the proof above, one has to keep in mind that $\sum A_\beta e^{\beta x} \beta^\mu$ was a complementary function, corresponding to $\frac{d^\mu}{dx^\mu}$ only if $\sum A_\beta e^{\beta x} = 0$. Expanding the exponentials into power series, Liouville obtained

$$\sum A_\beta e^{\beta x} = \sum A_\beta + \frac{x}{1} \sum A_\beta \beta + \frac{x^2}{1 \cdot 2} \sum A_\beta \beta^2 + \cdots = 0,$$

or

$$\sum A_\beta = 0$$
$$\sum A_\beta \beta = 0$$
$$\sum A_\beta \beta^2 = 0$$
$$\vdots \tag{35}$$

"essential conditions to which A_β is subject" [Liouville 1832b, p. 105]. The complementary function $\sum A_\beta e^{\beta x} \beta^\mu$ has the power series

$$\sum A_\beta e^{\beta x} \beta^\mu = \sum A_\beta \beta^\mu + \frac{x}{1} \sum A_\beta \beta^{\mu+1} + \cdots.$$

If μ is a positive integer, all the terms in this series are zero by (35), and thus the complementary functions are zero. If μ is a negative integer, all the terms from the $|\mu|$th are zero by (35), so that the complementary function is a polynomial of degree $|\mu| - 1$.

23. The above analysis of the complementary functions was contained in Liouville's second paper [1832b]. As I have already indicated, Liouville provided a new approach to the problem in his first paper in Crelle's *Journal* [1834a]. He used formula (32) for $n + \mu \neq -1, -2, \ldots$, interpreting $\Gamma(-n)$ to be infinite for $n \in \mathbf{N}$. Since the complementary function ψ corresponding to $\frac{d^{-\mu}}{dx^{-\mu}}$ has

$$\frac{d^\mu \psi}{dx^\mu} = 0, \tag{36}$$

Liouville wanted to determine the general solution to this equation. Suppose ψ is developed in a series

$$\psi = \sum \frac{A_n}{x^n}, \quad A_n \neq 0. \tag{37}$$

Then, according to (32),

$$\frac{d^\mu \psi}{dx^\mu} = (-1)^\mu \sum \frac{A_n \Gamma(n+\mu)}{\Gamma(n) x^{n+\mu}}.$$

If $\frac{d^\mu \psi}{dx^\mu}$ is zero, all the terms of the sum must be zero, so that

$$\frac{\Gamma(n+\mu)}{\Gamma(n)} = 0.$$

Since $\Gamma(x)$ is never zero, Liouville concluded that $\Gamma(n) = \infty$, which means that $n = 0$, or $n = -1$, or $n = -2$, or.... Therefore, the sum (37) is a polynomial Q.E.D. The proof does not apply to ordinary derivatives or integrals, because in these cases μ is an integer and some of the coefficients $\frac{\Gamma(n+\mu)}{\Gamma(n)}$ reduce to $\frac{\infty}{\infty}$.

24.   The complementary functions give Liouville's theory a peculiar undetermined character. However, he pointed out [1832b, p. 106] that in many applications the complementary functions can be omitted. For example, if it is known that the function $\frac{d^\mu f}{dx^\mu}$ vanishes at infinity and one has found a particular expression for $\frac{d^\mu f}{dx^\mu}$ with this property the complementary polynomial $c_0 + c_1 x + \cdots + c_n x^n$ must be zero. That is why Liouville found the correct solutions to the problems in his first paper, although he neglected the complementary functions completely. In the later application to Poisson's integral equation [Liouville 1835d], he was careful to introduce a complementary function and determine it by the conditions at infinity.

25.   Differentiation of trigonometric functions rival differentiation of polynomials in complexity, and Liouville devoted 24 pages of his second paper to this subject [1832b, pp. 119–143]. He first remarked that by Euler's formula (29) one finds

$$\frac{d^\mu \cos mx}{dx^\mu} = \frac{e^{mx\sqrt{-1}}(m\sqrt{-1})^\mu + e^{-mx\sqrt{-1}}(-m\sqrt{-1})^\mu}{2},$$

which can easily be reduced to

$$\frac{d^\mu \cos mx}{dx^\mu} = m^\mu \cos\left(mx + \frac{\mu\pi}{2}\right). \tag{38}$$

However, since $\cos mx = \cos(-mx)$, one may change m in the above formula into $-m$ so that

$$\frac{d^\mu \cos mx}{dx^\mu} = (-1)^\mu m^\mu \cos\left(mx - \frac{\mu\pi}{2}\right), \tag{39}$$

which is only identical to the first expression if μ is an integer. This shows that the derivative of $\cos mx$ has two values, and if we use $\cos mx = a \cos mx + (1 - a) \cos(-mx)$, we see that its derivative has infinitely many expressions.

Liouville gave several other arguments to show that the noninteger derivatives of the trigonometric functions are necessarily indeterminate. In the last of the arguments, he believed that he had found the true reason for this ambiguity. He determined, in two different ways, the derivatives of the product $e^{nx} \cos mx$ and found for $n > 0$:

$$\frac{d^{\mu}}{dx^{\mu}} e^{nx} \cos mx = e^{nx} \rho^{\mu} \cos(mx + \mu z), \tag{40}$$

and

$$\frac{d^{\mu}}{dx^{\mu}} e^{-nx} \cos mx = e^{-nx} (-1)^{\mu} \rho^{\mu} \cos(mx - \mu z), \tag{41}$$

where $\rho = \sqrt{m^2 + n^2}$ and $z = \arg\left(\frac{n+im}{\rho}\right)$. For n tending to zero, the first formula yields (38) and the second yields (39).

> One can see that this [formula 39] is the true one when $\cos mx$ is the limit of the quantity $e^{-nx} \cos mx$, which tends towards zero when the variable increases toward the positive infinity; and that [formula 38], on the contrary, must be used when $\cos mx$ is the limit of the quantity $e^{nx} \cos mx$, which becomes infinitely small when $x = -\infty$.
> [Liouville 1832b, p. 135]

This distinction corresponds to the distinction between formulas (2) and (4) or between the modern definitions of $_{\infty}D_x^{-\nu}$ and $_{-\infty}D_x^{-\nu}$, but Liouville made no remark to this effect.

26.   He did remark, however, that it made Fourier's integral formulas problematic. In [1832b], he stated:

> This indeterminacy does not prevent the derivatives we are talking about from being of a great use in analysis. Nevertheless, I must admit that it constitutes the main reason that has made me define, ..., the development in exponential series on which the new calculus is based. This shows why we have explicitly said that we shall not make use of a development in imaginary exponentials, that is in series of trigonometric curves such as those given by Fourier's theorem.
> [Liouville 1832b, p. 124]

In [Liouville 1835b], which was devoted particularly to that problem, Liouville more directly criticized Fourier's definition. Recall that Fourier [1822, p. 561] had defined

$$\frac{d^{\mu} f}{dx^{\mu}} = \frac{1}{2\pi} \int_{-\infty}^{\infty} d\alpha f(\alpha) \int_{-\infty}^{\infty} dp\, p^{\mu} \cos\left(px - p\alpha + \frac{\mu\pi}{2}\right), \tag{3}$$

which is the value obtained by differentiating under the integral sign in his formula[6]

$$f(x) = \frac{1}{2\pi} \int_{-\infty}^{\infty} d\alpha f(\alpha) \int_{-\infty}^{\infty} dp \cos(px - p\alpha) \tag{42}$$

and using the expression (38) for $\frac{d^\mu \cos mx}{dx^\mu}$. Liouville pointed out that this definition was based on an arbitrary choice, for if Fourier had used the formula

$$f(x) = \frac{1}{\pi} \int_{-\infty}^{\infty} d\alpha f(\alpha) \int_{0}^{\infty} dp \cos(px - p\alpha) \tag{43}$$

instead, he would have found

$$\frac{d^\mu f}{dx^\mu} = \frac{1}{\pi} \int_{-\infty}^{\infty} d\alpha \, f(\alpha) \int_{0}^{\infty} dp \, p^\mu \cos\left(px - p\alpha + \frac{\mu\pi}{2}\right), \tag{44}$$

which for $i \notin \mathbf{Z}$ does not generally give the same value as (3). For example, if $\mu = \frac{1}{2}$, (44) gives a purely real value if f is real, whereas Fourier's definition contains an imaginary part. Thus, even without using the indeterminacy of the derivatives of the trigonometric functions, inherent in his own definition, Liouville could point to an a priori arbitrariness in Fourier's definition that was not to be found in his own definition.

Liouville also pointed out that Fourier's definition always led to complex values of $\frac{d^{\frac{1}{2}}f}{dx^{\frac{1}{2}}}$ when f is real, whereas his own definition often gave real values of this derivative. Therefore, the two definitions are effectively different. Still, Liouville devoted the major part of his paper [1835b] to proving two formulas that would replace Fourier's (3) in his own theory. They are based on Fourier's two formulas:

$$f(x) = \frac{1}{\pi} \int_{-\infty}^{K} d\alpha \, f(\alpha) \int_{0}^{\infty} dp \cos p(x - \alpha) \quad \text{for} \quad x \in (-\infty, K) \tag{45}$$

and

$$f(x) = \frac{1}{\pi} \int_{-K}^{\infty} d\alpha \, f(\alpha) \int_{0}^{\infty} dp \cos p(x - \alpha) \quad \text{for} \quad x \in (-K, \infty). \tag{46}$$

He applied the first formula to functions $f(x) = \sum A_m e^{mx}$, where all exponents m have positive real parts and obtained:

$$\frac{d^\mu f(x)}{dx^\mu} = \frac{1}{\pi} \int_{-\infty}^{K} d\alpha \, f(\alpha) \int_{0}^{\infty} dp \, p^\mu \cos\left(px - p\alpha + \frac{\mu\pi}{2}\right), \, x \in (-\infty, K). \tag{47}$$

Functions with only negative real parts in the exponents in their exponential series can be integrated to $+\infty$, so in that case he used the last formula and obtained:

$$\frac{d^\mu f(x)}{dx^\mu} = \frac{(-1)^\mu}{\pi} \int_{-K}^{\infty} d\alpha \, f(\alpha) \int_{0}^{\infty} dp \, p^\mu \cos\left(px - p\alpha - \frac{\mu\pi}{2}\right), \, x \in (-\infty, K). \tag{48}$$

27.   Formulas (47) and (48) can be obtained by formally differentiating under the integral sign in (45) and (46) and using expression (38) in the first case and (39) in the second. That is in accordance with Liouville's remark quoted in §25, for since the integral of (45) goes to $-\infty$, one ought to consider $\cos mx$ as the limit of $e^{nx} \cos mx$ $(n > 0)$ and therefore (38) should be used, whereas (39) should be applied to (46) because its integral goes to $+\infty$.

However, Liouville did not derive (47) and (48) in this way because he had begun to doubt that it was permissible to differentiate fractionally under the integral sign (cf. [Liouville 1832b, p. 125, footnote]). He avoided this by applying formulas for the derivatives of arbitrary order, which generalized Cauchy's definition of a derivative as the limit of a difference quotient. These formulas (found in [Liouville 1832b, pp. 106–113]) are

$$\frac{d^\mu}{dx^\mu} f(x) = \frac{1}{h^\mu}\left(f(x) - \frac{\mu}{1}f(x-h) + \frac{\mu(\mu+1)}{2!}f(x-2h) - \cdots\right), \quad (49)$$

and

$$\frac{d^\mu}{dx^\mu} f(x) = \frac{(-1)^\mu}{h^\mu}\left(f(x) - \frac{\mu}{1}f(x+h) + \frac{\mu(\mu-1)}{2!}f(x+2h) - \cdots\right), \quad (50)$$

where h is infinitely small. The first formula applies to functions that have exponents m with positive real parts in their exponential series $\sum A_m e^{mx}$, whereas the last formula applies to functions with $\mathrm{Re}\, m < 0$. Liouville [1835b, pp. 224–227] calculated $\frac{d^\mu}{dx^\mu} f(x)$ by adding, according to (49), the double integrals of (45) corresponding to $f(x)$, $\frac{\mu}{1}f(x-h)$, etc., and interchanging the summation and integration (and implicitly the $\lim_{h\to 0}$ and integration also); thereby, he found (47). Similarly, from (46) and (50), he found (48).

28.   Before I finish this summary of the most important aspects of Liouville's foundations of his theory, I need to mention that he established the most common rules of differentiation in [1832b, pp. 113–119]. Among these we find the trivial rules

$$\frac{d^\nu}{dx^\nu}\left(\frac{d^\mu f(x)}{dx^\mu}\right) = \frac{d^{\mu+\nu} f(x)}{dx^{\mu+\nu}},$$

and

$$\frac{d^\mu(f+g)}{dx^\mu} = \frac{d^\mu f}{dx^\mu} + \frac{d^\mu g}{dx^\mu},$$

and also the more complicated rule

$$\frac{d^\mu \varphi\varphi_1}{dx^\mu} = \varphi_1 \frac{d^\mu \varphi}{dx^\mu} + \frac{\mu}{1}\frac{d\varphi_1}{dx}\frac{d^{\mu-1}\varphi}{dx^{\mu-1}} + \frac{\mu(\mu+1)}{1\cdot 2}\frac{d^2\varphi_1}{dx^2}\frac{d^{\mu-2}\varphi}{dx^{\mu-2}} + \cdots. \quad (51)$$

In a separate paper [1835c], he also discussed differentiation of composite functions $f(\omega(z))$. He studied the problem of expressing $\frac{d^\mu}{dz^\mu} f(\omega(z))$ in terms of $\frac{d^\nu}{dx^\nu} f(x)$ and $\frac{d^\rho}{dz^\rho}\omega(z)$, but admitted that he had not been able to solve this problem generally. He only solved it when $x = \omega(z) = z^n$, $n \in N$, or $z = \sqrt[n]{x}$. To begin with, he found four different formulas for $\frac{d^{\frac{1}{2}}y}{d(\sqrt{x})^{\frac{1}{2}}}$, for example,

$$\frac{d^{1/2}y}{d(\sqrt{x})^{1/2}} = \sqrt{2} \int^{1/4} x^{1/4} \frac{d^{3/4}y}{dx^{3/4}} \, dx^{1/4}. \tag{51a}$$

He proved this formula with two different proofs, one based on development of y in exponential series, the other using power series and an identity for the Γ-function. He then generalized the second proof and found $\frac{d^\mu y}{d(\sqrt{x})^\mu}$ and more generally $\frac{d^\mu y}{d(\sqrt[n]{x})^\mu}$. For the last quantity, he found $(n!)^2$ different expressions.

It is clear from Liouville's notes that he had discovered the second proof of (51a) in January 1832 [Ms 3615 (3), pp. 2r–4r]. It seems to have been in this connection that Liouville discovered the general applicability of formula (32) "if one attributes to the notation of the Γ-function all the extent that Legendre has given it" [Ms 3615 (3), p. 63v, July 1832, note 1].

29.   Liouville knew that there were alternative definitions of derivatives of arbitrary order. We have already seen how he compared his definition with Fourier's, and in [1834a, pp. 15–19] he compared it with Euler's method of fractional calculus. He quoted at length several pages of Euler's paper, *De progressionibus transcendentibus seu quarum termini generales algebraice dari nequeunt* [1730–1731], in which Euler, inspired by the ordinary formula

$$\frac{d^n (x^m)}{dx^n} = \frac{m!}{(m-n)!} x^{m-n}$$

defined

$$\frac{d^\mu (x^m)}{dx^\mu} = x^{m-\mu} \frac{\Gamma(m+1)}{\Gamma(m-\mu+1)}, \tag{52}$$

also for fractional values of μ.[7] Liouville pointed out that this formula clashes with his own formula (32), and he showed that the two methods gave entirely different results when applied to a fractional differential equation treated by Euler.

Liouville also realized that his own system of fractional calculus could be founded on other definitions than the one he chose in the first paper. For example in [1832b, p. 109], he observed that the generalized difference quotient formula (49) provided "a way to introduce even in the elements [of the curriculum] the idea of arbitrary differentials." However, partly because of problems with convergence of the series in (49), this alternative definition "does not seem to me [Liouville] to ensue from the true origin of these differentials" [1832b, p. 109]. Later, however, related definitions were suggested by A. K. Grunwald [1867] and recently by Butzer and Westfall [1975].

Similarly, Liouville pointed out that formula (32) could be used in an obvious way to define $\frac{d^\mu f}{dx^\mu}$ when f is expanded in the power series $f(x) = \sum \frac{A_n}{x^n}$. However, he preferred his original definition.

> This is not the place to go into the details of the very numerous reasons that have made us prefer our definition to the one provided by the equation (B) [32]
>
> [Liouville 1834a, p. 15][8]

Liouville was never very outspoken about these "very numerous reasons," but one finds hints to what he had in mind in the reflections following his first introduction of the definition:

> This definition is of great importance. It forms the crux of the difficulty of establishing the theory of the new calculus on a solid basis; for it does not suffice to create at random arbitrary notions and notations. It is necessary that these notions are in the nature of things, in order to make them useful and in order that the collection of operations which they give rise to merits the name of a calculus. The notation in particular is of a great importance since the entire analysis is, so to speak, a science of notations. It is therefore after a careful examination that I present the equation (1) as containing the true definition of differentials of arbitrary order, and these very differentials as an indispensable branch of mathematics, which it is impossible to set aside.
>
> [Liouville 1832a, p. 4]

Thus, he chose his definition so as to make the theory an applicable, connected calculus with a suitable notation. At the end of the paper, he indicated that Leibniz's ideas on the analogy between powers and differences also ensured him that his definition was the true one:

> I shall end this Memoir by observing that the definition [(1)], which is based on the exponential development of functions, contains the true key to what has been called the analogy between powers and differences, and at once makes the theorems very evident. Moreover, it was while working with this analogy, which he had discovered, that Leibniz was led to the differentials of arbitrary order, as is proved by his correspondence with Johann Bernoulli.
>
> [Liouville 1832a, p. 69]

Although Liouville never mentioned it explicitly, I believe that the applications of the fractional calculus convinced him that his version was better than Euler's. For, in Euler's fractional calculus, the derivatives of negative powers of x do not exist, and these were the most important in the applications to potential theory and electromagnetism.

There is no indication from the 1830s in either his publications or his notebooks that Liouville had any doubt that he had adopted the true definition. On the whole, there is no evidence that he ever seriously doubted the foundations of his new calculus. This was surely due to the fact, which I have tried to show in this section, that in spite of the peculiarities of Liouville's theory it was rather coherent and consistent.

Fractional Differential Equations

30.   Many applications of the fractional calculus lead to differential equations of fractional order. In [Liouville 1832a], most of the differential equations were of the simple form

$$\frac{d^\mu y}{dx^\mu} = f(x),$$

which can be solved by integrating both sides μ times (cf. (8), (14)). Others could be solved by equally simple means (cf. (11)). But, the last problem in [1832a, p. 67], led Liouville to the more complicated equation

$$y - nx\frac{d^{1/2}y}{dx^{1/2}} = 0, \tag{53}$$

where n is a complex constant.[9] To solve it, Liouville adapted the Laplace transform method, which he had learned from the *Théorie analytique des probabilité* [Laplace 1812, 3rd ed., p. 111], where Laplace used it to solve ordinary differential equations. Liouville wrote

$$y = \int_0^\infty e^{-\alpha^2 x}\psi(\alpha)\,d\alpha + A, \tag{54}$$

where A is a constant, and ψ is a function to be determined so that y fulfills (53). Differentiating under the integral sign, Liouville obtained

$$\frac{d^{1/2}y}{dx^{1/2}} = i\int_0^\infty e^{-\alpha^2 x}\alpha\psi(\alpha)\,d\alpha,$$

or, integrating by parts,

$$\frac{d^{1/2}y}{dx^{1/2}} = i\frac{\psi(0)}{2x} + \frac{i}{2x}\int_0^\infty e^{-\alpha^2 x}\,d\psi(\alpha). \tag{55}$$

Inserting these values of y and $\frac{d^{1/2}y}{dx^{1/2}}$ into (53) yields

$$\int_0^\infty e^{-\alpha^2 x}\psi(\alpha)\,d\alpha + A = \frac{1}{2}ni\,\psi(0) + \frac{1}{2}ni\int_0^\infty e^{-\alpha^2 x}\,d\,\psi(\alpha), \tag{56}$$

from which Liouville concluded that

$$A = \tfrac{1}{2}ni\,\psi(0) \quad \text{and} \quad \psi(\alpha)\,d\alpha = \tfrac{1}{2}ni\,d\psi(\alpha).$$

The last equation yields $\psi(\alpha) = ce^{(-2i/n)\alpha}$ so that $A = \frac{1}{2}nic$. When these values are applied to (54), one obtains the solution

$$y = c\left(\frac{1}{2}ni + \int_0^\infty e^{-\alpha^2 x} e^{-\frac{2i}{n}\alpha}\, d\alpha\right), \tag{57}$$

which, according to Liouville, is not the complete but only a particular integral.

31.   Five years later, Liouville published a paper [1837m] entirely devoted to the integration of differential equations of fractional order. This memoir contained three other methods of solution. However, it appears from his notebooks that he had discovered two of them already in 1832. Thus, after having found the complicated fractional differential equation for the generalized tautochrone in a resisting medium (cf. §13), he remarked in the 7th note from early January 1832:

> As one can see, one is more and more often led to the <u>necessity</u> of considering and integrating differential equations of fractional order. Therefore, it will be useful to work a little with this type of equation. One should, in particular, consider theories that might have some generality!
>
> [Ms 3615 (3), p. 9r]

As an example of how to find the complete solution, Liouville returned to equation (53). He differentiated the equation half a time and obtained

$$\frac{d^{1/2}y}{dx^{1/2}} - nx\frac{dy}{dx} - \frac{ny}{2} = \psi_{1/2}, \tag{58}$$

because by the product rule (51),

$$\frac{d^{1/2}}{dx^{1/2}}\left(x\frac{d^{1/2}y}{dx^{1/2}}\right) = x\frac{dy}{dx} + \frac{y}{2}. \tag{59}$$

Here $\psi_{1/2}$ denotes a complementary function. Eliminating $\frac{d^{1/2}y}{dx^{1/2}}$ from (53) and (58) leads to the linear first-order differential equation

$$\frac{dy}{dx} + y\left(\frac{1}{2x} - \frac{1}{n^2 x^2}\right) = -\frac{1}{nx}\psi_{1/2} \tag{60}$$

of which Liouville easily found the solution:

$$y = \frac{e^{-1/n^2 x}}{\sqrt{x}}\left(c - \frac{1}{n}\int \frac{\psi_{1/2}}{\sqrt{x}} e^{1/n^2 x}\, dx\right). \tag{61}$$

32.   Liouville immediately saw that this elimination of fractional derivatives by successive differentiation provided a general method for the solution of certain linear fractional equations, and in the next note, dated

January 3, 1832, he discussed its application to such equations with polynomial coefficients. He began with equations with constant coefficients. I have quoted this part of the argument in extenso in note 10, both in order to illustrate the use of the method and to show that Liouville's ideas, later published in [1837m], were indeed completely developed in 1832. The quote shows how the equation

$$Ay + B\frac{d^{1/3}y}{dx^{1/3}} + C\frac{d^{2/3}y}{dx^{2/3}} + D\frac{dy}{dx} = f(x)$$

can be reduced to a system of three ordinary differential equations by differentiating $\frac{1}{3}$ and $\frac{2}{3}$ times.

Liouville continued by showing that this method is applicable to general linear fractional differential equations with constant coefficients. That is obvious when the fractions are all reduced to a common denominator, as in note 10. Then, he proved by an example that the method is also applicable to linear equations with polynomial coefficients. The solution of (53) has already given us an example, so I shall not go into more detail, except to point out that the method works in this case because the series in the product rule (51) breaks off and becomes finite when one of the functions is a polynomial, as in (59). Liouville completed the note with a programme for further research:

> The method that we have explained for linear equations is, as one can see, very general, but it is necessary to extend it further and to show some applications both by integrating particular equations which depend on it and by solving some problems leading to these particular equations. It is also necessary to deduce a theorem concerning the number of particular integrals which are sufficient in order that the complete integral can be easily deduced.
>
> January 3, 1832, Joseph Liouville [Ms 3615 (3), p. 11r]

33.   During the following years, he used the method to solve some mechanical problems, but they were never published (see §13). The rest of the programme was never carried out, and the published account of the method in [1837m, pp. 71–83] was just an elaboration of the note, using other examples.

Liouville probably lost some of his enthusiasm for the method in February 1832, when in a series of notes [Ms 3615 (3), pp. 14v–23v] he compared and analyzed solution (61) of (53) and his earlier solution based on the Laplace transform. These confused notes show that at this point Liouville began to doubt both methods of solution, and indeed he had good reasons to be dubious.

It appears from the notes [Ms 3615 (3), p. 14v] that in the manuscript of his first paper, which he had sent to the *Journal de l'École Polytechnique*, he solved equation (53) by using a two-sided "Laplace transform":

$$y = \int_{-\infty}^{\infty} e^{-\alpha^2 x} \varphi(\alpha) \, d\alpha$$

instead of (54). By an argument similar to the one above, he found

$$y = C \int_{-\infty}^{\infty} e^{-\alpha^2 x} e^{-2i\alpha/n} \, d\alpha = C \int_{-\infty}^{\infty} e^{-\alpha^2 x} \cos \frac{2\alpha}{n} \, d\alpha,$$

which is different from (57).

In order to compare this solution with (61) [Ms 3615 (3), p. 15r], he transcribed it into

$$y = C \frac{e^{-1/n^2 x}}{\sqrt{x}} \tag{62}$$

which coincides with solution (61) when $\psi_{1/2} = 0$. This was clearly a nice corroboration of the two results, but soon Liouville began to have doubts. First [Ms 3615 (3), p. 15rv], he discovered that by taking

$$\frac{d^{1/2}}{dx^{1/2}} y = i \int_{-\infty}^{\infty} \alpha \, e^{-\alpha^2 x} \, d\alpha$$

as he had done, he had committed an error for the correct value is

$$\frac{d^{1/2}}{dx^{1/2}} y = i \int_{-\infty}^{\infty} \sqrt{\alpha^2} e^{-\alpha^2 x} \, d\alpha \,,$$

and if one decides to take the positive branch of $\sqrt{\alpha^2}$ this would give instead

$$\frac{d^{1/2}}{dx^{1/2}} y = 2i \int_{0}^{\infty} \alpha e^{-\alpha^2 x} \, d\alpha. \tag{62a}$$

He concluded these considerations with the words:

> Thus, our preceding method, which is nevertheless the one of our first memoir, is totally inadmissible. Yet, that does not mean that our result that is,
> $$y = \frac{Ce^{-\frac{1}{n^2 x}}}{\sqrt{x}}$$
> is itself incorrect.

[Ms 3615 (3), p. 15v]

To make sure that (62) was indeed a solution, Liouville expanded it in a series:

$$y = C \left(\frac{1}{\sqrt{x}} - \frac{1}{n^2 x \sqrt{x}} + \frac{1}{2n^4 x^2 \sqrt{x}} - \cdots \right)$$

and correspondingly found

$$\frac{d^{1/2}}{dx^{1/2}} y = C \frac{i}{\sqrt{\pi}} \left(\frac{1}{x} - \frac{2}{n^2 x^2} + \cdots \right),$$

and he was no doubt surprised to see that these series when substituted into (53) do not satisfy this equation (cf. [1837m, p. 80]). At first Liouville was confused:

In all these considerations, there is something obscure and contradictory.

[Ms 3615 (3), p. 16r]

Then, he was convinced that something was wrong:

Undoubtedly, the integral of our 1st memoir is completely incorrect.
[Ms 3615 (3), p. 17r]

Finally, he decided [Ms 3615 (3), p. 21v] that he had in fact made an
error in his memoir. Since the memoir was not printed until September
of that year, Liouville had time to correct the error by using the one-
sided Laplace transform instead as had been suggested by (62a). It is the
corrected solution I have summarized in §30.

34.   But, the discovery that $\frac{c}{\sqrt{x}}e^{-1/n^2x}$ is not a solution of (53) also showed
Liouville that solution (61) found by the other method is wrong, for $\psi_{1/2} =$
0 ought to be a possible complementary solution. This problem turned out
to be a genuine weakness of the method. Liouville discovered that because
of the successive differentiations performed in this method, the resulting
solution is too general. He compared the problem with the solution of the
equation

$$x\frac{dy}{dx} = y \tag{63}$$

[Ms 3615 (3), pp. 17v–21r]. For, if this equation is differentiated, one gets
the simple equation $\frac{d^2y}{dx^2} = 0$, the solution of which $A + Bx$ is too general
to be a solution of (63). It only fulfills (63) if $A = 0$. Therefore:

one must walk with [?] and caution on these new roads.
[Ms 3615 (3), p. 21v]

Thus, Liouville saw that his new method of integration yielded expressions that are too general and contain the solution of the equation as a
special case. He then set out to determine the limitations on the arbitrary
constant c and the arbitrary polynomial $\psi_{1/2}$ so that (61) would represent
the general solution of (53). This analysis, published for $n = 1$ in [Liouville
1837m, pp. 78–80] (beware of many errors of calculation) started by noting
that solution (61) can be written more conveniently:

$$y = y_1 + A_1x + A_2x^2 + \cdots + A_kx^k, \tag{64}$$

where

$$y_1 = \frac{e^{-1/n^2x}}{\sqrt{x}}\left(c + \frac{K}{n}\int\frac{1}{\sqrt{x}}e^{1/n^2x}\,dx\right). \tag{65}$$

This can be obtained by inserting an undetermined polynomial $a_0 + a_1x +$
$a_2x + \cdots + a_kx^k$ for $\psi_{\frac{1}{2}}$ in (61) and integrating by parts, or it can be seen

more elegantly as in [Liouville 1837m, p. 78]. Liouville then transformed the last integral in such a way that y_1 took the form

$$y_1 = C\frac{e^{-1/n^2x}}{\sqrt{x}} + 2K\left(1 - \frac{2}{n}i\int_0^\infty e^{-\alpha^2 x \exp(-2i\alpha/n)}\, d\alpha\right). \qquad (66)$$

The second term is precisely the solution (57) found by the Laplace transform method, and Liouville easily checked that in fact it does satisfy equation (53). On the other hand, Liouville found (§33) that the first term only satisfies (53) when $C = 0$. This, therefore, is the limitation on the arbitrary constants. Hence, the general solution to (58) is

$$K\left(1 + \frac{2}{n}i\int_0^\infty e^{-\alpha^2 x}e^{(2i/n)\alpha}\, d\alpha\right) + A_1 x + A_2 x^2 + \cdots + A_k x^k, \qquad (67)$$

where K, $A_1, A_2, \ldots, A_k$ are arbitrary constants, and k is an arbitrary integer.

In all the other examples where Liouville applied his new method of integration, he did not take the trouble to restrict the too general expressions arising from the successive differentiations (cf. [Liouville 1837m, pp. 80–82]).

35. When Liouville published the above method of solution, he referred to a paper by his former fellow student, Favre Rollin, in the first volume of his journal [Favre Rollin 1836], in which the author had used the same method to solve linear fractional differential equations with constant coefficients containing only one fractional derivative. He made clear that Favre Rollin had found the method without knowing Liouville's work. This was a kind gesture, for plagiarism could have been suspected since Liouville had already presented his memoir to the *Académie des Sciences* in February 1836, i.e., before the publication of Favre Rollin's paper. On the other hand, the public could have suspected that Liouville, being the editor of the journal, had stolen the idea from Favre Rollin before publishing Rollin's paper. Therefore, it is surprising that on this occasion Liouville did not point out that he had already developed the method in 1832. Most mathematicians at the time – and at any time – would have taken greater care to claim their own priority than to acknowledge the independent discoveries of their colleagues.

36. While investigating the solutions of (53) during February 1832, Liouville also discovered that he could use the method of successive approximations to solve this fractional differential equation. He included this method in his published paper [1837m, pp. 83–84], where he applied it to the more general equation

$$\frac{d^\mu y}{dx^\mu} = Py, \qquad (68)$$

where P is a function of x. By integration, this yields

$$y = \psi + \int^\mu P y \, dx^\mu,$$

where ψ is a complementary function. If y in the integral is replaced by $\psi + \int^\mu P y \, dx^\mu$, one gets

$$y = \psi + \int^\mu P\psi \, dx^\mu + \int^\mu P \, dx^\mu \int^\mu P y \, dx^\mu,$$

etc. Thus, Liouville concluded, if the repeated integral

$$\int^\mu P \, dx^\mu \int^\mu P \, dx^\mu \ldots \int^\mu P \, dx^\mu \int^\mu P y \, dx^\mu \tag{69}$$

tends to zero when the number of integrals tend to infinity, then the solution of (68) is represented by the convergent series

$$y = \psi + \int^\mu P\psi \, dx^\mu + \int^\mu P \, dx^\mu \int^\mu P\psi \, dx^\mu + \cdots. \tag{70}$$

37.   The fourth and simplest method of solution of fractional differential equations is not discussed in Liouville's notes. It was presented as the first method in the published memoir [1837m, pp. 62–71]. It applies to the linear differential equation

$$a\frac{d^\lambda y}{dx^\lambda} + b\frac{d^\mu y}{dx^\mu} + \cdots + c\frac{d^\nu y}{dx^\nu} = V \tag{71}$$

and is an easy generalization of the ordinary method of solving such equations when λ, μ, ν are natural numbers. When $V = 0$, one finds a solution by inserting $y = e^{mx}$ and solving the resulting equation

$$am^\lambda + bm^\mu + \cdots + cm^\nu = 0. \tag{72}$$

If $m_1, m_2, \ldots, m_i$ are the roots of this equation, the complete solution of the homogeneous equation is

$$y = c_1 e^{m_1 x} + c_2 e^{m_2 x} + \cdots + c_i e^{m_i x} + \psi, \tag{73}$$

where ψ is a complementary function. Liouville did not discuss in detail how to handle repeated roots of (72), but he showed how one could find a solution to the inhomogeneous equation, (71), from the complete solution, (73), of the homogeneous equation (cf. [1837m, pp. 65–71]).

38.   In a note of June 1832 [Ms 3615 (3), pp. 61v–62r], where Liouville for the first time planned a "Memoir on the integration of differential equations of arbitrary order" to be composed during his coming stay at Toul,

he counted the formulas for changes of variables among the subjects to be treated. He decided to compose a separate memoir on this subject [1835c], but in this paper the applications to solution of fractional differential equations were also central. For example, the equation

$$\frac{d^{1/2}y}{d(\sqrt{x})^{1/2}} + \frac{d^{1/2}y}{dx^{1/2}} = F(x) \tag{74}$$

can easily be solved by formula (51a). For substituting $\frac{d^{1/2}y}{d(\sqrt{x})^{1/2}}$ by its expression in (51a) leads to the equation

$$\sqrt{2} \int^{1/4} x^{1/4} \frac{d^{3/4}y}{dx^{3/4}} \, dx^{1/4} + \frac{d^{1/2}y}{dx^{1/2}} = F(x).$$

Liouville applied $\frac{d^{1/4}}{dx^{1/4}}$ to this equation and obtained

$$\sqrt{2} x^{1/4} \frac{d^{3/4}y}{dx^{3/4}} + \frac{d^{3/4}y}{dx^{3/4}} = \frac{d^{1/4}F(x)}{dx^{1/4}},$$

from which he found the solution

$$y = \int^{3/4} \frac{\frac{d^{1/4}F(x)}{dx^{1/4}} \, dx^{3/4}}{1 + \sqrt{2} x^{1/4}}.$$

This solution of (74) was sketched in [Liouville 1834b, p. 286], and in [1835c], Liouville treated several other examples.

He considered his ability to solve this type of equation to be a crucial step forward that would help convince other mathematicians of the utility of his new calculus:

> If one questions the necessity of our principles and the utility of the calculus of differentials of arbitrary order, it seems that one could pose as a challenge, to determine the unknown function φ, from one of the two following equations:

$$A \int_0^\infty \varphi(\sqrt{x^2 + \alpha^2}) \, d\alpha + \int_0^\infty \varphi(x + \alpha) \frac{d\alpha}{\sqrt{a}} = F(x),$$

and

$$Ax \int_0^\infty \varphi(\sqrt{x^2 + \alpha^2}) \, d\alpha + \int_0^\infty \varphi(x + \alpha) \frac{d\alpha}{\sqrt{\alpha}} = F(x),$$

> A being a constant, and $F(x)$ a given function.
> [Ms 3615 (3), p. 57r, note 4 of June 1832]

These two integral equations lead to equations similar to (74) for, as Liouville pointed out in the sixth note from June 1832 [Ms 3615 (3), pp. 57v–58v],

$$\int_0^\infty \varphi(x + \alpha) \frac{d\alpha}{\sqrt{\alpha}} = i\sqrt{\pi} \frac{d^{1/2}y}{dx^{1/2}},$$

and

$$\int_0^\infty \varphi(\sqrt{x^2 + a^2})\, dx = \frac{i\sqrt{\pi}}{2}\, \frac{d^{1/2}y}{d(x^2)^{1/2}}\,,$$

and the first expression can be transcribed by (52) into an expression involving $\frac{d^\nu y}{d(x^2)^\nu}$. In [1834b, pp. 286–287], Liouville posed the two integral equations as exercises to the reader.

39. In his notes, Liouville succeeded in solving many other fractional differential equations. Some of these occurred in connection with applications, in particular, to generalizations of the tautochrone problems, others were merely artificial equations that happened to be solvable by some special formula. To the latter category belongs the equations solved in [Liouville 1835c, p. 53]. Having studied them during October 1834, Liouville remarked in his notebook:

> It remains to be seen if these equations arise from any questions.
>
> [Ms 3615 (4), p. 63r]

When he wrote the paper [1835c], he had not yet found any problems leading to these equations, and this lack of applicability of the last section of this paper made Liouville conclude with the remark:

> But, that is enough about this subject. We shall no doubt have occasion to return to it at a later time. Moreover, one must know how to restrict oneself; for these studies of pure analysis end up being a bit vague if one pretends to push them to their last conclusion.
>
> [Liouville 1835c, p. 54]

Thus, Liouville finished his research on fractional differential equations with a feeling that he had pushed the theory so far away from applications that it began to be artificial and perhaps uninteresting.

40. Before concluding this chapter, I need to mention that Liouville also discovered that the fractional calculus could be useful in the solution of ordinary differential equations. In his third paper [1832c], he presented such an application to the hypergeometric equation:

$$(mx^2 + nx + p)\frac{d^2y}{dx^2} + (qx + r)\frac{dy}{dx} + sy = 0, \tag{75}$$

where m, n, p, q, r, and s are constants. He does not seem to have been aware that this equation had been the subject of intensive studies by Euler, Pfaff and Gauss (cf. [Gray 1986], pp. 1–14), but he did recall that Legendre had shown how to solve it when

$$(mx^2 + nx + p) = k(qx + r)^2,$$

where k is a new constant. In that case, it is easy to see that $y = (qx + r)^\alpha$ is a solution if α is a root of the equation

$$q^2 k\alpha(\alpha - 1) + q\alpha + s = 0.$$

Therefore, if α_1, α_2 denote two distinct roots,

$$y = c_1(qx + r)^{\alpha_1} + c_2(qx + r)^{\alpha_2}$$

is the complete solution of (75).

In order to solve the equation in its general form, Liouville substituted a new variable, z, defined by

$$y = \frac{d^\mu z}{dx^\mu} \tag{76}$$

into equation (75):

$$(mx^2 + nx + p)\frac{d^{\mu+2} z}{dx^\mu} + (qx + r)\frac{d^{\mu+1} z}{dx^{\mu+1}} + s\frac{d^\mu z}{dx^\mu} = 0. \tag{77}$$

He then integrated both sides μ times, and by the product rule (51), he obtained the ordinary differential equation

$$\begin{aligned}
(mx^2 + nx + p)\frac{d^2 z}{dx^2} \;&+\; ((q - 2m\mu)x + (r - n\mu))\frac{dz}{dx} \\
&+\; (m\mu(\mu + 1) - q\mu + s)z = \psi_\mu,
\end{aligned} \tag{78}$$

where ψ_μ is a complementary function. If μ is determined in such a way that the coefficient of z vanishes, z can easily be determined and by (76) also y. Liouville pointed out that in most cases one can let $\psi_\mu = 0$, for the two solutions μ_1, μ_2 of the equation

$$m\,\mu(\mu + 1) - q\mu + s = 0 \tag{79}$$

will usually lead to two independent solutions y_1, y_2 and then the complete solution is $y = c_1 y_1 + c_2 y_2$. Liouville gave many examples to illustrate the force of his method and also showed how to simplify the procedure when (79) has an integer μ as a solution.

The solution of equation (75) by way of the fractional calculus was later taken up by Holmgren [1867–1868]. He pointed out that special cases of this equation had occurred often in mathematics, astronomy, and physics, but he attributed to Liouville the first investigation of the general equation.

> Mr. Liouville is the first who has tackled the problem in all its generality. The ingenious method, which he used, was based on his celebrated "Differential calculus of arbitrary order." Yet, as a consequence of a certain indeterminacy that was still attached to the new calculus, the integrals given by Mr. Liouville could not be considered to be the definitive solution of the problem, though they give sufficient indications for finding them with the ordinary methods.
> [Holmgren 1867–1868]

Holmgren gave a complete solution "by taking Liouville's works as a starting point," but using his own definition (24) of the fractional derivative. As we have seen in Chapter V, Holmgren's interest in fractional calculus went back to the year 1853, when he visited Paris and became a good friend of Liouville's.

Rigor

41.   Liouville's early mathematics, and in particular his calculus of arbitrary order, is a peculiar mixture of an explicitly announced desire to follow the new standards of rigor and an inability to live up to this ideal.

He tried to give himself an image as a true follower of Cauchy when in [1832b, p. 77] he claimed that "the divergent series, which mostly lead to faulty results, must be completely banished from analysis." He therefore insisted that the exponential expansion $\sum A_m\,e^{mx}$ of a function f should only be considered for those x where it is convergent. He believed that if the series converges it will always converge to $f(x)$:

> The essential nature of our equation $y = \sum A_m\,e^{mx}$ is to be identical when the series converges whereas the periodic series of the theory of heat [Fourier series] express a function $F(x)$ only between given limits of x...

> [Liouville 1832b, p. 77]

As an example, Liouville mentioned that $\frac{1}{1-e^x}$ had to be represented by two different exponential series. One,

$$\frac{1}{1-e^x} = 1 + e^x + e^{2x} + e^{3x} + \cdots ,$$

is convergent for $x < 0$, and the other,

$$\frac{1}{1-e^x} = -(e^{-x} + e^{-2x} + e^{-3x} + \cdots),$$

is convergent for $x > 0$. However, he did not remark that even if the series $y = \sum A_m\,e^{mx}$ is convergent for some x its derivative $\frac{d^\mu y}{dx^\mu} = \sum A_m\,m^\mu\,e^{mx}$ need not be convergent. Moreover, his distinction between the exponential series and the Fourier series is not clearcut at all, for he often allowed complex exponents and in some cases even purely imaginary exponents, for example, in the development of the trigonometric functions.

42.   In spite of the Cauchy-inspired banishment of divergent series, Liouville did not keep strictly to Cauchy's concept of convergence and divergence. Sometimes he did not distinguish between convergence and absolute convergence. For example, in [Ms 3615 (3), p. 49r] he wrote that when $\sum A_n\,e^{-mx}$ is convergent for a real value of x then $\sum A_m\,e^{-m(x+iy)} = \sum A_m\,e^{-mx}\,(\cos my - i\sin my)$ is obviously convergent as well. Since A_m can be negative, this is only true for absolute convergence.

Liouville also often referred to series that are neither convergent nor divergent [1832b, pp. 113, 139]; [1835b, p. 225]. Such series, which do not exist in Cauchy's calculus, were introduced by Poisson [1823b, pp. 404–406]. They are the series $\sum_{i=1}^{\infty} a_i$, which are not convergent, but can be summed in the sense that $\sum_{i=1}^{\infty} e^{-ni}a_i$ is convergent for all $n > 0$ and its sum has a limit for n tending to zero. Poisson had used this summation in his investigation of Fourier series, and it was in the same connection that they entered Liouville's theory. For, when he tried to determine the derivatives of the trigonometric functions by way of generalized difference quotients (cf. (49), (50)) he was faced with series of the kind

$$1 - \frac{\mu}{1}\cos mh + \frac{\mu(\mu - 1)}{1 \cdot 2}\cos 2mh - \cdots . \tag{80}$$

He found its value as a Poisson sum:

$$1 + \lim_{n \downarrow 0}\sum_{i=1}^{\infty}(-1)^i\, e^{-nih}\frac{\mu(\mu - 1)\cdots(\mu - i - 1)}{1 \cdot 2 \cdots i}\cos imh$$

for $h > 0$. If $h < 0$, another formula must be used. This gives rise to the two different values for the derivative of $\cos x$.

43.   With respect to continuity of functions, Liouville also showed in a note from May 1832 [Ms 3615 (3), pp. 53v–54r], that he had not yet adopted Cauchy's ideas completely. There he had derived the equation

$$\int_0^{\infty}\int_0^{\infty} F'(x + \alpha^2 + \beta^2)\,d\alpha\,d\beta = -\frac{\pi}{4}F(x)$$

by way of the fractional calculus (cf. [Liouville 1834b, pp. 275–276]). He claimed that the formula is valid if (1) $F(\infty) = 0$ and (2) $F(x')$ is not infinite for x' greater than the x under consideration, and he added "and this function could even be discontinuous but not discontiguous." Here, Liouville used the distinction introduced by Arbogast [1791] between discontinuous and discontiguous functions. The former are presented by different expressions in different intervals, and the latter have graphs that are disconnected. It was the last class that Cauchy defined more precisely and called discontinuous. He later showed [Cauchy 1844a] that the old use of the word made no sense [cf. Lützen 1978].

44.   It is well-known that some weaknesses still remained in Cauchy's calculus. One of the well-known flaws was his inability to distinguish the order of different limit procedures, so that the difference between pointwise and uniform convergence (and continuity) was not pointed out (differing interpretations, however, have been suggested by Giusti [1984] and by Laugwitz [1987]). Liouville also generally interchanged the order of limit procedures. For example, he interchanged $\lim$ and $\sum$, although the "neither convergent nor divergent series" exemplified that this is not permissible. He also indiscriminately interchanged $\lim$ and $\sum$ with $\int$ despite some counterexamples he had himself discovered [1832b, pp. 84–85].[11] On the other hand, Liouville explicitly insisted that "differentiation of arbitrary order under the $\int$-sign is not always permissible" [Liouville 1832b, p. 125]. Therefore, in some proofs [1835b, pp. 224–227], he carefully avoided taking derivatives inside an integral sign by taking a limit inside the integral sign instead.

45. In his use of the limit concept, Liouville was not as consistent as Cauchy. For example, following the mathematicians of the 18th century, he sometimes wrote about the true value of $\frac{0}{0} = \frac{f(x+h)-f(x)}{h}$ when $h = 0$ instead of the limit for h tending to zero as Cauchy did. To be sure, Cauchy had introduced infinitesimals, in order to adjust to the curriculum at the *École Polytechnique* (cf. [Belhoste 1982, I, p. 80]), but, for example, in the treatment of the complementary functions (cf. §21–22), Liouville used infinitesimals in a way that broke with the spirit of Cauchy's calculus. Similarly, Cauchy would not have allowed the way in which Liouville got rid of an infinite integral by subtracting an infinite complementary function [Liouville 1832b, pp. 151–153].

46. Even the insistence on convergence of series is seemingly just lip service paid to Cauchy, for only infrequently did Liouville check the convergence of his series. He left it at one general remark, which put the burden on the reader:

> Therefore, if I sometimes happen to reason about series which are not always convergent, it is implied that one must restrict the consequences obtained in this way to within the limits where the convergence of the applied series is demonstrated.
>
> [Liouville 1832b, p. 77]

Similarly, Liouville often had to assume that his functions only have positive or negative real parts in the exponents of their exponential series, but he rarely mentioned this fact, except when two alternative formulas were found:

> All the formulas which I use in this Memoir are subject to restrictions similar to those of formula [(2)]; I have not considered it necessary to stress them. The way in which the theorems are demonstrated indicates well enough when and how they are exact.
>
> [Liouville 1832a, p. 8]

In [1832b], he excused himself for not always formulating his assumptions explicitly with the following remark:

> In this memoir, as well as in the preceding one, our only aim has been to give a general idea of the differential calculus of arbitrary order, and we have not intended to compose a complete treatise on this calculus. This is the reason why we have neglected certain details, and even, here and there, silently passed over the exceptions to which our rules are subject. It is sufficient that our readers are warned once and for all.
>
> [Liouville 1832b, p. 88]

As a whole, Liouville's view on exceptions to rules was in accordance with the 18th-century way of thinking. For example, having found that formula (32) does not apply when $n + \mu$ is zero or a negative integer, he remarked:

However, this particular case, which is otherwise very easy to treat directly, does not prevent the formula [(32)] from being the general formula, any more than the equivalent circumstance prevents the formula

$$\int x^m \, dx = \frac{x^{m+1}}{m+1}$$

from being the general formula for the integration of powers in the ordinary integral calculus.

[Liouville 1834a, pp. 8–9]

Thus, a rule is generally valid if there are only few exceptions to it.

We may conclude that Liouville was, in principle, sympathetic toward Cauchy's programme of rigor. However, he often uncritically took over the views of the 18th-century mathematicians whose works he had read. His main goal was not rigor but the development of a new theory.

47. Finally, I shall mention that Liouville had a somewhat peculiar view of the axiomatic deductive structure of mathematics. He often referred to "primordial truths" or "questions" in [1832b, p. 94] and [1834a, p. 2], as well as in his early papers. These seem to be something in between a definition and a theorem. In the physics papers [AC L Ms 5], it is mostly a truth that is forced upon mathematics by nature, and perhaps that is also the case with the primordial truth "$\frac{d^\mu \psi}{dx^\mu} = 0$," when ψ is a complementary function. Concerning this truth, Liouville remarked:

This property of complementary functions is not limited to the case where the number [μ] is a positive integer, and is not just a theorem that one must verify à posteriori, such as we have done above in this example. It is an almost evident primordial truth which applies to all the differentials, whatever the order μ of the differentiation may be.

[Liouville 1834a, p. 2]

Today, we believe, at least in principle, that a proof of a theorem is either valid or invalid, whereas Liouville seems to have followed, to some extent, his 18th-century predecessors, who believed that a proof could be more or less correct or convincing. For example, he gave several proofs that the derivatives of the trigonometric functions must be undetermined in order to dispense with all doubts [Liouville 1832b, p. 126]. Liouville also believed that one could ascribe a certain value to a "proof" using divergent series, in spite of their banishment from analysis, but he preferred a true "demonstration" using convergent series only. Thus, in connection with formulas (40) and (41) and the corresponding identities for sine, he remarked:

The mathematical considerations that I have used to demonstrate the formulas [C], [D], [E], [F] no doubt seem sufficient to most geometers. Still, one can object that the series which give $\cos \mu z$ and $\sin \mu z$ are not always convergent, ... I therefore think that it would be best to return to this subject and to give not just a simple proof

> of the formulas [C], [D], [E], [F], but a true demonstration which
> brings to light the unity of the properties of the differentials which
> we want to calculate and the intimate relation of the particular form
> that each of them takes.
>
> [Liouville 1832b, p. 135]

Thus, a "demonstration" carries more weight than a "proof." However, this seems to be an ad hoc distinction, which I have not found elsewhere in Liouville's writing.

Later in life, Liouville adopted more of the spirit of Cauchy but rigorization never achieved first priority in his writings (cf. Chapter III).

48.   Today, we may criticize Liouville for not being radical enough in his adaption of Cauchy's programme, but his immediate successor Peacock criticized him for being too radical:

> M. Liouville adopts an opinion, which has been <u>unfortunately</u> sanctioned by the authority of the great names of Poisson [sic] and Cauchy, that diverging series should be banished altogether from analysis, as generally leading to false results. (my underlining).
>
> [Peacock 1834, p. 219]

Much of what Peacock considered errors in Liouville's theory he attributed to this Cauchyan attitude and to a neglect of the so-called "principle of the permanence of equivalent forms." Peacock stated, as follows, this principle, which was central in most British mathematics at this time:

> If the conclusions thus obtained do not involve in their expression any condition that is essentially concerned with the specific values of the symbols, they may be at once transferred to symbolical algebra, and considered as true for all values of the symbols whatsoever.
>
> [Peacock 1834, p. 201]

As it stands, the principle is almost a tautology, but usually the principle was used to justify the insertion of any expression in algebraic formulas that are true for almost all real numbers.

> But, if we adopt the principle of the permanence of equivalent forms, we may safely conclude that the limitations of the formula's will be sufficiently expressed by means of those critical values which will at once suggest and require examination. The extreme multiplication of cases, which so remarkably characterizes M. Liouville's researches, and many of the errors which he has committed, may be principally attributed to his neglect of this important principle.
>
> [Peacock 1834, p. 219]

However, compared with much of the British fractional calculus based on this principle, Liouville's papers appear clear, precise, and rigorous.

Concluding Remarks

49. Liouville often called for the attention and participation of other mathematicians in the buildup of his new calculus. As I have mentioned, he began with the applications partly "in order to explain to geometers that this new calculus may one day be of great use and to urge them to pay attention to it" [Liouville 1832a, p. 2]. Similarly, he expressed the hope that [Liouville 1834b] would "urge some geometers to go the same way" [1834b, p. 287], because he would not alone be able to carry a whole new calculus far enough:

> I shall try to return later to this subject [of (1832b)] and to go deeper into this study. However, in such a vast area, one cannot hope for real progress without the collaboration of geometers.
> [Liouville 1832b, pp. 73–74]

In spite of these appeals, Liouville's calculus of arbitrary order had little if any impact on contemporary French mathematicians. It is even difficult to trace the reaction of the mathematics community. Liouville presented most of his papers on this subject to the *Académie des Sciences*: [1832a] and probably also [1832b] and [1832c] on June 18, 1832; [1835b] on February 17, 1834; [1835c] on February 2, 1835; [1835d] on June 29, 1835, and [1837m] on February 15, 1836. According to the *Procès-Verbaux*, committees were appointed to examine [1835b], (Lacroix, Poisson); [1835c], (Mathieu, Savary); and [1835d], (Lacroix, Poisson, Navier). The first three papers were probably also examined, but the examiners are not mentioned.

However, the *Procès-Verbaux* never mentioned the presentation of reports on any of these papers. This indicates that the examining committees were uneasy about the new theory and hesitated to support or criticize it openly.

50. In Liouville's notes, one can find a few indications that the fractional calculus had some opponents. I have already mentioned (§38) that Liouville designed special exercises for those who "question the necessity of our principles." In June 1832, he also refuted an objection of an imaginary opponent:

> At first it seems as though one can raise an objection to our great principle of complementary functions.
> [Ms 3615 (3), p. 61r]

The objection is that since $\frac{d^\mu}{dx^\mu} \sin x$ can be both $\sin(x + \frac{\mu\pi}{2})$ and $(-1)^\mu \sin(x - \frac{\mu\pi}{2})$ the difference

$$\psi = \sin\left(x + \frac{\mu\pi}{2}\right) - (-1)^\mu \sin\left(x - \frac{\mu\pi}{2}\right)$$

is the derivative of the function $0 = \sin x - \sin x$, and therefore it is a complementary function different from a polynomial. However, Liouville refuted this objection on the grounds that the first derivative is obtained when

$\sin x$ is considered as $e^{nx} \sin x$ for $n \downarrow 0$, whereas the second corresponds to $\sin x = \lim_{n \uparrow 0} e^{-nx} \sin x$.

Also, later in June 1832, Liouville wrote "in passing some remarks that are worth keeping in mind because they answer the completely invalid objections which I have rejected a thousand times." [Ms 3615 (3), p. 62v]. The reflections were to the effect that formula (51) "must be completed with a comple[mentary] function," for otherwise it is easy to see that the derivative of a polynomial is zero, which is not always the case.

Thus, apparently, Liouville had to fight for his new theory. However, it seems improbable that it aroused much public criticism in Parisian mathematical circles. Indeed, when Libri in 1839 launched a fierce attack on Liouville (Chapter II), he chose to direct it toward Liouville's theory of integration in finite terms instead, although both by modern and 19th-century standards, this was much more rigorous than his fractional calculus. Not until 1843 did the weakness of Liouville's fractional calculus catch the attention of Libri. In a new quarrel (Chapter III), he severely criticized Liouville's evasive remarks about divergent series and "exceptions" quoted in §46. At the same time, he referred to Peacock's criticism, even though it points in the opposite direction [CR 17 (1843), p. 440].

The small number of papers commenting on or continuing Liouville's theory before 1840 indicate that it was met mostly with indifference when it appeared. If so, this may well have been a reason why Liouville gradually abandoned fractional calculus during the 1830s.

51.   Soon after Liouville left the field, however, interest in fractional calculus arose, mainly in Great Britain. This happened in connection with the development of the principle of permanence of equivalent forms (cf. §48). If, according to this principle, differential operators are substituted for variables in an algebraic equation, one gets a formula in differential calculus. Operational calculus sprang from this principle, and it included as a special branch the study of derivatives of arbitrary order. From the start, this study was tied closely to Liouville's papers. In particular, people discussed if Euler's or Liouville's results for the derivative of x^n, $n \in R$, were correct.

Some mathematicians, like Greatheed [1839], used Liouville's definition; others, like Peacock [1834] (§48), criticized Liouville's theory severely, but whichever attitude was being advocated, Liouville's work was the point from which all further research in the field radiated [cf. Ross 1974, 1975, and 1977 for references to Liouville's successors].

Riemann is an exception in that he does not seem to have known Liouville's work when he wrote his paper on fractional calculus in 1847. He suggested an approach to fractional calculus based on Taylor's theorem. He, too, was led to the concept of a complementary function, but his complementary function, corresponding to $\frac{\partial^\nu}{\partial x^\nu}$ ($\nu < 0$) was of the form $\sum_{n=\infty}^{n=1} K_n x^{(-\nu-n)}$, where K_n are arbitrary constants. This is different from

Liouville's polynomials, and indeed his definition corresponds to the modern definition of $_cD_x^{-\nu}$ for c finite. Riemann's paper was not published until 1874, and therefore it had little impact.

Liouville's theory of fractional calculus was also discussed in 1844 in Italy by Tardy [Bottazzini 1978, 1983] and later by B. Tortolini (1808–1874) [1855]. These works have not found their way to Ross's bibliography. The only time Liouville entered the discussions for and against his fractional calculus was in an answer to Tortolini's paper, which the author had sent to Liouville. However, a remark in Liouville's answer suggests that, although he had not intervened earlier, he had followed the discussion of his definition of the fractional derivative:

> This definition, which is based on the consideration of series, has undoubtedly its disadvantages, which I know very well. One has and one can present other definitions and from them deduce results which are apparently different from mine, and I do not have the right nor the intent to be astonished. Perhaps, I shall myself one day return to this subject using an approach which is completely different from the one I first envisaged.
>
> [Liouville 1855a, p. 116]

One can feel a certain insecurity here as to the superiority of his definition over the alternative definitions, but in the paper, he continued to defend his definition as the one that most directly springs from Leibniz's "beautiful analogy between powers and differences." Thus, Liouville remained faithful to his own principles of fractional calculus, but he did not feel so strongly about them after 1840 that he bothered to enter the growing discussion in the field.

IX

Integration in Finite Terms

Introduction

1. While struggling with the theory of differentiation of arbitrary order, Liouville began to develop another theory for which he is usually hailed as the founder: the theory of integration in finite terms. He published 11 papers on the subject in the period from 1833 to 1841 and at some point had the idea of bringing the theory together in a book. However, this idea never materialized.

The main objective of Liouville's theory is to decide if a given indefinite integral can be expressed as a finite expression involving only algebraic, logarithmic, and exponential functions. To be more precise, we must introduce Liouville's classification of finite explicit functions. A function $y(x)$ is called *algebraic* if it is a root of a polynomial equation:

$$y^n + A_{n-1}(x)y^{n-1} + A_{n-2}(x)y^{n-2} + \cdots + A_1(x)y + A_0(x) = 0, \qquad (1)$$

where A_0, A_1, ... are rational (or entire) functions of x.

1. Logarithms and exponentials of algebraic functions are called (transcendental) monomials of the first kind. A function that is not algebraic is called a (transcendental) function of the first kind (or order) if it is an algebraic function of x and transcendental monomials of the first kind, i.e., if it is a solution of some polynomial equation, (1), with coefficients A_i, which are rational functions of x and transcendental monomials of the first kind. For example, $\log(1 + x) + \sqrt{1 + e^x}$ is a first-order transcendental.

2. Logarithms and exponentials of functions of the first kind are called (transcendental) monomials of the second kind, and algebraic functions of algebraic and first-order functions and transcendental monomials of the second kind are called functions of the second kind if they are not of the first kind or algebraic. For example, $\log(x + \log x)$ is of the second kind. Recursively, Liouville then defined transcendental functions of the nth kind. He called all functions defined in this way "finite explicit functions." I shall often leave out "explicit" or use Ritt's [1948] term "elementary functions." One must be careful not to be mislead by Liouville's use of the word "explicit," for it is only the transcendental operations log and exp that are restricted to act explicitly, whereas the algebraic functions can be defined by polynomial equations that are not solvable by radicals.

Liouville tried to devise methods by which he could decide if the integral of a given algebraic (or elementary) function was again elementary. For this purpose, he proved an important theorem, which is named after him. In its simplest form, it states:

Liouville's Theorem: Let y be an arbitrary algebraic function of x. If the integral $\int y\,dx$ is expressible in finite explicit form, it is always possible to write

$$\int y\,dx = t + A\log u + B\log v + \cdots + C\log w, \qquad ([2])$$

where $A, B, \ldots, C$ are constants and $t, u, v, \ldots, w$ are algebraic functions of x.

[Liouville 1834c, p. 42]

By means of this theorem and a generalization of it, Liouville succeeded in proving that a great number of integrals, in particular, certain elliptic integrals, are not expressible in finite terms. He also devoted his last three papers on the subject to the solution in finite terms of differential equations, such as the Ricatti equation.

Liouville's theory of integration in finite terms and its recent development have been the subjects of a good didactical survey paper by Toni Kasper [1980]. There is still need for a more detailed study of this aspect of Liouville's work, for Kasper does not consider the chronological development of Liouville's ideas, nor does he relate Liouville's ideas to those of his immediate precursors or successors. I shall try to add these aspects to the history by first discussing the historical setting for Liouville's theory, in particular, the work of Abel, which is discussed in a separate section. In the next two sections, I discuss Liouville's work on the evaluation of integrals in algebraic and in finite terms and then his theory of integration of differential equations in finite terms or by quadrature. Finally, the chapter concludes with a survey of the subsequent developments in this field.

Historical Background

2.   Everybody who pursues mathematical analysis has the feeling that of the infinitely many expressions representing a given function, some are more elementary than others. The singling out of the finite explicit functions, and the division of them into classes, is a way to make explicit this vague feeling about elementarity. It should be observed that this concept of elementarity has little to do with computational facility, for it is often easier to compute the values of a function from a series expansion or a successive approximation than from an "elementary" expression. The idea of elementarity is in part based on tradition, and as such it has a long history.

According to Pappus (fl. A. D. 300–350) [*Collection*, book III, proposition 5], the ancient Greeks divided problems into three classes characterized by the curves needed to solve them. This division corresponds to the following division of curves. The simplest were the "plane" curves: circles and straight lines. The second class consisted of the conic sections (the "spatial" curves), and all the rest were classified as "linear" curves. In *La Géométrie* [1637], Descartes further subdivided the third class. The curves that were graphs of polynomial equations, $f(x,y) = 0$, were called geometrical curves, and he divided them according to the degree of the corresponding polynomial. He classified all other curves as mechanical curves and banned them from mathematics all together.

These two classes were maintained after the introduction of the calculus under the names "algebraic" and "transcendental," but both Newton and Leibniz accepted transcendental curves as genuine mathematical objects. In fact, the power of their methods was most amply demonstrated by their ability to handle these curves. In transcendental problems, Newton mostly resorted to infinite series. Leibniz, on the other hand, while recognizing the generality of this method, preferred to reduce transcendental expressions to certain "simple" ˏand well-known forms. Thus, in two letters of March 30, 1675, and July 12, 1677 [Leibniz 1899, pp. 248–249], to Oldenburg, he discussed the possibility of reducing various quadratures to the quadrature of the hyperbola and the circle. He asserted, correctly, that Gregory was in error when he claimed that the rectification of the hyperbola could not be reduced to these quadratures. Gregory had made the same claim for the rectification of the ellipse, and in the last letter Leibniz asked if Newton, by using series, could decide if Gregory was right.

After the curve concept was replaced by the function concept in the 18th century, Leibniz's preference for the quadrature of the hyperbola and the circle was translated into a preference for the corresponding logarithmic and trigonometric functions and their inverses. The singling out of these transcendental functions as being "elementary" was probably also determined by the fact that they had proved important in fields outside the calculus, mainly in astronomy, and were accurately tabulated. Their excellence among the transcendental functions was firmly established in Euler's famous *Introductio* [1748], not only because they got their own names (log, exp, sin, cos, etc.), but also because the mathematical connection (Euler's formulas) between the exponential and the trigonometric functions was made explicit.

3.   In the degeometricized function language, Leibniz's question of reduction of quadratures became the problem of integration in finite terms. The first general theorem in this field was stated by Johann Bernoulli in the 1703 issue of *Acta Eruditorum*. He asserted that by the method of partial fractions the integral of any rational function can be expressed in terms of algebraic, logarithmic, and trigonometric functions. At first, the method of

partial fractions was questioned by Leibniz, but the purported counterexample he had found was soon refuted by Niclaus I. Bernoulli (1687–1759) (cf. [Kline 1972, pp. 411–412]). By the mid 18th century, Johann Bernoulli's theorem was an acknowledged fact, although a definitive proof could not be given before the fundamental theorem of algebra had been rigorously established by Gauss and Weierstrass.

The Bernoullis and their successors in the 18th century also devised many ingenious tricks to integrate irrational functions, but no general rules were found. However, two French mathematicians, Alexis Fontaine (1704–1771) and Marie-Jean Marquis de Condorcet (1743–1794), claimed to have found methods to solve the much more general problem of integration of any "solvable" differential equation in algebraic or finite terms. In his papers of [1764], Fontaine tabulated long lists of possible algebraic solutions to various types of differential equations, but as Condorcet [1765] pointed out, Fontaine did not delimit the possible solutions to a finite number of types, and therefore his determination of algebraic solutions became unmanageable. Liouville expressed this in stronger terms:

> The general method indicated by Fontaine is in reality nothing but a laborious groping whose least fault is its disheartening length. These gropings do not constitute a regular procedure for the reason that no circumstance tells us after how many attempts one will arrive at the desired result, that is the integral that one wants to calculate, so that if by chance this integral is impossible in the algebraic form that one a priori assigns to it, one will never be warned, however far one pushes these operations.
>
> [Liouville 1834c, pp. 37–38]

Condorcet claimed that he had solved this problem, even for transcendental solutions, by delimiting the possible solutions to a finite number of forms. Liouville, however, was also critical of Condorcet's programme:

> One must say that the writings of this illustrious author include some more or less fortunate views whose development will no doubt one day contribute to the progress of the integral calculus. However, most of his theorems lack demonstrations and some of them lack exactness.
>
> [Liouville 1834c, p. 38]

4.   Recently, Gilain [1988] has studied Condorcet's work on this subject in great detail, and he has found that in his unpublished works, in particular, in the unfinished manuscript *Traité du calcul intégral* from the last half of the 1770s, Condorcet anticipated several of Liouville's ideas. For example, he formulated a version of Liouville's theorem, but he did not prove it, nor did he succeed in deriving an integral that is not integrable in finite form. In Liouville's manuscripts, there is no evidence that he knew of this unpublished work. His only source was probably the book *Du calcul intégral* in which Condorcet's ideas are much less developed. To illustrate Liouville's judgment of his predecessor, I shall summarize how Condorcet

solved a first-order differential equation in this book. He considered [Condorcet 1765, chapter 2] the equation $f(x, y, dx, dy) = 0$, where f is an algebraic function of x, y and their differentials. Imagine that its solution $g(x, y, r) = 0$ contains a transcendental r. When the equation $g(x, y, r) = 0$ is differentiated, it will be a differential equation in x, y, r and their differentials. The given equation must be obtainable from $g(x, y, r) = 0$ and its derived equation by elimination of r and dr. Condorcet claimed that this is only possible if dr contains no transcendentals or only the transcendental r itself. In the first case, r must, according to Condorcet, be equal to $\log(h_1^{n_1} h_2^{n_2} \cdots h_k^{n_k})$, where $h_1, h_2, \ldots, h_k$ are algebraic functions and $n_1, n_2, \ldots, n_k$ are rationally independent numbers. In the second case, r must be of the form $r = e^{\phi(x,y)}$ or $r = e^{q \log \phi(x,y)}$, where ϕ is algebraic and q is irrational. Thus, there are three different types of transcendental solutions to a first-order algebraic differential equation, to which must be added possible algebraic solutions.

If the given algebraic differential equation is of the nth order, Condorcet solved it by successively lowering the order. Thus, according to the arguments above, one integration will lead to an $n - 1$-th order differential equation that contains at most the three mentioned transcendentals. One more integration will introduce new transcendentals as above, with the only modification that the h_i's, ϕ, and log may contain the transcendentals introduced in the first integration. This gives $9 = 3^2$ different types of transcendentals. After n integrations, the solution is reached, and it may therefore contain 3^n different types of transcendentals.

With similar arguments, Condorcet found limits on the degrees of the equations defining the algebraic functions in the possible solutions and the exponents $n_1, n_2, \ldots, n_n$ of the algebraic functions, but the argument summarized above is sufficient to judge the merits and weaknesses of Condorcet's theory. One of the "fortunate views" that Liouville referred to was probably Condorcet's recursive classification of transcendentals, which may well have inspired Liouville to his classification. Moreover, in the treatment of first-order equations, Condorcet delimited the possible forms of the solution to a small number of possibilities, just as Liouville did in the theorem named after him, and the special form, $\log(h_1^{n_1} h_2^{n_2} \cdots h_k^{n_k}) = n_1 \log h_1 + n_2 \log h_2 + \cdots + n_k \log h_k$, is of the form found in Liouville's theorem. However, Condorcet's classification is not identical to Liouville's since, for example, $e^{\pi \log x}$ is classified in the first class by Condorcet, but in the second class by Liouville. Moreover, strict proofs are missing in Condorcet's work, and the whole problem and solution is expressed in very vague terms, even by the standards of the 1760s. Finally, Condorcet's entire book rests on the fallacious belief that if a differential equation has a solution then it must be in finite terms. Thus, he ends the second chapter, "General method of finding the integral of a given differential equation," with the following conclusion:

> Those who have followed the spirit of this method can see that I have solved the problem I have set for myself in all its generality. In

fact, (1) there is no differential equation to which the method does not apply, (2) I give for each one the way to find all the forms to which the integral is subject, and I show that the number is finite. Therefore, the rest only depend on ordinary Algebra; thus I integrate every differential equation which is not absurd.

5.   Later Condorcet discovered that there might exist solutions to differential equations that are not in finite terms (cf. [Gilain 1988]), and it is in fact remarkable that d'Alembert and Bezout in their report of May 22, 1765, to the *Académie Royale des Sciences* (printed in [Condorcet 1765]) accepted Condorcet's conclusions, for it seems to have been the general belief that many integrals could not be expressed in elementary form. The simplest and most important class of apparently nonelementary integrals consisted of integrals of the form

$$\int \frac{P(x)}{\sqrt{R(x)}},$$

where P is a rational function and R is a polynomial of the third or of the fourth degree. Such integrals were called elliptic integrals because the rectification of the ellipse, discussed by Gregory, Leibniz, and the Bernoullis, had led to such an integral. I shall briefly recall two aspects of the history of elliptic integrals relevant to the present subject, referring to Chapter XIII, [Kline 1972, pp. 411–422, 644–655], [Houzel 1978] and [Fricke 1913] for more details.

   The first aspect is concerned with the so-called addition theorems. They were discovered by Euler, who, inspired by Fagnano, found that the sum of two similar integrals,

$$\int_a^x \frac{dx}{\sqrt{R(x)}} + \int_a^y \frac{dx}{\sqrt{R(x)}},$$

is equal to an integral of the same form:

$$\int_a^c \frac{dx}{\sqrt{R(x)}},$$

where c is an algebraic function of x, y, a and the values of $\sqrt{R(x)}$ at these points. Later, he proved a similar identity for arbitrary elliptic integrals, with the modification that he had to add to the one side of the equation a sum of a rational function and a linear combination of logarithms of rational functions. Thus, although an elliptic integral cannot generally be expressed in finite terms, a sum of similar integrals is expressed as a sum of rational and logarithmic terms if the limits of integration satisfy a certain algebraic equation. Formulated in this way, the addition theorem was generalized to integrals of arbitrary algebraic functions, so-called Abelian integrals, by Abel. I shall return to this generalization in the next section.

The second aspect from the history of elliptic functions to be mentioned is the reduction of elliptic integrals to a minimal number of standard forms. In this field, the main work was done by Adrien-Marie Legendre (1752–1833) who was the chief propagator of elliptic integrals from the 1780s to 1820. He collected his work in the *Traité des fonctions elliptiques* [1825–1828]. The main result of the book is that any elliptic integral can be written as a combination of algebraic and logarithmic functions and three canonical elliptic integrals of the first, second, and third kind:

$$\int \frac{dx}{\sqrt{(1-x^2)(1-k^2x^2)}}$$

$$\int \frac{x^2\,dx}{\sqrt{(1-x^2)(1-k^2x^2)}}$$

and

$$\int \frac{dx}{(x^2-a)\sqrt{(1-x^2)(1-k^2x^2)}}$$

or equivalently,

$$F(k,\phi) = \int_0^\phi \frac{d\phi}{\sqrt{1-k^2\sin^2\phi}}$$

$$E(k,\phi) = \int_0^\phi \sqrt{1-k^2\sin^2\phi}$$

and

$$\Pi(n,k,\phi) = \int_0^\phi \frac{d\phi}{(1+n\sin^2\phi)\sqrt{1-k^2\sin^2\phi}}$$

where $0 < k < 1$. When these integrals are tabulated, one can evaluate any elliptic integral.

6.   Both these trends in the history of elliptic integrals are ways to overcome the fact that elliptic integrals are not expressible in finite terms. However, the first mathematician who claimed to possess a proof of this fact was Pierre-Simon de Laplace (1749–1827). In the first book of his *Théorie analytique des probabilité*, he sketched a theory of elementary functions which is so remarkable that I shall follow Liouville [1833b, pp. 143–144] (almost the same extract as below) and quote the relevant passage. Having introduced the algebraic, exponential, and logarithmic functions, Laplace continued:

> These quantities are essentially distinct: the exponential a^x, for example, can never be identical with an algebraic function of x. In fact, every algebraic function is reducible to a descending series of the form $kx^n + k'x^{n-n'} + \cdots$; but it is easy to prove that when a is greater than unity and x is infinite a^x is infinitely greater than kx^n, however great k and n are supposed to be. Similarly it is easy to see that when x is infinite, x is infinitely greater than $k(\log x)^n$.

Thus, the exponential, algebraic and logarithmic functions cannot be reduced to one another; the algebraic quantities stand in the middle between the exponentials and the logarithms, in that, when the variable is infinite, we can consider the exponents as being infinite in the exponentials, finite in the algebraic and infinitely small in the logarithmic quantities.

Further, one can, in principle, establish that a radical function of a variable cannot be identical with a rational function of the same variable or with another radical function. Thus, $(1 + x^3)^{1/4}$ is essentially different from $(1 + x^3)^{1/3}$ and from $(1 + x)^{1/2}$.

These principles, which are based on the very nature of functions, can be of great use in analytic research because they indicate which form the functions we are searching for must necessarily have, and show that it does not exist in a great number of cases. However, one must then be absolutely sure not to omit any of the possible forms. Thus, since the differentiation lets the exponential and the radical quantities subsist and only makes the logarithmic quantities disappear when they are multiplied by constants, one may conclude that the integral of a differential function cannot include any other exponentials and radicals than those already included in this function. In this way, I have realized that one cannot obtain the integral $\int \frac{dx}{\sqrt{1 + \alpha x^2 + \beta x^4}}$ as a finite explicit or implicit function. Similarly, I have shown that the linear partial differential equations of the second order between three variables usually cannot be integrated in finite form, and this has led me to a general method of integrating them in this form when it is possible. In the other cases, one cannot obtain a finite integral except by means of definite integrals.

[Laplace 1812, pp. 4–5]

Laplace's sketch anticipated many of Liouville's discoveries, and if Laplace had published rigorous proofs of the theorems that he claimed to have found, he would certainly have been acknowledged as the founder of the theory of integration in finite form. However, the loose way in which Laplace formulated and proved that an integral can only contain the same radicals and exponentials as the integrand sounds more like Fontaine and Condorcet than like a rigorous statement of the 19th century. Therefore, I do not think Laplace was in fact in possession of anything approaching Liouville's theory. The only person who threatens Liouville's priority as the founder of the theory of integration in finite terms is Abel, whose works in this area we shall now consider.

Abel's Contributions

7.   Niels Henrik Abel (1802–1829), like Laplace, claimed that he had created a theory of integration in finite terms. In a footnote to theorem III in

his last major paper on elliptic functions *Précis d'une théorie des fonctions elliptiques*, he wrote:

> On this theorem, I have founded a new theory of integration of algebraic differential formulas, but the circumstances have not allowed me to publish it yet. This theory far surpasses the known results; its aim is to effect all the possible reductions of integrals of algebraic formulas using algebraic and logarithmic functions. In this way, one manages to reduce to the least possible number the integrals needed to represent in finite form all the integrals belonging to the same class.
>
> [Abel 1829, Oeuvres I, p. 550]

In Abel's case, we have good reasons to believe that he did possess the theory described in the quote, for not only is he known to have written a lost paper on the subject, we also have preserved elements of his theory in some of his other papers.

The lost paper was the first mathematical work Abel wrote in French. It is known from the minutes of the Collegium of Oslo University of March 22, 1823, here cited in Ore's translation:

> Professor Hansteen [Abel's supervisor] appeared before the Collegium and presented a manuscript by Student Abel constituting a memoir with the purpose of giving a general method to decide the integrability of any kind of differential formula.
>
> [Ore 1974, p. 63]

Abel had enquired whether the university would assist in the publication of the work, but knowing its importance, Hansteen found it more appropriate if it could be published abroad. Therefore, instead of supporting the publication directly, the university some years later decided to grant Abel a travel stipend so that he could present his paper to the Paris *Académie*. In the meantime, Abel became anxious about the fate of his great memoir and wrote in March the following year to Professor Degen in Copenhagen, who had proposed elliptic functions as a suitable field of enquiry for the young geometer:

> Since the time when I enjoyed the instructive association with the Herr Professor, I have occupied myself especially with the integral calculus, and I may venture to say, not without success. I have reviewed the first part of integration which I had developed before I came to Copenhagen, and have been fortunate enough to give it a very systematic form. I had hoped to get it printed here in Oslo at university expense, but nothing came out of it, since I was at the same time proposed for a travel grant and the Collegium was of the opinion that the publications of some academy of science would be a more suitable place for my memoir. God knows where I can get it printed! I wished it so dearly, for I believe that my own papers will be my best recommendation during my foreign travel, which I have reason to believe will occur in a year or thereabout.
>
> [Ore 1974, pp. 76,77]

However, the memoir was not printed before he set off, and during the journey through Europe to Paris, Abel discovered his celebrated theorem on Abelian integrals and so preferred to submit a paper on this subject to the *Académie des Sciences* instead. Abel's theorem is a generalization of Euler's addition formulas and states that if y and g are algebraic functions of x and $f(y, x)$ is a rational function of y and x, then there exists a fixed number p (the genus) depending only on y and not on f, such that any number of integrals,

$$\int_{g(a)}^{g(b_1)} f(y, x)\, dx + \int_{g(a)}^{g(b_2)} f(y, x)\, dx + \cdots + \int_{g(a)}^{g(b_m)} f(y, x)\, dx, \quad (3)$$

can be written in the form

$$\int_{g(a)}^{g(c_1)} f(y, x)\, dx + \cdots + \int_{g(a)}^{g(c_p)} f(y, x)\, dx + u + B_1 \log v_1 + \cdots + B_n \log v_n,$$
$$(4)$$

where $c_1, \ldots, c_p$ are algebraic functions of $b_1, \ldots, b_m$ and $u, v_1, v_2, \ldots, v_n$ are algebraic functions of x.

The paper, *Mémoire sur une propriété générale d'une classe très étendu de fonctions transcendantes*, in which Abel presented his theorem to the Paris *Académie* is famous not only for its content but also for its fate. It was mislaid by Cauchy and therefore did not reach publication until 1841, a long time after the death of its author. Yet, however sad were the circumstances surrounding the publication of this paper, the fate of the paper on integration was worse. According to Ore [1974, p. 63], it seems to be completely lost.[1]

8.   Still, some of Abel's results later appeared in two of his printed papers, *Sur l'intégration de la formule différentielle* $\frac{\rho\, dx}{\sqrt{R}} \ldots$ [1826a] and *Précis d'une théorie des fonctions elliptiques* [1829b], in a letter to Legendre published in [1830], and in a posthumously published memoir, *Théorie des transcendantes elliptiques* [1839]. Since these results did not appear in the order they were found by Abel, I shall discuss them not in chronological order but in logical order, starting with Abel's version of Liouville's theorem, and ending with Abel's application of these theorems to elliptic functions.

Abel's version of "Liouville's" theorem was announced in his letter of November 25, 1828, to Legendre in which he summarized the state of his mathematical research. Having explained his advances in the theory of elliptic functions, he continued:

> In addition to the elliptic functions there are two other branches of the analysis with which I have been occupied, namely the theory of integration of algebraic differential formulas, and the theory of equations. With the aid of a particular method, I have found many new results which are especially of very great generality. I have started from the following problem of the theory of integration:

"An arbitrary number of integrals $\int y \, \partial x, \int y_1 \, \partial x, \int y_2 \, \partial x$ etc. being given where $y, y_1, y_2, \ldots$ are arbitrary algebraic functions of x, to find all the possible relations between them that can be expressed by means of algebraic and logarithmic functions."

At first, I found that any relation must have the following form:

$$A \int y \, \partial x + A_1 \int y_1 \, \partial x + A_2 \int y_2 \, \partial x + \cdots = u + B_1 \log v_1 + B_2 \log v_2 + \cdots$$

$$(5)$$

where $A_1, A_2, \ldots, B_1, B_2, \ldots$ etc. are constants and $u, v_1, v_2, \ldots$ are algebraic functions of x. This theorem greatly facilitates the solution of the problem; but the most important theorem is the following:

"If an integral, $\int y \, \partial x$, where y is related to x by an arbitrary algebraic equation, can be expressed in some way *explicitly or implicitly* by way of algebraic and logarithmic functions, one can always suppose that: $\int y \, \partial x = u + A_1 \log v_1 + A_2 \log v_2 + \cdots + A_m \log v_m$, where $A_1, A_2, \ldots$ are constants and $u, v_1, v_2, \ldots, v_m$ are *rational* functions of x and y."

[Abel 1830, Oeuvres II, pp. 275–276]

At first sight, these two theorems may seem more restricted than Liouville's theorem (§1) in one sense: they do not mention the use of exponentials. Yet, in the last theorem, Abel allows the integral to be expressed implicitly by way of algebraic functions and logarithms, and since the equation $\log y = f(x)$ implicitly defines $y = e^{f(x)}$, Abel's theorems are at least as general as Liouville's later theorem. They surpass Liouville's theorem in three respects:

a. Abel's first theorem states that if F is an elementary function of $n + 1$ variables and if

$$F\left(x, \int y \, dx, \int y_1 \, dx, \ldots, \int y_{n-1} \, dx\right) = 0$$

for $y, y_1, \ldots, y_{n-1}$ algebraic, then some linear combination of these integrals is in the form prescribed by Liouville's theorem. A modern version of this theorem has recently been established by Prelle and Singer [1983] (Singer has written to me, "I did not know that Abel had thought about this").

With this extension of Liouville's theorem, Abel emphasized the connection to his theorem on Abelian integrals (§7). In fact, the integrals in (3) and (4) can be transcribed by the substitution

$$\int_{g(a)}^{g(x)} f(y, x) \, dx = \int_a^x f(y(g^{-1}(x)), g^{-1}(x))(g^{-1}(x))' \, dx$$

into integrals of different algebraic functions with the same bounds. Thus, the sum, (3), minus the integral in (4) is of the form appearing in the first of the theorems quoted above. This theorem therefore states that the most general elementary equation between the $m + p$ integrals in (3) and (4) is of the form (3) = (4). Thus, one cannot lower the number p of integrals needed in Abel's theorem by allowing a more complicated elementary relation between them.

This remark emphasizes the strong link there was between Abel's researches in the theory of integration in finite terms and in elliptic and Abelian integrals.

b. Abel's inclusion of implicitly given finite functions is also a real advantage over Liouville's statement, for, as Liouville proved [1837d], there exist implicitly defined finite functions that are not elementary. The solution $z(x)$ of Kepler's equation $x = z - h \sin z$ is an example [Liouville 1837d, p. 539].

It is not clear how general were the implicit functions Abel had in mind. He certainly thought of functions defined by an equation $f(x, y) = 0$, where f is elementary, but he may also have included functions defined by a system of elementary equations. His generalized version of Liouville's theorem is true in both cases. Ritt's paper [1923] supplies the proof in the first case, whereas the more general case has been established recently by Risch [1976]. I shall later return to this problem and to an unpublished proof by Liouville (§37–41).

c. In the last theorem quoted from Abel's letter to Legendre, Abel stated that the algebraic functions entering in Liouville's theorem can be chosen to be rational functions of x and y. This part of the statement was included as a separate theorem in *Précis d'une théorie des fonctions elliptiques* [Abel 1829b] and is now often called Abel's theorem, not to be confused with the theorem on Abelian integrals:

Abel's Theorem: If an integral of the form

$$\int (y_1 \, dx_1 + y_2 \, dx_2 + \cdots + y_\mu \, dx_\mu)$$

can be expressed by an algebraic and logarithmic function of the form

$$u + A_1 \log v_1 + A_2 \log v_2 + \cdots + A_\nu \log v_\nu,$$

one can always suppose that $u, v_1, v_2, \ldots, v_\nu$ are rational functions of $x_1, x_2, \ldots, x_\mu, y_1, y_2, \ldots, y_\mu$. Therefore, if one has an arbitrary algebraic equation, one can assume that u, v_1, v_2 etc. are rational functions of y and x.

[Abel 1829b, Oeuvres I, p. 550]

It was in a footnote to this theorem (quoted in §7) that Abel publicly announced his theory of integration in finite form.

9.   Abel's theorem is a clarification and a generalization of Laplace's statement (§6) that an integral can only contain the same radicals as the integrand. It is the only part of the theorems quoted from the letter to Legendre

of which we know Abel's proof. Since this gives a clue to the understanding of Abel's methods and also to his influence on Liouville, I shall outline this proof.

In the *Précis* [1829b], Abel's theorem appears only as a corollary to the following more general theorem:

Theorem II. If an arbitrary integral of the form

$$\int (y_1 \, dx_1 + y_2 \, dx_2 + \cdots + y_\mu \, dx_\mu),$$

where $y_1, y_2, \ldots, y_\mu$ are algebraic functions of $x_1, x_2, \ldots, x_\mu$ and where the latter are connected by an arbitrary number of algebraic equations, can be expressed by algebraic, logarithmic and elliptic functions such that one has

$$\int (y_1 \, dx_1 + y_2 \, dx_2 + \cdots + y_\mu \, dx_\mu) =$$

$$u + A_1 \log v_1 + A_2 \log v_2 + \cdots + A_\nu \log v_\nu$$

$$+\alpha_1 \cdot \psi_1 t_1 + \alpha_2 \cdot \psi_2 t_2 + \cdots + \alpha_n \cdot \psi_n t_n,$$

where $A_1, A_2, \ldots, \alpha_1, \alpha_2, \ldots$ are constants, $u, v_1, v_2, \ldots, t_1, t_2, \ldots$ are algebraic functions of $x_1, x_2, \ldots$, and $\psi_1, \psi_2, \ldots$ are arbitrary elliptic functions, then I say that one can always express the same integral in the following way:

$$\delta \int (y_1 \, dx_1 + y_2 \, dx_2 + \cdots + y_\mu \, dx_\mu) = r + A' \log \rho' + A'' \log \rho''$$

$$+ \cdots + A^{(k)} \log \rho^{(k)} + \alpha_1 \cdot \psi_1 \theta_1 + \alpha_2 \cdot \psi_2 \theta_2 + \cdots + \alpha_n \cdot \psi_n \theta_n,$$

where δ is an integer; $\alpha_1, \alpha_2, \ldots, \alpha_n$ are the same as in the given equation, $A', A,'' \ldots$ are constants, and $\theta_1, \Delta_1(\theta_1), \theta_2, \Delta_2(\theta_1), \ldots, \theta_n, \Delta_n$ $(\theta_n), r, \rho', \rho,'' \ldots, \rho^{(k)}$ are rational functions of the quantities

$$x_1, x_2, \ldots, x_\mu; y_1, y_2, \ldots, y_\mu.$$

[Abel 1829b, Oeuvres I, pp. 549–550]

Here, $\Delta_i(x) = \pm\sqrt{(1 - x^2)(1 - c_i^2 x^2)}$, where c_i is the modulus of ψ_i. However, in order to avoid too many technicalities, I shall follow Liouville and adapt Abel's proof to the simpler version of Abel's theorem stated in §8 with only one function y_1. Therefore, let

$$\int y \, dx = u + A_1 \log v_1 + A_2 \log v_2 + \cdots + A_\nu \log v_\nu, \tag{7}$$

where $y, u, v_1, v_2, \ldots, v_\nu$ are algebraic functions of x and $A_1, A_2, \ldots, A_\nu$ are complex constants. Abel first claimed without proof that there is an algebraic function λ, such that $u, v_1, v_2, \ldots, v_\nu$ can be expressed as rational functions of x and λ. In fact, this follows from Lagrange's important theorem [1770–1771] that if V_1 and V_2 are rational functions of the roots of a polynomial equation and if V_1 remains invariant under all the permutations of the roots leaving V_2 invariant, then V_2 is a rational function of V_1 and the coefficients of the polynomial equation. Indeed, let $f(x, u) = 0$, $g_1(x, v_1) = 0, \ldots, g_\nu(x, v_\nu) = 0$ be the irreducible polynomial equations defining $u, v_1, v_2, \ldots, v_\nu$, and let $u, v_1, v_2, \ldots, v_\nu, v_{\nu+1}, \ldots, v_m$ be all the roots of the following equation

$$f(x, y)g_1(x, y) \ldots g_\nu(x, y) = 0.$$

Consider $\lambda = \alpha u + \beta_1 v_1 + \beta_2 v_2 + \cdots + \beta_m v_m$, where $\alpha, \beta_1, \ldots, \beta_m$ are algebraically independent real numbers. Since λ is not invariant under any permutation of $u, v_1, v_2, \ldots, v_m$, it follows from Lagrange's theorem that $u, v_1, \ldots, v_\nu$ are rational functions of x and λ.

Since λ is algebraic, it is the solution of a polynomial equation with coefficients that are rational functions of x. Abel now chose to consider λ as a root of an irreducible polynomial equation whose coefficients are rational functions of x and y: $V(x, y, \lambda) = 0$.

If all the terms of (7) are placed on one side of the equality sign, one easily obtains by differentiation an equation $\varphi = 0$, where φ is a rational function of $x, y, u, v_1, v_2, \ldots, v_\nu$. Except for a constant, which can be considered as part of t, (7) is now fulfilled if and only if $\varphi = 0$. Introducing the function λ, φ becomes a rational function of x, y, and λ, which can easily be reduced to a polynomial in λ with rational coefficients in x and y. Since V is irreducible, all its roots, $\lambda, \lambda_1, \lambda_2, \ldots, \lambda_n$, are also roots of $\varphi(x, y, \lambda) = 0$. Thus, if $u, u_1, u_2, \ldots, u_n, v_i, v_{i1}, v_{i2}, \ldots, v_{in}$ are the functions one obtains by inserting $\lambda, \lambda_1, \lambda_2, \ldots, \lambda_n$ in the rational expressions of u and v_i, then the equation

$$\int y = u_j + A_1 \log v_{1j} + A_2 \log v_{2j} + \cdots + A_\nu \log v_{\nu j} \qquad (8)$$

holds for $j = 1, 2, \ldots, n$. Adding all these equations to (7), Abel obtained

$$(n+1) \int y \; = (u + u_1 + u_2 + \cdots + u_n)$$
$$+ A_1 \log(v_1 v_{12} v_{13} \ldots v_{1n})$$
$$+ \cdots$$
$$+ A_\nu \log(v_\nu v_{\nu 1} v_{\nu 2} \ldots v_{\nu n}).$$

The functions $(u + u_1 + u_2 + \cdots + u_n)$ and $(v_i v_{i1} v_{i2} \ldots v_{in})$ can be considered functions of x, y, and $\lambda, \lambda_1, \lambda_2, \ldots, \lambda_n$, and as such they are symmetric in the λ's. According to Lagrange's theory of symmetric functions, these functions are therefore rational functions of the coefficients of $V(x, y, \lambda) = 0$, which were in turn rational in x and y. Thus, (7) has been reduced to an expression of the same form where $u, v_1, v_2, \ldots, v_\nu$ are rational functions of x and y.   [Q.E.D.]

This ingenious proof has convinced several mathematicians, including Liouville, who cited it in [1834c] and Hardy, who referred to it in [1905]. However, Littlewood pointed out to the latter that Abel is not justified in concluding that all the roots of $V = 0$ are also roots of $\varphi = 0$, for the theorem on irreducible polynomials used here is only valid for functions $V(x, y, \lambda)$, where x, y are not algebraically related. The second edition of Hardy's book [1916] contained Littlewood's criticism as well as an alternative proof of Abel's theorem.

Rigorous or not, this proof shows clearly the strong connection between Abel's work in this field and his investigations of solvability of polynomial equations by radicals, for all the central theorems used in the proof have their origin in algebra of the time.

10.   Having explained this theorem in its general form (Theorem II, §9) to Legendre, Abel continued his letter discussing the difficulties he had in applying it:

> However, maintaining all the generality of the function y, I have been stopped by some difficulties which surpass my forces and which I shall never overcome. Therefore, I have been content with some particular cases, in particular with the one where y is of the form $\frac{r}{\sqrt{R}}$, r and R being two arbitrary rational functions of x. This is already very general. I have found that one can write the integral $\int \frac{r\,dx}{\sqrt{R}}$ in the form
>
> $$\int \frac{r\,dx}{\sqrt{R}} = p\sqrt{R} + A' \log\left(\frac{p' + \sqrt{R}}{p' - \sqrt{R}}\right) + A'' \log\left(\frac{p'' + \sqrt{R}}{p'' - \sqrt{R}}\right)$$
> $$+ \cdots + B_1 \Pi_1(y_1) + B_2 \Pi_2(y_2) + B_3 \Pi_3(y_3) + \cdots ,(9)$$
>
> where all the quantities, $y_1, y_2, y_3, \ldots, p, p', p.''\ldots$, are rational functions of the variable x.
>
> I have proved this theorem in the memoir on the elliptic functions, which has been printed in Crelle's journal.
>
> [Abel 1830, Oeuvres II, p. 278]

The last remark refers to the *Précis d'une théorie* [1829b]. The special case where r and R are entire functions was treated by Abel in the remarkable paper *Sur l'intégration de la formule différentielle* $\frac{\rho\,dx}{\sqrt{R}}$, R *et* ρ *étant des fonctions entières* [1826a]. In this paper, Abel solved the problem:

> Find all the differentials of the form $\frac{\rho\,dx}{\sqrt{R}}$, where ρ and R are entire functions of x, whose integrals can be expressed by a function of the form $\log \frac{p + q\sqrt{R}}{p - q\sqrt{R}}$.
>
> [Abel 1826a, Oeuvres I, p. 105]

Here, p and q are also supposed to be entire functions. In a final remark to the paper, Abel indicated that this seemingly special problem is in fact rather general:

When an integral of the form $\int \frac{\rho\,dx}{\sqrt{R}}$, where ρ and R are entire functions of x, is expressible by logarithms, one can always express it in the following way:

$$\int \frac{\rho\,dx}{\sqrt{R}} = A \log \frac{p + q\sqrt{R}}{p - q\sqrt{R}}, \qquad (10)$$

where A is a constant and p and q are entire functions of x. I shall demonstrate this theorem on another occasion.

[Abel 1826a, p. 144]

In fact, as remarked by Abel in the quote above (9), it follows easily from Liouville's and Abel's theorems that if the integral in question is elementary, then it must be of the form:

$$\int \frac{\rho\,dx}{\sqrt{R}} = \frac{p_0}{q_0}\sqrt{R} + A_1 \log\left(\frac{p_1 + q_1\sqrt{R}}{p_2 - q_1\sqrt{R}}\right) + \cdots + A_n \log\left(\frac{p_n + q_n\sqrt{R}}{p_n - q_n\sqrt{R}}\right).$$
$$(11)$$

Now it is easy to give an example where the algebraic term $\frac{p_0}{q_0}\sqrt{R}$ does not vanish[2]. Thus, by the phrase "expressible by logarithms" in the quote above, Abel clearly did not mean "in elementary form," but only referred to the logarithmic terms of (11), and he claimed that they reduce to one term. Abel died before he published the promised proof of his theorem, but it was later supplied by Chebyshev [1853].

11.   The aim of Abel's paper [1826a] is to decide for a given polynomial R, if there exists a polynomial ρ so that $\int \frac{\rho\,dx}{\sqrt{R}}$ is of the form (10) and to determine ρ, if it exists. He found and proved that the answer was to be found in the periodicity of the expansion of $\sqrt{R}$ as a continued fraction. More precisely:

When it is possible to find for ρ an entire function such that

$$\int \frac{\rho\,dx}{\sqrt{R}} = \log \frac{y + \sqrt{R}}{y - \sqrt{R}},$$

then the continued fraction resulting from $\sqrt{R}$ is periodic, and has the following form

$$\sqrt{R} = r + \cfrac{1}{2\mu + \cfrac{1}{2\mu_1 + \cdots + \cfrac{1}{2\mu_1 + \cfrac{1}{2\mu + \cfrac{1}{2r + \cfrac{1}{2\mu + \cfrac{1}{2\mu_1 + \cdots}}}}}}}$$

and conversely when the continued fraction resulting from $\sqrt{R}$ has this form it is always possible to find for ρ an entire function satisfying the equation

$$\int \frac{\rho\,dx}{\sqrt{R}} = \log \frac{y + \sqrt{R}}{y - \sqrt{R}}.$$

The function y is given by the following expression:

$$y = r + \cfrac{1}{2\mu + \cfrac{1}{2\mu_1 + \cfrac{1}{2\mu_2 + \cdots + \cfrac{1}{2\mu + \frac{1}{2r}}}}}.$$

[Abel 1826a, Oeuvres I, p. 136]

However surprising and ingenious this theorem is, it does not make it possible to prove the nonlogarithmic character of an integral of the given form, for however far one expands $\sqrt{R}$ as a continued fraction without finding a period there is no guarantee that an even longer period may not exist. This is probably what Liouville meant when he wrote about Abel's paper [1826a]:

> However, the ingenious method used by the author is not given by him as a general method. And in fact, if it were presented as such, one could raise the same objections as against Fontaine's method [cf. §3].

[Liouville 1834c, p. 38]

In order to be able to use Abel's theorem to prove the non logarithmic character of given integrals, one needs to determine a priori an upper bound for the period of the continued fraction for $\sqrt{R}$. This was done by Chebyshev [1860, 1861] in the case where R is a fourth-degree polynomial with rational coefficients.

12.   However, in the *Théorie des transcendantes elliptiques*, written in the period 1823–1825 and posthumously published [1839], Abel proved that the elliptic integral $\int \frac{x^2\,dx}{\sqrt{R}}$ is not expressible by logarithms when R is a general fourth-degree polynomial. This paper more generally treated Legendre's problem of reducing elliptic integrals to a small number of standard forms. In Chapter 1, Abel showed that, modulo algebraic functions, all elliptic functions,

$$\int \frac{\rho\,dx}{\sqrt{R}} \quad (\rho \text{ rational}, R \text{ polynomial of degree } 4),$$

can be reduced to the standard forms

$$\int \frac{dx}{\sqrt{R}}, \int \frac{x\,dx}{\sqrt{R}}, \int \frac{x^2\,dx}{\sqrt{R}}, \quad \text{and} \quad \int \frac{dx}{(x-a)\sqrt{R}},$$

which are in general irreducible modulo algebraic functions.

In Chapter 2, he discussed the relations that exist between these four types of integral modulo logarithmic expressions. Abel began this analysis by stating that "one can easily convince oneself" that the most general logarithmic expression for a linear combination of the four integrals must be of the form

$$\sum A_i + \log(P_i + Q_i\sqrt{R}),$$

where P_i and Q_i are entire functions and A_i are constants.

This follows from Abel's and Liouville's theorems. Then, he went on to establish several relations of this kind between the first two and the fourth kind of integral above, and proved that:

the integral $\int \frac{x^2\,dx}{\sqrt{R}}$ is irreducible in all the cases; thus, it constitutes a particular transcendental function.

This is the closest anybody before Liouville came to proving that a particular integral is not elementary. However, Abel only proved the irreducibility of $\int \frac{x^2\,dx}{\sqrt{R}}$ modulo algebraic functions and modulo logarithmic functions but not modulo both. Moreover, Abel's paper was not published until 5 years after Liouville had published his proof.

13.   To summarize Abel's contribution to the theory of integration in finite terms, Abel stated Liouville's and Abel's theorems and gave a proof of the latter, and he proved special theorems about expressibility of elliptic integrals in logarithmic terms. However, he never provided a proof of Liouville's theorem, a fact emphasized by Liouville [1834c, p. 38]. Moreover, Abel's ideas on this subject are mixed up in different works, consequences of theorems often being published before the theorems themselves (a fact I have tried to straighten out in the above summary). There seems little doubt that Abel had all or most of the ideas needed for a more systematic exposition of the theory of integration in finite terms, but because of his early death this job was left for Liouville.

Still, his colleague and rival, Jacobi, counted this subject among Abel's great achievements. Upon hearing the news of Abel's death, he wrote to Legendre [quoted from Ore 1974, pp. 233–234]:

A few days after my last letter I received the sad news that Abel was dead. Our government had called him to Berlin, but the invitation found him no longer alive; so I was bitterly disappointed in my hope to meet him in Berlin.

The wide scope of problems he proposed for himself, to find a necessary and sufficient condition for an algebraic equation to be solvable by radicals, for an arbitrary integral to be expressible in finite form, his wonderful discovery of a general property shared by all functions which are integrals of algebraic functions, etc., etc. — all these are questions of a form which is peculiar to him; no one before him had dared to propose them. He is departed, but he has left a great inspiration.

However, as we shall see in the next section, it was not the great inspiration left by Abel that made Liouville interested in this theory, but having learned of Abel's contributions he made ample use of them.

Integration in Algebraic Terms

14.   Liouville began his investigations of integration in finite terms with a less general question, namely, to decide if the integral of a given algebraic function is again *algebraic* and to find the integral when it is algebraic. Liouville solved this question completely in four published papers. The first two [1833b, 1833c] were revised versions of memoirs he had presented to the *Académie des Sciences* on December 17, 1832, and on February 4, 1833, and which are preserved in the files of the archives of the *Académie*. The third was a somewhat systematized summary of these two, written for Crelle's *Journal* [1833d], and the last, much later paper [1837l] contained simplifications of the procedures of the two first papers. Of additional material illuminating this work, we have Poisson's report to the *Académie* on Liouville's memoirs and a series of notes in Liouville's notebook [Ms 3615 (3)].

It appears from the notes that Liouville's interest in the subject developed from a desire to prove the transcendence of elliptic integrals. He might have learned about this question from Laplace's *Traité de probabilité* (see quote in §6) to which he referred in his first note. He seems to have been reminded of this problem in the middle of his work on differentiation of arbitrary order while working on the previous note [Ms 3615 (3), p. 70], which deals with expansion of algebraic functions in exponential series (cf. Chapter VIII, §19).

15.   The notes on integration in algebraic terms cover, with some omissions, the pages 72r–79v and show how Liouville gradually developed his theory. In order to illustrate Liouville at work, I shall consider this sequence in some detail.

The first appearance of the subject is in the third note of November 1832 [Ms 3615 (3), pp. 72r–73v]. It opens with the following project:

Problem. I propose to show that the quantity

$$\int_0^x \frac{dx}{\sqrt{1+x^4}}$$

is not an algebraic function of x.

[Ms 3615 (3), p. 72r]

Liouville went on to explain what he meant by an algebraic function:

One knows that a function is <u>algebraic</u> when the variable x only enters under the algebraic signs of operation, that is addition, subtraction, multiplication, division, raising to entire powers and extraction of roots.

[Ms 3615 (3), p. 72r]

Thus, Liouville initially only considered explicitly given algebraic functions as did his French precursors Condorcet and Laplace. This lasted a few months, and then, following Abel, he included implicitly given algebraic functions (cf. §§24–25).

Liouville assumed that $\int_0^x \frac{dx}{\sqrt{1+x^4}} = f(x)$, where $f(x)$ is an algebraic function and aimed for a contradiction. He first gave arguments for what we now recognize to be the relevant special case of Abel's theorem:

First, it is easy to see that $f(x)$ cannot contain any radical different from $\sqrt{1+x^4}$. This is evident when one considers the number of values of the functions. It is also a consequence of the rules of the differential calculus, as Laplace has observed. Thus, since $f(x)$ cannot contain radicals other than $\sqrt{1+x^4}$, it must be a rational function of x and $\sqrt{1+x^4}$, which can always take the form

$$\frac{M + P\sqrt{1+x^4}}{Q}, \qquad (12)$$

M, P, Q entire functions.

[Ms 3615 (3), p. 72r][3]

Here, Liouville indicates two proofs of this special case of Abel's theorem. The first, based on the number of values of the functions, probably ran approximately as follows: Since a square root has two different values, its integral must also have two different values. Since nth roots, for $n > 2$, have more than two values, $f(x)$ can only contain square roots, and since the different values of more than one square root would combine to more than two different values, there must be precisely one square root. It is not clear to me how Liouville meant to conclude, by this method, that the square root contained in f must be $\sqrt{1+x^4}$. This, however, is indicated by Laplace's argument (cf. §6) to which Liouville also referred. It is remarkable that Liouville initially accepted these two vague proofs of this theorem.

Continuing the note, Liouville set out to prove that the integral cannot be of the form (12). He first argued that $M = 0$:

This can be proved: 1° because the values must be equal and of opposite sign, 2° by the rule of differentiation.

[Ms 3615 (3), p. 72r]

Fortunately, Liouville clarified these hints in the next note [Ms 3615 (3), p. 75r]. There his argument for $M = 0$ ran as follows: Differentiation of the identity

$$\int_0^x \frac{dx}{\sqrt{1+x^4}} = \frac{M + P\sqrt{1+x^4}}{Q}$$

yields

$$\frac{1}{1+x^4}\sqrt{1+x^4} = \left(\frac{M}{Q}\right)' + \sqrt{1+x^4}\left(\frac{QP'-PQ'}{Q^2} + \frac{2x^3}{1+x^4}\frac{P}{Q}\right).$$

This gives $\left(\frac{M}{Q}\right)' = 0$, so that $\frac{M}{Q}$ is constant.

> Moreover since the integral must change sign with x one concludes
> that $R[=\frac{M}{Q}] = 0\ldots$

Here, Liouville has used that if S and T are rational functions and

$$S + T\sqrt{1+x^4} = 0,$$

then $S = T = 0$. He returned to a more general version of this theorem in a subsequent note (cf. §16).

The next step in Liouville's proof is to show that P is an odd and Q an even polynomial:

> One can prove without difficulty that $\int_0^x \frac{dx}{\sqrt{1+x^4}}$ will be reduced to x
> when x is very small, and will change sign with x. Thus P is an odd
> polynomial... and Q is an even polynomial.   [Ms 3615 (3) p. 73r]

After a page of fruitless calculations, due to a miscalculation, Liouville finished the argument on p. 74r. It is a result of the consideration above for x small that one can choose the coefficients to the lowest order term of P and Q to be 1, so Liouville set

$$P = x\left(1 + Ax^2 + \cdots + Gx^{2m-2} + Hx^{2m}\right) \text{ and}$$
$$Q = 1 + \alpha x^2 + \cdots + gx^{2n-2} + hx^{2n}.$$

> First, one can see that for $x = \infty$, $\frac{P\sqrt{1+x^4}}{Q}$ must be reduced to
> $\int_0^\infty \frac{dx}{\sqrt{1+x^4}}$ which is evidently a finite quantity $< \int_0^1 \frac{dx}{\sqrt{1+x^4}} + \int_1^\infty \frac{dx}{x^2}$.
> This gives $n = m + \frac{3}{2}$, which is absurd since n must be an integer.
>   [Ms 3615 (3), p. 74r]

Indeed, if $\frac{P\sqrt{1+x^4}}{Q}$ has a finite limit as x tends to infinity, $\deg P + 2 = \deg Q$, and hence $n = m + \frac{3}{2}$, which is certainly absurd, since m and n are integers. Thus, in his first note concerning integration in finite terms, Liouville succeeded in solving the problem he had given himself.

16.   He concluded the note by remarking that this method ought to be applicable to $\int \frac{dx}{\sqrt{R}}$, where R is a general fourth-degree polynomial and to many other integrals. However, before setting out on this generalization he began to doubt Laplace's argument for the exclusion of radicals other than $\sqrt{1+x^4}$ from $\int \frac{dx}{\sqrt{1+x^4}}$, and in the following note [note 4, November

1832, Ms 3615 (3), pp. 74v–75r], he sketched a proof of this fact. Assuming conversely that there are more radicals $\sqrt[n]{P}$, $\sqrt[m]{Q}$, etc., in $\int \frac{dx}{\sqrt{1+x^4}}$, one can remove irrational denominators and obtain

$$\int \frac{dx}{\sqrt{1+x^4}} = T_1 \sqrt{1+x^4} + A_1 \sqrt[n]{P} + B_1 \sqrt[m]{Q} + \cdots ,$$

where T_1, A_1, B_1, ... are rational functions. After differentiation of this equation, Liouville obtained an equation of the form

$$A \sqrt[n]{P} + B \sqrt[m]{Q} + \cdots = C \quad (A, B, \ldots, C \text{ rational}) ,$$

and he remarked that the question is solved if one can prove that $A = B = \ldots = C = 0$. First, he implicitly assumed that the functions under the root signs are rational functions and proceeded recursively according to the number of roots involved.

a. If there is only one radical, $A \sqrt[n]{P} = C$, this must be rational: $\sqrt[n]{P} = \frac{C}{A}$ and can therefore be removed.

b. Suppose there are two radicals $\sqrt[n]{P}$, $\sqrt[m]{Q}$, which are not "reducible one to the other," by which he meant that there exists no rational relation $\sqrt[n]{P} = R \sqrt[m]{Q}$ (R rational) between them. Differentiating the equation,

$$A \sqrt[n]{P} + B \sqrt[m]{Q} = C , \tag{13}$$

Liouville obtained[4]

$$\sqrt[n]{P} \left(A' + \frac{P'}{nP} \right) + \sqrt[m]{Q} \left(B' + \frac{Q'}{mQ} \right) = \frac{dC}{dx} ,$$

and thus

$$\sqrt[n]{P} \left(A \left(B' + \frac{Q'}{mQ} \right) - B \left(A' + \frac{P'}{nP} \right) \right) = \text{rational fct.,}$$

which according to a. implies that

$$A \left(B' + \frac{Q'}{mQ} \right) - B \left(A' + \frac{P'}{nP} \right) = 0.$$

It is now easy to deduce that if $A, B \neq 0$ we have

$$\frac{(A \sqrt[n]{P})'}{A \sqrt[n]{P}} = \frac{(B \sqrt[m]{Q})'}{B \sqrt[m]{Q}}$$

and hence that

$$\log A \sqrt[n]{P} = \log B \sqrt[m]{Q} + \text{constant} ,$$

so that

$$A \sqrt[n]{P} = k B \sqrt[m]{Q}.$$

This contradicts the assumed irreducibility of $\sqrt[n]{P}$ and $\sqrt[m]{Q}$.

c. For the more general cases, Liouville remarked, "Same thing for two radicals, three radicals, four radicals, etc." At this point, he explicitly mentioned that P, Q, ... were assumed to be rational above: "However, in all this it is only a question about simple radicals."

d. As for higher order roots, Liouville wrote:

> But, a double radical either is or is not reducible to simple radicals. If it is reducible, one reduces it. If it is not reducible, one carries out the above demonstration.
>
> [Ms 3615 (3), p. 75r]

Liouville also indicated that this method applies to higher order roots:

> This is based on the fact that in $\sqrt[n]{P}$, P is one degree less irrational than $\sqrt[n]{P}$. In this way, the thing seems to be arranged and the demonstration becomes completely rigorous.
>
> [Ms 3615 (3), p. 75r]

Indeed, the proof was made entirely rigorous by Liouville and published in [1833b, pp. 131–133 and 141–142].

17.   In the course of developing the above proof, Liouville must have realized the importance of classifying algebraic functions into functions of the first kind, where the root signs only operate on rational functions, of the second kind, where the root signs operate on functions of the first kind, etc. He also clearly realized that functions of the mth kind can be written in the form

$$P_0 + A_1 \sqrt[n]{P_1} + A_2 \sqrt[n_2]{P_2} + \cdots + A_i \sqrt[n_i]{P_i}, \qquad (14)$$

where the A_n's are rational functions and the P_n's are functions of the $(m-1)$st kind. Finally, Liouville also discovered the significance of supposing an algebraic expression to be given on the simplest possible form, i.e., being written as an expression of the lowest possible kind and with the smallest number of radicals in (14).

These are the basic ideas in the classification of algebraic functions put forward by Liouville in [1833b]. From this classification, it is only a small step to the classification of transcendental functions, which Liouville used in his later works (see §1). Abel had indicated a similar classification of explicit algebraic functions in his work on the unsolvability of the quintic [Abel 1824–1826] and Ritt suggested [1948 p. 47] that Liouville may have derived his classifications from him. That is clearly possible, but there is no cogent reason to assume an influence from Abel, for as we have seen the classification arose naturally in Liouville's own research.

18.   After this small detour, Liouville returned to the investigation of the algebraic expressibility of other integrals. In subsequent notes [Ms 3615 (3)], all dated November 1832, he gradually generalized the method applied in the first note to the following integrals:

$$
\begin{aligned}
&\text{note 6, p. 75v} &&: \int \frac{1}{\sqrt{1+x^2}} &&(x^2 \text{ is not a misprint!})\\
&\text{pp. 76r–76v} &&: \int \frac{P\,dx}{\sqrt{1+x^4}}, &&P \text{ polynomial}\\
&\text{note 7, pp. 76v–78r} &&: \int \frac{x^n}{\sqrt{1+x^m}}\\
&\text{pp. 78r–78v} &&: \int \frac{P\,dx}{\sqrt{Q}}, &&P,\,Q \text{ polynomials}\\
&\text{pp. 78v–79v} &&: \int \frac{P\,dx}{\sqrt[n]{Q}}, &&P,\,Q \text{ polynomials}\\
&\text{pp. 79v–80r} &&: \int \frac{P\,dx}{R\sqrt{Q}}, &&P,\,Q,\,R \text{ polynomials;}
\end{aligned}
$$

Q having only simple roots and no roots in common with R.

Except for the last step, this is a very systematic stepwise generalization. Liouville seems to have been very conscious of this gradual process and in particular strove not to make the steps in the process too big. For example, in note 6 he first began to investigate $\int \frac{P\,dx}{\sqrt{1+x^2}}$ for an arbitrary even polynomial P, but after two lines of calculation, he restricted P to be equal to one, so as to make an easier start. Similarly, after a few lines of calculations pertaining to $\int \frac{P\,dx}{\sqrt[n]{Q}}$ on p. 78r, he intercalated an investigation of the special case $n = 2$ before going on to the general problem.

This conscious attempt to generalize results and methods gradually was the pattern of discovery most applied by Liouville. It was this method that brought him his greatest triumphs in mathematics.

19.   Unfortunately, it would take up too much space to follow Liouville through all the successive generalizations. I shall therefore concentrate on the first example in note 7, where Liouville for the first time discovered that the problem of integration in algebraic terms can be reduced to the solution, in rational functions, of a linear rational ordinary differential equation. This insight is at the basis of all Liouville's subsequent research in this area.

In the opening remark of this note, Liouville pointed out that this example is well suited to illustrate his ideas at this stage:

However, here is a very general example that might be taken for a complete exposition of our method.

Problem. Determine when the integral

$$
\int \frac{x^n\,dx}{\sqrt{1+x^m}}
$$

is algebraic.

[Ms 3615 (3), p. 76v]

First, Liouville remarked that for $n \geq m$,

$$\int \frac{x^n\, dx}{\sqrt{1+x^m}} = \frac{2x^{n-m+1}\sqrt{1+x^m}}{2n-m+2} - \frac{n-m+1}{2n-m+2}\int \frac{x^{n-m}\, dx}{\sqrt{1+x^m}},$$

so that one can assume $n < m$. Assume that the integral is algebraic, i.e.; of the form

$$\int \frac{x^n\, dx}{\sqrt{1+x^m}} = \frac{P}{Q}\sqrt{1+x^m} + C, \tag{15}$$

where P, Q are mutually prime polynomials and C is a rational function. Differentiation of this equation yields

$$\frac{x^n}{1+x^m}\sqrt{1+x^m} = \left(\frac{QP'-PQ'}{Q^2} + \frac{P}{Q}\frac{mx^{m-1}}{2(1+x^m)}\right)\sqrt{1+x^m} + C',$$

which according to the analysis of §16 implies that

$$C' = 0$$

and

$$Qx^n = P'(1+x^m) - \frac{PQ'(1+x^m)}{Q} + \frac{mP}{2}x^{m-1}. \tag{16}$$

Hence, Q divides $PQ'(1+x^m)$. Since Q' and Q only share the multiple roots of Q and they have a smaller multiplicity in Q', and since P and Q are mutually prime, all the factors of Q are also factors of $(1+x^m)$. If these factors are $X, Y, \ldots, Z$, we can write

$$Q = X^\alpha Y^\beta \ldots Z^\gamma,$$

and we can assume $\alpha \geq \beta \geq \cdots \geq \gamma$.

Liouville then defined the polynomials M, N by

$$M = PY^{\alpha-\beta}\ldots Z^{\alpha-\gamma}$$

and

$$N = (1+x^m)^\alpha.$$

Since $\frac{P}{Q} = \frac{M}{N}$, one can substitute M for P and N for Q in (15) and therefore also in (16), which yields:

$$(1+x^m)^\alpha x^n = M'(1+x^m) + \frac{Mmx^{m-1}}{2}(1+2\alpha). \tag{17}$$

If $\alpha \neq 0$, this leads to a contradiction, because M is not divisible by $(1+x^m)$. Hence, $\alpha = 0$, so that Q is a constant, and equation (17) reduces to

$$x^n = M'(1+x^m) + \frac{Mmx^{m-1}}{2}. \tag{18}$$

Thus, Liouville reduced the original problem to the question of existence of a polynomial solution to (18). He continued to investigate this question and found easily that M must be a constant and $n = m-1$:

Thus, the integral

$$\int \frac{x^n \, dx}{\sqrt{1 + x^m}}$$

is not algebraic except when $n = m - 1$; this is certainly very remarkable although one can foresee it by the theory of binomial integrals.

[Ms 3615 (3), p. 78]

Binomial integrals are integrals of the form

$$\int x^n (1 + x^m)^{\frac{p}{q}} \, dx \qquad\qquad m, n, p, q \in \mathbb{Z}, \tag{19}$$

which for $\frac{p}{q} = -\frac{1}{2}$ reduces to Liouville's integral. It had been known since the childhood of the calculus that this integral was algebraic for $n = m - 1$ (see §58), but it had never been earlier established that this was the only case.

20. A comparison between Liouville's treatments of $\int \frac{dx}{\sqrt{1+x^4}}$ and $\int \frac{x^n \, dx}{\sqrt{1+x^m}}$ suggests, and the whole sequence of notes clearly demonstrates, that Liouville consciously altered the methods from being essentially analytic into being essentially algebraic. Thus, in the first few notes, the proof rested on the behavior of the polynomials or rational functions for particular values of x, in particular their growth at infinity and their alteration of sign at $x = 0$. In connection with the latter argument, continuity was implicitly used. Also, in the first note, he used the multivaluedness of the radicals involved.

In the note on $\int \frac{x^n \, dx}{\sqrt{1+x^m}}$ and in the subsequent notes on algebraic integrals on the other hand, the arguments only build on the algebraic theory of factorization of polynomials considered as formal expressions, without making use of their properties as functions. For example, x is never given a specified value in these arguments. We can illuminate Liouville's reason for preferring this twist of the theory by a later note dated February 1833, in which Liouville discussed the transcendental nature of e^x. He referred to the usual proof according to which any algebraic function grows as Ax^α for x tending to infinity, whereas e^x increases more rapidly (cf. Laplace's argument in §6) and continued:

> This proof requires the use of a development in series and is perhaps not absolutely exempt from difficulties, but in order to obtain the proposed goal, one can apply simpler and more rigorous considerations.

[Ms 3615 (3), p. 87r]

It was probably Cauchy's warning against divergent series that made Liouville doubt the rigor of the traditional argument. Since the growth of the functions in the argument about $\int \frac{1}{\sqrt{1+x^4}}$ also relied on developments in series, this seems to have been the reason why Liouville preferred the algebraic arguments.

21.   In the following notes, Liouville used the same algebraic methods to show that

$$\int \frac{P\,dx}{\sqrt[n]{Q}}$$

is algebraic if and only if the differential equation

$$P = M'Q + MQ' \left(\frac{n-1}{n}\right) \tag{20}$$

has a polynomial solution. It was well known at the time how to decide if an ordinary linear differential equation with polynomial coefficients has a polynomial solution. First, an upper bound for the degree of the solution was determined by a simple counting of the degree of the separate terms. This was formalized by the so-called "analytical parallelogram," which had been devised by Newton [1736 pp 9–10] to find the term of lowest degree in the power series solution of algebraic equations, and later applied by him to differential equations [Whiteside 1976, vol. 7, pp. 93–95, published in part in Wallis's *Collected Works*]. When an upper bound n of the degree of a possible solution has been determined, one only has to insert the general nth degree polynomial

$$a_0 + a_1 x + a_2 x^2 + \cdots + a_n x^n$$

into the equation and decide if the coefficients $a_0, a_1, \ldots, a_n$ can be determined so as to satisfy the equation identically. This can be done in a finite number of steps.

Thus, having reduced the algebraic integrability of $\int \frac{P\,dx}{\sqrt[n]{Q}}$ to the problem of finding a polynomial solution of (20), Liouville solved the problem completely. He even concluded this note with the words:

> The above seems to solve the problem of integration of irrational algebraic functions of the 1st kind, which seems to be a very beautiful result.
>
> [Ms 3615 (3), p. 79r]

Indeed, as pointed out in §16, any algebraic function of the first kind can be written in the form:

$$P_0 + A_1 \sqrt[n_1]{P_1} + A_2 \sqrt[n_2]{P_2} + \cdots + A_i \sqrt[n_i]{P_i},$$

where the A_i's are constants and the P_i's are rational functions and where one can assume that the function has no expression using fewer root signs (it is irreducible). It is easy to see (cf. [Liouville 1833b, pp. 139–140]) that this function has an algebraic integral if and only if all the integrals,

$$\int \sqrt[n_i]{P_i} = \int \frac{P_i}{\sqrt[n_i]{P_i^{n_i-1}}},$$

are algebraic, and this was the question that Liouville had just solved in the note above.

22.   At this point, Liouville systematized the whole theory of integration of algebraic functions of the first kind in algebraic form and worked it out as the *Premier mémoire sur la détermination des intégrales dont la valeur est algébrique* [1833b], which he presented to the *Académie des Sciences* between the 7th and 17th of December 1832, only a month after having started his investigations.

The general problem of integrating an algebraic function of higher order in algebraic terms was solved by Liouville in his second memoir presented to the *Académie* on February 4, 1833.

He began the second part of this paper by remarking that any algebraic function y satisfies an irreducible nth degree equation:

$$y^n + Ly^{n-1} + \cdots + My + N = 0,\tag{21}$$

where $L, \ldots, M, N$ are rational functions of x. He then showed that if $\int y\,dx$ is algebraic, it must be of the form

$$\int y\,dx = \alpha + \beta y + \gamma y^2 + \cdots + \lambda y^{n-1},\tag{22}$$

where $\alpha, \beta, \gamma, \ldots, \lambda$ are rational functions of x. Differentiating (22) and eliminating, by means of (21), $\frac{dy}{dx}$ and all powers of y higher than the $n-1$st, Liouville obtained an equation:

$$\left(\frac{d\alpha}{dx} + E\right) + \left(\frac{d\beta}{dx} + F\right)y + \left(\frac{d\gamma}{dx} + G\right)y^2 + \cdots + \left(\frac{d\lambda}{dx} + H\right)y^{n-1} = 0,\tag{23}$$

where $E, F, G, \ldots, H$ are known rational functions of $x, \beta, \gamma, \ldots, \lambda$. Since y was supposed to satisfy an irreducible equation, (21), of this kind of a higher degree, (23) can only be satisfied if

$$\frac{d\alpha}{dx} + E = 0$$
$$\frac{d\beta}{dx} + F = 0 \tag{24}$$
$$\frac{d\gamma}{dx} + G = 0$$
$$\vdots$$
$$\frac{d\lambda}{dx} + H = 0.$$

Thus, Liouville has reduced the problem to the question of finding rational solutions to a set of first-order linear rational differential equations, or equivalently, finding a rational solution to one nth order linear rational differential equation. It is the generalization of equation (16), considered as a differential equation in $\frac{P}{Q}$.

The general problem of finding rational solutions to a linear differential equation,

$$P\frac{d^n y}{dx^n} + Q\frac{d^{n-1}y}{dx^{n-1}} + \cdots + R\frac{dy}{dx} + Sy = T, \qquad (25)$$

where $P, Q, \ldots, R, S, T$ are polynomials, was solved in the first part of the second memoir. As in the special case of (16), it is enough to determine the denominator Y of $y = \frac{X}{Y}$, because then one can find X by the analytical parallelogram or prove that it does not exist. In fact, it is clearly enough to determine a polynomial L that is divisible by Y. Liouville obtained L by

a. proving that all roots of Y are to be found among the roots $a_1, a_2, \ldots, a_m$ of P, and

b. describing a procedure that determines upper bounds on the multiplicity $\alpha_1, \alpha_2, \ldots, \alpha_m$ of these roots.

Thus, one can choose $L = (x - a_1)^{\alpha_1} \cdots (x - a_m)^{\alpha_m}$.

It appears from the notebooks that Liouville began his investigations of rational solutions in December 1832 with a discussion of a first-order linear rational equation [Ms 3615 (3), pp. 80r–81r]. The same month he continued with higher order equations [Ms 3615 (3), pp. 81v–86v]. The method used in these notes seems to be entirely different from the one used in the published paper, and apparently did not lead to the desired result.

23.   The *Académie des Sciences* appointed a committee consisting of Poisson, Navier, and Lacroix to examine Liouville's paper and on February 18, 1833, they presented a report written by Poisson [1833b] to this learned society. The report was generally very positive, but also pointed out a few weaknesses primarily concerning the rational solution of rational differential equations. Here, Liouville had only investigated first- and second-order equations in detail, and although Poisson did not seriously doubt the generality of the method, he was worried about the increasing complexity of the calculations for higher order equations.

In the conclusion of the report, Poisson recommended that Liouville's two papers be approved by the *Académie* and printed in their *recueil des Savants étrangers*. This recommendation was followed in 1838, five years after they had appeared in the *Journal de l'École Polytechnique*. Before the publication, however, Liouville had "altered the composition at some points, but without changing anything in the principal theorems" [1833b p. 124].

I shall now analyze these alterations, because they shed light on Liouville's continued work with the theory and in particular on the influence of Abel on Liouville.

24.   This part of the story is complicated by the fact that Liouville presented two different versions of the first memoir to the *Académie*, both

preserved in the file of the meeting on December 17, 1832. The first of them is dated December 10th on the cover, later altered to December 18th (these is some confusion during the autumn of 1832 as to the dates of the meetings; Liouville himself [Liouville 1833c, p. 150] claimed that he submitted the paper on December 7th). The second version is dated February 4, 1833. Thus, when submitting the second memoir, Liouville included a new version of the first one.

The most conspicuous difference between the two is that Chapters III and IV (§§7–15) of the first version are left out in the second version. These chapters contain a number of examples of elliptic integrals which are shown not to be algebraic. The examples are a selection of the notes in the notebooks we have analyzed above, and even the methods are the same. Thus, analytical arguments relative to the increase of the functions at infinity are used. In §16 of the first version, Liouville began the purely algebraic approach with the words:

> By comparing the examples of deductions that we have given in the preceding sections, one can see that our method is fast and uniform and that it can be extended to a great number of algebraic functions. Nevertheless, it is not yet general enough, and it can with advantage be replaced by a new procedure, which we shall now explain.
>
> [AC L Ms 6, §16]

This remark corroborates the conclusions of §20. In the second version of the first memoir, the analytical approach is only mentioned in passing in the introduction just as in the printed version [1833b, pp. 124–125].

The published version differs from both of the manuscript versions in several ways. First, the discussion in the published version (§§7–12) of $\int \sqrt[4]{L}\,dx$ has replaced a similar discussion of the unnecessarily complicated expression $\int \frac{M\,dx}{N\sqrt[4]{L}}$, which successive generalizations had led Liouville to consider (cf. §18). More significantly, one does not find §17 and §18 of the printed memoir in the two manuscript versions of the first paper, but in §§22–24 in the manuscript version of the second paper, which is preserved in the files of the meeting of February 4, 1833, at the Archive of the *Académie des Sciences*. Why did Liouville transfer these sections to the first memoir?

25.   In order to answer this question, we must compare the second printed paper [1833c] with the manuscript version. The most significant difference is that while the manuscript only deals with explicit algebraic expressions the printed paper takes implicitly defined algebraic functions into account. In most places, for example, in the formulation of the main theorem in §1, this alteration has been carried out simply by intercalating the words "explicit or implicit" in suitable places. However, at the place where in the published paper Liouville cited Abel's proof of Abel's theorem, the manuscript version contained an entirely different argument. As in Abel's theorem, Liouville proved that if $y(x)$ is algebraic and if its integral is algebraic then it is a rational function of x and y, but in contrast to Abel, Liouville, without

saying so, only had explicit algebraic functions in mind. However, his proof did in fact allow y to be implicit, for he started out remarking that if y is algebraic it always satisfies an equation of the form (21), and from then on he based the argument on this equation. This allowing of implicit y's, however, does not seem to have been a conscious act, and when saying that $\int y\,dx$ is algebraic his method only covered the explicit case.

Indeed, Liouville began his proof by carrying over his classification of explicit algebraic functions to algebraic functions of two variables. Then, he first assumed that the integral is of the first kind in x and y, i.e., it is of the form

$$\int y\,dx = P + E\sqrt[m]{Q} + \cdots + H\sqrt[q]{S},$$

where P, E, Q, ..., H, S are rational functions in x and y. By differentiation, he obtained

$$y = P' + \frac{(E^m Q)'}{mE^m Q}E\sqrt[m]{Q} + \cdots + \frac{(H^q S)'}{qH^q S}H\sqrt[q]{S}.$$

From this equation, one can find $\sqrt[q]{S}$ as a rational function of x and y and the other roots, and inserting this value in the expression for $\int y\,dx$, the quantity $\sqrt[q]{S}$ has been eliminated. Similarly, all other radicals may successively be eliminated so that in the end y is expressed as a rational function of x and y.

If $\int y\,dx$ is expressed as a function of the second kind, it can be reduced to an expression of the first kind by the same argument, etc. From this theorem Liouville easily deduced (22).

It is this argument for Abel's theorem in the explicit case that Liouville removed from the manuscript version of the second memoir [AC L Ms 8 §22–24] and placed at the end of the printed version of the first memoir, where it came to serve as a commentary to the quotation from Laplace. The reason for this removal is clear. After having presented the second memoir to the *Académie* on February 4, 1833, Liouville became aware of the difference between explicit and implicit algebraic functions and realized the importance of taking the latter into account in his theory. He therefore had to replace the above-mentioned proof with that of Abel, and instead of discarding the first proof completely, he placed it at the end of the first paper in which he continued to deal only with explicit algebraic expressions.

Why then did Liouville change his approach suddenly by including implicit algebraic functions? There seems to be only one possible reason: he learned about Abel's work on the subject. To be sure, he did refer to Abel in the manuscript version of the second paper, but at that time he had "only known about it for a short while." From the absence of any mention of implicit algebraic functions, it is quite clear that Liouville had not yet digested Abel's ideas, and in particular that he had not yet appreciated the superiority of Abel's version of Abel's theorem over his own. Therefore, there is little doubt that Liouville did not know of Abel's contributions when he developed his theory of algebraic integrals. However, in his further research on integration he could draw on the ideas of his predecessor.

We can put an upper limit to the time when Liouville had completely digested the ideas he had read in Abel's work. For, in his third paper, dated March 1, 1833, he did subordinate the results of both his previous papers under the new approach. This paper was published in Crelle's *Journal* later the same year [1833d], headed by Poisson's report of the first two memoirs. In addition to the more general approach, Liouville supplied a proof, missing from the second paper, for the independence of the equations, (24), and finally he greatly simplified the algorithm for determining the algebraic integrability of functions of the first kind.

Thereby, Liouville had created and written a rather complete theory of integration in algebraic terms over a period of less than four months. He only returned to this subject in one short paper of 1837 in which he simplified the determination of rational solutions to (24) or (25), which Poisson had criticized in his report. In particular, he described without proof a simple method of determining, a priori, the denominator of the solution (cf. §60). Many of the ideas of this note had already been discovered in April 1834 (cf. [Ms 3615 (4), pp. 44r–45r]). Otherwise, from March 1833, he turned toward the general problem of integration of algebraic functions in *finite* terms.

Integration in Finite Terms

26.   From the very start of his investigations of algebraic integrals, Liouville probably considered them to be a first step toward a theory of integration in finite terms. Indeed, Laplace had discussed this problem in the passage (§6) to which Liouville referred in his first note on algebraic integrals. In any case, this continuation of his research on the calculus of integration was forcefully suggested by Abel's work and explicitly by Poisson. Thus, in his report on Liouville's papers on algebraic integrals, Poisson pointed out that

> by limiting his considerations to algebraic integrals of differential formulas, Mr. Liouville has therefore not completely solved the problem concerning the absolute possibility or impossibility of their integration in finite terms.
>
> [Poisson 1833b, p. 211]

Poisson even indicated how Liouville's methods could be extended to the general case (cf. §30).

Following either his own plan or Poisson's suggestions, in March 1833 Liouville began a study of integration in finite terms that resulted in his two most celebrated papers [1834c, 1835a] in this area, those presented to the *Académie* on June 24 and December 2, 1833, respectively. Like the previous papers, they are set into relief by an *Académie* report, this time written by Libri [1834], and by a collection of notes in Liouville's notebooks [Ms 3615 (3), pp. 91v–96r; Ms 3615 (4), pp. 18r–21v, 50r; Ms 3615 (5), pp. 60r–61v].

27.   The notes in [Ms 3615 (3)] show Liouville's first approach to the subject, and since it is again totally different from the published approach, I shall discuss the first note [Ms 3615 (3), pp. 91v–92v] of March 1833 in some detail.

Liouville first solved this problem:

Integrate $\frac{P\,dx}{\sqrt{Q}}$ in log. That is, in the form

$$\int \frac{P\,dx}{\sqrt{Q}} = \log f(x) + \text{Const.},$$

$f(x)$ being algebraic.

[Ms 3615 (3), p. 91v]

Possibly relying on Abel's theorem, but without reference to it, he immediately stated that $f(x)$ must be of the form $\alpha + \beta\sqrt{Q}$, α, β being rational functions, so that

$$\int \frac{P\,dx}{\sqrt{Q}} = \log(\alpha + \beta\sqrt{Q}) + C. \tag{26}$$

This is the problem that had been investigated by Abel in [1826a] (cf. §11), but instead of Abel's continued fraction method, Liouville employed his own much simpler ideas. He saw by differentiation of (26) that

$$\begin{cases} \beta P = \frac{d\alpha}{dx} \\ \alpha P = \frac{d\beta}{dx} + \frac{\beta Q'}{2Q} \end{cases} \cdot \tag{27}$$

Thus, he reduced the problem to the existence of rational solutions α, β to these differential equations, a problem that he had already solved.

Next, Liouville investigated the integration of $\frac{P\,dx}{\sqrt{Q}}$ "in an algebraic part + a logarithm." Liouville immediately claimed that then

$$\int \frac{P\,dx}{\sqrt{Q}} = \frac{\theta}{\sqrt{Q}} + \log(\alpha + \beta\sqrt{Q}) + \text{const.}, \tag{28}$$

where θ, α, β are rational functions. In fact, it is easy to see that the rational part γ in the general equation

$$\int \frac{P\,dx}{\sqrt{Q}} = \gamma + \frac{\theta}{\sqrt{Q}} + \log(\alpha + \beta\sqrt{Q}) \tag{29}$$

must vanish. For substituting $-\sqrt{Q}$ for $\sqrt{Q}$, one obtains

$$-\int \frac{P\,dx}{\sqrt{Q}} = \gamma - \frac{\theta}{\sqrt{Q}} + \log(\alpha - \beta\sqrt{Q}), \qquad (30)$$

and by subtraction of (30) from (29), one finds an expression for $\int \frac{P\,dx}{\sqrt{Q}}$ without a rational term.

Liouville observed that since (28) implies that

$$\int \frac{dx}{\sqrt{Q}} \left(\frac{2PQ - 2Q\theta' + \theta Q'}{2Q} \right) = \log(\alpha + \beta\sqrt{Q}) + \text{const.} \qquad (31)$$

this problem is reduced to the previous problem if one can determine θ. As a first step in this direction, he argued that θ must be a polynomial:

> One can prove that θ has no denominator, for if there was a divisor containing $(x + a)^m$, then, by putting $z = x + a$ and developing the two sides [of (28)] in powers of z, one finds the absurdity that the right-hand side contains a negative power of z which is stronger, disregarding the sign, than the left-hand side. Thus, θ is an entire polynomial.
>
> [Ms 3615 (3), p. 92r]

This argument was rephrased in the following, more comprehensible way in Liouville's next note [Ms 3615 (3), p. 93v]: If $(x - a)^m$ is a divisor in the denominator of θ and if $(x - a)^n$ is a divisor of Q, for $(x - a)$ small, the left-hand side of (28) grows as $(x - a)^{-(n/2)+1}$, and the right-hand side grows as $(x - a)^{-m-(n/2)}$. Since $-m \neq 1$, this is absurd. Thus, θ is a polynomial.

Considering the growth at infinity, Liouville similarly saw that θ must be of degree $n + 1$ where $n = \deg P$, so that

$$\theta = Ax^{n+1} + Bx^n + \cdots + Cx + D.$$

The problem is to find A, B, $\ldots$, C, D. Liouville first claimed:

> And by the same development, it is easy to determine A, B, &$_c$.
>
> [Ms 3615 (3), p. 92r]

However, he soon discovered that since the log terms can only be discarded by comparison with the strictly positive powers of x (for x tending to infinity), only the coefficients of $x^{n+1}, x^n, \ldots, x^{m+1}$ can be determined (where $2m$ is the degree of Q, which for convenience Liouville assumed to have even degree).

In order to determine the other coefficients, Liouville observed that the function in the parentheses in the integral, (31), must be a polynomial. For otherwise a factor $(x + a)$ of Q must remain in its denominator so that the integral would grow at least as fast as $(x + a)^{-(1/2)}$ for $(x + a)$ tending to zero, and that is absurd since it is equal to a logarithm. Let R be the polynomial equal to this parenthetical term. Then

$$2PQ - 2Q\theta' + \theta Q' = 2QR$$

from which it is apparent that Q must divide $\theta Q'$. At this point, Liouville assumed that Q has only simple roots, so that Q must divide θ. This requirement gives $2m$ conditions to determine the remaining $m+1$ coefficients of θ. Thus, θ is overdetermined and the problem is solved.

28.   Having tried, apparently without complete success, to extend this argument to the case where Q has multiple roots, he applied similar methods to show that "$\int \frac{dx}{\sqrt{R}}$ can never be expressed either algebraically or in logarithms." This time he allowed several logarithms to enter into the expression:

$$\int \frac{dx}{\sqrt{R}} = \alpha + \beta\sqrt{R} + A_1 \log(\alpha_1 + \beta_1\sqrt{R}) + \cdots + A_n \log(\alpha_n + \beta_n\sqrt{R}),\ (32)$$

where $\alpha, \beta, \alpha_1, \beta_1, \ldots, \alpha_n, \beta_n$ are rational functions. Liouville seems to have believed that his long argument [Ms 3615 (3), pp. 93r–94v] established the impossibility of such an expression and apparently attached a great deal of importance to it, for he carefully signed it "March 25th 1833 J. Liouville." However, the proof does contain some flaws, the last of which cannot, I believe, be easily repaired.[5]

At this point, Liouville believed that he possessed "almost the fundamental principles for the integration in finite quantities of all the irrational algebraic functions of the 1st kind," [Ms 3615 (3), p. 95r]. However, the subsequent attempt to solve this question broke down at the treatment of $\int \frac{P\,dx}{\sqrt{R}}$ where R has multiple roots. He never managed to solve the general case and in his published paper [1834c] one finds only the two following theorems concerning integration of algebraic functions of the first kind:

a. If P and R are polynomials of which R is of odd degree with only simple roots, then $\int \frac{P\,dx}{\sqrt{R}}$ is expressible in explicit finite form if and only if it is algebraic.

b. If P and R are polynomials of which R is of even degree with only simple roots and

$$\deg R > 2 \deg P + 2,$$

then $\int \frac{P\,dx}{\sqrt{R}}$ is not expressible in finite form.

Even though these two theorems fall short of Liouville's high goals, they are still remarkably general. Thus, in the case when $P = 1$ and $R = (1 - x^2)(1 - k^2 x^2)$, $(k \neq 1, 0)$, the last theorem implies that the elliptic integral of the first kind (§5) is not elementary. A slight modification of the proof also allowed Liouville to draw the same conclusion for the elliptic integrals of the second kind. He never discussed elliptic integrals of the third kind, and indeed they are expressible in finite form in some cases [Ritt 1948, p. 37].

29.   Although Liouville's published paper does not give a detailed treatment of a much larger class of integrals than his early notes, the published version represents an immense improvement over the manuscripts. For in [Liouville 1834c, pp. 63–81] the two theorems mentioned above are deduced from the equation

$$\int \frac{P\,dx}{\sqrt{R}} = \eta + \frac{\theta}{\sqrt{R}} + \sum_{i=1}^{n} A_i \log\left(\alpha_i + \beta_i \sqrt{R}\right) \tag{33}$$

($\eta, \theta, \alpha_i, \beta_i$ rational) by a careful algebraic study of its differentiated consequence. He consciously avoided all the arguments from the notes involving infinite series and the behavior of functions at infinity and relied entirely on factorization theory for polynomials.

Thus, Liouville's approach to integration in finite terms developed from being analytical into being algebraical, just as his earlier approach to algebraic integrals.

30.   However, Liouville's most remarkable achievement in [1834c] was, according to both him [1834c, p. 38, footnote] and his successors, the classification of explicit finite functions and the proof of Liouville's theorem (§1), which allowed him to conclude that $\int \frac{P\,dx}{\sqrt{R}}$ is of the form (33) if it is elementary. We saw that Abel had not published any proof of this theorem and Laplace had only given vague hints to a proof. The closest one can come to an argument for the theorem before Liouville is Poisson's suggestion in his report [1833b] that the methods Liouville had applied to radicals (cf. §26) also apply to exponentials and trigonometric functions. He wrote:

> One may add that a series of eliminations similar to the preceding one can also be used for making the exponentials and the trigonometric functions disappear, if there exist some of them in the integral of an algebraic differential; from this it follows that these transcendentals cannot enter in the integral of such a differential; but the same argument does not apply to logarithms.
>
> [Poisson 1833b, p. 212]

Liouville showed that Poisson was right, but the transfer of the methods summarized in §§14-25 to the transcendental case required a great deal of ingenuity. In order to illustrate the central ideas of Liouville's argument, I shall summarize the first few steps of the recursive proof of Liouville's theorem. It rests on two lemmas.

Lemma 1. *The derivative of a function of the nth kind is at most of the nth kind and the only nth-order monomials which can enter the derivative are those contained in the primitive function.*

Its proof is obvious. The second lemma is so central to the theory that Ritt [1948, p. 16] has named it after Liouville:

Lemma 2 (Liouville's principle). *If the number of monomials of the nth kind entering a function of the nth kind is reduced to a minimum, any algebraic relation in these monomials and functions of a lower kind must be a trivial identity, i.e., an equation that is fulfilled whatever is substituted instead of the monomials* (formulated explicitly in [Liouville 1837d, p. 76]).

Proof. If there existed such an algebraic relation which did not degenerate into a trivial identity, one could determine one of the monomials of the nth kind as an (implicit) algebraic function of the rest, in which case the number of monomials could be lowered by one, contrary to our assumption. Q.E.D.

31.   On the basis of these lemmas, Liouville proved his theorem, i.e., that if $\int y\, dx = \varphi(x)$ is a finite explicit function, then it is of the form

$$\int y\, dx = t + A \log u + B \log v + \cdots + C \log w,$$

where $t, u, v, \ldots, w$ are algebraic functions. He first assumed that φ is of the first kind, i.e., an algebraic function of some transcendental monomials, whose number is assumed to be reduced to a minimum. Liouville assumed that φ contains an exponential $\theta = e^u$, where u is algebraic, and then aimed for a contradiction.

Differentiating

$$\int y\, dx = \varphi(x, \theta),$$

he obtained

$$y = \varphi'_x(x, \theta) + \varphi'_\theta(x, \theta)\theta u'.$$

According to Lemma 1, this is an algebraic equation in the transcendental monomials contained in φ, and therefore, according to Lemma 2, it must be a trivial identity. Thus, when $\mu\theta$ is substituted for θ, the identity must hold so that

$$\varphi'_x(x, \theta) + \varphi'_\theta(x, \theta)\theta u' = \varphi'_x(x, \mu\theta) + \varphi'_{\mu\theta}(x, \mu\theta)\mu\theta u'.$$

Integrating this equation from b to x and assuming that $\theta(b) = a$, one gets

$$\varphi(x, \mu\theta) = \varphi(x, \theta) + \varphi(b, \mu a) - \varphi(b, a).$$

Again, this is an algebraic equation in the transcendental monomials contained in φ, and so it must hold if an arbitrary letter ζ is substituted for θ. Having differentiated the resulting equation and having set μ equal to one, Liouville found

$$\varphi'_\zeta(x, \zeta) = \frac{a\varphi'_\zeta(b, a)}{\zeta}.$$

Finally, he integrated this equation with respect to ζ from a fixed value ζ_0 and found

$$\varphi(x,\zeta) = a\varphi_a'(b,a)(\log\zeta - \log\zeta_0) + \varphi(x,\zeta_0).$$

When $\zeta = \theta = e^u$ is substituted for ζ, this equation yields

$$\varphi(x,\theta) = a\varphi_a'(b,a)(u - \log\zeta_0) + \varphi(x,\zeta_0).$$

Since the right-hand side does not contain the transcendental monomial θ, φ cannot contain any exponentials.

To finish the argument in the case where the integral is supposed to be of the first kind, Liouville must show that the logarithmic monomials can only enter linearly. To this end he took one of these monomials, θ, to be $\log u$, where u is an algebraic function, such that

$$\int y\,dx = \varphi(x,\theta).$$

Differentiating this, he obtained

$$y = \varphi_x'(x,\theta) + \varphi_\theta'(x,\theta)\frac{u'}{u}.$$

In this equation, one can, according to the lemmas, substitute $\mu + \theta$ for θ. Hence,

$$\varphi'(x,\mu+\theta) + \varphi_\theta'(x,\mu+\theta)\frac{u'}{u} = \varphi_x'(x,\theta) + \varphi_\theta'(x,\theta)\frac{u'}{u},$$

which yields by integration

$$\varphi(x,\mu+\theta) = \varphi(x,\theta) + \text{const.}$$

This implies that

$$\varphi_\theta'(x,\theta) = A,$$

where A is a constant. According to Lemma 2, we can substitute an arbitrary variable ζ for θ in this equation and integrating the resulting partial differential equation, one obtains

$$\varphi(x,\zeta) = A\zeta + \varphi(x,\zeta_0) - A\zeta_0,$$

where ζ_0 is an arbitrary fixed value of ζ. Thus, we have

$$\int y\,dx = \varphi(x,\theta) = A\theta + \varphi(x,\zeta_0) - A\zeta_0,$$

which makes it evident that $\theta = \log u$ can only enter the expression φ as a linear term.

This completes the proof of Liouville's theorem in the case where $\int y\,dx$ is assumed to be of the first kind. Liouville next showed that the integral cannot be of the second kind. First, he argued with an argument identical to the above that

$$\int y\,dx = t_1 + A\log u_1 + B\log v_1 + \cdots + C\log w_1,$$

where $t_1, u_1, v_1, \ldots, w_1$ are functions of the first kind. Then, he proved by contradiction that $u_1, v_1, \ldots, w_1$ contain neither exponentials nor logarithms, so that the integral cannot possibly be of the second kind. Exactly the same arguments show that the integral cannot be of a higher kind either. This ends Liouville's proof of his theorem.

32.   Because of their central role in Liouville's theory, these arguments require a closer analysis. The basic idea is, by differentiation of the given equation, to obtain an algebraic identity in which one can replace an exponential monomial with a multiple of itself and a logarithmic monomial with a constant plus itself. In a footnote in [1837d, p. 527], Liouville explained why different substitutions should be made in the two cases: "The reason for this difference stems from the very nature of the differential equations that serve to define these two types of functions."

Indeed, $z = \log x$ satisfies the equation $\frac{dz}{dx} = \frac{1}{x}$, whose complete solution is $z = \mu + \log x$, whereas $z = e^x$ satisfies $\frac{dz}{dx} = z$, whose complete solution is μe^x. Therefore, if $\theta = \log u$, the derivative

$$\frac{d\varphi(x,\theta)}{dx} = \frac{\partial\varphi(x,\theta)}{\partial x} + \frac{\partial\varphi(x,\theta)}{\partial\theta}\frac{u'}{u}$$

is transformed into the derivative of $\varphi(x, \mu + \theta)$:

$$\frac{d\varphi(x,\mu+\theta)}{dx} = \frac{\partial\varphi(x,\theta+\mu)}{\partial x} + \frac{\partial\varphi(x,\theta+\mu)}{\partial(\mu+\theta)}\frac{u'}{u}$$

simply by substituting $\theta + \mu$ for θ in the expression for $\frac{d\varphi}{dx}$. Similarly, changing θ to $\mu\theta$ in the expression $\frac{d\varphi(x,\theta)}{dx}$ when $\theta = e^u$ gives the expression for the derivative of $\varphi(x, \mu\theta)$. These are the properties of the substitutions used in the proofs.

Unfortunately, there are no sketches of this ingenious proof in Liouville's notebooks, so we cannot judge if Ritt was right when he conjectured [Ritt 1948, p. 27] that Liouville's proof was inspired by Abel's paper on the quintic [Abel 1824–1826].

33.   Only a few months after having begun the investigation of the integration of algebraic functions in finite terms, Liouville discovered that he could immediately transfer his ideas to integrals of transcendental functions of the form $\int e^x y\,dx$, where y is algebraic. Using the ideas from the proof of his theorem, he showed in May 1833 [Ms 3615 (4), pp. 5–7] that if such an integral is elementary, it must be of the form $e^x P$, where P is algebraic. More specifically, the integral must be of the form

$$\int e^x y\,dx = e^x(\alpha + \beta y + \gamma y^2 + \cdots + \lambda y^{\mu-1}), \tag{34}$$

where $\alpha, \beta, \gamma, \ldots, \lambda$ are rational functions. Liouville put this observation into a footnote of his second paper on algebraic integrals [1833c, pp. 192–193], in which he also claimed that on the basis of this theorem and the methods used in that paper, he had devised an elementary algorithm to find the integral $\int e^x y \, dx$, if it is elementary, or to show that it is not elementary.

He provided the details of this claim in a *Mémoire sur l'intégration d'une classe de fonctions transcendantes*, which he presented to the *Académie des Sciences* on December 2, 1833. He seemingly attached some importance to this paper, for in a letter of December 1833 to Crelle [Neuenschwander 1984b, II, 1], he explained that although he had promised to send a *Note sur l'usage de la formule de Fourier* (Liouville crossed out "l'usage") (this is surely [1835b]), he had chosen to finish this paper first. Having not heard from Crelle for two months, he wrote again because he feared that the manuscript might be lost:

> This uncertainty is the more disagreeable for me because I think it is important that my memoir be printed immediately. In fact it contains, I think, the most extensive method which has been published until now concerning the integration of explicit functions of one variable; and the principles that I develop, constitute the basis of a more complete theory... [Liouville to Crelle, February 9, 1834, Neuenschwander 1984b, II, 1] (cf. also Chapter III).

The paper appeared in Crelle's *Journal* in 1835 [1835a]. In this paper, the above theorem on $\int e^x y \, dx$ was derived from the following important generalization of Liouville's theorem:

Liouville's generalized theorem. *Let y and z, etc., be functions of x, which satisfy differential equations of the form*

$$\frac{dy}{dx} = p, \frac{dz}{dx} = q, \text{ etc.},$$

where p and q are algebraic functions of x, y, and z, etc. Further, let P be an algebraic function of x, y, and z, etc. If $\int P \, dx$ is a finite function of x, y, and z, etc., then

$$\int P \, dx = t + A \log u + B \log v + \cdots + C \log w,$$

where A, B, $\ldots$, C are constants and t, u, $\ldots$, w are algebraic functions of x, y, and z, etc. [1835a, pp. 98–99].

(An early version of this theorem involving only one equation $\frac{dy}{dx} = p$ is stated in [Ms 3615 (4), p. 12v], dated June 1833.)

Moreover, if P, p, q are *rational* functions of x, y, and z, etc., then t, u, v, $\ldots$, w can be chosen to be *rational* functions.

After a detailed proof of this theorem, Liouville showed how its consequence, (34), could be used to reduce the integration of $e^x y$ in finite form to the solvable problem of determining rational solutions to a set of linear rational differential equations, precisely as in the algebraic case. He even indicated that the same analysis applied to integrals of the form,

$$\int (e^p P + e^q Q + \cdots + e^r R)\, dx,$$

where P, Q, R, p, q, r, etc., are algebraic functions, and none of the functions p, q, r, etc., are constants. Thus, surprisingly enough, he found this question completely solvable in a finite number of steps, in contrast to the problem of the integration of purely algebraic functions (p, q, r, etc., constants) in finite terms, which he had only been able to solve in special cases.

In particular, Liouville proved that the integrals

$$\int_0^x \frac{e^x\, dx}{x} \quad \text{and} \quad \int_0^\infty \frac{\sin \alpha x}{1 + \alpha^2}\, d\alpha$$

are not elementary.

34. Three months after the *Académie* had received the paper on the integration of transcendental functions, a committee consisting of Libri, Lacroix, and Poisson presented a report written by Libri and signed by all three.[6] The report was generally positive. Thus, according to Libri [1834, p. 482], Liouville, "in this research, which concerns the most difficult questions of analysis, has shown evidence of a great knowledge and sagacity, as in his preceding works." However, Libri's summary of Liouville's paper is rather inaccurate in several points. For example, he claimed that the problem treated in this paper is "much more general" than that treated in his earlier papers. This is wrong, for as Liouville explicitly pointed out, his new theorems did not apply to integrals of algebraic functions. Libri also explained that the problem is reduced "in a very ingenious way to the *algebraic* integration of a system of differential equations...." In fact, the solutions must be rational, not only algebraic.

Libri also launched some criticism of Liouville's methods. In contrast to the observations made by Poisson in the earlier report, Libri's remarks say more about his poor understanding of the subject than about Liouville's work. In addition to pointing out that certain theorems, such as the transcendence of $\log x$, had not been established in all details, Libri had a fundamental criticism of the classification of explicit finite functions. In particular, he made complaints about the classification of the function $x^\alpha = e^{\alpha \ln x}$ when α is irrational:

> Mr. Liouville has classified these quantities among the transcendental functions of the second kind. However, it seems difficult to us to accept this classification; for according to the transformation, he has

indicated [$x^\alpha = e^{\alpha \log x}$] an arbitrary power of the variable can be classified among the transcendental functions of the second kind, and that is not admissible.

[Libri 1834, p. 483]

This criticism is based on a misunderstanding. It is clear that all explicit functions can be written in an infinity of different ways, e.g., $x = e^{\log x}$, $x^{p/q} = e^{(p/q) \log x}$, etc., which seem to fall into different classes in Liouville's classification. However, it is assumed that the class of a function is determined by its most elementary expression, that is, by the expression of lowest class. Thus, the function $x^{1/2}$ is an algebraic function, i.e., of the zeroth kind, even though it can be written in the form $e^{(1/2) \log x}$. The only problem in Liouville's classification of x^α is therefore to prove that it is not of a lower class than 2 when α is irrational. Libri, however, does not seem to have grasped this subtlety.

I have summarized Libri's report in such detail because it was to play a role in the discussions preceding Liouville's entry to the *Académie* (Chapter II). Also, in spite of the shallowness of the critical points, they may have convinced Liouville of the necessity of a very thorough discussion of his classification. This was published in [1837d].

35.   Libri's most constructive enterprise in the report was to urge Liouville to systematize his work in a connected treatise, which he knew Liouville had planned to write [Chapter IV]. This treatise on the subject never appeared, but in two notes in his notebooks, Liouville discussed its projected contents. The first note was written March 30, 1834, the day before Libri presented his report to the *Académie*. It says:

> We must begin to collect the material for a great work entitled *Essai sur la Théorie de l'intégration des formules différentielles en quantités finis.*
>
> The first chapter of this work, which should serve as an introduction, should be devoted to the definitions and the classification of functions.
>
> The second chapter will treat the differential formulas whose integral is of the same type as its differential. For example, consider $f(x, e^x) \, dx$, where f is algebraic in x and e^x, find $\int f(x, e^x) \, dx$ whenever the integral $= F(x, e^x)$, F also being algebraic etc. It is very important first to treat these two fundamental chapters thoroughly.
>
> [Ms 3615 (4), p. 44r]

At this point of the survey of his treatise, Liouville drifted off into an investigation of $\int f(x, e^x)$. However, early the following month, he returned to the planning and wrote:

> Our work will be divided into three parts or even into four. 1° Integration of explicit functions of one variable. 2° Integration of ordinary differential eq[uati]ons. 3° Classification of transcendentals;

this constitutes a theory similar to the one of elliptic functions but more general. 4° Historical studies concerning the questions treated in the 3 first parts. Each of these parts will be subdivided into books.

[Ms 3615 (4), p. 45v]

It would have been very interesting to see what Liouville had intended for these books. Unfortunately, he stopped the list of contents after the first book of the first part. This book was given the title "Integration in algebraic quantities" and was planned to consist of 10 chapters. I shall not quote Liouville's description of this book, as it corresponds closely to the contents of his three already published papers on the subject.

Clearly, the content of the papers on the integration of algebraic functions and functions of the form $e^x y$ (y algebraic) was also meant to go into the first part. I shall return to the possible content of the second part in the next section. Of the last projected section, nothing ever occurred, except for a few very loose critical comments on the work of some of his predecessors. On the other hand, Liouville wrote a paper *Sur la classification des transcendantes et sur l'impossibilité d'exprimer les racines de certaines équations en fonction finie explicite des coefficients* [1837d], which probably contains material he had intended for the third part of his Treatise (corresponding to the first chapter in the first list of contents).

36.   In this voluminous paper, Liouville analyzed in great detail the classification of elementary functions, which had been questioned by Libri. He provided a completely rigorous algebraic proof that $\log x$ cannot be expressed explicitly in algebraic and exponential terms, and conversely that e^x cannot be expressed explicitly by means of logarithms and algebraic functions (cf. the proof already mentioned in a note of February 1833 (§20)). Thus, all three kinds of functions are essential to the theory. He also rigorously determined the order of the ordinary transcendental functions: x^α, the trigonometric functions, and their inverses, and he proved that $\log \log \ldots \log x$ has an order equal to the number of log signs, so that there are functions of any order. Thereby, he answered Libri's criticism.

He obtained these results with the same methods he had used to prove Liouville's theorem. They exemplify the solution of the general problem:

Given a finite explicit function of x; how does one recognize in a certain way to what class this function belongs.

[Liouville 1837d, p. 60]

Concerning this question, Liouville rather vaguely explained:

Without trying to treat this problem in all its generality we shall show by selected examples which principles one must employ in order to solve it in each particular case.

[Liouville 1837d, p. 60]

By this slightly contradictory statement, Liouville probably meant that he was not able to devise a complete algorithm to solve the general problem, but he was convinced that skillful application of his methods would always solve any given particular question of this kind.

In the last section of the paper, Liouville described a method to determine whether a solution $y(x)$ of a transcendental equation $T(x, y) = 0$, where T is an explicit finite function, is expressible as an explicit finite function. The principles of the procedure were again similar to the ones used in the proof of Liouville's theorem, and again Liouville claimed that they would apply to all such equations although they "let the particular problems peculiar to each example subsist."

The examples he chose to illustrate the procedure were impressively general. For example, he proved [1837d, pp. 536–539] that the equation

$$\log y = F(x, y)$$

has no explicit finite solution when F is algebraic and $F'_x(x, y) \not\equiv 0$ and $F'_y(x, y) \not\equiv 0$. It easily implies that Kepler's equation

$$x = y - h \sin y$$

has no explicit finite solution $y(x)$. That had been supposed for some time, and Lambert had explicitly announced this theorem in [1767, p. 355]. However, as Liouville sarcastically remarked "it was easier to announce this theorem than to prove it" [1837d, p. 58].

37.   Thus, Liouville established that it is a real restriction to exclude implicit finite functions, i.e., functions defined by explicit transcendental equations. In his paper on the nonelementary character of elliptic integrals, he had already anticipated this in a remark:

> But, our goal in this Memoir is to prove that the quantity F [the elliptic integral of the first kind] cannot become equivalent to any finite explicit function; but, we do not assert that it would be absurd to regard it as a finite implicit function.
>
> [Liouville 1834c, p. 40]

It is obvious that if Liouville's theorem is proved in the generality with which Abel stated it (§8), it follows that any (possibly implicit) finite integral of an algebraic function is an explicit finite function, and therefore the elliptic integral of the first kind is not implicitly finite either. However, since Abel never published a proof of the theorem, Ritt [1923] is generally credited for being the first to show that if y is algebraic and if $\int y \, dx = z$ is a solution to a transcendental equation, $F(x, z) = 0$, where F is a finite explicit function, then $\int y \, dx$ is an explicit finite function.

However, Liouville had already established this theorem in two highly remarkable, but never published, notes dated July 1840 [Ms 3615 (5), pp. 60r–61v].

In an undated manuscript, Liouville even sketched the proof of the more general theorem that if $z = \int y\, dx$ (y algebraic) is defined implicitly by a number of equations,

$$
\begin{aligned}
F_1(x, z, z_1, z_2, \ldots, z_n) &= 0 \\
F_2(x, z, z_1, z_2, \ldots, z_n) &= 0 \\
&\vdots \\
F_{n+1}(x, z, z_1, z_2, \ldots, z_n) &= 0,
\end{aligned}
\tag{35}
$$

where F_i, $i = 1, 2, \ldots, n + 1$, are elementary functions, then z must be an explicit finite function.

Ritt conjectured this more general theorem in 1923 and wrote:

> It is natural to inquire as to whether an integral w of an elementary function is itself elementary if it is one of n functions, w, w_1, $\ldots$, w_{n-1}, which satisfy a system of elementary equations
>
> $$F_i(w, w_1, \ldots, w_{n-1}; z) = 0 \qquad (i = 1, 2, \ldots, n).$$
>
> While the formal elements of our proof, given in §IV, can be extended to settle this question affirmatively, we see no way of avoiding certain function-theoretic assumptions, which, though light in almost any other case, would be out of place in a problem of this kind. We shall therefore not write now on this question.
>
> [Ritt 1923, p. 212]

However, in 1948 he withdrew this statement:

> In the introduction to my paper, [Ritt 1923], I mentioned the problem of determining whether the integral of an elementary function is elementary if it is one of a set of functions which satisfy a set of elementary equations. I stated that the formal elements of the proof for one function could be carried over to answer this question affirmatively. I wish to withdraw this statement. I do not have the details now, if indeed I ever did.
>
> [Ritt 1948, p. 94]

The general problem was not solved in print until 1976 when Risch showed that Ritt's conjecture was true.

38. Let us now return to Liouville's proof more than a century earlier. His first note on the subject opened with a clear announcement of the simple theorem:

> Let $\int y\, dx = z$, y algebraic, z not expressible in explicit finite form. I say that z cannot be a finite implicit function either. Consider here the case where z is a root of one single eq[uati]on, $T = 0$.
>
> [Ms 3615 (5), p. 60r]

Here, $T(x, z)$ is an explicit finite function of x and z. Liouville assumed that z, defined by this equation, is not an explicit elementary function. Thus, $T(x, z)$ must contain at least one transcendental in z, when we consider x as a parameter. As usual, Liouville reduced T "to contain the least possible number of transcendentals of the highest degree, and this degree itself to be the smallest possible" [Ms 3615 (5), p. 60r].

In the first note [Ms 3615 (5), pp. 60r,v], he then continued with a proof that presupposed that T contains at least two transcendental monomials of the highest order. While reflecting on what to do when there is only one highest order monomial in T, he discovered a slightly different and "preferable" approach, which he then used in the subsequent note [Ms 3615 (5), pp. 61r,v].

With a slightly improved notation, he repeated the same argument in an undated manuscript of 11 pages. The manuscript has been classified among the papers of [Ms 3640] in the file dated 1877–1879, but to judge from the handwriting, it is of a much earlier date and was probably composed shortly after the two notes mentioned above, i.e., during the summer of 1840.

39.   I shall give a summary of this note and try to supply some of the details that Liouville left out. Assume first that there is a logarithm $\theta = \log f(x, z)$ among the highest order monomials in T. We can then isolate this logarithm and write $T(x, z) = 0$ in the form

$$\theta = \log f(x, z) = F(x, z), \tag{36}$$

where F is a finite function that does not contain θ. Differentiating (36) and keeping in mind that $\frac{dz}{dx} = y$, Liouville obtained the equation

$$\frac{f'_x(x, z) + y f'_z(x, z)}{f(x, z)} = F'_x(x, z) + y F'_z(x, z), \tag{37}$$

which contains one transcendental monomial (θ) less than (36). Since we have assumed that z cannot be defined by an equation with fewer monomials, (37) must be an identity in z, and thus we can substitute $(z + \mu)$ for z and get

$$\frac{f'_x(x, z + \mu) + y f'_z(x, z + \mu)}{f(x, z + \mu)} = F'_x(x, z + \mu) + y F'_z(x, z + \mu).$$

Liouville then integrated this equation between b and x and put $z(b) = a$. That gives

$$\log f(x, z + \mu) = F(x, z + \mu) + \log(b, a + \mu) - F(b, a + \mu).$$

Differentiating with respect to μ and setting $\mu = 0$ yields,

$$\frac{f'_z(x, z)}{f(x, z)} = F'_z(x, z) + \frac{f'_a(b, a)}{f(b, a)} - F'_a(b, a).$$

Again, this is an equation in z containing fewer transcendental monomials than (36), and therefore it is an identity in which we can replace z with an independent variable, i:

$$\frac{f_i'(x,i)}{f(x,i)} = F_i'(x,i) + m,$$

where m is the constant $\frac{f_a'(b,a)}{f(b,a)} - F_a'(b,a)$.

Integration of this equation with respect to i from i_0 to t yields

$$\log f(x,i) = \log f(x,i_0) + F(x,i) - F(x,i_0) + m(i - i_0),$$

in which we can reintroduce $i = z$; keeping in mind that $\log f(x,z) = F(x,z)$ we get

$$0 = \log f(x,i_0) - F(x,i_0) + m(z - i_0).$$

If $m \neq 0$, we can easily determine z as an explicit finite function of x contrary to our hypothesis. If, on the other hand, $m = 0$, we have $\log f(x,i_0) = F(x,i_0)$ for arbitrary i_0, and hence (36) cannot define z as a function of x. In this way, Liouville reached a contradiction.

Having seen all the details in the case where $\theta = \log f(x,z)$, it should be easy to follow the next quote from Liouville's proof in the case where $\theta = e^{f(x,z)}$:

$$2°\ \theta = e^u, \quad e^u \text{ or } e^{f(x,z)} = F(x,z),$$
$$e^{f(x,z)}(f_x'(x,z) + yf_z'(x,z)) = F_x'(x,z) + yF_z'(x,z),$$
$$f_x'(x,z) + yf_z'(x,z) = \frac{F_x'(x,z) + yF_z'(x,z)}{F(x,z)},$$
$$f_x'(x,z+\mu) + yf_z'(x,z+\mu) = \frac{F_x'(x,z+\mu) + yF_z'(x,z+\mu)}{F(x,z+\mu)},$$
$$f(x,z+\mu) = \log F(x,z+\mu) + f(b,a+\mu) - \log F(b,a+\mu).$$

Differentiate with respect to μ; then $\mu = 0$:

$$f_z'(x,z) = \frac{F_z'(x,z)}{F(x,z)} + m,$$
$$f_i'(x,i) = \frac{F_i'(x,i)}{F(x,i)} + m,$$
$$f(x,i) = f(x,i_0) + \log F(x,i) + m(i - i_0) - \log F(x,i_0);$$

let $i = z$,

$$0 = f(x,i_0) - \log F(x,i_0) + m(z - i_0),$$

but m different from zero allows one to solve with respect to z, and $m = 0$ gives

$$f(x,i_0) - \log F(x,i_0) = 0$$

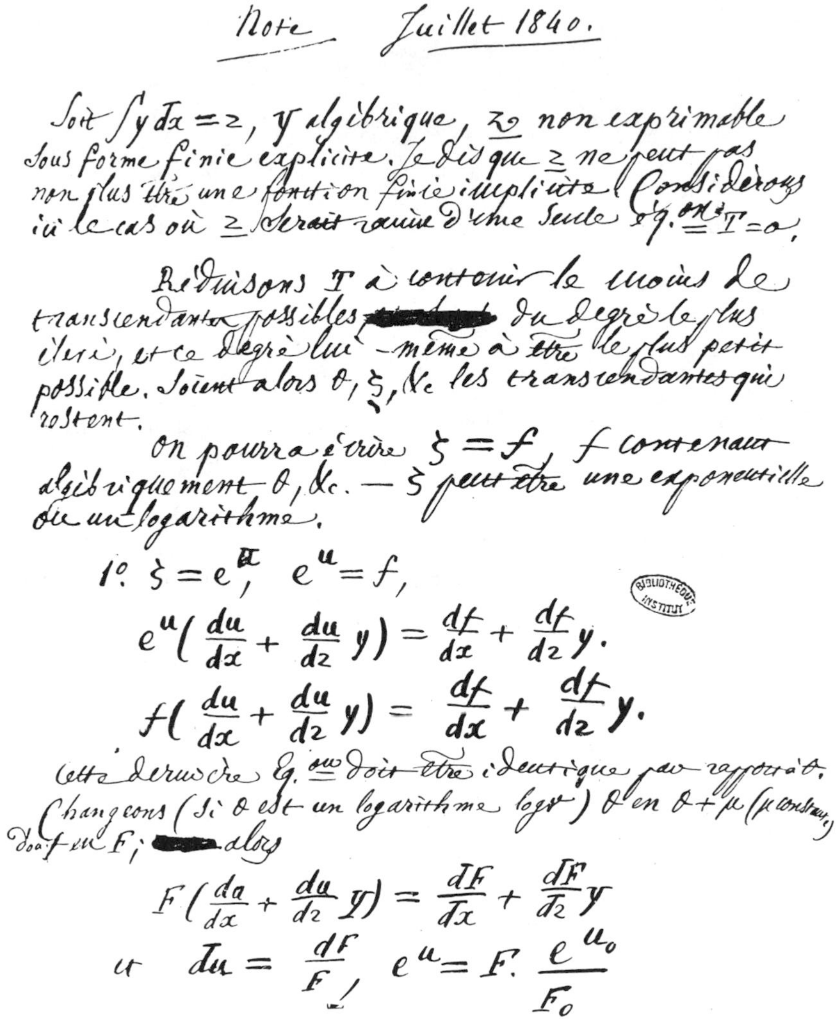

Note   Juillet 1840.

Soit $\int y\,dx = z$, y algébrique, z non exprimable
sous forme finie explicite. Je dis que z ne peut pas
non plus être une fonction finie implicite. Considérons
ici le cas où z serait racine d'une seule éq. ou $T=0$.

Réduisons T à contenir le moins de
transcendantes possibles ~~du degré~~ du degré le plus
élevé, et ce degré lui-même à être le plus petit
possible. Soient alors $\theta, \zeta, \&c$ les transcendantes qui
restent.

On pourra écrire $\zeta = f$, f contenant
algébriquement $\theta, \&c. - \zeta$ peut être une exponentielle
ou un logarithme.

$1^{\circ}.\ \zeta = e^{u},\ e^{u} = f,$

$$e^{u}\left(\frac{du}{dx} + \frac{du}{dz}\,y\right) = \frac{df}{dx} + \frac{df}{dz}\,y.$$

$$f\left(\frac{du}{dx} + \frac{du}{dz}\,y\right) = \frac{df}{dx} + \frac{df}{dz}\,y.$$

Cette dernière éq. ou doit être identique par rapport à θ.
Changeons (si θ est un logarithme $\log v$) θ en $\theta + \mu$ (μ constant)
$\ldots f$ en F; ~~alors~~ alors

$$F\left(\frac{da}{dx} + \frac{du}{dz}\,y\right) = \frac{dF}{dx} + \frac{dF}{dz}\,y$$

$$u\quad du = \frac{dF}{F},\ e^{u} = F.\frac{e^{u_{0}}}{F_{0}}$$

PLATE 25. Liouville's formulation of Ritt-Risch's theorem and the beginning of his proof. (Bibliothèque de l'Institut de France, Ms 3615 (5) p 60r)

or the identical eq. $U = 0[T = 0]$ which is absurd. Thus, $\int y\,dx$ [is] not an implicit function of the 1st class, that is, depending on the solution of one single equation.

[Ms 3640, p. 2 of a ms of 11 pages]

40.   In the undated manuscript, Liouville continued to ask "can it be of the 2nd class?," meaning can z be implicitly defined by two equations. He supposed z to be defined by $U = 0$, $V = 0$, or

$$\varphi(x, z; z_1) = 0, \quad \psi(x, z; z_1) = 0, \tag{38}$$

in which the highest order transcendentals in z are reduced to a minimum. He chose a highest order transcendental monomial θ in φ, which does not occur in ψ (one can always write φ and ψ so that this is possible) and assumed $\theta = \log f(x, z, z_1)$. The first equation can then be written

$$\log f(x, z, z_1) = F(x, z, z_1).$$

By differentiation, Liouville obtained

$$\frac{f'_x(x, z, z_1) + yf'_z(x, z, z_1) + \frac{dz_1}{dx} f'_z(x, z, z_1)}{f(x, z, z_1)}$$
$$= F'_x(x, z, z_1) + yF'_z(x, z, z_1) + \frac{dz_1}{dx} f'_{z_1}(x, z, z_1),$$

and

$$\psi'_x(x, z, z_1) + y\psi'_z(x, z, z_1) + \frac{dz_1}{dx}\psi'_{z_1}(x, z, z_1) = 0.$$

Since the equations, (38), are supposed to determine z, one can eliminate $\frac{dz_1}{dx}$ from the two equations above, and that would yield an equation that does not contain θ. Therefore, the differentiated equation and $\psi = 0$ cannot determine z, which means that for given x and z the two equations will lead to the same value of z_1. Now, alter the value of z to $z + \mu$, by which the common value of z_1 will change to $z_1(\mu)$ and integrate from a to z. One obtains

$$\psi(x, z + \mu, z_1(\mu)) = 0$$
$$\log f(x, z + \mu, z_1(\mu)) = \log f(b, a + \mu, A_1(\mu)) + F(x, z + \mu, z_1(\mu))$$
$$- F(b, a + \mu, A_1(\mu)),$$

where $z_1 = A_1$ for $z = a$. If these equations are differentiated with respect to μ and μ is put equal to zero, one obtains

$$\frac{f'_z(x, z, z_1) + f'_{z_1}(x, z, z_1)\frac{dz_1}{d\mu}}{f(x, z, z_1)} = m + F'_z(x, z, z_1) + F'_{z_1}(x, z, z_1)\frac{dz_1}{d\mu}, \tag{39}$$

and

$$\psi'_z(x, z, z_1) + \psi'_{z_1}(x, z, z_1)\frac{dz_1}{d\mu} = 0.$$

Again, if $\frac{dz_1}{d\mu}$ is eliminated, these two equations must yield the same value for z_1 as $\psi = 0$, so we can substitute an independent variable i for z. Keeping in mind that $\frac{dz_1}{d\mu} = \frac{dz_1}{dz} = \frac{dz_1}{di}$ for $\mu = 0$ and $z = i$, one can integrate equation (39) (for $z = i$) from i_0 to i, and one obtains

$$\begin{aligned}
\log f(x, i, z_1(i)) &= \log f(x, i_0, z_1(i_0)) + m(i - i_0) + F(x, i, z_1(i))\\
&- F(x, i_0, z_1(i_0)).
\end{aligned}$$

Liouville finally inserted $z = i$, and since $\log f(x, z, z_1) = F(x, z, z_1)$, he found

$$0 = \log f(x, i_0, z_1(i_0)) + m(z - i_0) - F(x, i_0, z_1(i_0)).$$

If $m \neq 0$, this yields an explicit expression for z in elementary terms, and if $m = 0$, the equation $\log(x, i_0, z_1(i_0)) = F(x, i_0, z_1(i_0))$ or $\varphi(x, i_0, z_1(i_0)) = 0$ is satisfied for all values of i_0, $z_1(i_0)$ that satisfy $\psi(x, i_0, z_1(i_0)) = 0$, and thus equation (38) cannot determine z. Hence, in both cases, Liouville reached a contradiction.

As for the case $\theta' = e^u$, he left it with the remark: "The exponentials will no doubt be very much easier." Thus, the integral $\int y\,dx = z$ cannot be an implicit finite function of the second class without being an explicit finite function.

41.   After an alternative proof on this fact, Liouville asked himself: "Can it be of the pth class?." I shall quote his sketch of the answer, which is not difficult to follow in the light of the above special case:

$$U = 0, \ V = 0, \ldots, \ W = 0$$

1st I display a transcendental which is not in any of the others and I solve with respect to this, so as to make it disappear by differ. After the limitations of the differentials of the auxiliary variables, I have an eq. which must enter into [be dependent on] the $(p-1)$ last. Thus, these $(p-1)$ last and this new subsist together when one replaces z by $z + \mu$ and when one modifies $z_1, \ldots$ accordingly. Then the diff. $dU = 0$ will also hold after these chang. and one has

$$U(\cdots, z + \mu, \ldots) = U(\cdots, a + \mu, \ldots)$$
$$V(\cdots, z + \mu, \ldots) = 0, W(\cdots, z + \mu, \ldots) = 0 \cdots$$

Diff. with respect to μ and setting $\mu = 0$, one has

$$\frac{dU}{dz} + \frac{dU}{dz_1}\frac{dz_1}{dz} + \cdots = m$$
$$\frac{dV}{dz} + \frac{dV}{dz_1}\frac{dz_1}{dz} + \cdots = 0$$
$$\cdots$$

equations, which according to what we have said above are again exact when one treats z as an indep. variable, provided that z_1, ... are determined by $V = 0, \ldots, W = 0$. Let's change z into i to clarify matters. Thus, $dU(i) = m$ so that $U(i) = m(i - i_0) + U(i_0)$ with $V(i) = 0, \ldots, W(i) = 0$. Let $i = z$.

Then $V = 0, \ldots, W = 0$ and $m(z - i_0) + U(i_0) = 0$; Thus, if m is not $= 0$, the first equation gives z, and if $m = 0$, it enters into [is a consequence of] the others — to be developed but good.

[Ms 3640, p. 8 of a ms of 11 pages]

The rest of the manuscript contains an alternative proof in the general case.

It is remarkable that Liouville produced a proof of this general theorem almost one and a half centuries before a proof was published by Risch. However, it must be admitted that in spite of its adequacy within Liouville's theory, it cannot pass as a proof in the modern sense, because the problem of the multivaluedness of the functions involved becomes more acute here than in the rest of Liouville's theory.

Solution of Differential Equations in Finite Terms

42.   According to the plans of April 1834, the second part of the great treatise on integration in finite terms was supposed to deal with "Integration of ordinary differential equations." Liouville had probably hoped to present a completely general and rigorous version of the loose ideas in Condorcet's treatise [1765]. However, the book never appeared, and Liouville only published three papers on special linear second-order differential equations. Although they fall short of Liouville's ambitious project, they still present highly important ideas and theorems. The three papers were presented to the *Académie des Sciences* on October 28, 1839 [1839e], January 6, 1840 [1840i], and November 9, 1840 [1841a], respectively.

The long period that elapsed between the treatise being projected and the publication of any results pertaining to part two reflects the complexity of the problems Liouville hit upon in this branch of the theory, compared to those he had found in the theory of elementary integrals. He clearly planned to conduct his investigations by stepwise generalization, as he had successfully done before, and therefore started out with the apparently simplest cases, namely, the first-order differential equation:

$$M(x, y)\, dx + N(x, y)\, dx = 0, \qquad (40)$$

where M, N are polynomials, and the second-order linear homogeneous differential equation

$$\frac{d^2y}{dx^2} + P(x)\frac{dy}{dx} + Q(x)y = 0, \qquad Q \not\equiv 0 \qquad (41)$$

with rational coefficients P and Q. However, even these first steps exceeded Liouville's forces, and therefore he had no intermediary results to publish when after a long time, he gave up his general plan and found instead even more special cases which the methods he had devised were able to solve. Thus, instead of the generalization Liouville had surely hoped for, he was in this case forced to invert the process.

43.   Liouville's many notes on first-order equations did not even result in a single published paper. His interest in this equation was clearly derived from Condorcet's work. Thus, his first note on integration of differential equations in finite form [Ms 3615 (4), pp. 2–5] established two theorems about the equation $M\,dx + N\,dy = 0$, due to Condorcet. The first theorem states that if this equation possesses a complete algebraic solution, $\varphi(x, y, a) = 0$, where φ is algebraic and a is an arbitrary constant, then the solution can be written in the form $f(x, y) = a$, where f is algebraic. This note is dated May 1833, that is, only two months after Liouville had begun the investigations of transcendental integrals, and almost a year before outlining the grand treatise.

Having generalized this theorem the following month to the equation

$$\frac{dy}{dx} = p(x, y)\,, \tag{42}$$

p being algebraic instead of rational as above [Ms 3615 (4), p. 8], he was led to his generalized theorem (cf. §33), the only published evidence of the research on first-order differential equations. Liouville returned to the subject on three later occasions in 1840 [Ms 3615 (5), p. 59r; Ms 3616 (3), pp. 37r–38v], 1842 [Ms 3617 (2), pp. 18v–19r, 40v–62r; Ms 3617 (3), pp. 29v–30r], and 1844 [Ms 3617 (5), pp. 10v–14r, 15v–17r, 21v–22r, 29v–30r], but these attempts were apparently almost as fruitless as his initial investigations. The basic idea in most of these notes is to discuss the connection between the solution to the equation and its integrating factors. It is obvious that if the equation $M\,dx + N\,dy = 0$ has a rational integrating factor λ, such that $\lambda M\,dx + \lambda N\,dy$ is an exact differential, $dF(x, y)$, then the complete solution, $F(x, y) = c$, to the equation is expressible in algebraic and logarithmic terms. Liouville's remarks are all developments of this idea.

For example, in a series of notes from July 1844, he tried to solve the special equation

$$r\,dx + (p + qy)\,dy = 0\,, \tag{43}$$

where p, q, and r are polynomials of x. The analysis rested on the claim that if equation (40) has no algebraic solutions but has a finitely expressible integrating factor then this factor must be of the form e^λ, where λ is a polynomial. He succeeded in finding λ from r, p, and q by a method that he found to be "not without elegance." However, returning to the general problem, he finally despaired over the unfruitfulness of the approach:

> Besides, this procedure is the one we have followed all the time; all this has made little progress since our first research.

[Ms 3617 (5), p. 164]

After one more abortive note written two days later on July 24, 1844, but subsequently crossed out, he apparently gave up and never returned to the subject of integration in finite terms.

44.   Because of their sketchy character, it is very difficult to detect possible good ideas or flaws in these notes, but from Liouville's own remarks, it is clear that he was generally not satisfied with his results. I shall, however, mention one remarkable instance where a careless use of his ingenious methods led Liouville to a completely fallacious result. I refer to the following theorem "found" in 1840:

> Theorem. When a differential equation of the first order $\frac{dy}{dx} = p$, where p is an algebraic function of x and y, has no particular algebraic integral, one can be certain that it has no particular integral of the form $T = 0$ either, where T is a function of x and y expressible by a limited number of algebraic, exponential, and logarithmic signs and even $\int$-signs, i.e. indefinite integrations [applied to exact differentials $f(x, y)\, dx + F(x, y)\, dy$].

[Ms 3616 (3), p. 38r]

The theorem is clearly false; for example, if $p = p(x)$, it says that an integral of an algebraic function is always algebraic if it is elementary.

The proof that there can be no logarithm in the solution runs as follows: Assume that a solution, $T(x, y)$, contains a logarithm $\theta(x, y) = \log u(x, y)$. Then, it can be written $\theta(x, y) = F(x, y)$, where F does not contain θ. By differentiation and by substitution of $p = \frac{dy}{dx}$, one gets

$$\frac{1}{u}\left(\frac{\partial u}{\partial x} + p\frac{\partial u}{\partial y}\right) = \frac{\partial F}{\partial x} + p\frac{\partial F}{\partial y}. \tag{43a}$$

Concerning this equation, Liouville wrote:

> If this eq[uation] is not an ident[ity], it must give the same value as the other [equation], which has not been simplified enough. If it is, $u = 0$ must give $\frac{du}{dx} + p\frac{du}{dy} = 0$ or that $u = 0$ is an integral.

[Ms 3616 (3), p. 37v]

Thus, according to Liouville, one can always eliminate a logarithm from the defining equation of $y(x)$. Similarly he argued that exponentials could be eliminated.

However, the last step in the proof is invalid. For if (43a) is an identity, its right-hand side may contain the same singularity as $\frac{1}{u}$ and so one cannot conclude that $u = 0$ is an integral. Consider, for example, the equation

$$\frac{dy}{dx} = \frac{1}{x},$$

for which $\theta = \log x$, $F = y$, $u = x$, and $p = \frac{1}{x}$. In this case, equation (43a) becomes

$$\frac{1}{x} \cdot 1 = \frac{1}{x},$$

and it does not follow that $u = 0$ is an integral.

Liouville's mistake is even more astonishing, because the same wrong theorem was derived in 1842, apparently using the integrating factor technique. The fact that Liouville never crossed out these errors indicates that despite many approaches to first-order equations he never reached the stage of critical revision caused by a planned publication.

45.   More fruitful were Liouville's researches on the linear second- order rational differential equation, (41), begun in June 1833, only a month after his initial note on first-order equations. The investigation of linear equations was inspired by his earlier successful determination of their rational solutions (cf. §22). Thus, his first note opens with the words:

> One can singularly generalize our method concerning rational integrals. For example, let it be proposed to find the integrals of linear equations with rational coefficients when they are no longer expressed by rational functions of x but more generally by functions of the form $y = \sqrt[m]{\varphi}$, φ being rational.

> [Ms 3615 (4), p. 114]

Liouville showed that if $\sqrt[m]{\varphi} = \sqrt[m]{\frac{M}{N}}$ (M, N polynomials) is a solution of

$$Py'' + Qy' + Ry = 0 \qquad (P, Q, R \text{ polynomials}) \qquad (44)$$

then any multiple root of N is also a root in P, and he claimed that the solution would easily follow.

He returned to the equation in February 1834, when he attempted to limit the possible forms of solutions of the first kind [Ms 3615 (4), pp. 42r–42v]. He first assumed that a solution $\varphi(x, \theta)$ of (44) contained a logarithm $\theta = \log u$. Then, according to Liouville's principle (cf. [1839e, pp. 436–437]) and the linearity, $B\varphi(x, \theta+A)$ is also a solution, when B and A are arbitrary constants. Liouville concluded that this must then be the complete integral, and therefore A and θ must enter in a linear way. Hence, $\varphi(x, \theta) = U + V\theta$, where U, V are independent of θ, and the complete integral must be

$$y = AV + B(U + V\theta).$$

It is easy to see that for this value of y

$$y'' + Py' + Qy = (V'' + PV' + QV)\theta + S, \qquad (45)$$

where S is independent of θ. Therefore, V is a solution to (44). Now, suppose V contained a second logarithm, θ_1. Then, it must be of the form $V = U_1 + V_1\theta$, where U_1 and V_1 do not contain θ_1, and where V_1 is a solution of (44). Thus, φ, V and V_1 are three independent solutions of (44). This is impossible, and therefore V cannot contain another logarithm.

Having treated exponential transcendentals in a similar way, Liouville concluded that the only possible forms of the solution of order ≤ 1 are:

Summary	$y = f$ algebraic	I
one logarithm	$y = AU + B(V + U \log u)$	II
one exponent	$y = AU + BVe^u$	III
two exponents	$y = AUe^u + BVe^v$	IV
one exp. one log.	$y = AVe^u + B(U + V \log v)$	V

[Ms 3615 (4), p. 42v]

Here A, B are constants and U, V, u, v are algebraic functions. However, convincing this argument is, it contains a flaw. For, it could happen that B and A in the expression $B\varphi(x, \theta + A)$ combine to one constant so that it does not represent the complete solution. Actually, in the exponential case, where the corresponding expression is $B\varphi(x, A\theta)$, Liouville later showed [1839e, p. 440], for a special family of such equations, that this is exactly what happens for $\varphi(x, \theta) = \psi(x)\theta^m$, so that $B\varphi(x, A\theta) = BA^m\psi(x)\theta^m$. In fact, Liouville showed for these equations that the only nonalgebraic elementary solution contains precisely one exponential, and he further showed that the complete integral is never expressible in finite form, as it is in II–V above [1839e, pp. 455–456]. Thus, the list is clearly deficient.

Liouville, however, did not discover this mistake at the time and continued to investigate the five possible forms. I shall briefly consider his remarks about algebraic solutions [Ms 3615 (4), p. 43r], for they contain interesting ideas and allow us to detect one of the major problems in the theory.

46.   Suppose y is an algebraic solution to the differential equation, (44). Then y is the root of an nth-degree irreducible polynomial equation $f(x, y) = 0$. It is easy to see [1839e, p. 432] that the n solutions, $y = y_1, y_2, \ldots, y_n$, must all be solutions of (44). Liouville distinguished two different cases:

1°. If the complete solution is not algebraic, all the functions $y_2, y_3, \ldots, y_n$, must be proportional to y, and therefore $y_1^n = y_1 \cdot y_2 \cdot \ldots \cdot y_n = u$, which is a symmetric function of the y's and therefore rational. Thus, the problem is to determine solutions of the form $\sqrt[n]{u}$ to equation (44), a problem he claimed to have solved in the note of June 1833.

2°. If the complete solution of (44) is algebraic, Liouville similarly concluded that

$$y_1 = \sqrt[n]{u + \sqrt{v}}, \quad y_2 = \sqrt[n]{u - \sqrt{v}},$$

where u, v are rational functions. (In a later note [Ms 3615 (4), pp. 54r–55r], he had second thoughts about this and arrived at a more general form of the solution.) He continued: "In this case the product equation [équation aux produits] has an integral of the form $\sqrt[n]{u^2 - v}$." It appears from a later note [Ms 3615 (4), p. 55v] that the *équation aux produits* is a particular third-order linear differential equation with rational coefficients that is constructed in such a way that it has as solutions any product $y_1 y_2$ of solutions to (44) (see also [1839e, p. 431, footnote]). Therefore, this case is also reduced to finding a solution of the form $\sqrt[n]{w}$ (w rational) to a linear rational equation.

It is neither indicated here nor in the earlier note how Liouville hoped to find the solutions of the form $\sqrt[n]{w}$ or to prove their nonexistence. It is possible however, that he had discovered that one can also determine a higher order linear rational differential equation that has as solutions all products of n solutions to the given differential equation. This would then have the rational solution w. However, in order to proceed in this way, one has to know n, or at least an upper bound for n. Liouville seems to refer to this problem in his final remark about algebraic solutions of (44):

> However, observe that one simply finds $\sqrt[n]{u^2 - v}$ without knowing n as a consequence of possible simplifications. It is necessary to dev[elop] that.
>
> [Ms 3615 (4), p. 43r]

The simplifications he had in mind may be the following trick indicated in the fourth note of July 1834:

> Thus, one takes
> $$y = \sqrt[n]{\varphi}$$
> and one looks for the rational values of φ. I observe that it gives
> $$y = e^{\frac{1}{n}\log \varphi} = e^{\frac{1}{n}\int \frac{\varphi' dx}{\varphi}} \qquad \left(\varphi' = \frac{d\varphi}{dx}\right).$$
> Thus, if one writes
> $$y = e^{\int t\, dx},$$
> the value of t will be rational.
>
> [Ms 3615 (4), p. 54r]

However, then one is left with the problem of determining rational solutions to the corresponding differential equation for t, which is not linear.

47.   In the July 1834 note, Liouville further remarked that equation (44) can easily be given in the form

$$\frac{d^2 y}{dx^2} = Py, \tag{46}$$

where P is another rational function. In fact, the substitution $z = y e^{-\frac{1}{2}\int \frac{Q}{P}\, dx}$ will do the trick. Therefore, in the following discussion, he always used this form of the equation. In that case, the equation for t, defined by $y = e^{\int t\, dt}$ is

$$\frac{dt}{dx} + t^2 = P. \tag{47}$$

Liouville's next (and last) series of notes deals with the algebraic solutions to this equation. It dates from April and May of 1838, i.e., four years after the previous notes [Ms 3616 (2), pp. 72r–75v]. He had by then struggled with this problem for some time and apparently attached such great importance to its solution that he grouped it with the reception of his first public honor:

> May 3, 1838. I received the official information about my nomination as a member of the *légion d'honeur...*
>
> I find the following solution of the problem which I have tried to find for a long time at Toul:
>
> $$\frac{dt}{dx} + t^2 = P = \text{fct. of } x,$$
>
> if there is a complete algebraic solution.

However, Liouville discovered that his solution was wrong and crossed out the note. A reason for the apparently scant success of these notes of May 1838 may be that the question he posed was not the most relevant one. He asked for a complete algebraic solution, which he later showed does not exist, instead of asking for particular solutions, which came to play a role in the later theory. Still, these notes contain some ideas that went into his first published paper on the integration of (46) in finite terms.

48.   This paper, *Mémoire sur l'intégration d'une classe d'équations différentielles du second ordre en quantités finies explicites* was presented to the *Académie des Sciences* on October 28, 1839, and its introduction was printed in the *Comptes Rendus*. The entire memoir appeared almost simultaneously in Liouville's *Journal* [1839e]. Liouville was only able to complete his research of the second-order equation,

$$\frac{d^2y}{dx^2} = P(x)y, \tag{46}$$

in the case where P was a polynomial, a restriction he had already made in his investigations the preceding year. Having summarized Liouville's long and frustrating approach, I shall give a short account of the final product both to do justice to Liouville and also to illustrate how some of the ideas from the notes were put to use.

In the first 8 sections of his paper, Liouville proved that the equation has no nontrivial algebraic integrals. He showed that for each natural number $\mu + 1$ there is a homogeneous linear differential equation of degree $\mu + 1$ with rational coefficients such that any product $y_1 y_2 \ldots y_\mu$ of integrals to (46) is a solution to the $(\mu + 1)$th degree equation. Certainly, for $\mu = 2$, and possibly for higher values of μ, this idea went back to his early notes from 1834 (cf. §46). The observation that if y is an algebraic solution to (46), then all the roots $y_1, y_2, \ldots, y_n$ of its irreducible equation satisfy the equation also stemmed from this time. Therefore,

$$y_1^\mu + y_2^\mu + \cdots + y_n^\mu$$

must satisfy the above $(\mu + 1)$th-degree equation. Since these expressions are symmetric in the y's, they are rational functions of x, and for $\mu = 1, 2, \ldots, n$ at least one of them must be different from zero, otherwise all the coefficients in the irreducible equation for the y's would vanish. Thus, at least one of the higher order equations should have a rational solution, but with the methods set forth in [1833c], he proved that this was not the case. Hence, y cannot be algebraic.

It is the investigation of rational solutions that does not work in general when P is a rational function. One can, of course, use Liouville's methods to investigate the $(\mu + 1)$-order equations for $\mu = 1, 2, \ldots, n$ and find out if they have rational solutions, but as long as there is no bound on n, this does not provide a general method. This problem was later solved by Fuchs [1876] (cf. §53). However, in certain cases, later studied by Liouville, it is possible to see by a general argument that none of the equations can have rational solutions.

Liouville remarked that the nonexistence of algebraic solutions to (46) could also be proved by the method of infinite series, which we saw him use in his early notes:

> Besides, one could have seen this in a very simple way by trying to represent y in a series ordered according to descending powers of x. But, although the use of series can be made rigorous, we thought that we should prefer the preceding method.
>
> [Liouville 1839c, p. 435]

So, also in his treatment of differential equations, he again deliberately chose the algebraic instead of the analytic methods, although the latter could be made rigorous. Watson [1922, p. 119] was later inspired by this hint.

49.   Having excluded algebraic integrals, Liouville then, by an ingenious use of his principle, proved that an elementary solution of (46) contains no logarithms and only one exponential of the form

$$y = e^{\int t \, dx},$$

where t is an algebraic function. Using the fact that the Wronskian of two solutions is a constant, Liouville then succeeded in proving that the irreducible equation for t is of degree 1, i.e., that t is a rational function. (In particular, the alternative proof in a footnote on pp. 447–448 is impressively simple.) Thus, t is a rational solution to the equation

$$\frac{dt}{dx} + t^2 = P. \tag{47}$$

Liouville determined that this equation has only rational solutions when P is of even degree, 2ν, and in that case the entire part of t is equal to the entire part of the square root of P, i.e., the polynomial Q defined such that

$$P = Q^2 + R \quad \text{where} \quad \deg R \leq (\nu - 1).$$

He further found that the fractional part was of the form

$$\sum_{i=1}^{k} \frac{1}{x - p_i}.$$

Thus, $e^{\int t\, dx}$ has the form $Y e^{\int \pm Q\, dx}$, where Y is a polynomial with only simple roots, and it is easy to see that if $+Q$ is used, Y must satisfy the equation

$$\frac{d^2 Y}{dx^2} + 2Q \frac{dY}{dx} + \left(\frac{dQ}{dx} - R \right) Y = 0.$$

When $-Q$ is used, a similar equation for Y is found. Liouville finally described the solution Y and showed that there is at most a polynomial solution to one of the two equations corresponding to Q and $-Q$, respectively. In this way, one can decide if the equation has an elementary integral, and it easily follows that the complete integral is never expressible in finite form.

50.   At the end of the paper, Liouville made the following significant observation:

> While trying to integrate equation (1) [46], we have always limited our analysis to functions that one obtains by combining the algebraic, exponential, and logarithmic signs a limited number of times; but this analysis can be generalized right away to the case where one joins to these three signs the sign $\int$, indicating an indefinite integral with respect to the variable x, that is, an integral whose upper limit is x and whose lower limit is a given or an arbitrary constant. In fact, the functions which are created by the use of the sign $\int$ possess, in this type of research, properties completely analogous to those of the logarithms and can be treated by the same methods.
> [Liouville 1839e, p. 456]

This remark was amplified in Liouville's next paper on this subject [1840i, p. 455] in a footnote where he pointed out that $\theta = \int v\, dx$ shares with $\log v$ the property that the derivative of $\varphi(x, \theta + \mu)$ is obtained by substituting $\theta + \mu$ for θ in the expression for $\varphi'(x, \theta)$. This is the fundamental property of $\log x$ used in all proofs (cf. §32). He also observed that when indefinite integration is allowed in the formation of elementary functions, the logarithm is superfluous since

$$\log f(x) = \int_a^x \frac{f'(x)\, dx}{f(x)} + \log f(a).$$

When a function can be written in terms of algebraic and exponential functions and integration, we shall say it is expressible by quadratures. It should be kept in mind that in these expressions only indefinite integration must be used, whereas expressions of the kind

$$g(x) = \int_0^a f(x, y)\, dy$$

are excluded. In fact, Liouville's *Mémoire sur les transcendantes elliptiques de première et de seconde espèce, considérées comme fonctions de leur module* [1840i] was devoted to two such integrals:

$$E(x) = \int_0^{\alpha_1} d\alpha \sqrt{1 - x^2 \sin^2 \alpha} \quad \text{and} \quad F(x) = \int_0^{\alpha_1} \frac{d\alpha}{\sqrt{1 - x^2 \sin^2 \alpha}},$$

about which Liouville proved that they were not expressible by quadratures for $\alpha_1 \neq 0$. These integrals are the two first of Legendre's elliptic integrals considered functions of the modulus x, instead of the amplitude α as in [1834c].

The results follow from an analysis of the differential equation

$$\frac{d^2 y}{dx^2} = -\frac{(1 + x^2)^2}{4(x - x^3)^2} y, \tag{48}$$

which is satisfied by the complete elliptic integral of the second kind (i.e. $\alpha_1 = \frac{\pi}{2}$). This equation is of the form $y'' = Py$, and although P is not a polynomial, many of the methods carry over from the previous paper. The major difficulty was to show that (48) has no algebraic solutions, but again Liouville succeeded in proving that the differential equations for products of solutions to (48) possess no rational solutions. The rest of the proof is identical to the previous, except for a slight twist of the discussion of rational solutions to (47), of which there are none in this case. The results for the general integrals E and F follow as easy consequences.

51.   Liouville presented a short summary of these results to the *Académie des Sceinces* on January 6, 1840. Having apparently read this summary in the *Comptes Rendus* [1840i], Dirichlet wrote in a letter of May 6, 1840, to Liouville:

> I have read with great interest the extension you have made of your research on elliptic functions and that you have succeeded in proving that these transcendentals, considered as functions of the modulus, cannot be expressed in finite form. I hope that your work will not be long in appearing in your journal, unless you have intended it for the *Académie* collection. In the latter case, I beg you to send me a separate copy when the printing is completed.
>
> [Tannery 1910, p. 7]

Dirichlet only had to wait half a year before he could read the entire paper in the October 1840 issue of Liouville's *Journal*. In his reply to Dirichlet of May 26, 1840, Liouville mentioned another work on a similar subject:

> I have not been able to work much this year. In addition to the Memoir on elliptic functions, I have composed a Note on the Riccati equation for which I find the cases of integrability and non-integrability directly. I will try to give you an idea about that in a letter which I will hand over to Mr. Riess.
>
> [Tannery 1910, pp. 11–12]

This last work of Liouville's on integration in finite terms was presented to the *Académie* on November 9, 1840, and appeared in Liouville's Journal the following January [1841a]. It provided the solution of an age-old problem about the Riccati equation:

$$\frac{dy}{dx} + ay^2 = b\,x^m \qquad (a, b, m \in \mathrm{R}). \tag{49}$$

It had been known since the research of the Bernoulli in the 1720s (see Chapter VII(§§35-36) that this equation could be solved by quadratures for particular values of the modulus m:

$$m = -\frac{4n}{2n \pm 1} \quad \text{for} \quad n \in \mathrm{N}.$$

Although the equation had been diligently studied by many of the greatest mathematicians of the 18th century and also by Liouville [1833a], no other integrable cases had been found. It was therefore generally believed that the Riccati equation could only be solved by quadratures for these values of m, but Liouville was the first to prove this conjecture in the memoir under consideration.

By taking the inverse step of the one that had originally led to the Riccati equation, he showed that (49) is equivalent to an equation of the form

$$\frac{d^2y}{dx^2} = Py,$$

where

$$P = \left(A(m) + \frac{B(m)}{x^2} \right) y.$$

He then derived the result by an analysis similar to those in the two preceding papers.

As a corollary, Liouville showed in a subsequent note [1841b] that the integral

$$\int_0^\pi \cos n(u - x \sin u)\, du \qquad (n \in \mathrm{N})$$

can never be expressed by quadratures. This integral presents itself in celestial mechanics.

The beautiful result on the Riccati equation and its corollary concluded Liouville's published work on integration in finite terms, and the work he continued to do in this field until 1844 (cf. §43) never resulted in published papers.

Further Developments

52.   There are two almost independent trends in the later history of the theory of integration in finite terms: one concerning integrals and the other

dealing with differential equations. Liouville's ideas on integrals expressible in finite form were barely kept alive for almost a century before they had a comeback during the 1940s. On the other hand, in the 1870s, the problem of solution of differential equations in finite form or by quadratures was incorporated in a dynamic algebraic theory of differential equations. The development of the latter theory is only outlined briefly here because it is rather loosely connected to Liouville's theory and because a closer study would fall outside the scope of this book. This sketch is followed (§56) by a more detailed survey of the slow development of the theory of algebraic and elementary integrals, which continued to be closely linked with Liouville's research.

53.   The only investigations of differential equations that sprang directly from those of Liouville were independently undertaken by P. T. Pépin [1863] (corrected in [1878]) and by Lazarus Fuchs [1876] and [1878]. They took up the major problem in Liouville's theory of second-order linear differential equations, (46), namely, that of determining the possible *algebraic* solutions. We recall that Liouville's problem had been reduced to determining an upper bound for the degree of the irreducible equation defining the possible algebraic solution. For if its degree is n then there exists a $\mu \leq n$ such that

$$y_1^\mu + y_2^\mu + \cdots + y_n^\mu \tag{50}$$

(defined in §48) is a *rational* solution of the "product equation" satisfied by all products of μ solutions to the original equation. Fuchs, whose papers were more influential than those of Pépin, approached the problem with a study of the general binary form, $F(u, v)$ (i.e., homogeneous entire function), of two independent solutions to (46). Liouville's expressions, (50), were particular examples of such forms of the μ'th-degree. Fuchs showed that if there is an algebraic solution to (46) then there exists a binary form $F(u, v)$, a so-called "Primform" of degree $N \leq 12$, which is an Lth root of a rational function, L being a known function of N. Thus, the determination of algebraic solutions to (46) was reduced to the determination of rational solutions to a finite number of "product equations," a problem that had been completely solved by Liouville.

Central to Fuchs's and Brioschi's later [1877] investigations of *Primformen* was invariant theory, the big revelation of the day. A more elegant application of the binary forms $F(u, v)$ in the search of algebraic solutions to (46) was initiated by Felix Klein in [1877a] and [1877b] and was generalized by Camille Jordan in [1878] to higher order equations. The problem of algebraic solutions continued to occupy such great mathematicians as Picard and Poincaré (cf. Vol. 3 of Poincaré's collected works), but I shall refer to Gray [1984 and 1986] and the historical introduction of Aug. Boulanger's thesis [1897] for more information.

In Klein's and especially Jordan's approaches, strong theorems about finite subgroups of the general linear group were used. Another much broader programme of application of group theory to differential equations was conceived by Sophus Lie (1842–1899). His idea was to study how differential equations behave under groups of transformations of their variables. One of his best known theorems about first-order partial differential equations states that the existence of a group of transformations, with certain specified properties, implies that the equation is solvable by quadratures [Lie 1888–1893, pp. 708–709]. Further accounts of Lie's methods can be found in [Demidov 1982] and [Vessiot 1915].

54.   Closer to Liouville's ideas was the adaption of Lie's methods to a regular "Galois theory" of ordinary differential equations. Galois had already indicated such an extension of his ideas at the end of his famous letter written to Auguste Chevalier the night before his fatal duel:

> You know, my dear Auguste, that these subjects are not the only ones I have investigated. For some time my principal meditations have been directed toward the application of the theory of ambiguity to transcendental analysis. It is a question of seeing a priori in a relation between transcendental quantities or functions which quantities one can substitute for the given quantities without the relation ceasing to hold. This makes it possible to recognize at once the impossibility of many expressions that one may look for. But, I do not have the time, and my ideas have not yet developed far into this immense area.
>
> [Galois 1832]

Libri attempted to apply algebraic ideas to differential equations [Libri 1839a] but the development of a coherent theory was left to Leo Königsberger (1837–1921) and in particular to Emile Picard (1856–1941) and Ernest Vessiot (1865–1952). Frobenius, and later Köningsberger, introduced the concept of irreducible differential equations in 1875 and in the early 1880s, respectively (see [Gray 1984, p. 5] and Köningsberger [1889]), and Picard defined the Galois group in [1883]. Picard's note was soon followed by a number of substantial memoirs by Picard himself, and from [1892] by Vessiot and later by Jules Drach (1871–1941) [1898] (see [Vessiot 1915, pp. 288–293] and [Picard 1908, p. 564] for further references). Of the results are connected to Liouville's theory, one should mention Vessiot's discovery in his thesis [1892] that it can be decided from the structure of the Galois group of a linear differential equation whether it is integrable by quadratures or in algebraic terms.

In the second quarter of this century, this field has been carried further under the name of "differential algebra" by Joseph Fels Ritt (1893–1951) (cf. his book [1950]). Following him, I. Kaplansky [1957] and E. R. Kolchin [1973] have added substantially to the theory, and in Pommaret's recent paper [1979], he announces a complete Galois theory for systems of partial differential equations. In the latter development of the theory, heavy use is made of algebraic geometry and differential geometry. Over the last decade,

interest in effective algorithms in differential algebra has grown considerably because of the new possibilities of making symbolic manipulations on
a computer. In this connection, one can consult the great number of papers
by M. F. Singer (for references see [Davenport and Singer 1986]).

55. In the preface to the second edition of his book on differential algebra,
Kaplansky wrote:

> There is another attractive chapter of differential algebra that is not
> represented in my book or in Kolchin's: the integration of functions
> in "elementary" terms (for instance, the proof that $\int e^{x^2}\,dx$ is not an
> elementary function). This is a kind of "pre-Galois" theory, in that
> only the basic properties of differential fields are involved. (In the
> same way, the theory of ruler and compass constructions precedes
> Galois theory in the study of ordinary fields.)

It is worth noting that this "pre-Galois" theory, which represents the second branch of the stream of mathematics evolving from Liouville's research
in this field, was not integrated into the more general differential Galois theory even by the 1960's, and historically the two branches developed almost
independently.

The classical work on the "pre-Galois" theory is the book *Integration in
Finite Terms; Liouville's Theory of Elementary Methods* [1948] by Joseph
Fels Ritt. As the subtitle suggests, little development had been made in this
domain since Liouville's work a century earlier. However, from Kasper's
historical survey [1980], we learn of some activity in the field in the first
half of the 20th century, and Hardy's bibliography of [1916] gives a hint of
some interest in integration in finite terms also in the 19th century after
Liouville had left the field. I shall now follow the thin thread leading from
Liouville to Ritt.

56. Much of this development took place in Russia (cf. [Demidov 1987]).
Only a few years after Liouville stopped his publications, Ostrogradsky
(1801–1862) published a couple of papers in this field. Youschkevitch [1970–
1980b] points out that he seems to have been inspired by Abel, whom he
may have met in Paris. On November 22, 1844, Ostrogradsky presented to
the St. Petersburg Academy a new method of integrating rational functions
(published [1845]). The advantage of Ostrogradsky's method over the older
procedure due to Johann Bernoulli is that one can determine the rational
part of the integral without developing the integrand in partial fractions.
Thereby, Ostrogradsky could decide if an integral of a rational function is
algebraic, for as he stated, without reference to Abel's theorem, such an
integral is algebraic only if it is rational.

In his *Sur les dérivées des fonctions algébriques* [1850a], Ostrogradsky
proved an interesting theorem on algebraic integrals. He defined the degree
of an algebraic function $y(x)$ as the value of α for which $\frac{y(x)}{x^\alpha}$ becomes finite
as x tends to infinity and showed that

> 1°, the degree of the derivative of an algebraic function is one single unit smaller than the function itself; at least when the latter is not zero.
>
> [Ostrogradsky 1850a, col 837]

and further that

> 2°, the degree of the derivative of a function of degree zero is at most one unit smaller than the highest power after zero which is in the function.
>
> [Ostrogradsky 1850a, col 837]

From these two theorems, he concluded:

> It follows that no algebraic function can have a function of degree −1 as a derivative. Thus, any function whose derivative has the negative unity as degree is necessarily transcendental.
>
> [Ostrodradsky 1850a, col 342]

As suggested by the concept of the degree of a function, Ostrogradsky's methods were analytical, like Liouville's first approach. The scope of his paper is much more limited than Liouville's. For example, Legendre's elliptic integrals are not treated, and integration in finite, nonalgebraic terms is not mentioned.

57.   Chebyshev, on the other hand, was mainly interested in the evaluation of elliptic integrals in finite terms. In his thesis, "On Integration by Means of Logarithms" [1847] (in Russian),[7] he solved some problems that according to Youschkevitch [1970–1980a], had been posed shortly before by Ostrogradsky. It is remarkable, however, that Chebyshev's methods were much more similar to Liouville's methods than to those of his compatriot. As I have remarked in Chapter V, an early contact between Chebyshev and Liouville in 1842 may explain this fact. We have seen how the two mathematicians became friends when Chebyshev visited Paris in 1852 and how Liouville instructed Chebyshev to write two papers on number theory for the 1852 volume of his journal and to revise his thesis for the 1853 volume.

58.   In the last of these papers, Chebyshev proved that Abel had been right when in [1826a] he had claimed that if $\int \frac{\rho \, dx}{\sqrt{R}}$, ρ and R being polynomials, is expressible by logarithms it can be written in the form

$$\int \frac{\rho \, dx}{\sqrt{R}} = A \log \frac{p + q\sqrt{R}}{p - q\sqrt{R}},$$

where p and q are polynomials and A is a constant (cf. §10). More generally, Chebyshev considered the integral

$$\int \frac{f \, dx}{\sqrt[m]{R}},$$

where f is only supposed to be rational but R is still a polynomial. He referred to Liouville and Abel for the fact than when expressible in finite form this integral must be of the form

$$U + A_0 \log V_0 + A_1 \log V_1 + \cdots + A_n \log V_n \,,$$

where $U, V_0, V_1, \ldots, V_n$ are rational functions of x and $\sqrt[m]{R}$. He then showed how to determine the algebraic part, U, thereby generalizing Ostrogradsky's method. After having subtracted this function on both sides, he was left with a function that must have a purely logarithmical integral, if it is integrable in finite terms at all. In the following theorem, Chebyshev then determined how many terms of the form $A_i \log V_i$ are needed:

> Theorem I. Let $\frac{fx}{Fx}$ be a rational fraction, θx a polynomial whose factors are raised to a degree less than m. If the value of the integral $\int \frac{fx}{Fx} \frac{dx}{\sqrt[m]{\theta x}}$ only contains logarithmic terms, the integral $\int \frac{fx}{Fx} \frac{dx}{\sqrt[m]{\theta x}}$ is equal to a sum of terms of the following form:
>
> $$A \log \left[\varphi \left(\sqrt[m]{\theta x} \right) \cdot \varphi^\alpha \left(\alpha \sqrt[m]{\theta x} \right) \cdot \varphi^{\alpha^2} \left(\alpha^2 \sqrt[m]{\theta x} \right) \cdots \varphi^{\alpha^{m-1}} \left(\alpha^{m-1} \sqrt[m]{\theta x} \right) \right]$$
>
> where $\varphi(\sqrt[m]{\theta x})$ is an entire function of x and $\sqrt[m]{\theta x}$, α is a primitive root of the equation
>
> $$x^m - 1 = 0 \,,$$
>
> and the number of these terms, sufficient to give the value of the integral $\int \frac{fx}{Fx} \frac{dx}{\sqrt[m]{\theta x}}$, does not surpass the degree of $F(x)$, if the function $\frac{fx}{Fx \sqrt[m]{\theta x}}$ is of a degree less than -1; if this is not the case, the number of terms does not surpass the degree of Fx by more than one unit.
>
> [Chebyshev 1853, p. 105]

In the special case of $m = 2$ and $F(x) = 1$, this theorem reduces to the one indicated by Abel. As a corollary, Chebyshev deduced that $\frac{\rho \, dx}{\sqrt[m]{R}}$ is not expressible in finite form if R has no roots of multiplicity greater than m and ρ is a polynomial of degree less than the degree of $\frac{\sqrt[m]{R}}{x}$. This includes Liouville's theorem about the elliptic integrals of the first kind. The most famous result of Chebyshev's paper, however, is related to the integral of a binomial differential,

$$\int x^{p-1}(1+x^q)^{\frac{m'}{m}} \, dx \,, \quad p,q,m,m' \in \mathbb{Z}, \quad q > 0,$$

whose algebraic integrability had also been discussed in Liouville's original version of his first memoir (§19). It had been known since Euler and Goldbach how to express this integral in elementary form if one of the three numbers $\frac{m'}{m}$, $\frac{p}{q}$ or $\frac{p}{q} + \frac{m'}{m}$ is an integer. Chebyshev proved that it followed from his theorems that these are the only cases where the binomial integral can be expressed in finite form.

59.   Chebyshev's 1853 paper ended with an enumeration of certain conditions which determine the logarithmic term of $\int \frac{\rho \, dx}{\sqrt[m]{R}}$. However, they do not

provide a method for deciding if the integral is expressible in finite form. He returned to this problem in [1857] in the particular case where $m = 2$ and R is a third- or fourth-degree polynomial. He showed how to reduce the general problem to the problem of deciding the integrability of

$$\int \frac{(x + A)}{\sqrt{R}}\, dx$$

in logarithms. In the paper of 1857, he discussed this problem by means of Abel's method of continuous fraction.

However, in [1860 and 1861], Chebyshev admitted that since Abel had given no upper bound on the length of the period of the continued fraction expansion of $\sqrt{R}$, his method did not provide a method for deciding whether the integral is of the desired form. Instead, he provided a finite algorithm to decide the question when the coefficients $\alpha, \beta, \gamma, \delta$ of $R = x^4 + \alpha x^3 + \beta x^2 + \gamma x + \delta$ are *rational*. This algorithm was indicated in the note of 1860 and described step by step in [1861]. Chebyshev also remarked that the integral $\int \frac{(x+A)}{\sqrt{r}}\, dx$ can be integrated in finite terms for at most one value of A; for if two such values of A existed, subtraction would yield a finite expression for $\int \frac{1}{\sqrt{R}}\, dx$, which is impossible according to his previous theorem.

In his last two papers on integration in finite terms [1865] and [1867], Chebyshev extended Abel's continued fraction method to integrals of the form

$$\int \frac{\rho}{\sqrt[3]{R}}\, dx\,,$$

where ρ and R are entire functions and $\sqrt[3]{R}$ has no rational factor.

Chebyshev published no proof of his algorithm for the integral of $\frac{x+A}{\sqrt{R}}$; this was provided by E. I. Zolotarev (1847–1878), who belonged to the famous St. Petersburg school, of which Chebyshev had been the main founder. The proof was published in Liouville's *Journal* [Zolotarev 1874]. Chebyshev and Zolotarev mainly used elementary methods, which resemble those of Liouville. However, they did not make a point of the pure algebraic nature of the theory; for example, they inserted specific values of the variable x into the functions under consideration.

60.   The elementary and slightly cumbersome methods of Chebyshev were criticized by Weierstrass in a paper [1857] written as a response to Chebyshev's paper of the same year. Weierstrass analyzed the question using Jacobi's theory of elliptic functions instead, which, he asserted, gave a "clearer and deeper insight into the essence of the matter." A similar criticism of the procedures of the Russian mathematicians was given by Ivan P. Dolbnia (1853–1902) in [1890]. He remarked that Zolotarev's "method of proof is very complicated, which can be explained by the fact that the author in this proof has used the old elliptic functions, with whose help the inversion of the elliptic integrals is obtained with a great difficulty."

Instead Dolbnia used Weierstrass' theory of elliptic functions to discuss "pseudo-elliptic integrals," that is integrals of the form $\int \frac{x+A}{\sqrt{R}} (\deg R \leq 4)$, which are expressible in finite form.

Pseudoelliptic integrals were studied in many connections toward the end of the last century. They obviously played a specific role in the many works on algebraic or elliptic relations between Abelian integrals; they were applied in dynamics [Greenhill 1894], and studied by Louis Raffy (1855– ?) with Liouville's methods. Raffy [1885] also provided proofs of the shortcuts Liouville had introduced in [1837l] in the theory of algebraic integrals.

A general criticism of Liouville's classification of transcendental functions was launched by Königsberger [1887]. He claimed that with the new theory of algebraic differential equations, and in particular his concept of irreducibility of such equations in mind, "it is not difficult to realize that Liouville has not formulated the question of the solvability of algebraic equations appropriately when it is to be carried over in a suitable way to transcendental functions." He seems to suggest that a function is best classified according to the order of its irreducible differential equation. He proved that an explicit function of the nth kind in Liouville's classification does satisfy an irreducible differential equation of order less than or equal to n. However, the fact that the order may be lower than n seemed to Königsberger to be an imperfection in Liouville's classification.

61.   Later, however, Godfrey Harold Hardy (1877–1947) defended Liouville's classification. In his "Properties of logarithmico-exponential functions" [1912], he stated:

> It would be unjustifiable to conclude from this [that the order of the function does not correspond to the order of its irreducible equation] (as Königsberger appears to have done) that Liouville's classification is in some sense illegitimate or trivial or uninteresting. It must not be forgotten that the particular is often more interesting than the general: and, in my opinion, the main interest of Liouville's classification lies in its application to two special problems - indefinite integration in finite terms on the one hand, orders of infinity on the other.

In the paper [1912] and in a previous book [1910], he investigated the second of these problems, by discussing how different functions behave as x tends to infinity. The first of the specific problems, namely, integration in finite terms, was the object of an earlier small book of his [1905–1916]. The book is much more function-theoretical (analytical) than Liouville's work and to some extent it centers around simpler problems than those discussed by Liouville. Still, many of Liouville's results are mentioned, and references are given to Liouville's original proofs. For example, in connection with Liouville's theorem, Hardy writes:

> But we would strongly recommend him [the reader] to study the exceedingly beautiful and ingenious proof of this proposition given by Liouville. We have unfortunately no space to insert it here.

[Hardy 1916, p. 45]

Hardy's book was the first work after Liouville to deal in full generality with the question of integration in finite terms. With its praise of Liouville, it probably recreated an interest in this almost forgotten field. Watson's [1922] summary of Liouville's work on the Riccati equation and Loria's biography [1936] of Liouville had similar effects.

At the same time, the Russian school began to add substantially to Liouville's theory. D. D. Mordukhai-Boltovskoi (1876–1952), who in 1906 had written on the integration of differential equations in finite terms of differential equations, published a book in 1913, *Integration of Transcendental Functions* (in Russian), and in 1946 Ostrowski also contributed to the theory (see [Kasper 1980] for references).

62.   However, it was Joseph Fels Ritt (1893–1951) who in [1948] wrote what is now considered the classical account of Liouville's methods. Comparing his work with that of Liouville, Ritt wrote:

> I should like, however, to say something in regard to the treatment given here of Liouville's work. Liouville's methods are ingenious and beautiful. From the formal standpoint, they are entirely sound. There are, however, certain questions connected with the many-valued character of the elementary functions which could be pressed back behind the symbols in Liouville's time but which have since learned to assert their rights.

[Ritt 1948, Preface]

For this reason, Ritt followed Hardy and stressed the function-theoretical aspects of the theory. However, apart from this, Ritt's theorems and methods are so close to Liouville's that one can, in a certain way, consider Ritt's book as the comprehensive work Liouville never published.

The algebraic aspects of Liouville's theory were again emphasized by Ostrowski [1946] and Rosenlicht [1968] and [1972]. Ostrowski made clear that the important property of the set of functions considered in Liouville's theory is that it constitutes a field of functions holomorphic in a certain region of the complex plane and that the derivative of a function in the field is again in the field. He called such a field a *corps liouvillien*. Rosenlicht went further and totally abandoned the set of functions as being of importance. He defined abstractly:

> A differential ring is a commutative ring R together with a derivation of R into itself, that is, a map $R \to R$ which, if denoted $x \to x'$, satisfies the two rules
>
> $$(x + y)' = x' + y'$$
> $$(xy)' = x'y + xy'.$$
>
> A differential field is a differential ring that is a field.

[Rosenlicht 1968, p. 153]

The set of elements with zero derivative are called constants. They constitute a subfield of the differential field. If a and b are elements of a differential field for which $b' = a'/a$, Rosenlicht calls a an exponential of b and b a logarithm of a. A differential field, G, is called an elementary extension of a differential field, F, if it can be obtained by successive adjunctions of logarithms, exponentials, or algebraic elements. In this abstract terminology, Rosenlicht generalized Liouville's theorem to the following:

> Theorem. Let F be a differential field of characteristic zero, $\alpha \in F$. If the equation $y' = \alpha$ has a solution in some elementary differential extension field of F having the same subfield of constants, then there exist constants $c_1, \ldots, c_n \in F$ and elements $u_1, \ldots, u_n, v \in F$ such that
>
> $$\alpha = \sum_{i=1}^{n} c_i \frac{u_i'}{u_i} + v'.$$

[Rosenlicht 1968, pp. 157–158]

Of course, it still requires some function theory to show that ordinary integration theory can be incorporated in the theory of differential fields, but the theory has become entirely algebraic. In a way, one can consider Rosenlicht's approach as being Liouville's algebraic approach pushed to its extremes, and in fact, several arguments are still close to Liouville's. The recent works of R. H. Risch, on the other hand, approach the problem of integration in finite terms with algebraic geometrical methods and seems rather far away from Liouville's way of thinking.

63. Thus, Liouville's theory has been generalized, it has been dressed in abstract terminology, and methods from other fields have been applied. What then are the new results? I have already mentioned (§37) that Ritt [1923] and Risch [1976] rediscovered the validity of Liouville's theorem for implicit finite functions, independently of Liouville's unpublished "proof."

The greatest triumph of the modern theory, however, is Risch's complete solution of Liouville's main problem: Given an explicit finite function, determine in a finite number of steps if its integral is also a finite function. He sketched the algorithm [1970] and promised to publish a complete programme that would allow a computer to determine the question. With his publications, Risch has thus rounded off Liouville's theory almost one and a half century after its creation.

Conclusion

64. With his penetrating investigations, Liouville deserves credit for being the founder of the theory of integration in finite terms, although Abel went further in this field than it is generally acknowledged. Liouville's early research on algebraic integrals were independent of Abel's, but when he

learned of Abel's work it made him introduce implicit algebraic functions and provided him with a proof of Abel's theorem. However, it was Liouville who gave the first proof of the most basic theorem in the whole theory, Liouville's theorem, and he showed how to apply it to prove the nonelementary nature of several important integrals. Liouville was also the first to give a thorough algebraic analysis of explicit finite functions, and he proved the impossibility of solving some important transcendental and second-order linear differential equations in explicit finite form. Liouville began his investigations using analytical techniques, but consciously changed them into the completely algebraic methods used in the published papers.

In spite of these admirable achievements, Liouville left many unsolved problems, of which the two most conspicuous were (1) design algorithms to decide (a) the order of an elementary function, (b) if an integral is elementary, (c) if the solution of an elementary transcendental equation is explicitly expressible; and (2) generalize the theories of second-order differential equations to cover all such equations and, more generally, all rational differential equations.

65.   Why did Liouville then discontinue this line of research in the early 1840s? I shall mention three possible reasons:

1°. He seems to have become convinced that the general algorithms were too difficult to find. On the other hand, his remarks about the problems a and c above suggest that he believed that his methods could relatively easily solve any particular new problem. Therefore, it became uninteresting to prolong the list of solutions to specific problems.

2°. In the study of elementary solutions to differential equations, Liouville had encountered so many technical problems that there seemed to be very little hope that he would ever find a complete treatment of first- or second-order equations, let alone find the general theory he had dreamed of.

3°. Liouville was probably discouraged from attempting to surmount these obstacles by the relatively little impact his theory had in his own time. Except for the *Académie* reports, I have only found two traces of interest in Liouville's papers on integration in finite terms. One is Dirichlet's remark quoted in §51, the other is a footnote in the edition of [1837e] in Liouville's Journal, where Liouville wrote:

> Some persons have urged me to reproduce this paper here which has already appeared in the *Compte•rendu de l'Académie des Sciences* (meeting of August 28. 1837).
>
> [Liouville 1837l, p. 20]

Whoever these persons were, they watched Liouville's work only passively, and nobody else seems to have been actively engaged in the field before Chebyshev began his work about one year after Liouville had stopped publishing on these matters.

Still, I do not believe Libri when he claimed in a critique [1839a, p.736] of Liouville's theory that "unfortunately the geometers have approved neither these methods nor these classifications." The general attitude was probably rather one of approval and indifference.

66.   Why did Liouville's theory not appeal to his contemporaries, but only to much later mathematicians? First of all, it did not really fit into the mathematical universe of the time. The *questions* and the *results* were closely related to the questions treated in the various *cours d'analyse*, and Liouville often expressed the hope that his theory would find its way into the graduate curriculum. However, the *methods* were clearly too difficult, and their algebraic nature fit poorly into such courses.

It was probably also the algebraic aspect that restrained research mathematicians from venturing into the field, for at the time, there were few mathematicians working on this type of algebra. Galois' work had not been published nor fully understood, and in Paris, the only person to continue some of his ideas was Libri, who, for political and personal reasons, chose to criticize Liouville instead of trying to contribute to his theory. It is probably not a coincidence that the first significant developments of Liouville's theory appeared after 1870 when Jordan made the first real additions to Galois' theory [Jordan 1870].

However, when the interest in the theory of integration in finite terms was being rekindled at the beginning of this century, the importance and ingenuity of Liouville's research were generally acknowledged, and because of the long period in which the theory virtually slept, Liouville became the immediate precursor of mathematicians one century later. The theory of integration in finite terms is therefore the theory where Liouville's research is closest to the mathematical research of today.

X

Sturm-Liouville Theory

Introduction

1. In a series of articles dating from 1836–1837, Sturm and Liouville created a whole new subject in mathematical analysis. The theory, later known as Sturm-Liouville theory, deals with the general linear second-order differential equation

$$(k(x)V'(x))' + (g(x)r - l(x))V(x) = 0 \quad \text{for } x \in (\alpha, \beta), \tag{1}$$

with the imposed boundary conditions

$$k(x)V'(x) - hV(x) = 0 \quad \text{for } x = \alpha, \tag{2}$$
$$k(x)V'(x) + HV(x) = 0 \quad \text{for } x = \beta. \tag{3}$$

Here, k, g, and l are given functions, h and H are given positive constants, and r is a parameter. The boundary-value problem only allows non-trivial solutions (eigenfunctions) for certain values (eigenvalues) of r, which can be considered as roots of a certain transcendental equation,

$$\Pi(r) = 0, \tag{4}$$

namely, the equation obtained by inserting the general solution of (1) and (2) into (3). The questions studied by Sturm and Liouville can roughly be divided into three groups:

1°. properties of the eigenvalues,

2°. qualitative behavior of the eigenfunctions, and

3°. expansion of arbitrary functions in an infinite series of eigenfunctions.

Of these, Sturm investigated 1° and 2°, and Liouville examined 3°, finding further results related to 1° and 2° in the process.

Before 1820, the only question taken up in the theory of differential equations had been: given a differential equation, find its solution as an analytic expression. For the general equation, (1), Sturm could not find such an expression, and the expression found by Liouville by successive approximation was unsuited for the investigation of the properties 1° − 3° above. Instead, they obtained the information about the properties of the solutions from the equation itself. This shows evidence of a new conception of the theory of differential equations characterized by a broader kind of question: given a differential equation, investigate some property of the solution.

Most conspicuous among the properties to be investigated in the early 19th century was existence. The existence theorem formulated and proved by Cauchy [1824–1981, 1835–1840] was the first to indicate the broader concept of differential equations.

The conceptual development in the field of differential equations ran parallel to the development in the field of algebraic equations. Here, the works of Abel and Galois shifted interest from the problem of finding solutions by radicals to a question of existence of such solutions and an investigation of their properties.

Since no workable explicit solutions to the general Sturm-Liouville problem could be found, the properties determined from the equation itself were necessarily qualitative in nature. Seen in this light, Sturm-Liouville theory was the first qualitative theory of differential equations, anticipating Poincaré's approach to non-linear differential equations developed at the end of the century. In addition, the Sturm-Liouville theory gave the first theorems on eigenvalue problems, and as such, it occupies a central place in the prehistory of functional analysis. But, the Sturm-Liouville theory was important not only as a herald of coming ideas; it was, and has remained until this day, of importance in the technical treatment of many concrete problems in pure and applied mathematics, and was, as such, of more than "just" conceptual importance.

2.   Of the two conceptual novelties in the theory of differential equations in early 19th century France, existence theorems have generally received more attention in the secondary literature than the Sturm-Liouville theory despite the fact that the latter presented a wider range of innovations. It even included the former in the sense that the first widely circulated existence proof was published in three of Liouville's papers on the Sturm-Liouville theory.

In two of the better surveys of the history of mathematics ([Kline 1972, pp. 715–717] and [Dieudonné 1978, pp. 140–142]), Sturm and Liouville's theory has received brief treatment. More information, particularly on the role of the Sturm-Liouville theory in the history of functional analysis, can be found in [Dieudonné 1981, pp. 16–21]. Richest on details are the two articles in the *Encyklopädie der mathematischen Wissenschaften* [Hilb & Szász 1922] and [Bôcher 1899–1916] and several of Bôcher's other works (e.g., [1911–1912, 1912, 1917]). However, it is rather difficult to extract a connected history from these older works since their primary goal was exposition of the mathematics, not of its history. [Bôcher 1911–1912] is an exception.

This chapter is an attempt to supply a comprehensive and coherent treatment of the emergence of this beautiful theory, taking all published as well as unpublished sources into account. All unpublished material from Sturm's hand seems to be lost, but some of Liouville's early memoirs presented to the *Académie des Sciences* and his notes have been preserved.

3. I have, as far as possible, unified, simplified, and clarified the notation. For example, letters have freely been replaced by others, the Lagrangean notation V' has been used instead of Sturm and Liouville's Leibnizian notation $\frac{dV}{dx}$, parenthesis have been inserted, and the variables have been introduced in the functions if clarity is gained (i.e., $V(x)$ instead of V). These changes do not alter the meaning at all.

I have freely used modern terms such as eigenfunctions, eigenvalues, spectrum, spectral theory, orthogonality of two functions, etc. Use of such modern terms abbreviates the discussion, but also presents the danger of over-interpretation. For example, the anachronistic term orthogonal suggests a geometric interpretation of functions as points in a Hilbert space. However, such a way of thinking was not introduced before the work of E. Schmidt [1907 and 1908].

The word "Fourier series" has throughout been used in its modern sense to describe the development in terms of any set of eigenfunctions of a Sturm-Liouville problem, whereas the Fourier series in the sense of the 19th century will be called trigonometric or ordinary Fourier series.

4. As I have explained in Chapter II, and Chapter V, §§4,5, Sturm and Liouville became devoted friends during the late 1920s and remained friends until Sturm's early death. They only wrote one joint paper on the theory named after them, but several remarks in their works bear witness to their collaboration. They always praised each other's achievements and even covered up each other's mistakes (see note 18). They discussed each other's papers before their publication, with the result that in some cases an elaboration of a certain discovery was published before the discovery itself.

Their works on linear differential equations fall into four periods. During the first period, 1829–1830, they formed and presented their initial ideas independently. In the middle of the period, 1831–1835, Sturm wrote his two large memoirs, which were eventually published simultaneously with Liouville's first famous memoir during the third period, 1836–1837. Liouville had begun his generalization of the theory to higher order equations in 1835, but his main work in this area fell into the last period, from 1838 to approximately 1840.

In the text below, I break this chronology by analyzing Sturm's work before Liouville's. This is justifiable since Liouville's definitive work drew heavily on Sturm's results, whereas Sturm only accidentally commented on Liouville's. To throw the work of the two friends into relief, the prehistory of Sturm-Liouville theory is recorded in §§5–15. Sturm's two impressive memoirs and their emergence are treated in the two following sections. In the last sections, I discuss Liouville's work on second-order and higher order equations. The first sections are based mainly on published sources, whereas the Liouville Nachlass has supplied valuable information on the subject treated in the last section.

The Roots of Sturm-Liouville Theory

5.   The following motivating considerations were presented in the opening phrases of Sturm's first large paper on Sturm-Liouville theory.

> The solution of most problems concerning the distribution of heat in bodies of various shapes and concerning small oscillatory motions of elastic solids, of flexible bodies, of elastic liquids and fluids, leads to linear differential equations of the second order...
>
>   [Sturm 1836a, p. 106]

In his second paper [1836b], he explained in more detail how the partial differential equations arising from the problems above can be solved by separating the variables, leading in general to a second-order ordinary differential equation with a parameter. The parameter must be chosen so that certain boundary conditions are satisfied.

As an example, he discussed heat conduction in an inhomogeneous thin bar. In this case, the temperature is governed by the equation

$$g\frac{\partial u}{\partial t} = \frac{\partial \left(k\frac{\partial u}{\partial x} \right)}{\partial x} - lu, \tag{5}$$

where $u(x, t)$ denotes the temperature at the point x and at the time t, and g, k, and l are positive functions of x. If the surroundings of the bar are maintained at zero degree, the temperature u must satisfy boundary conditions at the end points α and β:

$$k\frac{\partial u}{\partial x} - hu \; = \; 0 \quad \text{for } x = \alpha, \tag{6}$$

$$k\frac{\partial u}{\partial x} + Hu \; = \; 0 \quad \text{for } x = \beta, \tag{7}$$

where h and H are positive constants which may become infinite (implying $u = 0$). Sometimes the temperature is known when $t = 0$. That gives rise to the initial condition

$$u(x, 0) = f(x). \tag{8}$$

Ignoring (8), Sturm first looked for solutions to (5)-(7) of the form[1]

$$u = V(x)e^{-rt}. \tag{9}$$

When substituted into (5)-(7) the factors e^{-rt} cancel, leaving the boundary-value problem (1)-(3) for V. If $V_1, V_2, \ldots, V_n, \ldots$ are the eigenfunctions to

(1)-(3) corresponding to the eigenvalues $r_1, r_2, \ldots, r_n, \ldots$, the linear combination

$$u = \sum_n A_n V_n(x) e^{-r_n t}$$

is also a solution of (5)-(7). The initial condition, (8), thus poses the problem of determining the A_n's so that

$$\sum_n A_n V_n(x) = f(x). \tag{10}$$

This problem was taken up by Liouville.

THE PHYSICAL ORIGIN

6.   Eigenvalue problems of the form (1)-(3) had turned up in the early 18th century in the study of vibratory motions. In the papers of Brook Taylor on the vibrating string [1713] and of Johann Bernoulli [1728] on the hanging chain, the first eigenvalue was found, corresponding to the fundamental mode. The higher modes were discovered by Daniel Bernoulli (1700–1782) in his continuation of his father's research on the vibrating hanging homogeneous chain [1732–1733]. He derived the equation

$$\alpha \frac{d}{dx} \left(x \frac{dy}{dx} \right) + y = 0$$

for the shape $y(x)$ of the chain and found its solution as an infinite series that we would denote by

$$y = A J_0(2\sqrt{x/\alpha}),$$

J_0 being the zeroth-order Bessel function.[2] Daniel Bernoulli argued that there is an infinity of eigenvalues α satisfying $J_0(2\sqrt{l/\alpha}) = 0$, where l is the length of the chain, and investigated the distribution to the $n-1$ zeroes of the nth eigenfunction in the interval $(0, l)$. Later, he also discovered the possibility of superposing the eigenfunctions, and in connection with the vibrating string, he claimed that the general shape of the system could be obtained in this way.

   Taylor and the Bernoullis had derived the ordinary differential equation directly from physical principles. When d'Alembert and Euler from 1747 onward derived the partial differential equations describing vibrating strings, chains, and membranes, they obtained the eigenvalue problem by separating variables. Though they did not believe that they could get the complete solution by superposition of eigenfunctions, they investigated many specific cases of (5)-(7) using this technique (cf. [Truesdell 1960]). From 1807, separation of variables was widely used in the theory of heat, first by Fourier, and soon thereafter by almost all the younger French mathematicians. This vast complex of research presented ample motivation for Sturm and Liouville.

Before 1830, mathematicians almost exclusively studied such particular cases of (1)-(3) for which they could find an explicit solution either in finite form or in infinite series. Sturm and Liouville, however, could not find any manageable expression for the solution in the general case, and therefore they had to draw their conclusions directly from equations (1)-(3). This is the characteristic feature of the Sturm-Liouville theory. Because such a study of the equations had earlier been rendered unnecessary by the explicit knowledge of the solution, one can hardly find any anticipation of Sturm's and Liouville's methods and results. Nevertheless, some exceptional investigations pointing toward the Sturm-Liouville theory were made by d'Alembert, Fourier, and Poisson. I shall discuss their researches in chronological order below.

D'ALEMBERT'S CONTRIBUTION

7.   Jean le Rond d'Alembert (1717–1783), who had solved the problem of the vibrations of a homogeneous string in his famous paper [1747], turned to the nonhomogeneous string shortly after Euler had published his first investigations on this more difficult problem. In a letter of June 11, 1769 (later published as [1766]) to Lagrange, d'Alembert set up the differential equation governing the transversal amplitude $y(x,t)$:

$$\frac{\partial^2 y}{\partial x^2} = X \frac{\partial^2 y}{\partial t^2},\qquad(11)$$

where $X(x)$ is the distribution of mass along the string. Here and below, the physical constants contained in d'Alembert's equations have all been set equal to 1. D'Alembert sought solutions of the form

$$y = \zeta(x) \cos \lambda t,$$

for which the equation reduces to the form

$$\frac{d^2\zeta}{dx^2} = -X\lambda^2\zeta.\qquad(12)$$

After this separation of the variables, he let $\zeta = e^{\int p\,dx}$ and obtained for p the Riccati equation

$$dx = -\frac{dp}{p^2 + X\lambda^2}.\qquad(13)$$

Since he required the string to be fixed at the two endpoints,

$$y = 0 \quad \text{for} \quad x = 0 \quad \text{and} \quad y = 0 \quad \text{for} \quad x = a,\qquad(14)$$

he was faced with the question:

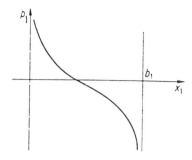

FIGURE 1

whether it is always possible to satisfy this double condition, the value of X being given; this is a question which nobody, as far as I know, has yet examined in general.

[d'Alembert 1766, p. 242]

In order to show it possible to determine such a value of λ, he considered the vibrations of a string of the uniform load $m = \min_{x \in [0,a]} X(x)$. It satisfies the equation

$$dx_1 = -\frac{dp_1}{p_1^2 + m\lambda^2} \tag{15}$$

corresponding to (13). He argued that if $\zeta_1 = e^{\int p_1 \, dx_1}$ is 0 at $x_1 = 0$, then p_1 must have vertical asymptotes at 0 and another point b_1, for which $\zeta_1(b_1) = 0$ (Fig. 1). A comparison of (13) and (15) shows that if $\zeta(0) = 0$, we must have $x < x_1$ at points where $p = p_1$. Therefore, p must also have a vertical asymptote at a point $b \leq b_1$ corresponding to $\zeta(b) = 0$ (Fig. 2). Finally, d'Alembert claimed it possible to choose λ in such a way that $b = a$; the two boundary conditions will then be fulfilled. The argument he had in mind, but did not write down, must have been the following:

One can prove that $b_1 \to 0(\infty)$ for $\lambda \to \infty(0)$, from which it follows that $b \to 0(\infty)$ for $\lambda \to \infty(0)$, and therefore λ can be chosen so as to make $b = a$. But, in this argument only the implication ($b_1 \to 0$ for $\lambda \to \infty$) $\Rightarrow$ ($b \to 0$ for $\lambda \to \infty$) follows from d'Alembert's inequality $b < b_1$. In order to get the other inequality, he could have compared equation (13) with the equation for a string of uniform load $M = \max X(x)$.

8.   D'Alembert established only the existence of one eigenvalue λ of the boundary-value problem, (11) and (14).[3] In spite of these shortcomings, d'Alembert's investigation was a remarkable anticipation of the Sturm-Liouville theory. Not only has the problem of existence of eigenvalues a central position in this theory, but also the method of basing the existence proof on a comparison with differential equations with constant coefficients

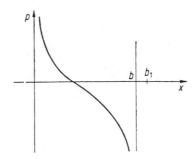

points directly to Sturm's comparison theorem (cf. §24). However, Sturm did not refer to d'Alembert's paper.

FOURIER'S CONTRIBUTION

9.   The work of Joseph Fourier (1768–1830), on the other hand, was well known to the young Sturm, who was his protegé. In his main work, *Théorie analytique de la chaleur* [1822], Fourier treated only heat conduction in homogeneous media, but was nevertheless led to differential equations with variable coefficients when he used spherical and cylindrical coordinates. In both cases, he succeeded in finding explicit formulas for the solutions of the separated equations. When using spherical coordinates, he found the solutions to be ordinary trigonometric functions so that the problem of finding the eigenvalues became a simple trigonometric problem. Heat conduction in an infinitely long homogeneous cylinder, on the other hand, posed problems similar to those Daniel Bernoulli had faced in his investigation of the hanging chain.

10.   After having set up the heat equation in cylindrical coordinates [1822, §§118–120], Fourier assumed as he usually did that the temperature is of the form $e^{-mt}u(x)$ and found [1822, §306] for $u(x)$ the equation

$$\frac{d^2u}{dx^2} + \frac{1}{x}\frac{du}{dx} + \frac{m}{k}u = 0, \qquad (16)$$

where k is a positive constant and x is the distance from the axis. Fourier imagined the cylinder to be immersed in a medium of constant temperature. Then, u must satisfy the boundary condition

$$hu + \frac{du}{dx} = 0 \quad \text{for} \quad x = \beta, \qquad (17)$$

where h is a constant and β is the radius of the cylinder.

Fourier found the solution to (16) expressed as an infinite series[4]:

$$u = 1 - \frac{mx^2}{k2^2} + \frac{m^2x^4}{k^22^2 \cdot 4^2} - \frac{m^3x^6}{k^32^2 \cdot 4^2 \cdot 6^2} + \cdots. \tag{18}$$

This is the Bessel function $J_0\left(x\sqrt{\frac{m}{k}}\right)$. Now, m must be chosen such that (17) is satisfied. Since Fourier wanted to represent any initial temperature $f(x)$ as a sum of solutions,

$$f(x) = a_1u_1(x) + a_2u_2(x) + \cdots + a_nu_n(x) + \cdots, \tag{19}$$

he needed to show that there is an infinite number of such values of m (eigenvalues). Setting $\theta = \frac{m\beta^2}{k2^2}$ and

$$f(\theta) = 1 - \theta + \frac{\theta^2}{2^2} - \frac{\theta^3}{2^23^2} + \cdots$$

makes it possible to write the boundary condition as follows:

$$\frac{h\beta}{2} + \theta\frac{f'(\theta)}{f(\theta)} = 0. \tag{20}$$

Fourier noted that $f(\theta)$ is a solution of the differential equation

$$y + \frac{dy}{d\theta} + \theta\frac{d^2y}{d\theta^2} = 0, \tag{21}$$

from which he deduced the existence of infinitely many roots of (20) in the following way. Successive differentiation of (21) yields

$$\frac{d^iy}{d\theta^i} + (i+1)\frac{d^{i+1}y}{d\theta^{i+1}} + \theta\frac{d^{i+2}y}{d\theta^{i+2}} = 0, \tag{22}$$

which shows that when $f^{(i+1)}$ has a root, $f^{(i)}$ and $f^{(i+2)}$ have opposite signs. Fourier claimed that if such a relation holds between the real roots of a function and its successive derivatives then the function has no imaginary roots. For polynomials, this theorem is valid, and it is closely related to Fourier's earlier investigation [1820] of the number of real roots of algebraic equations between given limits, which had resulted in the following theorem:

FOURIER'S THEOREM: Let $f(x) = 0$ be a real algebraic equation of the kth degree, let p denote the number of variations of signs in the sequence

$$f(\alpha), f'(\alpha), \ldots, f^{(k)}(\alpha),$$

and let q denote the number of variations of signs in the sequence

$$f(\beta), f'(\beta), \ldots, f^{(k)}(\beta).$$

Then, the number of real roots of $f(x)$ in (α, β) $(\alpha < \beta)$ is at most equal to $p - q$.

Fourier seems to have considered the function $f(\theta)$ above as a polynomial of infinite degree and concluded that since it has no imaginary roots it must have infinitely many real roots. A simple argument left out by Fourier then shows that (20) or (17) has infinitely many real roots. This result is correct, but as Poisson pointed out [1823a, p. 383, and 1830], Fourier's proof is inadequate because the theorem on the nonexistence of complex roots is not always true for transcendental functions. This criticism was met by Fourier [1831].

11. Having obtained an infinity of eigenfunctions, $u_1, u_2, \ldots$, to (16) and (17), Fourier desired to prove [1822, §§310–319] that any initial state $f(x)$ can be developed in a Fourier series, (19). As in all other cases of this kind, Fourier considered this to be proved if he could find a formula for the Fourier coefficients, a_i. After long calculations involving only equations (16) and (21), but not formula (18), for the solution, he found that

$$
\int_0^\beta x u_j(x) u_i(x) \, dx = \begin{cases} 0 & \text{for } i \neq j \\ \left[1 + \left(\frac{h\beta}{2\sqrt{\theta_i}} \right)^2 \right] \frac{\beta^2 u_i^2(\beta)}{2} & \text{for } i = j \end{cases} . \tag{23}
$$

From this statement of orthogonality and (19), it follows that

$$
a_i = \frac{2 \int_0^\beta x f(x) u_i(x)}{\beta^2 u_i(\beta) \left[1 + \frac{kh^2}{m_i} \right]} . \tag{24}
$$

Fourier's treatment of heat conduction in a cylinder anticipates the Sturm-Liouville theory in the sense that the conclusions are drawn directly from the differential equations, (21) and (16). However, since the deduction of (21) rested on the knowledge of the explicit expression (18) for the solution u, this equation is foreign to the Sturm-Liouville theory.

POISSON'S CONTRIBUTION

12. The only mathematician who proved general theorems in Sturm-Liouville theory before Sturm and Liouville was Siméon-Denis Poisson (1781–1840). He obtained his results in connection with his above-mentioned debate with Fourier over the reality of the roots of the transcendental equations determining the eigenvalues of various problems in heat conduction. Even in the paper [1823a] in which he first criticized Fourier's proof, he presented what he called an *a posteriori* proof of the reality of the eigenvalues for the boundary value problem describing heat conduction in a sphere consisting of two concentric homogeneous materials. He simply noted [1823a, p. 381] that by using two different methods [1823a1, §VII] and [1823a2,

§V] he could express the temperature in two ways as sums, $\sum_m u_m$, which were identical except that one sum ranged over all eigenvalues, whereas in the other only real eigenvalues were taken into account. Since the two expressions had to be equal, Poisson concluded that no complex eigenvalues existed.

13.   In [1823a, p. 382] he ascertained that no *a priori* method of proving the reality of the eigenvalues was known, but three years later, he presented such a method in a note read in the *Société Philomatique* [1826]. The proof was based on the orthogonality of the eigenfunctions. He had proved this relation in the case of the double layer sphere in [1823a p. 380], but he had not noticed that it implied the reality of the eigenfunctions as a simple consequence. In [1826] Poisson treated the particular case of (4):

$$\frac{\partial u}{\partial t} = \frac{\partial^2 u}{\partial x^2} + X(x)u \tag{25}$$

with the boundary conditions

$$\frac{du}{dx} \quad - \quad hu = 0 \quad \text{for} \quad x = \alpha, \tag{26}$$

$$\frac{du}{dx} \quad + \quad Hu = 0 \quad \text{for} \quad x = \beta. \tag{27}$$

As usual he noted that the function $y_\rho(x)e^{\rho t}$ solves this boundary-value problem if y_ρ is a solution to the ordinary differential equation,

$$\varrho y_\rho = \frac{d^2 y_\rho}{dx^2} + X y_\rho, \tag{28}$$

and satisfies boundary conditions similar to (26) and (27). In order to arrive at the orthogonality relations, he considered a particular eigenfunction $y_{\rho'}$ and the solution of the original problem, (25)-(27):

$$u = \sum y_\rho e^{\rho t}, \tag{29}$$

where the summation ranges over all the eigenvalues. Multiplying (25) with $y_{\rho'}$ and integrating over $[\alpha, \beta]$, he obtained

$$\frac{d \int_\alpha^\beta u y_{\rho'}\, dx}{dt} = \int_\alpha^\beta \frac{d^2 u}{dx^2}\, y_{\rho'}\, dx + \int_\alpha^\beta X u y_{\rho'}\, dx. \tag{30}$$

By partial integration, the first term on the right-hand side is transformed into

$$\int_\alpha^\beta \frac{d^2 u}{dx^2}\, y_{\rho'}\, dx = \int_\alpha^\beta u\, \frac{d^2 y_{\rho'}}{dx^2}\, dx, \tag{31}$$

because the boundary terms cancel when both u and $y_{\rho'}$ satisfy (26) and (27). Similarly, multiplication of (28) by u and integration over $[\alpha, \beta]$ yields

$$\varrho' \int_\alpha^\beta u y_{\rho'}\, dx = \int_\alpha^\beta u\, \frac{d^2 y_{\rho'}}{dx^2}\, dx + \int_\alpha^\beta X u y_{\rho'}\, dx. \tag{32}$$

From (30)-(32), Poisson deduced the equation

$$\frac{d \int_\alpha^\beta u y_{\rho'} \, dx}{dt} = \varrho' \int_\alpha^\beta u y_{\rho'} \, dx,$$

which can readily be integrated to give

$$\int_\beta^\alpha u y_{\rho'} \, dx = A e^{\rho' t}, \tag{33}$$

where A is an arbitrary constant. By reintroducing the particular form (29) of u into (33) and equating the coefficients of similar exponentials, Poisson demonstrated orthogonality:

$$\int_\alpha^\beta y_\rho y_{\rho'} \, dx = 0 \quad \text{for} \quad \varrho \neq \varrho'. \tag{34}$$

14.   Poisson then assumed ϱ to be a complex eigenvalue with the eigenfunction y_ρ. Since he implicitly assumed $X(x)$, h, and H to be real, he concluded that $\overline{\varrho}$ must also be an eigenvalue with the eigenfunction $\overline{v_\rho} = v_{\overline{\varrho}}$, and since $\varrho \neq \overline{\varrho}$, equation (34) implies

$$\int_\alpha^\beta |y_\rho|^2 \, dx = \int_\alpha^\beta y_\rho \overline{y_\rho} \, dx = 0. \tag{35}$$

Hence, $y_\rho \equiv 0$ in $[\alpha, \beta]$. This is a contradiction, and therefore Poisson concluded that all eigenvalues must be real.

In his *Théorie mathématique de la chaleur* [1835a], Poisson carried over this proof to the more general boundary-value problem

$$c(\overline{x}) \frac{\partial u}{\partial t} = \sum_{i=1}^{3} \frac{\partial k(\overline{x}) \left(\frac{\partial u_i}{\partial x_i} \right)}{\partial x_i} \quad \text{in } A, \tag{36}$$

$$k(\overline{\text{grad } u} \cdot \overline{n}) + pu = 0 \quad \text{on } \partial A, \tag{37}$$

where A is a domain in $\mathbf{R}^3$ and $\overline{n}$ is the outer normal of its boundary ∂A[5]. This is the three-dimensional analogue of Sturm's problem, (5)-(7). In this case, Poisson found the following orthogonality:

$$\int_A c(\overline{x}) P_i(\overline{x}) P_j(\overline{x}) \, d\overline{x} = 0 \quad \text{for } i \neq j. \tag{38}$$

15.   In the two works mentioned above, Poisson provided both theorems and methods of lasting value for the Sturm-Liouville theory. The orthogonalities (34) and (38), and the theorem on the reality of the eigenvalues were adopted with due credit in Sturm's papers, and the method used to obtain (33), particularly the application of partial integration, is still used, though in a slightly simplified form found by Sturm. Nonetheless, Poisson's research is of a limited scope compared with the gigantic advances in the field made by Sturm and Liouville within two years of the publication of Poisson's last results.

Sturm's First Memoir

ORIGINS

16. Sturm's mathematical masterpieces grew out of the blend of theorems on differential equations and roots of equations found by Fourier and Poisson. His famous algebraic theorem, improving Fourier's theorem on the determination of real roots, states

STURM'S THEOREM. Let $f(x) = 0$ be a real algebraic equation of arbitrary degree. Define $f_1(x) = f'(x)$ and $f_n(n \geq 2)$ successively as the negative of the remainder obtained by dividing f_{n-2} by f_{n-1}:

$$f_{n-2} = q_{n-1}(x)f_{n-1}(x) - f_n(x).$$

Let p be the number of variations of sign in the sequence

$$f(\alpha), f_1(\alpha), f_2(\alpha), \ldots, f_k(\alpha)$$

and let q be the number of variations of sign in the sequence

$$f(\beta), f_1(\beta), f_2(\beta), \ldots, f_k(\beta).$$

Then, the number of real roots of $f(x)$ in (α, β) $(\alpha < \beta)$ is precisely equal to $p - q$.

Sturm presented this theorem to the *Académie* on May 25, 1829, and during the following half year he presented a series of papers on transcendental equations and differential equations [1829b-f]. The papers are all lost, but, from the short summaries of some of them [c and d] in the *Bulletin de Férussac* and from Cauchy's report on [f], one sees that Sturm got most of his ideas on the Sturm-Liouville theory during this period. He proved that certain systems of differential equations [c] and algebraic equations [d] have real eigenvalues, he determined Fourier coefficients [c], he found a version of his oscillation theorem [f], and he applied the theorems to determine the temperature distribution in different bodies for large values of the time.

In the concluding remarks of the first of his large printed papers on the Sturm-Liouville theory, Sturm shed more light on his approach to both the algebraic and the analytical theorems:

> The theory explained in this memoir on linear differential equations of the form
>
> $$L\frac{d^2V}{dx^2} + M\frac{dV}{dx} + NV = 0 \qquad (39)$$
>
> corresponds to a completely analogous theory which I have previously made concerning linear second-order equations of finite differences of the form
>
> $$LU_{i+1} + MU_i + NU_{i-1} = 0 \qquad (40)$$

where i is a variable index replacing the continuous variable x; L, M, N are functions of this index i and of another undetermined quantity m which is subject to certain conditions. It is while studying the properties of a sequence of functions U_0, U_1, U_2, U_3, ... related by a system of equations similar to the above, that I found my theorem on the determination of the number of real roots of a numeric equation between two arbitrary limits, which is contained as a particular case in the theory which I just indicated here. It turns into the one which is the subject of this memoir by passing from finite differences to the infinitely small differences.

[Sturm 1836a, p. 186]

However, in the extant summaries of Sturm's early papers, there is no explicit mention of this theory of difference equations. To be sure, one of the papers [1829d] deals with systems of equations resembling (40), but the aim of the paper is so different from the one described in the quotations that it cannot be the work Sturm had in mind. Where, how, and why did Sturm then study the difference equation, (40)? A convincing answer to these questions has been given by Bôcher in his interesting paper [1911–1912]. Bôcher argues that the problem of Sturm's paper, *Sur la distribution de la chaleur dans un assemblage de vases* [1829e], of which only the title is known, lead Sturm to the difference equation (40), and he shows how an analysis of this equation can lead to Sturm's theorem as well as to a discrete version of the theorems found in Sturm's first published paper on the Sturm-Liouville theory. The reader is referred to Bôcher's paper for a further discussion of Sturm's unpublished papers on algebraic and differential equations.

CONTENTS

17.   The publication of a comprehensive version of Sturm's ideas was delayed until 1836 when Liouville urged Sturm [cf. Liouville 1855j] to publish in his newly founded journal a memoir presented to the *Académie* three years earlier[6] [Sturm 1836a]. A summary of Sturm's paper had appeared in the journal *L'Institut* [1833a],[7] but it contained only the main results, not the methods. Thanks to its conciseness, however, the summary displayed a clarity which the 80 pages of the memoir lacked. In the latter, every detail was proved, sometimes with several proofs, and this combined with the lack of emphasis on the important theorems created a rather unreadable paper. Sturm's reason for being so elaborate was probably the novelty of his methods.

The principle on which the theorems which I develop rests has never been used in analysis as fas as I know.

[Sturm 1836a, p. 107]

The paper dealt with the general second-order linear differential equation (39) in which the coefficient functions depend on a real parameter, r. For convenience, Sturm transcribed the equation into its self-adjoint form:

$$(K(r,x)V_r'(x))' + G(r,x)V_r(x) = 0, \quad x \in (\alpha, \beta), \tag{41}$$

which generalizes equation (1).[8] Here and in the rest of this chapter, $'$ as in f' always means differentiation with respect to x; δ will denote differentiation with respect to the parameter r.

In the first paper, Sturm did not discuss a boundary-value problem with two boundary conditions of the kind (2) and (3). He postponed the treatment of spectral theory to the second paper [Sturm 1836b] and imposed in the first paper only one boundary condition of this kind:

$$K(r,x)V_r'(x) - h(r)V_r(x) = 0 \quad \text{for} \quad x = \alpha, \tag{42*}$$

or equivalently,

$$\frac{K(r,x)V_r'(x)}{V_r(x)} = h(r) \quad \text{for} \quad x = \alpha, \tag{42}$$

in which he even allowed the "constant" h to vary with r. Sturm remarked that the following proposition secured the existence of a solution and its uniqueness, to within a multiplicative constant, to (41) and (42).[9]

Proposition A. *Suppose V_r is a solution to (41) and suppose $V_r(\alpha)$ and $V_r'(\alpha)$ are given. Then, V_r "has a determined and unique value for each value of x."*

By the 1830s, this basic theorem of existence and uniqueness had been generally believed for a century, but the first proof had only recently been provided by Liouville (see §32).

Thus, the problem (41), (42) leads to a continuous family of solutions V_r, one for each r. The aim of Sturm's first paper was to study the qualitative behavior of the solutions $V_r(x)$, and particularly how they varied with r. Sturm was primarily interested in the oscillatory properties of the $V_r(x)$'s, for example, their zeroes, their changes of sign, and their maxima and minima in the interval (α, β). He obtained this information "solely by considering the differential equations themselves without needing to solve them" [Sturm 1836a p. 107]. In this way, he made explicit the method that had been indicated by Fourier and used rather unconsciously by Poisson.

18.   Central in Sturm's paper is his investigation of the number of roots of V_r in (α, β), from which all the other results follow as easy corollaries. This investigation has two components:

1°. Proof that under variation of the parameter r a root $x(r)$ of V_r can appear or disappear from the interval (α, β) only if it crosses one of the boundaries ($x(r) = \alpha$ or $x(r) = \beta$).

2°. Determination of how the roots $x(r)$ of V_r move for varying r (also outside (α, β)), particularly how they enter the interval (α, β).

The first property is easily established, for if a root $x(r)$ appears or disappears without passing α or β, it gives rise to a double root (Fig. 3). However, this cannot happen according to Sturm's proposition:

Proposition B. *When V_r is a non-trivial solution to* (41), *then V_r and V_r' have no common roots.*

Sturm offered two proofs of Proposition B. First, he pointed out that if $V_r(c) = V_r'(c) = 0$ it follows from (41) that all the higher derivatives $V_r^{(n)}(c)$ will vanish in c, "and subsequently [by Taylor's theorem] V will be zero for all the values of x." This simple proof is followed by a second "more rigorous demonstration," which rests on the constancy of the Wronskian [Sturm 1836a, pp. 109–110]:

$$K(V_1 V_2' - V_2 V_1') = C,$$

where V_1 and V_2 are two arbitrary solutions to (41) and C is a constant. Indeed, if V_1 is a non-trivial solution, there is a point a such that $V_1(a) \neq 0$. Sturm then took a solution, V_2, with such values of $V_2(a)$ and $V_2'(a)$ that the Wronskian evaluated at the point a is different from zero. Then, the Wronskian is everywhere different from zero, and hence V_1 and V_1' cannot both be equal to zero.

Without comment, Sturm has here used the existence part of Proposition A when he chose V_2. On the other hand, Proposition B is a simple consequence of the uniqueness part of Proposition A. Sturm clearly did not see these relations between Propositions A and B, for he would probably consider the assertion of the existence of A as a stronger statement than the assertion of uniqueness.

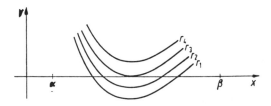

FIGURE 3

The two proofs reflect the radical changes taking place in the foundation of analysis during the early 19th century. The first proof reveals an author attached to the tradition of Lagrange, whereas the need to give an alternative proof would occur only to a mathematician influenced by the new standards of rigor. In particular, Cauchy's example [Cauchy 1829, 10. leçon] of a function whose Taylor series converges to a sum different from the function expanded shows that Sturm's first proof is invalid.

19.   Proposition B shows that it makes sense to follow a particular root x_r of V_r, when r varies, as is required in 2° above. In order to describe this variation, Sturm studied the behavior of the two fractions $\frac{V}{KV'}$ and $\frac{KV'}{V}$. For brevity, I shall discuss only the theorems concerning $\frac{KV'}{V}$, of which the first and most central stated [Sturm 1836a, p. 116]:

Proposition C. *If V is a solution to (41) and (42) and if*

$$
\begin{cases}
K > 0 & \forall r, \ \forall x \in [\alpha, \beta], \\
G \text{ is an increasing function of } r & \forall x \in [\alpha, \beta], \\
K \text{ is a decreasing function of } r & \forall x \in [\alpha, \beta], \\
\left[\frac{KV'}{V} \right]_{x=\alpha} = h(r) \text{ is a decreasing function of } r,
\end{cases}
$$
$$(43)$$

then $\frac{KV'}{V}$ is a decreasing function of r for all values of $x \in [\alpha, \beta]$.

Here "decreasing" is to be understood in the sense that when $V_r(x)$ for fixed x and variable r becomes zero the quotient $\frac{KV'}{V}$ jumps from $-\infty$ to ∞. Since both the proposition and its proof are fundamental in Sturm's theory, I shall consider the proof in detail.

Without bothering about differentiability,[10] Sturm differentiated equation (41) with respect to r and found:

$$
\frac{d\delta(KV')}{dx} + G\,\delta V + V\,\delta G = 0, \tag{44}
$$

where he has interchanged differentiation with respect to x (denoted d) and differentiation with respect to r (denoted δ). Multiplying (41) by $\delta V\,dx$ and subtracting (44) multiplied by $V\,dx$ yields

$$
\delta V\,d(KV') - V\,d\delta(KV') = V^2\delta G\,dx, \tag{45}
$$

which, integrated by parts between α and x, gives

$$
\delta V(KV') - V\delta(KV') = C + \int_\alpha^x V^2\delta G\,dx - \int_\alpha^x (V')^2\delta K\,dx, \tag{46}
$$

where C is the value of the left-hand side for $x = \alpha$. Since this left-hand side is equal to

$$
-V^2\,\delta\left(\frac{KV'}{V} \right),
$$

Sturm finally obtained

$$-V^2 \delta \left(\frac{KV'}{V}\right) = \left[-V^2 \delta \left(\frac{KV'}{V}\right)\right]_{x=\alpha} + \int_\alpha^x V^2 \delta G \, dx - \int_\alpha^x (V')^2 \delta K \, dx.$$
(47)

If the assumptions (43) of Proposition C are fulfilled, we have

$$\left[\delta \left(\frac{KV'}{V}\right)\right]_{x=\alpha} < 0$$

and

$$\delta G > 0 \quad \text{and} \quad \delta K < 0 \quad \text{for} \quad x \in [\alpha, \beta].$$
(48)

Hence, the left-hand side of (47) is positive, i.e.,

$$\delta \left(\frac{KV'}{V}\right) < 0 \quad \text{for} \quad x \in [\alpha, \beta],$$
(49)

which proves Proposition C.

20.   From Proposition C, Sturm easily obtained the desired description of the variation of a root $x(r)$ satisfying $V_r(x(r)) = 0$ when the conditions in (43) are fulfilled. When r increases to $r + dr$, x will change to $x + dx$ and V to the value $V_{r+dr}(x(r + dr)) = V_r(x(r)) + \frac{\partial V_r(x)}{\partial x} dx + \frac{\partial V_r(x)}{\partial r} dr$. Since $V_{r+dr}(x(r + dr)) = V_r(x(r)) = 0$ we have:

$$\frac{\partial V}{\partial x} dx + \frac{\partial V}{\partial r} dr = 0,$$

and

$$\frac{dr}{dx} = -\frac{\partial V}{\partial x} \bigg/ \frac{\partial V}{\partial r}.$$
(50)

Sturm then used (46) to see that $\frac{\partial V}{\partial r}$ (or δV) and $\frac{\partial V}{\partial x}$ (or V') have the same signs when $V = 0$. Therefore,

$$\frac{dr}{dx} < 0,$$

which implies the following proposition.

Proposition D. *If the assumptions* (43) *are satisfied, the roots* $x(r)$ *of the solution* $V_r(x)$ *to* (41) *and* (42) *are decreasing with* r.

Thus, new roots of V_r may enter the interval (α, β) through its right-hand end point β, but no roots can leave the interval through β. This means, according to property 1° (§18), that, if $V_r(\alpha) \neq 0$ for all r (i.e., $h(r) \neq \infty$ in (42) or (43)), the number of roots of V_r in (α, β) increases with r.

21.   Sturm could then easily deduce what he considered his main theorem:

Theorem E. *Let V_1 and V_2 denote solutions to the equations*

$$(K_i V_i')' + G_i V_i = 0, \quad \forall x \in (\alpha, \beta), \tag{51}$$

$$\frac{K_i V_i'}{V_i} = h_i \quad for \quad x = \alpha, \tag{52}$$

for $i = 1$ and $i = 2$, respectively. Suppose further that

$$G_2(x) \geq G_1(x), \quad K_2(x) \leq K_1(x), \quad \forall x \in [\alpha, \beta], \tag{53}$$

and

$$h_2 < h_1. \tag{54}$$

Then, V_2 vanishes and changes sign at least as many times as V_1 in (α, β); and if one lists the roots of V_1 and V_2 in increasing order from α, the roots of V_1 are larger than the roots of V_2 of the same order.

In order to deduce Theorem E from the remarks following Proposition D, Sturm "connected" the two situations ($i = 1$ and $i = 2$) by a continuous family of equations (41), (42) satisfying

$$G(x, r_i) = G_i(x), \quad K(x, r_i) = K_i(x), \quad i = 1, 2 \tag{55}$$

and

$$h(r_i) = h_i \tag{56}$$

in such a way as to satisfy the conditions (43). The possibility of selecting such a continuous family is secured by (53) and (54).

The introduction of the continuous family of equations also allowed Sturm to evaluate the difference Δ between the number of roots of V_1 and V_2 in (α, β). He showed that

Proposition F. $\Delta =$ *the number of roots of $V_r(\beta)$ for $r \in (r_1, r_2)$;*

that is, the number of zeroes which the solution V of the continuous family (41) and (42) assumes at the right-hand end point of the interval, when r varies from r_1 to r_2. This is the theorem which most clearly shows the connection between Sturm's theorem in algebra (cf. §16) and his analytical investigations.

22. The rest of the memoir consists of consequences, refinements, and variations of Theorem E and Proposition F. Of these, I shall, in the following, only discuss the most important, particularly those that where used in the subsequent development of the Sturm-Liouville theory. The first interesting consequence of Theorem E can be found in Section 16 of [Sturm 1836a]. It is

Sturm's comparison theorem. *If V_1 and V_2 satisfy (51) for $i = 1$ and $i = 2$, respectively, and G_i, K_i $(i = 1, 2)$ satisfy (53), the interval between two consecutive roots of V_1 will contain at least one root of V_2.*

Sections 19 to 35 investigate how

$$\left.\frac{KV' + HV}{V}\right|_{x=\beta} \tag{57}$$

behaves as a function of r (H is a constant) and how

$$KV' + p(x)V \tag{58}$$

behaves as a function of x and r when $V_r(x)$ is a solution to (41) and (42). As a result of this analysis, Sturm proved

Proposition G. *$[KV' + HV]_{x=\beta}$ considered as a function of r vanishes Δ or $\Delta + 1$ times for $r \in [r_1, r_2]$,*

where Δ has the meaning explained above Proposition F. A related consequence [Sturm 1836a, p. 141] of the analysis is

Sturm's oscillation theorem. *Let ϱ and ϱ' be two consecutive values of r that satisfy $KV'_r + HV_r = 0$ for $x = \beta$. Then, $V_{\varrho'}$ has one more root in (α, β) than V_ϱ.*

23.   In the last seven sections, Sturm gave methods of approximating the solutions V to (41) and their roots. During this part of the investigation, he transformed the general equation, (41), into the simpler form

$$V_r''(x) + \Gamma(x, r)V_r(x) = 0 \tag{59}$$

by altering either the independent variable x or the dependent variable V. He obtained the approximations by comparing equation (59) with the corresponding two differential equations, which emerge when Γ is replaced by its maximum and minimum in $[\alpha, \beta]$. The solution of these two differential equations with constant coefficients are trigonometric functions, and from their familiar properties Sturm could get approximations to V using the main theorem, E. This important method had, as we saw in §7, been perceived by d'Alembert.

By way of this method, Sturm also proved in passing [Sturm 1836a, §40]

Proposition H. *If $\Gamma(x, r) \to \infty$ for $r \to \infty$, $x \in [\alpha, \beta]$, if (43) is satisfied, and if $V_r(\alpha)$ has the same sign for all r, then*

$$[KV' + HV]_{x=\beta}$$

vanishes infinitely many times when $r \to \infty$.

An immediate consequence of Proposition H is that the boundary value problem (1)-(3) has an infinity of eigenvalues, but Sturm did not make this observation explicit here.

24.   I have treated Sturm's paper in such detail to give an impression of the wealth of new ideas and new results it presents. Of the Propositions A-H, only the first two had been known earlier, and even the idea of proving theorems like C-H was introduced by Sturm. The results were to constitute the basis of all the work of Sturm and Liouville in the theory named after them. The individual elements in Sturm's proofs, such as differentiation with respect to a parameter and partial integration, were all well known but Sturm combined them in an original way to obtain qualitative statements about the solutions.

As in all early 19th-century analysis, Sturm's methods do not meet the modern standards of rigor, but only minor alterations are needed to make them acceptable to a modern reader. Nevertheless, Sturm's exposition differs considerably from the presentation in modern textbooks. Today, the solutions V_r are thought of as vectors in a Hilbert space, and therefore their oscillation becomes uninteresting. If it is discussed at all, both boundary conditions (e.g. (2) and (3)) are usually introduced from the beginning, and so only the behavior of the eigenfunctions is studied, whereas the continuous family of equations and solutions V_r do not occur. In addition, the modern presentation generally concentrates on specific types of coefficients (e.g., as in (1)). Such limitations were applied by Sturm in his second memoir on Sturm-Liouville theory, devoted to spectral theory.

Sturm's Second Memoir

25.   Sturm's memoir on spectral theory was published in 1836 in Liouville's *Journal* [Sturm 1836b]. In its final form, it must have been composed during that year for it includes some comments on results obtained by Liouville toward the end of 1835 [Liouville 1836c and d]. However, it is clear both from Liouville's reference to the memoir in the papers mentioned above and from a one-page summary in *L'Institut* of 1833 [Sturm 1833b] that Sturm had written a preliminary version in 1833 in connection with the composition of the first paper [1836a]. As discussed in §16, the scanty evidence of Sturm's early works shows that he must have had some of the ideas even in 1829, but probably not in the polished form he gave them in the memoir of 1836.

26.   I have already summarized (§5) how Sturm, in the second memoir, deduced equation (1) with the boundary conditions (2)-(3) from a problem of heat conduction and how the eigenvalues can be considered as roots of a transcendental equation, $\Pi(r) = 0$ (4). This is the situation about which Sturm proved a number of results, the most important being propositions

J-N.

Proposition J. *There are infinitely many roots (eigenvalues) of the equation $\Pi(r) = 0$; they are all real and positive, and there are no multiple roots.*

Let $r_1 < r_2 < \cdots < r_n < \cdots$ be the eigenvalues[11] with the corresponding eigenfunctions $V_1, V_2, \ldots, V_n, \ldots$. In this notation, Sturm formulated the following results

Proposition K (orthogonality).

$$\int_\alpha^\beta g(x) V_m(x) V_n(x) \, dx = 0 \quad for \quad m \neq n. \tag{60}$$

Proposition L. *V_n never becomes infinite in $[\alpha, \beta]$ and has in (α, β) $n-1$ roots, at each of which it always changes sign.*

Proposition M. *Between two consecutive roots of V_m, there is precisely one root of V_{m-1}.*

Proposition N. *Let $m < n \in \mathbf{N}$ and $A_m, A_{m+1}, \ldots, A_n$ be constants that are not all zero. Define*

$$\psi(x) = A_m V_m(x) + A_{m+1} V_{m+1}(x) + \cdots + A_n V_n(x). \tag{61}$$

Then, $\psi(x)$ has at least $m - 1$ roots and at most $n - 1$ roots counted with multiplicity.

If h or H are infinite, the boundary conditions, (2) and (3), must be read $V(\alpha) = 0$ and $V(\beta) = 0$. Then, these roots must be given particular treatment in L, M, and N. Sturm painstakingly took care of these particular cases[12]. In the following brief discussion of the proofs, such details will be ignored, and so the principal ideas will become clearer.

27.   ad J and K: Sturm offered three proofs that the eigenvalues are real. The third was inspired by a method used in perturbation theory by Laplace in his *Mécanique Celeste* [1799–1825, book II, chapter 6] and the second was a revision of Poisson's proof, to which Sturm referred. Instead of Poisson's use of one solution to the ordinary differential equation (1) and one solution to the partial differential equation (5), Sturm started out at once with two different solutions, V_n, V_m, of the ordinary differential equation. This idea made the proof easier and independent of the partial differential equation. Following Poisson, he demonstrated orthogonality as an intermediate step.

ad J: Sturm's proof that there are infinitely many eigenvalues has now become standard. It rests upon comparing equation (1) with the equation

$$U_1'' + n(r)^2 U_1 = 0, \tag{62}$$

where $n(r)$ is a constant independent of x, so selected that

$$g(x)r - l(x) > n(r)^2 \sup_{x \in [\alpha, \beta]} k(x) \quad \text{for} \quad x \in [\alpha, \beta]$$

and that $n(r) \to \infty$ for $r \to \infty$.

The well-known solution, $C \sin(nx + c)$, of (62) has $\left[\frac{n(r)(\beta - \alpha)}{\pi}\right]$ roots in (α, β), and according to the comparison theorem of Sturm's first paper $V_r(x)$ has at least as many roots. Sturm had shown that $V_1(x)$ has no roots in (α, β); hence, it follows from Proposition G of the first paper that $K V_r'(\beta) + H V_r(\beta)$ has at least $\left[\frac{n(R)(\beta - \alpha)}{\pi}\right]$ roots when r runs through $[0, R]$. Therefore, there are infinitely many eigenvalues of (1)-(3). The proof is only a slight modification of the proof of Proposition H in the first memoir.

ad L and M: They are simple consequences of the oscillation theorem and the observation that $V_1(x)$ has no roots in (α, β).

ad N: Sturm obtained this result through careful investigation of the solution of the original boundary-value problem for the partial differential equation, (5)-(7):

$$C_m V_m(x)e^{-r_m t} + C_{m+1} V_{m+1}(x)e^{-r_{m+1} t} + \cdots + C_n V_n(x)e^{-r_n t}, \quad (63)$$

where $r_m < r_{m+1} < \cdots < r_n$. For large positive values of t, the solution (63) will be dominated by the first term, which has $m - 1$ roots, and for large negative values of t, it will be dominated by the last term, which has $n - 1$ roots. Sturm proved that for other values of t the number of roots lay between these extremes.

In connection with the argument above Sturm concluded that the temperature distribution in the bar,

$$C_1 V_1 e^{-r_1 t} + C_2 V_2 e^{-r_2 t} + \cdots + C_n V_n e^{-r_n t} + \cdots,$$

will eventually have m nodes, where m is the smallest value of i for which C_i is different from zero. This value of m can be found from the initial temperature distribution $f(x)$ since, in virtue of orthogonality, C_i is determined by[13]

$$C_i = \frac{\int_\alpha^\beta g(x) V_i(x) f(x)\, dx}{\int_\alpha^\beta g(x) V_i^2(x)\, dx}. \quad (64)$$

This result, which generalizes some of Fourier's and Poisson's theorems, had already been indicated by Sturm in [1829b].

28.   Few other papers in the history of mathematics can rival Sturm's two papers [1836a,b] for novelty of problem, methods, techniques, and results. Above, I have tried to summarize the theory as it appeared in 1833. When

preparing the last paper for publication, Sturm added a few remarks caused by Liouville's entrance on the scene (cf. §50). He returned to the subject only once, namely, in the following year with a paper written in collaboration with Liouville [Liouville and Sturm 1837a] (cf. §49). Otherwise, he left it to Liouville to extend his research on differential equations.

Liouville's Youthful Work on Heat Conduction

29. Liouville's work on the Sturm-Liouville theory concentrated on two major problems: expansion of functions in Fourier series of eigenfunctions and generalization of the theory to other types of differential equations. In his celebrated papers on these two problems, published during the years 1836–1838, he built directly on Sturm's research. His repeated reference to Sturm creates the impression that the work of his friend had been the starting point of his interest in these matters. However, as I indicate in Chapter VII, §18, Liouville's interests went as far back as Sturm's earliest investigations and had the same inspiration, namely, the study of Fourier's and especially of Poisson's work on the theory of heat. In Chapter VII, I discuss the more physical part of Liouville's juvenile research on heat conduction, so here only the mathematical questions are dealt with. As explained in Chapter I, Gergonne published this part of Liouville's work in his *Annales* while reproaching him, at the same time, for the bad style in which it was written [Liouville 1830].

It may have been the harsh criticism in Gergonne's footnote that restrained Liouville from referring to this paper in his later works. Otherwise, such references would have been appropriate since many of his most important ideas in the Sturm-Liouville theory can be found here in a preliminary or fully developed form. In order to give an account of these ideas I shall discuss in some detail those sections of the printed part of the memoir [Liouville 1830] where they were set forth. The bulk of the paper was devoted to a generalization of Poisson's theory of heat conduction in a thin metallic bar to the case where the surface is unequally polished and the material is inhomogeneous.

30. First, Liouville considered the homogeneous bar for which the temperature $u(x,t)$ must satisfy the following special case of $(5)^{14}$:

$$\frac{\partial u}{\partial t} = \frac{\partial^2 u}{\partial x^2} - f(x)u, \quad x \in (0,\beta), \tag{65}$$

where $f(x) > 0$ in $[0,\beta]$.

He assumed the temperature at the end points 0 and β of the bar to be fixed at θ and θ' degrees:

$$u(x,t) = \theta \quad \text{for} \quad x = 0, \tag{66}$$

$$u(x,t) = \theta' \quad \text{for} \quad x = \beta. \tag{67}$$

Since these boundary conditions, unlike Sturm's conditions, (6) and (7), are not homogeneous, the temperature distribution has a nontrivial stationary state. In the first part of [1830] Liouville described two different determinations of this stationary state u from the equation

$$u'' = f(x)u \tag{68}$$

and the boundary conditions (66) and (67).

The first method [§IV-§XI] consisted in approximating $f(x)$ by a polygon, solving the equation in this case, and letting the number of sides of the polygon tend to infinity. Liouville claimed, without proof, that the process would converge to a solution of the original equation. This polygon method, already developed in the memoir of June 1829 [Liouville 1830, p. 157], is of little interest to us, whereas the second method, invented in 1830, is very important.

31.   Liouville began the second method by observing that if $u_0, u_1, u_2, \ldots$ are solutions of the infinite system of differential equations,

$$\frac{d^2 u_0}{dx^2} = 0, \quad \frac{d^2 u_1}{dx^2} = f(x)u_0, \quad \frac{d^2 u_2}{dx^2} = f(x)u_1, \ldots, \tag{69}$$

then their sum

$$u = u_0 + u_1 + u_2 + \cdots \tag{70}$$

is a solution of (68). The solutions of (69) can be written

$$u_0(x) = A + Bx,$$
$$u_1(x) = \int_0^x dx \int_0^x (A + Bx)f(x)\,dx,$$
$$u_2(x) = \int_0^x dx \int_0^x f(x)\,dx \int_0^x dx \int_0^x (A + Bx)f(x)\,dx, \tag{71}$$

and so the general solution of (68) is of the form

$$A\left\{1 + \int_0^x dx \int_0^x f(x)\,dx + \int_0^x dx \int_0^x f(x) \int_0^x dx \int_0^x f(x)\,dx + \cdots\right\} +$$
$$B\left\{x + \int_0^x dx \int_0^x xf(x)\,dx + \int_0^x dx \int_0^x f(x) \int_0^x dx \int_0^x xf(x)\,dx + \cdots\right\}. \tag{72}$$

Substituting $M = \max_{x \in [0,\beta]} f(x)$ for $f(x)$ in the first series in (72), Liouville obtained the series

$$\left\{1 + \frac{Mx^2}{2!} + \frac{M^2 x^4}{4!} + \cdots\right\}. \tag{73}$$

Since this series converges, Liouville concluded that the first series of (72) converges as well. Similarly, he proved that the second series in (72) converges. The two arbitrary constants A and B can finally be determined from the boundary conditions (66) and (67).

32. In the above argument, Liouville used the method of successive approximation, later ascribed to E. Picard [1890], to show that a solution of the differential equation (68) such as to fulfill conditions (66) and (67), exists. Thus, Liouville's anticipation of Picard's proof of existence did not originate in [1836f and 1837c] as is generally believed, but was developed six years earlier.

It is now a well-established fact that Cauchy gave another type of existence proof for a general first-order differential equation in his lectures at the *École Polytechnique* during the 1820s [Cauchy 1824–1981] (publ. [1835–1840]), and it is often conjectured (e.g., by Kline [1972, p. 719] and by Birkhoff [1973, p. 243]) that Cauchy at that time knew also the proof of existence using successive approximation. The only argument in favor of that conjecture is Moigno's inclusion of the method in his *Leçons de calcul différentiel et de calcul intégral rédigées principalement d'après les méthodes de M. A.-L. Cauchy* [1844, pp. 702–707]. Recently, C. Gilain (cf. [Cauchy 1824–1981]) has argued convincingly that Cauchy did not apply successive approximation to prove existence, but only used this method to find approximate solutions. Still, Gilain is of the opinion that Moigno learned the method from Cauchy. However, it is striking that Moigno used the method of successive approximation to a second-order linear differential equation, just as Liouville did, whereas Cauchy proved existence for first-order equations and partial differential equations. This fact more than indicates that Moigno borrowed this part of his *Leçons* directly from Liouville (probably from [1837c]). Therefore, I think Liouville's proof in [1830] is his own original contribution to the theory of differential equations. If that is true, Liouville [1830] presented both the first published proof of the existence of a solution of a differential equation and the first application — published or unpublished — of the method of successive approximation for that purpose.

Admittedly, Liouville never formulated as explicitly as Cauchy the central question: Given a differential equation and certain boundary conditions or initial conditions; does a solution exist? In [1830] Liouville differed from Cauchy also by focusing on boundary-value problems. In [1836f,1837c] Liouville imposed Cauchy data at only one point.

33. After having thus found the stationary solution, hereafter called $V_0(x)$, of (65) with the boundary values of (66) and (67), Liouville supposed the general solution to (65)-(67) to be of the form

$$V_0(x) + C_1 V_1(x)e^{-r_1 t} + C_2 V_2(x)e^{-r_2 t} + C_3 V_3(x)e^{-r_3 t} \ldots, \qquad (74)$$

where each of the eigenfunctions must satisfy the special case of (1),

$$- rV_r = V_r'' - f(x)V_r, \qquad (75)$$

and the homogeneous boundary conditions,

$$V_r(x) = 0 \quad \text{for} \quad x = 0, \qquad (76)$$

$$V_r(x) \; = \; 0 \quad \text{for} \quad x = \beta. \tag{77}$$

Again, the method of successive approximation produces an expression for the solution as a sum of two series of which one vanishes if we take the boundary condition (76) into account (corresponding to $A = 0$ in (72)). In the remaining expression for V_r,

$$
\begin{aligned}
V_r \; = \; & x + \int_0^x dx \int_0^x x(f(x) - r) \, dx \\
& + \int_0^x dx \int_0^x (f(x) - r) \, dx \int_0^x dx \int_0^x x(f(x) - r) \, dx + \cdots, \tag{78}
\end{aligned}
$$

the eigenvalue r must be chosen so that the boundary condition (77) is satisfied. As usual, this gives rise to a transcendental equation (4) for r.

34.   Liouville went on to prove the orthogonality, (60) (for $g \equiv 1$), in essentially the same way as Sturm later did, that is, without using the partial differential equation as Poisson [1826] had done. Apparently, Liouville did not even know of Poisson's paper, for he used orthogonality only to determine the arbitrary constants C_i in (74) from a given initial temperature distribution $f(x) + V_0(x)$ (as in (64) with $g \equiv 1$), but used another method to prove the eigenvalues real:

> We will show $1°$ that the equation from which the values of $m[r]$ result, has all its roots real and positive; $2°$ we shall prove that the series forming the value of n [(74)] is a convergent series, which is necessary to complete the solution.
>
> [Liouville 1830, p. 164]

35.   Liouville's proof of $1°$ is both clumsy and unrigorous compared with Poisson's proof. It rests on the inspection of the transcendental equation $\Pi(r) = 0$ found by substituting (78) into (77) (in the following arguments he takes $\beta = 1$):

$$1 + \int_0^1 dx \int_0^x x(f(x) - r) \, dx + \tag{79}$$

$$\int_0^1 dx \int_0^x (f(x) - r) \, dx \int_0^x dx \int_0^x x(f(x) - r) \, dx + \cdots = 0.$$

Since $f(x) > 0$, it is clear that no negative value of r solves this equation. Liouville believed that he would exclude all nonreal eigenvalues if only he could prove the existence of an infinity of positive eigenvalues. In the most far-reaching of the two proofs he supplied for the latter, he substituted for $f(x)$ a loosely described "mean value," P_r, independent of x, reducing (79) to

$$1 - \frac{(r - P_r)}{3!} + \frac{(r - P_r)^3}{5!} - \frac{(r - P_r)^5}{7!} + \cdots = 0, \tag{80}$$

or

$$\frac{\sin \sqrt{r - P_r}}{\sqrt{r - P_r}} = 0. \tag{81}$$

In fact, this argument is not valid, since one cannot use the same "mean value" in the different terms of (79). But, taking this for granted and accepting Liouville's loose argument that P_r remains bounded as a function of r because $f(x)$ is bounded as a function of x, we can conclude with Liouville that there are infinitely many positive solutions of (79) or (81) of the form

$$r = P_r + n^2\pi^2, \quad n \in \mathbf{N}. \tag{82}$$

The argument even shows that for large values of n the eigenvalues are "very nearly" $r = n^2\pi^2$" in the sense that $|r_n - n^2\pi^2| \leq \max P_r < \infty$.

36.   This approximation of the eigenvalues was applied by Liouville in his subsequent proof of the convergence of the Fourier series (74) for $t = 0$, which will imply the convergence for $t > 0$. If we consider only large values of r (which suffices in a proof of convergence), $f(x)$ can be ignored in the expression (78) of V_r, leaving the approximate eigenfunction:

$$V_r = \frac{\sin \sqrt{r}x}{\sqrt{r}}. \tag{83}$$

Using this value of V_r in (74) and in formula (64) for the Fourier coefficients, Liouville was led to the approximate value of u for $t = 0$:

$$\sum_r \frac{\sin \sqrt{r}x}{\beta} \int_0^\beta f(x) \sin \sqrt{r}x \, dx. \tag{84}$$

If the "approximate" value $n^2\pi^2$ is substituted for r, (84) reduces to an ordinary Fourier series. Liouville believed that Fourier and Poisson had provided proofs of convergence for the ordinary Fourier series and concluded that since its terms coincided with the terms of (74) for large values of r, the latter, more general Fourier series must also converge.

   In the last sections [1830, §§24–27], Liouville generalized all these considerations to equation (5) for the heterogeneous bar.[15]

37.   In spite of the lack of rigor in the last arguments, Liouville's [1830] is of the greatest importance in the development of Sturm-Liouville theory because it constitutes the germ of Liouville's subsequent contributions to this theory. Together with Sturm's now lost memoirs from 1829 and Poisson's [1826], it presents the earliest advances in this branch of analysis. I have argued that Liouville did not know Poisson's work, and according to his own testimony in the introduction of [1830], he also did not know Sturm's work:

> Until now, these questions have not been treated by any geometer;
> they seem to offer almost insurmountable difficulties when one allows

the bodies to have their three dimensions [cf. Chapter XV, §12a]. If one sets aside the two of them, the problem is completely solved by my work.

There is also a striking difference between Sturm's and Liouville's approaches to the theory. Though they share some theorems in common, as, for example, the statement of orthogonality, the reality of eigenvalues, and the determination of the Fourier coefficients, the bulk of their papers have different goals, Sturm tending toward the qualitative behavior of the eigenfunctions, and Liouville toward expansion in Fourier series.

Liouville's Mature Papers on Second-Order Differential Equations. Expansion in Fourier Series

CONVERGENCE OF THE FOURIER SERIES

38.   Though Liouville's interests after 1830 turned to other fields, such as fractional differentiation and integration in finite terms, he continued to work on problems related to those treated in the paper of [1830] (cf. [Liouville 1836a, 1836b, 1836c]). However, these problems were all concerned with trigonometric series and he did not publish anything on the more general type of Fourier series until the three large *Mémoires sur le développement des fonctions ou parties de fonctions en séries dont les divers termes sont assujettis à satisfaire à une même équation différentielle du second ordre, contenant un paramètre variable* [1836f, 1837c, 1837i].

In these three important papers, Liouville was chiefly concerned with the questions he had treated in the memoir of [1830]. However, he now succeeded in rigorizing the theory considerably, partly by building on Sturm's much more detailed investigation of the eigenfunctions and partly by refining his own earlier arguments. He immediately turned to Sturm's general equation (1) with the boundary conditions (2) and (3), and repeated the solution [1830] by successive approximation, in this case for the Cauchy problem, (1) and (2). In the first memoir [1836f, p. 255], he merely wrote down the formulas similar to (72) and did not write one word on the question of convergence. However, in the second paper [1837c, pp. 19–22], he amply made up for this omission by supplying, in addition to the proof of convergence [1830], another proof based on conversion of the differential equation into an integral equation [cf. Lützen 1982b].

39.   Otherwise, Liouville devoted his second large memoir [1837c] to the other great question of convergence treated in the [1830] memoir, that of the Fourier series

$$F(x) = \sum_{n=1}^{\infty} \frac{V_n(x) \int_\alpha^\beta g(x) V_n(x) f(x)\, dx}{\int_\alpha^\beta g(x) V_n^2(x)\, dx}. \tag{85}$$

The central idea of the improved proof of convergence is the following uniform inequality:

> If one denotes by n a very large index, by u_n the absolute value of the nth term of the series [85] and by M a certain number independent of n, ... one has $u_n < \frac{M}{n^2}$.
>
> [Liouville 1837c, p. 18]

To obtain this conclusion, Liouville introduced the new dependent and independent variables z and U defined by

$$z = \int_\alpha^x \sqrt{\frac{g(x)}{k(x)}}\, dx, \tag{86}$$

$$V(x) = \theta(x)U(x), \quad \text{where} \quad \theta = \frac{1}{\sqrt[4]{g(x)k(x)}}, \tag{87}$$

and

$$r = \varrho^2.$$

Expressed in these new variables, the original problem (1)-(3) is reduced to the simpler problem

$$U''(z) + \varrho^2 U(z) = \lambda(z)U(z), \quad z \in [0, \gamma], \tag{88}$$

$$
\begin{aligned}
U'(z) - h'U(z) &= 0 \quad \text{for} \quad z = 0, \tag{89}\\
U'(z) + H'U(z) &= 0 \quad \text{for} \quad z = \gamma, \tag{90}
\end{aligned}
$$

where

$$\gamma = \int_\alpha^\beta \sqrt{\frac{g(x)}{k(x)}}\, dx,$$

$$\lambda = \frac{1}{\theta\sqrt{gk}}\left(l\sqrt{\frac{k}{g}}\theta - \frac{d\sqrt{gk}}{dz}\frac{d\theta}{dz} - \sqrt{gk}\frac{d^2\theta}{dz^2}\right), \tag{91}$$

and h', H' are constants that are not necessarily positive. The elegant transformation (86) and (87) is now called the Liouville transformation after its inventor. It combines Sturm's transformations of the dependent and independent variables (see §23) in such a way that the function k disappears in the transformed equation (88) at the same time as the coefficient to the undifferentiated term U retains its simple form.

40.   From the transformed equation (88)-(90) Liouville deduced the integral equation for U:

$$U(z) = \cos \varrho z + \frac{h' \sin \varrho z}{\varrho} + \frac{1}{\varrho}\int_0^z \lambda(z')U(z')\sin \varrho(z - z')\, dz'. \tag{92}$$

Here, he has fixed the arbitrary multiplicative constant in such a way that
$U(0) = 1$. Various estimates applied to this integral equation then lead to
the desired bounds of the numerator and denominator of the Fourier series,
(85):

$$\left| V_n \int_\alpha^\beta g(x) V_n(x) f(x)\, dx \right| < \frac{1}{r_n} K_1 \left(1 + \left(\frac{h'}{\varrho_n} \right)^2 \right), \qquad (93)$$

$$\int_\alpha^\beta g(x) V_n^2(x)\, dx > K_2 \left(1 + \left(\frac{h'}{\varrho_n} \right)^2 \right), \qquad (94)$$

K_1, K_2 being positive constants. Hence, the absolute value of the nth term
u_n of the Fourier series is bounded by

$$|u_n| < \frac{M}{r_n}. \qquad (95)$$

As in the proof of [1830], Liouville then needed only to show that r_n tends
to infinity fast enough when n tends to infinity. The idea of the proof is the
same as in [1830] but in [1837c] the trigonometric behavior of the eigenfunctions for large values of r was established from the integral equation, (92).
Using this asymptotic behavior of U_r and Sturm's oscillation theorem,[16]
Liouville rigorously established the asymptotic behavior of the eigenvalues:

$$\varrho_n = \sqrt{r_n} \sim \frac{(n-1)\pi}{\gamma} + \frac{P_0}{(n-1)\pi}, \qquad (96)$$

where P_0 is a constant. Combining this statement with (95), he finally
obtained the desired estimate:

$$|u_n| < \frac{M'}{n^2},$$

implying the convergence of the Fourier series, (85).

41. Liouville's deduction contained several innovations in addition to the
convergence theorem. I have mentioned the Liouville transformation, (86)
and (87). Just as important are the asymptotic expressions for the eigenvalues (96) and the corresponding approximate eigenfunctions to be found
from (92). According to Liouville, the latter complemented Sturm's methods of approximating the eigenfunctions (§23), which was manageable only
for small values of n.

   Finally, the ingenious application of equation (92) marks an important
instance in the early theory of integral equations (see [Lützen 1982b]).

   In the above proof of convergence, Liouville explicitly assumed that $g(x)$,
$k(x)$, $f(x)$ and their "first and second derivatives always conserves finite
values" ($g, k, f \in C^2[\alpha, \beta]$). Implicitly, he also assumed that $f(x)$ satisfies
the boundary conditions (2) and (3). Under these assumptions, Liouville's
proof even proves rigorously that the Fourier series converges uniformly.

However, Liouville could not appreciate this virtue of the proof since the difference between pointwise and uniform convergence had not been realized by then.

42.   In November of the same year, Liouville published a new proof of convergence [1837i] that did not make use of the too restrictive assumptions mentioned in the last section.

> Here I propose to get rid of these various restrictions as much as it is possible, in particular those concerning the function $f(x)$.
>
> [Liouville 1837i, p. 419]

In particular, he had discovered that it was unnecessary to impose boundary conditions (2) and (3) on f:

> these conditions, which I have inappropriately imposed on the function $f(x)$ in my first two memoirs, are unnecessary and must be left out.
>
> [Liouville 1837i, p. 421]

In place of them, Liouville assumed f to be continuous, and instead of $g, k, l \in C^2[\alpha, \beta]$, he assumed only that λ defined by (91) be absolutely integrable.[17]

An investigation of Liouville's proof reveals that he actually used more assumptions on g, k, l, and λ, such as differentiability of k and piecewise monotonicity of f. The last property is implicitly used in the proof [1837i, Sect. 4] of a strong version of Riemann's lemma:

$$\left| \int_0^z f(z) \sin \varrho z \, dz \right| < \frac{K}{\varrho}; \tag{97}$$

further it is presupposed at the end of the proof because Liouville referred to Dirichlet's proof of convergence [1829] for trigonometric Fourier series, which was explicitly carried through for piecewise monotone functions.

43.   In spite of these insufficiencies, Liouville's attitude toward the relation between assumption, proof, and theorem was very modern. He had realized that a theorem may be improved by relaxing its assumptions, and he had seen that the best assumptions can be found by examining the proof. For example, the assumption on λ clearly stems from the proof. In this respect, Liouville was more far-sighted than his leading compatriot Cauchy, who did not understand this interplay. Cauchy usually made the stereotypical assumption that the functions be continuous no matter what properties he actually used in his proofs. Related remarks have recently been made by Gilain (cf. [Cauchy 1824–1981, pp. XLI–XLIX]).

By 1837, the mathematician who had most explicitly put forward such proof-generated assumptions was Dirichlet in his paper on Fourier series [1829]. Liouville may have been influenced by his friend Dirichlet, but he took a step further by searching a new, and according to himself less elegant, proof of an already established theorem, with the sole aim of weakening the assumptions. Such an understanding of the interplay between assumptions and proofs did not catch on until the end of the 19th century.

44.   Clearly, Liouville could not use under the new assumptions the proof of convergence [1837i] with its uniform estimates. Instead, he produced a rigorized version of the proof [1830] by replacing the loose argument concerning the asymptotic behavior of the eigenfunctions with an intricate application of the integral equation (92), leading again to the asymptotic eigenvalues (96). In this way, he obtained the following expression for the terms of the Fourier series:

$$u_n(x) = \frac{2}{\gamma} \cos nz \int_0^\gamma F(z) \cos nz\, dz + \frac{\psi_1(x,n)}{n^2}, \qquad (98)$$

where $\psi_1(x,n)$ is bounded. From this expression, the convergence follows easily from the convergence of trigonometric Fourier series, for which Liouville this time referred to Cauchy's "and in particular the excellent Memoir of M. Lejeune Dirichlet" [1829].

DETERMINATION OF THE SUM OF THE FOURIER SERIES

45.   In his two proofs of convergence [1837c, 1837i], Liouville did not have to concern himself with finding the limit of the Fourier series, because he had already settled that problem in the first of the three large memoirs [1836f] presented to the *Académie* on November 30, 1835. Liouville claimed to prove "by a rigorous procedure" that the "value of the series" (85) is $f(x)$. The proof rests on two lemmas.

Lemma I. *Define inductively the functions $P_i^j(x)$ by*

$$P_i^1(x) = V_1(a_1)V_i(x) - V_i(a_1)V_1(x), \quad i = 2, 3, \ldots,$$
$$P_i^2(x) = P_2^1(a_2)P_i^1(x) - P_i^1(a_2)P_2^1(x), \quad i = 3, 4, \ldots,$$
$$\vdots$$
$$P_i^j(x) = P_j^{j-1}(a_j)P_i^{j-1}(x) - P_i^{j-1}(a_j)P_j^{j-1}(x), \quad i = j+1, j+2, \ldots,$$

where $a_1, a_2, \ldots, a_j, \ldots$ are different points of (α, β). Then $P_{j+1}^j(x)$ vanishes and changes sign at $a_1, a_2, \ldots, a_j, \ldots$, and it has no other roots.

It is easily verified that $a_1, a_2, \ldots, a_j$ are roots of P_{j+1}^j, and since P_{j+1}^j is of the form $\sum_{i=1}^{j+1} A_i V_i(x)$, it follows from Sturm's proposition N (§26) that it has at most j roots counted with multiplicity. Thus, P_{j+1}^j has precisely the simple roots $a_1, a_2, \ldots, a_j$.

Lemma II. *Let $\varphi(x)$ be a function of $x \in [\alpha, \beta]$. If*

$$\int_{\alpha}^{\beta} \varphi(x) V_n(x)\, dx = 0 \qquad (99)$$

for all eigenfunctions V_n of (1)-(3), then $\varphi \equiv 0$.

Liouville gave an indirect proof: Suppose φ changes sign j times, say, in $a_1, a_2, \ldots, a_j$. As in the first lemma, he constructs P_{j+1}^j corresponding to this series of roots. Since P_{j+1}^j is a linear combination of eigenfunctions, we have

$$\int_{\alpha}^{\beta} \varphi(x) P_{j+1}^j(x)\, dx = 0, \qquad (100)$$

which contradicts the fact that $\varphi(x) P_{j+1}^j(x)$ is not identically zero in $[\alpha, \beta]$, where it conserves its sign. Therefore, $\varphi(x)$ cannot change sign a finite number of times. Consequently, Liouville says, φ must be identically zero in $[\alpha, \beta]$.

It was an easy matter for Liouville to deduce the main theorem from this lemma. He multiplied both sides of (85) by $g(x) V_m(x)\, dx$ and integrated from α to β. Because of the orthogonality relations (60) only the mth term on the right-hand side survives:

$$\int_{\alpha}^{\beta} g(x) V_m(x) F(x)\, dx = \int_{\alpha}^{\beta} g(x) V_m(x) f(x)\, dx\,, \quad \forall m = 1, 2, \ldots,$$

or

$$\int_{\alpha}^{\beta} g(x)(F(x) - f(x)) V_m(x)\, dx = 0\,, \quad \forall m = 1, 2, \ldots. \qquad (101)$$

According to Lemma 2, $F(x) - f(x) = 0$, so that the sum in question equals $f(x)$. This completes Liouville's proof.

46.   In spite of its simplicity, this proof presents the most profound mistake in Sturm and Liouville's theory of second-order linear differential equations. I do not refer to the curious neglect of the problem of convergence, solved the following year,[18] nor to the term-by-term integration,[19] but to the last step in the proof of Lemma 2, where Liouville concludes that a function with infinitely many roots in $[\alpha, \beta]$ must be identically zero there. This would be true for an analytic function φ, but it is not easy to prove that F is analytic, even if f is analytic. In fact, one has to use a totally different approach to prove the theorem.

Liouville later came to realize at least a part of the problem. On March 28 (or 29), 1838, he wrote in his notebooks [Ms 3616 (2), p. 56v] that $\varphi(x)$ could be different from zero for isolated values of x, but if φ is continuous that cannot happen. In a note from September 7, 1840 [Ms 3616 (5), pp. 45r–46r], Liouville showed a deeper understanding of the problem. There, he mentioned the difficulties occurring if φ can "vanish an infinite number of times in each infinitely small interval lying between α and $\alpha + \varepsilon$, $\beta - \varepsilon$ and β" (Liouville does not use the letters α and β).

Furthermore, in a note [Ms 3617 (2)] (probably from 1842), Liouville wrote about a similar proof: "In all this I only see the objection of D regarding the functions changing sign an infinite number of times." Thus, at the latest in 1842, a certain D had explained the real problem in the above "proof." D is probably Dirichlet, who had taken great pains to exclude such infinitely often oscillating functions from his theorem on Fourier series.

47.   When Liouville repeated his insufficient argument in [1838h, pp. 603, 612], he dismissed the difficulty as a minor detail, but his subsequent attempts to supply another proof reveal that he was disturbed by the problem. Even in 1837, Liouville had devised a different proof of the expansion theorem in collaboration with Sturm [Liouville and Sturm 1837a]. They proved that $\varphi(x) \equiv F(x) - f(x)$ satisfies the equation

$$\int_\alpha^\beta \varphi(x) V_r(x)\, dx = 0, \quad \forall r \in \mathbf{R}, \tag{102}$$

for *all* solutions V_r of (1) satisfying only boundary condition (2).[20] However, in order to conclude that (102) implied $\varphi \equiv 0$, they referred to Liouville's original proof in [1836f], and so their argument was no more convincing than the original. Sturm and Liouville's proof was a small extract of a large memoir on Sturm-Liouville theory presented to the *Académie* on May 8, 1837. Unfortunately, the rest of the memoir is lost.

Liouville returned to this approach to the expansion theorem three years later. However, he left only an unfinished and insufficient sketch of a proof in a draft of a letter to an unnamed colleague [Ms 3616 (5), pp. 45r–46r].

Indications of other approaches can be found in Liouville's notes from March 8, 1838 [Ms 3616 (2)], August 21, 1839 [Ms 3616 (1)], and September 1839 [Ms 3616 (5)]. Of these notes, only the second, written in Bruxelles, led anywhere. The central idea of the proof was to make the coefficients dependent upon a new parameter, m.

At first, the new method did not work satisfactorily,[21] but after some revision, Liouville was so content with it that he sent the proof to Dirichlet in a letter dated by Tannery February 1841 [Tannery 1910, pp. 17–19]. He considered the simplified form (88)-(90) of the problem and chose $\lambda(z, m), h'(m), H'(m)$, so as to converge to zero as $m \to 0$. Thus, for $m = 0$, the eigenfunctions were simple trigonometric functions, and in that case Liouville knew that $\varphi(x, m) = F(x, m) - f(x) \equiv 0$. He then argued that if $\varphi(x, m) \equiv 0$ for some value of m it vanished necessarily in a whole neighborhood of m. Hence, he concluded that $\varphi(x, m) \equiv 0$ for all m[22].

Liouville did not notice that the size ε of the neighborhood $(m - \varepsilon, m + \varepsilon)$ was dependent on m. Therefore, he thought he could exhaust an interval $[0, M]$, with a finite number of neighborhoods. This part of Liouville's argument can be made rigorous if we note that in modern terms Liouville claimed that

$$\{m > 0 \mid \varphi(m, x) \equiv 0\}$$

is an open subset of $\mathbf{R}_+$. Under sufficiently strong conditions of regularity, it is clear that the set is also closed, and hence it is equal to $\mathbf{R}_+$.

However, there is also a grave irreparable mistake in the proof, namely, in the proof of convergence of a certain series, which he claimed to be "very easy" in the letter to Dirichlet. His faulty proof has been preserved in a note from the first half of 1840 [Ms 3616 (5), pp. 14v–15r].

The crucial mistake in this note occurred at the end, where Liouville stated that a series of the form

$$\sum_{i=1}^{\infty} \frac{1}{i} f(z, i) \cos iz$$

is convergent when $f(z, i)$ is a bounded function. He probably thought that this could be inferred from the behavior of the Fourier series:

$$\sum_{i=1}^{n} \frac{\cos iz}{i},$$

but that is not correct.

48.   Liouville's work on second-order linear differential equations almost exclusively dealt with Fourier expansion of "arbitrary" functions. However, in his second publication on Sturm-Liouville theory from 1836 [1836e], he gave a time-independent analysis leading to Sturm's proposition N (cf. §28). Using the techniques of Sturm, he showed that

$$A_m(r_m - r_1)V_m + A_{m+1}(r_{m+1} - r_1)V_{m+1} + \cdots + A_n(r_n - r_1)V_n \quad (103)$$

has at least as many roots in (α, β) as

$$A_m V_m + A_{m+1} V_{m+1} + \cdots + A_n V_n. \tag{61}$$

Repeated use of this observation shows that

$$A_m\left(\frac{r_m - r_1}{r_n - r_1}\right)^k V_m + A_{m+1}\left(\frac{r_{m+1} - r_1}{r_n - r_1}\right)^k + \cdots + A_n V_n \tag{104}$$

has at least as many roots as (61), and since $\left(\frac{r_{m+1}-r_1}{r_n-r_1}\right) < 1$, Liouville concluded that in the limit $k = \infty$ the number of roots of $V_n(= n - 1)$ is greater or equal to the number of roots of (61).[22] A similar argument gives the lower bound on the number of roots of (61). Inspired by Liouville, Sturm added a similar proof to his [1836b]. It was built on the observation that

$$A_m r_m V_m + A_{m+1} r_{m+1} V_{m+1} + \cdots + A_n r_n V_n \tag{105}$$

has at least as many roots as (61). His proof had the advantage of taking
into account possible roots at the end points, a problem Liouville had left
aside (as have I in §§26 and 27).

Liouville's new proof of proposition N can be viewed as the last step
in a process of establishing Sturm-Liouville theory as a self-contained subject, independent of the physical problems and partial differential equations
whence it had emerged. Sturm had taken the first step by freeing Poisson's
proof that the eigenvalues were real from the unnecessary use of a solution
of the partial differential equation. After [Liouville 1836e], all theorems in
Sturm-Liouville theory rested only on equations (1)-(3).

MINOR RESULTS

49. In addition to these papers on second-order linear differential equations, Liouville published some results that were inspired by specific ideas
in Sturm-Liouville theory.[23] For example, he proved that Lemma II is valid
when x^n is substituted for V_n [1837b] and concluded that the Fourier expansion of $f(x)$ on the orthogonal set of Legendre polynomials "have the
value" $f(x)$ [1837e]. The proof has the same weakness as the original deduction, and there is no proof of convergence. Three years later, Liouville
complemented Cauchy's convergence criteria of [1821] with a new *condition de convergence d'une classe générale de séries* [1840g]. The series in
question were power series of the form

$$\varphi(x) + \alpha\varphi_1(x) + \cdots + \alpha^n\varphi_n(x) + \cdots, \quad x \in [\alpha, \beta], \tag{106}$$

where φ_n is successively defined from the positive function $g(x)$ and the
arbitrary function $\varphi(x)$ by

$$\varphi_{n+1}(x) = \int_\alpha^x dx \int_\beta^x g(x)\varphi_n(x)\,dx. \tag{107}$$

He obtained the convergence criterion by remarking that the series (106) is
of the form $\varphi + \alpha s$, where s arises from solving, by successive approximation,
the boundary value problem

$$\frac{d^2s}{dx^2} - \alpha g s = g\varphi \tag{108}$$

with

$$s = 0 \quad \text{for} \quad x = \alpha, \quad \frac{ds}{dx} = 0 \quad \text{for} \quad x = \beta. \tag{109}$$

This application of the method of successive approximations to the case
where the boundary values are given at two points is an improvement over
the formulas from [1830] (cf. §§31-32).

Apart from these few short papers, Liouville stopped publishing on second-
order Sturm-Liouville theory in 1837. His further contributions to spectral

theory provided generalizations of Sturm-Liouville theory to different types of problems, particularly to higher order differential equations, to be discussed in the next chapter. Finally, Sturm-Liouville theory provided the background for Liouville's mostly unpublished work on spectral theory of integral operators, to which I shall return in Chapter XV.

Liouville's Generalization of Sturm-Liouville Theory to Higher-Order Equations

THIRD DEGREE; CONSTANT COEFFICIENTS

50. Sturm believed that he had exhausted the possibilities of his new methods:

> The principle on which the theorems which I develop rests has never been used in analysis as far as I know, and it does not seem to me to be susceptible of an extension to other differential equations.
>
> [Sturm 1836a, p. 107]

Liouville's greatest effort in Sturm-Liouville theory was an only partially successful attempt to disprove the last part of Sturm's conjecture by generalizing his and Sturm's theory to other types of equations. Thus, in Liouville's notebooks, the first recorded note inspired by Sturm's work, dating from April or May 1835 [Ms 3615 (4), pp. 79v–80r], dealt with the equation

$$\frac{du}{dt} = \frac{d^3u}{dx^3},\tag{110}$$

which, as we saw in Chapter VII, §33, had already been suggested by Poisson in 1828 as a difficult test case for Liouville's early methods in differential equations. He presented his new results to the *Académie* the following year (November 14, 1836) and had them published in [1837a]. In some of his notes (e.g., [Ms 3615 (5)]), Liouville attempted to apply Sturm's general methods to investigate the separated equation

$$V^{(3)}(x) + rV(x) = 0 \quad \text{for} \quad x \in (0,1)\tag{111}$$

with the boundary conditions

$$V(x) = V'(x) = 0 \quad \text{for} \quad x = 0,\tag{112}$$

$$V(x) = 0 \quad \text{for} \quad x = 1.\tag{113}$$

However, in the published account [1837a], he based his arguments on the explicit expression for the solution of (111) and (112):

$$V = \frac{1}{3\varrho^2}(e^{-x\rho} + \mu e^{-\mu x\rho} + \mu^2 e^{-\mu^2 x\rho}),$$

where $\varrho^3 = r$ and $\mu^3 = 1$, $\mu \neq 1$. He showed that all of Sturm's theorems were valid for the boundary-value problem (111)-(113) except for the orthogonality (60). This relation was replaced by the biorthogonality

$$\int_\alpha^\beta V_m(x)U_n(x)\,dx = 0 \quad \text{for} \quad m \neq n, \tag{114}$$

where $\alpha = 0$, $\beta = 1$, and U_n is the nth eigenfunction of the related boundary-value problem

$$U^{(3)}(x) - rU(x) = 0 \quad \text{for} \quad x \in (0,1), \tag{115}$$

$$U(x) = 0 \quad \text{for} \quad x = 0, \tag{116}$$

$$U(x) = U'(x) = 0 \quad \text{for} \quad x = 1. \tag{117}$$

Therefore, the Fourier series of a function $f(x)$ has the altered form

$$F(x) = \sum_{n=1}^\infty \frac{V_n \int_\alpha^\beta f(x)U_n(x)\,dx}{\int_\alpha^\beta V_n(x)U_n(x)\,dx}. \tag{118}$$

Liouville showed, with a proof similar to the one given in [1836f], that if the Fourier series converges it has the value $F(x) = f(x)$, but he did not prove convergence.

HIGHER DEGREE; VARIABLE COEFFICIENTS

51.   The boundary-value problem (115)-(117) is today called the adjoint of the original problem (111)-(113). At first, Liouville did not appreciate the profound difference between self-adjoint problems (e.g., (1)-(3)) and non-self-adjoint problems, like (111)-(113). For example, early in 1835, he apparently tried to prove ordinary orthogonality for the third-order differential equation (111).[24] He seems to have become aware of the importance of the adjoint equation during the winter of 1837–1838, when he taught his first course at the *Collège de France* as a substitute for Biot. This *Cours de Physique générale et Mathématique* was mainly a course òn differential equations and covered, for example, Lagrange's methods of integration using the adjoint equation [Liouville Ms 3615 (5), pp. 42v–54r]. It stimulated Liouville to take up his research from 1835 on higher order Sturm-Liouville theory, this time for equations with variable coefficients.

In [1838h], Liouville published the investigations of the most general of these equations:

$$(K(L\ldots(M(NV')')'\ldots)')' + rV = 0 \quad \text{for} \quad x \in (\alpha, \beta), \tag{119}$$

with the boundary conditions

$$V = A, NV' = B, \ldots, K(L\ldots(M(NV')')'\ldots)' = D \quad \text{for} \quad x = \alpha, \tag{120}$$

and

$$aU + b(NV') + \cdots + cK(L\ldots(M(NV')'\ldots)' = 0 \quad \text{for} \quad x = \beta, \quad (121)$$

where $K(x), L(x), \ldots, M(x), N(x) > 0$ for $x \in [\alpha, \beta]$ and
$A, B, \ldots, D, a, b, \ldots, c > 0$.

He found that the biorthogonality (114) would hold if U_n is the nth eigenfunction of the adjoint problem:

$$(N(M\ldots(L(KU')')'\ldots)')' + (-1)^\mu rU = 0 \quad \text{for} \quad x \in (\alpha, \beta), \quad (122)$$

$$DU - \cdots + (-1)^{\mu-2}B(M\ldots(KU')'\ldots)$$

$$+ (-1)^{\mu-1}AN(M\ldots(KU')'\ldots)' = 0 \quad \text{for} \quad x = \alpha,$$
(123)

$$U = c, \ldots, N(M\ldots(L(KU')')'\ldots)' = (-1)^{\mu-1}a \quad \text{for} \quad x = \beta, \quad (124)$$

where μ is the order of the differential equation. He proved, both for equations with constant coefficients [1837a, p. 102] and with variable coefficients [1838h, pp. 604–606], that the adjoint problem has the same eigenvalues as the original problem. Thus, Liouville extended Lagrange's concept of an adjoint (*conjugué*) differential equation to include the boundary values as well, though he did not introduce a term for the adjoint boundary values. During the spring of 1838, he found many other remarkable results pertaining to the general boundary-value problem (119)-(121) some of which he presented to his students at the *Collège de France* (cf. [Liouville 1838h, Introduction]).

52.   One can follow Liouville's successive progress with these questions in the approximately 200 pages of disorganized notes jotted down in his notebooks [Ms 3615 (5), Ms 3616 (2)] from February 1838 and later. In order to facilitate the understanding of the questions considered pell-mell in the notes, I shall discuss the ideas in their logical order, partly sacrificing the chronology.

Liouville's goal was still to expand arbitrary functions in series of eigenfunctions of the boundary value problem. In order to find the Fourier coefficients, he needed the biorthogonality (114). In some of his notes for example, [Ms 3616 (2)] from February 19, 1838, he experimented with suitable boundary conditions and corresponding adjoint conditions, but the problem does not seem to have caused him much trouble. Next, he wished Lemma II (§45) to hold so that he could show that the Fourier series (118) if convergent has the value $F(x) = f(x)$. The proof of this lemma could be taken over from [1836f] if only the problem would have an infinity of (positive) eigenvalues and if Sturms oscillation theorem would hold, i.e., if V_n has in (α, β) exactly $n - 1$ roots, all of which are simple. The existence of infinitely many positive eigenvalues is not dealt with in the notes. In the published paper [1838h, §7], Liouville based his proof on a comparison with the equation with constant coefficients, for which he could prove the theorem as he had done in [1837a]. However, since Liouville had nothing like Sturm's comparison theorem (§22) at his disposal, the proof necessarily differed from Sturm's, and it is in fact insufficient.

53.   The proof of Sturm's oscillation theorem caused Liouville the greatest troubles. The vast majority of the notes from the first half of 1838 are related to this problem. He indicated several ways to prove that the solutions of the equation with suitable boundary values at α (for example, (120)) had no multiple roots. Some proofs are wrong, for example, the second given on February 22, 1838 [Ms 3616 (2), pp. 21r–21v], whereas the first proof found that day [Ms 3616 (2), p. 18v], and repeated on April 14, 1838, is essentially right.

This proof, which he published in [1838h, §9], amounted to a rather simple accounting for successive roots of the quantities V, NV', ..., which are supposed to be positive for $x = \alpha$, (120). Such an investigation shows that no two of these quantities can vanish simultaneously; in particular, $V = 0$ and $V' = 0$ can have no roots in common. As Liouville indicated in a note from April or May 1848 [Ms 3616 (2), pp. 68v–77r] and published in [1838h, §10–11] such an argument also shows that the quantity on the left-hand side of (121) has a root between two consecutive roots of V. Therefore, following Liouville, we shall concentrate on problems where the boundary condition in β is of the form

$$V_r(x) = 0 \quad \text{for} \quad x = \beta. \tag{125}$$

For the treatment of the more general boundary condition, the reader can consult [Liouville 1838h].

Following Sturm (see §18), Liouville argued as follows [1838h §14]: Since V_r has no multiple roots, the number of roots of $V_r(x)$ in (α, β) can change only if a root passes one of the two endpoints, i.e., when r has a value for which $V_r(\alpha) = 0$ or $V_r(\beta) = 0$. Now, Liouville always imposed boundary conditions on V_r at α, which allow no root $V_r(\alpha) = 0$ (e.g. conditions (120)). Therefore, it is clear that if V_n and V_{n+1} are two consecutive eigenfunctions of (119), (120), and (125), the numbers of their roots in (α, β) differ by at most one. Since $V_1(x)$ can easily be seen to have no roots in (α, β), this implies that V_n has at most $n - 1$ roots in (α, β).

54.   Thus, we are led to the problem that troubled Liouville more than any of the other problems: to show that V_n has precisely $n - 1$ roots. In his notebooks, one can distinguish at least three different methods of proof.

Liouville called the first method "the Sturmian method" [Ms 3616 (2), p. 19v] because it was adapted from the method used in [Sturm 1836a].

Liouville had already applied this method with success in July 1836 [Ms 3615 (5), pp. 18v–20r] to an alternative treatment of equation (111) with constant coefficients. Therefore, it is natural that in his very first note on the equation with variable coefficients from February 1838 [Ms 3615 (5), pp. 30v–35r] he tried this method again. Recall (§19, 20) that Sturm's idea was to show that the roots of $V_r(x)$ decreased with increasing r. This fact would obviously complete Liouville's proof since it implies that V_{n+1} has one more root (and not one less) in (α, β) than V_n. Sturm had proved the decrease of the roots by showing that $\delta V_r(x) \frac{dV_r(x)}{dx} > 0$ when $V(x) = 0$, which is in turn implied by

$$\theta = \delta V \frac{dV}{dx} - V \frac{d\delta V}{dx} > 0, \tag{126}$$

or corresponding to (49),

$$\delta \left(\frac{V'}{V} \right) > 0 \tag{127}$$

(in Liouville's memoir $K(x)$ does not depend on r). Therefore, Liouville started out investigating the quantity θ defined by (126). After some pages of calculations, he believed he had established the inequality (126) and wrote enthusiastically:

> It is really proved by this procedure that Mr. Sturm's theory extends to linear equations of all orders by the consideration of Lagrange's conjugate [adjoint] equations that I have talked about in my lectures at the *Collège de France*. This result, which Sturm, according to the first lines of his Memoir [quoted in §50], must regard as completely unexpected, is at least very remarkable. However it is important to generalize it as much as possible[25].
>
> [Ms 3615 (5), p. 32r]

However, his continued calculations led to properties that he found strange, and after 10 pages, he finally arrived at a contradiction. He felt so uncomfortable with this result that he began to reflect upon the continuous dependence of $V_r(x)$ on r, which he everywhere else considered as self-evident:

> Is the explanation to be found in the lack of continuity of the function [V] with respect to r. That also seems absurd. Nevertheless, I cannot imagine another cause. The examples are favorable.
>
> [Ms 3615 (5), p. 35r]

In the end, he saw that the beginning of the note contained a simple miscalculation that invalidated the whole argument.

55.   Liouville did not lose courage after this initial failure. On the contrary, it seems to have become almost an obsession for him to make the Sturmian method work. Thus, during February, March, and April, he made over a dozen mostly fruitless attempt to apply the method to different higher order equations with variable coefficients (cf. [Ms 3615 (5), Ms 3616 (2)]). I shall

summarize the one partly successful application of the Sturmian method in order to show how far Liouville got with this approach. In this calculation, he showed that the first root of $V_r(x) = 0$ greater than α decreases with r when $V_r(x)$ is a solution of the equation

$$V_r^{(3)}(x) + g(x)rV_r(x) = 0, \tag{128}$$

satisfying suitable boundary conditions at $x = \alpha$. He proved it on February 10, 1838, in the case where $V_r(\alpha) = 0$ [Ms 3615 (5), pp. 38v–40r], and one month later [Ms 3616 (2), pp. 58v–61r] with the general boundary-value condition

$$V_r''(x) = aV_r'(x) - bV_r(x) \quad \text{for} \quad x = \alpha, \tag{129}$$

where $a, b > 0$, and $V_r(\alpha) \neq 0$, say, $V_r(\alpha) = c > 0$. After having arrived at the main lines of a proof in a messy way, on March 10, 1838, he immediately drew up a tidy version:

> Let us return to the previous method which we have so often tried to use and which suddenly turns out to provide an unexpected success.
> [Ms 3616 (2), p. 59v]

He combined (128) with the adjoint equation

$$U_r^{(3)}(x) - g(x)rU_r(x) = 0 \tag{130}$$

to find that

$$U_r(x)V_r''(x) - U_r'(x)V_r'(x) + U_r''(x)V_r(x) = \text{const.} \tag{131}$$

Under the conditions imposed on U_r,

$$U_r'(x) = aU(x), \quad U_r''(x) = bU(x), \quad \text{for} \quad x = \alpha, \tag{132}$$

the constant in (131) is zero. By differentiation of (128), he obtained

$$\delta V_r'''(x) + g(x)r\delta V_r(x) + g(x)V_r(x)\delta r = 0, \tag{133}$$

which combined with (130) yields

$$U_r(x)\delta V_r''(x) - U_r'(x)\delta V_r'(x) + U_r''(x)\delta V_r(x) = -\delta r \int_\alpha^x g(x)U_r(x)V_r(x)\, dx. \tag{134}$$

Liouville finally combined (131) and (134) and obtained

$$U_r(x)\theta_r'(x) - \theta_r(x)U_r'(x) = \delta r V_r(x) \int_\alpha^\beta g(x)U_r(x)V_r(x)\, dx, \tag{135}$$

or

$$\left(\frac{\theta_r(x)}{U_r(x)}\right)' = \frac{V_r(x)\delta r}{(U_r(x))^2} \int_\alpha^x g(x)U_r(x)V_r(x)\, dx. \tag{136}$$

If now we choose $U(\alpha) > 0$ and $r > 0$, it is easily seen that $U(x) > 0$ and $U'(x) > 0$ for $x \in [\alpha, \beta]$. If, further, $\delta r > 0$, (136) implies that $\theta_r(x)$ increases with x in the interval $[\alpha, x_1(r)]$, where $x_1(r)$ is the first root of $V_r(x) = 0$. Since $\theta_r(\alpha) = 0$, we have

$$\theta_r(x_1(r)) > 0; \tag{137}$$

hence, by the analysis above (126), $x_1(r)$ decreases when r increases.

Liouville tried to apply the method to the second root as well, as he had done for equation (111) in July 1836 [Ms 3615 (5), p. 19v], but this time he could not carry out the generalization. Thus, the unexpected success was very limited.

56.   On February 22, 1838, Liouville [Ms 3616 (2), p. 22–23r] believed that he had another success with the Sturmian method. He considered the equation

$$(K(x)(L(x)(V'(x))')' + g(x)rV(x) = 0, \tag{138}$$

where

$$K(x), L(x), g(x) > 0 \quad \text{for} \quad x \in [\alpha, \beta].$$

By successive differentiation, he obtained the differential equation

$$(L(K(L\theta)')')' = gr\theta, \tag{139}$$

which he integrated three times to give

$$\theta(x) = F(x) + \frac{r}{L(x)} \int_\alpha^x \frac{dx}{K(x)} \int_\alpha^x \frac{dx}{L(x)} \int_\alpha^x g(x)\theta(x)\,dx. \tag{140}$$

The function $F(x)$ depends on the initial conditions at $x = \alpha$. Liouville chose the conditions

$$LV' = a(r)V, \quad K(LV')' = b(r)LV', \quad \text{for} \quad x = \alpha, \tag{141}$$

and showed that if

$$a, b > 0, \quad \frac{da}{dr} < 0, \quad \frac{db}{dr} < 0, \tag{142}$$

then $F(x) > 0$. In that case, (140) implies that $\theta > 0$, which was the desired result.

However, his derivation of (139) contains an error, for in the beginning of the proof he chose V, V_1 to be two solutions to (138), with the same value r of the parameter. At the end of the proof, however, he set $V_1 = V + \delta V$ where $\delta V = \frac{dV}{dr}\delta r$. In that case, V_1 in fact satisfies (138) for the value $r + \delta r$, instead of the value r of the parameter.

Liouville did not cross out this proof as he usually did with erroneous calculations in his notebooks. That indicates that he did not discover the flaw. Yet, his continuous search for alternative derivations shows that he was not content with the result. The reason can be that the proof does not work in the most interesting case of constant boundary conditions. In fact, if $\frac{da}{dr} = \frac{db}{dr} = 0$, Liouville would get $F(x) \equiv 0$, and so nothing could be concluded from (140).

57.   In spite of these discouraging results, Liouville apparently continued to have confidence in the Sturmian method. It was clearly the method he liked the best. Thus, he continued to use it after he had found the more successful approach (February 18), which he later chose to publish [1838h].[26] He even returned to the method in two short notes as late as 1843–1844 and 1845–1846 [Ms 3617 (4), Ms 3618 (3)].

Before I discuss the published approach, I shall mention a third method, which is in a way a geometrical version of the Sturmian method. This method, thought out on February 19, 1838, and developed in three notes during the following few days (February 19–22 [Ms 3616 (2), pp. 3v–7r, 11v–12r]), consisted of a geometric investigation of the curves $V(r,x) = 0$. Recall that the main problem is to show it impossible that a root of $V(r,x)$ in (α, β) can leave the interval at β for increasing r. Liouville neatly argued that the only way that could happen was if a branch of the curve $V(r,x) = 0$ had the form shown in (Figure 4). But, that would mean that $V(r_1, x)$ and $V(r_2, x)$ has the same number of roots in (α, β), and Liouville considered that to be absurd. Liouville found this geometric argument "very clear but difficult to draw up" [Ms 3616 (2), February 19], but obviously the "absurdity" found at the end of the proof only arises when Sturm's oscillation theorem is used. Thus, he argued in a circle.

58.   One day before Liouville conceived the geometrical argument, he had thought out the central idea of the proof of Sturm's oscillation theorem that he eventually published in [1838h]. The inspiration clearly came from Sturm's proposition N (§26), more specifically from Sturm's alternative version of Liouville's proof of the theorem (§48). Thus, his aim was to

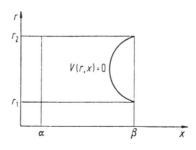

FIGURE 4

prove

Lemma III.

$$A_m r_m V_m + \cdots + A_n r_n V_n \tag{105}$$

has at least as many roots in (α, β) *as*

$$\psi = A_m V_m + \cdots + A_n V_n. \tag{61}$$

At first, Liouville had $m = 1$, but when he began his investigations of (146), he applied his results in the general case without comment.

In the note of February 18, 1838 [Ms 3616 (2), pp. 1r, v], he gave the following rigorous proof of this theorem when V_n, $n = 1, 2, \ldots$, are the eigenfunctions of equation (128) ($g(x) > 0$) with the boundary conditions

$$V' = aV, \quad V'' = bV \quad \text{for} \quad x = \alpha, \tag{143}$$

$$V'' + AV' + BV = 0 \quad \text{for} \quad x = \beta, \tag{144}$$

where $a, b, A, B > 0$.

Since ψ satisfies (144), either

1. both $\psi(\beta)$ and $\psi'(\beta)$ are of the opposite sign of $\psi''(\beta)$, or

2. only one of $\psi(\beta)$ and $\psi'(\beta)$ is of the opposite sign of $\psi''(\beta)$.

In the first case, Liouville argued as follows (Fig. 5). Let μ denote the number of roots of ψ in (α, β). Then, there are at least μ roots of ψ' in (α, β), namely, one between α and the first root of ψ, and at least one between each of the consecutive roots of ψ according to Rolle's theorem. Next, he showed that ψ'' has at least $\mu + 1$ roots in (α, β), namely, the $\mu - 1$ roots secured by Rolle's theorem plus one to the left of the first root of ψ' and one to the right of the last root of ψ'. A similar proof applies to the second case (Fig. 6). Therefore, according to Rolle's theorem, ψ''' has at least μ roots in (α, β), and by the differential equation (128) we have

$$\psi'''(x) = -g(x)(A_m r_m V_m + \cdots + A_n r_n V_n).$$

This establishes the theorem. As in [Sturm 1836b] and [Liouville 1836e], Liouville concluded that V_n has at least as many roots in (α, β) as ψ, which has in turn at least as many roots as V_m.

59.   Let us follow Liouville [1838h] and introduce the notation

$$p = \text{number of roots of } V_m,$$
$$q = \text{number of roots of } V_n,$$
$$\mu = \text{number of roots of } \psi.$$

Then, the above theorem states that

$$p \le \mu \le q. \tag{145}$$

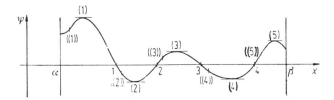

FIGURE 5

In the next note from the same day, Liouville considered the special case where $m = n - 1$ and $A_{n-1} = V_n(\alpha)$, $A_n = -V_{n-1}(\alpha)$[27]:

$$\psi(x) = V_n(\alpha)V_{n-1}(x) - V_{n-1}(\alpha)V_n(x). \tag{146}$$

According to (143), this particular function has $\psi''(\alpha) = 0$, and therefore ψ''' has an extra root between α and the first root of ψ'' in (α, β). Hence,

$$v(x) = V_n(\alpha)r_{n-1}V_{n-1}(x) - V_{n-1}(\alpha)r_n V_n(x)$$

has at least $\mu + 1$ roots. The second inequality of (145) applied to $v(x)$ thus yields $\mu + 1 \leq q$, and the first inequality applied to ψ yields $p \leq \mu$. Therefore, $p + 1 \leq q$. Since V_1 has no roots in (α, β), V_n has at least $n - 1$ roots in (α, β). On the other hand, Liouville knew that V_n has at most $n - 1$ roots in (α, β), and he had completed the desired proof of Sturm's oscillation theorem.

60. Four days after having designed the above proof, Liouville generalized it to equation (138) with variable coefficients and with the boundary conditions (143)-(144) and $V(\alpha) = 1$ [Ms 3616 (2), February 22]. In this general case, Liouville counted roots of U, U', and $(LU')'$ by reasoning as he had done a few days earlier. However, this time there is a flaw in the argument because one cannot conclude that $(LU')'$ has a root between α

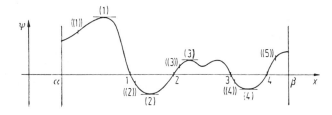

FIGURE 6

and the first root of U'. At first, Liouville did not see the mistake, but later he seems to have discovered it. At least he altered the boundary conditions in the published version of the proof [1838h] from the conditions (143), (144), and $V(\alpha) = 1$, which made V, V', V'' positive at α to the conditions (120) and (121), which made V, V', $(NV')'$ positive at α. Under the new conditions, Liouville's argument of February 22, 1838, is correct. Before, publishing the memoir [1838h] Liouville showed parts of it to Sturm, who suggested that the above simple proof of Lemma III by Rolle's theorem should be replaced by a more elegant argument resting on a generalization of Fourier's theorem on the number of roots of a function [Liouville 1838h, §17]. With this revision, the proof of February 18–22 of Sturm's oscillation theorem was published in [Liouville 1838h, §17].

61.   The only published result of Liouville's hard work during February, March, and April of 1838 in the field of higher order Sturm-Liouville theory was the paper *Premier Mémoire sur la Théorie des Équations différentielles linéaires et sur le développement des Fonctions en séries* from December [1838h] and a brief note published in May that year [1838c] announcing the theory later to appear. Let me briefly summarize the contents of [Liouville 1838h]. First, the solutions of (119) and (120) are found by successive approximation, and it is shown, by comparison with the equation with constant coefficients, that (119)-(121) have infinitely many positive eigenvalues. Then, it is shown that V_r has only simple roots, and the crucial proof of Sturm's oscillation theorem is given. Next, comes the argument for Lemma II (see §45) and the introduction of the dual problem (122)-(124) followed by the proof of the biorthogonality (114). Finally, it is proved that the Fourier series (118) of f has the value f if it converges, and Liouville argues that all eigenvalues are positive.

FURTHER GENERALIZATIONS

62.   The title of the memoir, *Premier Mémoire*, indicates that Liouville had planned a more comprehensive treatment of the generalized Sturm-Liouville theory. A gigantic project appears from the announcement of his new results in [1838c]:

> As soon as the abundance of material will allow me to occupy a sufficient space in this Journal, I shall hasten to publish the Memoir whose principal results I have indicated here, and which is, by the way, in my eyes only a small part of a very long work which I have embarked on concerning the general theory of differential equations and on the development of functions in series.
>
> [Liouville 1838c, p. 256]

In some of Liouville's notes, one can even get some idea of his general plan for this larger project. For example, on July 3, 1838, he began to reflect about the presentation of the material and remarked that in the first memoir the principal aim must be very clear; therefore, unnecessary difficulties, such as problematic boundary conditions, should be avoided. Here, he alluded to the boundary values mentioned in a note from March 31, where he suggested setting U or U' equal to zero at three different points. Such subtleties were apparently postponed to the later papers in the sequence. On August 5–10, 1838, Liouville drew up a few more points for his programme of research:

> Our studies of the general properties of integrals of linear equations gain more importance every day; it is necessary to deal carefully with
>
> 1° the convergence of the series to which our analysis leads us,
>
> 2° the study of simultaneous equations $\&_c\&_c$

$$\frac{dU}{dx} + rV = 0$$

$$\frac{dV}{dx} - rU = 0.$$

[Liouville Ms 3616 (2), p. 87v,[147]]

The question of convergence of the Fourier series (118), mentioned in 1°, was left open in both [1837a] and [1838h]. Liouville had already stressed its importance in a note from February 22, 1838, saying: " It is quite indispensable to deal very seriously with the convergence of these series" [Ms 3616 (2), p. 25r], but he left only one (or perhaps two) inconclusive notes on the question from February 23 and March 5, 1838, of which the first also contains an unsuccessful attempt to generalize his theory to the two dimensional Laplacian [Ms 3616 (2), pp. 32, 36]. In fact, U. Haagerup has pointed out that even for the boundary-value problem (111)-(113) with constant coefficients, the Fourier series converges only for very special functions f (cf. [Lützen 1984a, Appendix]).

The theory of simultaneous equations mentioned in 2° was studied in several particular cases in notes from March 11, 1838 (two second-order equations), August 5–10, 1838 (two and three first-order equations) [Ms 3616 (2), pp. 51v, 87v, 88] and one as late as March 1839 (three second-order equations) [Ms 3616 (5), pp. 8v–9v].

63.   On July 3, 1838, Liouville planned another "great extension" [3616 (2), p. 80v]. In the note, he seems to suggest that the expressions NV', $M(NV')'\ldots$ of (119) and (120) be replaced by

$$ll\nabla V = MV' + NV, \tag{148}$$

$$\nabla^2 V = M_1(\nabla V)' + N_1\nabla V - P_1 V, \tag{148}$$

$$\nabla^3 V = M_2(\nabla^2 U)' + N_2\nabla^2 U - P_2\nabla U \tag{148}$$

$$\vdots \tag{148}$$

Half a year later, on January 3, 1839, he began to study Sturm-Liouville theory for complex functions, and he intended to "follow this theory which may become very important, introduce into our functions a parameter; see the corresponding properties of the functions and their integrals in a contour, &$_c$&$_c$." [Ms 3616 (3), p. 6r].

Later the same month, he began a study of the eigenfunctions of the wave equation, or rather of the solutions of the equation:

$$\frac{d^2u}{dt^2} = a^2\frac{d^2u}{dx^2} + \lambda(x)u.$$

He found a solution by successive approximation:

$$u = \psi(t, x) + \psi_1(t, x) + \psi_2(t, x) + \cdots,$$

where

$$\frac{d^2\psi}{dt^2} = a^2\frac{d^2\psi}{dx^2}, \quad \frac{d^2\psi_1}{dt^2} = a\frac{d^2\psi_1}{dx^2} + \lambda\psi, \text{ etc.}$$

He chose $\psi = \frac{1}{2}(f(x + at) + f(x - at))$ and then determined

$$\psi_1 = \frac{1}{2a}\int_0^t dt' \int_{x-at'}^{x+at'} \lambda\psi(t - t', \gamma)\, d\gamma, \text{ etc.}$$

He even went so far as to prove the convergence of the series $\psi + \psi_1 + \psi_2 + \cdots$ by replacing λ with its maximum. "This analysis must apparently be useful in a great number of cases." Thus, Liouville concluded his note [Ms 3616 (3), pp. 10v–12r], and he never went on to study Sturm-Liouville-like problems for the wave operator.

However, a note, probably from 1842 [Ms 3617 (2), pp. 66v–67r], shows that Liouville continued to be occupied with Sturm-Liouville theory many years after 1838. This note contains a few results on "the properties of the powers or the products of the functions V of Mr. Sturm." Again, he put off further investigations though "all this seems to merit a serious study."

64.   In spite of his extensive research programme for the generalized Sturm-Liouville theory, Liouville never composed the subsequent memoirs in the series he had planned in 1838. What was the reason for his interruption of the series? An explanation cannot be found in a new, absorbing interest taking all of Liouville's time; he did turn to many other problems of analysis, algebra, mechanics, and celestial mechanics, but from his published papers and his notebooks it appears as if none of these interests was so strong that it could have taken him off the track. A more likely explanation will present itself after a view of the subsequent development of Sturm-Liouville theory. Such a view will be given in the concluding section.

Concluding Remarks

65.   The subsequent development of the Sturm-Liouville theory can be roughly divided into two mutually interacting categories: generalizations

and rigorizations. The following brief summary of the late 19th-century and early 20th-century advances in these two directions makes no claim to be complete. More details and references to primary sources can be found in the relevant articles of the *Encyklopädie der Mathematischen Wissenschaften* [Bôcher 1899–1916] and [Hilb and Szász 1922].

Around 1880, the problem of vibrating rods and plates led Lord Rayleigh [1877], G. Kirchhoff [1879], and others to develop theorems similar to Sturm's for higher order boundary-value problems. At the same time, the Sturm-Liouville theory of singular differential equations, where, for example, $k(x)$ in (1) has zeroes in the interval $[\alpha, \beta]$, was studied first in special cases such as the Bessel equation (16) (e.g. [Schläfli 1876]) and later in more generality by Bôcher and others. A third kind of generalization was undertaken by F. Klein [1881], who studied ordinary differential equations with several parameters. By allowing the solutions to have prescribed types of infinities at the boundary of the intervals considered, he was able to adapt his theory to the Lemé functions. Thereby, he combined the two hitherto entirely separate theories of boundary-value problems and polynomial solutions of ordinary differential equations. Finally, Poincaré initiated spectral theory of partial differential operators with his study of the Laplace operator [1894].

66.   The rigorization of Sturm-Liouville theory took place on two levels. To adjoin to Sturm's and Liouville's essentially correct arguments the lacking proofs of continuity, differentiability, and uniformity constituted the simpler task. It was undertaken piecewise in several papers from the 1890s and systematically carried through for the Sturmian theorems by Bôcher [1898, 1899]. Other mathematicians took up the much more difficult problem of finding a rigorous replacement for Liouville's essentially wrong proof that a Fourier series converges to the function that gives rise to it. H. Heine [1880] and U. Dini [1880] applied Cauchy's theorem of residues to this problem and Poincaré [1894] developed this idea in his proof of a general theorem on expansion in eigenfunctions for the Laplace operator. Poincaré also used complex function theory. An improvement of this method was carried over to the Sturm-Liouville problem by Stekloff [1898] to get the first rigorous proof that a twice differentiable function satisfying boundary conditions (2) and (3) could be expanded in a Fourier series. By combining the method of Poincaré and Stekloff, based on the theory of functions, with Liouville's original second proof of convergence [1837e] (see §44), A. Kneser [1904] succeeded in proving the expansion theorem for any piecewise continuous function of bounded variation (though taking the values $\frac{1}{2}(f(x+0)+f(x-0))$ at points of discontinuity). He here provided a parallel to the classical result of Dirichlet [1829] for trigonometric Fourier series.

67.   An entirely new approach to Sturm-Liouville theory was opened up by the rise of the general theory of integral equations, beginning at the end of the 1890s. The correspondence between spectral theory for differential

and integral equations is established by the Green's function. If r is not an eigenvalue of (1)-(3) and $\varphi(x) > 0$ in $[\alpha, \beta]$, there is a solution u of

$$(k(x)u'(x))' + (g(x)r - l(x))u(x) = -\varphi(x) \quad \text{for} \quad x \in (\alpha, \beta) \qquad (149)$$

satisfying boundary conditions (2) and (3). This solution can be expressed in the form

$$u(x) = \int_\alpha^\beta G(r, x, \xi)\varphi(\xi) \, d\xi. \qquad (150)$$

$G(r, x, \xi)$ is called Green's "function." If r_n is an eigenvalue to boundary-value problem (1)-(3) and if V_n is the corresponding eigenfunction, it follows from (149) and (150) that

$$V_n = (r_n - r) \int_\alpha^\beta g(\xi)G(r, x, \xi)V_n(\xi) \, d\xi.$$

Therefore, the eigenvalues and eigenfunctions of (1)-(3) correspond to eigenvalues and eigenfunctions of the integral operator

$$\varphi \to \int_\alpha^\beta g(\xi)G(r, x, \xi)\varphi(\xi). \qquad (151)$$

This idea is found in the work of Poincaré [1894] on the Laplace equation. However, its full importance was not revealed until Hilbert [1904/1910] and Schmidt [1907] proved that any continuous function of the form

$$f(x) = \int_\alpha^\beta g(\xi)G(r, x, \xi)\varphi(\xi)$$

can be expanded in a Fourier series of eigenfunctions to the integral operator (151). When applied to the Sturm-Liouville problem, this theorem almost immediately gives the expansion theorem for twice differentiable functions, which satisfies the boundary conditions. Thus, as long as mainly pointwise convergence was considered, Hilbert and Schmidt's method could not compete with Kneser's, but later in the 20th century when mathematicians became interested in L^2 convergence, the new methods proved more valuable. Today, the Sturm-Liouville theory is treated in close connection to the general theory of operators in Hilbert space, which developed from Hilbert and Schmidt's investigations.

68.   Of the generalizations listed in §65, only the study of higher order equations belongs to Liouville's research programme. He had also touched upon spectral theory for the Laplacian in his remark of 1830 about heat conduction in bodies of more than one dimension, cf. Chapter XV, §12a, but he had given it up because he could not even find the stationary temperature distribution explicitly. A further comparison of Liouville's research

programme with the late 19th century development is interesting because it may throw some light on Liouville's failure to carry out his plans.

What distinguishes Liouville from his successors is not so much the difference between their problems as their different motives for studying them. The generalizations of the Sturm-Liouville theory obtained in the late 19th-century were motivated by physics or by a desire to link up different trends of analysis. Liouville, on the other hand, appears in this case as a blind generalizer who wanted to extend his and his friend's new theory as far as possible just for the sake of generality. Of course, he may have had other motives, which he did not reveal to his readers or to his notebooks, but at least in the case of third-order equations his only motivation seems to have been Poisson's remark about the difficulty of the problem (§50). Thus, lack of genuine motivation may have been a reason why Liouville lost interest in the problems.

The missing physical inspiration had a serious consequence, for in analysis, and particularly in the type of analysis cultivated in early 19th century France, theorems and methods had generally been suggested by physical reality. For example, Sturm and Liouville's research on second-order differential equations had been guided by their knowledge of heat conduction and vibratory motion. In the extended research programme, however, Liouville did not have such physical guidance for what were the important questions and even what were the correct theorems to prove. A manifestation of Liouville's failure of intuition in the broader field was his belief in the convergence theorem for Fourier series.

To conclude, Liouville's lack of physical motivation, his resulting failing intuition, and more specifically, his inability to prove the central expansion theorem for higher order equations explain at least in part why the first chapter of the history of the Sturm-Liouville theory ended in 1838, only a few years after it had started. Yet, during these few years, Sturm and Liouville had advanced the theory to such a degree of completeness that no substantial additions were made during the next half century.

XI

Figures of Equilibrium of a Rotating Mass of Fluid

0. After Jacobi's surprising discovery in 1834 that the rotating triaxial ellipsoids of fluid could be in equilibrium, Liouville began a study of the properties of these figures and of the well-known equilibrium ellipsoids of revolution found by Maclaurin. He published six papers on the question, but only a small fraction of his most far-reaching investigations on the stability of the figures of equilibrium, made during the last months of 1842, appeared in print.

This chapter contains an almost complete reconstruction of Liouville's theory of stability in its historical context. It is based on two manuscripts published in [Lützen 1984b] and on numerous calculations in Liouville's notebooks. Liouville's idea is to determine whether the "force vive" (the kinetic energy) of a perturbed equilibrium figure has a maximum in the equilibrium state. That he did by expanding the perturbations in a series of Lamé functions.

Being largely unpublished, Liouville's theory of stability of equilibrium figures had little impact. Nevertheless, the published notes were known to one of his successors, Liapounoff, and the other successor, Poincaré, was indirectly influenced by Liouville's work on Lamé functions. Liouville had published this work without revealing that it had been developed as a tool for his investigations of the stability of rotating ellipsoids.

Prehistory; Maclaurin Ellipsoids

1. Which are the possible figures of equilibrium of a homogeneous mass of fluid rotating around an axis through its center of gravity under the influence of no forces beyond centrifugal force and the mutual gravitational attraction of its molecules when it is assumed that the whole fluid rotates as if it were solid? This is the question to which Liouville devoted much of his energy during the period 1834–1843. It had arisen in the late 17th century in connection with the problem of the figure of the earth. The history of this problem up to the time of Laplace has been treated in detail by Todhunter [1873] (see also [Szabó 1977]) and will therefore be mentioned only briefly here. Further interesting historical and scientific information can be found in [Oppenheim 1919], [Lichtenstein 1933], [Appell 1932–1937], and [Chandrasekhar 1969].

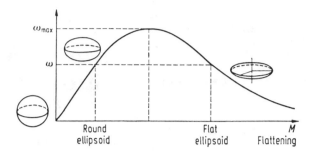

FIGURE 1

In the third book of the *Principia* [1687], propositions 18 and 19, Newton found the eccentricity of the earth under the assumption that its figure was a homogeneous ellipsoid of revolution, and in 1740 Maclaurin proved that such a figure would indeed be in equilibrium (detailed argument in [Maclaurin 1742]). Therefore, ellipsoids of revolution in equilibrium are known as Maclaurin ellipsoids. However, the theoretically obtained value for the eccentricity was not corroborated by measurements, and so in order to account for the discrepancies, Clairaut initiated the study of equilibrium figures of heterogeneous fluids [1743]. These investigations, important to both astronomy and geodesy, were carried further by Lagrange and Laplace, but for brevity, we shall restrict the discussion in this chapter to homogeneous fluids.

The latter part of the 18th century witnessed a clarification of the properties of the Maclaurin ellipsoids. First, d'Alembert [1773] pointed out that for a given mass of fluid and for each value of the angular velocity ω below a certain maximal value ω_{max} there exist precisely two Maclaurin ellipsoids, one "flat ellipsoid" with a large eccentricity e and one "round ellipsoid" with a small eccentricity. For $\omega = \omega_{max}$ the flat and the round ellipsoid coincide, and for $\omega > \omega_{max}$, a Maclaurin ellipsoid does not exist.

This strange behavior was explained by Laplace in the *Mécanique Céleste* [1799–1825, livre 3, §21, and earlier in 1784, sect. 13]. He observed that even though the angular velocity ω for the flat ellipsoids decreases as the values of the eccentricity increase, the moment of inertia I increases so rapidly that the angular momentum $M = \omega I$ increases to infinity (Fig. 1). Therefore, there is precisely one Maclaurin ellipsoid for each value of the physically more significant quantity M. Legendre [1784–1787] and Laplace [1782–1785, ch.4, 1799–1825, livre 3, §§25,26] also established that there are no almost spherical equilibrium figures other than the Maclaurin ellipsoids and in the *Mécanique Analytique*, 2nd ed., Lagrange [1811–1815, vol. 1, §§199–204] presented a "proof" that rotating ellipsoids in equilibrium must possess rotational symmetry.

Therefore, it came as a surprise when C. G. J. Jacobi on October 20, 1834, in a letter to the *Académie des Sciences* in Paris, announced that he had found a new family of ellipsoids of equilibrium having three different axes [Jacobi 1834].

Jacobi Ellipsoids

2. In the letter to the *Académie des Sciences*, Jacobi announced that he would "resume the *Mécanique celeste* in the pitiable state where Laplace has left it." Although Liouville was later to become the principal link between French and German mathematicians, he had strong patriotic feelings and regarded Jacobi's remarks as a challenge to French mathematics [Liouville 1834d, p. 290, footnote]. He immediately took up the challenge, and as soon as the following week he proved to the *Académie des Sciences* [Liouville 1834d] that Jacobi's theorem was only an easy consequence of formulas in Laplace's *Mécanique Céleste*. Liouville based his argument particularly on two of Laplace's results, the first being the fact that a mass of fluid is in equilibrium if and only if the effective force acting at the surface of the fluid is perpendicular to the surface [Laplace 1799–1825, livre 1, §§17 and 36]. (Huygens had used this idea as early as 1681 [Szabó 1977, pp. 225–226].) The second is the celebrated result [Laplace 1799–1825, livre 3, §§3–4] that inside the ellipsoid E, with the equation

$$\frac{x^2}{k_2^2} + \frac{y^2}{k_1^2} + \frac{z^2}{k^2} = 1 \tag{1}$$

having the axes $k < k_1 < k_2$, the gravitational force is of the form

$$\mathbf{F}_g = (A'x, B'y, C'z), \tag{2}$$

where A', B', C' are constants. They can be expressed in terms of

$$\lambda_1^2 = \frac{k_1^2 - k^2}{k^2}, \quad \lambda_2^2 = \frac{k_2^2 - k^2}{k^2}, \quad G^2 = (1 + \lambda_1^2 x^2)(1 + \lambda_2^2 x^2), \tag{3}$$

and the mass M of the ellipsoid as

$$A' = \frac{3M}{k^3} \int_0^1 \frac{x^2 \, dx}{(1 + \lambda_2^2 x^2)G},$$

$$B' = \frac{3M}{k^3} \int_0^1 \frac{x^2 \, dx}{(1 + \lambda_1^2 x^2)G},$$

$$C' = \frac{3M}{k^3} \int_0^1 \frac{x^2 \, dx}{G}. \tag{4}$$

Adding the centrifugal force under rotation around the z-axis, $\mathbf{F}_c = (-\omega^2 x, -\omega^2 y, 0)$, Liouville obtained the effective force

$$\mathbf{F}_e = ((A' - \omega^2)x, (B' - \omega^2)y, C'z), \tag{5}$$

which, in equilibrium, must be perpendicular to the surface. This condition gives two transcendental equations in k, k_1, k_2 and ω from which Liouville eliminated ω and got

$$(\lambda_2^2 - \lambda_1^2)\left[(1 + \lambda_1^2)(1 + \lambda_2^2)\int_0^1 \frac{x^4\,dx}{G^3} - \int_0^1 \frac{x^2\,dx}{G}\right] = 0. \tag{6}$$

Liouville observed that (6) can be satisfied in two ways. Either the first parenthesis is zero, in which case $k_1 = k_2$ and the ellipsoid is a Maclaurin ellipsoid of revolution, or the second parenthesis vanishes, in which case the elliptic equation

$$(1 + \lambda_1^2)(1 + \lambda_2^2)\int_0^1 \frac{x^4\,dx}{G^3} = \int_0^1 \frac{x^2\,dx}{G} \tag{7}$$

allows one to determine the least axis k for given arbitrary values of the two axes k_1, k_2 of the equatorial ellipse. Putting $x = \frac{k}{\sqrt{\alpha + k^2}}$, one finds

$$\int_0^1 \frac{x^2\,dx}{G} = \frac{k^3}{2}\int_0^\infty \frac{d\alpha}{(k^2 + \alpha)H}, \tag{8}$$

and

$$(1 + \lambda_1^2)(1 + \lambda_2^2)\int_0^1 \frac{x^4\,dx}{G^3} = \frac{k_1^2 k_2^2 k}{2}\int_0^\infty \frac{d\alpha}{(k_1^2 + \alpha)(k_2^2 + \alpha)H}, \tag{9}$$

where $H = \sqrt{(\alpha + k^2)(\alpha + k_1^2)(\alpha + k_2^2)}$, so that equation (7) can be written as

$$k_1^2 k_2^2 \int_0^\infty \frac{d\alpha}{(k_1^2 + \alpha)(k_2^2 + \alpha)H} = k^2 \int_0^\infty \frac{d\alpha}{(k^2 + \alpha)H}. \tag{10}$$

This is equivalent to the formula announced in Jacobi's letter. Written in the form

$$\int_0^\infty \frac{d\alpha}{\left(1 + \frac{\alpha}{k_1^2}\right)\left(1 + \frac{\alpha}{k_2^2}\right)H} = \int_0^\infty \frac{d\alpha}{\left(1 + \frac{\alpha}{k^2}\right)H},$$

this equation shows that $\frac{1}{k^2} > \frac{1}{k_1^2} + \frac{1}{k_2^2}$, which implies that the axis of rotation k is always the shortest axis. It also shows that the Jacobi ellipsoids are always far from spherical symmetry so that Laplaces's uniqueness theorem mentioned in §1 is not violated. Liouville also deduced Jacobi's formula for $\omega(k_1, k_2)$:

$$\omega^2 = 2\pi k k_1 k_2 \int_0^\infty \frac{\alpha\,d\alpha}{(k_1^2 + \alpha)(k_2^2 + \alpha)H}. \tag{11}$$

3.   While in 1834 Liouville tried to reestablish the virtues of Laplace's treatment of the figure of a rotating mass of fluid, three years later [1837g], he launched his own criticism of one of the two proofs given in the *Méca-*

nique Céleste [Laplace 1799–1825, livre 3, §26] for the fact that there are no almost spherical equilibrium ellipsoids other than the Maclaurin ellipsoids. With his thorough knowledge of integral equations [cf. Lützen 1982b], Liouville had detected a mistake in Laplace's treatment of such an equation, and succeeded in establishing an alternative proof on the basis of a theorem [Liouville 1837b] that had grown out of his work on the Sturm-Liouville theory (cf. Chapter X(49)). Later the same year, Poisson [1837b] showed how Laplace's mistake could be corrected with simpler means (cf. also [Wantzel 1839]).

In 1842, his discontent with Laplace's treatment of the motions of fluids had grown so strong that he wrote in his notebook:

> Laplace's analysis is very muddled.
>
> [Ms 3617 (3)]

While Liouville became increasingly more critical of Laplace's analysis, his esteem for Jacobi's discovery of the three axial ellipsoids gradually grew. In Chapter II, we saw that initially Liouville had expressed his doubts about the importance of Jacobi's results. In [1839b, p. 169], on the other hand, he described Jacobi's theorem as a "remarkable theorem" and in [1843–1984] as "Mr. Jacobi's beautiful discovery."

This revaluation was a result of the continued investigation of the Jacobi ellipsoids, which he undertook at Poisson's request after having presented his proof to the *Académie*. At first, however, Liouville did not find time enough to pursue the subject for long:

> However, since other works have not allowed me to treat it with all the care it merits, I felt forced to leave it.
>
> [Liouville 1839b, p. 169]

He would probably have kept his results to himself if it had not been for a paper of 1838 in which the British mathematician J. Ivory (1765–1842) reached results contradicting some of Liouville's "simple results." Ivory [1838] discussed the relation between the quantities $\tau = \lambda_2 - \lambda_1$ and $p = \lambda_1 \lambda_2$ (cf. (3)). In terms of these variables, Jacobi's determining equation, (7) or (10), takes the form

$$\int_0^1 \frac{x^2(1-x^2)(1-p^2x^2)\,dx}{[(1+px^2)^2 + \tau^2x^2]^{\frac{3}{2}}} = 0. \tag{12}$$

This transcendental equation has one solution p for each value of $\tau^2 \in [0, \infty]$, and Ivory claimed that p decreased from a maximal value p_0 to 1 when τ^2 increased from 0 to ∞. Liouville, on the other hand, reached the opposite conclusion that, when τ^2 increased from 0 to ∞, p increased from a minimal value p_0 to ∞. He published his proof in [1839b] and at the same time pointed out Ivory's error. Ivory immediately admitted his mistake and corrected it in a note [1839].[2]

4.   Neither Ivory nor Liouville discussed equation (11), determining the angular velocity. Jacobi saw this shortcoming and posed the problem of a simultaneous investigation of (10) and (11) to his student C. O. Meyer, who published his analysis in [1842] (see, in particular, p. 44).He found that for a given mass of fluid there is precisely one Jacobi ellipsoid for each value of the angular velocity ω less than or equal to a given ω'_{max}. For $\omega > \omega'_{max}$, no Jacobi ellipsoid exists. When $\omega = \omega'_{max}$, the Jacobi ellipsoid has $k_1 = k_2$, i.e., it is also a Maclaurin ellipsoid of revolution. When ω decreases, the Jacobi ellipsoid becomes flatter (k decreases) and the equatorial ellipse becomes more and more eccentric. When ω tends to zero, the axis of rotation k tends to zero and the medium axis k_1 becomes almost as small as $k \left(\frac{k_1}{k} \to 1 \right)$, whereas the major axis k_2 tends to infinity. In the limit $\omega = 0$, the Jacobi ellipsoid is therefore an infinitely long infinitely thin circular cylinder.

Since ω'_{max} is less than the maximal angular velocity of the Maclaurin ellipsoid, there are three equilibrium ellipsoids for $0 < \omega < \omega'_{max}$ (two of revolution and one with three different axes) and two equilibrium ellipsoids for $\omega'_{max} \leq \omega < \omega_{max}$ (both of revolution), whereas the fluid cannot form an equilibrium ellipsoid with an angular momentum greater than ω_{max}.

In a note read before the *Académie des Sciences* in January 1843, Liouville [1843e] carried further Meyer's investigation. He used the techniques of Meyer, whom he praised for his sagacity, to discuss the shape of the Jacobi ellipsoid as a function of its angular momentum. Here, he was following Laplace (cf. §1). He found that for a given mass of fluid ω is a decreasing function of M, which means that the Jacobi ellipsoid has a minimal value of M for which it has rotational symmetry and that it becomes long and cigar-shaped when M increases.[3] We can represent the results graphically (see Fig. 2):

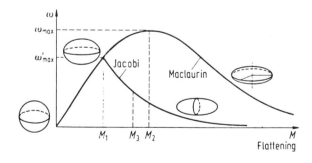

FIGURE 2

Stability of Equilibrium Figures

5.   These results concerning the Jacobi ellipsoid made only a small part of an investigation of rotating figures of equilibrium which Liouville had undertaken in the last half of 1842 during his yearly stay at his house in Toul. There he had taken up the new question of the stability of figures of equilibrium, that is, whether the figure after a small perturbation will remain close to the equilibrium state, like a rod hanging vertically from its upper end (stability), or whether it will diverge more and more from the equilibrium state, like a rod standing on its lower end (instability). The resulting theory was by far Liouville's most original and far-reaching contribution to the theory of figures of equilibrium; it clarified the physical principles and arrived, by new mathematical techniques, at new and surprising results. Still, only a very small part of it was ever published.

SOURCES

6.   The rest of this chapter deals with Liouville's investigation of stability, beginning with a discussion of the sources.
   Liouville informed the *Académie des Sciences* about his new results in two addresses, one given at the session of November 14, 1842, just after he had returned from Toul, and the second on February 13, 1843. Of these, the first, *Sur la stabilité de l'équilibre des mers*, was published in the *Comptes Rendus* [Liouville 1842h], whereas the second *Recherches sur la stabilité de l'équilibre des fluides* received only a brief mention [Liouville 1843f], which ended:

> We shall return to this communication when Mr. Liouville has presented his entire work to the *Académie*.

However, Liouville did not pursue the subject in the *Académie* and the *Comptes Rendus* never returned to it.
   The two published notes give only an incomplete presentation of Liouville's results. The first talk is mainly historical, setting the question in its proper context, and the second brief note mentions only one of Liouville's results, though admittedly the most striking one. However, the full text of Liouville's second speech [1843–1984], preserved in his *Nachlass* and printed in [Lützen 1984b Appendix B], contains a clear statement of his results.

7.   The talks indicate only key ideas in the method used by Liouville in the deduction of his results, but fortunately three other sources make possible a rather complete reconstruction of the proof. The first is the published paper *Formules générales relatives à la question de la Stabilité de l'équilibre d'une masse liquide homogène douée d'un mouvement de rotation autour d'un axe* [1852f], containing formulas for the kinetic energy of a slightly perturbed equilibrium ellipsoid.

According to Liouville [1852f, p. 164, footnote], this paper is an "extract of the first paragraph of the Memoir *Sur la stabilité de l'équilibre des mers*, communicated to the *Académie des Sciences* on November 14, 1842." Unfortunately, the rest of this memoir, which Liouville presented to the *Académie* with his first talk, seems to have been lost.

Instead, a preliminary version of the first address [Liouville 1842–1984] to the *Académie des Sciences*, dated Toul, November 2nd, has been preserved in Liouville's *Nachlass*. The reason why Liouville did not read this note to the *Académie* is explained by the remark he has written with pencil on the first page of the manuscript:

> Bad, with things that are even false, has been entirely redone.

Thus, Liouville discovered a mistake in his theory some time after November 2nd, but before November 14th, at which date he presented his corrected version to the *Académie*. In spite of the mistake, to which we shall return in §30, this first version of the talk is very interesting because, being more technical than the speech actually given on November 14th, it reveals a lot about Liouville's methods. One can also choose to say that it is interesting because of the mistake, since its shows something about the formation of the theory. For these reasons, I reproduced the note in [Lützen 1984b Appendix A].

8.   Being composed as an address to the *Académie*, Liouville's manuscript [1842–1984] naturally contains only the broad lines in the argument. The details must be filled in with the aid of various notes in Liouville's notebooks. Four of these, having the codes [Ms 3616 (3), 3617 (2), (3), and (4)], contain calculations and notes pertaining to the present subject. These notes show us Liouville at work. Unfortunately, a precise dating of the notes is impossible. For example, [Ms 3616 (3)] starts with a note dated January 1839 and contains towards the end a letter from 1854. The pages in between, which contain the notes we are interested in, are not dated. The notebook [Ms 3617 (2)] is dated Toul 1842, and [3617 (3)], Toul, July 1842, both on the first page, and it is clear from the contents that Liouville used them alternately. On the whole, the chronology can mostly be determined from the mathematical content.

The notes on stability of rotating masses of fluid are found mixed with notes on various other mathematical subjects, and in some cases it is hard to make out the result or the aim of a particular calculation. Hence, I may have overlooked notes pertaining to the problem of stability, but otherwise I have based the following account of Liouville's treatment of stability of masses of fluid on all the published and unpublished material mentioned.

RESULTS ON STABILITY

9.   In the spring of 1842, Liouville began a critical rereading of Laplace's *Mécanique Céleste*. He worked on the stability of a particular solution to

the three-body problem (results presented to the *Académie* in April of 1842 [1842g]), and over the summer and fall, he filled several notebooks with notes on, for example, the instability of Saturn's ring, considered as a fluid [Ms 3617 (2), pp. 21v–22r], on homogeneous nebulas [Ms 3617 (2), pp. 62v–66r], and on the stability of the equilibrium of the seas (for Liouville's other works in celestial mechanics see Chapter XVI §§9–17). Laplace's treatment of the last mentioned problem had originally been published in [1782–1785, ch. 5], and was reproduced with slight improvements in the fourth book of the *Mécanique Céleste*. There, Laplace had shown that the equilibrium figure of a thin layer of fluid on a rigid, almost spherical kernel is stable if and only if the density of the fluid D_f is less than the density of the kernel D_k.

> When I began my work [Liouville wrote], I first had as my only aim to simplify the rather long calculations of the *Mécanique celeste*; I think I have achieved this.
>
> [Liouville 1842h, p. 905]

Soon, however, Liouville widened the scope of the investigations to include the case where the kernel is far from spherical symmetry and the case of an entirely fluid ellipsoid of two or three unequal axes. "I have also treated these interesting questions, which I think are entirely new" [Liouville 1842h, p. 906].

In fact, these questions were not entirely new, for in [1773, §§27–31], d'Alembert had discussed the stability of the Maclaurin ellipsoids. He thought he had proved that the "round" ellipsoids were stable and the "flat" ellipsoids were unstable. However, his analysis is insufficient for two reasons. First, he considered only such perturbations as leave the fluid in the shape of an ellipsoid of revolution; second, he held constant the angular velocity ω rather than the angular momentum M. D'Alembert's investigation is discussed at great length in [Todhunter 1873, Vol. 1, §582] and in [Oppenheim 1919, §18].

10.   In the first address to the *Académie* [Liouville 1842h], Liouville mainly talked about the connection between Laplace's research and the subject matter of his new theory, and he postponed the announcement of his results to the second address [Liouville 1843–1984]. Then he affirmed Laplace's result on stability of liquid on an almost spherical kernel and stated that for kernels differing much from spherical symmetry, the condition $\frac{D_f}{D_k} < 1$ is still necessary but not sufficient. The quotient $\frac{D_f}{D_k}$ has to be less than a certain transcendental function in order to secure stability. However, "the most remarkable theorem," and the only one to be printed in the *Comptes Rendus* ([Liouville 1842h], from which this is quoted) "concerns Jacobi's ellipsoids of three different axes. The equilibrium state of these ellipsoids is always a stable state."

In fact, Liouville had, in the talk, given a complete solution to the problem of stability of entirely liquid ellipsoids, also including Maclaurin ellipsoids. The latter are stable when they are less flat than the Jacobi ellipsoid of revolution, which is at the branch point of the Maclaurin and Jacobi ellipsoids, and unstable if they are flatter than that figure. That means that there is precisely one stable ellipsoid for each value of the angular momentum M. When $M \leq M_1$ (Fig. 2), the Maclaurin ellipsoid is stable, and for $M > M_1$, the Jacobi ellipsoid is stable, and the Maclaurin ellipsoid is unstable. In §1, it was mentioned that Lagrange believed that there was precisely one equilibrium ellipsoid for each value of M, namely, the Maclaurin ellipsoid. This beautiful uniqueness theorem had been destroyed by Jacobi, but Liouville emphasized that he had reestablished it as far as stable ellipsoids are concerned.

Liouville's results were in surprising contradiction to the beliefs generally held until then. For example, in Laplace's discussion of the angular momentum of Maclaurin ellipsoids [1799–1825, livre 3 §21], he stated that any fluid mass, when left to itself, would after many oscillations attain the shape of the Maclaurin ellipsoid with angular momentum equal to the initial angular momentum of the fluid. Thus, Laplace was convinced that all Maclaurin ellipsoids were stable. Even after Jacobi's discovery of the tri-axial ellipsoid, it was believed that the stable figures were the Maclaurin ellipsoids. For example, Ivory ends his fallacious analysis of the Jacobi ellipsoids with the words:

> It would be superfluous to pursue this investigation further and a mere waste of labour to seek the easiest formulas for solving a problem which, it appears from what has been shown, can have no application in the theory of the figure of the planets. It is extremely probable that no such figures... will be found in nature. ... If the existence of such a figure can be supposed, would it be permanent? Would not the least action of other bodies of the system upon it be sufficient to destroy the exact conformation on which the equilibrium depends and leave the fluid to adjust its figure solely by the attraction and the centrifugal force of its particles?
>
> [Ivory 1838, pp. 65–66]

11.   Liouville's results raise several questions.

1°. Are they correct?

2°. How did they influence the later development?

3°. How were they obtained?

4°. Why did Liouville not publish them?

First, I shall answer 1°, and give a partial answer to 2°, by looking at the results of his successors. A more complete answer will follow after an analysis of Liouville's methods, i.e. after the answer to 3°. The last question will be discussed in the concluding section.

SUCCESSORS

12.   B. Riemann seems to have been the first person after Liouville who obtained results about stability of equilibrium ellipsoids. Following Dirichlet [1857] and [1858–1859], he considered motions of the fluid for which the positions of the particles at any fixed time are found by applying a linear transformation to the initial positions. Riemann [1860, §11] showed that under such motions the Jacobi ellipsoid is always stable, whereas the Maclaurin ellipsoid is stable only when its eccentricity is less than a certain limit. He made clear that the limits for stability might be narrower when arbitrary perturbations are allowed, and indeed Liouville's limit for the eccentricity of the Maclaurin ellipsoid is smaller than Riemann's.

In most of the paper, Riemann did not restrict his investigations to the study of figures of equilibrium, but studied arbitrary motions of Dirichlet's type, and his approach was rather different from Liouville's.

Sir W. Thomson (Lord Kelvin), on the other hand, asked himself exactly the same questions as Liouville had asked. In the first edition of his and Tait's famous *Treatise on Natural Philosophy* [Thomson and Tait 1867], the Maclaurin and the Jacobi ellipsoids received only a brief mention, but in the second edition [1879–1883, vol. 2 §778'], a discussion of stability was added. It was introduced as follows:

> During the fifteen years which have passed since the publication of our first ed. we have never abandoned the problem of the equilibrium of a finite mass of rotating incompressible fluid. Year after year questions of the multiplicity of possible figures of equilibrium have been almost incessantly before us, and yet it is only now under the compulsion of finishing this second edition... that we have succeeded in finding anything approaching to full light on the subject.

In the above quotation, "we" and "us" refer solely to Thomson, for he also published word for word the same results under his name alone [Thomson 1882]. He announced that when one allows only perturbations that leave the fluid in the shape of an ellipsoid of revolution, all Maclaurin ellipsoids are stable. If the ellipsoids are restricted to remain ellipsoids but not necessarily having rotational symmetry, those Maclaurin ellipsoids that are less flat than the Jacobi ellipsoid of revolution remain stable, and the rest become unstable. Under that restriction, all Jacobi ellipsoids are stable. If no restrictions are laid on the perturbations, the Jacobi ellipsoids are unstable when they have a high angular momentum.

Thomson offered no proofs but supported the last statement with the intuitive argument that if a long, cigar-shaped ellipsoid is being made thinner in the middle and thicker toward its ends, its potential energy will diminish. He conjectured that Jacobi's ellipsoids close to rotational symmetry were in stable equilibrium, but he was not able to prove it.

Thomson's incomplete analysis was carried further by H. Poincaré in a long paper in the *Acta Mathematica* of 1885 and in a series of lectures

[1902]. Poincaré's more general approach confirmed Thomson's results and established moreover that under arbitrary perturbations the Maclaurin ellipsoids less flat than the Jacobi ellipsoid of revolution (i.e., with $M < M_1$ (Fig. 2)) are still stable, as are the Jacobi ellipsoids with angular momenta less than a certain M_3. For angular momenta above these limits, there is instability, at least when the viscosity is nonzero (secular instability). Exactly the same results had been obtained one year earlier by the Russian mathematician Alexander Liapounoff in his doctoral thesis [1884], which was translated into French in 1904.

Thus, Liouville's results about the Maclaurin ellipsoids and about the Jacobi ellipsoids for $M < M_3$ were confirmed in the 1880s (and by Chandrasekhar [1969]). For higher momenta, however, contrary to Liouville's claim, the Jacobi ellipsoids are not stable. After I have discussed Liouville's methods, I shall, in §32, be able to point out why he missed the boundary M_3.

13.   The only one of these successors to mention Liouville's work on stability of fluid ellipsoids was Liapounoff. It is understandable that Riemann did not know of the results of his predecessor, but Thomson's and Poincaré's ignorance is surprising. Thomson had met Liouville in Paris in 1845, and they developed a long lasting friendship (cf. Chapter III). However, it is most unlikely that they discussed the question of fluid ellipsoids. For if they had, Thomson would probably have asked Liouville about his results instead of speculating about them for 15 years, and he would no doubt have been happy to cite Liouville.

It is less surprising that Poincaré did not learn about Liouville's theory from its creator, for when Poincaré began his studies in Paris, Liouville was old and weak. However, it is remarkable that nobody referred Poincaré to Liouville's few published notes. It indicates that by 1885 the scientific community in Paris had forgotten Liouville's two addresses to the *Académie*.

Poincaré's ignorance of Liouville's work is set into relief by the extensive knowledge of Liouville's published notes possessed by Liapounoff, whose teacher Chebyshev had worked on this problem as well. As we have seen in Chapter V, Chebyshev began to collaborate with Liouville in 1842 and the two mathematicians became good friends at the latest in 1852. It is therefore most likely that Chebyshev told Liapounoff about Liouville's work:

> Liouville has dealt with the question 30 years ago. Unfortunately, he has not published his works.
>
> [Liapounoff 1884, p. 6]

Liapounoff referred to all Liouville's published papers and notes on the subject and concluded:

> I have not been able to find any other indications of Liouville's research in this area.
>
> [Liapounoff 1884, p. 7]

Liapounoff also quoted the most important parts of [Liouville 1842h and 1843f] and indicated why Liouville had obtained the incomplete result about the Jacobi ellipsoids:

> As the reader will see below, the results concerning Mr. Jacobi's ellipsoids were probably obtained by Liouville by a particular hypothesis concerning the disturbances.
>
> [Liapounoff 1884, p. 7]

Liapounoff probably believed that Liouville had limited his discussion to perturbations for which the fluid maintains an ellipsoidal shape. However, as we shall see later (§§32–35), this is true only in a very subtle way.

To conclude, Liouville's few published notes on stability of fluid ellipsoids inspired only Liapounoff directly, and none of Liouville's successors knew about the unpublished material. However, after we have analyzed Liouville's methods, an indirect influence on Poincaré will come to light (§36).

Methods and Proofs

FORMULAS FOR THE FORCE VIVE

14.   The type of stability Liouville was interested in is called relative stability, because it is stability relative to a coordinate system moving with the fluid. A mass of fluid is in relative "static"[4] stable equilibrium if the relative potential energy E_{pot} has a minimum. When viscosity is disregarded, the total energy of the fluid is preserved, and so the configuration is stable when the live force or twice the kinetic energy E_{kin} has a maximum. The latter property is the one used by Liouville. In his first talk, he presented it as follows:

> One knows that in order to decide if the equilibrium of a system is stable or unstable one must move the system a little away from the state of rest, determine the value of the *force vive* after an arbitrary time, and see if it becomes a maximum by supposing that the system goes through the equilibrium position again. However, I have obtained an expression for this *force vive* which is as simple as possible... for a kernel of arbitrary shape and even for an entirely liquid system.
>
> [Liouville 1842h, pp. 905–906]

Liouville published this expression for the *force vive* and its deduction 10 years later [Liouville 1852f]. His starting point was the equations of motion for an inviscid fluid, which Laplace had also used. He expressed the equations in a coordinate system (x, y, z) rotating around the z-axis with the angular velocity $\omega(t)$. For small values of the velocity field $\mathbf{v} = (u, v, w)$, the total live force $\sum m|\mathbf{v}|^2$ can then be expressed as

$$2E_{\text{kin}} = 2 \int dt \iiint \left(\operatorname{grad} \varphi \cdot \mathbf{v} + (uy - vx)\frac{d\omega}{dt} \right) dx\, dy\, dz, \qquad (13)$$

where the space integral is to be taken over the entire fluid and where φ is the effective potential

$$\varphi = V + \frac{1}{2}(x^2 + y^2)\omega^2 - p, \qquad (14)$$

and p is the pressure. Here, I have followed Liouville and the other early 19th-century mechanicians who defined the potential or force function V as a function whose gradient equals the Newtonian force (here from the particles of the fluid). Thus, it is minus the quantity we call the potential today. Moreover, I have set the density of the fluid equal to one.

The last term of the integral, (13), vanishes in two interesting cases:

1° If

$$\frac{d\omega}{dt} = 0,$$

i.e., if the coordinate system rotates with constant angular velocity, and

2° if

$$\iiint (uy - vx)\, dx\, dy\, dz = 0, \qquad (15)$$

i.e., if the total angular momentum of the fluid around the z-axis is zero in the rotating coordinate system.

15.   In both cases, Liouville was interested in finding the kinetic energy of a figure close to an equilibrium ellipsoid. He denoted the perpendicular distance from the equilibrium surface to the perturbed surface ζ, counted positively outward. The quantity ζ is then a function of the time t and the point (x, y, z) on the surface of the ellipsoid. He also used the expression ζ' for the same quantity considered as a function of (x', y', z') and the symbol Δ for the distance $\sqrt{(x - x')^2 + (y - y')^2 + (z - z')^2}$ between the two points (x, y, z) and (x', y', z'). Moreover, he let g denote the effective attractive force from the equilibrium ellipsoid at its surface, that is, the gravitational plus the centrifugal force. With these symbols, he found, after long calculations, that in the first case the *force vive* could be expressed as

$$2E_{\text{kin}} = \iiiint \frac{\zeta\zeta'}{\Delta}\, d\sigma\, d\sigma' - \iint g\zeta^2\, d\sigma + \text{const} \quad (\omega \text{ const}), \qquad (16)$$

where the integrals are taken over the surface ∂E of the equilibrium ellipsoid.

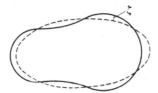

In the second case, where the coordinate system rotates in such a way that the angular momentum with respect to it is zero, the angular velocity $\omega(t)$ will vary because the moment of inertia varies. Taking that into account, Liouville found

$$2E_{\text{kin}} \;=\; \iiiint \frac{\zeta\zeta'}{\Delta}\, d\sigma\, d\sigma' - \iint g\zeta^2\, d\sigma$$

$$-\frac{\omega^2}{I_0}\left[\iint \zeta(x^2+y^2)\, d\sigma\right]^2 + \text{const} \quad (M=0), \qquad (17)$$

where I_0 is the moment of inertia of the equilibrium ellipsoid. In (17), the integrals must be taken over an imaginary surface ∂E, namely, the solidified equilibrium surface rotating with the variable angular velocity $\omega(t)$ of the coordinate system. However, g still represents the effective gravitation at the surface of the equilibrium ellipsoid, when it rotates with its constant angular velocity.[5]

As pointed out by Liouville in the passage quoted above (§14), the fluid is in stable equilibrium if and only if the kinetic energy has a maximum in the equilibrium state, where $\zeta = 0$. Formulas (16) and (17) therefore lead to the stability conditions

$$\iint g\zeta^2\, d\sigma - \iiiint \frac{\zeta\zeta'}{\Delta}\, d\sigma\, d\sigma' > 0 \quad (\omega \text{ const}), \qquad (18)$$

$$\iint g\zeta^2\, d\sigma - \iiiint \frac{\zeta\zeta'}{\Delta}\, d\sigma\, d\sigma' + \frac{\omega^2}{I_0}\left(\iint \zeta(x^2+y^2)d\sigma\right)^2 > 0 \quad (M=0)$$
$$(19)$$

for all sufficiently small values of ζ. In [Ms 3616 (3), pp. 38v–39r, 44r], Liouville found an expression for the live force of a fluid mass relative to a coordinate system that rotates in an arbitrary way around the origin. He ended his calculations with the words "Verify this formula and place it at the end of the memoir." However, no such note is preserved in [Liouville 1852f].

16. In [1852f] Liouville explained that the first condition (18) was suited for the investigation of the stability of a thin layer of fluid on a solid kernel,

spherical or not. For in that case, the total angular momentum is almost exclusively carried by the kernel, and the variation of the angular momentum of the fluid (called M above) has a negligible effect on $\omega(t)$. When there is no kernel, the angular momentum M of the fluid is conserved, so in that case the second condition (19) must be used.

In the second talk to the *Académie des Sciences* [1843–1984], Liouville explained what goes wrong if one applies formula (18) in the case of an entirely fluid ellipsoid: if, for example, a Jacobi ellipsoid is perturbed into another Jacobi ellipsoid with slightly smaller angular velocity, the new ellipsoid will keep its shape and therefore continue to have a shape close to the original ellipsoid. We must therefore conclude that under such a perturbation the motion is stable. However, since the angular velocity of the perturbed ellipsoid is smaller than that of the original ellipsoid, the major axis of the two will be perpendicular to each other after a finite time. Therefore, if one looked at the perturbed ellipsoid from a coordinate system with constant angular velocity equal to the angular velocity of the original ellipsoid, one would find considerable fluctuations in the x and y coordinates of the surface of the perturbed ellipsoid, and one would wrongly conclude that the system was unstable. Describing the motion with respect to a coordinate system in which $M = 0$ overcomes the above problem.

Liouville had good reasons for giving a particularly illuminating explanation of this point, because he himself initially obtained wrong results by applying formula (18) to Jacobi ellipsoids. He had also applied (18) to Maclaurin ellipsoids, and as the above argument suggests, he obtained the correct results in that case (see §30).

LIQUID ON AN ALMOST SPHERICAL KERNEL

17.   In the published papers, Liouville gave only vague hints about how he employed the conditions of stability. Concerning the problem of stability of fluid on an almost spherical kernel, he revealed that he had confirmed Laplace's results using two different methods:

> The first of these methods is based, as is the one due to Laplace, on a certain development in series which is incessantly used in the theory of attraction of spheroids. The second, which is independent of this type of development, is based on a remarkable method of minimum and seems to be subject to a great extension.
>
> [Liouville 1842h, p. 906]

The first version of his talk [Liouville 1842–1984, §§3, 4] throws light on the two methods.

18.   When the kernel is almost spherical with mean radius 1 and density D_k, the attractive force is approximately $g = \frac{4}{3}\pi D_k$ (the density D_f of the

fluid is still assumed to be 1). Therefore, the condition of stability, (18), can be written as

$$\frac{4}{3}\pi D_k \iint \zeta^2 \, d\sigma - \iiiint \frac{\zeta\zeta'}{\Delta} \, d\sigma \, d\sigma' > 0. \tag{20}$$

Liouville, then, followed Laplace and developed ζ in an infinite series of spherical harmonics:

$$\zeta = Y_1 + Y_2 + Y_3 + \cdots.$$

When this expression is inserted into (20), it reduces, by well-known formulas for spherical harmonics, to

$$\left(\frac{D_k}{3} - \frac{1}{3}\right) Y_1^2 + \left(\frac{D_k}{3} - \frac{1}{5}\right) Y_2^2 + \cdots > 0. \tag{21}$$

This condition is satisfied for arbitrary values of ζ only if all the coefficients of Y_i^2 are positive. Therefore, the motion is stable if and only if $D_k > 1$.

The basic idea of this proof goes back to Laplace, but in accordance with his initial aim (§9), Liouville obtained some simplifications and clarifications. Most striking is his simplification of the condition of stability. Laplace had developed the vibrations of the molecules of the liquid in a series of exponentials $e^{\lambda_i t}$ and had taken as the criterion for stability the requirement that all the numbers λ_i be negative (Oppenheim [1919] calls this kinetic stability).

In the original paper [1782–1785], Laplace did not mention the "force vive" in his analysis, but in the *Mécanique Céleste* [1799–1825], he did remark that a certain complicated expression in the spherical harmonics was the live force. Liouville apparently discovered that rendering this expression maximal was the crucial point in the proof of relative stability, just as in the case of ordinary stability. Yet, admittedly, Liouville never defined the meaning of such stability as the maximal kinetic energy might imply, and even today the physical problem of relative stability is a very difficult one.

19.   The second method mentioned in the quotation in §17 is explained in [Liouville 1842/1984 §4]. It is inspired by Gauss's *Allgemeine Lehrsätze in Beziehung auf die im verkehrten Verhältnisse des Quadrats der Entfernung wirkenden Anziehungs- und Abstossungs-Kräfte* [1840], which Liouville had just published in a French translation in the August 1842 issue of his journal, (cf. Chapter XV, §§4–7).

Stability is secured if the minimal value of the left-hand side of (18) exists and is positive. Liouville first looked at perturbations ζ such that

$$\iint g\zeta^2 \, d\sigma = H > 0, \tag{22}$$

where H is a positive constant. Naturally, conservation of mass gives the further restriction

$$\iint \zeta \, d\sigma = 0. \tag{23}$$

Liouville claimed that under these conditions it is easily shown that the left-hand side of (18) has a minimum.

In Liouville's time, it was generally believed that a minimum existed if a lower bound could be found (cf., however, Chapter XV, §§45–48). The proof of the existence of a lower bound is indicated in a note [3617 (2), p. 78r] that can be interpreted as follows.[6] Since there are k, $K \in \mathbf{R}^+$, such that $0 < k \leq g(x, y, z) \leq K < \infty$, it is enough to show that the quantity

$$A \iint g\zeta^2 \, d\sigma - \iiiint \frac{gg'\zeta\zeta'}{\Delta} \, d\sigma \, d\sigma' \tag{24}$$

has a lower bound when A is a constant. Determine a negative constant B such that

$$A - B \geq \iint g' \frac{d\sigma'}{\Delta} \quad \text{for all} \quad (x, y, z) \text{ on } \partial E.$$

The existence of such a constant follows from [Gauss 1840, §12] because the integral is the potential due to a mass distribution g on ∂E. Then,

$$\frac{1}{2} \iiiint gg' \frac{(\zeta - \zeta')^2}{\Delta} \, d\sigma \, d\sigma' \geq 0$$

$\Downarrow$

$$\frac{1}{2} \iiiint gg' \frac{\zeta^2 + \zeta'^2}{\Delta} \, d\sigma \, d\sigma' - \iiiint gg' \frac{\zeta\zeta'}{\Delta} \, d\sigma \, d\sigma' \geq 0$$

$\Downarrow$

$$\iint \left(\iint g' \frac{d\sigma'}{\Delta} \right) g\zeta^2 \, d\sigma - \iiiint gg' \frac{\zeta\zeta'}{\Delta} \, d\sigma \, d\sigma' \geq 0$$

$\Downarrow$

$$(A - B) \iint g\zeta^2 \, d\sigma - \iiiint gg' \frac{\zeta\zeta'}{\Delta} \, d\sigma \, d\sigma' \geq 0$$

$\Downarrow$

$$A \iint g\zeta^2 \, d\sigma - \iiiint gg' \frac{\zeta\zeta'}{\Delta} \, d\sigma \, d\sigma' \geq B \iint g\zeta^2 \, d\sigma = BH.$$

Hence, expression (24) is bounded below by BH. Therefore, the minimum of the left-hand side of (18) exists and can be determined by taking its variation, subject to conditions (22) and (23). According to Lagrange's method there is a multiplyer $(a - 1)$ such that at the minimum

$$\delta \left[(a - 1) \iint g\zeta^2 \, d\sigma + \left(\iint g\zeta^2 \, d\zeta - \iiiint \frac{\zeta\zeta'}{\Delta} \, d\sigma \, d\sigma' \right) \right] = 0,$$

or

$$\iint \left(ag\zeta - \iint \frac{\zeta'}{\Delta} \, d\sigma' \right) \delta\zeta \, d\sigma = 0, \tag{25}$$

where the variation $\delta\zeta$ is now only subject to the restriction (23).

20. Using an argument due to Gauss [1840, §31] (cf. Chapter XV, §§6,7), Liouville concluded that there is a second constant a' such that

$$W = ag\zeta - \iint \frac{\zeta'}{\Delta} d\sigma' = a'. \tag{26}$$

Indeed, if W is not constant, let D be a "mean" value between its maximal and minimal value. Let P and Q denote the parts of the surface of the ellipsoid where W is larger than D and smaller than D, respectively, and determine a variation $\partial\zeta$, subject to (23), which is negative in a portion of P and positive in a portion of Q and zero elsewhere. Then,

$$\iint W \,\delta\zeta \,d\sigma = \iint (W - D)\delta\zeta \,d\sigma < 0 \tag{27}$$

contrary to (25).

At the minimum, the right-hand side of (18) is equal to $(1 - a)H$ because

$$\zeta^2 \,d\sigma \;-\; \iiiint \frac{\zeta\zeta'}{\Delta} d\sigma \,d\sigma'$$
$$= \;(1 - a)H + \iint \zeta \left(ag\zeta - \iint \frac{\zeta'}{\Delta} d\sigma' \right) d\sigma$$
$$= \;(1 - a)H + a' \iint \zeta \,d\sigma = (1 - a)H.$$

Thus, the motion is stable if and only if this quantity is positive for all values of $H > 0$, that is, if the largest possible value of a is less than 1.

In the case of a thin layer of liquid on a kernel of density D_k, (26) becomes

$$a\frac{4}{3}\pi D_k \zeta - \iint \frac{\zeta'}{\Delta} d\sigma' = a'. \tag{28}$$

When this equation is multiplied by the associated Legendre polynomial P_n^m $(n > 0)$ and integrated over the sphere, one gets[7]

$$\left(\frac{4}{3}\pi D_k a - \frac{4\pi}{2n + 1} \right) \iint P_n^m \zeta \,d\sigma = 0, \quad \forall n > 1. \tag{29}$$

If the second factor is zero for all $n > 1$ and if we take (21) into account, the completeness of the P_n^m's implies $\zeta \equiv 0$, in contradiction with (22). Therefore, the first factor must vanish for a value of n, which means that $a = \frac{3}{(2n+1)D_k}$. The largest possible value of a is therefore $a = \frac{1}{D_k}$, and so the condition for stability[8] is $\frac{1}{D_k} < 1$ or $D_k > 1$.

Liouville emphasized the advantage of this method because it does not rely on arbitrary functions having expansions in spherical harmonics. Nevertheless, it uses the completeness of these functions in a different way. In Chapter XV, we shall see how further speculations along these lines led Liouville to a remarkable and mostly unpublished approach to potential theory of arbitrary surfaces.

LAMÉ FUNCTIONS

21.   Liouville remarked that when there is no kernel the above methods still work, and prove the stability of the almost spherical Maclaurin ellipsoids [1842–1984, §4].

When the fluid is no longer almost spherical, spherical harmonics cannot be used; instead, Liouville got the idea of using another system of functions, and immediately jotted down the following programmatic note [Ms 3617 (2), p. 72v]:

> In order to treat $\iiiint \frac{\zeta \zeta' \, d\omega \, d\omega'}{\Delta} [d\omega = d\sigma]$, it would be best to study first $\iint \frac{\zeta' \, d\omega'}{\Delta}$ and to see if the terms of the develop[ment] of ζ given by Lamé follow from it. Moreover, follow all that takes place for the sphere. This merits being generalized and extended at least to the ellipsoid. It seems almost impossible that such an extension should not hold.

The idea of using the Lamé functions was a fortunate one and indeed it led Liouville to the solution of the problem. It was also a rather natural idea; Lamé had introduced his functions as an aid in finding the steady-state temperature distribution in a tri-axial ellipsoid. Since the equation governing stationary heat conductions is the Laplace equation, which is also the fundamental equation for the potential, it is not surprising that Liouville was so confident of success, even before he had begun his calculations.

22.   The Lamé functions are defined in ellipsoidal coordinates, which are tailored to problems concerning ellipsoids, just as spherical coordinates are well adapted to spheres. They were defined as follows by Lamé [1837 and 1839].

Let $0 < b < c$ be given real numbers. The ellipsoidal coordinates of a point $P = (x, y, z)$ are defined as the roots ρ, μ, ν of the equations:

$$\frac{x^2}{\rho^2} + \frac{y^2}{\rho^2 - b^2} + \frac{z^2}{\rho^2 - c^2} = 1, \tag{30}$$

$$\frac{x^2}{\mu^2} + \frac{y^2}{\mu^2 - b^2} - \frac{z^2}{c^2 - \mu^2} = 1, \tag{31}$$

$$\frac{x^2}{\nu^2} - \frac{y^2}{b^2 - \nu^2} - \frac{z^2}{c^2 - \nu^2} = 1. \tag{32}$$

It is supposed that $0 < \nu^2 < b^2 < \mu^2 < c^2 < \rho^2$. The surface $\rho = \text{const}$, $\mu = \text{const}$, $\nu = \text{const}$ are ellipsoids and hyperboloids of one and two sheets, respectively, which cut each other orthogonally. Liouville took (30) to be the equation of the surface of equilibrium. It would be more correct to let the equilibrium ellipsoid be determined by

$$\frac{x^2}{\varrho_0^2} + \frac{y^2}{\varrho_0^2 - b^2} + \frac{z^2}{\varrho_0^2 - c^2} = 1,$$

so as to distinguish between the variable ρ and the fixed first coordinate ρ_0 on the equilibrium surface. For notational convenience, I shall follow Liouville and disregard this distinction since it does not cause any confusion. The points on the ellipsoid are determined by the coordinates μ, ν, i.e., $\zeta = \zeta(\mu, \nu)$.

23.   Lamé [1839] had considered the equation

$$(\rho^2 - b^2)(\rho^2 - c^2)\frac{d^2R}{d\rho^2} + (2\rho^3 - (b^2 + c^2)\rho)\frac{dR}{d\rho} = (n(n+1)\rho^2 - B)R \quad (33)$$

and had introduced the particular solutions $R(\rho)$, which are polynomials in ρ^2, or such polynomials multiplied by one, two, or three of the factors ρ, $\sqrt{\rho^2 - b^2}$, $\sqrt{\rho^2 - c^2}$. In (33), $n \in \mathbf{N}$ is the total degree of $R(\rho)$ (counting each of the factors ρ, $\sqrt{\rho^2 - b^2}$, $\sqrt{\rho^2 - c^2}$ as having degree one), and B is a number for which a solution $R(\rho)$ exists. Lamé had shown that such values of B exist and are all real. Liouville observed that this followed easily from Sturm's method of determining roots in transcendental equations [1842–1984, §5 and 1846e, p. 221] and later Poincaré [1885, §8] showed, with due reference to Liouville, how this could indeed be done.

To give an impression of the character of the Lamé functions, I shall reproduce Liouville's list of the R's corresponding to $n = 0, 1, 2$ as he wrote them down in his notebook [Ms 3617 (2), p. 77r] (in the list he wrote Lamé's symbol E instead of R; I have added the indices on R according to Liouville's principle discussed in §27):

$n = 0 \quad R_0 = 1,$

$n = 1 \begin{cases} B = b^2 + c^2, & R_1 = \rho, \\ B = c^2, & R_2 = \sqrt{\rho^2 - b^2}, \\ B = b^2, & R_3 = \sqrt{\rho^2 - c^2}, \end{cases}$

$n = 2 \begin{cases} B = \frac{2b^2c^2}{l^2}, \; l^2 = \frac{1}{3}(b^2 + c^2 - \sqrt{b^4 + c^4 - b^2c^2}), & R_4 = \rho^2 - l^2, \\ B = \frac{2b^2c^2}{l'^2}, \; l'^2 = \frac{1}{3}(b^2 + c^2 + \sqrt{b^4 + c^4 - b^2c^2}), & R_8 = \rho^2 - l'^2, \\ B = b^2 + 4c^2, & R_5 = \rho\sqrt{(\rho^2 - b^2)}, \\ B = c^2 + 4b^2, & R_6 = \rho\sqrt{(\rho^2 - c^2)}, \\ B = b^2 + c^2, & R_7 = \sqrt{(\rho^2 - b^2)(\rho^2 - c^2)}. \end{cases}$

The Lamé function R is always normalized so that $\frac{R(\rho)}{\rho^n} \to 1$ for $\rho \to \infty$.

Along with the functions R, Lamé had introduced the functions $M(\mu)$ and $N(\nu)$, which are found from $R(\rho)$ by changing ρ into μ and ν and writing $\sqrt{c^2 - \mu^2}$ instead of $\sqrt{\mu^2 - c^2}$, and $\sqrt{c^2 - \nu^2}$, $\sqrt{b^2 - \nu^2}$ instead of $\sqrt{\nu^2 - c^2}$, $\sqrt{\nu^2 - b^2}$. The product $R(\rho)M(\mu)N(\nu)$ of corresponding Lamé functions is then a solution of Laplace's equation; in fact, (33) is precisely the equation obtained by separating variables in Laplace's equation when it is written in ellipsoidal coordinates.

Lamé had also shown that a product $M_i N_i$ of corresponding Lamé functions and a different pair $M_j N_j$ satisfy the orthogonality relation [Lamé 1839, p. 158]:

$$\iint h M_i N_i M_j N_j \, d\sigma = 0 \,, \qquad (34)$$

where the integral ranges over the surface of the ellipsoid, (30), and h is defined by

$$h(\rho, \mu, \nu) = \frac{\sqrt{\rho^2 - b^2}\sqrt{\rho^2 - c^2}}{\sqrt{\rho^2 - \mu^2}\sqrt{\rho^2 - \nu^2}} \,. \qquad (35)$$

"I have realized," Liouville wrote, "that it is important in addition to consider a function S, which Mr. Lamé does not talk about and which satisfies the same equation as R" [Liouville 1842–1984, §5]. The function in question is the Lamé function of the second kind:

$$S(\rho) = (2n + 1) R(\rho) \int_\rho^\infty \frac{d\rho'}{R(\rho')\sqrt{(\rho'^2 - b^2)(\rho'^2 - c^2)}} \,. \qquad (36)$$

Heine introduced this function at the same time [Heine 1845]. In his standard work, *Theorie der Kugelfunktionen* [1878–1881], Heine makes a strong claim of priority. Having recalled that Liouville presented his new function to the *Académie des Sciences* on May 12 and June 2, 1845 (cf. §36), he continued:

> As the signature shows, I submitted my work *Beitrag zur Theorie der Anziehung und der Wärme* to the founder and editor of Crelle's Journal on April 19, 1844. It appeared in 1845 in the third issue of the 29th volume of this Journal (which, as far as I have learned, was sent off at the latest on May 20th of the same year).
> [Heine 1878–1881, vol. 1, p. 384]

In the parentheses Heine seems to suggest that Liouville might have stolen the idea from him. However, as we shall see, Liouville made extensive use of this function as early as 1842, so we can establish his priority unambiguously.

STABILITY OF FLUID ELLIPSOIDS

24.   Just as Liouville - following Laplace - had developed ζ in a series of spherical harmonics in the almost spherical case (§18), so in the case of an arbitrary ellipsoid Liouville developed ζ in a series of Lamé functions:

$$\zeta(\mu, \nu) = h(\rho, \mu, \nu) \sum_i A_i M_i(\mu) N_i(\nu) \,, \qquad (37)$$

where h is defined by (35), the A_i's are real coefficients, and M_i, N_i are pairs of corresponding Lamé functions numbered conveniently. In [1842–1984], Liouville promised to supply a proof that "arbitrary" functions can be developed in this way, and indeed, in [1846k] he supplied the proof (cf. Chapter XV, §§18–19). His aim was to determine a simple expression for the right-hand side of (18) in terms of the coefficients A_i.

We shall analyze the two terms of (18) separately, starting with the last term.

25.   Liouville's calculation of

$$\iiiint \frac{\zeta \zeta'}{\Delta} \, d\sigma \, d\sigma' \tag{38}$$

is not preserved, but it can easily be reconstructed by following the procedure suggested in his programmatic note quoted in §21. There, his idea was to calculate

$$\iint \frac{\zeta' \, d\sigma'}{\Delta} = \sum_j A_j \iint h' \frac{M_j' N_j'}{\Delta} \, d\sigma'. \tag{39}$$

This must have led him to the discovery of the important formula

$$\iint h' \frac{M'N'}{\Delta} \, d\sigma' = \frac{4\pi}{2n+1} RSMN \sqrt{(\rho^2 - b^2)(\rho^2 - c^2)}, \tag{40}$$

where n is the degree of the Lamé functions M and N entering their defining equation, (33) [Liouville 1842/1984, §6 and 1845d, 1846e,f] (cf. Chapter XV, §20).

With this formula at hand, one can write (38) in the form

$$4\pi \sum_{ij} A_i A_j \frac{R_j S_j \sqrt{(\rho^2 - b^2)(\rho^2 - c^2)}}{2n+1} \iint h M_i N_i M_j N_j \, d\sigma. \tag{41}$$

By the orthogonality relation (34) only the terms for which $i = j$ survive, and so finally we have

$$\iiiint \frac{\zeta \zeta'}{\Delta} \, d\sigma \, d\sigma' = 4\pi \sum_i A_i^2 \frac{R_i S_i \sqrt{(\rho^2 - b^2)(\rho^2 - c^2)}}{2n+1} \iint h M_i^2 N_i^2 \, d\sigma. \tag{42}$$

26.   In order to find a similar expression for the first term in (18), Liouville first had to determine $g(\mu, \nu)$; for when the ellipsoid is not almost spherical, this quantity is no longer a constant. His determination of g can be found in [1843b] and on page 2 of a bundle of papers put into [Ms 3617 (2)].

The potential that creates the effective gravitation (5), is expressed by

$$V = \frac{1}{2}(A' - \omega^2)x^2 + \frac{1}{2}(B' - \omega^2)y^2 + \frac{1}{2}C'z^2. \tag{43}$$

When the ellipsoid is in equilibrium, its surface, expressed by (1), must be an equipotential surface. Thus, the effective potential V is proportional to

$$\frac{x^2}{k_2^2} + \frac{y^2}{k_1^2} + \frac{z^2}{k^2}. \tag{44}$$

The force $g = |\operatorname{grad} V|$ is therefore proportional to

$$\sqrt{\frac{x^2}{k_2^4} + \frac{y^2}{k_1^4} + \frac{z^2}{k^4}} \tag{45}$$

which is inversely proportional to the perpendicular distance from the center of the ellipsoid to the tangent plane through (x, y, z). This neat geometrical observation was published separately in [Liouville 1843f].

He also saw that this perpendicular distance is equal to $h\rho$ so that [Liouville 1842–1984, §7; Ms 3617 (2), loose sheets, p. 2]

$$g = \frac{2C}{h\rho} = \frac{2C}{hk_2}. \tag{46}$$

The last equality is obtained by comparing the expressions for the axis in the two different expressions (1) and (30) for the equilibrium ellipsoid:

$$\rho^2 = k_2^2, \quad \rho^2 - b^2 = k_1^2, \quad \rho^2 - c^2 = k^2. \tag{47}$$

In order to determine the constant C, Liouville looked at the "North Pole" $(0, 0, k)$ [Ms 3617 (2), loose sheets, p. 2], where the perpendicular distance $h\rho$ is equal to k. The gravitational force g_N at this point by (4) and (5) is

$$g_N = \frac{2C}{h\rho} = \frac{2C}{k} = k\frac{3M}{k^3}\int_0^1 \frac{x^2}{G}\,dx\,, \tag{48}$$

or, using identity (8) and inserting the mass $M = \frac{4}{3}\pi k k_1 k_2 D_k$, we have

$$g_N = \frac{2C}{k} = 2\pi D_k k^2 k_1 k_2 \int_0^\infty \frac{d\alpha}{(\alpha + k^2)H}. \tag{49}$$

Therefore,

$$2C = 2\pi D_k k_1 k_2 k \int_0^\infty \frac{k^2\,d\alpha}{(\alpha + k^2)H}, \tag{50}$$

so that

$$g = \frac{2C}{h\rho} = \frac{2C}{hk_2} = 2\pi D_k \frac{k_1 k}{h}\int_0^\infty \frac{k^2\,d\alpha}{(\alpha + k^2)H}. \tag{51}$$

Now, Liouville could easily determine the expression for the first term of (18) when $\zeta = h\sum A_i M_i N_i$:

$$\iint g\zeta^2\,d\sigma = \iint \frac{2C}{h\rho}\left(\sum_i hA_iM_iN_i\right)\left(\sum_j hA_jM_jN_j\right)d\sigma$$

$$= \frac{2C}{\rho}\iint h\left(\sum_i A_i^2M_i^2N_i^2 + 2\sum_{i\neq j} A_iA_jM_iM_jN_iN_j\right)d\sigma$$

$$= \frac{2C}{\rho}\sum_i A_i^2 \iint h\,M_i^2N_i^2\,d\sigma \tag{52}$$

because by (34) the mixed terms cancel.

27.   Thus, the criterion (18) for stability can be expressed by a combination of (42) and (52):

$$\sum_i \left(\frac{2C}{\rho} - 4\pi \frac{R_i S_i \sqrt{(\rho^2 - b^2)(\rho^2 - c^2)}}{2n+1} \right) A_i^2 \iint h M_i^2 N_i^2 \, d\sigma > 0 \quad (53)$$

[Liouville 1842–1984]. This criterion is fulfilled for all perturbations, i.e., for all sequences A_i, if and only if for all $i \in N$

$$\frac{2C}{\rho} - 4\pi \frac{R_i S_i \sqrt{(\rho^2 - b^2)(\rho^2 - c^2)}}{2n+1} > 0. \quad (54)$$

Altering the terminology of Poincaré slightly, I shall use the term "stability coefficient" for the left-hand side of (54). Since the equilibrium is stable when the smallest stability coefficient is positive, the question is to find the smallest stability coefficient. To this end, Liouville applied an ordering principle for Lamé functions, which he claimed to follow from "a method due to Mr. Sturm." The principle states the following.
    [**Principle** W.]

$$\frac{R_i S_i \sqrt{(\rho^2 - b^2)(\rho^2 - c^2)}}{2n+1} \quad (55)$$

is smaller the greater n is, and... for the same value of n, it increases with the root B.   [Liouville 1842–1984, §5]

    This principle is wrong. Poincaré [1885 §10] and Liapounoff [1884, ch. 4] have discussed what information Sturm's method can give about the relative size of $\frac{R_i S_i}{2n+1}$, and the conclusion is considerably more complicated than Liouville's Principle W. We shall return to a discussion of this mistake and its consequences in §§32, 34, but for the moment we shall follow Liouville's argument on the basis of the principle.
    If the Lamé functions are arranged according to decreasing values of (55), as I have done in the list in §22, the stability condition (54) will, according to Principle W, be satisfied for all Lamé functions if it is satisfied for the first. The first Lamé function is $R_0 = 1$ $(n = 0)$ corresponding to $M_0 N_0 = 1$. However, this term cannot be allowed in the expansion of ζ since $\zeta = A_0 M_0 N_0 = A_0$ would violate conservation of the volume of the fluid. Thus, when there is a solid kernel, the crucial term is the term corresponding to $R = \rho$ $(n = 1, B = b^2 + c^2)$. It gives rise to the stability condition

$$\frac{2C}{\rho} > 4\pi \int_\rho^\infty \frac{\rho^2 \sqrt{(\rho^2 - b^2)(\rho^2 - c^2)}}{\rho'^2 \sqrt{(\rho'^2 - b^2)(\rho'^2 - c^2)}} \, d\rho'. \quad (56)$$

Since C contains the density D_k of the kernel as a factor (50), this gives a transcendental inequality for D_k, which Liouville mentioned in his second talk [1843–1984].

28.

However, if it is a matter of an entirely liquid mass (and that is the most interesting case), the center of gravity of this mass must remain fixed so that one will have $A_1 = 0$, $A_2 = 0$, $A_3 = 0$.

[Liouville 1842–1984, §7]

In fact, $M_1 N_1 = \mu\nu$ is proportional to x and an addition of $\zeta = Ax$ would move the center of gravity along the x-axis. Similarly, $M_2 N_2 = \sqrt{\mu^2 - b^2}\sqrt{b^2 - \nu^2}$ and $M_3 N_3 = \sqrt{c^2 - \mu^2}\sqrt{c^2 - \nu^2}$ leads to a shift of the center of gravity along the y-axis and the z-axis, respectively. The crucial stability coefficient therefore corresponds to $n = 2$ and $B = \frac{2b^2c^2}{l^2}$, which Liouville easily showed to be the largest value of B for $n = 2$ [Ms 3617 (2), p. 77r,v]. The criterion (54) of stability of the Jacobi ellipsoid is then the inequality

$$\frac{2C}{\rho} - 4\pi \frac{R_4 S_4 \sqrt{(\rho^2 - b^2)(\rho^2 - c^2)}}{5} , \tag{57}$$

where the density D_k in (50) equals 1.

Liouville's calculation of the sign of this stability coefficient for a Jacobi ellipsoid can be found in [Ms 3617 (3), pp. 36r–37r]. In this note, he used expression (51) for $g = \frac{2C}{\hbar\rho}$, and for the second term of (57), he used the equivalent expression in k, k_1, k_2, to be found in footnote 9. In this terminology, the stability coefficient (57) has the following appearance:

$$2\pi k k_1 \int_0^\infty \frac{d\alpha}{H} \left(\frac{k^2}{\alpha + k} - \frac{(k_2^2 - l^2)^2}{(\alpha + k_2^2 - l^2)^2} \right). \tag{58}$$

Liouville transformed the first term of (58) using the equilibrium condition (10) for the Jacobi ellipsoid and obtained the expression

$$2k k_1 \int_0^\infty \frac{d\alpha}{H} \left(\frac{k_1^2 k_2^2}{(\alpha + k_1^2)(\alpha + k_2^2)} - \frac{k_2^2 - l^2}{(\alpha + k_2^2 - l^2)^2} \right) \tag{59}$$

for the stability coefficient. The parenthesis can be written in the form

$$\frac{A\alpha^2 + B\alpha + C}{(\alpha + k_1^2)(\alpha + k_2^2)(\alpha + k_2^2 - l^2)}. \tag{60}$$

Liouville then calculated the coefficients [Ms 3617 (3), pp. 36r–37r] and showed that $C = 0$, $A < 0$, $B < 0$. Therefore, the stability coefficient (59) or (57) is negative, "which implies the instability of Jacobi's liquid ellipsoid" [Ms 3617 (3), p. 37r].

29.   Liouville claimed [1842/1984] that he had proved by the same method that the Maclaurin ellipsoids are stable when they are less flat than the Jacobi ellipsoid of revolution and unstable otherwise. I have not found Liouville's calculations relative to the Maclaurin ellipsoids; one procedure,

which Liouville may have used, was found by Poincaré [1885, §11]. Liouville also indicated in [1842/1984, §8] that the results about the Jacobi and Maclaurin ellipsoids could be obtained by using the minimum method described in §§19–20. Indeed, the argument leading to (26) is equally valid in this case. By multiplying both sides of (26) by $hM_iN_i(i \geq 1)$ and integrating over the equilibrium ellipsoid, the right-hand side vanishes because of the orthogonality relation (34), and one has

$$\iint hag\zeta M_i N_i d\sigma - \iiiint h\zeta' \frac{M_i N_i}{\Delta} d\sigma \, d\sigma' = 0, \qquad (61)$$

or by (40)

$$\iint hag\zeta M_i N_i d\sigma - \iint \zeta' \frac{4\pi R_i S_i \sqrt{(\rho^2 - b^2)(\rho^2 - c^2)}}{2n+1} M_i' N_i' \, d\sigma = 0.$$

Since gh is a constant on the surface of the ellipsoid according to (46),

$$\left(agh - \frac{4\pi R_i S_i \sqrt{(\rho^2 - b^2)(\rho^2 - c^2)}}{2n+1} \right) \iint M_i N_i \zeta = 0.$$

By an argument similar to the one used in the spherical case (§20) the completeness of the Lamé functions implies the stability condition (54).

A CORRECTED AND AN UNCORRECTED ERROR

30.   Thus, on the basis of the above calculations, Liouville concluded, in the manuscript [1842–1984, §7] from November 2, 1842, that the Jacobi ellipsoids are always unstable. However, before he informed the *Académie des Sciences* about his discoveries 12 days later, he found an error in the argument. Namely, he realized that in the discussion of the Jacobi ellipsoids it is not ω but M that is constant, and therefore the stability condition (18) does not apply. Immediately, Liouville corrected his memoir according to his new insight and entirely rewrote the talk [1842h] with which he presented his memoir to the *Académie des Sciences* on November 4, 1842. We have already seen how he derived the alternative stability condition (19) in the first section of the memoir published in [1852f] (§14).

Having found this new stability condition, Liouville studied its implications for the Jacobi ellipsoids. His investigations were not published, but enough notes are left in [Ms 3617 (2), pp. 81r–90r, and loose sheets, pp. 1–6; Ms 3617 (3), pp. 40v–42v] to allow a partial reconstruction. The only difference between the old and the new stability condition is the third term of (19). Liouville calculated this term by noting [Ms 3617 (2), p. 81v] that

$$\begin{aligned} (x^2 + y^2) &= a_4(\mu^2 - l^2)(\nu^2 - l^2) + a_8(\mu^2 - l_1^2)(\nu^2 - l_1)^2 \\ &= a_4 M_4 N_4 + a_8 M_8 N_8, \end{aligned} \qquad (62)$$

where a_4 and a_8 are suitable constants.

When $\zeta = \sum h A_i M_i N_i$ is substituted into the additional term of (19) and the orthogonality relation is used, one gets

$$\frac{\omega^2}{I_0}\left(\iint (x^2+y^2)\zeta\, d\sigma\right)^2 = \frac{\omega^2}{I_0}\left(A_4 a_4 \iint h M_4^2 N_4^2\, d\sigma + A_8 a_8 \iint h M_8^2 N_8^2\, d\sigma\right)^2,$$

(63)

which Liouville wrote in the form

$$\frac{\omega^2}{I_0}(Q_4 A_4 + Q_8 A_8)^2,$$

(64)

where

$$Q_4 = a_4 \iint h M_4^2 N_4^2\, d\sigma,$$

$$Q_8 = a_8 \iint h M_8^2 N_8^2\, d\sigma.$$

(65)

Similarly, he called the coefficients in the old stability condition (53) P_i:

$$P_i = \left(\frac{2C}{\rho} - 4\pi \frac{R_i S_i \sqrt{(\rho^2 - b^2)(\rho^2 - c^2)}}{2n + 1}\right) \iint h M_i^2 N_i^2\, d\sigma.$$

(66)

With this notation, the new stability condition (19) takes on the form

$$\sum_i P_i A_i^2 + \frac{\omega^2}{I_0}(Q_4 A_4 + Q_8 A_8)^2 > 0, \quad \forall A_i \neq 0.$$

(67)

31.   Comparison with the former investigation (§§27–28) reveals that only the two terms corresponding to $M_4 N_4$ and $M_8 N_8$ are altered. The first of these terms was precisely the criminal term that had a negative stability coefficient (57). Now, a positive term must be added, so Liouville realized that there was new hope for the stability of the Jacobi ellipsoids and set out to calculate the sign of the part of (67) corresponding to

$$\zeta = A_4 h M_4 N_4 + A_8 h M_8 N_8.$$

(68)

This time the effect of $M_4 N_4$ and $M_8 N_8$ cannot be separated because the quadratic form (67) does not decompose into quadratic factors. Thus, Liouville wished to prove that for all $A_4, A_8 \neq 0$,

$$P_4 A_4^2 + P_8 A_8^2 + \frac{\omega^2}{I_0}(Q_4 A_4 + Q_8 A_8)^2 > 0.$$

(69)

He knew that $P_8 > 0$. I have not found the argument, but either he calculated P_8 just as he calculated P_4 (§28) or he concluded it from Principle

W (see §32). In any case, he knew that the quadratic form (69) is positive for $A_4 = 0$. Therefore, it is positive-definite if its discriminant is negative (cf. Fig. 4 of [Lützen 1984b]), that is, if

$$P_4 P_8 + \frac{\omega^2}{I_0}(P_4 Q_8^2 + P_8 Q_4^2) > 0. \tag{70}$$

In long and complicated calculations, Liouville then estimated the quantities in (70) for large values of k_2 when the ellipsoid is almost cylindrical. He determined a_1 and a_2 [Ms 3617 (2), pp. 85v–86r; Ms 3617 (3), p. 42r], $\iint h M_4^2 N_4^2$ and $\iint M_8^2 N_8^2$ [Ms 3617 (2), pp. 83v–85v], $\frac{\omega^2}{I_0}$ [Ms 3617 (3), pp. 41v–42r] and finally P_4 and P_8 or rather the quantity (54) [Ms 3617 (2), pp. 82v–83r, and loose sheets, pp. 1–3; Ms 3617 (3), pp. 40v–41v]. After having corrected some miscalculations (for example, [Ms 3617 (3), p. 40v]), he finally found that in this limit inequality (70) is indeed satisfied, and he triumphantly wrote "Stability!" [Ms 3617 (2), loose sheets, p. 5; Ms 3617 (3), p. 42r].

I do not know what made him conclude that when (70) is satisfied in the limit where k_2 is large it must also be satisfied for smaller values of k_2. There are also traces of a calculation of the sign of (70) in the other limit, where the Jacobi ellipsoid is almost an ellipsoid of revolution [Ms 3617 (2), pp. 88v–89v], but it is incomplete, and I cannot judge if it had any place in the argument. Complete lack of sources forces me to leave this minor hole in the reconstruction of Liouville's proof.

Whatever argument Liouville may have used, his conclusion that (70) is positive is correct, so that the addition of the extra term (64) does "rescue" the negative stability coefficient P_4.

32.   However, the overall conclusion, stated after the calculation and in [1843a] and [1843c], that this implies the stability of the Jacobi ellipsoids is not correct. For Liouville's argument takes care only of the perturbations of the form $A_4 M_4 N_4 + A_8 M_8 N_8$, and it does not ensure that none of the higher coefficients of stability are negative. Indeed, Poincaré [1885] and Liapounoff [1884] have shown that when the Jacobi ellipsoid has a high angular momentum, certain of the higher coefficients of stability do become negative. When the Jacobi ellipsoid is close to rotational symmetry, i.e., has a small angular momentum $\sim M_1$ (Fig. 2), all the coefficients of stability are positive, and for increasing M, the first coefficient of stability to vanish is the coefficient corresponding to

$$R_{15} = \frac{1}{5}\rho(5\rho^2 - 2(b^2 + c^2) + \sqrt{4b^4 - 7b^2 c^2 + 4c^4}). \tag{71}$$

When M is less than the vanishing value M_3 of R_{15}, the Jacobi ellipsoids are therefore stable, whereas they are unstable for $M > M_3$. Why did Liouville neglect the examination of the higher coefficients of stability? Was he perhaps so delighted to get rid of the negative P_4 that he completely

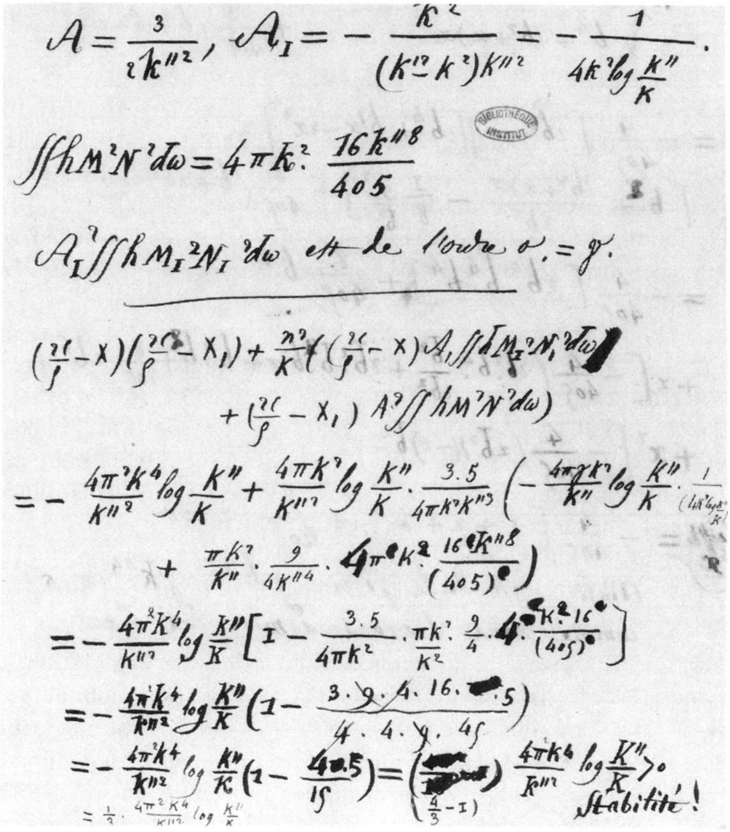

PLATE 26. Liouville concluded in his long calculation that the Jacobi Ellipsoids are stable. (Bibliothèque de l'Institut de France, Ms 3617 (3) p 42r)

forgot about the rest? Since there is no explicit mention of the problem in his notes, one cannot exclude this possibility, but his thoroughness makes it seem highly improbable. Instead, I shall argue that the problem can be traced back to the fallacious Principle W in §27, and that granted this principle, Liouville had a rational argument to infer that all the higher stability coefficients should be positive.

Indeed, if the negative P_4 is saved by the extra term (19), the critical coefficient of stability is the one corresponding to R_5. In a rather early note [Ms 3617 (2), pp. 79v–81r] to be studied in the next section, Liouville had discovered that this coefficient was identically zero, and he had also convinced himself that this fact had no influence on the stability. Thus, Principle W has the important, but wrong, consequence that all the higher stability coefficients are positive.

33.   It is of interest to see in which context Liouville discovered the disturbing fact that one of the coefficients of stability was identically zero.

In [Ms 3617 (2), pp. 79v–81r], he studied the quantity

$$
-g\zeta + \iint \frac{\zeta' \, d\sigma'}{\Delta} \;=\; -\frac{2C}{\rho}\sum_i A_i M_i N_i
$$
$$
+ \sum_i A_i \frac{4\pi R_i S_i M_i N_i \sqrt{(\rho^2 - b^2)(\rho^2 - c^2)}}{2n+1}. \tag{72}
$$

This is the difference between the potential of a point on the surface of the equilibrium ellipsoid and the potential at the corresponding point of the perturbed surface. The potential on the equilibrium surface is a constant, and so if quantity (72) is constant, the perturbed figure is also an equilibrium surface. Liouville discovered that this was the case for a particular value of $\zeta = \sum_i h A_i M_i N_i$:

However, one has

$$
\frac{2C}{\rho} - \frac{4RS\sqrt{(\rho^2 - b^2)(\rho^2 - c^2)}}{2i+1} = 0 \tag{73}
$$

for a particular value of (i, B). Thus, confining oneself to this term, new figures of equilibrium. It is $\zeta = Ahxy$: this is very curious.

[Ms 3617 (2), p. 79v]

In fact, one can easily show (see (51) and note 9, formula ∗) that

$$
\frac{2C}{\rho} = \frac{4\pi R_3 S_3 \sqrt{(\rho^2 - b^2)(\rho^2 - c^2)}}{3}. \tag{74}
$$

This identity is of little consequence for the term $M_3 N_3$ does not enter in the development of ζ (§28). However, by the stability condition (10) (and note 9, equation ∗ ∗ ∗),[10] the right-hand side of (74) can be transformed to yield

$$\frac{2C}{\rho} = \frac{4\pi R_5 S_5 \sqrt{(\rho^2 - b^2)(\rho^2 - c^2)}}{5}.$$ (75)

The term corresponding to R_5 in the development of ζ is

$$Ah\,\mu\sqrt{\mu^2 - b^2}\nu\sqrt{b^2 - \nu^2}$$

which is proportional to $Ahxy$, the term mentioned by Liouville above.

Clearly, Liouville was disturbed by this fact, for that means that one can deform the Jacobi ellipsoid into another equilibrium ellipsoid corresponding to $\zeta = Ahxy$ at no cost in energy. Intuitively, this indicates that the Jacobi ellipsoids cannot be stable. In fact quantity (75) is precisely the coefficient of stability.

On the next three pages in his notebook [Ms 3617 (2), pp. 80r–81r], Liouville solved the problem by calculating the equation of the figure that results from a small perturbation of the Jacobi ellipsoid of the form $Ahxy$. He concluded:

> One deduces this new figure from the original one by rotating it by a finite quantity once and for all.
>
> [Ms 3617 (2), p. 81r]

Therefore, it is only natural that the two figures have the same energy and that the stability coefficient corresponding to R_5 vanishes identically. These facts have no influence on the stability.

It is interesting to note that here Liouville implicitly formulates the principle that when a stability coefficient vanishes there is another equilibrium figure close to the original. Since he found no vanishing coefficients of stability except for the one corresponding to R_5, he found no figures of equilibrium close to the Jacobi ellipsoids, except those ellipsoids rotated through a fixed angle. On the other hand, when Poincaré later used the principle on the stability coefficient corresponding to R_{15} (71), he found a family of pear-formed states of equilibria that look somewhat like Figure 3. The question of stability of the pear-shaped configurations was debated by Liapounoff and Darwin and was finally settled by Jeans [1919, pp. 87–102]. He proved that they are unstable, so that there is no stable continuation of the Jacobi ellipsoids for $M > M_3$. [11]

34.   Liouville pursued the question of equilibrium states close to equilibrium ellipsoids in the notebook [Ms 3617 (4)] that is dated 1843–1844, and thereby he almost came to realize the invalidity of Principle W. He first tried with limited success to prove that a homogeneous liquid in absolute equilibrium must be a sphere, a cylinder, or a plane [Ms 3617 (4), pp. 15–23]. On page 28, he then returned to the revolving ellipsoids and explicitly derived the identity

$$\frac{R_3 S_3}{3} = \frac{R_5 S_5}{5},$$ (76)

which he had implicitly used in the earlier note [Ms 3617 (2), pp. 79v–81r]
(§33). First, he remarked "There is an absurdity here" [Ms 3617 (4), p. 31],
but then he discovered that (76) "is the equation determining ρ. We have
seen it already in the past!!!" On the following page, he continues:

> The essential thing is to recognize properly that here is a condi-
> tion equation. It is not less true that there is here an equality that
> hold between two quantities $\frac{RS}{2n+1}$ and that from the beginning it is
> necessary to indicate this possibility. The rest is easy, and one con-
> cludes that the ellipsoids alone give poss. [ible] eq.[uilibrium] figures
> differing slightly from an ellipsoid.

Thus, Liouville realized that (76) represented an exception from Principle
W. If he had kept in mind that $\frac{R_4 S_4}{5} > \frac{R_3 S_3}{3}$ (§28), the contradiction would
have been even clearer. However, the end of the quoted passage clearly
shows that Liouville in general believed in the principle and even considered
its proof easy. Still, a slightly earlier note shows that he had verified the
principle only in special cases and had not yet taken the trouble to write
down the general proof: "One sees the law of the problem; but one must
write down the general proof" [Ms 3617 (4), p. 11].

I have not found any sign that Liouville later attempted to give such a
proof, nor have I found further indication that he realized his Principle W
was basically wrong.

Concluding Remarks

35.   In the last sections, I have tried to answer, as completely as the unpublished sources allow, the third question in §11: How did Liouville obtain
his results? We have seen that he did so by clarifying Laplace's ideas from
the *Mécanique Céleste*, particularly 1° by finding an explicit formula for
the kinetic energy also in the case where the angular momentum and not
the angular velocity is conserved and 2° by using Lamé functions instead
of spherical harmonics.

Hereby, it has also become clear that Liouville did not get exactly the
correct result because he did not investigate stability coefficients of sufficiently high order, because of a wrong theorem about Lamé functions. In
fact, Liouville studied only terms containing Lamé functions with $n \leq 2$.
The perturbed figures corresponding to these terms are also ellipsoids, so
Liouville's result agrees with Thomson's later observation that the Jacobi
ellipsoids are stable when the perturbations are subject to the restriction
always to yield ellipsoids. However, as we have seen, Liouville's method was
capable of treating arbitrary perturbations, and he thought he had covered
the arbitrary case.

Thus, Liapounoff was wrong when he guessed that Liouville's results
were obtained "by a particular hypothesis concerning the disturbances"

(§13). Still, there is one particular note [3616 (3), p. 47r] in which Liouville proposed to study the stability of the Maclaurin ellipsoids under "ellipsoidal perturbations."

He first wrote down the force components and the potential V according to Laplace. He then suggested "augment λ^2 and dim. λ'^2 [λ_1, λ_2 in (3)] with the same quantity ($\delta M = 0$) in the ellipsoid of revolution." He then calculated the variation δV of the potential due to this variation and ends: "the calculation will be easy: carry it out... and then the calculations for the instability of Maclaurin's ellipsoids can be made directly, I think." Liouville's idea was obviously to find, directly from the formulas for the ellipsoids, that the effective potential energy had a negative variation for the Maclaurin ellipsoids flatter than the Jacobi ellipsoid. However, he does not seem to have carried the idea through.

36.   The method that Liouville did follow is almost the same as Liapounoff's and Poincaré's. Anyone who reads their papers, after having read the above summary of Liouville's line of procedure, will be struck by the similarity. It is most conspicuous in the case of Poincaré, who did not refer to Liouville's notes on the subject. One may get the impression that Poincaré had the *Nachlass* of the recently deceased Liouville in front of him. However, the similarity of Poincaré's and Liouville's approaches can be explained in other ways.

Mathematical necessity often explains similarity between two mathematical theories and it does in this case as well, but as Riemann's and Thomson's different approaches show, it does not explain everything. More is explained by the fact that both Liouville and Poincaré found their points of departure in Laplace's *Mécanique Céleste*. Finally, Poincaré did use some of Liouville's theory of stability without knowing it, namely, his formulas for Lamé functions.

Liouville had presented some of these formulas to the *Académie des Sciences* in two addresses [1845b, 1845c] (reprinted in [1845d]) and developed his ideas the following year in two long published letters to P. H. Blanchet [1846e,f]. In Chapter XV, I shall return to the content of these letters and Liouville's remarkable generalization of some of the ideas presented in them.

These two letters were the sources of Poincaré's knowledge about Lamé functions, and he frequently referred to them in [1885]. In the letters, Liouville indicated that his formulas stemmed from 1842 when he had read Gauss's work on potential theory, but his remarks about the origin of the theorems would not give Poincaré any reason to guess that they had been obtained as a means of solving the very problem to which Poincaré applied them.

37.   Having stressed the similarity of Liouville's approach to those of his two successors, I shall briefly indicate a few differences. Both Poincaré and Liapounoff presented more thorough investigations of the Lamé functions to

replace the erroneous Principle W, and both avoided the complicated direct calculation of the sign of the quadratic form (69), Liapounoff by referring to a direct examination of "ellipsoidal perturbations" and Poincaré by an elegant application of a theorem in analysis situs, now called the Poincaré-Hopf Index Theorem.

Moreover, Liapounoff's paper bore the mark of the higher level of rigor of the 1880s as compared to the 1840s, and Poincaré's work was highly influenced by his global theory of differential equations. For example, Poincaré formulated the principle that when a stability coefficient vanishes in a family of equilibrium figures this family must meet another family of equilibrium figures. This is a global version of Liouville's idea explained in §§33 and 34.

38.   A discussion of the influence of Liouville's theory of stability of fluid ellipsoids would be incomplete without a remark about its influence upon Liouville's own later research and publications.

As we have already seen (§36), the fluid ellipsoids, and in particular the use of ellipsoidal coordinates and Lamé functions, were central in Liouville's general studies of potential theory in 1842 and therefore constitute the background of many of the ideas we shall discuss in Chapter XV. Furthermore, the Jacobi ellipsoids inspired Liouville to a study of different aspects of the geometry of ellipsoids such as their geodesics and lines of curvatures and therefore were a great source of inspiration for his further work on differential geometry (cf. Chapter XVII). Finally, the ellipsoidal coordinates were of great importance in Liouville's two largest works on rational mechanics from 1846 and 1847 (cf. Chapter XVI).

In his biography of Liouville in the Dictionary of Scientific Biography, René Taton writes: "Despite its great diversity, [Liouville's] literary output is marked by a limited number of major themes, the majority of which were evident in his first publications" [Taton 1973, p. 383]. The theory of equilibrium of rotating masses of fluid is one of these major themes; it brings coherence to much of Liouville's production in the 1840s.

39.   Why did Liouville not publish his theory of stability of fluid ellipsoids? I cannot give a satisfactory answer to this last question of §11, but I shall conclude by giving some tentative remarks.

The reason does not seem to have been purely mathematical. To be sure, Liouville's research contained an error, but obviously this would have kept him from publishing only if he had discovered it. As pointed out in §34, he does not seem to have seriously questioned the wrong principle and even if he had discovered the mistake, he could rather easily have repaired it. Indeed, Poincaré's later analysis was based on nothing but the formulas and ideas explicit in Liouville's papers [1846e,f]. Remarks in [1843f] and in the notebooks reveal that early in 1843 Liouville had the idea of publishing his theory, but he postponed it indefinitely and stopped this line of research.

The only reason I can find for this postponement and final abandonment of the publication is lack of time. As we saw in §3, Liouville had already felt obliged to leave the area in the 1830s in favor of other occupations, and from his letters and his notes, it appears that the pressure of time grew over the years (cf. Chapter III). Still, Liouville did find time to publish many papers over the next years, but they were generally short notes, often linked directly to the publications of other people. The composition of such notes was clearly more suited for the spare moments left between the many duties of the busy and scrupulous professor who also edited a prestigious journal. It was in such small notes that Liouville revealed the formulas for Lamé functions, although he expanded on the subject at the request of Blanchet. As Liouville explained to Blanchet, the publication of the entire theory of fluid ellipsoids and related topics would have required a rather long book, which he apparently did not find the time to compose.

Liouville's theory of stability of fluid ellipsoids thereby shared the fate of several of Liouville's other most important ideas from the 1840s and 1850s, such as the related spectral theory of integral operators in potential theory, Galois theory, various results in mechanics and doubly periodic functions, of which only the latter became known during his life time.

XII

Transcendental Numbers

1.   In 1844, Liouville showed that transcendental numbers exist, i.e., numbers that are not root in any algebraic equation:

$$f(x) = a_n x^n + a_{n-1} x^{n-1} + \cdots + a_1 x + a_0 = 0 \,, \tag{1}$$

where $a_n, a_{n-1}, \ldots, a_1, a_0$ are integers. Unlike many of Liouville's deeper ideas, the importance of this discovery was immediately recognized by his contemporaries, and it has remained one of his most celebrated results. In this chapter, I shall analyze its historical background, Liouville's gradual approach to the question, and the methods he used. For a more comprehensive treatment of the history of transcendental numbers, see [Waldschmidt 1983].

As I have mentioned in the brief summary of this result in Chapter III, Liouville's interest in transcendental numbers had two origins: 1° a correspondence between Christian Goldbach and Daniel Bernoulli on series representing transcendental numbers and 2° the more well-known investigations made during the 18th century of the nature of specific numbers, in particular, e and π.

Historical Background

2.   The Goldbach-Bernoulli discussion was but a small offshoot of a long correspondence on infinite series. It began on April 28, 1729, when Bernoulli, in passing, claimed about the series

$$\log \frac{m+n}{m} = \frac{n}{m} - \frac{n^2}{2m^2} + \frac{n^3}{3m^3} - \frac{n^4}{4m^4} + \cdots \tag{2}$$

that "not only can it not be expressed in rational numbers but it cannot be expressed in radical or irrational numbers either" [Fuss 1843, II, p. 301]. Goldbach understood the importance of this theorem and asked for "a clear and precise demonstration" [Fuss 1843, II, p. 306]. Confronted with this direct question Bernoulli beat a retreat:

> As an answer to that, I have the honor of telling you that one does not know how to demonstrate it, for perhaps it is not impossible as such but only to the mind of the geometers.
>
> [Fuss 1843, II, p. 310]

In his answer of August 18, 1729, Goldbach pointed out that it was still an open question if

$$\sum_{n=1}^{\infty} \frac{1}{n^2 + \frac{p}{q}n}$$

can be reduced to a root of a rational number.

> What is more, I doubt that anybody has ever produced a series of rational numbers about which he has proved that its sum cannot be expressed in either rational or irrational numbers; however, this is not impossible, and I can give an infinity of such series whose sum cannot be the root with a rational power of a rational number.
>
> [Fuss 1843, II, p. 313]

Now, it was Bernoulli's turn to ask for more explanation "to enlighten me on this matter" [Fuss 1843, II, p. 319], and on October 20th, Goldbach sent him an example:

> Here follows a series of fractions like the ones you have asked me for whose sum is neither rational nor the root of any rational number:
>
> $$\frac{1}{10} + \frac{1}{100} + \frac{1}{10000} + \frac{1}{100000000} + \text{etc.} \qquad [3]$$
>
> The general term is $1/10^{2^{x-1}}$.
>
> [Fuss 1843, II, p. 326]

Goldbach offered no proof, but in his answer Bernoulli sketched one in order to show that he had understood Goldbach:

> because each fraction when transformed into a decimal fraction will have a certain period of the digits which the sum of this series
>
> $$\frac{1}{10} + \frac{1}{100} + \frac{1}{10000} + \frac{1}{100000000} + \text{etc.}$$
>
> does not have, no more than an arbitrary power of an arbitrary number, so that this sum cannot be a root of any rational number.
>
> [Fuss 1843, II, p. 329]

Bernoulli's, and probably Goldbach's, idea seems to be that if the interval between the nonzero digits of an infinite decimal fraction increases fast enough no power of it will be periodic, i.e., rational. However, it is not clear to me how he proved this, for the number above and the further correspondence gives no clue.

It is remarkable that Goldbach and Bernoulli almost insensibly changed the subject during the correspondence. Initially, they discussed if certain series could be *expressed* by rational numbers and roots of rational numbers, that is, by explicit algebraic numbers. Finally, however, they did not go beyond simple rational numbers or roots taken separately, ignoring expressions like $\sqrt[5]{1 + \sqrt{2}}$. Thus, even if they had provided proofs, they would still have been far from the existence of transcendental numbers. (In 1938, Kuz'min showed that Goldbach's number [3] is transcendental).

In his second letter on $\log \frac{m+n}{m}$, Bernoulli remarked that if it could be expressed algebraically "one would have the quadrature of the hyperbola which the geometers have searched for with the same care as that of the circle" [Fuss 1843, II, p. 310]. This leads us directly to the second root of Liouville's research.

3.   The famous classical problem of the quadrature of the circle states: is it possible by ruler and compass alone to construct a square with the same area as a given circle? During the 17th and 18th centuries, this question was reformulated in algebraic terms: can π be written as a finite expression using only rational numbers, rational operations, and square roots? The 18th-century mathematicians generally believed that this was not the case and even that π was transcendental. Similarly, it was believed that e was transcendental. This is related to the quadrature of the hyperbola discussed above.

Still, in the 18th century, only the irrationality of these numbers was proved. First, Euler [1737] showed that e and e^2 are irrational, and in [1766,1767a] Lambert proved with a similar method that π is also irrational. In Lambert's papers, the irrationality of e and π are easy consequences of a more general theorem:

Theorem *If x is a rational number different from zero, then e^x and $tg\, x$ are irrational.*

Indeed, according to this theorem, $e^1 = e$ is irrational, and since $tg\frac{\pi}{4} = 1$, Lambert concluded that $\frac{\pi}{4}$ and thus π are not rational. He derived the general theorem from the following expansions in continued fraction:

$$\frac{e^x+1}{2} = \cfrac{1}{1-\cfrac{1}{\frac{2}{x}+\cfrac{1}{\frac{6}{x}-\cfrac{1}{\frac{10}{x}+\cdots}}}},$$

or

$$= \cfrac{x}{1-\cfrac{x^2}{2+\cfrac{x^2}{6-\cfrac{x^2}{10+\cdots}}}},$$

and

$$\tan x = \cfrac{1}{\frac{1}{x}-\cfrac{1}{\frac{3}{x}-\cfrac{1}{\frac{5}{x}-\cfrac{1}{\frac{7}{x}-\cdots}}}},$$

or

$$= \cfrac{x}{\cfrac{x}{1-\cfrac{x^2}{3-\frac{x^2}{7-\cdots}}}}.$$

If x is a unit fraction, $\frac{1}{\mu}$, $\mu \in N$, these continued fractions immediately show that Euclid's algorithm used on $\frac{e^x+1}{2}$ or $\frac{\tan x}{1}$ never finishes so they cannot be rational numbers. In the case of a mixed fraction $x = \frac{\nu}{\mu}$, however, Lambert's proofs were rather long-winded, and in [1794] Legendre gave a condensed but complete proof of the $\tan x$-part, including a beautifully short demonstration of the

Theorem *The continued fraction*

$$\cfrac{m}{n + \cfrac{m_1}{n_1 + \cfrac{m_2}{n_2 + \cfrac{m_3}{n_3 + \cdots}}}},$$

where $m_i, n_i \in Z \setminus \{0\}$, converges to an irrational number if $\frac{m_i}{n_i} < 1$ for all i larger than some i_0.

From this, Lambert's theorem follows directly.

A simpler proof of the irrationality of e, based on its series expansion

$$e = 1 + \frac{1}{1!} + \frac{1}{2!} + \frac{1}{3!} + \cdots, \tag{4}$$

was later given by Fourier (cf. [Rudio 1892, pp. 57–58]). He observed that if $e = \frac{p}{q}$ $(p, q \in N)$ then by multiplication of the series (4) by $q!$ one finds that

$$\frac{1}{(q+1)} + \frac{1}{(q+1)(q+2)} + \cdots \tag{5}$$

is a natural number. On the other hand, (5) is smaller than

$$\frac{1}{(q+1)} + \frac{1}{(q+1)^2} + \frac{1}{(q+1)^3} + \cdots,$$

which is in turn equal to $\frac{1}{q}$, and thus it cannot be equal to a natural number.

As for the transcendence of π (and e), Legendre probably spoke on behalf of all the leading mathematicians when he declared it a difficult problem:

> It is probable that the number π is not even comprised among the algebraic irrationals, that is, it cannot be the root of an algebraic equation of a finite number of terms whose coefficients are rational, but it seems very difficult to prove this proposition rigorously.
>
> [Legendre 1794, Note 4]

In this note, Legendre himself took one small step in this direction by showing that π^2 was not rational.

Liouville on the Transcendence of e (1840)

4.   Thus, when Liouville entered the scene, it was known that e, e^2, π, and π^2 were irrational numbers. Liouville's aim was to show the transcendence of e, but he only had partial success. In a one-page note in the May 1840

issue of his *Journal* [Liouville 1840b], he used Fourier's method to show that e is not a root of a quadratic equation with rational coefficients, i.e., that there do not exist integers a, b, c ($a > 0$) such that

$$ae + \frac{b}{e} = c. \tag{6}$$

For, inserting the infinite series

$$e = 1 + \frac{1}{1!} + \frac{1}{2!} + \frac{1}{3!} + \cdots \tag{7}$$

and

$$e^{-1} = 1 - \frac{1}{1!} + \frac{1}{2!} - \frac{1}{3!} + \cdots \tag{8}$$

into (6) and multiplying by $n!$, he obtained an equation of the form

$$\frac{a}{n+1}\left(1 + \frac{1}{n+2} + \cdots\right) \pm \frac{b}{n+1}\left(1 - \frac{1}{n+2} + \cdots\right) = \mu, \tag{9}$$

where μ is an integer. Choosing n odd if $b > 0$ and n even if $b < 0$, he made sure that $\pm\frac{b}{n+1}$ is always positive. Furthermore, if n is large enough, the left-hand side of (6) is less than one, and since it is strictly positive, it cannot equal the integer μ. That completes the proof.

In the following monthly issue of his *Journal*, Liouville [1840c] extended the result and showed that e^2 cannot satisfy a second-degree equation either, i.e., that an identity

$$ae^2 + be^{-2} = c \tag{10}$$

cannot be satisfied for $a_1 \in Zh_+$ and $b, c \in Z$. In order to estimate the terms of the series expansion

$$e^x = 1 + \frac{x}{1} + \frac{x^2}{2!} + \cdots + \frac{x^n}{n!}\left(1 + \frac{xe^{\theta x}}{n+1}\right) \qquad \theta \in [0,1] \tag{11}$$

with the Lagrangian remainder for $x = \pm 2$, Liouville observed that 2 divides $(m!)$ less than m times and precisely $(m-1)$ or $(m-2)$ times when $m = 2^i$ or $m = 2^i + 1$. Thus,

$$\frac{2^m}{m!} = \frac{2^{\alpha_m}}{p_m}, \qquad 2 \nmid p_m,$$

where $\alpha_m = 1$ when $m = 2^i$ and $\alpha_m = 2$ when $m = 2^i + 1$. Moreover, $p_m \mid p_n$ when $n \geq m$. Introducing these expressions into (11) for $x = \pm 2$, Liouville obtained

$$e^2 = 1 + \cdots + \frac{2^{\alpha_m}}{p_m} + \cdots + \frac{2^{\alpha_n}}{p_n}(1 + 2\beta) \tag{12}$$

$$e^{-2} = 1 - \cdots \pm \left(\frac{2^{\alpha_m}}{p_m}\right) \pm \cdots \pm \frac{2^{\alpha_n}}{p_n}(1 - 2\gamma), \tag{13}$$

where β and γ tend to zero when n tends to infinity. He finally inserted (12) and (13) into (10), multiplied the equation by p_n, and found

$$a2^{\alpha_n} \cdot 2\beta \pm b2^{\alpha_n} \cdot 2\gamma = \mu,$$

where μ is an integer. This is absurd, for, by taking $n = 2^i$ when $b < 0$ and $n = 2^i + 1$ when $b > 0$, we have $\alpha_n \leq 2$ and the left-hand side is positive and becomes smaller than 1 for sufficiently large values of i.

5.   Hurwitz later pointed out that it is also possible to prove with elementary means that e is not a root of a third-degree equation with rational coefficients. However, Liouville only published one more note on e [1840d] in which he improved the proof of the formula:

$$\lim_{m \to \infty} \left(1 + \frac{1}{m}\right)^m = e.$$

In Cauchy's *Cours d'Analyse* [1821, Chapter VI, p. 167], the ideal of rigor of those days, this was obtained as a simple corollary of the binomial series

$$\left(1 + \frac{1}{m}\right)^m = 1 + \frac{1}{1} + \frac{1}{2!}\left(1 - \frac{1}{m}\right) + \frac{1}{3!}\left(1 - \frac{1}{m}\right)\left(1 - \frac{2}{m}\right) + \cdots$$
$$+ \frac{1}{n!}\left(1 - \frac{1}{m}\right)\left(1 - \frac{2}{m}\right)\cdots\left(1 - \frac{n-1}{m}\right) + \cdots . \qquad (14)$$

Letting m tend to infinity, Cauchy obtained

$$\lim\left(1 + \frac{1}{m}\right)^m = 1 + \frac{1}{1} + \frac{1}{2!} + \frac{1}{3!} + \cdots + \frac{1}{n!} + \cdots .$$

Thus, Cauchy uncritically interchanged the order of the limit and summation, a weakness repeated several times in the *Cours d'Analyse*. If we call the terms of series (14) $f_n(m)$, we would allow the reversal,

$$\lim_{m \to \infty} \sum_{n=0}^{\infty} f_n(m) = \sum_{n=0}^{\infty} \lim_{m \to \infty} f_n(m),$$

either if the sum is uniformly convergent on an interval $[K, \infty)$ or if the family $f_n(x)$ is equicontinuous at $x = \infty$. In modern terms, Liouville detected the lack of equicontinuity and then implicitly used the uniformity in his proof. However, neither of these concepts were developed in 1840, so Liouville was more vague when he stated that the traditional proof:

> is not sufficiently rigorous; in fact, it assumes that for $m = \infty$ the product $(1 - \frac{1}{m})(1 - \frac{2}{m})\ldots(1 - \frac{n-1}{m})$ is reduced to unity, which is correct when n is a fixed number independent of m, but is not correct when one has, for example, $n = m$ or $n = m - 1$.
>
> [Liouville 1840d]

He mended the weakness by focusing on the finite series with the remainder:

$$e = 1 + \frac{1}{1} + \frac{1}{2!} + \cdots + \frac{1}{n!}\left(1 + \frac{\theta}{n}\right); \quad 0 < \theta < 1$$

$$\left(1 + \frac{1}{m}\right)^m = 1 + \frac{1}{1} + \cdots +$$

$$\frac{1}{n!}\left(1 - \frac{1}{m}\right) \cdot \left(1 - \frac{2}{m}\right) \cdots \left(1 - \frac{n-1}{m}\right) \cdot \left(1 + \frac{\eta}{n}\right); 0 < \eta < 1.$$

From the last formula, he concluded that

$$\left(1 + \frac{1}{m}\right)^m = 1 + \frac{1}{1} + \frac{1}{2!} + \cdots + \frac{1}{m!}\left(1 + \frac{\zeta}{n}\right) + \varepsilon_n(m),$$

where $0 < \zeta < 1$ and $\varepsilon_n(m) \to 0$ for $m \to \infty$ for all n. Thus,

$$\left| e - \left(1 + \frac{1}{m}\right)^m \right| = \left| \frac{1}{n!}\frac{\theta - \zeta}{n} - \varepsilon_n(m) \right|,$$

"and one can see that it can be made smaller than any given number, by *first* taking n very large, and *then* letting m increase over all limits" (my italics) [Liouville 1840d].

As we saw in Chapter III, this meticulous attitude to rigor was unusual for Liouville and for his day, and was probably suggested to Liouville by Dirichlet, to whom Liouville attributed the general idea. Liouville developed his improved proof on February 22, 1838 [Ms 3616 (2), pp. 12v–13v], but he did not publish it until 1840, after his two other notes on e.

Construction of Transcendental Numbers (1844)

6.   Another four years elapsed before Liouville returned to the study of transcendental numbers. His interest in the subject seems to have been renewed by reading the correspondence between Goldbach and Bernoulli, which was published by P. H. Fuss in 1843. Two pieces of evidence point in this direction, the first being a reference in his notebook [Ms 3617 (4)] dating from 1843–1844 to "Euler (corresp. T.1, p. 523)." This shows that Liouville at this time knew at least volume 1 of Fuss's edition, which contained the Goldbach-Bernoulli correspondence in volume 2. Second, Liouville actually referred to the Goldbach correspondence in connection with his own transcendental number $\sum 10^{-n!}$, but by 1851 when this happened his memory of it was vague:

> I seem to remember that one can find a theorem of this sort announced in a letter from Goldbach to Euler; but to my knowledge the demonstration has never been given.
> [Liouville 1851e, p. 140]

Liouville must have misremembered the letter for in the correspondence between Goldbach and Euler there is no "theorem of this sort" [Fuss 1843, I, Euler-Goldbach 1965, cf. pp. 160–161]. This also suggests that many years had elapsed since he read the relevant letters in Fuss's book and corroborates the conjecture that it was Goldbach and Daniel Bernoulli's remarks which soon after their publication inspired Liouville to his new and most important work on transcendental numbers.

7. Liouville presented his results first to the *Bureau des Longitudes* on March 27, 1844 (cf. Chapter III) and then to the *Académie des Sciences* on May 13th and 20th. The latter communications were printed in the *Comptes Rendus* as *remarques relatives à des classes très-étendus de quantités dont la valeur n'est ni rationnelle ni même réductible à des irrationnelles algébriques* [Liouville 1844e,f]. The transcendental numbers presented in these notes were not given by infinite series as were those suggested by Goldbach and Daniel Bernoulli and his own research on *e*, but by continued fractions, as in Lambert's approach. Thus, Liouville constructed continued fractions that do not satisfy any algebraic equation:

$$f(x) = a_n x^n + a_{n-1} x^{n-1} + \cdots + a_1 x + a_0 = 0. \tag{15}$$

Before I can describe Liouville's method, I must introduce some notation. Let x be a real number and let $b_0 = [x]$ be the greatest integer not exceeding x. Then, x is of the form $x = b_0 + \frac{1}{x_1}$, where $x_1 > 1$. Similarly, define $b_1 = [x_1]$ and $x_1 = b_1 + \frac{1}{x_2}$, etc. After $m+1$ steps, one has the continued fraction

$$x = b_0 + \cfrac{1}{b_1 + \cfrac{1}{b_2 + \cfrac{1}{b_3 + \cdots \cfrac{1}{b_m + \frac{1}{x_{m+1}}}}}}, \tag{16}$$

where $b_1, b_2, \ldots, b_m \in \mathbf{N}$ and $x_{m+1} > 1$. If x is rational, the process will end after a finite number of steps. Otherwise, the continued fraction is infinite. Liouville called x_{m+1} the "complete quotient" and the partial numerators b_i the "incomplete quotients." The (reduced) convergents $\frac{p_i}{q_i}$ are defined by

$$\frac{p_i}{q_i} = b_0 + \cfrac{1}{b_1 + \cfrac{1}{b_2 + \cdots + \frac{1}{b_i}}} \qquad p_i, q_i \in \mathbf{N}, (p_i, q_i) = 1. \tag{17}$$

Now, we can understand Liouville's first basic proposition pertaining to the continued fraction (16) of a real root of equation (15):

The incomplete quotient $\mu[= b_{n+1}]$ which comes after the convergent $\frac{p}{q} \left[= \frac{p_m}{q_m}\right]$ and which serves in forming the following convergent, will (according to a formula of Lagrange, see the *Mémoires de Berlin* year 1768) for very large values of q end up being constantly smaller than

$$\pm \frac{df(p,q)}{q f(p,q) \, dp}, \tag{[18]}$$

which is an essentially positive expression, where one supposes that

$$f(p,q) = q^n f\left(\frac{p}{q}\right) = a_n p^n + a_{n-1} p^{n-1} q + \cdots + a_0 q^n \qquad ([19])$$

[Liouville 1844e, pp. 883–884; he has $a, b, \ldots, h$ instead of $a_n, a_{n-1}, \ldots, a_0$.]

Liouville did not prove this proposition, but his notes [Ms 3617 (7), pp. 2r–8r] make a reconstruction of his proof possible and, what is even more interesting, they shed light on his gradual approach to transcendental numbers.

8.   The notes, probably written immediately before March 27, 1844, suggest that Liouville's primary aim was still to prove the transcendence of e. Since he contemplated using the continued fraction approach, he needed some characteristics of continued fraction expansions of algebraic numbers. To this end, he naturally turned to Lagrange's three large memoirs [1769, 1770a, and 1770b], to which he referred in the quote above. In these memoirs, Lagrange developed a method of (approximately) solving polynomial equations using continued fractions. He first described a means of finding an interval, $[b_0, b_0 + 1)$, containing a real root of $f(x)$, (15), if it has real roots at all. He then inserted $x = b_0 + \frac{1}{x_1}$ into $f(x)$, and multiplying through by x_1^n, he obtained a similar equation in x_1. Repeating this procedure $m+1$ times, he arrived at the continued fraction, (16). It will be clear from what follows that Liouville paid special attention to the following formula in [Lagrange 1770a, §62]:

$$u = x - \frac{p_m}{q_m} = (-1)^m \frac{1}{q_m(x_{m+1} q_m + q_{m-1})}. \qquad (20)$$

We do not know what else he learned from Lagrange's papers, for the notes he took while reading them are lost. The notebook [Ms 3617 (7)] only contains the notes taken during his subsequent careful study of Vincent's *Note sur la résolution des Équations numériques* [1833–1838] that Liouville had reprinted in the first and third volumes of his journal. In this paper, Vincent proved that if the original equation (15) has a real root then after a number of applications of Lagrange's procedure the equation in x_{n+1} has only one change of signs in the sequence of coefficients. However, after the notes pertaining to Vincent's paper, Liouville remarked:

> We do not see that these results, curious as they may be, can be of use in our current research.

He then threw himself upon the nature of $e^y (y \in)$ using its series expansion

$$\begin{aligned}
e^y &= 1 + \frac{1}{1}y + \frac{1}{2!}y^2 + \cdots + \frac{1}{n!}y^n + \frac{1}{(n+1)!}(y+\theta)y^{n+1} \\
&= \frac{p}{q} + \frac{(1+\theta)y^{n+1}}{(n+1)q},
\end{aligned} \qquad (21)$$

where $q = n!$ (there is an insignificant error in the remainder). He then assumed that e^y is a root of $f(x)$ (15) and used the Taylor series for f to obtain an estimate of the remainder:

$$f(x) = f\left(\frac{p}{q}\right) + u\frac{df\left(\frac{p}{q}\right)}{d\left(\frac{p}{q}\right)} + \cdots, \tag{22}$$

where $x = e^y$ and $x - \frac{p}{q} = u = \frac{(1+\theta)y^{n+1}}{(n+1)q}$. In fact, Liouville only considered the first two terms of the series explicitly written down above, but since he assumed that $e^y = x$ is a simple root of f so that $f'(x) \neq 0$, the inclusion of higher order terms would only have resulted in an extra factor $(1 + \eta)$, where $\eta \to 0$ for n (or q) $\to \infty$, which will not alter the conclusions. Thus, when $f(x) = 0$ (22) yields

$$u = \frac{(1+\theta)y^{n+1}}{(n+1)q} = -\frac{f\left(\frac{p}{q}\right)}{f'\left(\frac{p}{q}\right)} = -\frac{f(p,q)}{q\frac{d}{dp}f(p,q)}, \tag{23}$$

where

$$f(p,q) = q^n f\left(\frac{p}{q}\right) = a_n p^n + a_{n+1} p^{n-1}q + \cdots + a_0 q^n. \tag{24}$$

From (23) Liouville concluded that

$$\frac{(1+\theta)y^{n+1}}{n+1} = -\frac{f(p,q)}{\frac{d}{dp}f(p,q)}, \tag{25}$$

and he remarked that when n increases "the left hand side increases and the right hand side decreases indefinitely" [Ms 3617 (7), p. 4v].

For a moment, he believed that he had thus established the transcendence of e^x, but soon he discovered that this hasty conclusion pertaining to the right-hand side of (25) was wrong, and the attempts made on the following couple of pages to repair the proof were unavailing.

9. Still, his reflections on e^x proved of the utmost importance, for, although they failed to establish the transcendence of e, they led Liouville to a construction of other transcendental numbers. Indeed, on the following pages of his notebook [Ms 3617 (7), pp. 6r–8r], he jotted down the note that he read with some modifications to the *Académie* on May 13th [1844e]. This is the talk of which I have quoted the beginning in §7. The notebooks do not contain any proof of the central proposition of the quote ((18) and above), but Liouville's discussion of e^x suggests that he obtained it by combining the above reflections with Lagrange's formula (20) as follows. Formula (23) remains valid when $\frac{p}{q}$ is replaced by the convergent $\frac{p_m}{q_m}$

of a root x of $f(x)$ and $x - \frac{p_m}{q_m} = u$. Thus, from (20) and (23), one finds that

$$\frac{1}{q_m(x_{m+1}q_m + q_{m-1})} = \pm \frac{f(p_m, q_m)}{q_m \frac{d}{dp_m} f(p_m, q_m)}, \tag{26}$$

or

$$x_{m+1}q_m^2 + q_m q_{m-1} = \pm \frac{q_m \frac{d}{dp_m} f(p_m, q_m)}{f(p_m, q_m)}, \tag{27}$$

such that

$$x_{m+1} < \pm \frac{\frac{d}{dp_m} f(p_m, q_m)}{q_m f(p_m, q_m)}. \tag{28}$$

This immediately implies Liouville's proposition ((18) and above), because $\mu = b_{m+1} \le x_{m+1}$.

Liouville continued his talk by remarking that if $f(x)$ is irreducible we know that $f(p_m, q_m) = q_m f\left(\frac{p_m}{q_m}\right)$ is an integer different from zero, so that (18) implies

$$b_{m+1} < \left| \frac{df(p_m, q_m)}{q_m \, dp_m} \right| = q_m^{n-2} \left| f'\left(\frac{p_m}{q_m}\right) \right|. \tag{29}$$

When $\frac{p}{q}$ approaches x, $f'\left(\frac{p}{q}\right)$ is bounded, and so there exists a constant A such that

$$b_{m+1} < A \, q_m^{n-2}. \tag{30}$$

This is the fundamental inequality that must be satisfied by the continued fraction of a root of an nth-degree equation.

It was then easy for Liouville to construct a continued fraction that does not satisfy an inequality like (30) for any value of n. For example, he started with an arbitrary $b_1 \ge 2$ and chose the following b's successively, such that

$$b_{m+1} = q_m^m. \tag{31}$$

This continued fraction represents the first transcendental number in the history of mathematics. As another example of a transcendental number, Liouville mentioned without proof

$$\frac{1}{l} + \frac{1}{l^{2!}} + \frac{1}{l^{3!}} + \cdots + \frac{1}{l^{n!}} + \cdots, \tag{32}$$

where l is a natural number.

10.   The following week, Liouville inserted in the *Comptes Rendus* a more direct proof of (30) [1844f]. Let x be a real root of $f(x)$ (15) and let the other roots be $x_1, x_2, \ldots, x_{n-1}$. Then,

$$f\left(\frac{p}{q}\right) = a\left(\frac{p}{q} - x\right)\left(\frac{p}{q} - x_1\right) \cdots \left(\frac{p}{q} - x_{n-1}\right), \tag{33}$$

such that

$$\frac{p}{q} - x = \frac{f\left(\frac{p}{q}\right)}{a\left(\frac{p}{q} - x_1\right)\dots\left(\frac{p}{q} - x_{n-1}\right)} = \frac{f(p,q)}{q^n a\left(\frac{p}{q} - x_1\right)\dots\left(\frac{p}{q} - x_{n-1}\right)}. \quad (34)$$

Since $a\left(\frac{p}{q} - x_1\right)\dots\left(\frac{p}{q} - x_{n-1}\right)$ converges to a finite quantity when $\frac{p}{q}$ tends to x, it has a maximum A and hence

$$\left|\frac{p}{q} - x\right| > \frac{1}{Aq^n}. \quad (35)$$

Inserting expression (20) for $\frac{p_m}{q_m} - x$, he finally obtained, as in (26)-(28), that

$$x_{m+1} < Aq_m^{n-2}, \quad (36)$$

which implies (30).

11.   After these two notes, another 7 years passed before Liouville, for the last time, returned in print to transcendental numbers in the April 1851 issue of his *Journal*. In this paper, he quoted his two earlier notes and added some new remarks, the most important being that continued fractions were in fact irrelevant to the question. His notebook [3617 (5), pp. 114v–116r] reveals that already in 1844 or shortly thereafter he became aware that the proof of (35) does not presuppose that $\frac{p}{q}$ is a convergent of x, but only that it is sufficiently close to x. Indeed, in these notes, one can find the first formulation of the famous Liouville theorem (or inequality), which he later published in 1851.

Liouville's Theorem *If x is an algebraic number of degree n (i.e., root of an irreducible nth-degree equation with rational coefficients), then there exists a positive number A such that for all rational numbers $\frac{p}{q}$ with $q > 0$ and $\frac{p}{q} \neq x$ the following inequality holds:*

$$\left|x - \frac{p}{q}\right| > \frac{1}{Aq^n}. \quad (35)$$

In fact, the proof above only works when x is known to be irrational (i.e., $n > 1$), but in [1851e] Liouville gave a simple argument for the case $n = 1$. (According to Waldschmidt [1983, p. 99], Dirichlet had already proved (35) for $n = 1$ in 1842).

Liouville's theorem implies that if we can find a sequence $\frac{p_i}{q_i}$ converging to x and such that no A exists that will make (35) hold true for all i, then x is not a root of an nth-degree equation, and clearly not a root of an equation of smaller degree either (because $\frac{1}{Aq^n} < \frac{1}{Aq^m}$ for $q > 1$ and $n > m$). Hence, if (35) is false for all A and all n, then x must be transcendental.

As an example, Liouville [1851e] repeated the number:

$$x = \frac{1}{l} + \frac{1}{l^{2!}} + \frac{1}{l^{3!}} + \cdots + \frac{1}{l^{n!}} + \cdots. \tag{32}$$

Taking $\frac{p_i}{q_i}$ equal to the sum of the first i terms, one has $q_i = \ell^{i!}$ and

$$x - \frac{p_i}{q_i} = \frac{1}{l^{(i+1)!}} + \frac{1}{l^{(i+2)!}} + \cdots \leq \frac{2}{l^{(i+1)!}} \leq \frac{2}{q_i^{i+1}}, \tag{37}$$

and so (35) cannot be satisfied by any fixed A and n. Thus, Liouville concluded that (32) is transcendental. For $\ell = 10$, this is the often quoted Liouvillian number

$$0.1100010000\ldots,$$

with 1's in the i!th place after the decimal point and zeros otherwise.

The reason why continued fractions worked so well in the initial investigations is that they converge quicker than any other sequence of $\frac{p_i}{q_i}$. Thus, if one can find a sequence that converges quickly enough to x to falsify (35) for all A and n's, the convergents of x's continued fraction will a fortiori falsify it.

Liouville finished his paper by giving other examples of transcendental numbers and generalizing the procedure to polynomials with complex integers as coefficients.

The Impact of Liouville's Discovery

12.   Considering the importance of Liouville's discovery of transcendental numbers, its announcement in 1844 was very modest. Liouville's verbal communication of May 13th was published on less than two pages of the *Comptes Rendus*, and the *Nouvelle démonstration d'un théorème sur les irrationnelles* only took up one page of the *Comptes Rendus* of the following week. Liouville even blurred the content of the verbal communication by presenting at the same time a remark on a passage in Newton's *Principia*. The combination of these two notes indicates that in 1844 Liouville did not appreciate the importance of his transcendental numbers, whereas his 1851 paper may testify a later recognition of their importance.

Transcendental numbers of the Liouvillian type have later been studied in great detail (cf. [Maillet 1906 and Schneider 1957]), but Liouville himself was probably disappointed that he was not able to show the transcendence of e. In fact, in 1845 Liouville returned to this problem, and he even thought that his methods provided a proof of the transcendence of π. His note [Ms 3618 (5), pp. 9v–10r] ends:

Thus, π is not the root of any algebraic equation with rational coefficients.

However, he soon discovered a mistake and crossed out the note.

It may have been a consolation for Liouville that it was one of his former protégés, Hermite, who solved the problem concerning e. In his work [1873], Hermite even stated that some of his methods were similar to Liouville's ideas in [1844e,f], and he also referred to [Liouville 1840b,c].

On the basis of Hermite's ideas, Lindemann [1882a] finally showed that e^x is irrational when x is a real or complex algebraic number and concluded that since $e^{i\pi} = -1$ the number π must be transcendental. Thereby he had settled the problem of the quadrature of the circle. He presented his solution to this classical problem to the Berlin Academy on June 22, 1882, and on July 10th Hermite broke the news to the Paris Academy [Lindemann 1882b]. Thus, there is a good chance that Liouville also learned about it before his death in September of that year.

Eight years earlier, Cantor had shown that the algebraic numbers are countable, whereas the real numbers are not, and thereby he had given

> a new proof of the theorem, first proved by Liouville, that in each given interval $(\alpha, \ldots, \beta)$ there exist infinitely many transcendental numbers.

Thus, Cantor wrote in his first paper [1874] on set theory. In fact, it is easy to modify Liouville's number $\sum 10^{-n!}$ so that it ends up in a given interval. The new proof of Liouville's result was therefore one of the first triumphs of Cantor's works on cardinality, but Liouville's methods have continued to fascinate number theorists, as can be seen from Roth's inaugural lecture [1962].

XIII

Doubly Periodic Functions

General Introduction; Diffusion of Liouville's Ideas

1. In the mathematical community of today, Liouville's name is primarily attached to the theory of complex functions. The reason is that among the many mathematical theorems named after him, the most celebrated is the one stating that a bounded holomorphic function boldmath $C \rightarrow$ boldmath C is necessarily constant. It is less well known that Liouville conceived of this theorem as the basis of an elegant approach to elliptic functions. The basic element of this approach is to deduce the properties of elliptic functions from their double periodicity, rather than from their representation as inverses of elliptic integrals. After Liouville this has become the standard approach in elementary textbooks to the extent that an elliptic function is nowadays usually defined as a doubly periodic meromorphic function. In this chapter, however, I shall follow Liouville and call such functions doubly periodic, whereas the term "elliptic function" will be used in the Jacobian sense to denote the inverse of an elliptic integral.

Considering the great impact of Liouville's ideas on the theory of doubly periodic functions, a biography of Liouville could hardly be considered complete if this subject was left out. Therefore, I have included this chapter despite the fact that Jeanne Peiffer has recently published an excellent paper on this subject [Peiffer 1983]. Another justification for this chapter is that I shall emphasize other aspects of the story than did Jeanne Peiffer, such as Liouville's early paper on the division of elliptic functions, the chronological development of his ideas on doubly periodic functions, and the essential elements of his final approach. In particular, I shall reconstruct Liouville's first proof of Liouville's theorem, a proof which she only implicitly mentioned. On the other hand, I shall gloss over Liouville's ideas on the foundation of complex analysis and their relation to the ideas of his contemporaries, an intricate question discussed in detail by Peiffer.

2. Jeanne Peiffer was the first to illustrate the great use one can make of Liouville's notebooks in the study of his life and mathematical work. In the case of doubly periodic functions, these notes are particularly valuable because Liouville himself published very little on the subject. These notebooks reveal that Liouville developed his new approach to elliptic functions during the fall of 1844. On December 9th, he informed the *Académie* about it in a short note [Liouville 1844g], which in the *Compte Rendu* of the meeting takes up less than one page. It contains a formulation of Liou-

ville's theorem for doubly periodic functions; the idea that doubly periodic functions are characterized solely by their periods, poles, and zeros; and an indication that Jacobi's theorem on division of elliptic functions can be solved on the basis of these observations. Thus, the readers of the *Comptes Rendus* would hardly guess the importance of Liouville's approach; and to make it worse, the note was not a separate publication, only an appendix to a remark on a paper of Chasles, which dealt with other aspects of elliptic functions.

Liouville did not return in print to doubly periodic functions until 1851 in a slightly longer note in the *Comptes Rendus*. Here, he included an index of a set of notes taken by Borchardt during a private course Liouville had given him and Joachimsthal in 1847. In 1855, Liouville republished the two previous notes in his *Journal* together with a few additional comments, and thereafter he did not return to the subject. Two years before Liouville's death, Borchardt published the lecture notes from 1847 in Crelle's Journal which he then edited. By then, however, Liouville's ideas had become known to a wide circle of mathematicians through courses given by Liouville at the *Collège de France* during 1851 and 1859 and privately (in addition to the course mentioned above, he lectured to Chebyshev [cf. Chebyshev 1852, pp. 8–9]). Moreover, Borchardt's lecture notes were circulated among a wide circle of French and German mathematicians. In fact, Liouville had allowed Borchardt to show the lecture notes to Jacobi, and on September 28, 1848, Borchardt informed Liouville about the positive reception his ideas had received from the master of elliptic functions:

> Mr. Jacobi writes to me that he has read your theory with great pleasure, that he has drawn much information from it, that he strongly wishes to see it published soon, and that he plans to write to you then.

Liouville quoted this passage in [1855i] "at the risk of being accused of vanity." Dirichlet also got hold of a copy of the notes, and in this way the Berlin group was informed of Liouville's theory.

Briot and Bouquet were among the "select audience" who followed Liouville's course during 1851 at the *Collège de France* [Bertrand 1867, p. 3]. In their first memoir, they referred t Collo'eville's "beautiful theory," and in their widely read book, *Théorie des fonctions elliptiques*, from 1859, they openly admitted their debts to their former teacher. Through this book, Liouville's approach became known to the entire mathematical world.

3.   Liouville's reluctance to publish his theory may give the impression that he did not care for the priority, but that impression is quite wrong. In fact, most of his more or less public utterances about doubly periodic functions were meant to establish his priority. Thus, the 1851 note was triggered off by a review of Hermite's work on this subject written by Cauchy, and it was followed by a rather sharp discussion between Cauchy and Liouville. The course held the same year at the *Collège de France* grew out of this episode. Furthermore, it was clearly not by accident that Liouville's republication of

the *Compte Rendu* notes [1855i] appeared only a few months after Briot and Bouquet's first paper had been presented to the *Académie* (during February 1855) or that the last course at the *Collège de France* took place the very year Briot and Bouquet's book [1859] was published. During his last years, Liouville, in his notebooks, bitterly accused his former students of having stolen his theory. This opinion was shared by Weierstrass [Biermann 1960]. However, it is not quite fair to Briot and Bouquet, for their work was much more than a reproduction of Liouville's ideas. They placed these ideas in a broader framework of a general theory of complex functions, including also Cauchy and Hermite's ideas.

As has been pointed out by Peiffer [1983, pp. 235–240], Liouville did not develop such a general theory, and only on a few occasions did he use Cauchy's results of complex integration. This may seem strange at first sight, but one has to keep in mind that although Cauchy had published widely on complex integration and the calculus of residues he had not by 1844 produced anything approaching a connected theory of complex functions. In particular, he had only in passing mentioned the possibility of a geometric representation of complex numbers as points in a plane despite the fact that Argand had advanced the idea in 1817 and Gauss had expressed his approval in 1831 (cf. [Bottazzini 1986, p. 139], Caspar Wessel's Danish publication from 1797 went unnoticed at the time).

In this respect, Liouville was ahead of Cauchy, for in 1836 he and Sturm had explicitly drawn contours in the complex plane, in a paper on *un théorème de M. Cauchy relatif aux racines imaginaires des équations*, which was undoubtedly known to Cauchy. Although Liouville's arguments did not rely on the idea of a complex plane, his vocabulary shows that he did introduce it in his lectures on elliptic functions. In particular, the term *period parallelogram* seems to be due to Liouville (cf. [Osgood 1901]). Of course, skeptics may argue that the use of this term in the published lecture notes [Liouville 1880, p. 278] could be a later addition due to Borchardt, but in fact Liouville also referred to a *rectangle des périodes* in his notes from 1844 (cf. [Ms 3617 (5), p. 92v]).

Division of the Lemniscate

4. The paper by Sturm and Liouville shows that Liouville had thought about questions related to complex analysis before 1844. He was also well informed about the theory of elliptic functions. In the 1830s, he had shown that the elliptic integrals of the first and second kind cannot be expressed in explicit finite form, neither when considered as a function of the upper limit in the integral [Liouville 1834c] nor as a function of the modulus [Liouville 1840 a] (cf. Chapter IX; for a definition and a classification of elliptic integrals, the reader is referred to §5 of that chapter). This research

was mostly based on the results of Legendre, but it inspired Liouville to read Abel and later Jacobi (cf. Chapter IX, §25). Thereby he became acquainted with their new approach to the subject, which rested on two ingenious novelties: 1° to extend the theory to complex values of the variables and 2° instead of the integrals to emphasize the inverse functions. Both Abel and Jacobi showed that these inverse complex functions have two periods in the same way as the inverse of the integral $\int_0^t (1 - x^2)^{\frac{1}{2}} \, dx$, that is, $\sin t$, has one period.

This double periodicity, which was to become the foundation of Liouville's theory, was explicitly underscored in a paper written by Liouville in the autumn of 1843 [1843k], that dealt with the division of the lemniscate. I shall summarize the context and content of this paper, because it shows some important elements of Abel's theory, and illustrates Liouville's knowledge of the work of his greatest hero.

5.   Abel introduced elliptic functions in his masterpiece *Recherches sur les fonctions elliptiques* [1827–1828]. He considered an elliptic integral of the first kind:

$$\alpha = \int_0^x \frac{dx}{\sqrt{(1 - c^2 x^2)(1 + e^2 x^2)}} \tag{1}$$

and defined its inverse function

$$\varphi(\alpha) = x. \tag{2}$$

For the sake of brevity, he also introduced two other functions of α, namely,

$$f(\alpha) = \sqrt{1 - c^2 \varphi^2(\alpha)}, \qquad F(\alpha) = \sqrt{1 + e^2 \varphi^2(\alpha)}. \tag{3}$$

The elliptic function φ satisfies the fundamental addition formula:

$$\varphi(\alpha + \beta) = \frac{\varphi(\alpha) f(\beta) F(\beta) + \varphi(\beta) f(\alpha) F(\alpha)}{1 + e^2 c^2 \varphi^2(\alpha) \varphi^2(\beta)}. \tag{4}$$

Initially, Abel only defined φ for real values of α, but he formally inserted $i\alpha$ in (1) and found that $i^{-1}\varphi(\alpha i)$ is changed into $\varphi(\alpha)$ when c and e are interchanged. Thereby $\varphi(\alpha i)$ is defined; f and F have similar continuations to the imaginary axis. Abel then defined φ of an arbitrary complex number $\alpha + i\beta$ by the addition formula, (4). It is remarkable that he did not need complex integration in order to extend $\varphi(\alpha)$ analytically.

Abel showed that the function φ has two periods, 2ω and $2i\omega'$, where

$$\frac{\omega}{2} = \int_0^{(1/c)} \frac{dx}{\sqrt{(1 - c^2 x^2)(1 + e^2 x^2)}}; \qquad \frac{\omega'}{2} = \int_0^{(1/e)} \frac{dx}{\sqrt{(1 - e^2 x^2)(1 + c^2 x^2)}}. \tag{5}$$

In fact, $\omega + i\omega'$ is also a period; Abel found that φ has zeros at the values $m\omega + ni\omega'$, $(m, n \in \mathbf{Z})$ and infinities (poles) at the points $(m + \frac{1}{2})\omega + (n + \frac{1}{2})i\omega'$. Moreover, φ satisfies the symmetry relation

$$\varphi(m\omega + ni\omega' \pm \alpha) = \pm(-1)^{m+n} \varphi(\alpha), \qquad (m, n \in \mathbf{Z}). \tag{6}$$

By repeated use of the addition theorem (4), Abel easily solved the problem of multiplication of φ which is to express $\varphi(m\alpha)$ in terms of $\varphi(\alpha)$. He showed that for $n \in N$

$$
\begin{aligned}
\varphi(2n\alpha) &= \varphi(\alpha)f(\alpha)F(\alpha)R_1 , \\
\varphi((2n+1)\alpha) &= \varphi(\alpha)R' ,
\end{aligned}
\tag{7}
$$

where R_1 and R' are rational functions of $\varphi^2(\alpha)$. The primary aim of the *Recherches* was to solve the inverse problem of division, that is, given $\varphi(m\alpha)$, find $\varphi(\alpha)$. This can be obtained by solving (7) considered as equations in $\varphi(\alpha)$. It was Abel's great merit to show that these equations can be solved by radicals in terms of $\varphi(m\alpha)$ and the quantities e, c, an mth root of unity, $\varphi(\frac{\omega}{m})$, and $\varphi\left(\frac{i\omega'}{m}\right)$.

6.   The question therefore arises: can $\varphi(\frac{\omega}{m})$ and $\varphi(\frac{i\omega'}{m})$ be expressed by radicals in terms of the moduli e, c, and some roots of unity? Since $\varphi(m\cdot\frac{\omega}{m})$ $= \varphi(m \cdot \frac{i\omega'}{m}) = 0$, this corresponds to solving (7) for vanishing left-hand sides. It is called the division problem for complete elliptic functions.

Abel [1827–1828, §22] remarked that this problem cannot in general be solved by radicals, but he claimed that it was possible in a number of cases, for example, for $e = c$. In that case, integral (1) can be changed by a simple substitution of variables into the simple lemniscatic integral ($e = c = 1$):

$$
\alpha = \int_0^x \frac{dx}{\sqrt{1 - x^4}} .
\tag{8}
$$

(It takes its name because if $AMBN$ is a lemniscate and if x denotes the chord AM, then $\alpha = $ the arc AM is given by (8).)

The problem of finding $\varphi(\frac{\omega}{m})$ corresponds in this case to the division of the lemniscatic arc $AMBNA$ in m equal arcs (Fig. 1). For, according to (5), the half arc AMB, corresponding to $x = 1$, is equal to $\frac{\omega}{2}$, and so the entire arc is ω. The chord corresponding to the mth part of the arc $AMBNA$ is therefore $x = \varphi(\frac{\omega}{m})$.

In *Disquisitiones Arithmeticae*, Gauss claimed that he could treat this problem with the same methods he used in the division of the circle [Gauss

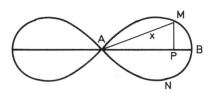

FIGURE 1

1801, §335], and after his death, his notebooks revealed that in his investigations of this problem he had anticipated many of Abel's ideas [cf. Houzel 1978, p. 17]. Abel corroborated Gauss's assertion by showing that the division of the lemniscate by m is possible by ruler and compass (i.e., $\varphi(\frac{\omega}{m})$ is expressible by square roots), if m is of the form $m = 2^n \cdot p_1 \cdots p_k$, where $p_1, p_2, \ldots, p_k$ are different Fermat primes, that is, primes of the form $2^{n_i}+1$, [Abel 1827–1828, §40]. As pointed out by Abel, this is the same requirement that Gauss had found for the construction of the regular polygon of m sides [Abel 1827–1828, §22; Gauss 1801].

The division of the lemniscate by radicals, which Abel claimed he could solve, is not treated in its full generality in the *Recherches* nor in any of his later writings, but he came most of the way. Indeed, as in the circular case, it is enough to treat division by prime numbers, and as explained above, the division by 2 is simple. Abel further showed that division by an odd prime of the form $4\mu + 1$ is possible by radicals. What separates these primes from those of the form $4\mu + 3$ is that the former can be written as the sum of two squares or as the product of two conjugate Gaussian integers:

$$4\nu + 1 = \alpha^2 + \beta^2 = (\alpha + i\beta)(\alpha - i\beta). \tag{9}$$

Abel used this fact. He first showed that $\varphi(\frac{m\omega}{\alpha+i\beta})$ can be obtained by radicals. By changing i into $-i$, he obtained $\varphi(\frac{m\omega}{\alpha-i\beta})$, and by the addition theorems, he could express the quantities:

$$\varphi\left(\frac{2m\alpha\omega}{4\nu + 1}\right) = \varphi\left(\frac{m\omega}{\alpha + i\beta} + \frac{m\omega}{\alpha - i\beta}\right)$$

by radicals. Finally, since 2α and $4\nu + 1$ are relatively prime, he could choose $m, t \in \mathbb{Z}$, such that $1 = 2m\alpha - (4\nu + 1)t$, and then obtain

$$\varphi\left(\frac{2m\alpha\omega}{4\nu + 1}\right) = \varphi\left(\frac{\omega}{4\nu + 1} + t\omega\right) = (-1)^t \varphi\left(\frac{\omega}{4\nu + 1}\right).$$

Now, the reader will of course ask: if Abel can solve the problem of division by a Gaussian integer, $\alpha + i\beta$, has he not then a fortiori solved the division by a real integer? The answer is yes, but Abel did not treat the division by an arbitrary Gaussian integer. In the proof, he used that one of the numbers, α or β, is odd and the other is even, that α and β are mutually prime, and that $\alpha^2 + \beta^2$ is a prime. Therefore, Abel's proof cannot easily be extended to an arbitrary Gaussian integer, and he never published a proof for a prime of the form $4\mu + 3$, let alone for an arbitrary Gaussian integer.

7.   This is precisely the problem solved by Liouville in [1843k]. Abel had treated the even and odd cases separately because of the two different expressions in (7). Liouville simplified the discussion by remarking that they both state that $\varphi^2(m\alpha)$ is a rational function $\frac{T}{S}$ of $\varphi^2(\alpha)$, where T and S are irreducible entire functions. With his somewhat more liberal

use of complex numbers, Liouville saw that this also holds when m is a Gaussian integer. In order to divide the complete lemniscatic function, where $\varphi(m \cdot \alpha) = 0$, Liouville had to solve

$$\frac{T(\varphi^2(\alpha))}{S(\varphi^2(\alpha))} = 0 \quad \text{or} \quad T(\varphi^2(\alpha)) = 0 \tag{10}$$

for $x = \varphi^2(\alpha)$. Since φ has zeros in the points $r\omega$, where r is a Gaussian integer, we have

$$\frac{T(\varphi^2(\frac{r\omega}{m}))}{S(\varphi^2(\frac{r\omega}{m}))} = \varphi^2(m(\frac{r\omega}{m})) = \varphi^2(r\omega) = 0. \tag{11}$$

Thus, the roots of $T(x) = 0$ are of the form $\varphi^2(\frac{r\omega}{m})$. Abel had also made use of these transcendental expressions for the roots, but for his method to work, he had been forced to list a complete set of values of r for which $\varphi^2(\frac{r\omega}{m})$ runs through the different roots precisely once. That led to differences in the proofs for odd and even m and for the specific Gaussian integers he treated.

Liouville overcame these problems by using a theorem of Abel's published a year after the *Recherches* [Abel 1829a]:

Given a polynomial equation whose roots can all be expressed rationally in terms of one of them, x, say, $x_i = \theta_i(x)$, the equation is solvable by radicals if

$$\theta_i\theta_j(x) = \theta_j\theta_i(x). \tag{12}$$

The requirement (12) is equivalent to saying that the Galois group is Abelian, but of course Abel did not know that.

This theorem can be applied to the equation $\frac{T(x)}{S(x)} = 0$ above, for according to the multiplication formula, all the roots $\varphi^2(r\frac{\omega}{m})$ are rational functions of one of them, namely, $\varphi^2(\frac{\omega}{m})$. Moreover, let θ_1, θ_2 be any two of these functions defined by

$$\theta_1(\varphi^2(\frac{\omega}{m})) = \varphi^2(r_1\frac{\omega}{m})$$

$$\theta_2(\varphi^2(\frac{\omega}{m})) = \varphi^2(r_2\frac{\omega}{m}).$$

Then

$$\theta_1(\theta_2(\varphi^2(\frac{\omega}{m}))) = \theta_1(\varphi^2(r_2\frac{\omega}{m})) = \varphi^2(r_1r_2\frac{\omega}{m})$$

$$= \varphi^2(r_2r_1\frac{\omega}{m}) = \theta_2(\theta_1(\varphi^2(\frac{\omega}{m}))).$$

Thus, according to Abel's theorem, the equation relative to the division of the lemniscate by a Gausian integer is solvable by radicals.

8. This proof surpassed Abel's, both in generality and in elegance. But, as Liouville pointed out, "the above analysis... is... implicitly contained in Abel's works." Indeed, in *Recherches*, Abel announced:

In all these cases, the equation $P_n = 0[T = 0]$ can be solved by one and the same uniform method, which is applicable to an infinity of other equations of all degrees. I shall explain this method in a separate memoir.

[Abel 1827–1828, §34]

This "uniform method" is no doubt the one expounded in [Abel 1829a] and used by Liouville. In the introduction to that paper, Abel even announced that "after having explained this theory in general, I shall apply it to the trigonometric and elliptic functions" [Abel 1829a], and the last section does indeed contain an application of the theorem above to the Gaussian problem of division of the circle. However, one looks in vain for the promised application to elliptic integrals. As pointed out by Liouville, the reason is probably that "death prevented him from carrying out his plan." Whether Abel had planned to use his algebraic theorem in the complex case is doubtful; at any rate, he only treated division of the trigonometric functions by ordinary integers. Still, the project is so strongly indicated by Abel, that any good mathematician who had studied his works closely could have supplied the details.

Liouville's paper [1843k] therefore provides us with the conclusion that by 1843 Liouville mastered Abel's ideas on elliptic functions and on the theory of equations; he was probably the only French mathematician apart from Hermite who did that. In particular, Liouville had fully appreciated the importance of doubly periodicity:

The above analysis, [he wrote in the conclusion] is based on the transcendental expression of the roots deduced from the consideration of the two periods of elliptic functions. Without this preliminary knowledge, it seems very difficult indeed to reach a satisfactory result concerning the required algebraic solution.

[Liouville 1843k]

The Discovery of Liouville's Theorem

9. The paper discussed above was provoked by Libri's unfounded claim that he had proved the theorem on the division of the lemniscate before Abel. The quarrel between Libri and Liouville on this point had in turn arisen as a result of the latter's review of Hermite's first great paper on the division of the Abelian functions.

In fact, Liouville's new interest in elliptic and Abelian functions was probably kindled by his discussions with this promising young student. It was also nourished by his investigations on geodesics on ellipsoids [Liouville 1844c], a subject that Jacobi had already shown would lead to Abelian equations. Indeed, this geometrical approach to Abelian functions was what made Liouville respond to Chasles' geometrical work on elliptic functions, and gave him the occasion to announce his theory of doubly periodic functions [1844g]. Finally, in 1843, Liouville had begun to penetrate into the works of Galois, an enterprise that surely made him receptive to the problem of solving the equation of division of elliptic functions by radicals.

Now, we shall answer the question: how was Liouville led to his own approach to elliptic functions and in particular to Liouville's theorem? This question can be answered in some detail by considering his notebook [Ms 3617 (5)]. The first notes of the book are carefully dated July 15, 1844, etc., but after August 1st, when the theory really began to take off, Liouville was too absorbed to care for any dating. To judge from the intensity of the ideas and the notes, they may well have been completed in a month or so, and probably no later than December 1844, when he presented his theory to the *Académie*. It is a special attraction of these notes that they contain Liouville's first derivation of Liouville's theorem. Until Peiffer dug out traces of this proof together with Liouville's other proofs, it had been a secret how Liouville had originally argued for his theorem. For when Borchardt published the lecture notes, he chose to replace Liouville's proof with another one, giving only vague hints of a proof he had learned from Liouville, and which is different from the original one. For these reasons, I shall summarize the content of this notebook.

10.   The first notes from July 1844 contain little of interest. Among other things, Liouville continued to study Holmboe's 1839 edition of Abel's works, and he made several notes concerning various aspects of elliptic functions. On August 1st, the notes take a new turn:

> Mr. Hermite would, he says, use the development of a function $f(x)$ in sine- and cosine-series to prove the impossibility of the existence of two real and incommensurable periods of a function $f(x)$ of one variable.
>
> [Ms 3617 (5), p. 35v]

Jacobi [1828–1834, 2nd part, §1] had proved that if a nontrivial holomorphic function has two periods then they are linearly independent over R. Hermite's idea was to reprove this in a special case using trigonometric series. Since "any" function with just one real period can be expanded in Fourier series, this is a natural suggestion, and that was precisely Liouville's immediate reaction: "Nothing seems to me to be simpler." The subsequent proof confirmed this:

Assume, with Liouville, that f has periods $\alpha, \beta \in$ R. Develop f in a trigonometric series according to the period α:

$$f(x) = A_0 + \sum_{i=1}^{\infty} A_i \cos \frac{2i\pi x}{\alpha} + \sum_{i=1}^{\infty} B_i \sin \frac{2i\pi x}{\alpha} \tag{13}$$

$$= A_0 + \sum_{i=1}^{\infty} A_i \cos \frac{2i\pi(x+\beta)}{\alpha} + \sum_{i=1}^{\infty} B_i \sin \frac{2i\pi(x+\beta)}{\alpha}, \tag{14}$$

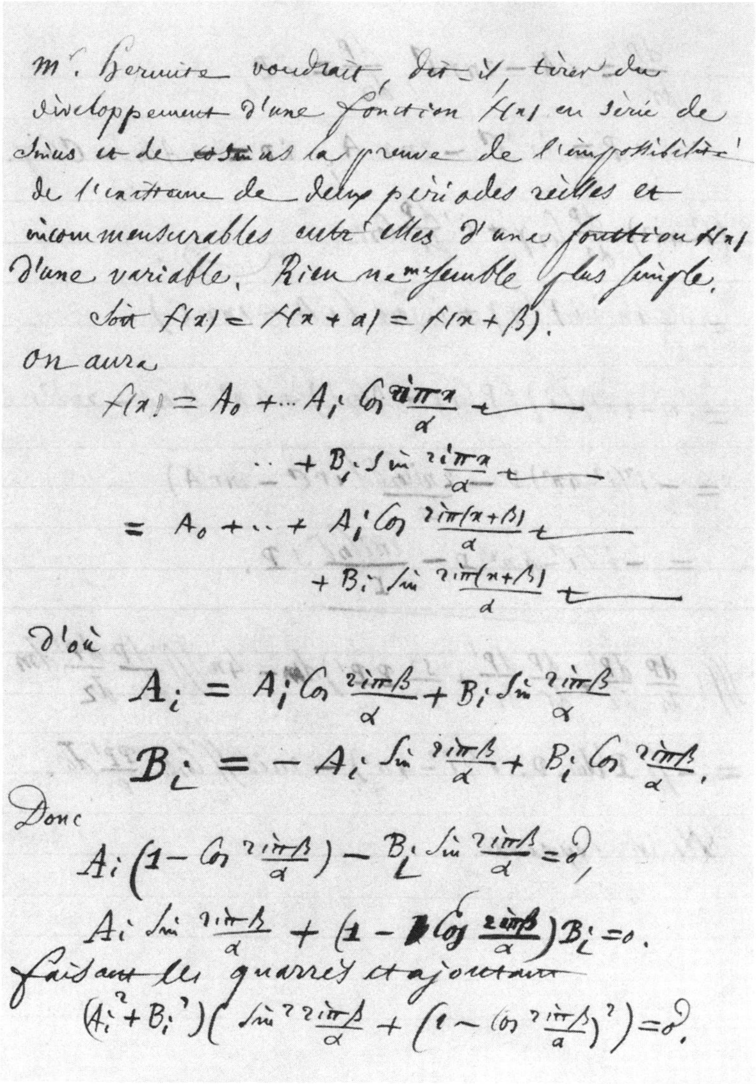

M. Hermite voudrait, dit-il, tirer du diveloppement d'une fonction $f(x)$ en série de sinus et de cosinus la preuve de l'impossibilité de l'existence de deux périodes réelles et incommensurables entre elles d'une fonction $f(x)$ d'une variable. Rien ne me semble plus simple.

Soit $f(x) = f(x + a) = f(x + \beta)$.

On aura

$$f(x) = A_0 + \cdots A_i \cos \frac{2 i \pi a}{a} +$$

$$\cdots + B_i \sin \frac{2 i \pi a}{a} +$$

$$= A_0 + \cdots + A_i' \cos \frac{2 i \pi (x + \beta)}{a} +$$

$$+ B_i' \sin \frac{2 i \pi (x + \beta)}{a} +$$

d'où

$$A_i = A_i' \cos \frac{2 i \pi \beta}{a} + B_i' \sin \frac{2 i \pi \beta}{a}$$

$$B_i = - A_i' \sin \frac{2 i \pi \beta}{a} + B_i' \cos \frac{2 i \pi \beta}{a},$$

Donc

$$A_i \left(1 - \cos \frac{2 i \pi \beta}{a} \right) - B_i \sin \frac{2 i \pi \beta}{a} = 0,$$

$$A_i \sin \frac{2 i \pi \beta}{a} + \left(1 - \cos \frac{2 i \pi \beta}{a} \right) B_i = 0.$$

faisant les quarrés et ajoutant

$$(A_i^2 + B_i^2) \left(\sin^2 \frac{2 i \pi \beta}{a} + \left(1 - \cos \frac{2 i \pi \beta}{a} \right)^2 \right) = 0.$$

PLATE 27. The beginning of Liouville's notes on doubly periodic functions. The inspiration from Hermite is mentioned. (Bibliothèque de l'Institut de France, Ms 3617 (5) p 76v)

the last equality resulting from $f(x) = f(x + \beta)$. By equating the ith coefficients and adding the squares of the resulting equations, Liouville obtained for $i \geq 1$:

$$(A_i^2 + B_i^2)\left(\sin^2\frac{2i\pi\beta}{\alpha} + \left(1 - \cos\frac{2i\pi\beta}{\alpha}\right)^2\right) = 0. \qquad (15)$$

The last factor is zero only for $\frac{2i\pi\beta}{\alpha} = 2m\pi$, but if $\frac{\alpha}{\beta} \notin \mathbf{Q}$, this implies that $i = m = 0$. Thus, for $i \neq 0$, the first factor is zero, i.e., $A_i = B_i = 0$, and so "the pretended function $f(x)$ is reduced to a simple constant" (the note is quoted in extenso in [Peiffer 1983, pp. 241–242]). At first, Liouville was apparently of the opinion that this proof was a bit farfetched, for he concluded the argument:

> But isn't this to look for noon at two o'clock?

11.   However, he soon discovered the fruitfulness of the method, for it provided him with the key to his theorem. He formulated it for the first time on page 76v in the case of one real and one imaginary period:

> If $\frac{\varphi(x)}{\psi(x)}$ preserves the two periods 2ω, $2\overline{\omega}\sqrt{-1}$ and does not become infinite for real or imaginary values of x, it must be a simple constant. The development in sine- and cosine-series proves it.
> $$[\text{Ms 3617 (5), p. 76v}]$$

To avoid confusion, I shall use Liouville's own term, "Liouville's [my] principle," to denote the theorem that a doubly periodic function without poles is identically constant. As usual, "Liouville's theorem" shall refer to the more general theorem stating that a bounded holomorphic function on boldmath C is a constant.

How did Liouville prove his principle by way of trigonometric series? Jeanne Peiffer, who discusses the most interesting notes in this notebook concerning the development in trigonometric series, concludes: "We still do not know how Liouville deduced from this that the function is constant" [Peiffer 1983, p. 226], after which she gives an argument of Liouville's from the 1850s. Certainly, at first sight, one does not find a particular note in [Ms 3617 (5)] containing Liouville's proof, but I shall now argue that by combining several notes it is possible to reconstruct with reasonable certainty the general lines of the argument.

12.   First, let me quote a simplified proof of the above theorem concerning the impossibility of two incommensurable real periods. It can be found in Liouville's notebook 20 pages after the first proof:

> A periodic function of x can be developed in a series of the form:
> $$f(x) = \sum A_i \cos\left(\frac{2i\pi x}{\alpha} + \epsilon_i\right), \qquad \alpha \text{ being a period.}$$

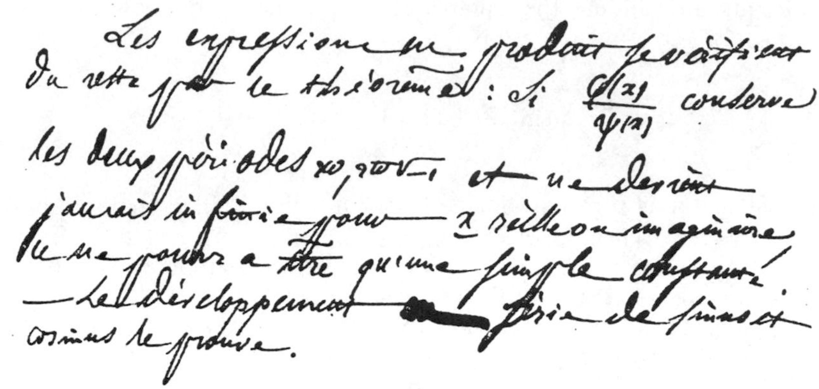

PLATE 28. The first appearance in Liouville's notebooks of his famous theorem on meromorphic functions. (Bibliothèque de l'Institut de France, Ms 3617 (5) p 76v)

In order that another period α' can exist, it is necessary that:

$$A_i \cos\left(\frac{2i\pi x}{\alpha} + \epsilon_i\right) = A_i \cos\left(\frac{2i\pi x}{\alpha} + \epsilon_i + \frac{2i\pi\alpha'}{\alpha}\right).$$

Thus, $A_i = 0$ or $\frac{2i\pi\alpha'}{\alpha} = 2m\pi$, where m is an integer, positive, zero or negative.

If the ratio of α' to α is incommensurable, the last equation can only hold if we take $m = 0$, $i = 0$, so that the pretended function is a simple constant.

If the ratio $\frac{\alpha'}{\alpha} = \frac{p}{q}$,

$$i\frac{p}{q} = m \text{ gives } m = \mu p, \ i = \mu q,$$

and the terms of $f(x)$ are expressed by the formula

$$A_{\mu q} \cos\left(\frac{2\mu q\pi x}{\alpha} + \epsilon_i\right),$$

which has the period $\frac{\alpha}{q} = \frac{\alpha'}{p}$ where α et α' are just multiples, so that the two given periods have been reduced to only one.

[Ms 3617 (5), p. 45v, quoted by Peiffer 1983, p. 242]

It is worth noting that this and the earlier proof does in fact establish Liouville's principle in the case of two real periods. My guess is that at a certain point Liouville discovered that it could be used for complex periods as well. Indeed, assume $f(z)$ is a complex function that has an expansion in trigonometric series:

$$f(z) = \sum_{i=0}^{\infty} A_i \cos\left(\frac{2i\pi z}{\alpha} + \epsilon_i\right). \tag{16}$$

If α' is another real or complex period, the argument above can be copied verbatim to show that if α and α' are incommensurable all the coefficients A_i, except for A_0, vanish. One just has to utilize that even as a complex function cosine has the periods $2m\pi$ only.

This explains why Liouville did not need to write up a "proof of my principle." He probably realized that any doubly periodic function can be transformed into the form (16), and he could then take over his earlier proof by considering x, A_i, and β to be complex.

13.   How then did Liouville deduce the expansion (16) in the complex case? This can be seen in the following quote from his notebook:

> Thus: 1° A well-determined doubly periodic function which never becomes infinite is reduced to a simple constant.
>
> . . .
>
> Let $\Psi(t)$ be a function with two periods, $a + b\sqrt{-1}$ being one of them. Let $\frac{t}{a+b\sqrt{-1}} = \frac{z}{2\pi}$ and the function of z, ψz which follows [from this substitution], will have the period 2π and another period, which is necessarily imaginary, $2\omega + 2\overline{\omega}\sqrt{-1}$.
>
> If $\psi(z)$ does not become infinite we let $z = x + y\sqrt{-1}$, x real. By the theory of sine- and cosine-developments
>
> $$\psi(x + y\sqrt{-1}) = \sum (P_m \cos mx + Q_m \sin mx),$$
>
> P_m and Q_m functions of y. Let h be real. Then
>
> $$\psi(x + h + y\sqrt{-1}) = \sum \begin{cases} (P_m \cos mx + Q_m \sin mx)\cos mh \\ +(Q_m \cos mx - P_m \sin mx)\sin mh \end{cases},$$
>
> from which
>
> $$P_m \cos mx + Q_m \sin mx = \frac{1}{\pi} \int \psi(x + y\sqrt{-1} + h)\cos mh \, dh.$$
>
> Therefore, $P_m \cos mx + Q_m \sin mx$ is a function of $x + y\sqrt{-1}$, and thus it must be of the form
>
> $$A_m \cos m(x + y\sqrt{-1}) + B_m \sin m(x + y\sqrt{-1}).$$
>
> where A_m et B_m are constants.
> Consequently,
>
> $$\psi(x + y\sqrt{-1}) = \sum A_m \cos(x + y\sqrt{-1}) + B_m \sin(x + y\sqrt{-1})$$

or

$$\psi(z) = \sum H_m \cos m(x + \epsilon_m) \text{ [error for } \cos m(z + \epsilon_m)],$$

and our <u>principium</u> is deduced from this in the simplest way.

[Ms 3617 (5), pp. 113v–114r]

This "way" is probably the one I have quoted in §12. There are two steps in the above deduction. First, a substitution is given that makes one of the periods real (equal to 2π). This means that along every line $y = $ constant, ψ has a Fourier expansion, whose Fourier coefficients are functions of y. This is an easy step. The interesting argument is the last one, where Liouville shows that this y dependence of the Fourier coefficients can be accounted for simply by substituting z for x in the trigonometric functions. Phrased differently, Liouville shows that the analytic continuation of the real Fourier series (13) can in this case be found by continuing each term analytically. He first develops $\psi(x + h + y\sqrt{-1})$ as a trigonometric series in h, and expresses its Fourier coefficients by the usual Fourier formula. This step is all right, since ψ is holomorphic so that the Fourier series converges uniformly. From the resulting formula,

$$P_m(y) \cos mx + Q_m(y) \sin mx = \frac{1}{\pi} \int_0^{2\pi} \psi(x + y\sqrt{-1} + h) \cos mh \, dh \,,$$

(16a)

he concluded that $P_m(y) \cos mx + Q_m(y) \sin mx$ is holomorphic and therefore equal to

$$P_m(0) \cos m(x + y\sqrt{-1}) + Q_m(0) \sin m(x + y\sqrt{-1}).$$

This step is also acceptable to a modern reader, for according to Lebesgue's theorem, we can differentiate with respect to the parameters x and y under the integral sign on the right-hand side of (16a), and therefore this integral satisfies the Cauchy-Riemann equations because ψ does.

It is remarkable that Liouville saw the need for this latter argument, and that he discovered a proof that can be made valid even by modern standards. Yet, despite its apparent simplicity, Liouville's proof of his principle does require so many deep theorems about Fourier series and analytic continuation that it is hardly suitable for an introductory course in complex analysis.

14.   However, the above proof is written more than 100 pages after Liouville formulated his principle for the first time, so it is entirely possible that he was initially led to the analytic continuation in a different way. It is of course possible that following many of his 18th-century precursers he naively let x turn complex in the series (13) or the corresponding cosine series in §12, and only later discovered that this was problematic. Still, another possibility is strongly suggested in a note 24 pages after the note

quoted in §12. It deals with Abel's elliptic function, φ, of the first kind with the periods 2ω and $2i\overline{\omega}$ (cf. §5) and begins: "Let $\varphi(x + y\sqrt{-1})$ and x be arbitrary, $y^2 < (\overline{\omega}/2)^2$. The function does not become infinite" [Ms 3617 (5), pp. 57v–58r]. Thus, Liouville considered the function in a band around the x-axis, where it has no poles. He developed the function in a Fourier series according to its real period, 2ω, and included the y dependence in the coefficients (it is the first note in which he extended the trigonometric series to the complex plane):

$$\varphi(x + y\sqrt{-1}) = \sum A_n(y) \cos \frac{n\pi x}{\omega} + B_n(y) \sin \frac{n\pi x}{\omega}. \qquad (17)$$

In order to determine $A(y)$ and $B(y)$, he applied the Cauchy-Riemann equations,

$$i \frac{\partial f(x + iy)}{\partial x} = \frac{\partial f(x + iy)}{\partial y} \qquad (18)$$

term by term to the identity (17) (this is suspect) and obtained

$$
\begin{aligned}
i\varphi'(x + iy) &= i \sum -\frac{n\pi}{\omega} A_n(y) \sin \frac{n\pi x}{\omega} + \frac{n\pi}{\omega} B_n(y) \cos \frac{n\pi x}{\omega} \\
&= \sum \frac{dA_n(y)}{dy} \cos \frac{n\pi x}{\omega} + \frac{dB_n(y)}{dy} \sin \frac{n\pi x}{\omega}.
\end{aligned}
$$

By equating coefficients, Liouville found

$$\frac{dA_n(y)}{dy} = i \frac{n\pi}{\omega} B_n(y), \quad \frac{dB_n(y)}{dx} = -i \frac{n\pi}{\omega} A_n(y). \qquad (19)$$

Now, it was an easy task for Liouville to determine the complete solution to this set of differential equations:

$$
\begin{aligned}
B_n(y) &= a_n e^{(n\pi y/\omega)} + b_n e^{-(n\pi y/\omega)} \\
A_n(y) &= i \left(a_n e^{(n\pi y/\omega)} - b_n e^{-(n\pi y/\omega)} \right).
\end{aligned} \qquad (20)
$$

At this point, he utilized the specific properties of Abel's elliptic function to find that $a_n = b_n$: "But how to find $\underline{a}$? Perhaps by looking for the derivative for $y = 0$" [Liouville Ms 3617 (5), pp. 57v–58r]. (The entire note is quoted in [Peiffer 1983, pp. 242–243]).

If Liouville, in his continued search for a_n, tried to insert $B_n(y)$, $A_n(y)$ from (20) into the original series (17) he would only need a little calculation with complex exponentials to find that

$$\varphi(x + iy) = \sum i(a_n - b_n) \cos \frac{n\pi}{\omega}(x + iy) + (a_n + b_n) \sin \frac{n\pi}{\omega}(x + iy), \qquad (21)$$

i.e., that the y variation can be included in the trigonometric functions. In the derivation of (21), we have only used that φ is a doubly periodic function with a real period and with no poles in the domain. If φ has no singularities at all, (21) is valid in the entire complex plane. Liouville was probably led to the analytic continuation of the Fourier series in this way and then discovered his principle by combining it with the argument in §12. Indeed, as pointed out by Peiffer [1983, p. 226], after the note summarized above, Liouville suddenly considered his principle as an established fact. The above argument only works when one of the periods is real, and in fact, Liouville seems to have discovered this special case first, for when he formulated his principle explicitly for the first time (cf. quote in §11), he assumed one period to be real. After all, the step to arbitrary periods is small, as we saw in §13.

Thus, the proof inspired by Hermite's remark turned out to be the way to Liouville's famous principle.

15.   Liouville changed his proof several times. Borchardt's notes show that in 1847 he proved the analytic continuation property using the *postulatum* that if a function $f(x + y\sqrt{-1})$ is independent of x then it is a constant. As pointed out by Peiffer [1983, p. 227], he later (1850) deduced this property from the Cauchy-Riemann equations. The rest of the 1847 proof is similar to the arguments quoted in §10 and §12, except that Liouville used complex exponential notation.

In the 1850–1851 lectures at the *Collège de France*, Liouville gave an entirely different "local" proof. Let $f(z)$ be a doubly periodic function without poles so that $|f(z)|$ is bounded and takes on its maximum A in a point, say, z_0. Assuming that f is not constant, Liouville used the mean value theorem (Cauchy's integral formula), to the effect that the mean value of f around a circle with center in z_0 is equal to $f(z_0)$ to arrive at the following contradiction:

$$A = |f(z_0)| = \left| \frac{1}{2\pi} \int_0^{2\pi} f(z_0 + re^{i\theta})\, d\theta \right| < \frac{1}{2\pi} \int_0^{2\pi} |f(z_0 + re^{i\theta})|\, d\theta \leq A\,.$$

A more detailed analysis of these proofs and references to the original manuscripts can be found in [Peiffer 1983].

16.   Liouville's last proof, which does not use the double periodicity in an essential way, leads us to the question: when was the general "Liouville's theorem" on bounded holomorphic functions formulated? In public, this happened on December 16, 1844, the week after Liouville had summarized his new ideas on doubly periodic functions in the *Académie*. Eager to secure his priority in this field, Cauchy reminded the *Académie* of his earlier works on complex integration [Cauchy 1844b]. In particular, he emphasized that the previous year he had applied his calculus of residues to establish all of Jacobi's results on elliptic functions and some new theorems. Moreover, Cauchy analyzed the relation between the theorem announced by Liouville at the previous meeting and some of his own earlier results, in particular, the following one from 1843:

> If for each real or imaginary value of the variable z, the function $f(z)$ always preserves a unique and determined value; if, moreover, it is

reduced to a certain constant $\mathcal{F}$ for all infinite values of z, then it will be reduced to this very constant when the variable z takes on an arbitrary finite value.

[Cauchy 1844b]

From the particular case $\mathcal{F} = I$, Cauchy easily deduced the general Liouville theorem for a function $f(z)$, which is "always continuous." According to his definition, this implies that $f(z)$ is never infinite, and without saying so explicitly, he meant this to hold also at infinity so that f is bounded. Moreover, he implicitly assumed continuity to mean (complex) differentiability, a mistake which is often repeated in his papers. With that in mind the ratio

$$\frac{f(z) - f(a)}{z - a} \qquad \text{for some fixed} \qquad a \in C$$

is a finite continuous function tending to zero at infinity. According to Cauchy's earlier theorem, it is therefore constantly zero, so that $f(z) = f(a)$. Thus, Cauchy had found

a proposition which is apparently more general, namely:

3rd Theorem. If a function $f(z)$ of the real or imaginary variable z always remain continuous, and consequently always finite, it is reduced to a simple constant.

[Cauchy 1844b]

Thus, Cauchy was the first to announce and prove the general form of Liouville's theorem. In 1851, he strained his claim of priority:

There is more to it. As I remarked in 1844...one of these formulas [the theorem from 1843] provides the fundamental theorem involved by Mr. Liouville concerning doubly periodic functions and

[Cauchy 1851b, p.453]

17. Still, I think Liouville justly deserves the honor of having his name attached to the theorem for the following three reasons. 1° Liouville was the first to publish the theorem in the doubly periodic case, Cauchy's 1843 theorems being clearly different from, although closely related to, Liouville's theorem. 2° Liouville was the first to discover the fundamental importance of the theorem. 3° Liouville probably had arrived at the general form of Liouville's theorem before Cauchy. This appears from the following note:

Let $f(z)$ be a well-determined function of z. If the moduli of $f(z) = f(x + y\sqrt{-1})$ do not exceed M, one has $f(z) = $ constant.
    For if $\varphi(x)$ is the elliptic function sinam (x), $f(\varphi(x))$ does not become infinite and has two periods. Therefore &$_c$.

[Ms 3617 (5), p. 94r]

Peiffer, who quotes a slightly different passage from Ms 3617 (5) [Peiffer 1983, p. 226], says that Liouville had an "intuition" of this general version of the theorem and calls the appended proof the "only element of a proof."

In fact, however, this quote not only gives the precise formulation of the general theorem, it also provides a completely adequate proof in the case where f is a bounded holomorphic function on the entire Riemann sphere. The trick is to reduce the general theorem to Liouville's principle on doubly periodic functions. To this end, Liouville composed f with Jacobi's elliptic function sinam z, which is the same as Abel's function $\varphi(z)$, (cf. (1) and (2)) for $c = 1$. Then, the meromorphic function f (sinam z) is doubly periodic, because sinam is; and it is bounded because f is bounded, and so it has only removable singularities. Thus, f(sinam z) satisfies the assumptions in Liouville's principle. Considering this to be known, Liouville concluded that f(sinam z) is constant, and whence that f itself is constant. In the last step, it is important that sinam maps its period parallelogram onto the entire complex plane.

Did Liouville formulate this general version before Cauchy, that is, before he summarized his theory in the *Académie*? I shall discuss this problem while following Liouville's notes chronologically (cf. §21).

The Gradual Development of the General Theory

18.   Strangely enough the first application Liouville made of his principle was an elegant proof of the fundamental theorem of algebra:

> Note. It is very remarkable that our general principle on doubly periodic functions proves that every algebraic Eq$\underline{\text{on}}$ has a root. For if
>
> $$\frac{1}{x^m + Ax^{m-1} + \&_c}$$
>
> does not become infinite, the same holds for
>
> $$\frac{1}{\varphi^m(x) + A\varphi^{m-1}(x) + \&_c},$$
>
> where $\varphi(x)$ is the inverse elliptic function of the first kind. Consequently this last quantity is reduced to a simple constant and that is absurd."

[Ms 3617 (5), p. 85r]

The composition with φ was necessary for Liouville, because he had not yet discovered the general theorem; except for that this elegant proof is still standard today; it was probably rediscovered independently later by another mathematician, for Liouville did not publish this note.

19.   Liouville went on to apply his principle to the study of elliptic functions, beginning with the problem of division. Let φ be Abel's elliptic function (cf. §5) and let α be an nth root of unity ($\alpha^n = 1$, $\alpha \neq 1$). Liouville then defined $\psi(x)$ by

$$\psi(x) = \varphi(x) + \varphi\left(x + \frac{2\omega}{n}\right)\alpha + \cdots + \varphi\left(x + \frac{2(n-1)\omega}{n}\right)\alpha^{n-1} \quad (22)$$

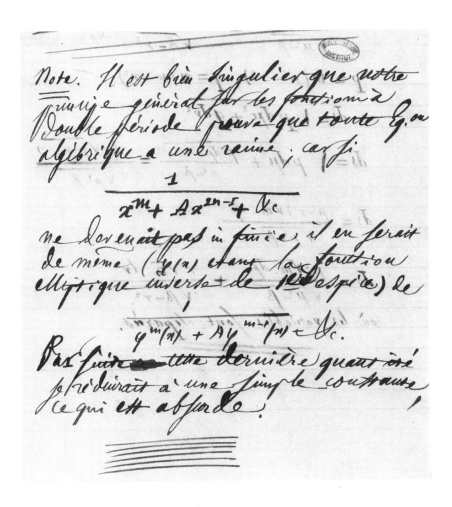

PLATE 29. Liouville's deduction of the fundamental theorem of Algebra from Liouville's principle. (Bibliothèque de l'Institut de France, Ms 3617 (5) p 85r)

and continued:

Consider the product

$$\psi(x)\psi\left(x + \frac{2\overline{\omega}}{n}\right)\cdots\psi\left(x + \frac{2(n+1)\overline{\omega}}{n}\right),\qquad [23]$$

$2\overline{\omega}$ being the second period conjugate to 2ω. This product becomes infinite for the same values as $\varphi(nx)$ and has the same periods $\frac{2\omega}{n}$, $\frac{2\overline{\omega}}{n}$. Thus its difference from $A\varphi(nx)$ does not become infinite if A is chosen conveniently. Thus B being another constant, the <u>principium</u> gives

$$\psi(x)\psi\left(x + \frac{2\overline{\omega}}{n}\right)\cdots\psi\left(x + \frac{2(n-1)\overline{\omega}}{n}\right) = A\varphi(nx) + B.\qquad [24]$$

[Ms 3617 (5), p. 89v]

How did Liouville arrive at this conclusion? In his lecture notes, he first concluded from his principle that a doubly periodic function has to have at least two poles, counted with multiplicity in the period parallelogram, if it is not constant. If A is chosen such that the residue of [23] and $A\varphi(nx)$ is equal in one of their common poles, their difference has at most one pole and is therefore equal to a constant B. However, the later notes seem to indicate that Liouville had not yet excluded nontrivial doubly periodic functions with one pole, and if that is so he must have had another argument showing that the above choice of A would also make the residues in the other pole equal. I have not found traces of this argument.

Liouville continued: "However, it is very important to prove that $B = 0$." Using some of Abel's theorems, he succeeded:

In other words, $B = 0$ and

$$\psi(x)\psi\left(x + \frac{2\overline{\omega}}{n}\right)\cdots\psi\left(x + \frac{2(n-1)\overline{\omega}}{n}\right) = A\varphi(nx).\qquad [25]$$

Thus one can see that the roots of

$$\psi(x) = 0,\ \psi\left(x + \frac{2\omega}{n}\right) = 0,\ldots$$

are nothing but groups of the roots of $\varphi(nx) = 0$. - This argument is certainly favorable to the <u>principium</u>. But we must go through all the details with care.

[Ms 3617 (5), p. 90v]

Note Liouville's use of Galois' term group! He was clearly aware that Abel had used this conclusion in his proof of the solvability of the equation of division by radicals, i.e., the expression of $\varphi(x)$ in terms of $\varphi(nx)$. He then went on to show that "Jacobi's formula for the roots can be deduced very easily from this."

20.   Until this point, Liouville had mainly used his principles as a means of discussing rather technical matters concerning the special Abelian and Jacobian elliptic functions. On page 92, however, he began to generalize these ideas to general doubly periodic functions:

> But, we have always known that the <u>Principium</u> is much more extensive. <u>Example</u>
>   <u>Theorem.</u> Let $\mathcal{F}(x)$ be a doubly periodic function and [α being an nth root of unity]
>
> $$\mathcal{F}(x) + \alpha \mathcal{F}\left(x + \frac{2\omega}{n}\right) + \cdots + \alpha^{n-1} \mathcal{F}\left(x + \frac{2(n-1)\omega}{n}\right) = \psi(x, \alpha).$$
>
> Let β be another root of unity and $2\overline{\omega}$ the second period conjugate to 2ω.
>   The two products
>
> $$\psi(x, \alpha)\psi\left(x + \frac{2\overline{\omega}}{n}, \alpha\right) \cdots \psi\left(x + \frac{2(n-1)\overline{\omega}}{n}, \alpha\right)$$
>
> $$\psi(x, \beta)\psi\left(x + \frac{2\overline{\omega}}{n}, \beta\right) \cdots \psi\left(x + \frac{2(n-1)\overline{\omega}}{n}, \beta\right)$$
>
> become infinite together with each of the terms $\mathcal{F}(x)$, $\mathcal{F}(x + \frac{2\omega}{n}m + \frac{2\overline{\omega}}{n}m')$, but if the infinities of $\mathcal{F}(x)$ are of the first order (that is as those of the elliptic function $\varphi(x)$) the difference will not be infinite; this difference is therefore constant.
>   In general our theory extends all we have said about elliptic functions to arbitrary functions $\mathcal{F}(x)$.
>
> [Ms 3617 (5), p. 92r]

This last remark reflects Liouville's growing awareness that he was on the track of a general theory of doubly periodic functions. However, he soon discovered that it should be taken with a grain of salt. Indeed, in the following note, he first found that if $\mathcal{F}$ is as above, and if ψ is defined by

$$\psi(x) = \prod_{p=1}^{n-1} \prod_{q=1}^{n-1} \mathcal{F}\left(x + \frac{2\omega}{n}p + \frac{2\overline{\omega}}{n}q\right),$$

then

$$\psi(x) = \text{constant} \cdot \mathcal{F}(nx).$$

In the proof, he had implicitly assumed that $\mathcal{F}$ has its poles at the points $(p + \frac{1}{2})\omega + (q + \frac{1}{2})\overline{\omega}$, just as Abel's elliptic function φ. When he discovered this hole in the argument, he crossed out the note with the words: "Error. It is necessary to have a relation between the roots and the periods - otherwise correct" [Ms 3617 (5), p. 92v]. Before he crossed out the result, he wrote: "This answers the question which Mr. Hermite has suggested to me" [Ms 3617 (5), p. 93v], revealing thereby that he continued to be in contact with Hermite while developing his theory.

21.   At this point, Liouville made the generalization of his theorem to arbitrary holomorphic functions that I discussed in §17 . There, I left open the question whether this generalization predates Cauchy's formulation of the general version or, equivalently, Liouville's first *Académie* note. Unfortunately, I cannot find a conclusive answer. One can account for most of the content of Liouville's *Académie* note in the manuscripts I have discussed above. He stated his principle and indicated that he had used it to deduce the "known theorems concerning both the multiplication and the transformation of elliptic functions and their development in series." Moreover, he explained that he had found Jacobi's formulas, from which the solvability of the equation relative to division could be proved. Finally, he described beautifully how in his method "the integrals which have given rise to the elliptic functions and even the moduli disappear in a way leaving only the periods and the points for which the functions become zero or infinite" [Liouville 1844g].

To be sure, in the manuscripts discused above, Liouville did not treat the problem of transformation of one elliptic function into another, but he could have guessed that it would work out in the same way as the related problem of multiplication. His claim that his principle "leads to numerous and useful consequences in other parts of analysis"[Liouville 1844g] however, does indicate a knowledge of its general validity for holomorphic functions, if it is not a bold guess on the basis of the proof of the fundamental theorem of algebra. As a whole, the note does seem to reveal a more mature understanding of doubly periodic functions than he had before generalizing his principle, so I believe that Liouville did prove his theorem before Cauchy published it.

22.   Having generalized his principle, Liouville returned to doubly periodic functions [Ms 3617 (5), pp. 94v–95r]. As a replacement for the incorrect notes on pages 92v–93v, he now considered the two symmetric expressions:

$$P(x) = \mathcal{F}(x)\mathcal{F}\left(x + \frac{2\omega}{n}\right) \cdots \mathcal{F}\left(x + \frac{2(n-1)\omega}{n}\right) \qquad (26)$$

and

$$\mathcal{F}(x) + \mathcal{F}\left(x + \frac{2\omega}{n}\right) + \cdots + \mathcal{F}\left(x + \frac{2(n-1)\omega}{n}\right), \qquad (27)$$

where $\mathcal{F}$ has the two periods 2ω and $2\overline{\omega}$ and the simple poles $\alpha, \beta, \ldots, \gamma$ with residue 1, which Liouville expressed with the phrase "I suppose as $\frac{1}{x-\alpha}$." Since the two functions have the same poles, Liouville concluded that

$$\mathcal{F}(x) + \mathcal{F}\left(x + \frac{2\omega}{n}\right) + \cdots + \mathcal{F}\left(x + \frac{2(n-1)\omega}{n}\right) = AP(x) + B.$$

Moreover, he defined the quantities

$$R(x) = P(x)P\left(x + \frac{2\overline{\omega}}{n}\right) \cdots P\left(x + \frac{2(n-1)\overline{\omega}}{n}\right)$$

and

$$Q(x) = \psi(x) \cdot \psi\left(x + \frac{2\overline{\omega}}{n}\right) \cdots \psi\left(x + \frac{2(n-1)\overline{\omega}}{n}\right)$$

where

$$\psi(x) = \mathcal{F}(x) + \alpha\mathcal{F}\left(x + \frac{2\omega}{n}\right) + \cdots + \alpha^{n-1}\mathcal{F}\left(x + \frac{2(n-1)\omega}{n}\right),$$

and concluded as above that

$$Q(x) = CR(x) + D.$$

"When $B = 0$ [he meant $D = 0$], we see that $R(x)$ is proportional to $Q(x)$ and that gives a remarkable decomposition into factors" [Ms 3617 (5), p. 95r].

23.   These beautiful generalizations of Jacobi's and Abel's results encouraged Liouville to continue with a general study of functions for which "$\varphi(\alpha) = \infty$ only," that is, having only one simple pole in a period parallelogram. On page 101v, he discovered that φ must be constant, and on the following pages, he simplified the proof.

A couple of pages further on, Liouville [Ms 3617 (5), pp. 102v–105r] made a very important discovery, namely, that if φ has two simple poles, α and β, then there exists a constant D such that $u = (\varphi(\alpha+x)-D)+(\varphi(\alpha-x)-D)$ is a solution to a second-order differential equation of the form:

$$\left(\frac{du}{dx}\right)^2 = a + bu + cu^2 + du^3 + eu^4, \tag{28}$$

so that $u(x)$ is the inverse of the elliptic integral:

$$x = \int \frac{du}{\sqrt{a + bu + cu^2 + du^3 + eu^4}}. \tag{29}$$

By a suitable choice of parity and two constants, Liouville was led to "the usual functions sinam(x)" [Ms 3617 (5), p. 113r] and then summed up his discoveries:

Thus:

1° A well-defined, doubly periodic function which never becomes infinite is reduced to a simple constant.

2° Such a function cannot become infinite once and of the first order only.

3° If it becomes infinite twice, it boils down to elliptic functions.
[Ms 3617 (5) p. 113v]

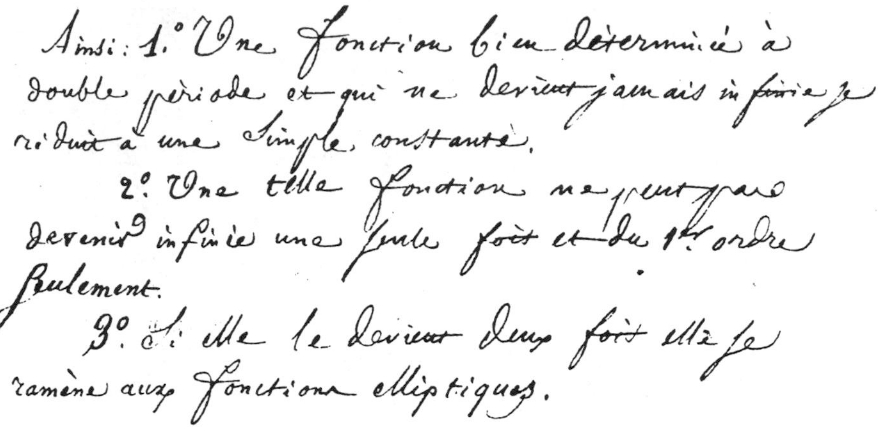
Ainsi: 1.° Une fonction bien déterminée à double période et qui ne devient jamais infinie se réduit à une simple constante.
2.° Une telle fonction ne peut pas devenir infinie une seule fois et du 1er ordre seulement.
3.° Si elle le devient deux fois elle se ramène aux fonctions elliptiques.

PLATE 30. Liouville's three main theorems concerning doubly periodic functions as they appear in his notes from 1844. (Bibliothèque de l'Institut de France, Ms 3617 (5) p 113v)

This statement is followed by the proof of Liouville's principle, discussed in §13.

At this point Liouville had discovered the essential elements of his theory. In the rest of [Ms 3617 (5)] and in countless other notes (e.g., in [Ms 3618 (6), (9); 3619 (4)]), he reorganized, polished, and extended his original ideas. In contrast to the initial discoveries, which were probably the result of a few weeks of inspired creativity, the subsequent reorganization may have lasted months or years. Yet, when Liouville gave his private lecture in 1847 to Borchardt and Joachimsthal, his theory had reached a final form. This can be seen when comparing Borchardt's published notes [Liouville 1880] with the notes from the 1851 course [Ms 3640 (1846–1851)] and from the 1859 course [Ms 3624 (10)], the latter being in Berthelot's hand. Instead of following this stage of cleaning up the theory, I shall leave it at a brief summary of the published lecture notes of 1847.

The Final Form of Liouville's Theory

24. Whereas the notes summarized above had begun with highly technical investigations on the multiplication and division of elliptic functions, and only gradually had uncovered the importance of a general theory, in the lectures of 1847, Liouville started with six sections on *Théorie générale* and

then demonstrated its power in the last three sections on *Applications* by deriving some of Abel's and Jacobi's results in a simple way. Having proved his principle in §1, he used it in §2 to derive that there does not exist a doubly periodic function with one simple pole. I shall summarize Liouville's proof of this theorem in order to illustrate his elegant and simple methods.

As in his notes, Liouville implicitly assumed that complex functions were meromophic. Let φ be such a function with the two periods 2ω and $2\omega'$, and with the poles $\alpha_1, \alpha_2, \ldots, \alpha_n$ in the period parallelogram

$$P_{z_0} = \{z = z_0 + u\omega + u'\omega' | u, u' \in [-1, 1)\}. \tag{30}$$

If the poles are all simple, there exist constants $G_1, G_2, \ldots, G_n$ such that

$$\varphi(z) - \left(\frac{G_1}{z - \alpha_1} + \frac{G_2}{z - \alpha_2} + \cdots + \frac{G_n}{z - \alpha_n} \right) \tag{31}$$

is finite in P_{z_0}. This is true also if some of the poles (e.g., α_k) are multiple as long as the corresponding fractions $\frac{G_k}{z - \alpha_k}$ are replaced by

$$\frac{G_k}{z - \alpha_k} + \frac{G_k'}{(z - \alpha_k)^2} + \cdots + \frac{G_k^{(i)}}{(z - \alpha_k)^i}, \tag{32}$$

where i is the multiplicity. Liouville called the sum of these fractions "the fractional part of $\varphi(z)$" in P_{z_0} and denoted it $[\varphi(z)]^{z_0}$. Thus,

$$\varphi(z) - [\varphi(z)]^{z_0} \tag{33}$$

is finite in P_{z_0}. The numbers G_j are what Cauchy had called the residues of φ in α_j, and Liouville did admit that $[\varphi(z)]$ "plays an important part in the calculus of residues." When Liouville presented Borchardt's notes from his 1847 lectures to the *Académie* in 1851, Cauchy immediately spotted that $[\varphi(z)]$ was a central tool in Liouville's theory, and he did not hesitate to claim priority for this idea. In the very issue of the *Comptes Rendus* in which Borchardt's index appeared, Cauchy wrote:

> As for the method of exhaustion employed by Liouville which consists of subtracting successively from a given function $f(z)$ other functions that become infinite at the same rate as the given function for certain systems of values of the variable z, so as to obtain as final remainder a function $\widetilde{\omega}(z)$ which has an always finite or even constant value for finite values of z; that is precisely the method which I have used in the first volume of the *Exercices de Mathématique* to establish the fundamental principles of the calculus of residues.
>
> [Cauchy 1851b, p. 453]

25.   In terms of "the fractional part," Liouville rephrased his principle: a doubly periodic function having $[\varphi(z)]^{z_0} = 0$ is a constant. Now assume that φ has only one simple pole α in P_{z_0} so that

$$[\varphi(z)]^{z_0} = \frac{G}{z - \alpha}, \tag{34}$$

where

$$G = \lim_{z \to \alpha} (z - \alpha)\varphi(z). \tag{35}$$

For simplicity, let $z_0 = \alpha$ and $z - \alpha = t$ such that

$$[\varphi(z)]^{\alpha} = [\varphi(\alpha + t)]^0 = \frac{G}{t}, \tag{36}$$

where

$$G = \lim_{t \to 0} t\varphi(\alpha + t). \tag{37}$$

Substituting $-t$ for t, Liouville found

$$[\varphi(\alpha - t)]^0 = -\frac{G}{t}, \tag{38}$$

and by adding (36) and (38) he concluded that

$$[\varphi(\alpha + t) + \varphi(\alpha - t)]^0 = 0, \tag{39}$$

so that by Liouville's principle

$$\varphi(\alpha + t) + \varphi(\alpha - t) = 2C, \tag{40}$$

where $2C$ is a constant. Defining $f(t)$ by

$$f(t) = \varphi(\alpha + t) - C, \tag{41}$$

we have by (40)

$$f(t) = -f(-t). \tag{42}$$

Combining this with the equations

$$f(t + 2\omega) = f(t), \qquad f(t + 2\omega') = f(t)$$

it is easy to see that

$$f(\omega) = f(\omega') = f(\omega + \omega') = 0. \tag{43}$$

Liouville then defined the function

$$F(t) = f(t)f(t + \omega), \tag{44}$$

which has singularities at most at the poles of its two factors, i.e., at the points

$$2m\omega + 2m'\omega' \qquad \text{and} \qquad \omega + 2m\omega + 2m'\omega',$$

m, m' being even integers. But, at these points, where one of the factors of (44) has a pole, the other is zero, according to (43), and since the poles are assumed to be simple, the singularities of $F(t)$ are removable. Hence, $F(t)$ is a doubly periodic function without poles, and so it is a constant:

$$F(t) = f(t)f(t + \omega) = k. \tag{45}$$

In the same way, Liouville found that

$$f(t)f(t + \omega') = k', \tag{46}$$

and

$$f(t)f(t + \omega + \omega') = k'',$$

or substituting $t + \omega$ for t:

$$f(t + \omega)f(t + \omega') = k''. \tag{47}$$

Multiplying (45) and (46), we have

$$(f(t))^2\, f(t + \omega)f(t + \omega') = kk',$$

and so by (47)

$$(f(t))^2 = \frac{kk'}{k''}, \tag{48}$$

which is contrary to the assumption that f has a pole at $t = 0$. This completes Liouville's proof that a doubly periodic function with only one simple pole in a period parallelogram does not exist.

26.   In the following section (§3), Liouville first demonstrated that there exists doubly periodic functions with two poles (in the rest of this chapter the phrase "n the period parallelogram" will usually be left out). He did so by constructing the simple example

$$\varphi(z) = \sum_{i=-\infty}^{\infty} f(z + 2i\omega'), \tag{49}$$

where

$$f(z) = \frac{1}{\cos \frac{\pi}{\omega}(z - h) - \cos \frac{\pi}{\omega} h'}. \tag{50}$$

Indeed, the function φ has the periods 2ω and $2\omega'$ and the poles $\alpha = h + h'$ and $\beta = h - h'$. Further, he showed that given one function φ with the periods 2ω and $2\omega'$ and two simple poles, then any other function of this kind is a rational function of φ, in particular, if φ_1 has the same simple poles as φ, it is of the form

$$\varphi_1 = C\varphi + C_1, \tag{51}$$

where C and C_1 are constants.

Liouville's theorems about the relationship between poles and zeros of doubly periodic functions are among his most beautiful. They are discussed in §3 for two poles and generally in §4. The first theorem says that the number of poles is equal to the number of zeros counted with multiplicity. It is a simple consequence of this theorem that the function assumes any value (including ∞) the same number of times. In [*Encyclopedic Dict. of Math.*, 1977, p. 483], this is called Liouville's third theorem, and the number of times any value is assumed is called the order of the function. Liouville also proved that the sum of the zeros minus the sum of the poles is a period (Liouville's fourth theorem). In [*Encyclopedic Dict. of Math.* 1977, p. 483], "Liouville's second theorem" is also mentioned, to the effect that the sum of the residues of a doubly periodic function is zero. As far as I can see, Liouville only discovered this theorem in the case of two poles. According to [Houzel 1978, p. 22], the general theorem is due to Hermite (1848; cf. [Cauchy 1851a]), who deduced it from the simple observation that the integral of the function around the edge of a period parallelogram is zero.

27.   From these theorems, Liouville deduced various expansions of doubly periodic functions in series and products, and he showed in two ways that an arbitrary doubly periodic function can be reduced to functions of order two: 1° [Liouville 1880, p. 293]: a doubly periodic function of order n can be written as a product of $n - 1$ functions of order two. 2° [Liouville 1880, §6]: given a doubly periodic function $\psi(z)$ of any order and another $\varphi(z)$ of order two and with the same periods, then there exist entire functions L, M, and N of $\varphi(z)$ such that

$$\psi(z) = \frac{M + N\varphi'(z)}{L}. \tag{52}$$

When applied to the doubly periodic function $\psi(z) = (\varphi'(z))^2$, Liouville easily deduced that $L = 1$, $N = 0$, and M is of degree four, so that

$$(\varphi'(z))^2 = A\varphi^4(z) + B\varphi^3(z) + C\varphi^2(z) + D\varphi(z) + E,$$

where $A = 0$ when the two poles of φ coincide. As remarked in §23, this implies that φ is the inverse of an elliptic integral of the first kind.

The latter result is to be found in the second part of Liouville's paper [1880, pp. 298–99]; it continues with the derivation of general addition theorems and a characterization of the particular function sinam through its parity, poles, and zeros. The last two sections are devoted to the transformation of the sinus amplitudinus. Let $\varphi(\omega, \omega', z) = \operatorname{sinam} \epsilon z$, where $\epsilon = \varphi'(0)$, be the sinam function with periods 2ω and $2\omega'$. Moreover, let $\varphi_1 = \varphi(\omega_1, \omega'_1, z)$ and assume that there exists a constant H such that

$$\frac{\omega'_1}{\omega_1} = n\frac{\omega'}{\omega},$$

where n is an odd number, and define

$$\psi(z) = \varphi_1(\frac{z}{M}),$$

where $M = \frac{\omega'}{\omega'_1}$. Then, Liouville showed that there exists a constant H such that

$$\psi(z) = H\{\varphi(z) + \varphi(z + \frac{2\omega}{n}) + \cdots + \varphi(z + \frac{2(n-1)\omega}{n})\}, \qquad (53)$$

and that conversely $\varphi(z)$ can be expressed in terms of ψ by an expression involving only radicals. When applied to the case $\varphi_1 = \varphi$, we recover formulas for multiplication and division similar to those Liouville had found in his notebooks.

Conclusion

28.   The novelty of Liouville's approach to elliptic functions is striking. To be sure, Abel and Jacobi had made great use of the double periodicity, but their work always took the elliptic integrals as the point of departure. Liouville turned the theory upsidedown by starting with a study of general doubly periodic functions only to apply them to the inverses of the particular integrals afterward. This is an example of a method commonly used by mathematicians, namely, to select just those elements of a branch of mathematics that will preserve an essential class of its theorems. This process is particularly forceful either if the new theory is vastly more general than its origin or if it turns out to be almost the same as its origin, in which case the selected elements characterize the theory. Liouville's singling out of the double periodicity is an example of the latter. For as he proved in his lectures, any doubly periodic function can be written simply in terms of a doubly periodic function of order two, and the latter is in turn the inverse of an elliptic integral.

One year after Liouville, Eisenstein [1844 and 1847] independently proposed a direct approach to elliptic functions circumventing the elliptic integrals. He had discovered that Abel and Jacobi's introduction of these functions as inverses of elliptic integrals was not rigorous. Indeed, since these pioneers did not have a concept of complex integration, their formal extension of elliptic functions to the complex plane was somewhat artificial (cf. §5). In particular, Eisenstein pointed out that according to the ordinary concept of integration the elliptic integrals possess only one value, and so the periodicity of the inverse functions is a mystery. For this reason, he relegated this traditional approach to the history books and replaced it with a definition of elliptic functions as infinite products. Liouville's general theory of doubly periodic functions is a more radical answer to the problem pointed out by Eisenstein, but it is uncertain if Liouville thought of it in this way.

In fact, Eisenstein's problem was quickly solved in a direct way by the interpretation of the elliptic integrals as integrals in the complex domain. Cauchy suggested this way out in a brief note [1846b] in which he tried to extend his ideas of complex integrals to multiple valued functions. He pointed out that elliptic or Abelian integrals do not depend only on the limits but also on the way in which the path of integration winds around the ramification points where the radicals vanish. This programme, which Cauchy only sketched in general terms, was carried out by Puiseux [1850] and further by Riemann, in whose hands it led to the introduction of the Riemann surfaces.

Thus, the importance of Liouville's theory does not consist in its rigorization of the previous approach but in its simplification and generalization; and these virtues explain why it has survived until this day. Still, for all the similarities between Liouville's approach and modern introductions to elliptic functions, there are also striking differences. Liouville, like Cauchy at this time, had rather imprecise ideas about what kind of functions his theory dealt with. He did use the Cauchy-Riemann equations, but there is no evidence that he thought of them as a defining property of what he called "well defined functions." Another great difference between Liouville's and our modern approach is that Liouville did not use complex integration, even for his single-valued doubly periodic functions for which Cauchy's ideas were applicable. Yet, it was only a few years until Hermite began to merge Cauchy's methods with Liouville's theory, and in Briot and Bouquet's book, many of Liouville's proofs based on Liouville's principle were replaced by Cauchian proofs. Thus, in the 1850s, both Jacobi's and Abel's original approach and Liouville's approach to elliptic functions became an integrated part of the steadily developing theory of complex functions.

29.   As can be seen from his later courses and his notebooks, Liouville did not lose interest in elliptic functions after 1844. He continued to work and publish on various geometric and mechanical applications of elliptic and Abelian functions (cf. Chapter XVI). Moreover, in 1849, seemingly in great haste, he wrote a paper on differential equations "on the occasion of a Memoir of Mr. Jacobi on some elliptic series" [Liouville 1849b]; similar problems were the topic of two small notes [Besge 1846a] and [Liouville 1854b].

There is no doubt that Liouville knew that his general approach to doubly periodic functions was more novel and important than these later works on the application of specific aspects of elliptic functions. So we are faced with the question why did Liouville not publish his general theory? A similar question can be asked about some of his other far-reaching ideas, but this particular case presents an intriguing circumstance. In Weierstrass's copy of the Borchardt notes (which are preserved in a copy by H.A. Schwartz), Weierstrass remarked that the typeset proof sheets of these notes meant for Liouville's *Journal* were found among Dirichlet's papers after his death in 1859, [Peiffer 1983, p. 232]. When were they set, and why were they not published until 1880 — and then in Borchardt's Journal? Peiffer [1983, p. 232] suggests the following answer: having not been able to produce his projected large treatise on doubly periodic functions toward the end of the 1850s and feeling that competitors were quickly entering into this domain, Liouville decided to publish the lecture notes taken by Borchardt. However, when Briot and Bouquet's book, *Théorie des fonctions doublement périodique...*, was published in 1859, he interrupted the printing process. This gives a plausible explanation of the proof sheets, but it leaves the problem of why Liouville never got around to writing a treatise on this subject before 1859. I cannot find any explanation other than Liouville's usual lack of time to write comprehensive works, and neither could Weierstrass in his *Wahlvorschlag* (cf. [Biermann 1960]), where he spoke of "reasons unknown to us."

Weierstrass regarded Liouville's unpublished work in this area as his most important:

> In particular we must mention a very important work by Liouville which has had a curious fate, namely his theory of elliptic functions in which all the properties of these transcendentals are derived from their doubly periodicity. This theory, of which the essentials are completely developed [by Liouville], is distinguished by the originality of the basic ideas.
>
> [Weierstrass 1876 in Biermann 1960, p. 48]

This appreciation by one of the greatest masters of elliptic functions is an appropriate conclusion to this chapter.

XIV

Galois Theory

1. The last letter Galois wrote to his friend Auguste Chevalier the night before the fatal duel ends with the words:

> After that, I hope some men will find it profitable to sort out this mess.

Liouville was the person to do this job. The great importance of this event in the history of mathematics justifies a closer analysis of Liouville's merits in this field.

In Chapter III and Chapter V, I told how Liouville began to read Galois's unpublished papers in 1842, how he announced their profundity to the *Académie* in September of 1843 and at the same occasion promised to publish them together with a commentary, and how he prepared the proof sheets of the most important unpublished memoir for the December 1843 issue of his *Journal*, but withdrew it at the last minute only to publish it three years later. This publication also contained another unpublished fragment, the letter to Chevalier, and Galois's previously published papers, and was prefaced by an *Avertissement* in which Liouville repeated his never to be fulfilled promise to publish a commentary. We have also seen that Liouville at an unknown point gave a course to Serret and others in order to understand Galois better, that he did prepare some unpublished notes on Galois's *Mémoire*, and that his continued intention to publish a commentary prevented Serret from including Galois theory in the two first editions of his *Cours d'Algèbre*. In this chapter, I shall analyze some obscure points in this story: How and from whom did Liouville obtain Galois's papers? Why did he postpone their publication for three years? Why did he never publish the commentary? How well did he understand Galois? What did his notes contribute mathematically? What was his influence on later developments?

The "Avertissement"

2. Some of these questions are elucidated in the three-page *Avertissement* with which Liouville prefaced his edition of Galois's work. The *Avertissement* had a twofold purpose: 1° to bring out the genius of the young Galois and 2° to justify the conduct of Liouville's predecessors and colleagues. As to the first point, Liouville affirmed the correctness of Galois's results and the profundity of his methods while emphasizing that these ideas were

obtained by a mathematician who did not live to see his 21st birthday and who spent most of his last two years "fruitlessly in political agitations, in clubs, or in the prison of Sainte-Pélagie."

The second point was more delicate. In 1832, Auguste Chevalier had published an obituary of his late friend Galois together with the letter Galois had written the night before the duel [Chevalier 1832]. Here, Chevalier almost accused the academicians of having killed Galois with their repeated rejections of his work. Liouville, on the other hand, defended their conduct by pointing out that the commissions had not rejected Galois's ideas, but had found his presentation of them obscure. Liouville agreed with this judgment: "An exaggerated wish to be concise was the cause of this defect which one must try to avoid particularly when treating the abstract and mysterious subjects of pure algebra" [Liouville 1846j, p. 382]. Liouville hinted that Galois's last judges, Poisson and Lacroix, entertained the hope that Galois's results were correct. He did so by quoting a passage from the sixth edition of Lacroix's algebra book [1831] where Lacroix mentioned one of the central theorems in Galois's last *Mémoire* while adding "but this *Mémoire* seemed nearly incomprehensible to the members of the commission in charge of its examination." Liouville went so far as to accentuate the noble motives of Galois's judges:

> we can understand that some illustrious geometers might have thought it good to try to lead this ingenious but inexperienced novice on the right track by the severity of their wise advice.
>
> [Liouville 1846j, p. 382]

With a view to the 1840s and the future, Liouville continued:

> But, now all has changed. Galois is no more! Let us guard ourselves from pursuing futile criticism; let us leave the defects aside and exhibit the merits.... I have therefore considered it my only aim to investigate, to unravel, what it is that is new in this work in order to bring it out the best I can.
>
> [Liouville 1846j, p. 382]

It would have been easy for Liouville to draw a picture of himself as an outstanding examiner of Galois surrounded by nitwits. But, instead he modestly assigned to himself the role as the first examiner who had no hopes of a more comprehensive formulation and who was therefore forced to make sense out of the given material. Thereby he presented a version of the Galois legend that is more sober and closer to modern interpretations than Chevalier's and Bell's [1937] melodramatic stories [cf. Kiernan 1971–1972 and Rothman 1982].

Galois's Friends?

3.   We shall now address the question: who persuaded Liouville to take on the time-consuming job of an unofficial examiner of Galois's work? Liouville

twice shed some light on this problem in the *Avertissement*:

> ...yielding to pleas of *some friends* of Evariste, I have devoted my-
> self, so to speak, under the gaze of his brother, to a careful study of
> all the printed or handwritten papers which he has left behind.
>
> As for us, who have neither known nor even seen this ill-fated
> young man, we limit ourselves to our role as a geometer; the ob-
> servations we can allow ourselves, publishing these works under the
> instigation of *his family*, concerns only the mathematics.
>
> [Liouville 1846j, p. 382]

These quotes leave the question: Did the ill-defined group of "friends of Evariste" in the first quote include other than Galois's family, in particular, the brother Alfred? This question has been analyzed by Kiernan [1971–1972]. His analysis, though very interesting, is not entirely conclusive or convincing. As for the obvious candidate, Auguste Chevalier, Kiernan states: "Liouville does not explicitly state from whom he received the papers. It seems implied in his comments that he had no direct contact with Chevalier...." To be sure, this is the impression one gets from Liouville's *Avertissement*, where he explicitly rejects the biased view of Galois's life to be found in Chevalier's obituary, while pardoning its author on the grounds of his affectionate friendship with Galois. Still, in his first announcement to the *Académie* [Liouville 1843i, p. 448] of the importance of Galois's papers, Liouville wrote that "These manuscripts have been entrusted to me by M. Auguste Chevalier." He repeated this in a note that he inserted after Galois's letter to Chevalier:

> Inserting the above letter in their journal, the editors of the *Re-
> vue encyclopédique* announced that they would shortly publish the
> manuscripts left behind by Galois. But, this promise has not been
> kept. However, Mr. Auguste Chevalier has prepared the work. He
> has handed it over to me, and one will find in the following pages
>
> 1° An entire *Mémoire*...
>
> 2° A fragment of a second *Mémoire*...
>
> We have preserved the majority of the notes which Mr. Auguste
> Chevalier has added to the *Mémoires* mentioned above.
>
> [Liouville 1846j, pp. 415–416]

Thus, although Chevalier may not have been in possession of all of Galois's original manuscripts and although he may not have been among the persons to contact Liouville in the first place, he was the one that provided Liouville with the manuscripts and the annotated fair copies that were eventually printed.

In passing, it must, however, be remarked that when Liouville, in the continuation of the above quote, says that the notes marked A. Ch. are due to Chevalier this must be taken with a grain of salt. In fact, a comparison of Chevalier's fair copy and Liouville's printed text show that Liouville changed Chevalier's notes to Proposition II and III. These are the two theorems that Galois altered, probably the night before the duel. Chevalier had remarked that the originals were different, but it was Liouville who added the original versions in the two footnotes. We can infer that Liouville had Galois's original papers at his disposal and carefully compared them with Chevalier's fair copy. This remark does not change the fact that Chevalier is the only person, in addition to Alfred Galois and Liouville, who we positively know was involved in the publication of Galois's collected works.

4.   Kiernan conjectures that Sturm, Hermite and Richard were the "friends of Evariste," who convinced Liouville that he should examine Galois's papers [Kiernan 1971–1972, p. 103]. Before arguing against this point of view, I shall outline the relationship of these three mathematicians with Galois and Liouville.

Richard, professor at the *Collège Louis le Grand*, was highly praised by Liouville in his *Avertissement*:

> [Galois's] great faculties were encouraged by an excellent professor, by an excellent man, M. Richard. [And added in a footnote]: M. Le Verrier, M. Hermite and other distinguished savants have followed M. Richard's classes. The good students are the glory of the master.
>
> [Liouville 1846j, p. 381]

At the latest, Liouville probably made the acquaintance of Richard during 1834, when he too taught at the *Collège Louis le Grand*. There he may also have met Chevalier, who taught at this institution around the same time (cf. [Kiernan 1971–1972, p. 102]).

Liouville's good friend Sturm had met Galois personally and had, according to one of Galois's notes, been receptive to Galois's ideas. Moreover, as an editor of Ferrussac's *Bulletin*, he might have been influential in the publication of three of Galois's papers in this journal. For these reasons, Kiernan writes "It may well be true that Sturm is the man most responsible for the world's knowing today that Galois ever existed." [Kiernan 1971–1972, p. 74].

As a student of both Richard and Liouville, Hermite qualifies in Kiernan's opinion as a third of the "friends of Evariste." Moreover, Hermite had written a paper on the solvability of the quintic as early as 1842, and later he received some of Galois's classroom papers (subsequently given by his son-in-law Emile Picard to the *Bibliothèque de l'Institut*). Still, Hermite's first reference to Galois appeared in a letter to Jacobi written between the summer of 1847 and November 1848 [Hermite 1850] (dating is possible from remarks at the beginning of the first and the fourth paper, the latter referring to [Jacobi 1848]). Kiernan [1971–1972, p. 101] thinks that this reference and the general direction of Hermite's research in the 1850s indicate "a familiarity with Galois's results and concepts at a curiously early time, too early for it to have arisen after their publication in 1846.... Hermite's independence of his contemporaries tends to confirm him as a direct link from Galois to Jordan, or more correctly, from Richard to Jordan." This idea is interesting, but I find the argument rather weak. In fact, in the letter to Jacobi, Hermite merely quoted one theorem on the division of elliptic functions from Galois's letter to Chevalier, and not until the early 1850s did he show any real insight into Galois's ideas. Thus, he would have had ample time to penetrate into Galois's works, even if he saw them for the first time after their publication in 1846, or, what I find more plausible, during Liouville's course, which may have taken place around 1844.

To conclude, it is possible, but not documented, that Richard, Sturm, Hermite, and even other students of Richard's, such as J. A. Serret and his friend Bertrand, had heard of Galois's papers and joined Alfred Galois, and perhaps Chevalier, to persuade Liouville to have a look at them. But, I find it hard to believe that these mathematicians should have had any knowledge or understanding of the contents of Galois's papers. Had one of them understood Galois's ideas, he would undoubtedly have explained this to Liouville, and considering how readily Liouville, during this period of his life, acknowledged and propagandized for all of these friends (Richard, only in the *Avertissement*), it is inconceivable that Liouville should not have mentioned their efforts. But, as it was, Liouville mentioned only Galois's family, and in particular, the brother Alfred. I find it most plausible that they, and possibly Chevalier, and no mathematicians, were the friends of Evariste who persuaded Liouville to read Galois's papers.

5. If this conjecture is true, the "friends" clearly had to find a benevolent and clever mathematician in order to establish Evariste's mathematical repute. As one of the two most famous French mathematicians, Liouville was an obvious choice. The other, Cauchy, was clearly excluded because he was among the academicians whom Galois in the *Préface* had accused of having lost his earlier memoirs [Galois 1962, pp. 3–11]. Moreover, as a republican (although less radical than Galois), Liouville would presumably be sympathetic with the activities of Galois; and last, but not least, Liouville was the publisher of a journal where Galois's posthumous papers could appear.

I therefore find that Liouville, and not Sturm, is the mathematician most responsible for the world's knowing today that Galois even existed. The other central figure in the publication of Galois's works is Galois's brother. This is asserted not only by Liouville's *Avertissement*, but also by a letter in which Galois's works were presented to Jacobi. The letter is signed by Alfred, who thereby stands out as the man behind the project. Still, the handwriting of the original draft reveals that Liouville was its author. This draft is kept among Galois's papers (the many corrections show that it is indeed the original). It runs as follows:

November 7, 1847

Monsieur

I have the honor of sending you... a copy of the first part of the mathematical works of my brother.

It appeared in Mr. Liouville's *Journal* almost a year ago, and, if I have not sent it to you earlier, it is because I continued to wish to be able to send you some day the complete work, whose publication has been delayed by various circumstances. Still, this first part includes the most important papers which my unfortunate Evariste has left behind, and we shall hardly need to add more than a few fragments selected from the disorder of his papers. Thus, one has not found anything concerning the theory of elliptic and Abelian functions; one can only see that he has studied your Works profoundly with pen in hand. As for the theory of equations Mr. Liouville and other geometers I have consulted affirm that his Memoir, which was rejected very strongly by Mr. Poisson, contains the basis of a very fruitful doctrine and a first important application of this doctrine. "This work," they tell me "assures a place for your brother forever in the history of Mathematics..."

[Galois 1962, p. 521]

This letter elucidates Liouville's understanding of Galois theory and his plan to continue the publication of Galois's posthumous works, but before we analyze Liouville's publication, we must briefly consider the mathematics.

Galois Theory According to Galois

6.   Harold Edwards [1984] has devoted an entire book to a comprehensive and comprehensible presentation of the ideas found in Galois's *Mémoire sur les conditions de résolubilité des équations par radicaux*, so it would be presumptuous to try and do so in a few pages. Therefore, this section is confined to a summary of some basic ideas from this memoir [Galois 1831–1846], necessary for the understanding of Liouville's notes. For the rest, the reader is referred to Edward's admirable book.

Consider a polynomial equation with coefficients in a number field (modern term) of rationally known quantities (Galois's term). The rationally known quantities can be the rational numbers or an extension obtained by *adjoining* (Galois's term) one or more quantities. Galois also considered eqations with literal coefficients; in this case the known quantities are obtained by adjoining these letters: Let a, b, ... be the roots of this equation. Following Cauchy [1815], Galois denoted an arrangement of these letters a *permutation* and a passage (mapping) from one permutation to another is called a *substitution* (this is our modern permutation). He represented a substitution by writing the two permutations one above the other. By writing several permutations under each other, one can represent a set of substitutions, namely, the mappings from one fixed row, e.g,. the first, to itself and to all the other rows. In this situation, Galois spoke of a group of substitutions if one gets the same set of substitutions irrespectively of which row one maps from. This is in fact equivalent to our modern definition. However, Galois was not always clear as to whether it was the permutations or the substitutions that were being grouped.

Already Lagrange [1770–1771] had emphasized the idea of considering the number of values obtained by a function of the roots when these roots are permuted. Abel had claimed [1829b, Chapter 2, §1], and Galois independently indicated in Lemma II, that there exists a rational function $V = \theta(a, b, \ldots)$ that obtains $k!$ values under the permutation of the k roots. This function is sometimes called the Galois resolvent. It follows from a theorem of Lagrange [1770–1771 §104] that the roots are all rational functions of $V : a = \phi(V), b = \phi_1(V), \ldots$. Galois sketched another proof of this important fact in Lemma III.

7.   Now we can define the Galois group of the given equation. The Galois resolvent V is a root of a polynomial with known coefficients, namely,

$$\Pi_{T \in S_k}[x - \theta(T(a), T(b), \ldots)],$$

where S_k is the full symmetric group of the k roots. This polynomial may be reducible over the known quantities, but it has a unique irreducible factor of which V is a root:

$$F(X) = \Pi_{T \in G}[x - \theta(T(a), T(b), \ldots)]. \tag{1}$$

The substitutions in G form a group called the Galois group - Galois called it the group of the equation. Let $V, V', \ldots, V^{m-1}$ denote the roots of this irreducible factor $(V^{(i)} = \theta(T_i(a), T_i(b), \ldots)$ for $T_i \in G)$. Then, Galois showed that each line in the following scheme was a permutation of the roots $a = \phi(V), b = \phi_1(V)$, etc.:

$$
\begin{array}{c|cccc}
(V) & \phi(V) & \phi_1(V) & \cdots & \phi_{k-1}(V) \\
(V') & \phi(V') & \phi_1(V') & \cdots & \phi_{k-1}(V') \\
(V'') & \phi(V'') & \phi_1(V'') & \cdots & \phi_{k-1}(V'') \\
\vdots & & & & \\
(V^{m-1}) & \phi(V^{m-1}) & \phi_1(V^{m-1}) & \cdots & \phi_{k-1}(V^{m-1}).
\end{array}
\tag{2}
$$

The substitutions of the scheme are a presentation of the Galois group. According to Galois's Proposition I [Galois 1831–1846], this group has the following properties: each function of the roots invariant under the substitutions of the Galois group will be known rationally, and conversely every function of the roots that can be determined rationally will be invariant under these substitutions.

8.   Proposition II explains what is today thought of as the central idea of Galois theory, namely, the interplay between the rationally known quantities and the Galois group:

Proposition II. Theorem.

If one adjoins to a given equation the root r of an auxiliary irreducible equation

(1) one of two things will happen: either the group of the equation will not be changed; or it will be partitioned into p groups, each belonging to the given equation respectively when one adjoins each of the roots of the auxiliary equation;

(2) these groups will have the remarkable property that one will pass from one to the other in applying the same substitution of letters to all the permutations of the first.

[Galois 1831–1846, translation by Edwards 1984]

In the original formulation of this theorem, Galois had assumed that the auxiliary equation was "of prime degree p." He crossed out this phrase, probably the night before the duel. Note that in Proposition II Galois spoke about groups of permutations instead of groups of substitutions. The idea is that table (2) is divided into p blocks of $\frac{m}{p}$ permutations. In modern terms, these blocks can define sets of substitutions in two ways. 1° Consider the substitutions taking the first row *in a block* into all the rows of that block. In this way, one will get the Galois group of the original equation when one of the roots of the auxiliary equation is adjoined. 2° Consider the substitutions sending the first row of the entire scheme (2) into the rows of one particular block. In this way, only the first block corresponds to the Galois group of the original equation when one of the roots of the auxiliary equation is adjoined, and the other sets of substitutions are its left cosets. The latter property follows from the second statement in Galois's Proposition II.

In Proposition III, Galois remarked that if all of the roots of the auxiliary equation are rational functions of one of these roots, all the subgroups of substitutions defined in the first interpretation above are identical. This means in modern terms that the left and right cosets of the first of the subgroups are identical, so that it is normal.

9.   Galois then discussed solutions of polynomial equations by radicals. The main idea is to reduce the Galois group of the equation successively until it contains only the identity. For when that is obtained, all the roots are rationally known according to Proposition I. If this can be done by successively adjoining radical quantities, the equation is said to be solvable by radicals, or just solvable.

First Galois assumed that the equation was solvable by radicals. One can obviously assume that all the radicals are pth roots where p is a prime. Consider the smallest p for which a pth root, say $\sqrt[p]{A}$ of the known quantities reduces the Galois group G to a subgroup G_1. According to Gauss's results on cyclotomic equations [Gauss 1801], the pth roots of unity are expressible by qth roots for $q < p$. Therefore, one may assume that the pth roots of unity have been adjoined, for that does not change the Galois group. In that case, all the roots of the irreducible equation $X^p = A$ are rational functions of one of them, and thus G_1 satisfies both Propositions II and III (G_1 is normal in $G : G \triangleright G_1$). From Galois's *original* version of Proposition II, it follows that the index of G_1 in G is equal to p. Continuing in this way, we see that if an equation is solvable by radicals then there is a tower of subgroups, $G_1, G_2, \ldots, G_k$ of its Galois group such that

$$G \triangleright G_1 \triangleright G_2 \triangleright \ldots \triangleright I, \tag{3}$$

and such that the index of one group in the preceding one is a prime. Conversely, Galois showed that the existence of such a tower ensures that the given equation is solvable by radicals.

10. In the final section of [Galois 1831–1846], "Application to irreducible equations of prime degree," Galois used this criterion to establish (Proposition VII) that an irreducible equation of prime degree is solvable by radicals if and only if there exists a cyclic order of the roots $x_0, x_1, \ldots, x_{p-1}$, where the index is counted modulo p, such that its Galois group consists exclusively of linear substitutions of the form

$$x_i \rightarrow x_{ai+b}$$

for fixed $a = 1, 2, \ldots, p-1$ and $b = 0, 1, \ldots, p-1$.

In Proposition VIII, this is used to describe the solvability of the irreducible equations of prime degree in a way that does not depend on the notion of a Galois group. It is shown that in order for such an equation to be solvable by radicals, it is necessary and sufficient that once any two of the roots are known, the others can be deduced from them rationally. This is the result Liouville reported to the *Académie* in 1843 [Liouville 1843i].

Galois's *Mémoire* is concluded by a strange list of permutations headed by the words: "let $n = 5$. The group will be the following one." In fact, this list represents the $4 \times 5 = 20$ substitutions $x_k \mapsto x_{ak+b}$ which constitute the largest possible Galois group of a solvable quintic. The idea is apparently to show that this group is different from the full symmetric group of order 5! of the five roots, which Galois without proof had indicated would be the Galois group of the general quintic [remark after Proposition II]. Thus, he had proved that the general quintic is not solvable by radicals. This proof is more elegant than Abel's proof of the same result from 1826 [Abel 1826].

Liouville's Commentaries

11. Bound together with Galois's manuscripts at the *Bibliothèque de l'Institut de France* are a collection of notes or commentaries written by Liouville [Ms 2108]. They are numbered fol. 36 and fol. 260–285. Fol. 36, which is attached to the 1843 proof sheets of Galois's *Mémoire*, has been published by Tannery [Galois 1906], and by Bourgne and Azra [Galois 1962, p. 492; a ϕ is missing here in the first line]. The last sheets are likewise commentaries on Galois's great *Mémoire*. Fol. 260–265, carrying the title *Notes sur Galois* and consecutively paginated (1–6), is a commentary on

Propositions II–VIII. On the following sheets, Liouville went over the same material three more times. First Fol. 268–271 (268 recto and verso has been reversed in the binding) has the headline: "Galois" and paginations 1–4, Fol. 272–277 (pp. 1–6) likewise on "Galois" and third Fol. 278–284 (pp. 1–6) without headline.

The two versions on Fol. 268–271 and Fol. 278–284 give a fuller account of Proposition II than Fol. 260–265. I therefore conjecture that they are the first versions, the other two being later condensed versions. However, one cannot exclude that conversely Fol. 260–265 or Fol. 268–271 is the first version, the two others being later extensions. In Appendix II, the reader will find Fol. 260–265 quoted in full together with those parts of Fol. 278–284, which are more detailed than the corresponding sections of Fol. 260–265. Here, I shall summarize the commentaries making use of the four versions indiscriminately.

PROPOSITION II

12.   All Liouville's commentaries start with a proof of Galois's Proposition II. Galois had remarked that if V's irreducible equation $F(X) = 0$ (over the field K) remains irreducible after the adjunction of a root r of the auxiliary equation then the Galois group naturally remains unchanged. If however it becomes reducible after the adjunction (i.e., in $K(r)$), Galois claimed that it can be written as a product:

$$F(X) = f(X, r) \cdot f(X, r_1) \dots f(X, r_{q-1}), \qquad (4)$$

where $r = r_0, r_1, \dots, r_{q-1}$ are some of the n roots of the auxiliary equation, and where f is a polynomial of two variables. (Galois and Liouville use V to denote both the "variable" X and one of the roots of $F(X) = 0$. I shall not adopt this somewhat confusing convention.) From this decomposition, it is easily seen that the Galois group (2) splits into q smaller "groups" of an equal number of permutations, each group corresponding to the roots $V^{(j)}$ of one of the factors. These are then the Galois groups when $r = r_0, \dots, r_{q-1}$, respectively, are adjoined. But, it is not obvious how to obtain the decomposition, (4); after all, the decomposition of F when r is adjoined gives another decomposition:

$$F(X) = f(X, r) f_1(X, r) \dots f_{q-1}(X, r). \qquad (5)$$

Liouville showed how (4) can be obtained from (5). He first observed that since all the roots of F are rational functions over K of each other, their irreducible equations over $K(r)$ must have the same degree so that all the f_j's have the same degree, say, s. Thus $s \cdot q = m$, m being the degree of F. Then, Liouville claimed that (5) implies that

$$F(X) = f(X, r_i) f_1(X, r_i) \dots f_{q-1}(X, r_i), \qquad (6)$$

where r_i is any of the n roots of the auxiliary equation. In fact, this follows from

Lemma A. *If $P_1(X,r) = P_2(X,r)P_3(X,r)$, where P_1, P_2, and P_3 are rational functions over K in two variables and polynomials in X (so that $P_1(X,r)$, $P_2(X,r)$, and $P_3(X,r)$ are polynomials in X over $K(r)$), then $P_1(X,r_i) = P_2(X,r_i)P_3(X,r_i)$ when r_i is a root in the irreducible equation having r as a root.*

This Lemma in turn is a simple consequence of Galois's Lemma I, which Liouville used repeatedly and which Abel also formulated as Theorem I in [1829a]:

Lemma I. *If a rational equation, $f(x) = 0$, has one root in common with an irreducible polynomial, $p(x)$, then all the roots of $p(x)$ are roots of $f(x) = 0$, and so $p(x)$ divides $f(x)$.*

Multiplying all the equations (6) for $i = 0, 1, \ldots, n-1$, Liouville obtained

$$
\begin{aligned}
F(X)^n \ &= \ f(X,r)f(X,r_1)\ldots f(X,r_{n-1}) \\
&\times \ f_1(X,r)f_1(X,r_1)\ldots f_1(X,r_{n-1}) \\
&\times \ f_{q-1}(X,r)f_{q-1}(X,r_1)\ldots f_{q-1}(X,r_{n-1}).
\end{aligned} \tag{7}
$$

But, since every substitution of $r_1, \ldots, r_{n-1}$ leaves each line in this product invariant, these lines are, according to Lagrange, polynomials in $K[X]$. Therefore, $f_i(X,r)f_i(X,r_1)\ldots f_i(X,r_{n-1})$, for $i = 0, 1, 2, \ldots, q-1$, are all divisors of $F(X)^n$ in $K[X]$ and therefore equal to a power of F (and since they all have the same degree, this power is the same for all i):

$$
F(X)^\mu = f_i(X,r)f_i(X,r_1)\ldots f_i(X,r_{n-1}), \quad i = 0, 1, \ldots, q-1. \tag{8}
$$

Counting degrees on both sides, we find

$$
m \cdot \mu = s \cdot n,
$$

and since $m = s \cdot q$, this implies that

$$
q \cdot \mu = n. \tag{9}
$$

Now, Liouville distinguished two cases: n prime and n composite. In the first case, (9) implies that $q = n$ and $\mu = 1$. Thus, for n prime,

$$
\begin{aligned}
F(X) \ &= \ f(X,r)f(X,r_1)\ldots f(X,r_{n-1}) \\
&= \ f(X,r)f_1(X,r)\ldots f_{n-1}(X,r),
\end{aligned} \tag{10}
$$

where the first decomposition agrees with Galois's decomposition (4) and includes all the roots of the auxiliary equation. This is the only case treated by Liouville in the 1843 note published by Bourgne and Azra [Galois 1962].

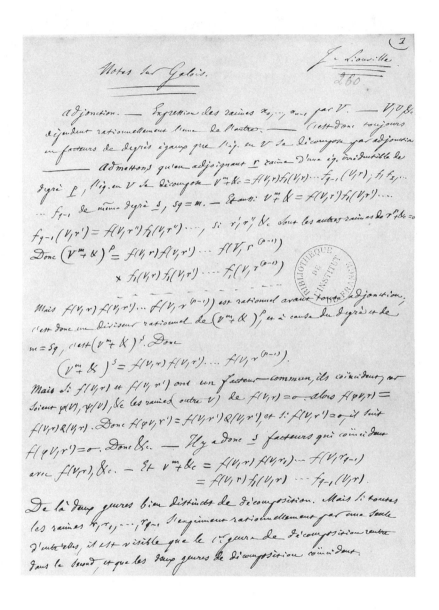

PLATE 31. The first page of Liouville's notes on Galois theory. (Bibliothèque de l'Institut de France, Ms 2108 Fol 260)

In the later notes, Liouville also analyzed what happens when n is a composite number. In this case μ, may be different from one, so that the roots of (8) are μ-fold. Using Lemma A, Liouville easily showed that if two factors, $f_i(X, r_j)$ and $f_i(X, r_l)$, have a common root, then they are identical. Thus, there are μ factors equal to $f_i(X, r)$, μ factors equal to $f_i(X, r_1)$, etc. Having grouped them together and possibly changed the indexing of the r's, one can extract the μth root of (8) (for example for $i = 0$), obtaining:

$$
\begin{aligned}
F(X) &= f(X, r)f(X, r_1)\ldots f(X, r_{q-1}) \\
&= f(X, r)f_1(X, r)\ldots f_{q-1}(X, r).
\end{aligned}
\tag{11}
$$

Again, the first line is Galois's decomposition (4), but now only q of the roots of the auxiliary equation are used, where q is a divisor in the degree n of the auxiliary equation.

13.   Here, I shall intercalate a warning to those who read Galois's *Mémoire* in Edwards' translation. Edwards has incorrectly translated " $r, r', r,''$ ... [i.e., $r, r_1, \ldots, r_{q-1}$] etant d'autres valeurs de r," into " $r, r', r,''$ ... being *the* other values of r." Moreover, in the preface, Edwards writes: "In fact, the Proposition, as stated is false. The index of the subgroup need not be 1 or $p[n]$ when p is not prime - it must simply be a divisor of p" [Edwards 1984, p. x]. In my opinion, Galois very carefully avoided this mistake when rephrasing Proposition II. Originally, he had assumed that the auxiliary equation be of "prime degree p," and in the proof he had written: " $r, r', r,''$ ... [$r, r_1, r_2, \ldots, r_{q-1}$ of (4)] etant les divers valeurs de r." This is the correct version for n prime $= p$. When altering the theorem, he maintained the phrase that the group would be partitioned into one or p groups, but it is not stated that p is the degree of the auxiliary equation. On the contrary, his alteration of "les divers valeurs de " into the indefinite "d'autres valeurs de r" shows unambiguously that he was aware that the decomposition (4) need not involve all the roots of the auxiliary equation, and consequently, that p need not be its degree. Thus, Galois's formulation of Proposition II is completely correct, although admittedly incomplete since he forgot to add that p is a divisor of the degree of the auxiliary equation.

14.   The two different factorizations of F in (11) correspond to different divisions of the roots of F, i.e., to different partitionings of the Galois group (2) (over K). They were analyzed by Liouville in Fol. 279–281. He first considered Galois's factorization

$$
F(X) = f(X, r)f(X, r_1)\ldots f(X, r_{q-1}).
\tag{12}
$$

All the roots of $f(X, r)$ are rational functions (over K) in V. Let them be denoted by

$$
V, \varphi_1(V), \ldots, \varphi_{s-1}(V).
\tag{13}
$$

Liouville showed that if $V_j = \omega_j(V)$ is a root of $f(X, r_j)$ then $\varphi_1(V_j), \ldots, \varphi_{s-1}(V_j)$ are its other roots. For the rational function, $f(\varphi_i(X), r)$ has the root V, and so, according to Lemma I, it is divisible by $f(X, r)$:

$$f(\varphi_i(X), r) = f(X, r)Q_i(X, r).$$

Lemma A then implies that

$$f(\varphi_i(X), r_j) = f(X, r_j)Q_i(X, r_j), \quad i = 0, 1, \ldots, s - 1,$$

and consequently

$$f(\varphi_i(V_j), r_j) = f(V_j, r_j)Q_i(V_j, r_j) = 0, \quad i = 0, 1, \ldots, s - 1.$$

Therefore, one passes from the permutations of the first "group," characterized by $V, \varphi_1(V), \ldots, \varphi_{q-1}(V)$, to the jth group, characterized by $V_j = \omega_j(V), \varphi_1(V_j), \ldots, \varphi_{q-1}(V_j)$, by changing V into V_j; or put differently, by performing the same substitution to all the permutations of the first group. This proof is an elaboration of Galois's proof of the last statement in Proposition II.

Liouville's subsequent analysis of the partitioning of the Galois group corresponding to the second factorization in (11) has no parallel in Galois's *Mémoire*.

Let

$$F(X) = f(X, r)f_1(X, r) \ldots f_{q-1}(X, r) \tag{14}$$

and assume as above that (13) is a list of the roots of $f(X, r)$. Assume moreover, that $V_j = \omega_j(V)$ is a root of $f_j(X, r)$, that is, that V is a root of $f_j(\omega_j(X), r)$. As above, Lemma I and Lemma A show that

$$\omega_j(V), \omega_j\varphi_1(V), \ldots, \omega_j\varphi_{s-1}(V)$$

is a list of the roots of $f_j(X, r)$ (that they are different also follows from Lemma I). Thus, the substitution taking the first permutation in the first group into the ith permutation in the first group (and taking V into $\varphi_i(V)$) also carries the first permutation of the jth group into the ith permutation of that "group" (i.e., takes $\omega_j(V)$ into $\omega_j\varphi_i(V)$). Therefore, all the "groups" in the partitioning according to (14) contain the same substitutions.

15.   Liouville [Fol. 268] remarked that this analysis gives the key to what Galois said about "the two modes of decomposition of the total group." Since the first type of decomposition is the only one discussed in Galois's *Mémoire*, Liouville must here refer to Galois's letter to Chevalier, in which he discussed right and left cosets in a rather modern way:

> In other words, when a group G contains another H the group G can be partitioned into groups which are each obtained by operating on the permutations of H with one fixed substitution, such that
>
> $$G = H + HS + HS' + \cdots.$$

And it can also be partitioned into groups which all have the same substitutions, such that

$$G = H + TH + T'H + \cdots.$$

These two types of decomposition generally do not coincide. When they coincide, the decomposition is said to be "propre."

[Galois 1962]

In fact, on Fol. 283, Liouville similarly spoke of a "propre" decomposition in the case where we would speak of a normal subgroup.

In particular, if all the roots, $r_1, r_2, \ldots, r_{n-1}$, of the auxiliary equation depend rationally on one of them, then $f(X, r_i)$ is of the form $f_i(X, r)$, and thus the two types of decomposition in (11) coincide as stated in Galois's Proposition III. With these remarks, Liouville completed the proofs of Proposition II and III about which Galois had written: "There is something that needs completing in this demonstration. I haven't the time" and "One will find the proof."

Referring to the 1843 proof of Proposition II, Edwards [1984, p. x] wrote that "Liouville found it necessary to circumvent Galois's proof entirely," and continued "I believe now that the proof given in §44 is very close to what Galois had in mind." The above summary, however, has shown that Liouville's proof is in full accordance with the sketch given by Galois, and in fact, it is not very different from the proof given by Edwards. (I showed a sharper version of this statement and those of §13 to Harold Edwards, and we almost agreed on the formulations given here, except that Edwards still maintains that it sounds as if p in Galois's Proposition II is the degree of the auxiliary equation (cf. [Edwards 1989])).

PROPOSITIONS VI–VIII

16.   Liouville rather quickly passed over Proposition V (cf. §9), which is proved in unusually great detail in Galois's paper, and so arrived at the application of the general theory to the equation of prime degree. In Galois's manuscript, this section begins with the following Proposition VI:

Lemma. An irreducible equation of prime degree cannot become reducible by the adjunction of a radical.

This statement is clearly false, and so in the published version Liouville added:

whose index [i.e., p in $\sqrt[p]{A}$] is different from the degree of the equation.

He did not mention that this was an addition of his own. The new statement is only correct for p prime. Indeed, an nth root, where n is divisible by the order of the equation, may of course do as well. Galois's sketch of a proof as elaborated by Liouville goes as follows: Let $\mathcal{F}(\S)$ be the given irreducible equation that becomes reducible after the adjunction of $\sqrt[p]{A} = r$. According to the analysis above we have

$$\mathcal{F}(x) = \mathcal{G}(x, r)\mathcal{G}(x, r_1) \ldots \mathcal{G}(x, r_{p-1}),$$

where $r_1, \ldots, r_{p-1}$ are the other roots of $r^p = A$. When the degree of $\mathcal{F}$ is prime, this is only possible when $p = \deg \mathcal{F}$ and $\deg \mathcal{G} = 1$.

In this proof, Galois used the fact that if $r^p = A$ does not have a root among the known quantities K then it is irreducible over K. This is not established by Galois, but Liouville provided a proof in the special case that interested him, namely, when p is the smallest prime for which the Galois group of a given polynomial in $K[X]$ decomposes by the adjunction of a pth root, say, $\sqrt[p]{A}$. Indeed, as remarked by Galois (cf. §9), we may assume that the pth roots of unity are in K. Now, assume that $r^p - A$ decomposes over K, and let $h(x)$ be a factor of degree q, where $1 < q < p$. It must be of the form

$$h(x) = (x - \alpha^{\nu_1} r) \ldots (x - \alpha^{\nu_q} r),$$

where α is a pth root of unity. Its constant term H is equal to $\alpha^{\nu_1 + \nu_2 + \cdots + \nu_q} r^q$, which lies in K, and since we have assumed that $\alpha \in K$, we conclude that $r^q = H' \in K$. Liouville has $H = H'$, which is obviously a slip of the pen. Thus, $r = \sqrt[q]{H'}$, contrary to the assumption that p was the smallest prime for which a pth root would partition the Galois group. The modern general proof of the irreducibility of $r^p = A$ over K when $\sqrt[p]{A} \notin K$ is but a small elaboration of this proof.

17.   In Fol. 261–263, Liouville continued to analyze Galois's Proposition VII, which deals with the Galois group of an irreducible solvable equation, $\mathcal{F}(\S) = \prime$ of prime degree p. As discussed in §9, the solution of such an equation can take place by successively adjoining pth roots for increasing primes p. The preceding proposition shows that the given equation $\mathcal{F}(x) = 0$ remains irreducible during this process until the last adjunction - which must be the adjunction of a pth root - after which it decomposes into first-degree factors, which means that the roots, and thus the Galois resolvent V, are known, and the Galois group is reduced to one element. Consequently, before the adjunction of the final pth root, the Galois group must have consisted of p elements.

At this point of the argument, Galois referred to Cauchy [1815] for the theorem that a group of order p must be cyclic. To be sure, Cauchy did not speak of a group, but of the substitutions of $x_1, \ldots, x_p$, leaving a given function $K(x_1, \ldots, x_p)$ unchanged. Still, for a genius like Galois, it was a triviality to prove the theorem on the basis of Cauchy's concept of the order (Cauchy called it degree) of a substitution and Cauchy's analysis of the cyclic subgroup generated by one substitution. (Cauchy spoke of a circle of permutations.) In modern terms, the proof runs as follows: Consider the cyclic subgroup $\langle a \rangle$ of G generated by an arbitrary element of G different from the identity. Its index in G divides the order p of G and is therefore 1, so that $\langle a \rangle = G$.

Liouville provided another argument for this fact: Let K denote the field of known quantities before the adjunction of $\sqrt[p]{A}$. Then, Liouville claimed that the roots of V's irreducible polynomial over K must be of the form $V, \varphi(V), \varphi^2(V), \ldots, \varphi^{p-1}(V)$, where φ is a rational function over K. He probably knew this fact from [Abel 1829a, §1]. Before any adjunction, we have $x_0 = \lambda(V)$, λ being rational over the originally known quantities, and by applying Lemma 1, one can see that $\lambda(V), \lambda(\varphi(V)) \ldots \lambda(\varphi^{p-1}(V))$ are different roots of $\mathcal{F}(\S)$, so that one can assume

$$x_0 = \lambda(V), x_1 = \lambda(\varphi(V)) \ldots x_{p-1} = \lambda(\varphi^{p-1}(V)).$$

The substitution $V \to \varphi^k(V)$ therefore yields the cyclic substitution $x_i \mapsto x_{i+k}$ of the roots, where the indices are counted modulo p.

It is worth remarking that here Liouville preferred an argument based on rational functions to Cauchy's general ideas about permutation groups. In fact, it is characteristic of Liouville's arguments that they tuned down the importance of the permutation groups and emphasized the rational manipulations with the Galois resolvent.

18.   This is certainly true of his subsequent two proofs of the fact that before the last but second adjunction in the above case the permutations in the Galois group are all of the form

$$x_i \mapsto x_{ai+b} \qquad \begin{aligned} a &= 1, 2, \ldots, p-1 \\ b &= 0, 1, \ldots, p. \end{aligned} \tag{15}$$

I refer the reader to Liouville's own proofs in Appendix II. A rather simple extension of the above argument shows that even before the adjunction of any radicals the permutations of the Galois group must be of the form (15). This argument is not found in Liouville's notes; neither is an elaboration of Galois's sketchy proof of the converse statement of Proposition VII (cf. §10).

Instead, Liouville continued to deduce Proposition VIII (cf. §10) from this characterization of the Galois group of a solvable irreducible equation of prime degree. About this long and correct proof (cf. Fol. 263–265), I shall merely remark that it follows Galois's indication of a proof that Edwards described as being "difficult to complete" [Edwards 1984, p. 98, Exercise 4].

When Tannery [Galois 1906] published Liouville's brief 1843 note, he expressed the view that it was "a touching trace of the care and scrupulousness which Liouville displayed in his publication" and further, that "Certainly while composing this note Liouville stuck to the recommendation to be 'transcendentally clear' which he has given in the *Avertissement* of the *Oeuvres mathématique of Galois*." This characterization may be extended to most of Liouville's other notes on Galois theory.

Liouville's Publication of the Works of Galois

19.   We can now turn to the questions: why did Liouville postpone the publication of Galois's works three years, and why did he never publish his commentaries? Being unable to give a definitive answer, I shall suggest a number of possible reasons.

In his *Avertissement* to Galois's collected works, Liouville described how he "enjoyed a vivid pleasure at the moment where, after having filled in some small holes, I realized the complete exactness of the methods with which Galois proves [Proposition VIII]." This moment of excitement occurred before September 1843, for when Liouville announced his discovery of Galois's papers to the *Académie* [Liouville 1843i, p. 448], he explicitly stated that Galois's analysis was "exact and profound." However, that does not quite exclude internal mathematical reasons for the delay in publishing Galois's works; Liouville may have had doubts about the correctness of some details of Galois's *Mémoire*. Indeed, on the proof sheets from the December 1843 issue of his *Journal*, Liouville maintained the phrase in Proposition II that Galois had struck out, to the effect that the auxiliary equation be of prime degree. Moreover, as we saw above, Liouville's 1843 proof of Proposition II only applied to this case. Such a restriction does not affect the correctness of Proposition VIII, which is concerned solely with adjunction of pth roots where p is a prime. Still, it is possible that the general version of Proposition II and other details puzzled Liouville so much that it made him defer the planned December 1843 publication. As we saw above, Liouville later extended his proof to cover auxiliary equations of composite degree.

Problems with finding space for Galois's work in the issues of the *Journal* may also account for the delay. Postponing the papers of an author long since dead probably caused less problems than putting off a recent paper. More specifically, Liouville may have found it more urgent to publish the paper of his young protegé J. A. Serret, in the December 1843 issue, than a paper ten years old.

Finally, it is possible that as early as 1843 Liouville decided to extend the project from publishing merely Galois's great *Mémoire*, the only one typeset in the 1843 proof sheets, to a complete publication of all the published and unpublished works of Galois. In that case, he also had to postpone the great *Mémoire*, if only to rewrite the *Avertissement* and to sort out the remaining *Nachlass*. A combination of these reasons together with his constant lack of time, may account for the delay in Liouville's publication of Galois's works.

20.   As for Liouville's commentaries, the four different versions indicated some indecision as to the formulation, and indeed none of the versions were carried to a publishable state. Still, as we saw in the preceding section, it would not have been but a small work to combine and rephrase these notes into a highly informative commentary. It would not have become

completely exhaustive, for it does not elucidate the first part of Galois's paper or the converse of Proposition VII. However, the methods used in Liouville's comments would suffice to elucidate these parts of the *Mémoire* as well, except, perhaps, for the assertion that the general equation of degree m has S_m as its Galois group. To these problems of edition may have come a desire to comment on other manuscripts by Galois.

Indeed, the publication of Galois's other unpublished manuscripts may provide the key to the indefinite postponement of the commentary. In the 1846 volume of his *Journal*, Liouville merely published the letter to Auguste Chevalier, the great *Mémoire*, and a fragment on solvability of "primitive equations," in addition to the previously published papers. But, in the note following Galois's letter to Chevalier, Liouville expressed his plan to publish more of Galois's unpublished notes:

> We shall complete this publication with some fragments extracted from Galois's papers, and which, though without being of great importance, may still be read with interest by geometers.
>
> [Liouville 1846j, p. 416]

However, these fragments never appeared, and that was probably responsible for the nonappearance of the commentary, for as he explained in the *Avertissement*, Liouville had planned to publish the commentary last:

> We reproduce first the different articles published by Galois ....
> Then come the unpublished pieces, and finally a commentary in which we intend to complete certain passages and clarify some delicate points.
>
> [Liouville 1846j, p. 383]

This does not answer the question of why Liouville did not publish a commentary, but it replaces it with a better question: why did Liouville not fulfill his promise to publish more fragments from Galois's *Nachlass* and a commentary? As we saw in §5, Liouville referred to this problem in the letter to Jacobi written in November 1847 for Alfred Galois. However, the letter throws little light on the reasons, explaining merely that the publication of the fragments "has been delayed by various ["diverses"] circumstances." These circumstances probably included an attempt to annotate the rather incomprehensible published fragment on "primitive" equations. We do not know if there were other circumstances, but it is a fact that the fragments not included in Liouville's 1846 edition remained unpublished until Tannery edited the *Manuscrits et Papiers inédits de Galois* in 1906–1907 and 1908 [Galois 1906].

Liouville's Understanding of Galois Theory

21. The extent to which Liouville understood Galois has been questioned [see, e.g., Kiernan 1971–1972 and Hirano 1984]. Kiernan goes as far as

to state that "Liouville's surviving notes indicate only a few notational amendments to Galois's text." To be sure, Liouville did make a few notational improvements, such as a less confusing indexing (Tannery in [Galois 1906] lists these changes), and he did unsuccessfully suggest the name *permutation tournante* instead of Galois's *substitution circulaire* (our "element of a cyclic group") (cf. [Fol. 261]), but as the above summary has shown, Liouville went much further. The main merit of his commentary is its clarification of the most obscure places in Galois's great *Mémoire*. It shows that by 1846 Liouville completely understood this paper as well as the algebraic parts of the letter to Chevalier. It is less certain whether he was ever able to comprehend all the details in Galois's other unpublished works.

In his interesting book on the history of the group concept, Wussing [1984–1969] expressed the view that Liouville did not understand the scope of Galois's work, and offered as an example that "Liouville regarded as Galois's central achievement the theorem that an irreducible equation of prime degree is solvable by radicals if and only if its roots are rational functions of any two of them." Indeed, Liouville explicitly mentioned this theorem, both in his 1843 announcement to the *Académie* and in his *Avertissement*, but there he merely followed Galois [cf. Galois's preface to 1831–1846]. At the time, one could not have introduced Galois's work better than by quoting this theorem, which was the only one that did not involve unfamiliar concepts. Thus, it is by no means certain that Liouville considered this proposition as Galois's central achievement. In fact, Liouville often emphasized Galois's methods rather than his results. In [Liouville 1843i], he spoke of a "profound solution," in the *Avertissement*, of "the bases of a general theory which he applies in detail to equations whose degree is a prime number," and in the letter to Jacobi he stated that the "memoir... contains the basis of a very fruitful doctrine and a first important application of this doctrine." It is my impression that the doctrine Liouville had in mind was the interplay between the adjunctions to the known quantities and the splitting of the Galois group as expressed by the factorization of the irreducible equation of the Galois resolvent V.

Still, I agree with Wussing that Liouville did not grasp the group theoretic core of Galois's paper. As pointed out in §17, Liouville's comments largely avoid the general ideas of permutation groups as developed by Cauchy, even in cases where Galois referred to them. Instead, he relied on arguments involving rational functions and the theory of factorization of polynomials, which is suggested by Galois's definition of the Galois group by way of the Galois resolvent V and its irreducible equation. This preference is probably influenced by Liouville's familiarity with the works of Abel, who had used similar arguments in [1829a]. So, although Liouville's commentaries are probably closer to Galois than many modern commentaries with their use of abstract group theory, it must be admitted that they were somewhat conservative and not very far-sighted. They certify a thorough understanding of Galois's methods, but no attempts to carry the ideas further.

Liouville's Impact

22.   It is obvious that Liouville had a great impact on the later development of algebra, for the simple reason that all the subsequent developments of Galois theory were based on his edition of Galois's works. It is not so easy to evaluate the impact of Liouville's unpublished analysis of Galois's works, in particular his course. It probably alerted several of the young generation (certainly Serret and probably Bertrand and Hermite) to the beauty of Galois's ideas, and as argued earlier, it may have inspired Cauchy to his last papers on permutation groups. Still, his somewhat conservative conception of Galois theory may have retarded the development. Even worse, with his expressed intention to publish a commentary, he monopolized the field in France for several years. In Chapter V, we have seen how the fear of Liouville's reaction kept Serret away from the subject for 20 years. Even the Italian mathematician Enrico Betti found it necessary to refer to Liouville's plan and to argue why he thought it proper to publish a commentary of his own, even if Liouville's paper had not appeared. Having pointed out the unintelligible conciseness of Galois's *Mémoire*, Betti wrote in 1851:

> In a note in his Treatise on higher algebra, however, Serret has announced Liouville's intention to publish some day certain developments that will clarify and complete Galois's work. Nevertheless, since Malmsten has already made known to science Abel's theorem, I believe that it will be both proper and useful to show, using theories already developed in Algebra, how it can be transformed into Galois's. In calling attention to this achievement, which the young geometer had the time to obtain almost entirely on the basis of his own profound views, I wish that the great Liouville would not deprive the public any longer of the results of the study that he made on the same, for they will undoubtedly provide the stimulus for remarkable progress in Algebra.
>
> [Betti 1851, Opere, p. 18, transl. by Wussing 1984–1969]

Liouville did not follow this invitation, and so one year later, Betti published his own commentary [Betti 1852]. This first published paper on Galois theory went over much the same material as Liouville's unpublished notes, but put more emphasis on the group theoretic aspects.

The only indication that Liouville continued to work on his commentary is eight pages of notes in Ms 3619 (6), which stem from 1851–1852. They contain no mathematical novelties, but a remark "Lire Betti," shows that Liouville was aware of the work of his Italian colleague. Thus, Liouville's positive influence on Galois theory seems to have ended in 1846. But, by then he had served the development of algebra in a way that can hardly be overestimated. He had been the first mathematician to understand Galois's profound ideas, and he had spread these ideas in the first course given on this subject, and not least, he had published Galois's works. Had he published his commentaries, his influence might have been even greater, and his effort would undoubtedly have been esteemed more highly by later generations.

XV

Potential Theory

1. Liouville is not usually mentioned in historical accounts of potential theory despite the fact that he discussed the relation between harmonic and holomorphic functions in the plane and the uniqueness of an equilibrium distribution on given conductors in a more elegant way than his contemporaries and contributed substantially to the theory of ellipsoidal harmonics. It is the main purpose of this chapter to remedy this neglect and in particular to present a most remarkable but unfinished and unpublished contribution by Liouville to general potential theory. In this work, Liouville anticipated many of the central ideas and theorems concerning integral operators, in particular, their spectral theory, including the Rayleigh-Ritz method for determining eigenvalues. As background material, I shall first sketch the state of affairs in potential theory around 1840 and discuss Liouville's published contributions.

The Genesis of Potential Theory

2. Potential theory emerged as the theory of gravitational forces. In the *Principia*, Newton had argued that any two masses attract each other along the line connecting them with a force inversely proportional to the square of their distance and proportional to their masses. From this universal law of gravitation, he concluded, in the notation of his successors, that a mass distribution of density $\rho(\overline{x}')$ attracts a unit mass at a point $\overline{x}$ with a force

$$\overline{F}(\overline{x}) = \int \frac{\rho(\overline{x}')(\overline{x} - \overline{x}')}{|\overline{x} - \overline{x}'|^3} \, d\overline{x}'. \tag{1}$$

Newton succeeded in showing geometrically that the force from a homogeneous sphere on a point outside the sphere is equal to the force that would be from the center of the sphere if the entire mass was concentrated there. Moreover, he showed that a homogeneous body bounded by two similar ellipsoids, having their axes along the same lines (in particular two concentric spheres) exercises no attraction on bodies in the cavity.

The complicated expression (1) impeded further progress until Lagrange [1774] had the ingenious idea of introducing the potential, that is, the scalar function

$$V(\overline{x}) = \int \frac{\rho(\overline{x}')}{|\overline{x} - \overline{x}'|} \, d\overline{x}' \tag{2}$$

from which the force can be deduced simply by taking minus the gradient. Laplace [1782] took up this idea and proved that at points outside the mass distribution the potential satisfies

$$\Delta V = 0. \tag{3}$$

This is called Laplace's equation. A function satisfying Laplace's equation was later called harmonic by William Thomson. At points carrying mass, the potential V satisfies the equation

$$\Delta V = -4\pi\rho. \tag{4}$$

This equation was found by Poisson [1813], after whom it is named; it was rigorously proved by Gauss [1840, §§9–11]. Poisson found his results in connection with electrostatics, where, according to Coulomb, the interactions are similar to Newtonian gravitation with the only difference that charges (i.e., ρ above) can be negative. Since by (4) conductors only carry charges on their surface, it was natural to introduce surface distributions: a distribution μ on a surface S gives rise to the potential

$$V(x) = \int_S \frac{\mu(\overline{x}')}{|\overline{x} - \overline{x}'|} \, d\omega, \tag{5}$$

which satisfies Laplace's equation outside the surface. It is continuous across the surface, it has tangential derivatives on the surface, and its normal derivative from the outside and inside differs by $4\pi\mu(\overline{x})$:

$$\frac{\partial V}{\partial n_-} - \frac{\partial V}{\partial n_+} = 4\pi\mu. \tag{6}$$

Here and in the following, n denotes the outward normal and $\frac{\partial V}{\partial n_-}$, $\frac{\partial V}{\partial n_+}$ denote the limit of the derivative of V in this direction as the point tends to a point on the surface along the normal from the inside and from the outside, respectively. These facts were known to Poisson (cf. [Poisson 1812]).

Here, I must confess that in order not to overload this account the exact regularity conditions needed for the theorems to be correct will be mentioned only in so far as they are important for the story. The rigorous statement of the classical theorems can be found, for example, in [Kellogg 1929].

After Fourier, the theory of heat became a third source for potential theory. The reason is that when a heat flow in a homogeneous medium is stationary the temperature satisfies Laplace's equation.

3.   During the 1830s and 1840s, great advances were made both in general potential theory and in the search for potentials caused by concrete charge distributions and harmonic functions with given boundary values on given rather simple surfaces. The last, more technical advances can in turn

be divided into two types: one concerns the determination in finite form (or by quadrature) of the potential for a few simple mass distributions. The homogeneous ellipsoid, in particular, attracted the interest of Laplace, Maclaurin, Ivory, Green, Gauss, and Chasles (cf. [Burkhardt and Meyer 1899–1916, p. 483]). Second, more complicated potentials were expressed as infinite series of special functions. The latter method was initiated by Legendre and Laplace's treatment of almost spherical mass distributions. In this connection, they introduced spherical harmonics. This approach was continued by Lamé in his studies of the stationary temperature distribution in ellipsoids. Thereby, he was led to the ellipsoidal harmonics or Lamé functions (cf. Chapter XI, §§21–23).

4. The general theorems of potential theory were greatly advanced in two classics: *An Essay on the Application of Mathematical Analysis to the Theories of Electricity and Magnetism* by George Green and the *Allgemeine Lehrsätze in Beziehung auf die im verkehrten Verhältnisse des Quadrats der Entfernung wirkenden Anziehungs-und Abstossungs-Kräfte* by Gauss. Though published in 1828, Green's essay was neglected until 1845 when William Thomson called attention to it (cf. Chapter III). Therefore, Gauss's independent *Allgemeine Lehrsätze* from 1840 was the first of the two to influence the mathematical community, including Liouville.

In the first part of the paper, Gauss presented rigorous proofs of the theorems due to his French colleagues - without ever mentioning any of their names. Then, he developed his new results, the first being the *reciprocity theorem* [§19]: If two mass distributions μ and ν on a surface S give rise to potentials U and V, respectively, then

$$\int_S V\mu\,d\omega = \int_S U\nu\,d\omega. \qquad (6a)$$

5. From this result Gauss deduced a theorem of which I shall quote a special case. Consider a spherical surface S of radius R and let V be the potential of a mass distribution outside the sphere. Then, the value V_0 of the potential at the center of the sphere is the arithmetical mean of its values on S:

$$4\pi R^2 V_0 = \int_S V\,d\omega. \qquad (6b)$$

This so-called Gaussian theorem of the arithmetic mean has as an immediate consequence that a potential cannot have a maximum or a minimum outside the mass distribution; however, Gauss did not make this observation.

In §23, he then proved the so-called *Gauss theorem*: If V is a potential due to some mass distribution and if M denotes the total mass inside a closed surface S, then

$$\int_S \frac{\partial V}{\partial n}\,d\omega = -4\pi M. \qquad (6c)$$

Finally, Gauss showed [§26 and §27] that if V is the potential due to a charge distribution μ on a surface S and if V has the constant value V_0 on S then V is everywhere zero if and only if the total charge is zero and it is numerically smaller than V_0 outside of S if and only if the total charge is different from zero.

6.   These theorems may all be considered as introductions to the existence and uniqueness theorems of §§29–34. In these sections, which had the greatest influence on Liouville, Gauss simultaneously discussed two problems concerning the charge distribution on a closed surface S: 1° Charge the surface with a given total charge; show that there exists a unique equilibrium distribution of the charge on the surface. 2° Given a function U on the surface, show that there exists a unique charge distribution on the surface giving rise to the potential U on the surface.

Gauss's argument is a mathematization of the following physical argument. If the mass distribution $\mu(\overline{x})$ on S gives rise to the potential $V(\overline{x})$, then the potential energy stored in the distribution can be written

$$E = \int_S V(\overline{x})\mu(\overline{x})\,d\omega. \tag{7}$$

The equilibrium distribution of a given total charge M must minimize E among all distributions having

$$\int_S \mu(\overline{x})\,d\omega = M. \tag{8}$$

In order to include the second problem above, Gauss more generally showed that if U is a given function on S and if $\mu(\overline{x})$ minimizes the integral

$$\Omega = \int_S (V(\overline{x}) - 2U(\overline{x}))\mu(\overline{x})\,d\omega \tag{9}$$

under condition (8), then $V(\overline{x}) - U(\overline{x})$ is a constant on S. The proof is simple and well known from the calculus of variations: Let $\Omega + \delta\Omega$ and $V + \delta V$ be the values of Ω and V, respectively, when $\mu + \delta\mu$ is substituted for μ. Then,

$$\delta\Omega = \int \delta V \mu\,d\omega + \int (V - 2U)\delta\mu\,d\omega. \tag{10}$$

According to the reciprocity theorem (6a) we have

$$\int \delta V \mu\,d\omega = \int V \delta\mu\,d\omega, \tag{11}$$

so that at a minimum

$$0 = \delta\Omega = \int 2(V - U)\delta\mu\,d\omega. \tag{12}$$

Now, $\delta\mu$ is arbitrary, except that

$$\int \delta\mu \, d\omega = 0, \tag{13}$$

and so Gauss concluded that $(V - U) = $ constant. For, if $(V - U)$ varies over S, Gauss selected a "mean" value A and two regions P, Q (of measure > 0) such that $V - U$ is larger than A in P and smaller than A in Q. He then selected a $\delta\mu$ that is positive in P, negative in Q, and satisfies (13) and found that

$$\begin{aligned}
\delta\Omega &= \int 2(V - U)\delta\mu \, d\omega = \int 2(V - U)\delta\mu \, d\omega - \int 2A\delta\mu \, d\omega \\
&= \int 2\left[(V - U) - A\right]\delta\mu \, d\omega < 0, \tag{14}
\end{aligned}$$

contradicting the assumption that Ω had a minimum.

In particular, if $U = 0$, we have $V = $ constant on S, so that μ is an equilibrium distribution. We also remark that if we minimize Ω without assuming that the total charge is a constant $V - U$ must satisfy (12) for arbitrary $\delta\mu$, and thus we conclude that $V = U$. Gauss showed the existence of μ, making $V = U$ in a different way by considering a linear combination of the equilibrium distribution and the distribution found above corresponding to U.

7.   These existence theorems require that Ω has a minimum. Gauss first tried to establish that this minimum exists among distributions that are everywhere nonnegative. In this case, it is obvious from (5) that $V(x) > \frac{M}{R}$, where R is the greatest distance between two points in S (the diameter of S). Therefore, by (7), $E > \frac{M^2}{R}$, and if U_0 is the minimum of U, we have $\Omega > \frac{M^2}{R} - 2U_0M$. From the existence of this lower bound, Gauss concluded that a minimizing distribution among the everywhere nonnegative μ's exists.

In fact, in Gauss's paper, the argument in §6 above is only carried out for nonnegative μ's for which he could only conclude that $U - V$ is constant in places where the minimizing μ is different from zero. In regions where μ is zero, $U - V$ is greater or equal to this constant.

In the case of the equilibrium distribution, Gauss showed that the minimizing, everywhere nonnegative distribution is everywhere positive, and thus its potential is constant over the entire surface. It is also the only equilibrium distribution with the given total charge. Indeed, if there were two equilibrium distributions V, V', with the charge distributions μ and μ', then their difference $V - V'$, which is constant on S, must be a potential of the charge distribution $\mu - \mu'$, which has total charge zero. Therefore, $\mu - \mu'$ must be everywhere zero and $V = V'$ according to the theorem at the end of §5.

For $U \neq 0$ and for charge distributions of varying sign, Gauss could not find a lower bound for Ω and so instead he established the general existence of μ, making $U - V$ constant by reducing it to the case of one sign. This rather unelegant twist to the argument was soon circumvented by Thomson and Dirichlet, who replaced Gauss's problem by the so-called Dirichlet problem (cf. §9), and thereby replaced Ω by another integral which evidently has a lower bound.

However, it took much longer to spot the basic weakness in Gauss's proof. The problem is that although one can prove the existence of a (greater) lower bound for Ω it is not certain that one can find a distribution μ for which Ω attains this greatest lower bound. One only knows that there exists a sequence of μ's that makes Ω tend to its greatest lower bound. Another criticism was raised later by de La Vallée-Poussin [1962]. I shall return to these problems and their history after summarizing Green's results.

8.   Green's analysis is based on a number of theorems relating volume and surface integrals, the most important being

$$\int_\Omega (U\,\Delta V - V\,\Delta U)\,dv = \int_{\partial\Omega} \left(U\frac{\partial V}{\partial n} - V\frac{\partial U}{\partial n} \right) d\omega, \tag{15}$$

where Ω is a bounded region in space with boundary $\partial\Omega$ and where n is the outward normal. This formula is called Green's (second) theorem. It was independently found the same year by the Russian mathematician Ostrogradsky, who also found Gauss's theorem, or the divergence theorem, independently of Gauss (cf. also [Liouville 1854d]). As a whole, the technique used in these proofs - componentwise partial differentiation — was so familiar in the 19th century that any mathematician working in potential theory stumbled on some theorems relating volume and surface integrals. Therefore, the history of Green's, Gauss's, and Stokes's theorems are full of independent discoveries. By way of example, Liouville in 1842 derived Green's first theorem

$$\int_\Omega \nabla\varphi \cdot \nabla\psi\,dv + \int_\Omega \varphi\Delta\psi\,dv = \int_{\partial\Omega} \varphi\frac{\partial\psi}{\partial n}\,d\omega \tag{16}$$

(where $\nabla\varphi = \operatorname{grad}\varphi$) in the case where $\varphi = \psi$. Since Liouville's proof works just as well for $\varphi \neq \psi$ and since Green's second theorem is a trivial consequence of (16), we can conclude that Liouville was in possession of Green's first two theorems in 1842, that is, before he learned of Green's *Essay*. Moreover, in his [1846e,f], the content of which dated back to 1842, Liouville derived Gauss's theorem [1846e, pp. 233–235] and Green's second identity [1846f, pp. 265–267] for special functions U, V, but with a general proof. A few pages further on, however [1846f, p. 271], he declared it unnecessary to prove virtually the same formula, not only because he had already "carried out a similar transformation" but also because such calculations were "very familiar to geometers." Earlier, Lamé and Duhamel had derived similar formulas (cf. [Burkhardt and Meyer 1889–1916, p. 478]). Clearly, these theorems were in the air. What was lacking was an agreement as to which of them should be selected as the fundamental ones.

9. After this digression about the theorems of vector calculus, we shall return to Green's *Essay*. In Green's identity (15) U and V are supposed to be "regular" (e.g., $\mathcal{C}^2(\Omega)$). If $U(\overline{x})$ is regular except at a point $\overline{x}_1$ in Ω where it is "sensibly equal to" $\frac{1}{|\overline{x}-\overline{x}_1|}$, then, according to Green (15) it must be replaced by

$$\int_\Omega (U\Delta V - V\Delta U)\, dv = \int_{\partial\Omega} \left(U\frac{\partial V}{\partial n} - V\frac{\partial U}{\partial n}\right) d\omega - 4\pi V(\overline{x}_1), \quad (17)$$

where ΔU is put equal to zero at $\overline{x}_1$. From this identity, Green first showed that if V and V' are harmonic inside and outside of $\partial\Omega = S$, respectively (V' tending to zero at infinity), and if their boundary values $\overline{V}$ and $\overline{V}'$ at S coincide, then they can be considered the potential inside and outside of S of the same surface distribution on S, namely, the one having density μ defined by (6). In fact, this argument is only correct if the normal derivatives in formula (6) exist.

Second, he observed that if the function U in (17) is chosen such that it is harmonic in Ω outside the singular point $\overline{x}_1$ and such that it has zero boundary values on $\partial\Omega$, and if furthermore V is harmonic in Ω, then by (17),

$$V(x_1) = \frac{1}{4\pi} \int_{\partial\Omega} \overline{V}\frac{\partial U}{\partial n}\, d\omega , \quad (18)$$

where $\overline{V}$ denotes the boundary values of V on $\partial\Omega$. This formula yields an explicit solution of the so-called Dirichlet problem: Given a function $\overline{V}$ on $\partial\Omega$, determine a harmonic function V in Ω that has the boundary values $\overline{V}$.

Green and later Dirichlet argued that this problem was equivalent to the problem posed by Gauss: Given a function $\overline{V}$ (or U in formulas (9)-(14)) on $\partial\Omega$ (or S), determine a charge distribution μ on $\partial\Omega$ that has the potential $\overline{V}$ on $\partial\Omega$. Indeed, it is clear that if we have found a μ satisfying the conditions in Gauss's problem the potential (5) of this distribution is a solution of Dirichlet's problem. Conversely, suppose V and V' are solutions of the Dirichlet problem in Ω and its complement $\mathcal{C}\Omega$, which has the same boundary as Ω (the Dirichlet problem implicitly requires the solution in the unbounded component, say, $\mathcal{C}\Omega$, to tend to zero at infinity). Then, according to Green's observation following (17) above, V and V' are potentials of a distribution μ on $\partial\Omega$. This is evidently the solution of Gauss's problem.

However, as pointed out above, Green's observation is only valid if $\left(\frac{\partial V}{\partial n_+} - \frac{\partial V}{\partial n_-}\right)$ exist at the surface, and here it is not certain that this is the case. In fact, de la Vallée Poussin [1962, p. 325] pointed out that there exist Dirichlet problems whose solutions are not Newtonian potentials obtained from a surface distribution. Thus, the Dirichlet problem has a larger class of solutions than Gauss's problem; still, one should keep in mind that the mathematicians of the 19th century believed that the two problems were equivalent.

10.   The function U in (18), which is in fact a function of the two variables $\overline{x}$ and $\overline{x}_1$, is the celebrated Green's function. Once it has been found for a given surface, formula (18) gives the solution of the Dirichlet problem for "any" function on the boundary. This idea was later generalized to other operators than the Laplacian. The problem of finding the Green's function, or even of proving its existence, is often difficult. Green himself gave a physical existence argument by pointing out that U is the potential due to a point charge in $\overline{x}_1$ and the charges it induces on $\partial\Omega$ considered as an earthed conducting surface.

In a later paper [Green 1833–1835, §4], Green "proved" the existence of a solution to the Dirichlet problem by a variational technique. This was the first occurrence of the so-called Dirichlet principle, which is similar to Gauss's existence argument discussed above. The principle was later published by Thomson [1847b] and used by Dirichlet in his lectures. The argument runs as follows: Consider all (C^2) functions V on Ω having the given boundary values $\overline{V}$ on $\partial\Omega$. Since the Dirichlet integral

$$D(V) = \int_\Omega \left[\left(\frac{\partial V}{\partial x}\right)^2 + \left(\frac{\partial V}{\partial y}\right)^2 + \left(\frac{\partial V}{\partial z}\right)^2 \right] dv \qquad (19)$$

is nonnegative, there exists at least one function V of this kind minimizing $D(V)$. This V is the solution of the Dirichlet problem. In order to see that, let W be an arbitrary C^2 function on Ω having boundary values 0 and let $U = V + hW$. Then,

$$D(U) = D(V) + 2h \int \nabla V \cdot \nabla W \, dv + h^2 D(W), \qquad (20)$$

and since D has a minimum in V, we have

$$\int \nabla V \cdot \nabla U \, dv = 0, \qquad (21)$$

and thus by Green's first identity (16),

$$\int_\Omega \Delta V \, W \, dv = 0. \qquad (22)$$

This must hold for all functions W, and so $\Delta V = 0$. Moreover, the way the minimizing function was selected ensures that V has the correct boundary value. The above argument follows Dirichlet [1856–1857/1876] and differs from Green and Thomson only in the details. Dirichlet applied this existence theorem in order to prove the existence of a solution of Gauss's problem making use of the (incorrect) equivalence of the two problems pointed out in §9. With this twist to the argument, he elegantly overcame the problem Gauss had with charge distributions of varying sign. Yet Green's, Thomson's, and Dirichlet's arguments share with Gauss's the confusion of a greater lower bound and a minimum.

11. In order to throw Liouville's contributions into relief, I shall briefly sketch the traditional "comedy of errors" as Monna [1975] has called the history of the Dirichlet principle. Its name stems from Dirichlet's student Riemann. He used it in a generalized form in the plane to prove the existence of analytic functions on a Riemann surface with prescribed boundary values and prescribed singularities. This approach works because any harmonic function in the plane is the real part of an analytic function and vice versa (cf. §12). However, in 1870, Weierstrass pointed out the weakness of the principle by displaying another variational problem whose greatest lower bound cannot be obtained among the admissible functions. This criticism discredited the principle for several decades until Hilbert in 1899 showed that under proper conditions on the boundary curve and on the boundary value any minimizing sequence of the Dirichlet integral (19) has a subsequence converging toward an admissible function (i.e., C^2 with the given boundary values), which is then a solution to the Dirichlet problem. This is the so-called direct method in the calculus of variations.

In the meantime, H. A. Schwarz, Carl Neumann, and Poincaré had shown the existence under rather general conditions of a solution to the Dirichlet problem with other techniques. I shall return to some of these methods in §43.

Liouville's Published Contributions

12a. Already in his early investigations concerning the theory of heat, Liouville was faced with problems of potential theory. In Chapter X, §30, we saw how he treated heat conduction in a one-dimensional metallic rod. In his published paper [Liouville 1830, §1], he stated that the problem offers "almost insurmountable difficulties" if we consider a two-dimensional plate. In the original memoir [AC L Ms 5], Liouville indicated what kind of problems were involved in determining the stationary distribution of temperature in a metal plate kept at zero degrees along its two horizontal edges and radiating heat into a medium of varying temperature along its left, unequally polished, and vertical edge.

The mathematical formulation of this problem is the boundary-value problem

$$\frac{d^2u}{dx^2} + \frac{d^2u}{dy^2} = 0, \tag{a}$$

$$u = 0 \quad \text{for} \quad x = \pm\frac{\pi}{2}, \tag{b}$$

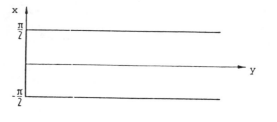

$$\frac{\partial u}{\partial y} - uf(x) + F(x) = 0 \quad \text{for } y = 0 \tag{c}$$

(in fact, in the 1830 memoir, Liouville kept the left edge at a constant temperature and therefore he had $f(x) = F(x)$). Under the conditions that $f(x) = f(-x)$ and $F(x) = F(-x)$, Liouville solved the first two equations by separating variables and inserted the result into the last equation. He obtained

$$\sum_{m=0}^{\infty} mA_m \cos mx + f(x) \sum_{m=0}^{\infty} A_m \cos mx = F(x), \tag{22a}$$

which he could not solve for A_m. He also tried to find what happened if the two horizontal edges are situated at $x = \pm\infty$ instead of at $x = \pm\frac{\pi}{2}$. As usual, he obtained an equation similar to the above, but having Fourier integrals instead of Fourier series. Concerning this equation, he wrote to Le Verrier in 1833:

> Now it has been almost six years since I began in vain to try to solve a problem that Mr. Poisson had posed to me in the theory of heat. The simplest case comes down to this problem of analysis: Let $f(x)$ and $F(x)$ be two given functions of the variable x which never become infinite and the first of which is everywhere positive; to find $\varphi(x)$ such that one has (x always being real):
>
> $$\int_0^{\infty} \alpha\varphi(\alpha)\cos\alpha x\, d\alpha + f(x)\int_0^{\infty} \varphi(\alpha)\cos\alpha x\, d\alpha = F(x).$$
>
> I have not been able to solve this question even by a reasonable method of approximation, and I have again wasted the short time of work during this holiday on it.
>
> [Neuenschwander 1984a, III, 1]

Liouville's notebooks [Ms 3615 (3,4)] also bear witness to his fruitless work with these problems. During the winter of 1833–1834, he finally found a partial solution that he presented to the *Académie* on March 17, 1834, and published two years later [Liouville 1836b]. By developing the functions f and F in Fourier series, he arrived at a system of integro-differential equations from whose solutions one can find A_m or φ above. However, he was only able to solve the integro-differential equations for particular classes of functions f, F. In a note from December 1835 [Ms 3615 (4), pp. 87v–90v], Liouville returned to the solution of (22a), but this time he substituted eigenfunctions V_m into a general Sturm-Liouville problem for the trigonometric functions $\cos my$. The note ends with the words "This cannot be done."

12.   After this Fourier type approach to potential theory, eight years passed before Liouville returned to harmonic functions in the plane, and now he considered them from a more Gaussian point of view. In this short paper [1843a], he first showed that if u is a solution of the equation

$$\frac{\partial^2 \varphi}{\partial x^2} + \frac{\partial^2 \varphi}{\partial y^2} = 0, \tag{23}$$

i.e., if u is harmonic, then $\frac{\partial u}{\partial x}\,dy - \frac{\partial u}{\partial y}\,dx$ is an exact differential and

$$v = \int \left(\frac{\partial u}{\partial x}\,dy - \frac{\partial u}{\partial y}\,dx \right) \tag{24}$$

satisfies

$$\frac{\partial v}{\partial y} = \frac{\partial u}{\partial x} \quad , \quad \frac{\partial v}{\partial x} = -\frac{\partial u}{\partial y}. \tag{25}$$

From (25), it follows that v is another solution of (23). Moreover, Liouville pointed out that the level curves $u = \text{const}$ and $v = \text{const}$ cut each other orthogonally so that one can define a new set of orthogonal coordinates (α, β) in the plane by setting $\alpha = u$ and $\beta = v$. Liouville showed that if φ satisfies (23) then it also satisfies

$$\frac{\partial^2 \varphi}{\partial \alpha^2} + \frac{\partial^2 \varphi}{\partial \beta^2} = 0. \tag{26}$$

In particular, x and y satisfy (23), and so we have

$$\frac{\partial^2 x}{\partial \alpha^2} + \frac{\partial^2 x}{\partial \beta^2} = 0, \quad \frac{\partial^2 y}{\partial \alpha^2} + \frac{\partial^2 y}{\partial \beta^2} = 0. \tag{27}$$

In the concluding passage, Liouville pointed out that this reciprocity between (x, y) and (α, β) is also a consequence of the remark that "$u + iv$ is a simple [analytic] function of $x + iy$." Indeed, equations (25) are nothing but the Cauchy-Riemann equations. "However," Liouville concluded "it seems more convenient to avoid the use of imaginaries."

d'Alembert [1752] had observed that if $u\,dx - v\,dy$ and $u\,dy + v\,dx$ are both exact then $u + iv$ is a holomorphic function. Moreover, Lamé had shown that if two solutions, u, v, of (23) have orthogonal level curves then (23) can be transformed into (26). Finally, Gauss [1822–1825] had implicitly stated that conformal mappings of the plane are "holomorphic," so the ideas of Liouville's paper were in the air. Still, I do not think they had been expressed more clearly earlier. In particular, Liouville's proof that any harmonic function is the real part of a holomorphic function seems to be remarkable for the time. It points toward Riemann's use of harmonic functions in complex function theory.

13.   Liouville's contributions to potential theory in space began with a small note [1838i] where he provided an analytic proof of a formula that Poisson had found synthetically. Of much greater importance and at the origin of many of his later ideas were the works on rotating masses of fluids, (cf. Chapter XI). These works culminated in 1842, the very year Liouville published his first paper on general potential theory. This later paper was occasioned by a discrepancy between a remark by Chasles and a proof in Gauss's *Allgemeine Lehrsätze*, which Liouville was about to publish in his *Journal*.

Chasles's remark had appeared in the *Comptes Rendus* of February 11, 1839. It began with the following beautiful theorem:

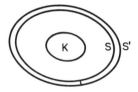

<small>FIGURE 1</small>

Let K be an arbitrary bounded solid and consider one of its equipotential surfaces S outside K and another equipotential surface S' infinitely close to it (Fig.1). Equip S with a charge distribution inversely proportional to the normal distance between S and S' (that is proportional to $\frac{\partial V}{\partial n}$). Then, this distribution will exert no attraction on a point inside S, and its attraction on exterior points will be proportional to the attraction exerted by K and directed along the same line.

By analogy (cf. Chapter III), Chasles stated the following theorem which mixes the theories of heat and electricity in a peculiar way: Consider a homogeneous body bounded by two closed surfaces S and S_1 each kept at a constant temperature. When temperature equilibrium has been reached, one takes any isothermal surface S_α between S and S_1 and equips it with a charge distribution proportional to the heat flux through the surface at each point of the surface. Then, this distribution will not attract points inside S_α, and the force with which it attracts an external point is independent of the choice of isothermal S_α.

Chasles knew quite well that the equilibrium state of a charge distribution on the surface S satisfies the property that it does not attract points inside S, and so he asked himself the question: is this equilibrium state uniquely determined by this property? He claimed that the last-mentioned theorem answers this question in the negative:

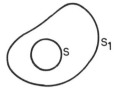

FIGURE 2

> A given surface can always carry an infinity of infinitely thin layers
> possessing the property of exercising no action on any interior points.
>
> [Chasles 1839, p. 211]

Apparently, Chasles thought it physically plausible that a given surface
can be an isothermal for several heat conduction problems having different
(i.e., not proportional) heat fluxes through the surface. But, as he realized
three years later, he had in this case been misled by the analogy:

> These considerations on heat seemed to me to indicate that one could
> form, on the same surface, several infinitely thin layers possessing
> the property of exercising no action on the interior points. Since
> then, I have recognized that a first outline had led me in error. The
> opposite proposition is demonstrated for an arbitrary body and even
> for several bodies in a Note by M. Liouville following this Memoir.
>
> [Chasles, 1842]

14.   This quote stems from a footnote in a paper by Chasles that was
printed in the *Connaissance des Temps* at Liouville's request. The *Procès
Verbaux* of the *Bureau des Longitudes* of June 1, 1842, recounts:

> Mr Liouville presents a memoir by Mr Châle entitled *Théorèmes
> géneraux sur l'attraction des corps* and asks that it be printed in the
> *connaissance des temps*. The *Bureau* adopts this proposal.

In this paper, Chasles proved the general theorems he had announced
in 1839 and was careful to point out that he had priority over Gauss,
whose *Allgemeine Lehrsätze* he had just recently heard of. Still, it is hardly
a coincidence that Chasles wrote the long memoir at the same time as
Liouville had Gauss's paper translated for his *Journal*. It is highly probable
that the two friends discussed Gauss's work with each other and that this
made Chasles include the general version of Gauss's theorem in a footnote
[Chasles 1842, p. 26] and made him realize the error he had made in the
1839 note.

15.   In any case, in the subsequent note in the *Connaissance des Temps*,
Liouville [1842i] explicitly declared that he had modeled his proof after
Gauss's paper. Indeed, Liouville's argument is a slightly generalized and

more elegant version of Gauss's uniqueness proof of the equilibrium distribution. I shall summarize it here because it came to play an important role in Liouville's later research.

Let $A, B, \ldots, C$ be conducting bodies, each charged with a given total charge. Assume that there exist two distributions of this charge over the surfaces that give rise to a constant potential inside each of these bodies. Then, the linearity of the problem implies that the difference between these distributions also gives a constant potential inside the bodies and this difference distribution has total charge zero on each of the conductors. Liouville's aim is to show that such a distribution does not exist.

To this end, let Ω denote the space outside the conductors and let V be the potential due to the assumed charge distribution. By partial integration, Liouville deduced Green's first formula (16) for $\varphi = \psi = V$:

$$\int_{\Omega} \left[\left(\frac{\partial V}{\partial x} \right)^2 + \left(\frac{\partial V}{\partial y} \right)^2 + \left(\frac{\partial V}{\partial z} \right)^2 \right] dv + \int_{\Omega} V \Delta V \, dv = \int_{\partial \Omega} V \frac{\partial V}{\partial n} \, d\omega.$$

(28)

By Laplace's equation, the second integral is zero. Moreover, since V is constant over each of the conductors, the last integral can be written

$$\int_{\partial \Omega} V \frac{\partial V}{\partial n} \, d\omega = \quad V(A) \int_{\partial A} \frac{\partial V}{\partial n} \, d\omega$$

$$+ V(B) \int_{\partial B} \frac{\partial V}{\partial n} \, d\omega + \cdots + V(C) \int_{\partial C} \frac{\partial V}{\partial n} \, d\omega, \quad (29)$$

where $\frac{\partial V}{\partial n}$ denotes the derivative $-\frac{\partial V}{\partial n_+}$, which according to (6) (and the constancy of V inside $A, B, \ldots, C$) is equal to -4π times the density of the charge distribution at the given point. Thus, the integrals on the right-hand side of (29) represent the total charge on $A, B, \ldots$ and C, which we had assumed to be zero. Therefore, Liouville concluded that the first integral in (28) is zero, so that

$$\left(\frac{\partial V}{\partial x} \right)^2 + \left(\frac{\partial V}{\partial y} \right)^2 + \left(\frac{\partial V}{\partial z} \right)^2 = 0 \quad \text{in } \Omega.$$

(30)

This implies that V is also constant outside the conductors, and by (6) we see that the charge distribution is zero at each point of the surface of $A, B, \ldots, C$. This was what Liouville wanted to establish.

It is worth noting that although this uniqueness theorem uses many of the same transformations as the existence proof by way of Dirichlet's principle, it is not subject to the same objections.

16.   In 1845, Liouville's interest in potential theory was aroused anew when William Thomson gave him a copy of Green's *Essay*, which had remained unknown until that year. In Chapter III, I have told how Liouville took great pains to spread the news of Green's work. It is therefore not surprising

that when in 1846 Liouville contributed a paper to the first volume of Thomson's *Cambridge and Dublin Mathematical Journal* he chose a subject from potential theory. He considered the situation in Figure 1 (§13) and showed that the center of mass of the solid K is the same as the center of mass of the distribution $(\frac{\partial V}{\partial n})$ on the equipotential surface S. For the fact that this is an equilibrium distribution Liouville did not refer to Chasles but to Green, who had already proved it in his *Essay*. Liouville's elegant proof of the theorem on the center of mass uses methods similar to Green's or rather similar to those used in [Liouville 1842i].

Except for one insignificant note [Liouville 1848b], this was Liouville's last publication on general potential theory. However, Green's *Essay* made Liouville resume the work on potentials from ellipsoids, which he had begun in 1842 under the double inspiration from 1° his work on rotating fluids and 2° Gauss's *Allgemeine Lehrsätze*.

Liouville on Potentials of Ellipsoids

17.   On May 12 and June 2, 1845, Liouville read two papers to the *Académie des Sciences* in which he presented several theorems on potential theory and Lamé functions without proof [Liouville 1845b,c]. The *Comptes Rendus* notes were reproduced in the June 1845 issue of Liouville's *Journal* [Liouville 1845d]. The following year, Liouville published the proofs in the form of two long letters to P. H. Blanchet, who had asked Liouville to elaborate his sketchy ideas [Liouville 1846e,f]. As pointed out by Belhoste [1982, p. 210], Blanchet was a talented mathematician who had worked on the theory of light during the period 1830–1842, partly on the instigation of Liouville. Liouville had published several of Blanchet's papers (cf. in particular [Blanchet 1840]) and had refereed some of his works for the *Académie*.

In passing, I shall mention that Liouville had contacts with other Blanchets. At the *École Polytechnique*, he had been the classmate of Jos. Bernard Achille Blanchet, later a professor at the *Collège Royal* in Avignon [Fourcy 1828], and the remark "Blanchet, manufacturer of pianos" in Liouville's notebook [Ms 3617 (4)] indicates that he was also in contact with the famous piano builder Pierre-Armand-Charles Blanchet, likewise a former Polytechnician. I do not know the relation between these three Blanchets.

At the end of the second letter to P. H. Blanchet, Liouville admitted that they had been written in great haste, and he expressed his fear that they may therefore have become both long and obscure. To be sure, the composition of the letters is somewhat messy but their content places them among Liouville's most important papers. Moreover, Liouville claimed that all the essential ideas in the letters dated back to the last four months of 1842 when he had just read Gauss's *Allgemeine Lehrsätze*; the notes on the equilibrium of fluid masses shows that this claim is justified (cf. Chapter XI, §19).

18.   The *Comptes Rendus* notes and the letters to Blanchet treated various "questions of analysis and mathematical physics concerning the ellipsoid" [title of Liouville 1846e,f]. The first [1846e] contains an introduction of ellipsoidal coordinates and the Lamé functions, including Liouville's new Lamé function S of the second kind. I have already summarized the content of this part in Chapter XI, and pointed out that as late as the 1880s Poincaré still used it as a comprehensive introduction to these functions. Second, Liouville discussed the relation between the Lamé functions and Laplace's spherical harmonics Y_n, and in that connection he proved the completeness of the system of Lamé functions $M_i(\mu)N_i(\nu)$ on the surface of the ellipsoid $\varrho = $ const. (cf. [Liouville 1845b,d, pp. 223–224; 1846e, p. 220; 1846f, pp. 273–280]).

To this end he considered what Ivory had called corresponding points on the unit sphere (x_1, y_1, z_1) and on the ellipsoid (x, y, z) related by the formulas

$$x = \varrho x_1, \quad y = \sqrt{\varrho^2 - b^2}\, y_1, \quad z = \sqrt{\varrho^2 - c^2}\, z_1 \tag{31}$$

(for the notation see Chapter XI, §22–23; in particular, I shall often adopt the single indexing of the Lamé functions (N_i) instead of the double indexing $(N_{n,B})$). By transforming the differential equation of $M_{n,B}(\mu)N_{n,B}(\nu)$ into the new coordinates and further into spherical coordinates (φ, θ), Liouville found the differential equation defining the spherical harmonics Y_n. The details of this argument were carried out in a later note [Liouville 1846k]. The equation of the spherical harmonics is the same for all the $2n + 1$ values of B corresponding to a fixed value of n, and therefore any linear combination $\sum_B A_B M_{n,B} N_{n,B}$ when considered as a function of the (x_1, y_1, z_1) corresponding to (μ, ν) is a spherical harmonic Y_n; and since Y_n contains exactly $2n + 1$ arbitrary constants, Liouville concluded that it is always of the form $\sum_B A_B M_{n,B} N_{n,B}$.

19.   Now, Liouville could easily prove the completeness of the system $M_{n,B} N_{n,B}$ using the completeness of the spherical harmonics, which had been established by Dirichlet. Indeed, let $\overline{V}$ be a function on the ellipsoid $\varrho = \varrho_0$. Express it in terms of the corresponding (x_1, y_1, z_1) on the unit sphere and expand it in terms of the Y_n's: $\overline{V} = \sum A_n Y_n$. When the Y_n is written as $\sum_B A_B M_{n,B} N_{n,B}$ and the ellipsoidal coordinates are reintroduced, $\overline{V}$ is expressed as a convergent series:

$$\overline{V} = \sum_{n,B} A_{n,B}\, M_{n,B}(\mu)N_{n,B}(\nu), \tag{32}$$

or with the single index convention,

$$\overline{V} = \sum_i A_i\, M_i(\mu)N_i(\nu). \tag{33}$$

This establishes the completeness. Since $R_i\,M_i\,N_i$ and $S_i\,M_i\,N_i$ are harmonic functions for $\varrho \neq \infty$ and for $\varrho \neq 0$, respectively, it is obvious that

$$V = \sum_i A_i \frac{R_i(\varrho)M_i(\mu)N_i(\nu)}{R_i(\varrho_0)} \tag{34}$$

and

$$V = \sum_i A_i \frac{S_i(\varrho)M_i(\mu)N_i(\nu)}{S_i(\varrho_0)} \tag{35}$$

formally solve the Dirichlet problem inside and outside the ellipsoid with $\overline{V}$ given on the boundary of the ellipsoid $\varrho = \varrho_0$.

In [1845d], Liouville proved that these sums also converge outside the ellipsoidal surface. The following year, he generalized the argument to any infinite series representing a harmonic function in an arbitrary domain. I shall now quote this elegant unpublished note. It begins with a proof of the maximum principle, which Gauss had left unnoticed. In the published special case [1845d], Liouville had given a physical argument for this principle: If u is harmonic in a closed domain, it can be considered as the temperature arising from a stationary conduction of heat. If u had a maximum in an interior point, it would evidently be cooled by the surrounding medium, and so the temperature could not be stationary. In the unpublished note, Liouville replaced this argument with a purely mathematical one:

Heat

$$\frac{d^2V}{dx^2} + \frac{d^2V}{dy^2} + \frac{d^2V}{dz^2} = 0,$$

$$V = F \text{ at the surface.}$$

In the interior there is neither a maximum nor a minimum: for $\iint \frac{dV}{ds}\,d\omega = 0$ and so $\frac{dV}{ds}$ [i.e. $\frac{\partial V}{\partial n}$] cannot be constantly of the same sign on a very small surface around a point.

— Thus, the maximum and the minimum of V is attained at the surface, and if F is a small quantity, the same holds for V. — Let $F = F_0 + F_1 + F_2 + \cdots$ and in general V_m corresponding to F_m is $> F_m$ if only the maxima are considered, and $V_{n+1} + V_{n+2} + \cdots + V_{n+m} <$ the corresponding $F_m + \cdots + F_{m+n}$. Therefore, if the series $\sum F_m$ is convergent, the series $\sum V_m$ is convergent as well. This will be the value of V. — The application to the ellipsoid is easy"

[Ms 3618(5), p. 25r]

This argument requires that $\sum F_m$ be uniformly convergent on the boundary surface, but, of course, Liouville did not notice that.

20.   The central theme of Liouville's papers [1845c,d, and 1846e,f] is the proof and the application of the three fundamental formulas,

$$\iint_E \frac{l'M'N'\,d\omega'}{\Delta} = \frac{4\pi RS}{2n+1}MN \quad \text{for } \varrho = \varrho', \tag{36}$$

$$\iint_E \frac{l'M'N'\,d\omega'}{\Delta} = \frac{4\pi R S'}{2n+1} MN \quad \text{for } \varrho < \varrho', \qquad (37)$$

$$\iint_E \frac{l'M'N'\,d\omega'}{\Delta} = \frac{4\pi R' S}{2n+1} MN \quad \text{for } \varrho > \varrho'. \qquad (38)$$

Liouville's compact notation requires some comments. Let $\bar{x}$ and $\bar{x}'$ be two points with ellipsoidal coordinates (ϱ, μ, ν) and (ϱ', μ', ν'), respectively. Then, the unprimed functions are evaluated at $\bar{x}$ (e.g., $M = M(\mu)$), and the primed letters represent the same functions evaluated at $\bar{x}'$ (e.g., $M' = M(\mu')$). Moreover, Δ is an abbreviation for the distance $|\bar{x} - \bar{x}'|$, and the Lamé functions all correspond to the same values of B and n (the n occurring on the right-hand side). Finally, E is the ellipsoidal surface corresponding to ϱ', $d\omega'$ is its surface element, and l is the function

$$l = \frac{1}{\sqrt{\varrho^2 - \mu^2}\sqrt{\varrho^2 - \nu^2}}, \qquad (39)$$

which is closely related to the function h in Chapter XI, §23.

Physically, the integral

$$V = \iint_E \frac{\lambda'}{\Delta}\,d\omega' = \iint_E \frac{\lambda(\bar{x}')}{|\bar{x} - \bar{x}'|}\,d\omega' \qquad (40)$$

is the potential due to a surface charge distribution λ on E. Thus, (36)-(38) give the potential of a charge distribution lMN on E; (36) gives the potential on the ellipsoidal surface E itself, (37) and (38) give it inside and outside of E, respectively. Liouville's first proof of (36)-(38) used this fact elegantly [Liouville 1846e, pp. 226–230]: according to Gauss there exists a unique charge distribution λ on E that has the potential MN on E; that is,

$$\iint_E \frac{\lambda'\,d\omega'}{\Delta} = MN \quad \text{on } E. \qquad (41)$$

The problem is to determine λ. According to the above arguments, the function

$$u = \iint_E \frac{\lambda'\,d\omega'}{\Delta} - \frac{RMN}{R'} \qquad (42)$$

(where R' is the constant $R(\varrho')$) is harmonic in the interior of E and has the boundary value 0 on E. Therefore, according to the uniqueness of the solution to the Dirichlet problem (for which Liouville referred to Green), u is everywhere zero, so that the potential inside E is given by

$$\iint_E \frac{\lambda'\,d\omega'}{\Delta} = \frac{RMN}{R'} \quad \text{for } \varrho < \varrho'. \qquad (43)$$

Similarly, since $S \to 0$ for $\rho \to \infty$, we have

$$\iint_E \frac{\lambda'\,d\omega'}{\Delta} = \frac{SMN}{S'} \quad \text{for } \varrho > \varrho'. \qquad (44)$$

In the argument for (44), Liouville at first forgot to make sure that S had no singularities outside E, but he filled this hole in [1846f, p. 261–264]. Having thus found the potential inside and outside of E, Liouville determined λ by way of (6) as the jump in the normal derivative at E. He found

$$\lambda = \frac{(2n+1)lMN}{4\pi RS},\tag{45}$$

which substituted into (41), (43), and (44) yields (36)-(38) (here one must remember that R and S are constant on the surface E).

In [1846f, pp. 265–269], Liouville presented an alternative proof that was less elegant but had the virtue of being independent of the existence theorem used above.

21. Liouville proved other interesting formulas, e.g., the development of $\frac{1}{\Delta}$ as a series of Lamé functions [Liouville 1846f, p. 286], and showed how formulas (36)-(38) could yield easy proofs of some well-known theorems on attraction of ellipsoids and of the orthogonality relation

$$\iint_E lMN\,M_1N_1\,d\omega = 0\,,\tag{46}$$

when MN and N_1N_1 correspond to different pairs (n, B).

Finally, he used formula (36) to derive the solution of Gauss's problem: Given a function Q on the ellipsoidal surface E, determine the charge distribution λ on E, which has the potential Q on E. Liouville's solution is simple:

$$\lambda = \frac{l}{4\pi}\sum_{n,B}\frac{(2n+1)MN}{RS}\frac{\iint_E lQMN\,d\omega}{\iint_E lM^2N^2\,d\omega}.\tag{47}$$

His last proof [1846f, p. 284] is noteworthy because it points to a very important generalization. Liouville remarked that in (36) the factor, $\frac{4\pi RS}{2n+1}$ is a constant on E. Thus, if we let ζ_i denote the product $M_i\,N_i$, we have the fundamental formula

$$\iint_E \frac{l'\zeta_i'\,d\omega'}{\Delta} = m_i\zeta_i\,,\tag{48}$$

from which Gauss's problem can easily be solved.

Indeed, since $\zeta_i = M_iN_i$ is a complete system, we can develop

$$Q = \sum_i A_i\zeta_i.\tag{49}$$

As usual, in Sturm-Liouville theory, A_i can be determined by multiplying (49) by $l\zeta_j$, integrating over E, interchanging summation and integration (which Liouville did without a second thought), and using the orthogonality relation (46) $\iint_E l\zeta_i\zeta_j = 0$ for $i \neq j$. The result is

$$A_i = \frac{\iint_E lQ\zeta_i\,d\omega}{\iint_E l\zeta_i^2\,d\omega} = \frac{\iint_E lQM_iN_i\,d\omega}{\iint_E lM_i^2N_i^2\,d\omega}.\tag{50}$$

Now, let us calculate the potential on E due to the charge distribution:

$$\lambda = l \sum_i \frac{A_i}{m_i} \zeta_i. \tag{51}$$

It is equal to

$$\iint_E \frac{\lambda'}{\Delta} d\omega' = \iint_E \frac{l' \sum_i \frac{A_i}{m_i} \zeta'}{\Delta} d\omega'$$

$$= \sum_i \frac{A_i}{m_i} \iint_E \frac{l' \zeta'}{\Delta} d\omega' \overset{*}{=} \sum_i \frac{A_i}{m_i} \cdot m_i \zeta_i = \sum A_i \zeta_i = Q.$$

At *, formula (48) has been used. Thus, (51) is the solution of Gauss's problem, and if the values of A_i and ζ_i are inserted, we find (47).

22.   In [1845d, pp. 227–228], Liouville remarked that this way of solving Gauss's problem could be generalized to "arbitrary surfaces." What did he mean by that? He meant that if S is a closed surface in space and if it is possible to find a complete system of functions ζ_i defined on S satisfying

$$\iint_S \frac{l' \zeta_i' d\omega'}{\Delta} = m_i \zeta_i \tag{52}$$

then if we expand a given function Q on S, as in (49), (51) solves Gauss's problem. In order to make this statement precise, I must explain how Liou-ville defined the function l for a general surface. Liouville's papers contain two interesting remarks concerning l. First, he showed that by the Ivory correspondence (31), the ordinary surface element $d\sigma$ on the sphere corresponds to $l\, d\omega$ on the ellipsoid, or as we would say, the Lebesgue measure $d\sigma$ on the sphere corresponds to the measure $l\, d\omega$ on the ellipsoid, where $d\omega$ is its Lebesgue measure. In fact, Liouville strongly emphasized the importance of this measure on the ellipsoidal surface:

> In general, the introduction of the quantity $l\, d\omega$ simplifies and clari-
> fies the formulas very much; M. Lamé who has not used it was bound,
> for this reason alone, to run into more complicated calculations than
> ours.
>
> [Liouville 1846f, p. 287]

This underlines the importance of finding a measure on a general surface S corresponding to $l\, d\omega$, but it does not help find it. The key to the problem lies in the fundamental formula (36) calculated for the first Lamé function, $M_0 = N_0 = 1$:

$$\iint_E \frac{l'\, d\omega'}{\Delta} = \text{constant} \qquad (\text{on } S). \tag{53}$$

Indeed, this shows that l is the equilibrium distribution of charge on the ellipsoidal surface. Liouville concluded his second *Comptes Rendus* note with the following sketch of such a generalization of his theory:

Let us add that the product $E_1 E_2$ (or rather $lE_1 E_2$) $[E_1 E_2 = MN]$ has very curious properties, several of which are related to maxima and minima. Besides, these properties are not limited to the ellipsoid, and even less are they confined to the coordinates we have used; they can be extended to arbitrary surfaces. One can deduce them by considering certain functions ζ or ζ' related to the elements $d\omega$ or $d\omega'$ of the surface one is dealing with. Denote by l and l' the density of the distribution of an electric layer in equilibrium on the surface, such that, when Δ is the distance from $d\omega$ to $d\omega'$, one has

$$\iint \frac{l' \, d\omega'}{\Delta} = \text{constant}. \tag{54}$$

The functions ζ, mentioned above, are then defined by equations of the form

$$\iint \frac{l' \zeta' \, d\omega'}{\Delta} = m\zeta, \tag{55}$$

where m is a constant that changes when one passes from one function to another.

After having studied the matter I do not hesitate to regard the functions ζ as being of the utmost importance in analysis. But, it is not adequate to treat this difficult subject incidentally. Therefore, I propose to return to this matter soon in a separate communication.
[Liouville 1845c]

Liouville did not keep his last promise, and so one does not find in his published papers any hint as to why the functions ζ exist at all, how one can find the numbers m, and many other questions raised by the quote above. However, I have found a long series of notes in Liouville's notebooks pertaining to these problems. They show that Liouville's ideas were indeed of the utmost importance.

Liouville's Unpublished Notes on Spectral Theory of Integral Operators in Potential Theory

INTERPRETATION OF THE MAJOR RESULT

23. Approximately 100 pages of notes in Liouville's notebooks deal with the general ideas sketched above, namely, [Ms 3618 (1), pp. 15v–26r, 40v–55v and 68v–72r] from 1845–1846; [Ms 3618 (5), pp. 32v–39v] written between October 1846 and March 1847; a few pages of [Ms 3619 (5)] from 1851; and [Ms 3622 (2), pp. 106r–114v] from 1857. On the very first of these pages, Liouville summarized his most interesting results as follows (the content of the quote will be explained below):

1°. Render $\iiiint \frac{\lambda \lambda' \, d\omega \, d\omega'}{\Delta}$ a minimum with $\iint \lambda \, d\omega = \text{const.}$; and you have a function l for which $\iint \frac{l' \, d\omega'}{\Delta} = \text{const.} = 1$, say; that corre-

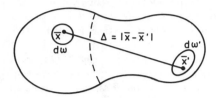

FIGURE 3

sponds to the equilibrium distribution of electricity.

2°. Render <u>maximum</u>

Let $\iint \lambda \, d\omega = 0$ and $\iint \frac{\lambda^2 \, d\omega}{l} = $ const. and you find a second function $l\zeta_1$ such that

$$\iint \frac{l' \zeta_1' \, d\omega'}{\Delta} = m_1 \zeta_1 \quad ; \quad m_1 < 1 \quad \text{if} \quad \iint \frac{l' \, d\omega'}{\Delta} = 1.$$

3°. Let $\iint \lambda \, d\omega = 0$, $\iint \lambda \zeta_1 \, d\omega = 0$ and $\iint \frac{\lambda^2 \, d\omega}{l} = $ const. You find the function $l\zeta_2$ such that

$$\iint \frac{l' \zeta_2' \, d\omega'}{\Delta} = m_2 \zeta_2 \quad , \quad m_2 < m_1$$

and so on.

And a function Q can be expressed as

$$Q = l(A_0 + A_1\zeta_1 + A_2\zeta_2 + \cdots + A_n\zeta_n + \cdots) \cdot$$
$$\iint l\zeta_m \zeta_n \, d\omega = 0, \quad A_n = \frac{\iint Q\zeta_n \, d\omega}{\iint [l]\zeta_n^2 \, d\omega}$$

That is what we found a long time ago.

> [Ms 3618 (1), pp. 15v–16r;
> Liouville has forgotten the l in the last denominator]

Here, the geometrical situation is a generalization of the one in §20 above: S is a closed surface in space (Fig. 3), $d\omega$ and $d\omega'$ are two surface elements of S at $\overline{x}$ and $\overline{x}'$, whose distance in space is denoted by Δ. If λ is a function of $\overline{x}$ ($\lambda = \lambda(x)$), λ' denotes the same function evaluated at $\overline{x}'$ ($\lambda' = \lambda(x')$). If we think of λ as a charge distribution on S, its total charge is

$$\text{total charge} = \iint_S \lambda \, d\omega \, ; \tag{56}$$

its potential at x is (cf. (40))

$$V = \iint_S \frac{\lambda'}{\Delta} \, d\omega' \, ; \tag{57}$$

and so its potential energy is (cf. (7)):

$$\text{potential energy} = \iint_S \iint_S \frac{\lambda \lambda'}{\Delta}\, d\omega\, d\omega'. \tag{58}$$

Thus, in the first step of the above quote, Liouville finds the equilibrium distribution l on S by minimizing the potential energy among all charge distributions with a constant total charge. This is a variant of Gauss's argument (§6) (I shall return to the differences in §31). The distribution l will have a constant potential (57) on S, and by changing its total charge, one can by linearity make

$$\iint_S \frac{l'}{\Delta}\, d\omega' = 1 \quad \text{everywhere on } S. \tag{59}$$

Thereby, the function l — and thus the important expression (measure) $l\, d\omega$ — has been fixed.

24.   Before I explain the rest of the quote, I shall introduce some shorthand notation. It will serve two purposes: 1° it will abbreviate the calculations, and 2° it will suggest a modern framework in which one can study the correctness of Liouville's theorems. I shall introduce a Hilbert space of functions on S and some of the associated operations. In order that the reader shall not over-interpret Liouville, I wish to stress that the Hilbert space — i.e., the domain of the operations — does not have a counterpart in Liouville's notes (he just speaks of (arbitrary) functions on S), whereas the operations are direct translations of Liouville's terms and formulas.

Consider the Hilbert space $L^2(S, l\, d\omega)$ of square integrable functions on S with respect to the measure $l\, d\omega$. By $\langle \xi, \eta \rangle$, we denote the inner product of two functions $\xi, \eta \in L^2(S, l\, d\omega)$:

$$\langle \xi, \eta \rangle = \iint_S \xi \eta l\, d\omega.$$

As usual $\|\cdot\|$ denotes the $L^2(S, l\, d\omega)$ norm: $\|\xi\|^2 = \langle \xi, \xi \rangle$. Finally, we define the operator A on $L^2(S, l\, d\omega)$ by

$$A(\zeta) = \iint_S \frac{\zeta'}{\Delta} l'\, d\omega'.$$

Thus, A maps a function ζ on S onto the potential on S due to the charge distribution ζl on S. I shall return to the admissibility of this definition in §49.

With this notation, we can write (59) as

$$A(1) = 1. \tag{60}$$

Thus in step 1° Liouville has chosen l such that the constant function 1 is an eigenfunction of A with the eigenvalue 1.

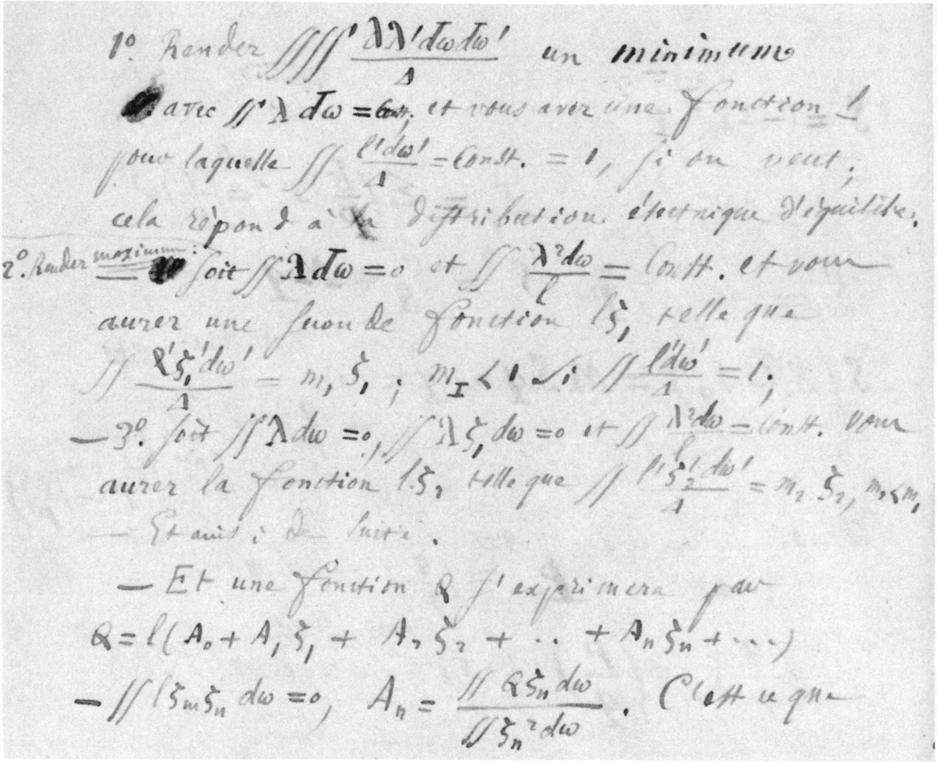

PLATE 32. Liouville's first note concerning the spectral theory of his integral operators in potential theory. The eigenvalues are determined by the "Rayleigh Ritz Method". (Bibliothèque de l'Institut de France, Ms 3618 (1) p. 15v)

25. In the following steps, he considers a function λ that is finally made equal to $l\zeta$. I shall make this substitution from the start. Liouville writes: "Render maximum," but he does not write which expression he intends to maximize. The subsequent notes show that he has the expression (58) in mind in all the steps. With our abbreviated notation, this expression can be written:

$$\iint_S \iint_S \frac{\lambda \lambda'}{\Delta} \, d\omega \, d\omega' = \iint_S \iint_S \frac{\zeta \zeta'}{\Delta} l \, d\omega l' \, d\omega' = \langle A\zeta, \zeta \rangle. \qquad (61)$$

This quantity is maximized among all functions on S having

$$\iint_S \lambda \, d\omega = \iint_S \zeta l \, d\omega = \langle 1, \zeta \rangle = 0 \qquad (62)$$

and

$$\iint_S \frac{\lambda^2 \, d\omega}{l} = \iint_S \zeta^2 l \, d\omega = \langle \zeta, \zeta \rangle = \|\zeta\|^2 = \text{const.} \qquad (63)$$

Calling the maximizing function ζ_1, Liouville claims that it satisfies

$$\iint_S \frac{\zeta_1'}{\Delta} l' \, d\omega' = A\zeta_1 = m_1 \zeta_1, \qquad (64)$$

$$m_1 < 1, \qquad (65)$$

i.e., that ζ_1 is an eigenfunction of A with an eigenvalue m_1 less than 1. In the third step, Liouville maximizes $\langle A\zeta, \zeta \rangle$ under the conditions:

$$\langle 1, \zeta \rangle = \langle \zeta_1, \zeta \rangle = 0 \quad \text{and} \quad \|\zeta\|^2 = \text{const.}, \qquad (66)$$

and he claims that the maximizing function ζ_2 is an eigenfunction of A with an eigenvalue m_2, which is less than the previous eigenvalue

$$A\zeta_2 = m_2 \zeta_2, \quad m_2 < m_1. \qquad (67)$$

Now, the process is easily continued. In the nth step, the eigenfunction ζ_n is determined as the function of a given norm making $\langle A\zeta, \zeta \rangle$ a maximum in the orthogonal complement of the preceding eigenfunctions.

Liouville concludes the note by claiming that the generated set of eigenfunctions is complete in the sense that any function on S can be developed in a series of ζ's, and he indicates the usual expression of the Fourier coefficients A_n.

26. The procedure described above is the so-called Rayleigh-Ritz method of finding eigenvalues to an operator, A. It is remarkable that Liouville discovered the method 30 years before Rayleigh and 60 years before Ritz. However, the note raises a number of questions:

1° How and when was Liouville led to these results?

2° How much of a proof did he possess?

3° Are the results correct?

4° What place does Liouville's notes play in the history of mathematics?

I shall discuss these questions in the rest of this chapter.

A POSTERIORI MOTIVATION

27.   The notebook containing the above note carries the date March 24, 1845, on page 3, and on page 70, there is a reference to the letters to Blanchet published in the summer of 1846. In fact, it seems probable that the notes on spectral theory of the operator A were composed in the early summer of 1845 in connection with the presentation of the two *Comptes Rendus* notes in which he announced the importance of the matter (cf. §22). This dating is corroborated by the fact that Liouville published a few of the theorems in a more generalized version in the August 1845 issue of his *Journal* (cf. §40).

Yet, according to the end of the above quote, Liouville had found the results much earlier. This could mean that he had discovered these general theorems in 1842, at the same time as many of the results in the letters to Blanchet. Still, Liouville's enthusiasm in the 1845 *Comptes Rendus* notes seems to indicate that they stem from the spring of that year. "A long time" could just mean a few weeks or months.

At any rate, it is evident from the quote in §22 that the general ideas contained in the notes had their source in a wish to generalize formula (36) to arbitrary surfaces. The question addressed in Liouville's unpublished notes is the following: How do we find the eigenfunctions and eigenvalues satisfying (55) (or $A\zeta = m\zeta$)?

28.   How did Liouville arrive at the answer to this question, that is, at the Rayleigh-Ritz method? Since we do not possess his earliest notes on this problem, we cannot tell for sure. Still, it seems probable that the argument following the quote in §23 shows the reasoning that originally lead Liouville to his new insight. The argument is a posteriori, that is, Liouville assumed that any function Q on the surface S can be expanded as

$$Q = A_0 + A_1\zeta_1 + A_2\zeta_2 \ldots , \tag{68}$$

where the ζ_i's are eigenfunctions satisfying (64), (67), etc., and $1 > m_1 > m_2 > \cdots$, and he then deduced that the ζ_i's can be found by the above variational method. In fact, at first [Ms 3618 (1), pp. 16r–16v] he assumed in accordance with the quote in §23 that any arbitrary function Q_1 can be written in the form

$$Q_1 = l(A_0 + A_1\zeta_1 + A_2\zeta_2 + \cdots) , \tag{69}$$

but soon [Ms 3618 (1), p. 40v] he made the substitution $Q_1 = lQ$, and so he was led to (68). I shall make this substitution from the start, as I did for $\lambda = l\zeta$ above; in fact, Q in this section corresponds to ζ in the quote in §23.

Liouville [Ms 3618 (1), pp. 16r–16v] first calculated the two quantities $\langle AQ, Q \rangle$ and $\|\zeta\|^2$:

$$
\begin{aligned}
\langle AQ, Q \rangle &= \iint_S \iint_S \frac{QQ'}{\Delta} l \, d\omega \, l' \, d\omega' \\
&= \iint_S Q l \, d\omega (A_0 + m_1 A_1 \zeta_1 + m_2 A_2 \zeta_2 + \cdots) \\
&= A_0^2 \iint_S l \, d\omega + A_1^2 m_1 \iint_S l \zeta_1^2 \, d\omega + A_2^2 m_2 \iint_S l \zeta_2^2 \, d\omega + \cdots \\
&= A_0^2 \|1\|^2 + A_1^2 m_1 \|\zeta_1\|^2 + A_2^2 m_2 \|\zeta_2\|^2 + \cdots.
\end{aligned}
\tag{70}
$$

$$
\begin{aligned}
\|Q\|^2 &= \iint_S Q^2 l \, d\omega \\
&= A_0^2 \iint_S l \, d\omega + A_1^2 \iint_S l \zeta_1^2 \, d\omega + A_2^2 \iint_S l \zeta_2^2 \, d\omega + \cdots \\
&= A_0^2 \|1\|^2 + A_1^2 \|\zeta_1\|^2 + A_2^2 \|\zeta_2\|^2 + \cdots.
\end{aligned}
\tag{71}
$$

Here, Liouville has used the orthogonality of the eigenfunctions ζ_i. He proved this fact later (cf. §40).

In the first step above (§23), Liouville minimized $\langle AQ, Q \rangle$ (or $\langle A\zeta, \zeta \rangle$) under the assumption that

$$
\iint Q l \, d\omega = \langle 1, Q \rangle = A_0
\tag{72}
$$

is a constant. Expression (70) demonstrates that this minimum is attained when $A_1 = A_2 = \ldots = 0$, that is, when Q is a constant A_0. This corroborates that we have chosen l to be the equilibrium distribution.

In order to prove the second step, we shall consider functions Q for which

$$
\langle Q, 1 \rangle = A_0 = 0,
\tag{73}
$$

and

$$
\|Q\|^2 = A_1^2 \|\zeta_1\|^2 + A_2 \|\zeta_2\|^2 + A_3 \|\zeta_3\|^2 + \cdots = \text{const} = B.
\tag{74}
$$

Liouville formed the difference

$$
\begin{aligned}
\langle AQ, Q \rangle - m_1 B &= \langle AQ, Q \rangle - m_1 \|Q\|^2 \\
&= A_2^2 (m_2 - m_1) \|\zeta_2\|^2 + A_3^2 (m_3 - m_1) \|\zeta_3\|^2 + \cdots,
\end{aligned}
\tag{75}
$$

so that

$$
\langle AQ, Q \rangle = m_1 B - A_2^2 (m_1 - m_2) \|\zeta_2\|^2 - A_3^2 (m_1 - m_3) \|\zeta_3\|^2 + \cdots.
\tag{76}
$$

Since $m_1 > m_2 > m_3 \ldots$, this quantity has its maximum for $A_2 = A_3 = \ldots = 0$, that is, for $Q = A_1 \zeta_1$, or, if the constant B is chosen equal to $\|\zeta_1\|$,

we have the maximum for $Q = \zeta_1$. This is what Liouville claimed in 2°
above (§23).

29.   At this point, Liouville saw how the proof continued and so his original
note stopped. However, in a later note [Ms 3618 (5), pp. 34v–35r], he wrote
down the general argument for ζ_i, which I shall now quote (here, Q is
replaced by λ):

> <u>Max.</u> $\iiiint \frac{ll'\lambda\lambda'\,d\omega\,d\omega'}{\Delta}$ admitting the series $\lambda = \sum A\zeta$, that is $=$
> $\sum A^2 m \iint l\zeta^2\,d\omega$. — Impose the following conditions $\iint l\lambda\,d\omega =$
> 0, $\iint l\zeta\lambda\,d\omega = 0, \ldots \iint l\zeta_{i-1}\lambda\,d\omega = 0$, and $\iint l\lambda^2\,d\omega = C$, that is
> $A_0 = 0, \ldots, A_{i-1} = 0$ and
>
> $$A_i^2 \iint l\zeta_i^2\,d\omega + A_{i+1}^2 \iint l\zeta_{i+1}^2\,d\omega + \cdots = C.$$
>
> Thus,
>
> $$\iiiint \frac{ll'\lambda\lambda'\,d\omega\,d\omega'}{\Delta} = m_i(C - A_{i+1}^2 \iint l\zeta_{i+1}^2\,d\omega - \cdots)$$
> $$+ m_{i+1}A_{i+1}^2 \iint l\zeta_{i+1}\,d\omega + \cdots$$
> $$= m_iC - (m_i - m_{i+1})A_{i+1}^2 \iint l\zeta_{i+1}^2\,d\omega - \cdots \le m_iC.$$
>
> Therefore, the <u>maximum</u> is m_iC corresponding to $A_{i+1} = 0$, $A_{i+2} =$
> $0 \ldots$ or to $\lambda = A_i\zeta_i$, A_i depending on $A_i^2 \iint l\zeta_i^2\,d\omega = C$. — Thus,
> the functions ζ_i can be defined successively by the properties of max-
> imum or of minimum: for one has a minimum in the following way.
> Let $\iint l\lambda = 0, \ldots \iint l\lambda\zeta_{i-1} = 0$,
>
> $$\iiiint \frac{ll'\lambda\lambda'\,d\omega\,d\omega'}{\Delta} = C = A_i^2 m_i \iint l\zeta_i^2 + \&_c.$$
>
> Render $\iint \lambda^2\,d\omega$ a minimum.

[Ms 3618 (5), pp. 34v–35r]

In the last line, Liouville clearly meant $\iint \lambda^2 l\,d\omega$. This mistake is corrected in the following proof, which is a simple recasting of the above
argument.

I believe that the a posteriori argument summarized in this and the
previous sections was the one that suggested the Rayleigh-Ritz method to
Liouville.

RECONSTRUCTION OF THE A PRIORI PROOF

30.   As an introduction to the above quote, Liouville wrote:

> <u>Remark</u>: The functions $\zeta_1, \zeta_2, \ldots$ corresponding to the roots
> $m_1, m_2, \ldots$ satisfy <u>max.</u> and <u>min.</u> properties which follow easily once

the series have been assumed, but which, considered <u>a priori</u>, demonstrate the existence of these functions and the roots m.

[Liouville Ms 3618 (5), p. 34v]

Moreover, he underlined this idea in the conclusion of the quote:

> Now, one can prove <u>a priori</u> the existence of these <u>maxima</u> and <u>minima</u> and thus of m and ζ by a method similar to that of Gauss for the electric layer.

[Ms 3618 (5), p. 35r]

The majority of the notes mentioned at the beginning af §23 are devoted to carrying out this a priori proof of the existence of a complete system of eigenfunctions for the operator A. The various elements of the proof are written pell-mell among each other and among alternative and abortive attempts and erroneous ideas. In the following, I shall first give a reconstruction of the proof in its logical order, making clear exactly how much of the proof can be found in the notes; this will point out two holes in the argument. In doing so, I shall have to reorder the propositions and to pass over all the alternative ideas, to which I shall then return in §39.

31.   At first sight, Liouville's a priori proof of step $1°$ in the quote in §23, which leads to l, seems to be circular. One might expect that he would need to prove that $\iint_S \iint_S \frac{\lambda\lambda' \, d\omega \, d\omega'}{\Delta}$ is bounded from below, and indeed he does prove that this integral is positive or zero for all λ. However, in this proof, he uses the existence of an equilibrium state l, which is the function he is after. A more careful study of the notes reveal that the argument is not circular after all. On the contrary, it is an elegant application of Gauss's existence theorem (the numbering of the following propositions is mine):

Proposition 1 [Ms 3618 (5), p. 32v]. *On S, there exists an equilibrium distribution of charge l. It is strictly positive (if its total charge is positive), and it minimizes*

$$\iint_S \iint_S \frac{\lambda\lambda' \, d\omega \, d\omega'}{\Delta} \, d\omega \, d\omega' \tag{77}$$

among all nonnegative functions λ on S with $\int \lambda \, d\omega = C$. It yields a constant potential on S, and in the subsequent we shall normalize it such that

$$\iint_S \frac{l' \, d\omega'}{\Delta} = m_0 = 1 \text{ on } S. \tag{78}$$

Liouville accepted Gauss's proof of this proposition (cf. §§6,7). Indeed, he headed the statement of the minimal properties with the words "Proposition demonstrated by Gauss." What is the difference between this proposition and Liouville's first step in the quote in §23? The difference is that Liouville wanted to minimize (77) among all functions irrespective of their sign. Therefore, it became acutely important to demonstrate the following theorem, which I shall quote directly from Liouville:

Proposition 2 *Let us develop the following theorem separately:*

The integral $\iiiint \dfrac{\lambda\lambda'\,d\omega\,d\omega'}{\Delta}$ is essentially positive.

Let $\iint \frac{l'\,d\omega'}{\Delta} = 1$, $\iint l\,d\omega = C$; *Multiplying* λ *with a convenient constant factor, and thus also* λ', *one does not change the sign of the proposed quantity, and one can assume that* $\iint \lambda\,d\omega = C$ *[the problem with* $\iint \lambda\,d\omega = 0$ *is discussed below].*

Let $\lambda - l = lq$ *so that* $\iint lq\,d\omega = C - C = 0$. *Thus, if* μ *is a* <u>*small constant*</u>, *and if one considers the quantity* $\varphi = l(1 + \mu q)$, *it gives* $\iint \varphi\,d\omega = \iint l\,d\omega = C$, *and moreover it will be essentially positive [the compactness of* S *is used implicitly here]; according to a theorem of Gauss [Proposition 1], we therefore have*

$$\iiiint \frac{\varphi\varphi'\,d\omega\,d\omega'}{\Delta} > \iiiint \frac{ll'\,d\omega\,d\omega'}{\Delta}$$

[here and below he probably meant $\geq$]. *But,*

$$\iiiint \frac{\varphi\varphi'\,d\omega\,d\omega'}{\Delta} = \iiiint \frac{ll'\,d\omega\,d\omega'}{\Delta} + \mu \iiiint \frac{ll'q\,d\omega\,d\omega'}{\Delta}$$
$$+ \mu \iiiint \frac{ll'q'\,d\omega\,d\omega'}{\Delta} + \mu^2 \iiiint \frac{ll'qq'\,d\omega\,d\omega'}{\Delta}$$

But, $\iiiint \frac{ll'q\,d\omega'\,d\omega}{\Delta} = \iint lq\,d\omega = 0$. *The same holds for the* 3rd *term. Hence,*

$$\iiiint \frac{ll'qq'\,d\omega\,d\omega'}{\Delta} \quad \text{is} > 0. \tag{79}$$

But, $\lambda = l + lq$, *and so*

$$\iiiint \frac{\lambda\lambda'\,d\omega\,d\omega'}{\Delta} = \iiiint \frac{ll'\,d\omega\,d\omega'}{\Delta} + \iiiint \frac{ll'qq'\,d\omega\,d\omega'}{\Delta}$$
$$= C + \iiiint \frac{ll'qq'\,d\omega\,d\omega'}{\Delta}. \tag{80}$$

Therefore, this quantity is essentially positive QED.

[Ms 3618 (1), p. 42v–43r]

This proof fails if $\iint_S \lambda\,d\omega = 0$. Later, Liouville returned to this case and even made it the basis of a more elegant version of the above proof:

$$\iiiint \frac{qq'\,d\omega\,d\omega'}{\Delta} \quad , \quad \iint q\,d\omega = 0$$

cannot be neg.

For one must have, at least for $l \mp \mu q > 0$:

$$\iiiint \frac{(l \mp \mu q)(l' \mp \mu q')\,d\omega\,d\omega'}{\Delta} > \iiiint \frac{ll'\,d\omega\,d\omega'}{\Delta},$$

and it is $= \displaystyle\iiiint \frac{ll'\,d\omega\,d\omega'}{\Delta} + \mu^2 \iiiint \frac{qq'\,d\omega\,d\omega'}{\Delta}.$ [81]

[Here Liouville has used that $\iint_S \iint_S \frac{lq'\,d\omega\,d\omega'}{\Delta} = \iint_S q'\,d\omega = 0$; he now goes on with the general case:]

$$\iiiint \frac{\zeta\zeta'\,d\omega\,d\omega'}{\Delta} \qquad \text{cannot be negat.}$$

For one can put $\zeta = a + q$, $\iint q\,d\omega = 0$, $\iint \zeta\,d\omega = a\int d\omega$. And thus, the expression becomes:

$$a^2 + \iiiint \frac{qq'\,d\omega\,d\omega'}{\Delta} = + \text{ and } + \text{ not negative.}$$

[Ms 3622 (2), p. 113v]

After the a^2 in the last line, Liouville has left out a positive integral but that, of course, does not alter the correctness of this elegant proof.

Proposition 3 *The equilibrium state l (in Proposition 1) can be found by minimizing (77) over all functions λ on S.*

Liouville stated this theorem in step 1° of the quote in §23. I have not found a proof of it in Liouville's notes but it follows easily from Proposition 2, and even more easily from identities [79], [80], and [81] above. Indeed, they show that for any λ the energy integral (77) is larger than its value for l. I feel rather sure that this is a correct reconstruction of Liouville's proof of Proposition 3. To be sure, this proposition is stated before Proposition 2, but if Liouville had had a proof of Proposition 2, which did not use Proposition 3, he would no doubt have stated the latter as a trivial consequence of the former. But, as we saw above, Liouville was careful to base his proof of Proposition 2 exclusively on Gauss's property, i.e., that l minimizes (77) among all *nonnegative* functions λ. One must admire Liouville's way of proving the general minimum property by first using Gauss's weaker property to prove the positivity of (77) and then applying this to the general situation.

32.   We now proceed to a discussion of the eigenfunction ζ_1.

Proposition 4 [Ms 3618 (1) p. 20r]. *The quantity*

$$\langle A\zeta, \zeta \rangle = \iint_S \iint_S \frac{ll'\zeta\zeta'}{\Delta}\,d\omega\,d\omega' \qquad (82)$$

has an upper bound (Liouville writes a maximum) among functions ζ for which

$$\|\zeta\|^2 = \iint_S l\zeta^2\,d\omega = constant = A$$

Proof [also Ms 3618 (1), p. 20r.]

$$
\begin{aligned}
\langle A\zeta, \zeta \rangle &= \iint_S \iint_S \frac{ll'\zeta\zeta'}{\Delta}\, d\omega\, d\omega' \\
&= \frac{1}{2} \iint_S \iint_S \frac{ll'(\zeta^2 + \zeta'^2)}{\Delta}\, d\omega\, d\omega' - \frac{1}{2} \iint_S \iint_S \frac{ll'(\zeta - \zeta')^2}{\Delta}\, d\omega\, d\omega' \\
&\le \iint_S \iint_S \frac{ll'\zeta^2\, d\omega\, d\omega'}{\Delta} = \iint_S l\zeta^2 \left(\iint \frac{l'\, d\omega'}{\Delta} \right) d\omega \\
&= m_0 \iint_S l\zeta^2\, d\omega = m_0 \|\zeta\|^2 = m_0 A
\end{aligned}
$$

(we have chosen $m_0 = 1$).   (83)

□

Proposition 5 (step 2 in quote in §23). *If ζ_1 maximizes $\langle A\zeta, \zeta \rangle$ (82) under the assumption that*

$$\langle \zeta, 1 \rangle = \iint_S l\zeta\, d\omega = 0 \tag{84}$$

and

$$\|\zeta\|^2 = \iint_S l\zeta^2\, d\omega = \text{constant}, \tag{85}$$

then there exists a constant m_1 such that

$$A\zeta_1 = \iint_S \frac{l'\zeta_1'}{\Delta}\, d\omega' = m_1\zeta_1.$$

Proof [Ms 3622 (2), p. 109r.] (it is a generalization of Gauss's variational proof, cf. §6).

At the maximum, we must have

$$
\begin{aligned}
\delta\langle A\zeta, \zeta \rangle = \delta \iint_S \iint_S \frac{ll'\zeta\zeta'\, d\omega\, d\omega'}{\Delta} &= 2 \iint_S l\delta\zeta \left(\iint_S \frac{l'\zeta'}{\Delta}\, d\omega' \right) d\omega \\
&= 2 \iint_S l\delta\zeta A\zeta\, d\omega = 0, \tag{86}
\end{aligned}
$$

where the variation $\delta\zeta$ is restricted by (84) and (85), i.e., where

$$\delta\langle \zeta, 1 \rangle = \iint_S l\delta\zeta\, d\omega = 0 \tag{87}$$

and

$$\frac{1}{2}\delta\|\zeta\|^2 = \iint_S l\zeta\delta\zeta\, d\omega = 0. \tag{88}$$

Now, if $A\zeta = \text{const} = C$, equality (84) holds because of (87), and if $A\zeta = m\zeta$, then (84) holds because of (88). Thus, (84) is satisfied if

$$A\zeta = m\zeta + C. \tag{89}$$

Conversely, this is the only case where (84) holds for all $\delta\zeta$ satisfying (87) and (88). However, $C = 0$ because

$$0 = \iint_S l'\zeta' \, d\omega' = \iint_S \iint_S \frac{ll'\zeta' \, d\omega \, d\omega'}{\Delta} = \iint_S l(m\zeta + C) \, d\omega = C \iint_S l \, d\omega. \tag{90}$$

$\square$

33. REMARK. In modern operator theory, it is common to state that ζ_1 maximizes the Rayleigh quotient

$$\frac{\langle A\zeta, \zeta \rangle}{\|\zeta\|^2} \tag{91}$$

under condition (84). Since A is linear, this is, of course, equivalent to Liouville's requirement.

Proposition 6 [Ms 3618 (1), p. 17r; Ms 3618 (5), p. 34v; Ms 3622 (2), p. 109].
   Let ζ_1 and m_1 be as in Proposition 5. Then,

$$\iint_S \iint_S \frac{ll'\zeta_1\zeta_1' \, d\omega \, d\omega'}{\Delta} = m_1 \iint l\zeta_1^2 \, d\omega \tag{92}$$

or

$$\langle A\zeta_1, \zeta_1 \rangle = m_1 \|\zeta_1\|^2. \tag{93}$$

Thus, m_1 is the maximal value of the Rayleigh quotient (91).

Proof [same Ms as above].

$$\iint_S \iint_S \frac{ll'\zeta_1\zeta_1'}{\Delta} \, d\omega \, d\omega' = \iint l\zeta_1 \left(\frac{l'\zeta_1' \, d\omega'}{\Delta} \right) d\omega$$
$$= m_1\zeta_1$$

$\square$

Proposition 7 (step 2° in quote in §23)

$$1 = m_0 \geq m_1.$$

REMARK. Liouville actually wrote $1 > m_1$ or, more generally, $1 = m_0 > m_1 > m_2 > m_3 > \cdots$. Still, he knew perfectly well that these eigenvalues may be multiple, i.e., they may have an eigenspace of more than one dimension, such that some of the m's may be equal. Indeed, in [Liouville 1846f, pp. 273–278], he showed that for an ellipsoid of revolution the Lamé functions $M_i N_i$ come in pairs with the same eigenvalue, and for a sphere, the eigenvalues have a $2n+1$-fold multiplicity. There is therefore no doubt that Liouville meant $\geq$.

I have not found an explicit proof of Proposition 7 in Liouville's notes, but it is an immediate consequence of the proof of Proposition 4.

Proof of Proposition 7. According to (83), we have $\langle A\zeta, \zeta \rangle \leq m_0 \|\zeta\|^2$. Thus,

$$m_0\|\zeta_1\|^2 \geq \langle A\zeta_1, \zeta_1 \rangle = \langle m_1\zeta_1, \zeta_1 \rangle = m_1\|\zeta_1\|^2$$

and so $m_0 \geq m_1$. □

34. The subsequent eigenfunctions can now be determined in the same way:

Proposition 8 (step $3°$ in quote in §23). *Let ζ_2 denote the function that maximizes $\langle A\zeta, \zeta \rangle$ under these conditions:*

$1°$: $\langle \zeta, 1 \rangle = 0$,

$2°$: $\langle \zeta, \zeta_1 \rangle = 0$, and

$3°$: $\|\zeta\|^2 = \text{const.}$;

then $A\zeta_2 = m_2\zeta_2$, $\langle A\zeta_2, \zeta_2 \rangle = m_2\|\zeta_2\|^2$, and $m_1 \geq m_2$.

Proof Liouville did not bother to write down this proof, which is a simple generalization of the above proofs. The inequality $m_1 \geq m_2$ follows from the fact that in finding ζ_2 we maximize over a smaller set of functions than when we determined ζ_1. □

Proposition 9 (cf. quote in §23). *If we continue this process, we get a sequence, $\zeta_1, \zeta_2, \zeta_3, \ldots$, such that*

$$A\zeta_i = m_i\zeta_i$$

and such that

$$1 = m_0 \geq m_1 \geq m_2 \geq \cdots$$

The eigenfunctions are orthogonal:

$$\langle \zeta_i, \zeta_j \rangle = \iint_S l\zeta_i\zeta_j \, d\omega = 0 \quad \text{for} \quad i \neq j. \tag{94}$$

35. Following Liouville we shall now investigate the sequence of eigenvalues.

Proposition 10 [Ms 3618 (1), pp. 21v–22r; Ms 3618 (1), p. 41v; Ms 3622 (2), p. 113v].

All the eigenvalues m_i are greater than zero.

Proof This follows easily from Proposition 2 which states that $\langle A\zeta, \zeta \rangle$ is always positive. Indeed, in [Ms 3622 (2), p. 113v], Liouville stated the non-negativity of the eigenvalues as a corollary to Proposition 2. However, in [Ms 3618 (1), pp. 41v–42r] which was written before he formulated Proposition 2, Liouville gave the following independent proof of Proposition 10.

Assume that $A\zeta = -n\zeta$, where $n > 0$, and let $\lambda = l(1 + \mu\zeta)$, where $\mu \ll 1$. Then, we have

$$\iint_S \iint_S \frac{\lambda\lambda'\, d\omega\, d\omega'}{\Delta} = \iint_S \iint_S \frac{ll'(1+\mu\zeta)(1+\mu\zeta')\, d\omega\, d\omega'}{\Delta}$$

$$= \iint_S \iint_S \frac{ll'\, d\omega\, d\omega'}{\Delta} + 2\mu \iint_S \iint_S \frac{ll'\zeta'\, d\omega\, d\omega'}{\Delta} + \mu^2 \iint_S \frac{ll'\zeta\zeta'\, d\omega\, d\omega'}{\Delta}$$

$$= \iint_S \iint_S \frac{ll'\, d\omega\, d\omega'}{\Delta} - 2\mu n \iint_S l\zeta\, d\omega + \mu^2(-n) \iint_S l\zeta^2\, d\omega.$$

However, for $\zeta = \zeta_i$, the middle term is zero, and thus,

$$\iint_S \iint_S \frac{\lambda\lambda'\, d\omega\, d\omega'}{\Delta} \leq \iint_S \iint_S \frac{ll'\, d\omega\, d\omega'}{\Delta},$$

contrary to Gauss's theorem.   □

REMARK. Although he did not say so explicitly, Liouville choose μ small, so as to make $\lambda = l(1 + \mu\zeta)$ positive, so that Gauss's Proposition 1 applies. There is no doubt that the above proof suggested the proof of Proposition 2, which is in fact formulated on the following page of Liouville's notebook.

Proposition 11 *The sequence of eigenvalues m_i found in Proposition 9 converges to a limit in the interval $[0, 1]$.*

REMARK. Though this is an immediate consequence of Proposition 9 and 10, it is not stated explicitly in Liouville's notes. In fact, Liouville was convinced that the sequence m_i tends to zero, but as far as I can see, he did not find a satisfactory argument for this. In some of the following theorems concerning the completeness of the system ζ_i, Liouville explicitly wrote "at least when $m_i = 0$ for $i = \infty$" [Ms 3618 (1), p. 41r], but in other places, he simply admitted this implicitly. In order to spot how crucial this assumption is, I shall mention explicitly when it is used in the subsequent propositions.

36. Let us now turn to the "Fourier" series (68):

Proposition 12 (quote in §23 [Ms 3618 (5), p. 33r]). *If* $Q = \sum_i A_i \zeta_i$, *then the Fourier coefficients* A_i *can be determined as*

$$A_i = \frac{\iint_S lQ\zeta_i d\omega}{\iint_S l\zeta_i^2\, d\omega} = \frac{\langle Q, \zeta_i \rangle}{\|\zeta_i\|^2}. \tag{95}$$

Proof Liouville was so familiar with this theorem from his works on Sturm-Liouville theory that he only hinted at the following proof: Multiply Q with $l\zeta_j$ and integrate over S. Then, according to the orthogonality relation (94) we have

$$\iint_S lQ\zeta_i d\omega = \iint_S l\sum_i A_i \zeta_i \zeta_j\, d\omega$$

$$= \sum_i A_i \iint_S l\zeta_i \zeta_j\, d\omega = A_j \iint_S l\zeta_j^2\, d\omega,$$

from which (95) is an immediate consequence. With our present-day standards of rigor, we know that the interchange of summation and integration is problematic, but Liouville had no such scruples (cf. §21). □

Proposition 13 [Ms 3618 (5), p. 37r].

$$\text{Let } Q = A_0 + A_1 \zeta_1 + \cdots + A_{n-1}\zeta_{n-1} + R_n\,, \tag{96}$$

where A_i *is determined by (95). Then,*

$$\langle R_n, \zeta_i \rangle = \iint_S lR_n \zeta_i d\omega = 0 \quad \text{for} \quad i = 1, 2, \ldots, n-1\,. \tag{97}$$

Proof Let i be one of the numbers $1, 2, \ldots, n-1$. Then,

$$\langle R_n, \zeta_i \rangle = \langle Q - (A_0 + A_1 \zeta_1 + \cdots + A_{n-1}\zeta_{n-1}), \zeta_i \rangle$$

$$= \langle Q, \zeta_i \rangle - A_i \|\zeta_i\|^2 = 0.$$

□

REMARK. After this theorem, Liouville remarked: "This theorem is useful. The following is an immediate consequence" [Ms 3618 (5), p. 37r], and he went on to formulate

Proposition 14 [Ms 3618 (5), p. 37r]. *If* Q *and* R_n *are as in Proposition 13, we have*

$$\iint_S lQ^2 = A_0^2 \iint_S l\, d\omega + A_1^2 \iint_S l\zeta_1^2\, d\omega + \cdots + A_{n-1}^2 \iint_S l\zeta_{n-1}^2\, d\omega + \iint_S lR_n^2\, d\omega$$

or

$$\|Q\|^2 = A_0^2 \|1\|^2 + A_1^2 \|\zeta_1\|^2 + \cdots + A_{n-1}^2 \|\zeta_{n-1}\|^2 + \|R_n\|^2. \tag{98}$$

Proof As Liouville remarked, it is an immediate consequence of Proposition 13. Liouville went on with the following, even more immediate, consequence: $\square$

Proposition 15 (Bessel's inequality) [Ms 3618 (5), p. 37r].   Thus, the series of which $A_i^2 \iint l\zeta_i^2 \, d\omega$ is the general term is convergent and cannot surpass $\iint lQ^2 \, d\omega$.

Or, in modern notation: $\sum_i A_i \|\zeta_i\|^2$ *is convergent with a sum less than or equal to* $\|Q\|^2$.

REMARK. After the theorem quoted above, Liouville remarked: "(in fact it is equal to it)." This is Parseval's equality. However, Liouville gave no proof at this point, and so it is more an evidence of his convictions than a theorem.

We observe that Proposition 15 implies that $\|R_n\|^2$ is decreasing with n, but we have no proof as yet that it tends to zero.

37.   Liouville then went on to prove the following theorem:

Proposition 16 [Ms 3618 (5), p. 37r]. *If Q and R_n are as in Proposition 13, we have*

$$\iint_S \iint_S \frac{ll'QQ' \, d\omega \, d\omega'}{\Delta} = A_0^2 \iint_S l \, d\omega + \cdots \tag{99}$$

$$+A_{n-1}^2 m_{n-1} \iint l\zeta_{n-1}^2 \, d\omega + \iint_S \iint_S \frac{ll'R_n R_n' \, d\omega \, d\omega'}{\Delta}$$
$$(99)$$

or

$$\langle AQ, Q \rangle = \sum_{i=0}^{n-1} A_i^2 m_i \|\zeta_i\|^2 + \langle AR_n, R_n \rangle. \tag{100}$$

Proof Let me just give Liouville's proof in modern notation:

$$\langle AQ, Q \rangle = \left\langle \sum_{i=0}^{n-1} A_i m_i \zeta_i + AR_n, \sum_{i=0}^{n-1} A_i \zeta_i + R_n \right\rangle$$

$$= \sum_{i=0}^{n-1} A_i^2 m_i \|\zeta_i\|^2 + \sum_{i=0}^{n-1} A_i m_i \langle \zeta_i, R_n \rangle + \sum_{i=0}^{n-1} A_i \langle AR_n, \zeta_i \rangle + \langle AR_n, R_n \rangle$$

$$= \sum_{i=0}^{n-1} A_i^2 m_i \|\zeta_i\|^2 + \langle AR_n, R_n \rangle.$$

The last equality holds because for $i < n$ we have $\langle \zeta_i, R_n \rangle = 0$ according to Proposition 13, and similarly

$$\langle AR_n, \zeta_i \rangle = \langle R_n, A\zeta_i \rangle = A_i \langle R_n, \zeta_i \rangle = 0.$$

Here, I have used A as formally self-adjoint. Of course, Liouville did not state this fact as a theorem, but he used the symmetry of the kernel in a way corresponding to the self-adjointness.   □

Proposition 17 [Ms 3618 (1), p. 17r,v; Ms 3618 (5), p. 37r].   Thus, the series of which $A_i^2 \, m_i \iint l\zeta_i^2 \, d\omega$ is the general term converges also (this is evident) and cannot surpass $\iiiint \frac{QQ'll' \, d\omega \, d\omega'}{\Delta}$,, or in modern notation: If A_i is determined by (95), $\sum_{i=0}^{\infty} A_i^2 m_i \|\zeta_i\|^2$ is convergent with a sum less than or equal to $\langle AQ, Q \rangle$.

Proof Since A is positive definite (cf. Proposition 2 and 10), the quantity $\langle AR_n, R_n \rangle$ in (100) is nonnegative, and since $m_i > 0$, the sequence in question has positive terms, and by (100) it is bounded by $\langle AQ, Q \rangle$.   □

38. REMARK: Propositions 12–17 and their proofs are valid if ζ_i is just some orthogonal set of eigenfunctions of the operator A. Next, Liouville showed that the particular system arising from the Rayleigh-Ritz method (Proposition 9) is complete if the eigenvalues tend to zero.

Proposition 18 [Ms 3618 (1), p. 41r; Ms 3618 (5), p. 37v]. *Let the eigenfunctions ζ_i be constructed as in Proposition 9, let Q and R_n be as in Proposition 13, and assume that $m_i \to 0$ for $m \to \infty$. Then,*

$$\langle AR_n, R_n \rangle = \iint_S \iint_S \frac{ll' R_n R_n' \, d\omega \, d\omega'}{\Delta} \to 0 \quad \text{for} \quad n \to \infty$$

Proof The following elegant proof can be found in [Ms 3618 (5), p. 37v]. According to Proposition 13, we have $\langle R_n, \zeta_i \rangle = 0$ for $i = 1, 2, \ldots, n-1$. Now, ζ_n is the function that, under these conditions and for $\|\zeta\|^2 = \text{const.}$, maximizes $\langle A\zeta, \zeta \rangle$. Thus, if B is a constant such that $\|B^{-1} R_n\|^2 = \|\zeta_n\|^2$ or $\|R_n\|^2 = B^2 \|\zeta_n\|^2$, we must have

$$0 \leq \langle AR_n, R_n \rangle \leq B^2 \langle A\zeta_n, \zeta_n \rangle = B^2 m_n \|\zeta_n\|^2 = m_n \|R_n\|^2 .$$

According to Proposition 15 (cf. the remark), $\|R_n\|^2$ is decreasing, in particular, it is bounded, and if $\lim_{n \to \infty} m_n = 0$, we conclude that

$$lim_{n \to \infty} \langle AR_n, R_n \rangle = 0.$$

□

Theorem 19 [Ms 3618 (5), p. 37v]. *Under the assumption in Proposition 18, we have*

$$\iint_S \iint_S \frac{ll'QQ'\, d\omega\, d\omega'}{\Delta} = \sum_{i=0}^{\infty} \left(A_i^2 m_i \iint_S l\zeta_i^2\, d\omega \right)$$

or

$$\langle AQ, Q \rangle = \sum_{i=1}^{\infty} A_i^2 m_i \|\zeta_i\|^2.$$

Proof It is a consequence of Proposition 16, formula (100), and Proposition 18.   □

From this main theorem, Liouville immediately deduced two other completeness properties:

Corollary[20 [Ms 3618 (5), p. 38r].] *Under the assumptions of Proposition 18, we have*

$$\iint lQ\zeta_i = 0 \quad \text{for all} \quad i = 0, 1, 2, \ldots \Rightarrow Q \equiv 0,$$

or phrased differently: the only function that is orthogonal to all the eigenfunctions ζ_i is the zero function.

Proof If $\langle Q, \zeta_i \rangle = 0$ for all i, Theorem 19 implies that

$$\langle AQ, Q \rangle = 0.$$

But, since A is positive definite (cf. Propositions 2 and 10), this is only possible if $Q \equiv 0$.   □

Theorem 21 [Ms 3618 (5), p. 38r].   If

$$u = \sum \frac{\zeta_i \iint lQ\zeta_i d\omega}{\iint l\zeta_i^2\, d\omega},$$

where Q is an arbitrary function, converges, one has $u = Q$.

Or, in terms of the A_i's (95), if the Fourier series of Q: $\sum_{i=0}^{\infty} A_i \zeta_i$ converges, then it has Q as its limit.

Proof Liouville simply wrote:

"For evidently $\iint l(u-Q)\zeta_i d\omega = 0$ for whatever i you choose." In modern terms, we can write this as follows:

$$\left\langle Q - \sum_{j=0}^{\infty} A_j \zeta_j, \zeta_i \right\rangle = \langle Q, \zeta_i \rangle - A_i \|\zeta_i\|^2 = 0$$

according to (95).   □

39. Having thus established this theorem, Liouville remarked about the Fourier series:

> It remains to be established that it converges. The difficulty undoubtedly stems from the fact that the function Q may not become infinite. Let us try to express this condition. Meanwhile, these are already some very rigorous proofs of some beautiful propositions.
>
>   [Ms 3618 (5), p. 38r]

Liouville's theorems and their proofs are beautiful indeed. The entire structure and even the details of the proofs are so similar to most modern treatments of spectral theory of compact operators that it is almost unbelievable that these notes were written more than half a century before Fredholm and Hilbert published the first works on such questions.

There are several other outbursts of enthusiasm scattered among Liouville's notes. Having proved Corollary 20, he wrote "Courage; we shall attain the end, and it will be a model of elegance" [Ms 3618 (1), p. 44v]. When he returned to these ideas in 1856, he once more contemplated publishing them: "Summary. We find nothing but our old conclusions (which, perhaps, we had better publish)" [Ms 3622 (2), p. 112r]. However, he only published one short note about this research.

THE PAPER ON THE GENERAL SPECTRAL THEORY OF SYMMETRIC INTEGRAL OPERATORS

40. The origin of this note can be found in [Ms 3618 (5), pp. 32v–33r]. Here, Liouville analyzed what can be said about the eigenfunctions and eigenvalues of the operator A without assuming that they have been found by the above variational method. First, Liouville established the orthogonality relation:

> Theorem. $\iint l\zeta_i\zeta_j\,d\omega = 0$ if i and j are different (or rather m_i and m_j):

$$\iint \frac{l'\zeta'\,d\omega}{\Delta} = m\zeta, \qquad \iint \frac{l'\zeta_1\,d\omega_1'}{\Delta} = m_1\zeta_1$$

$$m \iint l\zeta\zeta_1\,d\omega = \int \frac{ll'\zeta'\zeta_1\,d\omega'\,d\omega}{\Delta}$$

$$= \iint l'\zeta'\,d\omega' \int \frac{l\zeta_1\,d\omega}{\Delta} = m_1 \int l'\zeta'\,d\omega\zeta_1' = m_1 \iint l\zeta\zeta_1\,d\omega$$

$$[101]$$

Thus, etc.

                                                    [Ms 3618 (5), pp. 32v-33r]

Probably because of its familiarity from Sturm-Liouville theory, Liouville omitted the end of the argument, which runs: From (101), we conclude that

$$(m - m_1) \iint l\zeta\zeta_1 \, d\omega = 0 \,,$$

and so $\iint l\zeta\zeta_1 \, d\omega = 0$ if $m \neq m_1$.

From this theorem, Liouville deduced two corollaries. The first is identical to Proposition 12, and the second states that the eigenvalues of A are all real. The proof, which is only sketched (again it is similar to Sturm-Liouville theory), runs as follows: If $\zeta = P + Q\sqrt{-1}$ has the eigenvalue $p + q\sqrt{-1}$, it is easy to see that $\overline{\zeta} = P - Q\sqrt{-1}$ has the eigenvalue $p - q\sqrt{-1}$. If $q \neq 0$, the two eigenvalues are different and thus the eigenfunctions are orthogonal:

$$0 = \iint l\zeta\overline{\zeta} \, d\omega = \iint l(P^2 + Q^2) \, d\omega$$

and so $P = Q = 0$, or $\zeta = \overline{\zeta} = 0$.

In the summer of 1845, Liouville discovered that the only property of the kernel $\frac{1}{\Delta}$ used in this proof is its symmetry, and so he generalized the results accordingly and published them in *Sur une propriété générale d'une classe de fonctions* [Liouville 1845h]. He considered the general eigenvalue equation

$$\int_D l' T\zeta' \, dx_1 \, dx_2 \dots dx_n = m\zeta \,, \tag{102}$$

where D is a subset of $\boldsymbol{R}^n$ (or perhaps some n-dimensional manifold), l is any given function on D, and $T(\overline{x}, \overline{x}')$ is a symmetric function on $D \times D$. The orthogonality relation is derived as above and so is the reality of the eigenvalues m when "l is always positive or always negative" and when T is real — the latter assumption is not mentioned by Liouville.

Liouville concluded his published note by the following prophetic remark:

> Moreover, one can easily see that instead of the left-hand side of the equation A [(102)] one can substitute more complicated integrations or even operations of another kind without the theorem (conveniently modified, if necessary, to suit these new operations) and even the proof ceasing to be correct; for the statements above rely essentially on a certain symmetry which it will be sufficient to retain. Let us add, that there are analogous results for some classes of simultaneous equations. Besides, as is well known, the solution of most questions of mathematical physics rely on equations of the form $B[\int_D l' T\zeta' \, dx_1 \, dx_2 \dots dx_n = 0]$.
>
> [Liouville 1845h]

With our present-day knowledge of symmetric operators and their applications, we must admire Liouville's foresight. Still, for Liouville's contemporaries, these remarks must have appeared very vague, so it is hardly surprising that this unique publication before Volterra on the general theory of integral operators has been overlooked by later mathematicians.

Unsolved Problems, Alternative Methods

41.   Why did Liouville hesitate to publish his other results in this area? Generally, we have seen that Liouville had a peculiar reluctance to publish some of his most interesting results. In this case, in particular, there were additional shortcomings in the theory. First, he was well aware (cf. §39) that he had not proved the convergence of the Fourier series. He had only proved that if it converges it has the "correct" sum. This situation is quite the reverse of the situation in Sturm-Liouville theory (cf. Chapter X, §§45–47). Second, we have remarked that the proofs of the completeness of the system ζ_i assumed in an essential way that the eigenvalues tend to zero. Though Liouville often assumes this implicitly, he was probably aware that he had not supplied a rigorous proof. Thus, his notes in [Ms 3618 (1)] end with the note "Examine m_i and show that it tends to the limit 0" [Ms 3618 (1), p. 44v].

42.   Probably in an attempt to overcome these problems Liouville tried a new approach in [Ms 3618 (5) p. 35v–36v and 38v–39v]:

> Note. Instead of demonstrating a priori the existence of the functions ζ and the related series for an arbitrary surface, one can start from a surface (from the ellipsoid, for example) for which their existence and their properties are known by other means. Afterward one varies a parameter, and one passes to another surface.

The idea is to perturb a simple surface and follow the eigenfunctions ζ_i. First, Liouville considered the equilibrium distribution l on an ellipsoid. It satisfies $\iint_E \frac{l' \, d\omega'}{\Delta} = 1$.

Let us pass to another neighboring surface:

$$\iint \frac{\delta l' \, d\omega'}{\Delta} = - \iint l' \delta \left(\frac{d\omega'}{\Delta} \right) . \qquad [103]$$

> The right-hand side is known. In order to find $\delta l'$ it is therefore enough to solve Gauss's problem on the ellipsoid. — And generally on the surface from which one passes to the following.
>
> [Ms 3618 (5), p. 35v]

Thus, if we can solve Gauss's problem: "given a function; find the charge distribution having this function as its potential" on some surface, then we can pass from its equilibrium distribution to the equilibrium distribution of a neighboring surface. We can do the same with the eigenfunctions ζ_i. Assume that

$$\iint_S \frac{l' \zeta_j' \, d\omega'}{\Delta} = m_j \zeta_j , \qquad \iint_S l \zeta_j^2 \, d\omega = 1 , \qquad (104)$$

and that on a neighboring surface S' we have $\zeta'_j = \zeta_j + \delta\zeta_j$ at "corresponding points", then

$$\iint \frac{l'\delta\zeta'_j \, d\omega'}{\Delta} = m_j\delta\zeta_j + \zeta_j\delta m_j - \iint \zeta'_j\delta\left(\frac{l' \, d\omega'}{\Delta}\right). \tag{105}$$

The last term is known, and Liouville assumed it could be written as $\sum_i B_i\zeta_i$; similarly, he assumed that the unknown $\delta\zeta_j$ could be written $\sum_i A_i\zeta_i$. Then, the problem is to determine A_i from the equation

$$\sum_i A_i m_i\zeta_i = m_j\sum_i A_i\zeta_i + \zeta_j\delta m_j + \sum_i B_i\zeta_i. \tag{106}$$

For $i \neq j$, we have $A_i(m_i - m_j) = B_i$; the jth term gives $\delta m_j = -B_j$. Finally, from (104), Liouville concluded that $\iint l\zeta\delta\zeta \, d\omega = 0$, and so according to the orthogonality relation $A_j = 0$:

> Thus the functions ζ, $\iint \frac{l'\zeta' \, d\omega}{\Delta} = m\zeta$ exist also for the other surface, each one corresponding to another. Therefore, they also have the property $\iint l\zeta_1\zeta_2 \, d\omega = 0$.
>
> [Ms 3618 (5), p. 36r]

In this argument, Liouville had assumed that the system ζ_i on the initial surface is complete. His next move therefore was to establish that this property carries over from the simple surface to the neighboring one.

Now, for every surface let,

$$u = \sum \frac{\zeta_i \iint l\zeta_i \, d\omega Q}{\iint l\zeta_i^2 \, d\omega}.$$

One gets

$$\iint (u - Q)l\zeta_i \, d\omega = 0.$$

Let us pass from a surface to another, allowing the function Q to be the same at corresponding points. One has:

$$\iint \delta u l\zeta_i \, d\omega + \iint (u + Q)\delta(l\zeta_i \, d\omega) = 0.$$

But, for the first surface (e.g. the ellipsoid) one has $u = Q$; hence, $\iint \delta u l\zeta_i \, d\omega = 0$ and thus $\delta u = 0$. Thus, for the second surface, the series is again $= Q$. — Consequently one can solve Gauss's problem for that surface $\&_c\&_c\ldots$ going from surface to surface.

> [Ms 3618 (5), p. 36r]

Despite the ingenuity of this approach, it raises several questions. What are corresponding points? Also, more seriously, may singularities arise when this method is iterated? Liouville may have been aware of these problems, and so after two pages, he returned to the "direct and absolute method."

43.   Before returning to the old approach, Liouville discovered a new method of determining the eigenvalues m_i [Ms 3618 (5), p. 36v]. Like his first proof, it is a posteriori and so assumes the existence of a complete system of eigenfunctions ζ_i.

<u>Direct method for ζ.</u> Let Q be an arbitrary function and

$$Q - \frac{\iint lQ\,d\omega}{\iint l\,d\omega} = P.$$

We have

$$P = A_1\zeta_1 + A_2\zeta_2 + \cdots .$$

Thus,

$$\iint \frac{l'P'\,d\omega'}{\Delta} = P_1 = A_1 m_1 \zeta_1 + A_2 m_2 \zeta_2 + \cdots$$

and similarly,

$$P_2 = \iint \frac{l'P_1'\,d\omega'}{\Delta} = A_1 m_1^2 \zeta_1 + A_2 m_2^2 \zeta_2 + \cdots$$

and

$$P_s = A_1 m_1^s \zeta_1 + A_2 m_2^s \zeta_2 + \cdots .$$

Thus

$$\frac{P_{s+1}}{P_s} = m_1 + \varepsilon \text{ because } \left(\frac{m_2}{m_1}\right)^s \text{ is very small.}$$

m_1 is therefore $= \lim \frac{P_{s+1}}{P_s}$; and ζ_1 is proportional to $\lim \frac{P_s}{m_1^s}$. Similarly for the others

[Ms 3618 (5), p. 36v]

Here, Liouville has implicitly assumed that $1 > m_1 > m_2 > \cdots$. His result can be translated thus: Let Q be an arbitrary function and assume that ζ_1, $\ldots$, ζ_{n-1} are known. Then, let P_0 be defined by

$$P_0 = Q - \sum_{i=0}^{n-1} \frac{\langle Q, \zeta_i \rangle}{\|\zeta_i\|},$$

and let P_s be defined successively by

$$P_s = A(P_{s-1}).$$

Then,

$$m_n = \lim_{s \to \infty} \frac{P_{s+1}}{P_s}, \tag{107}$$

and

$$k\zeta_n = \lim_{s \to \infty} \frac{P_s}{m_n}.$$

Of course, this argument fails if $\langle Q, \zeta_n \rangle = 0$.

44. A few pages further on, Liouville designed another argument of the same type. It relied on a development of the quantity $\frac{1}{\Delta} = \frac{1}{|\overline{x} - \overline{x}'|}$. I shall quote the beginning of Liouville's note, the formal elements of which are easy to follow:

$\frac{1}{(\Delta)} = \sum A\zeta'$, (Δ) being the distance from x', y', z' on the surface to (x, y, z) outside, therefore

$$\frac{1}{(\Delta)} = \sum \frac{\zeta' \iint \frac{l'\zeta' d\omega'}{\Delta}}{\iint l\zeta^2 d\omega};$$

and consequently, on the surf. $\frac{1}{\Delta} = \sum \frac{m \zeta \zeta'}{\iint l\zeta^2 d\omega}$. Hence

$$\iint \frac{l' d\omega'}{\Delta^2} = \sum \frac{m^2 \zeta^2}{\iint l\zeta^2 d\omega},$$

and

$$\iiiint \frac{ll' d\omega d\omega'}{\Delta^2} = \sum (m^2).$$

But, the argument relies on a rather delicate series"

[Ms 3618 (5), p. 38v]

Indeed, in [1846f], Liouville had written down the development of $\frac{1}{\Delta}$ in the system of Lamé functions, and he had explicitly pointed out that the development only works when (x, y, z) is not on the ellipsoidal surface. So, he was clearly aware of the problematic singularity in $\frac{1}{\Delta}$. Still, he continued to show in a formal way that

$$\iint_S \iint_S \iint_S \frac{ll'l'' d\omega d\omega' d\omega''}{\Delta(0,1)\Delta(1,2)\Delta(2,0)} = \sum_i (m_i)^3,$$

where $\Delta(k, l)$ is the distance between $\overline{x}_k$ and $\overline{x}_l$ and $l^{(k)}$, $d\omega^{(k)}$ has the obvious meaning. Continuing this way, "one has in general $\sum (m^i)$."

Liouville denoted this sum S_i and concluded that

$$S_i - S_{i+1} = m_1^i(1 - m_1) + m_2^i(1 - m_2) + \cdots,$$

and thus,

$$\lim_{i \to \infty} \frac{S_{i+1} - S_{i+2}}{S_i - S_{i+1}} = m_1 \tag{108}$$

Liouville did not indicate how the method goes on, but it is obvious how m_2 can be found once m_1 has been determined, etc. He simply concluded "Practicable or not, this procedure is elegant." Finally, he designed a method to determine ζ_i along these lines, but it seems neither elegant nor practicable.

Anticipation of Weierstrass's criticism of the Dirichlet principle

45.   After this digression, I shall return to the unsolved problems in Liouville's approach to spectral theory via the Rayleigh-Ritz method. In addition to the question of m_i tending to zero and the question of convergence of the Fourier series, there is the fundamental problem of the existence of a minimum or maximum in the variational procedure (cf. §23). We have seen that Liouville was content to establish that the integral was bounded (from below in the first step and from above in the following steps). This is not surprising, for as I have mentioned in §§6–10 Gauss, Green, Thomson, and Dirichlet likewise confused a greater lower bound with a minimum. It is unlikely that Liouville was aware of this problem in 1845 when he began his research, but he spotted the weakness at the latest in 1857 when he wrote the following note:

> Notice that <u>the existence</u> of the electric layer in equilibrium is in fact proved for an infinity of surfaces where it is known. And no objections can be raised against the proof of Gauss, which shows that it is unique.
>
> [Ms 3622 (2), p. 113r]

This quote shows that Liouville had objections to Gauss's existence proof, and a few pages further he revealed the nature of his objections:

> On the ellipsoid, the electric layer corresponds in fact to a <u>minimum</u> for which there is also a <u>well-determined</u> state and not a mere indefinite tendency toward the goal
>
> [Ms 3622 (2), p. 114v]

Thus, Liouville had become aware of the problem that although a variational integral like (58) has a lower bound there need not be a minimizing function but only a sequence that makes the integral tend toward the "goal," i.e., toward the greatest lower bound. He had spotted the weak point in the Dirichlet principle, but how?

46.   I think this insight had its origin in an attempt in his earliest notes to determine the eigenvalues and eigenfunctions from below. Recall that the Rayleigh-Ritz method of §23 successively determines the eigenvalues $1 \geq m_1 \geq m_2 \geq m_3 \geq \cdots$ from above by maximizing the $\langle A\zeta, \zeta \rangle$ for $\|\zeta\|^2$ constant. Having established that $\langle A\zeta, \zeta \rangle$ is nonnegative (cf. Proposition 2), Liouville thought that by minimizing $\langle A\zeta, \zeta \rangle$ for $\|\zeta\|^2$ constant he would find the "smallest" eigenvalue and its corresponding eigenfunction. However, he was also aware that m_i decrease toward zero and that zero is not an eigenvalue. His notes [Ms 3618 (1), pp. 16r, 20r–22r, 43v–44r] show his initial confusion and slowly emerging awareness that these ideas clash with each other. On page 16r, he simply states that one can take minima instead of maxima in the Rayleigh-Ritz method. Having formally found an eigenvalue μ_1 in this way, Liouville remarked:

but, it seems to me that μ_1 cannot help being infinitely small. This is at once clear if one cannot give it a negative value. Let us therefore see if a negative value is possible.

[Ms 3618 (1), p. 20v]

Having answered this question in the negative, he concluded:

The minimum is zero and one does not attain it except for functions changing sign incessantly.

[Ms 3618 (1), p. 22r]

Finally, Liouville gave up finding the m_i's and the ζ_i's from below, but since he wanted to have a minimum method, as well as a maximum method, he minimized $\|\zeta\|^2$, holding $\langle A\zeta, \zeta \rangle$ constant (cf. the quote in §29). This is, of course, equivalent to the original method, but it permitted him to continue to write (e.g., §30) that the eigenfunctions satisfy max. and min. properties.

Thus, Liouville discovered that although the integral $\langle A\zeta, \zeta \rangle$ is bounded from below by zero there is no ζ with $\|\zeta\|^2 = $ const. corresponding to its greater lower bound but only a sequence of more and more wildly oscillating functions that make it "tend toward the goal." Could something similar happen when we take a minimum in the first step in the variational procedure (§23) (where we find l) or when we take a maximum in the following steps (where we find ζ_n)? This, I believe, was the question that made Liouville doubt Gauss's existence proof, or equivalently, the Dirichlet principle, and consequently, the entire Rayleigh-Ritz method.

47.   Still, this weakness does not seem to have been a serious problem for Liouville, for as the continuation of the last quote of the previous section shows, he believed that one could avoid such pathological behavior by selecting nice surfaces:

Thus, in general this state will also be well-determined, and we shall exclude the cases where it turns out not to be determined. It is therefore a general theory that we develop here. Afterward the particular exceptional cases.

[Ms 3622 (2), p. 114v]

Here, Liouville betrayed his 18th-century conception of mathematical rigor and reduced his own important discovery to a pathology of little importance. Where we would formulate Liouville's discovery as follows: the Dirichlet principle and similar variational principles, are "generally" incorrect, Liouville wrote that they are "generally" correct because they work in the one and only case where he knew the solution explicitly.

Unfortunately, I have not found any notes where Liouville discusses which surfaces fall under the "general theory" and which surfaces are "particular exceptional cases," and it is questionable if he contemplated the matter in any detail at all. In all probability, he did not anticipate Hilbert by half a century by giving an existence proof for a general class of surfaces.

Yet, one cannot exclude the possibility that he discovered a surface for which the existence theorems do not hold. In fact, the counterexample suggested by Lebesgue [1913] (Lebesgue's spine) is rather intuitive in the sense that a charge distribution on the surface will send out an electric spark from the needle and thus cannot be in equilibrium. Moreover, the integrals considered by Liouville diverge in this case, which also suggests that something is wrong.

48.   Liouville's discovery of the weakness of the Dirichlet principle at least 13 and probably 25 years before Weierstrass's published criticism is highly remarkable. However, it should be remarked that around the same time Riemann seems to have been aware of the problem as well. Indeed, Felix Klein [1894] claimed that Riemann only used Dirichlet's principle as a "heuristic principle." It has often been pointed out that Klein sometimes read things into the work of his hero which were not there, but in this case, Klein's testimony is corroborated by Riemann's own dissertation [1851, publ. 1867], in which he used the principle as the basis for his theory of complex functions. A careful reading of this passage reveals that Riemann tried to establish a kind of "compactness property" for the set of admissible functions and a kind of "continuity property" of the Dirichlet integral. To be sure, Riemann was not entirely successful, but his attempt shows that he was aware of the problems encountered in the naïve use of the Dirichlet principle. The delayed publication of Riemann's dissertation excludes an influence on Liouville.

ARE LIOUVILLE'S RESULTS CORRECT?

49.   After having recited all the weaknesses of Liouville's theory, one may wonder whether his results are in fact correct, that is, does the operator A have a complete set of eigenfunctions? Do the eigenvalues tend to zero? Can the eigenfunctions and eigenvalues be found by the variational method (the Rayleigh-Ritz method)? From modern operator theory, it is known that all these questions are answered in the affirmative if A is a compact operator on $L^2(S, l\, d\omega)$. Whether this will be the case clearly depends on the surface. Indeed, as remarked above, even for continuous functions ζ, the integral defining $A\zeta$ will diverge if the surface has a point that cannot be reached by a cone from both sides of the surface. One might therefore hope that the operator would be Hilbert-Schmidt (and therefore compact) for smooth surfaces, but even this is not the case. For A to be Hilbert-Schmidt, its kernel $\frac{1}{\Delta} = \frac{1}{|\bar{x} - \bar{x}'|}$ should be an $L^2(S, l\, d\omega)$ function in each of its variables, but it is only an $L^{2-\varepsilon}$ function. However, for a smooth surface S, for which l is a continuous and hence bounded function, the compactness follows since A has a so-called weak singularity [Mikhlin 1957]. Christian Berg, who kindly called my attention to this result, has further utilized refined ideas of potential theory and interpolation theory to establish the

compactness of A for a large class of nonsmooth surfaces S, namely those enclosing a domain Ω that is not thin in any point of S (cf. [Berg and Lützen 1990]). For such surfaces, therefore, Liouville's results are entirely correct. However, they were so far ahead of their time that his immediate successor Poincaré gave up proving them — or similar results — rigorously.

Poincaré's Fundamental Functions

50. It is well known that Poincaré invented the ingenious *méthode de balayage*, or sweeping out method, to establish the existence of a solution to the Dirichlet problem. However, as he admitted, this method is not suitable for finding the solution. To this end, Carl Neumann's method of the arithmetic mean is preferable. Neumann [1870, 1877] had only proved the convergence of his method for convex boundary surfaces, but Poincaré [1895, 1896] extended the result to more general surfaces. In the first four chapters of that work, Poincaré presented his polished rigorous proof, and in the last two chapters he revealed "the general ideas which have initially led me to guess the results" [Poincaré 1896, p. 142]. These last chapters, which are of main concern to us, begin with the words:

> Up to this point, I have tried to be perfectly rigorous. Now, I think it will be useful to connect the above with other considerations and therefore to enter into a domain which I have not investigated properly and where I must content myself with a simple outline
> [Poincaré 1896, pp. 118–119]

51. These considerations concern what Poincaré called the "fundamental functions." Like Liouville's functions ζ_i, the fundamental functions arise from single-layer distributions on a surface S, but contrary to Liouville, Poincaré used their potential W on and outside the surface as the basic quantity. Let S be the boundary of a domain Ω and consider the Dirichlet integral inside and outside of D:

$$D_-(W) \;=\; \iiint_\Omega \left(\left(\frac{\partial W}{\partial x}\right)^2 + \left(\frac{\partial W}{\partial y}\right)^2 + \left(\frac{\partial W}{\partial z}\right)^2 \right) dv, \quad (109)$$

$$D_+(W) \;=\; \iiint_{c\bar\Omega} \left(\left(\frac{\partial W}{\partial x}\right)^2 + \left(\frac{\partial W}{\partial y}\right)^2 + \left(\frac{\partial W}{\partial z}\right)^2 \right) dv. \quad (110)$$

As a first step, Poincaré minimized $D_-(W)$ under the assumption that $D_+(W) = 1$. This evidently gives the potential ϕ_0 due to the equilibrium distribution on S. In fact, $D_-(\phi_0) = 0$. If W is the potential of another single-layer distribution on S, then Green's second identity (15) implies that

$$\iint_S W \frac{\partial \phi_0}{\partial n_+} \, d\omega = \iint_S \phi_0 \frac{\partial W}{\partial n_+} \, d\omega \,,$$

and similarly for ∂n_- (for notation see §2). In the second step, Poincaré then minimized $D_-(W)$ for $D_+(W) = 1$ and under the additional requirement that

$$\iint_S W \frac{\partial \phi_0}{\partial n_+} \, d\omega = \iint_S \phi_0 \frac{\partial W}{\partial n_+} \, d\omega = 0. \tag{111}$$

With the ordinary methods of the calculus of variations, he showed that the normal derivatives of the minimizing function ϕ_1 satisfy:

$$\frac{\partial \phi_1}{\partial n_-} = -\lambda_1 \frac{\partial \phi_1}{\partial n_+} \qquad \text{on} \quad S, \tag{112}$$

where λ_1 is a constant equal to $D_-(\phi_1)$. (The details of this proof can be found in [Poincaré 1899].)

Continuing this process, Poincaré found a sequence of fundamental functions ϕ_i satisfying

$$\frac{\partial \phi_i}{\partial n_-} = -\lambda_i \frac{\partial \phi_i}{\partial n_+}, \tag{113}$$

where the λ_i's are positive constants:

$$0 = \lambda_0 < \lambda_1 \le \lambda_2 \le \cdots. \tag{114}$$

The fundamental functions satisfy the "orthogonality relation"

$$\iint_S \phi_i \frac{\partial \phi_j}{\partial n_+} \, d\omega = \iint_S \phi_j \frac{\partial \phi_i}{\partial n_+} \, d\omega = 0 \quad \text{for} \quad i \ne j \tag{115}$$

and therefore also,

$$\iint_S \phi_i \frac{\partial \phi_j}{\partial n_-} \, d\omega = \iint_S \phi_j \frac{\partial \phi_i}{\partial n_-} \, d\omega = 0 \qquad \text{for} \quad i \ne j \tag{116}$$

Moreover, the fundamental functions are normalized such that

$$D_+(\phi_i) = 1 \quad \text{and} \quad D_-(\phi_i) = \lambda_i. \tag{117}$$

52.   Poincaré went on to show that if ϕ is a function satisfying an equation like (113), that is,

$$\frac{\partial \phi}{\partial n_-} = -\lambda \frac{\partial \phi}{\partial n_+}, \tag{118}$$

then λ is one of the numbers λ_i above and ϕ is proportional to the corresponding ϕ_i. In the process, he used a method similar to Liouville's to prove that two functions ϕ, ψ satisfying (118) for two different values of λ must satisfy an "orthogonality relation" like (115); as a corollary, he deduced that λ in (118) must be real. It was then a simple matter for Poincaré to show that λ in (118) can neither be negative nor lie between two consecutive λ_i's. It remained to be seen that λ cannot be larger than all the λ_i's, but Poincaré only presented an unrigorous argument to this effect. The argument was based on the completeness of the system of fundamental functions, and as he admitted in the introduction to the section on "development in series," this was nothing but a bold conjecture:

Let F be an arbitrary function of the coordinates of a point on S; various analogies make me believe that F can be developed in a series of fundamental functions

[Poincaré 1896, p. 124]

Taking this completeness for granted, Poincaré determined the Fourier coefficients in the usual way and went on to show how these series expansions can be used in solving Neumann's generalization of the Dirichlet problem. These considerations are analogous to Liouville's solution of Gauss's problem by way of the ζ_i's (§22). Poincaré was well aware of the problems involved in this method — the existence of the minimizing functions as well as the completeness — and so he explained his motives for including it as follows:

> By drawing consequences which I have succeeded in proving rigorously in another way from considerations which rest solely on a shaky basis, I simply wanted to explain the course of my thoughts and to show how I was led to the result.
>
> However, one can pose the problem in another way; can one use the propositions established at the beginning of this work to prove the existence of the fundamental functions?
>
> I have not yet succeeded, but it is evident that one can try to do it by a process analogous to the one I have employed in my memoir on the equations of mathematical physics, published in the Rendiconti del Circolo di Palermo (1894).
>
> [Poincaré 1896, pp. 129–30]

The paper mentioned at the end of the quote contains Poincaré's discussion of the spectral theory of the Laplacian with given boundaries (the vibrating membrane) (cf. Chapter X, §65).

53.   Edouard Le Roy followed these hints of Poincaré, and two years later he was able to give a rigorous proof of the existence of the fundamental functions. More generally, he proved that if φ is a function on a connected analytic surface S, so that $\alpha < \varphi < \beta$ (α, β are two positive constants), there exist a sequence of positive constants $\xi_1 \leq \xi_2 \leq \xi_3 \leq \cdots$ tending to infinity and a corresponding sequence of single-layer potentials W_1, W_2, W_3, ... such that

$$W_i = \frac{\xi_n}{4\pi} \iint_S \frac{\varphi' W_i'}{\Delta} \, d\omega \tag{119}$$

and

$$\iint_S \varphi \, W_i W_j \, d\omega = 0, \quad m \neq n. \tag{120}$$

Le Roy found these functions by minimizing $D_+(W) + D_-(W)$, keeping $\iint_S \varphi W^2 \, d\omega = 1$. Moreover, Le Roy proved that any C^∞ function on S can be developed on the system W_i. In particular, he observed that if φ is the equilibrium distribution, the W_i are equal to the fundamental functions ϕ_i conjectured by Poincaré.

54. A comparison between (119), (120), Liouville's (64) (and similar equations for ζ_i), and (94) shows that Liouville's eigenvalues m_i are equal to Poincaré-Le Roy's $\frac{4\pi}{\xi_i}$ and that the eigenfunctions ζ_i are precisely the restriction to S of Poincaré's fundamental functions ϕ_i. Thus, Le Roy's argument implicitly showed that Liouville's results were correct, at least for connected analytic surfaces S. Le Roy's methods can be generalized to other surfaces, but Christian Berg's modern proofs are more direct and much more general.

55.   The identity of Poincaré's fundamental functions and Liouville's ζ_i's makes one wonder if Poincaré or Le Roy were inspired by Liouville. Poincaré [1896] did not refer to Liouville, and it is also unlikely that the old and weak Liouville should have told the young Poincaré about his unpublished notes. Still, from Poincaré's paper on rotating fluids [1885], we know that he was familiar with Liouville's work on Lamé functions, and when Le Roy showed that the fundamental functions for an ellipsoid were the Lamé functions, he also referred to Liouville's results. Therefore, it is not impossible that the two authors had read the short note [1846c] in which Liouville sketched the properties of the functions ζ_i for a general surface (cf. §22). This may consciously or unconsciously have inspired them.

Later Developments

56.   To my knowledge, Poincaré was the only mathematician who continued directly in Liouville's footsteps, but just after the turn of the century, the general development of potential theory, integral operators, and spectral theory had developed so far that the central ingredients in Liouville's method were rediscovered and appreciated at their true value. Here, I shall summarize just those aspects of these developments that relate to Liouville's notes.

The idea of using integral equations in the solution of Dirichlet's problem is so natural that it was taken up by other mathematicians soon after Liouville. Indeed, when presented in its Gaussian version the problem asks for a function λ on a surface S such that

$$\iint_S \frac{\lambda(\overline{x}')}{|\overline{x} - \overline{x}'|} \, d\omega' = f(x), \tag{121}$$

where $f(x)$ is a given function. This is an integral equation for λ of the first kind. Beer [1856] had the idea of solving this equation, or rather its plane logarithmic equivalent, by the method of successive approximations, which, as we saw in Chapter X, had been invented by Liouville. Beer's idea was generalized by C. Neumann [1870, 1877], who, contrary to Beer, proved the convergence of the procedure for a well-defined class of boundary surfaces and functions f. Since Neumann sought the solution of the Dirichlet problem as the potential from a double layer on the surface, his integral equation was different from (106) (cf. [Burkhardt und Meyer 1899–1916]).

Compared to these procedures, Liouville's method is highly complicated, but it has the great advantage that once the eigenfunctions ζ_i have been found for a given surface S the solution of the Dirichlet problem is in principle easy for any given boundary value f. With Beer and Neumann's method, one must do the calculations over again for each new function f. Moreover, as we saw above, Poincaré showed how functions similar to the ζ_i could lead to a generalization of Neumann's procedure.

57.   The first works on general integral equations were published by Volterra [1896] and carried further by Fredholm [1903]. Both mathematicians were inspired by the Dirichlet principle; indeed, Fredholm's first work carried the title *Sur la nouvelle méthode pour la résolution du problème de Dirichlet* [Fredholm 1900]. Spectral theory is not dealt with in any of these early works, except perhaps implicitly in Fredholm's paper. It was Hilbert who first pointed out the importance of eigenvalues and eigenfunctions and extended the previous isolated observations on integral equations into a connected large-scale theory, which became the starting point of modern operator theory and an important source of inspiration for the development of all of functional analysis [Bernkopf 1966–1967].

Hilbert was probably the first mathematician whose understanding of spectral theory went beyond that of Liouville. Like Liouville, he was of the opinion that the possibility of expanding "arbitrary" functions on a set of eigenfunctions was among the most important properties of analysis. In his papers from 1904 to 1910 — later collected in a monograph (1912) — he showed in several ways that large classes of integral operators have a complete system of eigenfunctions. Several of Hilbert's methods are the same as Liouville's. In particular, he established the existence of the eigenfunctions by the very same variational method Liouville had used. Yet, Hilbert's use of the method was more satisfactory than that of his predecessor, in that he singled out the property of the operator — compactness or *vollstetigkeit* of the corresponding quadratic form — which allowed him to conclude that the successive greater lower bounds are attained and that the method will yield a sequence of eigenfunctions tending to zero.

Hilbert's method was inspired by his work on the Dirichlet principle and the direct method in the variational calculus. In fact, the idea of finding eigenfunctions in this way had been suggested by Holmgren one year earlier than Hilbert [Holmgren 1906] (for a more detailed summary of the early history of integral operators see [Hellinger and Toeplitz 1923–1927]).

In modern textbooks, the method of determining eigenvalues in this way is often called the Rayleigh-Ritz method. Rayleigh's name is attached to the method because in his *Theory of Sound* [1877] he used a variational technique to determine the first few eigenvibrations of a plate. In fact, Weber [1869] had made a similar application of the method. Since these applications concern differential equations and not integral equations, they are more closely attached to Sturm-Liouville theory than to the ideas dealt with in this section. The association of Ritz's name with this method seems to be more farfetched. He developed the numerical aspect of Hilbert's direct method in the calculus of variations in [1909], but he did not discuss eigenvalues. Thus, Poincaré, Holmgren, and Hilbert seem to have been the first to use the method in the same elegant way as Liouville. As I have pointed out in Chapter X, §67, Hilbert went on to apply his theory of integral operators to Sturm-Liouville theory, uniting thereby the two trends in Liouville's work on spectral theory.

58.   In §56, we left potential theory around 1900, and in fact, integral operators from then on played a smaller role in the development of this theory. In recent years, however, this approach has been rehabilitated, and Richter [1977] has discussed the logarithmic potential, in particular, the numerical aspects, with a method that is somewhat similar to Liouville's treatment. There may also be a connection to the so-called Galerkin method, but I have not investigated its history.

59.   This leads us to the question of the place of Liouville's theory in the history of mathematics. First of all, with the possible exception of Poincaré, there is no trace of any influence on later mathematicians. Liouville may have elaborated on his vague hints in his *Academie* talks to his friends, for example, Sturm and Chasles. However, even if this happened, it is not certain that they would have paid much attention, for the importance of spectral theory was not appreciated by anybody else at this period, and the application of the method to potential theory is problematic. Indeed, as had been pointed out by Chasles, the mere determination of the equilibrium distribution l surpassed the forces of analysis except in very simple cases, and so it would seem almost hopeless to find the eigenfunctions ζ_i. Of course, a more qualitative approach to potential theory was beginning to emerge in the works of Gauss and Green, but the ultimate goal was still to find explicit solutions to Laplace's equation, and Liouville's nonconstructive ideas do not seem promising in this respect. This fact may also have discouraged Liouville from publishing them. Still, in his work on Sturm-Liouville theory, Liouville had shown his awareness of qualitative general theorems in spectral theory, and his enthusiasm for integral operators revealed by his notes seems to be genuine.

One may wonder what would have happened if Liouville had published his spectral theoretic approach to potential theory. For the reasons mentioned above, I think that his ideas would have had little immediate impact. Indeed, Sturm-Liouville theory lay dormant for half a century before its qualitative results were extended, and in this case, at least the differential equations were familiar objects. In the theory discussed in this chapter, almost all the ingredients were unfamiliar: the ideas of eigenfunctions, the qualitative nonconstructive character, and the notion and importance of integral operators or integral equations. Indeed, as I have explained [Lützen 1982b], expect for Abel, Liouville was the only person who had studied such equations. They were central in his fractional calculus (cf. Chapter VIII) and in Sturm-Liouville theory (cf. Chapter X, §40–41), but nobody else dealt with this subject. In fact, his small published note on the spectral theory of a symmetric integral operator [Liouville 1845h] (cf. §40) seems to have been entirely overlooked by his contemporaries and successors.

Thus, Liouville's spectral theory of integral operators and its application to potential theory was more than 50 years ahead of his time, and so would probably have been overlooked by his contemporaries if it had been published. However, there is no doubt that if that had happened Liouville would have appeared in the histories of mathematics as the inventor of the "fundamental functions" of Poincaré, and he would have figured along with Fredholm and Hilbert as the creators of spectral theory of integral operators. Finally, the Rayleigh-Ritz method might have been called the Liouville method.

XVI

Mechanics

1. In the mid-19th century, when Liouville made his contributions, mechanics had become the most highly mathematized of all the sciences. In fact, rational or analytical mechanics was considered an integral part of mathematics.

After some tentative efforts in the middle ages, this mathematization was begun by Galileo and reached a high point in Newton's *Principia* [1687]. Newton's successors, Euler and Lagrange, turned the basic laws of point mechanics into precise statements of mathematical analysis: differential equations and calculus of variations. This tendency was expressed with great clarity by Lagrange in his famous preface to the second edition of the *Mecanique Analytique*:

> One will not find any figures in this Work. The methods I present need neither constructions, nor geometric or mechanical arguments but only algebraic operations subject to a regular and uniform procedure. Those who love the Analysis will be pleased to see the Mechanics becoming a new branch of it and will be grateful to me for having extended its domain.
>
> [Lagrange 1788, 2nd ed. 1811, *Avertissement*]

Mechanics was not mechanical any more! But, despite the mathematical clarification of the foundations, many dynamical problems remained unsolved. In particular, planetary motion, which Newton had made a problem of mechanics, defied exact solution. The two-body problem was solved by Newton, but it is well known that even today we can only approximately solve the equations of motion for three-point masses.

Faced with these problems, mathematicians went in two different directions. They devised highly complicated methods for approximating planetary motions, and they created an elegant general theory of the transformation and integration of the equations of motion. This corresponds to the division into celestial and rational (or analytical) mechanics. Liouville contributed to both these disciplines; to celestial mechanics in the period 1836–1840 and to rational mechanics around 1846 and in the period 1853–1856.

In the history of mechanics before Liouville, the two branches were closely connected, making the division somewhat artificial. In particular, the development of the theory of perturbations and the method of variation of the arbitrary constants enriched the general theory as well as the details of planetary calculations. I shall start out with a summary of this development because it is a precondition for appreciating Liouville's work in celestial, as well as in rational, mechanics.

The Theory of Perturbation by Variation of the Arbitrary Constants

2.   Lagrange briefly sketched the method later to be called the variation of the arbitrary constants in a paper [1776], the bulk of which was devoted to something completely different. He considered a differential equation, for example,

$$x^{(n)}(t) = P\left(t, x, x^2, \ldots, x^{(n-1)}\right). \tag{1}$$

and assumed a complete solution $F(t, a_1, \ldots, a_n)$ to be known. Here, the a_i's are arbitrary constants. Let equation (1) be perturbed by adding a function $Q(t, x, x^2, \ldots, x^{(n-1)})$ to P:

$$x^{(n)}(t) = P + Q. \tag{2}$$

Then, Lagrange sought solutions $y(t)$ to (2) of the form $y(t) = F(t, a_1(t), \ldots, a_n(t))$ where the a's, which were constants before, are now allowed to vary. He argued that $y(t)$ satisfies (2) if $\frac{da_i}{dt}$ satisfies the following linear system of equations:

$$\frac{\partial F}{\partial a_1}\frac{da_1}{dt} + \frac{\partial F}{\partial a_2}\frac{da_2}{dt} + \cdots + \frac{\partial F}{\partial a_n}dt = 0$$

$$\frac{\partial F'}{\partial a_1}\frac{da_1}{dt} + \frac{\partial F'}{\partial a_2}\frac{da_2}{dt} + \cdots + \frac{\partial F'}{\partial a_n}\frac{da_n}{dt} = 0$$

$$\vdots$$

$$\frac{\partial F^{(n-2)}}{\partial a_1}\frac{da_1}{dt} + \frac{\partial F^{(n-2)}}{\partial a_2}\frac{da_2}{dt} + \cdots + \frac{\partial F^{(n-2)}}{\partial a_n}\frac{da_n}{dt} = 0$$

$$\frac{\partial F^{(n-1)}}{\partial a_1}\frac{da_1}{dt} + \frac{\partial F^{(n-1)}}{\partial a_2}\frac{da_2}{dt} + \cdots + \frac{\partial F^{(n-1)}}{\partial a_n}\frac{da_n}{dt} = Q, \tag{3}$$

where $F' = \frac{\partial F}{\partial t}, F'' = \frac{\partial F'}{\partial t}$, etc.

Indeed, the first $n - 1$ equations imply that $y'(t) = F'$, $y''(t) = F'' \ldots$ $y^{(n-1)} = F^{(n-1)}$, and the last equation implies that

$$y^{(n)} = \frac{dy^{(n-1)}}{dt} = \frac{\partial F^{(n-1)}}{\partial t} + Q = P + Q,$$

so that y is a solution of (2). From (3), one can find $\frac{da_i}{dt}$ by way of Cramer's method. The resulting system of differential equations can be solved approximately, and if the original equation (1) is linear, Lagrange could even find an exact solution $a_i(t)$ that inserted into $F(t, a_1(t), \ldots, a_i(t))$, yields the solution of the perturbed equation, (2).

3. Lagrange [1776] indicated that this method was of interest in celestial mechanics, and indeed, Euler, Laplace, and Lagrange himself had for several years used specialized versions of it in the study of planetary motion.

It is well known that one isolated planet will move around the sun in an ellipse according to Kepler's laws. Such a solution to the two-body problem is characterized by six constants $\gamma, a, e, \alpha, \omega$, and ϵ. I shall define these so-called elliptic elements in the next section; for the moment, it is sufficient to consider them as arbitrary constants, corresponding to a_i above. The presence of a second planet will perturb the first planet by introducing in its equations of motion an extra term corresponding to Q above, and as above we can try to describe the perturbed motion of the first planet by varying the elliptic elements of its unperturbed elliptic motion. The problem is to determine the variation of the elements as functions of time.

Lagrange [1781–1782] simplified this method: A single planet of mass m moving around the sun M is governed by Newton's second law:

$$\frac{d^2 r}{dt^2} + (M + m)\frac{r}{|r|^3} = 0, \tag{4}$$

r being the radius vector of the planet from the center of the sun. If the planet is influenced by perturbing forces described by a potential R, then its equation of motion becomes

$$\frac{d^2 r}{dt^2} + (M + m)\frac{r}{|r|^3} + \operatorname{div} R = 0. \tag{5}$$

In particular, if the perturbation is caused by another planet with mass m' and radius vector r', the perturbing function R has the form

$$R = \frac{m' r \cdot r'}{|r'|^3} - \frac{m'}{|r - r'|}. \tag{6}$$

This function can be expressed as a function of time and the elliptic elements of the planets. The great result in Lagrange's papers says that the variation of the elements of the perturbed planet can be determined from the derivatives of R. For example,

$$\frac{da}{dt} = -\frac{2}{an}\frac{\partial R}{\partial \epsilon} \tag{7}$$

(n is defined in §8).

These ideas were developed by Laplace and incorporated into his great *Mécanique Celeste* [1799–1825, Part 1, Livre II, §22, pp. 48 and 49]. In fact, the notation R for the perturbation function is due to Laplace.

Most of Liouville's work in celestial mechanics deal with the integration of (7) and similar equations for the other elements.

4. When the method of variation of the arbitrary constants made its next great theoretical progress, it happened within the Lagrangean formalism, so I shall take the opportunity to discuss this improvement of rational mechanics, which shaped all subsequent development, including Liouville's work.

The Lagrange formalism was the answer to a problem that rarely appears in celestial mechanics, but often turns up when treating terrestrial systems, namely, how to deal with point masses that do not move freely. More specifically, Lagrange considered systems with holonomic constraints, that is, systems of m point masses whose configuration must satisfy a number of equations of the form

$$f_j(x_1, y_1, z_1, x_2, y_2, z_2, \ldots, z_m) = 0 \qquad j = 1, 2, \ldots, k, \tag{8}$$

where x_i, y_i, z_i are the Cartesian coordinates of the ith particle. Lagrange devised two ways of solving the problem. One is to use his famous multipliers, and the other is to reduce the number of variables to $n = m - k$ independent parameters, nowadays called generalized coordinates and denoted q_i. Lagrange gave the equations of motion a form valid for any set of generalized coordinates, and not just for the usual orthogonal coordinates, as in Newton's second law.

Let T denote the kinetic energy,

$$T = \frac{1}{2} \sum_{i=1}^{m} m_i(\dot{x}_i^2 + \dot{y}_i^2 + \dot{z}_i^2), \tag{9}$$

of the system, where we denote the time derivative by a dot. Express x_i, y_i, z_i in terms of the q_i's and insert them into T in order to get a function

$$T = T(q_1, \ldots, q_n, \dot{q}_1, \ldots, \dot{q}_n). \tag{10}$$

Assume that the system is conservative so that the forces can be derived as $- \operatorname{grad} U$, where $U = U(x_1, \ldots, z_m)$. Substitute also the variables q_i in U such that

$$U = U(q_1, \ldots, q_n). \tag{11}$$

Then, the equations of motion of the system can be written

$$\frac{d}{dt} \frac{\partial T}{\partial \dot{q}_i} = \frac{\partial T}{\partial q_i} - \frac{\partial U}{\partial q_i}, \tag{12}$$

or

$$\frac{d}{dt} \frac{\partial L}{\partial \dot{q}_i} = \frac{\partial L}{\partial q_i}, \tag{13}$$

where we have introduced the Lagrangian

$$L = T - U. \tag{14}$$

These are Lagrange's equations. They were derived in the *Mécanique Analytique* [Lagrange 1788, 1811–1815, Part 2, Sect. IV].

5. In [1808] Lagrange returned to the variations of the elements of the planets after having participated in the commission evaluating a memoir on this subject read by Poisson to the *Institut de France* on June 20th the same year [Poisson 1809a]. In this paper, Lagrange discussed only perturbations of planetary motions, but half a year later [Lagrange 1809], he generalized his new approach to "all the problemes of mechanics."

Let $q_1, \ldots, q_n$ be a complete set of solutions of Lagrange's equations (12) containing $2n$ arbitrary constants $a_1, \ldots, a_{2n}$. Lagrange then perturbed the potential with a quantity $-\Omega(q_1, \ldots, q_n)$, obtaining the new equations

$$\frac{d}{dt}\frac{\partial T}{\partial \dot{q}_i} = \frac{\partial T}{\partial q_i} - \frac{\partial U}{\partial q_i} + \frac{\partial \Omega}{\partial q_i}. \tag{15}$$

By inserting $q_1, \ldots, q_n$ as functions of the a_i's, Ω can be considered a function of these arbitrary constants. He then showed that $q_i(t, a_1(t), \ldots, a_{2n}(t))$ is a solution of (15) if the $\frac{da_i}{dt}$'s satisfy the equations

$$\frac{\partial \Omega}{\partial a_i} = \sum_j [a_i, a_j]\frac{da_j}{dt}, \tag{16}$$

where $[a_i, a_j]$ is a quantity called the Lagrange bracket. In the first of the papers [1808], Lagrange defined it as follows for one free particle with orthogonal coordinates (x_1, x_2, x_3):

$$[a, b] = \sum_{i=1}^{3} \frac{\partial x_i}{\partial a}\frac{\partial^2 x_i}{\partial t \partial b} - \frac{\partial x_i}{\partial b}\frac{\partial^2 x_i}{\partial t \partial a}, \tag{17}$$

"when one removes all the terms that contain t after the substitution of the partial differences [derivatives] of the values x, y, z [i.e., x_1, x_2, x_3]." In the general case, his expression for $[a, b]$ was appalling [Lagrange 1809, §16], but in an *Addition* to the printed paper, he simplified it by introducing what we now call the generalized momentum

$$p_i = \frac{\partial T}{\partial \dot{q}_i} = \frac{\partial L}{\partial \dot{q}_i}. \tag{18}$$

Using these variables, Lagrange wrote

$$[a, b] = \sum_{i=1}^{n} \left(\frac{\partial q_i}{\partial a}\frac{\partial p_i}{\partial b} - \frac{\partial q_i}{\partial b}\frac{\partial p_i}{\partial a}\right). \tag{19}$$

It is most probable that Lagrange's *Addition* was inspired by Poisson's next memoir, read to the first class of the *Institut* on October 16, 1809. It is characteristic, however, that now when Poisson had actually surpassed him, Lagrange did not give him a word of mention. In his second paper on the same subject, however, he could not help referring to his young colleague [Lagrange 1810].

6.   Poisson characterized his own formulas as the inverse of Lagrange's. Having defined p_i as in (18), he first pointed out that Lagrange's equations could be written

$$\frac{dp_i}{dt} = \frac{\partial L}{\partial q_j}. \tag{18a}$$

As a comment on these equations, which were to become half of Hamilton's (53), Poisson wrote: "In this way the equations of motion are transformed to the simplest form that one can give them" [Poisson 1809b, p. 273].

Then, Poisson introduced the so-called Poisson bracket of two functions $a(t, q_1, \ldots, q_n, p_1, \ldots, p_n)$ and $b(t, q_1, \ldots, q_n, p_1, \ldots, p_n)$:

$$(a, b) = \sum_{i=1}^{n} \frac{\partial a}{\partial p_i} \frac{\partial b}{\partial q_i} - \frac{\partial a}{\partial q_i} \frac{\partial b}{\partial p_i} \tag{19a}$$

and showed that if a and b are integrals of the system then the bracket is also independent of time. Moreover, he addressed the perturbed equation (15), writing it in the form

$$\frac{dp_i}{dt} = \frac{\partial L}{\partial q_i} + \frac{\partial \Omega}{\partial q_i}. \tag{15a}$$

As a "dual" of Lagrange's complete solution, $q_i(t, a_1, \ldots, a_{2n})$, Poisson assumed $a_i(t, q_1, \ldots, q_n, p_1, \ldots, p_n)$, $i = 1, 2, \ldots, 2n$, to be a complete set of integrals of the original equation (18a). He then proved that $q_i(t, a_1(t), \ldots, a_{2n}(t))$ is a solution of (15) or (15a) if $a_i(t)$ satisfies the equations

$$\frac{da_i}{dt} = \sum_{j=1}^{n} \frac{\partial \Omega}{\partial a_j}(a_j, a_i). \tag{16a}$$

As remarked by Poisson, this is the "inverse" of Lagrange's formulas (16). In fact, we see that the matrix $\{(a_i, a_j)\}$ is the inverse of $\{[a_i, a_j]\}$. This also shows that the Lagrange brackets as defined by (19) are independent of time; the quote below formula (17) indicates that Lagrange had not discovered this initially. Poisson's equation (18) is the natural generalization of equations (7), etc., for planetary motion.

I shall consistently denote Lagrange's bracket with a bracket and Poisson's with a parenthesis. Thereby, I follow the tradition begun by Poisson and still used by Whittaker [1937]. Lagrange used the opposite convention, as do most modern textbooks that usually only discuss the Poisson bracket.

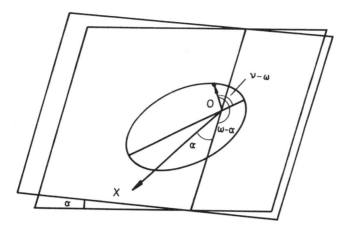

FIGURE 1

Celestial Mechanics

7.   Liouville's contributions to celestial mechanics were all additions and perfections of methods and results in Laplace's *Mécanique Celeste*. In particular, he improved the standard techniques for solving equations (7) and the like for the variation of the elliptic elements. In order to understand these technicalities, we must define the elements characterizing the elliptic orbit of a single planet.

The standard reference frame (Fig. 1) is a coordinate system with its origin 0 in the center of the sun, its (x, y) plane coinciding with the ecliptic, and the x axis pointing in the direction of the vernal equinox. The orbit plane of the planet is then determined by its inclination γ with the ecliptic and the longitude α of the ascending node, measured from the x axis.

The size of the orbit is characterized by its semimajor axis a and its eccentricity e; its position on the orbit plane is fixed by an angle ω, where $\omega - \alpha$ is the angle between the ascending node and the perihelion. The elements $\gamma, \alpha, a, e,$ and ω completely determine the orbit. Finally, we shall introduce an element ϵ to determine the position of the planet at time $t = 0$.

To this end, we must introduce the mean anomaly. Consider a circle (Fig. 2) in the orbit plane with radius a and concentric with the ellipse. Project the position of the planet R onto the circle (A) by a projection orthogonal to the major axis. Let $\nu - \omega$ be the angle between the radius vector of the planet and its perihelion (Figs. 1 and 2), and let the so-called eccentric anomaly $< \text{PCA}$ (Fig. 2) be denoted u. The connection between ν and u

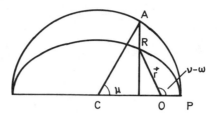

FIGURE 2

is given by

$$\tan \tfrac{1}{2}(\nu - \omega) = \sqrt{\tfrac{1+e}{1-e}} \tan \tfrac{1}{2} u, \tag{19b}$$

and the solar distance $|\vec{r}| = r$ of the planet is determined by

$$r = a(1 - e \cos u). \tag{20}$$

A simple geometric consideration shows that the area POR covered by the radius vector is proportional to

$$\zeta = u - e \sin u, \tag{21}$$

the so-called mean anomaly. According to Kepler's second law, this area, and therefore ζ, increases with a constant velocity:

$$\zeta = nt + \epsilon - \omega. \tag{22}$$

This defines ϵ and n.

It is easily seen that ζ can be interpreted as an arc of a circle with its center in the sun, namely, the arc that the planet would have described if it had encircled the sun with constant angular velocity and its actual period. Thus, at a given time t, we can determine ζ from (22) and then find u from equation (21) (called Kepler's equation, see Chapter IX, §36). Then, the position of the planet can be determined from (19b) and (20). This is all we need to know about positional astronomy in order to understand Liouville's work.

8.   Let us now return to the perturbation problem. Lagrange and Laplace had found equations (7) etc., for the variation of the elliptic elements, but two problems remained: 1° determine the right-hand side of (7) and 2° solve the differential equation (7). These were the problems addressed by Liouville.

When the planet is perturbed by another planet (with primed elements and variables), the standard method of solving the first problem, used by

Lagrange and Laplace, is to develop the perturbing function R (6) in a series of the form

$$R = \sum_{i,i' \in Z} M_{ii'} \sin(i\zeta - i'\zeta') + N_{ii'} \cos(i\zeta - i'\zeta'), \tag{23}$$

where the M's and the N's are functions of the first five elements, but where ϵ only enters through $\zeta = nt + \epsilon - \omega$. Therefore,

$$\frac{\partial R}{\partial \epsilon} = \frac{\partial R}{\partial \zeta} = \sum_{ii' \in Z} i\, M_{ii'} \cos(i\zeta - i'\zeta') - i\, N_{ii'} \sin(i\zeta - i'\zeta'). \tag{24}$$

It is now possible to determine a from (7) to the first order by reducing the elliptic elements on the right-hand side to their constant values. In that case, t only enters through $i\zeta - i'\zeta' = (in - i'n')t + i\epsilon - i'\epsilon - \omega + \omega'$. Hence, from (6) and (8):

$$a = a_0 - \sum_{i,i' \in Z} \frac{2i}{an(in - i'n')}(M_{ii'} \sin(i\zeta - i'\zeta') + N_{ii'} \cos(i\zeta - i'\zeta')). \tag{25}$$

The exceptional term $i = i' = 0$ gives rise to a term $k \cdot t$, called a secular term. The other terms give rise to periodic inequalities. Because of the denominator $in - i'n'$ in (25), those terms for which $in - i'n'$ are approximatively zero stand out as particularly important; they are called long periodic. For example, if we consider the perturbation of Saturn (n) produced by Jupiter (n'), the numbers $n = 120.''45504$ and $n' = 299.''12838$ have a ratio close to $2 : 5$. Therefore, the term corresponding to $5n - 2n'$ is the leading one.

TWO PUBLICATIONS BY LIOUVILLE ON PLANETARY PERTURBATIONS

9.   Liouville's first paper on celestial mechanics [1836d] dealt with the periodic inequalities, more precisely, with the determination of the coefficients $M_{ii'}$ and $N_{ii'}$. The standard method was to expand them as infinite series of ascending powers of the excentricities e, e' and of $\sin\frac{I}{2}$, where I is the angle between the orbit planes of the two planets. In [1822] Poisson had hinted at the possibility of expressing the coefficients $M_{ii'}$ and $N_{ii'}$ by double integrals analogous to the recently discovered Fourier coefficients:

$$M_{ii'} = \frac{1}{2\pi^2} \int_0^{2\pi} \int_0^{2\pi} R \sin(i\zeta - i'\zeta')\, d\zeta\, d\zeta'$$

and similarly for $N_{ii'}$. After P. A. Hansen had made use of this method in his prize essay on the mutual perturbations of Jupiter and Saturn [1831], Poison spelled it out in more detail in [1833a].

Liouville's aim was to show how to replace Poisson's double integrals by single integrals. I shall follow Liouville and illustrate the method by the great inequality of Saturn produced by Jupiter. Abbreviating $5\zeta - 2\zeta'$ by θ, Liouville wrote the expansion (23) in the form:

$$R(\zeta, \zeta') = A + \sum_{p=1}^{\infty} A_p \cos p\theta + \sum_{p=1}^{\infty} B_p \sin p\theta + \phi(\zeta, \zeta'), \qquad (26)$$

where $\phi(\zeta, \zeta')$ has no terms depending on θ or its multiples. Then, he introduced the new variables (σ, θ) defined by $\zeta = 2\sigma$, $\zeta' = 5\sigma - \frac{\theta}{2}$ by means of which $R(\zeta, \zeta')$ can be written

$$R\left(2\sigma, 5\sigma - \frac{\theta}{2}\right) = A + \sum_{p=1}^{\infty} A_p \cos p\theta + \sum_{p=1}^{\infty} B_p \sin p\theta + \phi\left(2\sigma, 5\sigma - \frac{\theta}{2}\right).$$
$$(27)$$

The last function contains only terms having an integer times σ inside a trigonometric function, so if integrated from $\sigma = 0$ to $\sigma = 2\pi$, they all disappear. Thus, integrating (27), Liouville found

$$A + \sum_{p=1}^{\infty} A_p \cos p\theta + \sum_{p=1}^{\infty} B_p \sin p\theta = \psi(\theta), \qquad (28)$$

where

$$\psi(\theta) = \frac{1}{2\pi} \int_0^{2\pi} R\left(2\sigma, 5\sigma - \frac{\theta}{2}\right) d\sigma. \qquad (29)$$

Replacing θ with $-\theta$ in (28),

$$A + \sum_{p=1}^{\infty} A_p \cos p\theta - \sum_{p=1}^{\infty} B_p \sin p\theta = \psi(-\theta), \qquad (30)$$

and defining

$$\varphi(\theta) = \frac{1}{2}(\psi(\theta) + \psi(-\theta)) \quad \text{and} \quad \Phi(\theta) = \frac{1}{2}(\psi(\theta) - \psi(-\theta)), \qquad (31)$$

he found by (28) and (30)

$$A + \sum_{p=1}^{\infty} A_p \cos p\theta \;=\; \varphi(\theta) \qquad (32)$$

$$=\; \sum_{p=1}^{\infty} A_p \sin p\theta \;=\; \Phi(\theta) \qquad (33)$$

As a first approximation, he put $A_2 = A_3 = \cdots = B_2 = B_3 = \cdots = 0$, so that

$$A + A_1 \cos \theta = \varphi(\theta) \qquad \text{and} \qquad B_1 \sin \theta = \Phi(\theta). \qquad (34)$$

Setting $\theta = 0$ and $\theta = \pi$ successively in the first of these equations, he got, by adding and subtracting, the result

$$A = \tfrac{1}{2}(\varphi(0) + \varphi(\pi)) = \tfrac{1}{2}(\psi(0) + \psi(\pi)) \qquad (35)$$

and

$$A_1 = \tfrac{1}{2}(\varphi(0) - \varphi(\pi)) = \tfrac{1}{2}(\psi(0) - \psi(\pi)), \qquad (36)$$

where he has used the fact that ψ is periodic with period 2π. Similarly, by setting $\theta = \frac{\pi}{2}$ in the last equation in (34), he found:

$$B_1 = \Phi\left(\frac{\pi}{2}\right) = \frac{1}{2}\left(\psi\left(\frac{\pi}{2}\right) - \psi\left(\frac{3\pi}{2}\right)\right). \qquad (37)$$

Thus, Liouville had expressed the coefficients pertaining to the great inequality of Saturn due to Jupiter in terms of the four simple integrals $\psi(0), \psi(\frac{\pi}{2}), \psi(\pi)$, and $\psi(\frac{3\pi}{2})$ defined by (29). He argued that formulas (35)–(37) are in fact correct to the ninth order in the eccentricities.

In the rest of the paper, Liouville considered the general term $i\zeta - i'\zeta'$ and showed how to determine the corresponding A's and B's when $A, A_1, A_2, \ldots, A_q, B_1, B_2, \ldots, B_q$ are considered different from zero. In that case, it is necessary to find ψ at $2q - 1$ points, i.e., to evaluate $2q - 1$ simple integrals. Liouville admitted that when q becomes too large ($q > 4$) the work involved in finding these integrals becomes at least as hard as using Poisson's double integrals.

10. A few months after publishing this simplification of Laplace's and Poisson's work, Liouville made an alternative improvement of the method of perturbations [1836g]. As remarked by Liouville [1836d], the inequalities caused by the first term of (6) are easy to calculate. Therefore, it is enough to deal with the last term

$$R_1 = \frac{m'}{|r - r'|} = \frac{m'}{\sqrt{r^2 + r'^2 - -2rr' \cos\varphi}}, \qquad (38)$$

where φ is the angle between the two radius vectors r and r', and r, r' are their lengths. If the planets describe almost circular orbits lying near the ecliptic, we can replace φ with the difference in their mean anomalies:

$$\varphi = \zeta - \zeta' = (n - n')t + \epsilon - \epsilon' - (\omega - \omega'),$$

and we can substitute for r and r' the semimajor axes a, a':

$$R_1 \sim \frac{m'}{\sqrt{a^2 + a'^2 - -2aa' \cos((n - n')t + \epsilon - \epsilon' - (\omega - \omega'))}}. \qquad (39)$$

In order to determine the variation of a from (7), we need to integrate the function

$$\frac{\partial R_1}{\partial \epsilon} = \frac{2m' \, aa' \sin((n - n')t + \epsilon - \epsilon' - (\omega - \omega'))}{[a^2 + a'^2 - 2aa' \cos((n - n')t + \epsilon - \epsilon' - (\omega - \omega'))]^{3/2}} \qquad (40)$$

with respect to t.

More generally, the variation of all the elliptic elements can, with a suitable choice of unit of time, be reduced to integrals of the form

$$\int Q(t) \cos gt \, dt, \quad \int Q(t) \sin gt \, dt$$

$$\int \left(\int Q(t) \cos gt \, dt \right) dt, \quad \int \left(\int Q(t) \sin gt \, dt \right) dt , \qquad (41)$$

where Q is of the form

$$Q(t) = \frac{1}{[A + A' \cos t]^{\mu/2}} , \qquad (42)$$

μ being an odd number. Liouville's aim in [1836g] was to show how these integrals could be estimated using elliptic integrals.

Writing

$$\cos gt = \cos qt \cos \ell t - \sin t \sin \ell t , \qquad (43)$$

where $g = q + \ell$, $\ell \in \left(-\frac{1}{2}, \frac{1}{2}\right]$ and similarly for $\sin gt$, Liouville reduced the problem to finding the simple and double integrals of the following quantities:

$$Q(t) \cos qt \cos \ell t, \quad Q(t) \sin qt \cos \ell t, \quad Q(t) \sin qt \sin \ell t, \quad Q(t) \sin qt \cos \ell t. \qquad (44)$$

To exemplify his method, I shall consider the simple integral of the first of these quantities. The standard method called for developing $P(t) = Q(t) \cos qt$ into a cosine series:

$$P(t) = \sum_{p=1}^{\infty} H_p \cos pt. \qquad (45)$$

According to Fourier's formulas, the coefficients H_p can be found by

$$H_p = \int P(t) \cos pt \, dt = \int \frac{\cos qt \cos pt}{(A + A' \cos t)^{\mu/2}} dt , \qquad (46)$$

where the integrals are taken from 0 to 2π. That these integrals, including $\int P(t) \, dt$, can be reduced to elliptic integrals, had already been pointed out by Legendre "in his work" (cf. [Liouville 1836g, p. 449]), he probably referred to the *Traité des Fonctions elliptiques*, but I have not been able to locate the exact place]. Having thus estimated the first $i - 1$ coefficients H_p and writing $P' = \sum_{p=1}^{i-1} H_p \cos pt$ and $P'' = \sum_{p=i}^{\infty} H_p \cos pt$, it is easy to find the first term on the right-hand side in the equation:

$$\int_0^t P(t) \cos \ell t = \int_0^t P'(t) \cos \ell t \, dt + \int_0^t P''(t) \cos \ell t \, dt . \qquad (47)$$

This gives a good approximation to the integral on the left-hand side, which was the one we wanted to find. In his paper, Liouville showed that for i sufficiently large one can also estimate the remainder term $\int P''(t) \cos \ell t \, dt$ by elliptic integrals. Indeed,

$$
\begin{aligned}
\int_0^t P'' \cos \ell t \, dt &= \int_0^t \sum_{p=i}^{\infty} H_p \cos pt \\
&= \cos \ell t \sum_{p=i}^{\infty} \frac{p H_p \sin pt}{p^2 - \ell^2} - \sin \ell t \sum_{p=i}^{\infty} \frac{\ell H_p \cos pt}{p^2 - \ell^2}. \quad (48)
\end{aligned}
$$

Since $\ell^2 \leq \frac{1}{4}$ and p is large (larger than i), Liouville approximated

$$
\frac{p}{p^2 - \ell^2} \approx \frac{1}{p} \quad \text{and} \quad \frac{\ell}{p^2 - \ell^2} \approx 0, \quad (49)
$$

so that

$$
\int_0^t P'' \cos \ell t \, dt \approx \cos \ell t \sum_{p=i}^{\infty} \frac{H_p \sin pt}{p} = \cos \ell t \int_0^t P''(t) \, dt. \quad (50)
$$

Replacing $P''(t)$ by $P(t) - P'(t)$, the last integral can be reduced to elliptic integrals, as remarked in connection with (46). This was what Liouville wanted to prove.

He admitted that if i is only around 10 the estimate of the remainder is not always sufficiently exact. He solved this problem by developing $\frac{p}{p^2 - \ell^2}$ and $\frac{\ell}{p^2 - \ell^2}$ in a power series in ℓ. In this case, his method also led exclusively to elliptic integrals.

THE CHRONOLOGICAL DEVELOPMENT OF LIOUVILLE'S IDEAS

11. Already as a student, Liouville's list of "works to read" contained Lagrange's *Mécanique Analytique* and Laplace's *Mécanique Celeste* [Ms 3615 (1)]. However, the first sign of an emerging interest in celestial mechanics to be found in Liouville's notebooks is a note on the attraction from a spherical shell dated June 1834 [Ms 3615 (4), p. 52v]. In October of the same year, he turned to a critical investigation of the method of least squares, concluding that "The method by which Laplace proves that the method of least squares is the one which ought to be used seems completely incorrect to us" [Ms 3615 (4), p. 61 r,v]. In this respect, he was in agreement with Gauss, who had infuriated Legendre maintaining that his own proof of the method was the first rigorous one.

These scattered notes indicate that Liouville began to read the classics in celestial mechanics, in particular, Laplace's *Mécanique Céleste* in the summer of 1834. On April 7, 1835 [Ms 3615 (4), pp. 67v–74v, 80v–82r], he had penetrated so far into the methods that he began to apply them to a new problem. He had noticed that the orbit planes of Jupiter, Saturn, and Uranus cut each other almost along one line, and he asked himself whether this would continue to be the case. He set up differential equations, similar to (7), for the angles between the lines of intersection of each pair of planes and showed that to the first order one of these angles will oscillate between certain limits, whereas the two others will increase beyond all limits. Thus, the present-day configuration of the orbit planes of the three planets will change drastically in the future. In these calculations, Liouville showed a great familiarity with the methods of the *Mécanique Céleste* and also with Pontecoulant's newly published *Théorie Analytique du Système du Monde* [1829–1843], from which he took the numerical values. He was later to criticize the latter treatise severely.

The computations needed in order to solve this particular problem were the only extended numerical calculations that Liouville ever published, but his notebooks contain several unpublished applications of the methods of celestial mechanics to particular concrete problems. In fact, at first Liouville did not publish the work on Jupiter, Saturn, and Uranus either. It remained in his notebooks until 1839, when he dug it up and published it [1839f] as a part of his campaign for the seat in the astronomy section of the *Académie des Sciences*.

12.   After a short note on the three-body problem in June 1835 [Ms 3615 (4), pp. 82v–83v], Liouville seems to have taken a break from celestial mechanics until January the following year, when he returned to the "Perturbations of planets" [Ms 3615 (5), pp. 1r–16v]. He tried out his theoretical considerations on the perturbations of Vesta and Jupiter and during February was finally led to the simple-integral method discussed above (§9). He immediately presented this result to the *Académie des Sciences* on February 29th. It could not originally be applied to the terms for which $i = i'$, but in a supplementary note, included as §9 in the published version [1836d], Liouville extended the method to cover these cases as well.

The note with its supplement was well received by the *Académie* commission consisting of Poisson, Mathieu, and Damoiseau. Poisson, whose results Liouville had improved, acknowledged in the report that:

> Mr. Liouville has been able to simplify it [the determination of the coefficients] greatly by giving a method, which has remained unknown until now, of reducing the approximate value of a double integral to a simple integral.
>
> [Poisson 1836, p. 394]

Therefore, the *Académie* approved of Liouville's paper and decided to print it in its *Mémoires des Savants Étrangers*. However, the decision was not carried out, possibly because Liouville published the note in his *Journal* during the summer of 1836. In the report, Poisson suggested that Liouville apply his method to an inequality in the motion of the earth produced by

Venus, which had been discussed by Airy. This suggestion did not come as a surprise to Liouville, who had already concluded an early draft of [1836d] in his notebooks with the words:

> This very well developed theory may lead to some useful consequences; I must apply it to the inequality produced by the action of Venus on the motion of the Earth, an inequality which Mr. Airy has discovered recently and which has been calculated by Pontécoulant in one of the volumes of the *connaissance des temps.*
>
> [Ms 3615 (5), p. 12r]

Still, Liouville does not seem to have used his method on examples other than the long periodic inequality in the motion of Saturn caused by Jupiter, the example he published in [1836d]. Instead, he turned in March 1836 to the "perturbations produced by Mars in the motion of Vesta" [Ms 3615 (5), pp. 13v–14v], discovering in the process the importance of elliptic integrals. Having approached the perturbations of Ceres with the same methods, he concluded:

> These terms can be expressed, the first of them exactly and the others nearly, by elliptic integrals.
>
> [Ms 3615 (5), p. 15r]

This is the idea he presented to the *Académie* on July 11th and published in December of the same year [1836g]. As pointed out in §10, Liouville's considerations only work for almost circular orbits close to the ecliptic. However, in continuation of the above quote, Liouville claimed:

> One could also take the eccentricity into account.
>
> [Ms 3615 (5), p. 15r]

He returned to this claim in public in 1839, when he summarized his achievements in celestial mechanics; he specified that his method could be generalized at least if the perturbing planet has a small eccentricity.

> It is in fact easy to show that by means of certain particular tables, which one must first construct, they can help calculate immediately, and without successive quadratures, the perturbations produced after an arbitrary given time.
>
> The difficulties of calculation which still have to be overcome depend uniquely on the construction of these tables, which one must work out once and for all.
>
> [Liouville 1839i, p. 698]

Liouville never constructed these tables nor did he explain his generalization in detail. Indeed, after June 1836, Liouville's notebooks reveal only little activity in this area (cf. [3615 (5), pp. 25v–26r]), before he suddenly launched the campaign for the astronomy seat left vacant by the death of Lefrançois Lalande in 1839. Liouville's only publication in this three-year period relating to perturbation theory was an improvement of Lagrange's method of variations of the arbitrary constants. This result is of such importance that it will be analyzed in a separate section (cf. §16).

13.   After Liouville entered the astronomy section of the *Académie des Sciences*, he only published one more note on the theory of perturbations. It dealt with the question of convergence of the series obtained by the various approximation procedures. His investigations on the use of elliptic integrals implicitly touched on this problem, for it provided a method for estimating the remainder. In fact, in his notebooks, Liouville had used it to show the convergence of series pertaining to Ceres and Vesta [Ms 3615 (5), pp. 15r–16v, March or April 1836]. It was a major drawback to the ordinary series expansions that their terms sometimes contained the time variable in algebraic expressions other than the trigonometric functions. These terms will grow in time making an initially convergent series divergent. Lagrange and Laplace (see references in Liouville [1840n]) had devised various ways out of this difficulty, but when Cauchy, about 1840, scrupulously reconsidered the convergence question, he pointed out that their methods were vague and unrigorous. Two weeks after Cauchy had voiced this criticism in the *Comptes Rendus* on August 3, 1840 [Cauchy 1840], Liouville presented a note to the *Académie* in which he pretended to show that:

> even if Lagrange's and Laplace's methods are not rigorous they can at least be made rigorous by means of very slight modifications, which alter neither the procedure nor its essential character, so that by adding some details, perhaps omitted by mistake by the illustrious authors but easy to supply, one entirely satisfies the need for geometric rigor.
>
> [Liouville 1840n, p. 252]

In the introduction to the note, Liouville pointed out that these results also stemmed from the period 1835–1836:

> While working on these very questions four years ago I have, for my part experienced some difficulty in giving an exact account of the mentioned passages from Lagrange and Laplace. However, it seemed to me that after a careful examination I managed to solve the objections one can raise against them. However, being diverted from this work by other studies, I did not finish the edition of it. Mr. Cauchy's observations naturally forced me to take it up again, and I have had the satisfaction of seeing that I could confirm the exactness of all the results contained in my old notes.
>
> [1840n, pp. 252–253]

In the *Comptes Rendus* note, Liouville indicated only briefly how to repair specific problematic points in the *Mécanique Céleste*. For a fuller account, he referred to "my Memoir." This memoir is lost, if it was ever composed. In fact, it is rather unclear from Liouville's note whether the memoir had been completed or only projected. If the latter is the case, it was in all probability never completed, because of unforeseen problems. Indeed, as shown by Poincaré [1892–1899], the series do diverge, but can be ascribed a sum considered an asymptotic series (cf. [Kline 1972, pp. 1103–1109]) and [Schlissel 1976–1977].

On November 2, 1840, Liouville presented an *Addition* to this memoir, but the *Comptes Rendus* does not even contain a summary of it [CR 10, p. 678].

14.   The first result of his own, which Liouville read to the *Bureau des Longitudes* after his election, dealt with planetary motion, namely, with the integrable case of the three-body problem, which he had already studied in 1834.

> Mr. Liouville has examined the motion of a moon which Mr. Laplace has placed in such a way that it always remains in opposition to the sun.
>
>   According to the general result which Mr. Liouville has found, the resulting motion in question (position in a straight line of the sun, the moon, and the earth) will not be stable. The least perturbation will give considerable and very fast disturbances of the relative positions.
>
> [P.V. Bur. Long. September 29, 1841][1]

This result had theological implications. For in the *Exposition du Système du Monde* [1835], Laplace had criticized the "supporters of final causes" for maintaining that the moon was created in order to shine on the earth by night. If that was the aim, Laplace argued, the moon would have been placed in the orbit described above, for one can easily calculate that the shadow of the earth would not reach so far that it could eclipse the moon, and therefore the moon would shine whenever the sun had set. By pointing to the instability of the configuration, Liouville invalidated Laplace's argument. However, in the published version of the paper [1842g], Liouville did not draw any philosophical conclusions, but let the quote from Laplace speak for itself.

It may have been these theological overtones that had provoked a dissertation published in Rome in 1825, entitled *Paucis expenditur cl. Laplace opinio de illorum sententia qui lunam conditam dicunt ut noctu tellurem illuminet* (I have not been able to locate this paper). At the meeting when Liouville presented his ideas to the *Bureau des Longitudes*, one of its members called Liouville's attention to this work, but at the following meeting Liouville informed the *Bureau* that it did not deal with stability at all [PV. Bur. Long., October 6, 1841]. Indeed, in the published paper, Liouville stated that the Roman author had only made the trivial remark that the other planets would perturb Laplace's moon [Liouville 1842g, p. 5].

Liouville's result on Laplace's moon bears witness to his growing interest in questions of stability, which manifested itself in the most interesting way in his investigations of another classical Laplacian theme of celestial mechanics: the shape of the planets. These mostly unpublished works from the period 1834–1842 are discussed in Chapter XI.

15.   When Liouville was elected member of the *Bureau des Longitudes*, one of the arguments in favor of his candidacy was that he could supervise the calculations necessary for producing the astronomical tables in the *Connaissance des Temps* (cf. [Bigourdan 1930, p. A12]). With this in mind, it is remarkable that Liouville did not publish one single paper on planetary motion or perturbation theory after this note on Laplace's moon. Moreover, his oral communications to the *Bureau* only rarely touched these questions. In 1845, he announced that "Mr. Jacobi has discovered a very simple method of calculating secular inequalities" [P.V. Bur. Long., November 5, 1845], and the following year, he discussed the libration of the moon modeled on a pendulum:

> Mr. Liouville communicates the results of a calculation concerning the increase of amplitude which occurs in the very small oscillations of a pendulum whose length continuously diminishes. By applying these formulas to the Moon, with the aim of inquiring if (as one has thought) the libration of this star can in the long run become perceptible as a result of an alteration of the dimensions of this star produced by a decrease of its temperature, Mr. Liouville finds that by assuming an exaggerated value of the given quantities and after millions of years the libration will not have increased by more than a small fraction of its present value, which is imperceptible.
>
> [P.V. Bur. Long., June 3, 1846][2]

After this period, the *Procès Verbaux* reports only a few casual remarks from Liouville on the theory of perturbations. Thus, discussing the old appearances of Halley's comet:

> Mr. Liouville thinks that a calculation of perturbations, which cannot be very long if it is conducted conveniently, will considerably increase the certainty of the obtained results.
>
> [P.V. Bur. Long., June 3, 1846][3]

Finally, in his clashes with Le Verrier in 1852, Liouville announced some alleged mistakes in Le Verrier's planetary theories (Chapter V).

In the *Académie*, Liouville regularly dealt with questions of planetary theory in the prize committees as well as in his reports or discussions of the works of Le Verrier, Delaunay, Pontécoulant, and others. Moreover, his notebooks bear witness to some activity in the field until the mid-1850s. There are notes from 1840 [Ms 3615 (5), p. 62r, p. 69 ff; 3616 (5), p. 39r ff] (lunar theory etc.), 1842 [Ms 3617 (2)], 1848 [Ms 3618 (8)] (on the precession of the equinox), 1850 [Ms 3619 (3)], 1852–1853 [Ms 3620 (1)], 1854–1855 [Ms 3621 (2)] and 1861 [3627 (2)] (planetary theory according to Ptolemaios).

Finally, Liouville also took up astronomical subjects at the *Collège de France* and at the *Faculté des Sciences*. During his first year as a teacher of applied mathematics at the *Collège*, he included the subjects *Pertur-bations planétaires* and *réfractions astronomique*, and in connection with his election to the *Bureau des Longitudes*, he lectured for two semesters (1840–1841) on *Mécanique céleste* (cf. lecture notes [Ms 3616 (5), p. 39r ff, and 3618 (2)]). As a professor of pure mathematics, he returned to the subject once, in 1857–1858 (2), in a course on the *Méthodes d'analyse dont on peut faire usage dans les problèmes de mécanique céleste.*

This activity, however, cannot hide the conspicuous fact that after having been placed in the ranks among professional astronomers in the *Académie* and in the *Bureau des Longitudes*, Liouville almost stopped research on the core of celestial mechanics. His later and more peripherical contributions to astronomy were in rational mechanics and differential geometry and in a work on refraction.

ATMOSPHERICAL REFRACTION

16.   On various occasions during his life, Liouville occupied himself with astronomical refraction, but we only have one publication on this sub-ject from his pen. The *Démonstration d'un théorème de M. Biot sur les réfractions astronomiques près de l'horizon* was published in 1842, after Liouville had lectured on the subject at the *Collège de France*, first during the late 1830s and then during the first semester of 1841–1842 [Ms 3640]. In these lectures and in the published paper, Liouville presented a purely mathematical investigation of refraction in spite of his own observation, written in a review of a paper of Ritter on *Recherches analytique sur le problème des réfractions astronomique*, to the effect that "the subsequent progress of this part of science depends on experiment and observation rather than on the calculus [or calculations]" [Liouville 1839l].

During 1841–1842, he lectured on Laplace's treatment of refraction, as found in the fourth volume of the *Mécanique Céleste*, connecting it to the earlier investigations of Euler, Lagrange, and the later results of Jacobi. The variation of temperature in the atmosphere is only taken into account in the last lectures. Liouville did not flinch from long calculations, but he emphasized the derivation and evaluation of the definite integrals entering this field. In particular, in the 9th lecture, he derived from an integral in the *Mécanique Céleste* a differential equation for the refraction R:

$$i\, r \cos\theta \, \frac{\partial R}{\partial r} = \sin\theta \, \frac{\partial(ir)}{\partial r} \frac{\partial R}{\partial\theta} + r \sin\theta \, \frac{\partial i}{\partial r} , \qquad (51)$$

where r is the distance from the observer to the center of the earth, i is the refractive index of the atmosphere at the place where the observation is made, and θ is the apparent zenith distance of the ray. For a ray close to the horizon, $\theta = \frac{\pi}{2}$, and then the formula is reduced to

$$\frac{\partial R}{\partial \theta} = -\frac{rdi}{d(ir)} ,\tag{52}$$

which is equivalent to a formula Biot had derived in 1839 [Biot 1839]. Liouville published formula (51) in the paper mentioned above, not only because it generalized Biot's argument but also because he believed that his arguments disclosed its "physical origin."

The physical principles used by Liouville stemmed from a particle model of light: (1) The speed of light depends only on the refractive index. (2) Assuming that the atmosphere is built up from shells concentric with the earth, the refracting forces on the light particles will act only in the radial direction, and thus their angular momenta around the center of the earth are preserved, or as Liouville expressed it: the *principe des aires* is valid. Thus, he did not subscribe to Fresnel's new wave theory (cf. Chapter VII, §28).

During the summer of 1850, Liouville returned to astronomical refraction in a long series of notes [3619 (3)], the aim of which he indicated to the *Bureau des Longitudes* on July 25th, shortly after having begun these inquiries:

> One talks about the calculation of astronomical refraction where geometers are in the habit of supposing that the atmosphere is in a state of mean stability. Mr. Liouville announces to the *Bureau* that he is engaged in a work which may shed some light on the question in the case of an atmosphere in motion.
>
> [P.V. Bur. Long., July 25, 1850][4]

Four months later, Liouville announced that his research had led to the desired result:

> Mr. Arago speaks about a memoir on astronomical refraction presented to the *Académie* by Mr. Robert Lefebre. On this occasion, Mr. Liouville announces that he has found a new proof of Mr. Biot's theorem on refractions close to the horizon and a partial differential equation, which he has already published and which contains this theorem as a particular case. The new proof does not suppose that one knows the expression for the refraction. The method can be extended with suitable modifications to an arbitrary atmosphere. One has even results of the same kind for all problems of mechanics where the integral of *force vive* holds.
>
> [P.V. Bur. Long., November 13, 1850][5]

Less than a month later, Liouville discovered a generalization of these ideas:

> Mr. Liouville announces that the method which he explained to the *Bureau* at a previous meeting is even more general than he thought;

it is not necessary that the integral of *force vive* holds, in order that it be applicable. It can be extended to all cases.

[P.V. Bur. Long., December 4, 1850][6]

These indications of Liouville's unpublished research on refraction show that by 1850 this question had become a part of his general ideas on mechanics, to which I shall return in Section 34.

CONCLUSION

17. Almost all of Liouville's works on celestial mechanics stem from a brief period during 1835 and 1836. They have not found a lasting place in the history of the field and would have been passed over here if it had not been for the great influence they exerted on Liouville's career. They made him formally qualified for the astronomy section of the *Académie* and the *Bureau des Longitudes* and contributed to his election, although his greater achievements in other fields were decisive.

Liouville's published and unpublished works bear witness to his great familiarity with the classical and more recent methods in celestial mechanics and to his ability to combine them with other fields, such as elliptic functions, differential geometry, and of course, rational mechanics. In this way celestial mechanics was an important and integral part of Liouville's scientific activity.

Liouville's Theorem "on the Volume in Phase Space"

18. Among physicists, Liouville is celebrated for the important theorem on the constancy of the volume in phase space. In its modern formulation, this theorem deals with physical systems with generalized coordinates $q_1(t), q_2(t), \ldots, q_n(t)$ and generalized momenta $p_1(t), p_2(t), \ldots, p_n(t)$, satisfying Hamilton's equations:

$$\frac{dq_i}{dt} = \frac{\partial H}{\partial p_i}, \quad \frac{dp_i}{dt} = -\frac{\partial H}{\partial q_i}, \quad i = 1, 2, \ldots, n, \tag{53}$$

where the Hamiltonian H is a function of $q_1, q_2, \ldots, q_n, p_1, p_2, \ldots, p_n$. The phase flow is defined to be the one parameter group of transformations of phase space:

$$g^t : (q_1(0), q_2(0), \ldots, p_1(0), p_2(0), \ldots) \to (q_1(t), q_2(t), \ldots, p_1(t), p_2(t), \ldots) \tag{54}$$

where $q_1(t), q_2(t), \ldots, p_1(t), p_2(t), \ldots$ are solutions of Hamilton's equations. Now we can formulate

[Liouville's theorem] The phase flow preserves volume, i.e., for any region D in phase space, we have

$$\text{volume of } g^t(D) = \text{volume of } D.$$

This is the formulation in [Arnold 1978] of a theorem whose fame rests on its great importance in statistical mechanics.

In this section, I shall first explain what Liouville had to do withthis theorem; more precisely, how and when Liouville formulated his version of the theorem and what motivated him. Second, I shall explain the mathematical relation between Liouville's original formulation and the modern one, and finally, I shall analyze how the importance of the theorem in statistical mechanics was discovered and how Liouville's name was attached to it.

19.   In one of Liouville's notebooks [Ms 3615 (5), pp. 28r–29v], one finds a note dated January 1838: *Sur les méthodes d'approximation*. It begins:

Let $\frac{d^3x}{dt^3} = P$ be a given equation where P is of the form $f(t, x, x', x'')$, where $x' = \frac{dx}{dt}$, $x'' = \frac{dx'}{dt}$ &c. The integral will be $x = F(t, a, b, c)$. Suppose

$$
\begin{aligned}
u = \ & \frac{dx}{da} \cdot \frac{dx'}{db} \cdot \frac{dx''}{dc} - \frac{dx}{da} \cdot \frac{dx'}{dc} \cdot \frac{dx''}{db} + \frac{dx}{dc} \cdot \frac{dx'}{da} \cdot \frac{dx''}{db} \\
& - \frac{dx}{db} \cdot \frac{dx'}{da} \cdot \frac{dx''}{dc} + \frac{dx}{db} \cdot \frac{dx'}{dc} \cdot \frac{dx''}{da} - \frac{dx}{dc} \cdot \frac{dx'}{db} \cdot \frac{dx''}{da}.
\end{aligned}
$$

In order to find the differential $\frac{1}{dt}\, du$ one must successively differentiate x, x', x''. However, the differentiation of x leads to replacing x by x' and gives zero as a result; similarly the differentiation of x' gives zero. Thus, one needs only to differentiate x'' or to replace it by P. Therefore,

$$
\begin{aligned}
\frac{du}{dt} = \ & \frac{dx}{da} \cdot \frac{dx}{db} \cdot \frac{dP}{dc} - \frac{dx}{da} \cdot \frac{dx'}{dc} \cdot \frac{dP}{db} + \frac{dx}{dc} \cdot \frac{dx'}{da} \cdot \frac{dP}{db} \\
& - \frac{dx}{db} \cdot \frac{dx'}{da} \cdot \frac{dP}{dc} + \frac{dx}{db} \cdot \frac{dx'}{dc} \cdot \frac{dP}{da} - \frac{dx}{dc} \cdot \frac{dx'}{db} \cdot \frac{dP}{da}.
\end{aligned}
$$

If P is differentiated, one has

$$\frac{dP}{da} = \frac{dP}{dx} \cdot \frac{dx}{da} + \frac{dP}{dx'} \cdot \frac{dx'}{da} + \frac{dP}{dx''} \cdot \frac{dx''}{da},$$

and as a result

$$\frac{du}{dt} = \frac{dP}{dx''} \left(\frac{dx}{da} \cdot \frac{dx'}{db} \cdot \frac{dx''}{dc} - \cdots \right) = u \frac{dP}{dx''}.$$

This quantity is zero if P does not contain x'', and then u is independent of time t.

[Ms 3615 (5), p. 28r]

Thus, Liouville assumed that he knew a complete solution $x = F(t, a, b, c)$ to the differential equation

$$x^{(3)} = P(t, x, x', x''),\tag{55}$$

where a, b, c are arbitrary constants, and he showed that the determinant

$$u = \begin{vmatrix} \frac{\partial x}{\partial a} & \frac{\partial x}{\partial b} & \frac{\partial x}{\partial c} \\ \frac{\partial x'}{\partial a} & \frac{\partial x'}{\partial b} & \frac{\partial x'}{\partial c} \\ \frac{\partial x''}{\partial a} & \frac{\partial x''}{\partial b} & \frac{\partial x''}{\partial c} \end{vmatrix}\tag{56}$$

satisfies

$$\frac{du}{dt} = u \frac{\partial P}{\partial x''}.\tag{57}$$

In particular, $\frac{du}{dt}$ is zero if P does not contain x'', and then u is a constant.

This note shows that Liouville was familiar with the properties of determinants, in particular, that they are multilinear and that they are zero when two rows or two columns are equal. In the published version of the note, Liouville used the Jacobian symbol

$$\sum \left(\pm \frac{dx_1}{da} \frac{dx_2}{db} \cdots \frac{dx_n}{dc} \right)$$

to denote determinants, and in the 1850s he employed Cayleyan double bars (as in (56)) in [1854d] and in his notes [Ms 3637 (13), p. 41v].

20. What motivated Liouville to prove this theorem? The answer can be found on the following page of his notebook, where he treats the differential equation

$$x^{(3)} = P + Q(t)\tag{57}$$

with the Lagrangean method of variation of the arbitrary constants as in §2. The determination of $\frac{da_1}{dt}, \frac{da_2}{dt}, \frac{da_3}{dt}$, or as Liouville has it, $\frac{da}{dt}, \frac{db}{dt}, \frac{dc}{dt}$ from (3) by Cramer's method (Liouville attributed it to Laplace, who had applied it to differential equations) calls for the determinant of (3). This is precisely the quantity u, (56). Therefore, it is of interest in perturbation theory to know how u behaves, in particular, that it is a constant if P does not contain x''. This was the reason why Liouville derived the theorem.

21. Liouville immediately generalized the theorem to a differential equation of arbitrary (nth) order,

$$X^{(n)} = P(t, x, x', \ldots, x^{(n-1)}).\tag{58}$$

and sent it to the *Bureau des Longitudes* on January 17, 1838 (cf. [Ms 3615 (5)]). Shortly thereafter, he discovered that he could apply the same idea to a system of first-order equations:

$$\frac{dx_i}{dt} = P_i(t, x_1, x_2, \ldots, x_m) \qquad i = 1, 2, \ldots, m.\tag{59}$$

If $x_i = x_i(t, a_1, a_2, \ldots, a_m), i = 1, 2, \ldots, m$, is a complete system of solutions, the determinant

$$u(t) = \left| \frac{\partial x_i}{\partial a_j} \right| \tag{60}$$

satisfies

$$\frac{du}{dt} = \left(\sum_{i=1}^{m} \frac{\partial P_i}{\partial x_i} \right) u. \tag{61}$$

Thus, u is a constant if $\sum_{i=1}^{m} \frac{\partial P_i}{\partial x_i}$ is zero. Liouville further remarked that if the arbitrary constants a_i are chosen to be the value of x_i at a particular time, say, $t = 0$, i.e., $a_i = x_i(0)$, so that we have a solution of a Cauchy problem, then

$$u(t) = \left| \frac{\partial x_i(t)}{\partial x_i(0)} \right| = 1 \tag{62}$$

if $\sum_{i=1}^{m} \frac{\partial P_i}{\partial x_i} = 0$, for $u(0)$ is clearly equal to $|I| = 1$, and since $u(t)$ is a constant, it must remain equal to one. Liouville published these theorems the same year in his *Journal* [1838e].

22.   In order to show what this has to do with the theorem on the volume in phase space, we shall apply Liouville's idea to Hamilton's equations (53), which are a system of type (59). If we define

$$\begin{aligned} x_i &= q_i & \text{for} & \quad i = 1, 2, \ldots, n\,, \\ x_{i+n} &= p_i & \text{for} & \quad i = 1, 2, \ldots, n, \end{aligned}$$

Hamilton's equations are transformed into

$$\begin{aligned} \frac{dx_i}{dt} &= \frac{\partial H}{\partial x_{i+n}}, & i = 1, 2, \ldots, n\,, \\ \frac{dx_i}{dt} &= -\frac{\partial H}{\partial x_{i-n}}, & i = n+1, n+2, \ldots, 2n. \end{aligned}$$

Thus,

$$\sum_{i=1}^{2n} \frac{\partial P_i}{\partial x_i} = \sum_{i=1}^{n} \frac{\partial^2 H}{\partial x_i \partial x_{i+n}} - \sum_{i=1}^{n} \frac{\partial^2 H}{\partial x_{i+n} \partial x_i} = 0\,,$$

and therefore according to Liouville's observation, we have

$$u(t) = \left| \frac{\partial x(t)}{\partial x_i(0)} \right| = 1. \tag{62}$$

But, the determinant (62) is precisely the Jacobian determining the ratio of volume in phase space:

$$dx_1(t)\, dx_2(t) \ldots dx_{2n}(t) = \left| \frac{\partial x_i(t)}{\partial x_i(0)} \right| dx_1(0)\, dx_2(0) \ldots dx_{2n}(0). \tag{63}$$

Thus, by integration, we find the modern formulation of Liouville's theorem (cf. §18).

23.   Did Liouville himself discover this formulation of his general theorem? This question cannot be answered affirmatively on the basis of his publications for neither in [1838e] nor in his later publications on rational mechanics did he apply his theorems (61) and (62) to Hamilton's equations. However, in a manuscript from around 1856, he did relate his theorem to Hamilton's equations in two-dimensional phase space (cf. §53), and it seems likely that he was also aware, of its importance in the general case. For not only had Hamilton [1834, 1835; see Prange 1933, pp. 609–611] and Jacobi [1837b] emphasized the canonical form of the equations of motion in connection with the theory of perturbations, Jacobi had also explicitly used Liouville's theorem to prove his theorem on the last multiplier.

In a specialized case, this theorem deals with a system of differential equations

$$\frac{dx_i}{dt} = X_i(t, x_1, \ldots, x_n), \qquad i = 1, 2, \ldots, n. \tag{64}$$

Assume that the expression

$$\sum_{i=1}^{n} \frac{\partial X_i}{\partial x_i} \tag{65}$$

can be transformed into a total derivative of a function with respect to t (e.g., it may be zero). Moreover, assume that $n - 1$ integrals,

$$\varphi_i(t, x_1, \ldots, x_n) = a_i, \qquad i = 2, 3, \ldots, n, \tag{66}$$

are known. Use these integrals to express $x_2, \ldots, x_n$ as functions of t, x_1, and the φ's and insert these expressions into X_1. Since the φ_i's are kept constant equal to a_i, there remains one equation to be solved, namely,

$$dx_1 - X_1 \, dt = 0. \tag{67}$$

Jacobi's theorem shows how to determine an integrating factor - a multiplier - of this equation, that is, a function $N = N(t, x_1)$ such that

$$N \, dx_1 - N X_1 \, dt = 0 \tag{68}$$

is a total differential. I shall paraphrase Jacobi's analysis [Jacobi 1844, §8 and 1866, lecture 12]:

Let $\varphi_1(t, x_1, x_2, \ldots, x_n) = a_1$ denote the last integral. If again $x_2, \ldots x_n$ are expressed as functions of t, x_1, and the constants $\varphi_2, \varphi_3, \ldots, \varphi_n$, then

$$d\varphi_1 = \left(\frac{\partial \varphi_1}{\partial x_1}\right) dx_1 + \left(\frac{\partial \varphi_1}{\partial t}\right) dt = 0,$$

which shows by comparing the first term with (68) that $\left(\frac{\partial \varphi_1}{\partial x_1}\right)$ is an integrating factor of (67). Here, the parentheses announce that the partial derivatives are taken while considering $x_1, t, \varphi_2 \cdots \varphi_n$ as independent variables.

Of course, we do not know φ_1, so this does not help unless we can determine $\left(\frac{\partial \varphi_1}{\partial x_1}\right)$ in another way. To this end, we evaluate the "volume element" $dx_1 \, d\varphi_2 \cdots d\varphi_n$ in two different ways. First, relate it to the original volume element $dx_1 \, dx_2 \cdots dx_n$:

$$dx_1 \, d\varphi_2 \cdots d\varphi_n = \sum \pm \frac{\partial \varphi_2}{\partial x_2} \frac{\partial \varphi_3}{\partial x_3} \cdots \frac{\partial \varphi_n}{\partial x_n} \cdot dx_1 \, dx_2 \cdots dx_n, \qquad (69)$$

where the Jacobian on the right is written in Jacobi's notation. Next, relate it to the volume element $d\varphi_1 \, d\varphi_2 \cdots d\varphi_n$:

$$d\varphi_1 \, d\varphi_2 \cdots d\varphi_n = \left(\frac{\partial \varphi_1}{\partial x_1}\right) dx_1 \, d\varphi_2 \cdots d\varphi_n. \qquad (70)$$

Combining (69) and (70) with the identity

$$d\varphi_1 \, d\varphi_2 \cdots d\varphi_n = \left|\frac{\partial \varphi_i}{\partial x_j}\right| dx_1 \, dx_2 \cdots dx_n, \qquad (71)$$

we get

$$\left(\frac{\partial \varphi_1}{\partial x_1}\right) \sum \pm \frac{\partial \varphi_2}{\partial x_2} \cdots \frac{\partial \varphi_n}{\partial x_n} = \left|\frac{\partial \varphi_i}{\partial x_j}\right|. \qquad (72)$$

According to Liouville's theorem (61), the right-hand side of (72) is equal to

$$u = \left|\frac{\partial \varphi_i}{\partial x_j}\right| = \exp \int \sum_{i=1}^{n} \frac{\partial X_i}{\partial x_i} \, dt,$$

which we assumed to be known. We also assumed $\varphi_2, \ldots, \varphi_n$ and therefore $\sum \pm \frac{\partial \varphi_2}{\partial x_2} \cdots \frac{\partial \varphi_n}{\partial x_n}$ to be known, and consequently, we have determined the integrating factor

$$\left(\frac{\partial \varphi_1}{\partial x_1}\right) = \left(\sum \pm \frac{\partial \varphi_2}{\partial x_2} \cdots \frac{\partial \varphi_n}{\partial x_n}\right)^{-1} \exp \int \sum_{i=1}^{n} \frac{\partial X_i}{\partial x_i} \, dt. \qquad (73)$$

Both in his first published proof [Jacobi 1844, §7] and in his lectures [Jacobi 1866, lecture 12], Jacobi referred to the original paper by "Cl$^\mathrm{O}$ Liouville."

It is almost unthinkable that Liouville should not have heard of this application of his theorem. Indeed, the first time Jacobi pronounced his principle was in a speech given before the *Académie des Sciences* during a visit to Paris in the summer of 1842 [Jacobi 1842]. Considering Liouville's great admiration for Jacobi and his works in rational mechanics (to which I shall return in the next section), it is highly probable that Liouville showed Jacobi his usual hospitality, and discussed these new ideas with him. This would explain how Jacobi became aware of Liouville's short note, and also why around this time Liouville repeated his theorem in one of his note-books, adding the comment "Revise this" [Ms 3617 (2), p. 32v].

24.   The time has come to remove a widespread misunderstanding ex-pressed as follows in *Mayer's Encyclopädisches Lexikon* (Mannheim 1975) "Liouville...studied among other things problems of...statistical mechan-ics."

This idea has arisen because Liouville's theorem is of the greatest im-portance in statistical mechanics. Still, the idea that Liouville should have contributed to this theory is absurd. It is indeed unlikely that he ever heard of this new approach to the theory of heat, which was being con-ceived by Maxwell, Gibbs, and Boltzmann toward the end of Liouville's life. For Liouville, the theorem was originally just a lemma in the theory of perturbations, and if or when he realized its applicability to Hamilton's equations, he probably never thought of it as dealing with volume in phase space, let alone with the theory of heat. Thus, Liouville never came to realize what we today consider the true importance of his theorem.

25.   This raises the historical question: How did it happen that Liouville's rather pedestrian observation on perturbation theory was seen to have im-plications in statistical mechanics?

As could have been expected, Liouville's paper had fallen into oblivion when Maxwell and Boltzmann began their work, and therefore they redis-covered Liouville's theorem independently. In special cases, it was derived by Maxwell in his second discussion of the statistical distribution of the velocities of monatomic gas molecules in thermal equilibrium [1868] and by Boltzmann for other physical systems [1868]. Here, I shall illustrate the use of Liouville's theorem in statistical mechanics with a summary of its first general appearance in Boltzmann's *Über das Wärmegleichgewicht zwischen mehratomigen Gasmolekülen* [1871a]. He described the r material points of a molecule by their coordinates $x_1, x_2, \ldots, x_{3r}$ and their velocities $x_{3r+1}, x_{3r+2}, \ldots, x_{6r}$. In the subsequent discussion, it will be convenient to denote the space R^{6r}, to which $(x_1, \ldots, x_{6r})$ belongs, the phase space, although Boltzmann does not introduce such a concept explicitly, and al-though in modern texts phase space is a coordinate-momentum space.

Let

$$dN = f(t, x_1(t), \ldots, x_{6r}(t)) \, dx_1(t) \cdots dx_{6r}(t) \qquad (74)$$

denote the number of molecules having

$$x_i \in [x_i(t), x_i(t) + dx_i(t)], \qquad i = 1, 2, \ldots, 6r, \qquad (75)$$

at time t, where f is proportional to the probability distribution in the phase space R^{6r}. Boltzmann wanted to determine this distribution in ther-mal equilibrium, that is, in a situation where f does not depend explicitly on t.

At time $t + \delta t$, the same dN molecules will occupy the following volume in phase space:

$$\{\overline{x} \in R_{6r} | x_i \in [x_i(t + \delta t), x_i(t + \delta t) + dx_i(t + \delta t)]\},$$

and therefore

$$
\begin{aligned}
dN &= f(t, x_1(t), x_2(t), \ldots, x_{6n}(t)) \, dx_1(t) \, dx_2(t) \cdots dx_{6r}(t) \\
&= f(t + \delta t, x_1(t + \delta t), \ldots, x_{6r}(t + \delta t)) \, dx_1(t + \delta t) \cdots dx_{6r}(t + \delta t).
\end{aligned}
\tag{76}
$$

Thus, the behavior of f is determined by the Jacobian

$$\frac{dx_1(t + \delta t) \cdots dx_{6r}(t + \delta t)}{dx_1(t) \cdots dx_{6r}(t)} = \left| \frac{\partial x_i(t + \delta t)}{\partial x_j(t)} \right|. \tag{77}$$

First, Boltzmann assumed that the molecules did not interact, so that the evolution of the x_i's is governed by Hamiltonian equations of the form

$$\frac{dx_i}{dt} = \kappa_i, \qquad i = 1, 2, \ldots, 6r, \tag{78}$$

where the first half of the κ_i's are only functions of the last haft of the x_i's (the velocities), and the last half of the κ_i's are only functions of the first half of the x_i's (the positions). Writing

$$x_i(t + \delta t) = x_i(t) + \delta x_i,$$

he developed the Jacobian to the first order in the δx_i's and found

$$\left| \frac{\partial x_i(t + \delta t)}{\partial x_j(t)} \right| = 1 + \sum_{i=1}^{6r} \frac{\partial \delta x_i}{\partial x_i}, \tag{79}$$

and since by (78) $\delta x_i = \kappa_i \delta t$, we have

$$\left| \frac{\partial x_i(t + \delta t)}{\partial x_j(t)} \right| = 1 + \sum_{i=1}^{6r} \frac{\partial \kappa_i}{\partial x_i} \delta t = 1, \tag{80}$$

because κ_i does not depend on x_i. By (80), the "volume" in phase space is conserved from time t to time $t + \delta t$ and thus forever. This is Liouville's theorem.

By (76), this implies that

$$f(t_1, x_1(t_1), \ldots, x_{6r}(t_1)) = f(t, x_1(t), \ldots, x_{6r}(t)),$$

and thus in thermal equilibrium,

$$f(x_1(t_1), \ldots, x_{6r}(t_1)) = f(x_1(t), \ldots, x_{6r}(t)). \tag{81}$$

Thus, the probability distribution is constant along the trajectory of any molecule. This was an important step in the determination of f. It is outside the scope of this book to follow Boltzmann's determination of f when the molecules interact.

26.   Boltzmann's derivation of Liouville's theorem was less satisfactory than Liouville's original theorem, because Boltzmann did not show that the higher order terms, discarded in (79), do not add up in the final result, and second, because equations (78) were less general than Liouville's. However, in *Einige allgemeine Sätze über Wärmegleichgewicht* from later the same year [1871b], Boltzmann derived the general theorem (61), and, moreover, showed that it was equivalent to Jacobi's principle of the last multiplyer.

René Dugas [1959] has analyzed Boltzmann's work on Liouville's theorem in more detail. However, the reader must be warned that when Dugas claims that Boltzmann saw in 1868 that the theorem (80) was "a special case of Liouville's theorem"[Dugas 1959, p. 147] and that Boltzmann extended Liouville's theorem [Dugas 1959, pp. 151–152] he refers to the modern version of Liouville's theorem and not to Liouville's original paper, which does not even appear in the bibliography. In fact, Boltzmann did not generalize Liouville's theorem further than (61), and he did not refer to Liouville in his 1868 paper, neither did Maxwell [1879], when he clarified Boltzmann's ideas.

27.   Still, as soon as the connection between Liouville's theorem and Jacobi's last multiplyer had been established, the realization of Liouville's priority was not far away. Indeed, in Kirchhoff's *Vorlesungen über die Theorie der Wärme*, held five times during the years 1876–1884 in Berlin, and published posthumously by Max Planck in 1894, it is remarked that Jacobi in his *Vorlesungen über Dynamik* attributed the theorem to Liouville (cf. §23). Four years later, in his second part of the *Vorlesungen über Gastheorie*, Boltzmann [1896–1898] baptized the theorem "Liouville's theorem."

To conclude, it was thanks to a reference in Jacobi's famous *Vorlesungen über Dynamik*, that Liouville's obscure paper became known to statistical mechanicians. If it had not been for this reference, the theorem would probably have been known as Boltzmann's theorem.

Rational Mechanics

28.   Liouville is still remembered in rational mechanics for two discoveries: (1) Liouville's theorem stating that the knowledge of n independent integrals in involution is enough to solve Hamilton's equations of motion in $2n$ dimensional phase space by quadrature (cf. [Arnold 1978, p. 271 ff]) and (2) a particular class of potentials for which the equations of motion can be integrated, called "Liouville's case of complete integrability" [Campbell 1971, p. 87]. Liouville discussed the latter discovery in three voluminous

papers from 1846 and 1847 [1846i, 1847h,i] in which he drew a great many
geometrical conclusions and in a note [1848a]. Liouville's theorem, on the
other hand, was published in a brief note in 1855. It was one of a series
of notes dealing with fundamental questions in rational mechanics and differential equations encompassing the three long memoirs in time. The first
note appeared in 1840 [1840f] (see also [1844c]), but the majority were published in the mid-1850s [1855c,d,h, 1856f,g and k, cf. also 1856p (composed
shortly after 1840) and 1857q] as a spinoff from a lecture he held at the
Collège de France during the summer of 1853.

I shall first discuss the shorter notes, in particular, Liouville's theorem, a
geometrization of the principle of least action, and an unpublished generalization of the methods of mechanics, and end with an analysis of Liouville's
longer memoirs. However, before discussing Liouville's contributions, I shall
summarize the advances made prior to his entry on the scene, in particular,
the ideas of Jacobi, which were at the origin of all of Liouville's investigations in this area.

The Hamilton-Jacobi Formalism

29.   The formation of the basic notions and principles of mechanics has
been a central theme for historians of physics [cf. Szabó 1977]. Historians of
mathematics, on the other hand, have been attracted by the mathematical
clarifications of mechanics, in particular, its importance to the development of differential equations [e.g., Demidov 1980, 1981]. The borderland
between the history of physics and the history of mathematics is surveyed
in Dugas's fine *Histoire de la Mécanique* [1950], but since it covers the
whole period from the ancient Greeks to quantum-mechanics, only a little
space is devoted to Hamilton-Jacobi theory. Thus, here is a lacuna to be
filled. To be sure, Prange's great contribution [1933] to the *Encyclopädie
der Mathematischen Wissenschaften* and Whittaker's book [1937] contain
a wealth of historical information, but since their aims are purely scientific,
they do not tell a connected story. It is my plan to let this be my next
research project. Until then, I shall leave it as a summary of those aspects
of the development that directly influenced Liouville's contributions.

30.   Inspired by his works in optics [1828], William Rowan Hamilton (1805–
1865) gave mechanics a new mathematical twist in his two papers, *On a
general method in dynamics; by which the study of the motions of all free
systems of attracting or repelling points is reduced to the search and differentiation of one central relation or characteristic function* [1834,1835]. It
is most convenient to start with a summary of the second of these papers.
Being of the opinion that constraints were going out of fashion [Hamilton
1834, Introduction and §8], Hamilton limited his theory to free particles.
Since he used generalized coordinates and stated [Hamilton 1834, §8] that

his ideas were also applicable when holonomic constraints were acting, I shall at once use this general approach.

Consider a conservative mechanical system as in §4. Hamilton defined the generalized momenta p_i as Poisson had done earlier (18) and considered T as a function of the generalized coordinates and momenta. Having introduced the Hamiltonian

$$H(p_1, \ldots, p_n, q_1, \ldots, q_n) = T + U, \qquad (82)$$

he showed that the equations of motion (12) or (13) take the form

$$\dot{q}_i = \frac{\partial H}{\partial p_i}, \quad \dot{p}_i = -\frac{\partial H}{\partial q_i}. \qquad (83)$$

We would say that Hamilton took the Legendre transform of the Lagrangian, viewed as a function of $\dot{q}$ [cf. Arnold 1978, pp. 61–65].

Here, and in the subsequent development, I shall use U in the modern sense introduced by Lagrange (who called it V). Hamilton and his successors, Jacobi and Liouville, however, used a potential function whose gradient equals the total force on the free system. It therefore has the opposite sign of the U in this section. According to our convention, U denotes the potential energy so that H represents the total energy of the system, which Hamilton assumed to be constant.

The form (82) of the equations of motion was later called "canonical" by Jacobi [1837c, p135]. Poisson, as we saw in §6, had already derived half of these equations and Lagrange [1788, 1811–1815, vol. 1, Part 2, 5th sect., §14] had expressed the fundamental equations of perturbation theory in this form (cf. [Prange 1933, note 90] and [Whittaker 1937, p. 280] for more references to early appearances of Hamilton's equations). Still, it was through Hamilton's paper that "Hamilton's equations" (83) became known. Hamilton himself did not value this canonical form particularly highly [Conway and Mc Connell's introduction to Hamilton's *Mathematical Papers* vol. 2., p. x]. The main purpose of his papers was, as indicated in the title, to reduce their solution to the determination of one solution of two partial differential equations.

31.   To this end, he let $q_1^o, \ldots, q_n^o$ and $q_1, \ldots, q_n$ be two configurations of the system; consider the integral

$$\int_0^t (T - U)\, dt = \int_0^t L\, dt$$

along any virtual path taking the system from the first configuration at time 0 to the second at time t. Varying the virtual path for fixed t, the vanishing of the first variation yields the Lagrange equations (12) or (13). Thus, the actual motion of the system makes (84) an extremum. Although often called Hamilton's principle, this variational principle was known before Hamilton (cf. [Voss, 1901, nr. 42, note 243]). Hamilton, however, was the first to investigate the extremal integral (i.e., the integral along the actual path) as a function of t, $q_1^o, \ldots, q_n^o$, and $q_1, \ldots, q_n$:

$$S(t, q_1, \ldots, q_n, q_1^o, \ldots, q_n^o) = \int_0^t (T - U)\, dt = \int_0^t L\, dt. \tag{84}$$

Denoting the corresponding generalized momenta by p_i^o and p_i, respectively, Hamilton showed that the total derivative of S is of the form

$$dS = \sum_{i=1}^n (p_i\, dq_i - p_i^o\, dq_i^o) - H\, dt. \tag{85}$$

Hamilton called this the law of varying action. It implies that

$$\frac{\partial S}{\partial q_i} = p_i, \tag{86}$$

$$\frac{\partial S}{\partial q_i^o} = -p_i^o, \tag{87}$$

and finally

$$\frac{\partial S}{\partial t} = -H. \tag{88}$$

Hamilton remarked that given S equations (86) and (87) are $2n$ equations to be satisfied by q_i and p_i, and since they contain the $2n$ arbitrary constants q_i^o and p_i^o, they must be a complete set of integrals to the equations of motion (83). Thus, integration of Hamilton's equations is equivalent to the determination of S.

A priori S was defined as an integral along the path of the system, and thus by its definition presupposes that we have integrated (83). However, Hamilton discovered that by (86), (87), and (88) S satisfies the two partial differential equations

$$\frac{\partial S}{\partial t} + H\left(\frac{\partial S}{\partial q_1}, \ldots, \frac{\partial S}{\partial q_n}, q_1, \ldots, q_n\right) = 0 \tag{89}$$

and

$$\left.\frac{\partial S}{\partial t}\right|_o + H\left(-\frac{\partial S}{\partial q_1^o}, \ldots, -\frac{\partial S}{\partial q_n^o}, q_1^o, \ldots, q_n^o\right) = 0 \tag{90}$$

Thus, the solution of system (83) is equivalent to solving (89) and (90). Hamilton called S the principal function. He discovered it while finishing off the first of his papers [Hamilton 1834, §23] and developed the idea in the second paper.

32.   In the first paper, he had instead considered the so-called characteristic function

$$V(h, q_1, \ldots, q_n, q_1^o, \ldots, q_n^o) = \int_0^t T\, dt, \tag{91}$$

where h is the total energy H. It is the Legendre transform of S:

$$V = t\,H + S. \tag{92}$$

According to the celebrated principle of least action, vaguely indicated by Maupertuis (1698–1759) and clarified by Euler and Lagrange [Szabó 1977 p. 92–107], the actual path of the system minimizes the integral (91), when the virtual path between $\overline{q_i^o}$ and $\overline{q_i}$ and the time t is varied but the total energy h is kept fixed. The characteristic function is this extremum. It has properties analogous to S: it satisfies the two partial differential equations

$$H\left(\frac{\partial V}{\partial q_1}, \ldots, \frac{\partial V}{\partial q_n}, q_1, \ldots, q_n\right) = h \tag{93}$$

and

$$H\left(-\frac{\partial V}{\partial q_1^o}, \ldots, \frac{\partial V}{\partial q_n^o}, q_1, \ldots, q_n\right) = h. \tag{94}$$

When V is a solution of (93) and (94), the following equations for q_i, p_i yield a complete set of integrals to the equations of motion (83):

$$\frac{\partial V}{\partial q_i} = p_i\,, \tag{95}$$

and

$$\frac{\partial V}{\partial q_i^o} = -p_i^o. \tag{96}$$

Moreover, the equation

$$\frac{\partial V}{\partial h} = t \tag{97}$$

determines how the system progresses along its path in time.

In his two papers, Hamilton showed how this new approach could facilitate perturbation theory.

33.   Inspired by Lagrange and Pfaff, Jacobi had in his youth [1827a,b] occupied himself with the reduction of the solution of systems of first-order ordinary differential equations. For that reason, he found Hamilton's work highly interesting, but he took exception to the need for two partial differential equations, (89)-(90) and (93)-(94). Indeed, he showed [Jacobi 1837b] that if $S(t, q_1, q_2, \ldots, q_n, a_1, \ldots, a_n)$ is any complete solution to the first equation (89) containing, in addition to the obvious additive constant, n other arbitrary constants $a_1, \ldots, a_n$ then the equations

$$\frac{\partial S}{\partial a_i} = b_i, \quad i = 1, 2, \ldots, n\,, \tag{98}$$

and

$$\frac{\partial S}{\partial q_i} = p_i, \quad i = 1, 2, \ldots, n\,, \tag{99}$$

where $b_1, \ldots, b_n$ are other arbitrary constants, are a complete system of integrals to the equations of motion. If $a_i, \ldots, a_n$ are chosen to be the initial values of $q_1(t), \ldots, q_n(t)$, S is Hamilton's principal function (84), but as Jabobi's theorem shows, any complete integral of (89) will do as well.

Similarly, if $V(h, q_1, q_2, \ldots, a_1, \ldots, a_n)$ is a complete solution to (93) containing, in addition to the obvious additive constant, n arbitrary constants $a_1, a_2, \ldots, a_n$, then

$$\frac{\partial V}{\partial a_i} = b_i, \qquad i = 1, 2, \ldots, n \tag{100}$$

and

$$\frac{\partial V}{\partial x_i} = p_i, \qquad i = 1, 2, \ldots, n \tag{101}$$

are a complete system of integrals of the equations of motion. The first-order partial differential equation (89) is usually called the Hamilton-Jacobi equation.

Moreover, Jacobi extended the method to potentials and thus Hamiltonians, depending explicitly on time, as, for example, in the famous Jacobian three-body problem: one (small) mass attached to two moving (large) masses, when the motion of the two latter are known. In this case the Hamilton-Jacobi equation is

$$\frac{\partial S}{\partial t} + H\left(\frac{\partial S}{\partial q_1}, \ldots, \frac{\partial S}{\partial q_n}, q_1, \ldots, q_n, t\right) = 0. \tag{102}$$

Most often, Jacobi discussed n free particles and thus wrote the equation on the form:

$$\frac{\partial S}{\partial t} + \frac{1}{2} \sum \frac{1}{m_i}\left[\left(\frac{\partial S}{\partial x_i}\right)^2 + \left(\frac{\partial S}{\partial y_i}\right)^2 + \left(\frac{\partial S}{\partial z_i}\right)^2\right] + U = 0. \tag{103}$$

In fact, the general equation (102) and the like did not occur in the early development of the Hamilton-Jacobi formalism, so the transformation of (103) into coordinates other than the Cartesian ones became a constant struggle both for Jacobi and Liouville. In the subsequent discussion, however, I shall often play down this aspect of the development.

Liouville's Theorem and a Precursor

34.   At this point, I shall interrupt the chronology in order to explain Liouville's theorem. For with the Hamilton-Jacobi formalism at hand, we are able to understand it; and by introducing it now, I can focus the continuation of the story on those aspects that throw Liouville's achievement into perspective.

In a one-page communication to the *Bureau des Longitudes* on June 29, 1853, republished in his *Journal* two years later [1855d], Liouville stated

[Liouville's Theorem] Consider Hamilton's equations

$$\dot{q}_i = \frac{\partial H}{\partial p_i}, \quad \dot{p}_i = -\frac{\partial H}{\partial q_i}, \qquad i = 1, \ldots, n, \tag{83}$$

where H is a function of t and the q_i's and the p_i's. Assume that we have found half of their integrals:

$$\varphi_i(t, p_1, \ldots, p_n, q_1, \ldots, q_n) = a_i, \qquad i = 1, 2, \ldots, n, \tag{104}$$

and assume that the Poisson bracket of any two of them is zero:

$$(\varphi_i, \varphi_j) = \sum_{k=1}^{n} \frac{\partial \varphi_i}{\partial p_k} \frac{\partial \varphi_j}{\partial q_k} - \frac{\partial \varphi_i}{\partial q_k} \frac{\partial \varphi_j}{\partial p_k} = 0 \qquad \text{for} \quad i \neq j \tag{105}$$

(nowadays we say that φ_i and φ_j are in involution). Finally, assume that the integrals are independent in the sense that they can be solved (theoretically) to yield p_i as functions of t, the q_i's and the constants a_i.

Then, the quantity

$$\sum_{i=1}^{n} p_i \, dq_i - H \, dt \tag{106}$$

is an exact differential form. If $S(t, q_1, q_2, \ldots, q_n, a_1, a_2, \ldots, a_n)$ denotes its integral, the following $2n$ equations produce a complete set of integrals to Hamilton's equations:

$$\frac{\partial S}{\partial q_i} = p_i \quad , \quad \frac{\partial S}{\partial a_i} = b_i. \tag{107}$$

Thus, if half of the integrals in involution are known, the equations of motion can be completely solved by integrating an exact differential, i.e., by quadratures. By an exact form, Liouville and his contemporaries meant a form $\sum y_i \, dx_i$ for which

$$\frac{\partial y_i}{\partial x_j} = \frac{\partial y_j}{\partial x_i}.$$

Problems with surfaces, that are not simply connected, had not yet crept into mathematics. In modern language, a_i and b_i are called the action-angle coordinates of the system [Arnold 1978, pp. 279–285].

This theorem clearly grew out of the Hamilton-Jacobi formalism. For, the function S, defined by

$$dS = \sum_{i=1}^{n} p_i \, dq_i - H \, dt, \tag{108}$$

is the Hamilton-Jacobi function; it satisfies $\frac{\partial S}{\partial t} = H$ and $\frac{\partial S}{\partial q_i} = p_i$ and therefore satisfies the Hamilton-Jacobi equation (102). Thus, equations (107) are nothing but Jacobi's complete integrals (98) and (99).

What then are the principal components in Liouville's discovery? First, the realization that in order to determine S it is enough to know n integrals, which makes $\sum_i p_i \, dq_i - H \, dt$ exact. Second, that if n integrals are in involution, they possess this property. I shall show below that the first of these ideas had been published earlier by Poisson and that the second was in the air, and had even published by Donkin in the period between Liouville's presentation of the note to the *Bureau des Longitudes* and its publication.

35.   First, however, I shall summarize Liouville's very first published note on rational mechanics [1840f], which contained his theorem in a special case. He considered a system with two degrees of freedom and a Hamiltonian of the special form

$$H = \lambda(t) H_1(p_1, p_2, q_1, q_2). \tag{109}$$

In this case, $H_1 = a_1$ is obviously an integral of Hamilton's equations. Moreover, from a theorem of Lagrange, Liouville deduced that $\varphi(p_1, p_2, q_1, q_2) = a_2$ is a time-independent integral to these equations if and only if

$$(H_1, \varphi) = 0 \tag{110}$$

(or equivalently if $(H, \varphi) = 0$). I do not know if this is the first appearance of this important fact.

Liouville now showed that if he used the two integrals, H and φ, to express p_1 and p_2, as functions of q_1, q_2, a_1 and a_2, then $p_1 \, dq_1 + p_2 \, dq_2$ is an exact differential whose integral he denoted θ:

$$\theta = \int p_1 \, dq_1 + p_2 \, dq_2 \,.$$

The two remaining integrals are then

$$\frac{\partial \theta}{\partial a_2} = b_2 \quad \text{and} \quad \frac{\partial \theta}{\partial a_1} = \int \lambda(t) \, dt + b_1 \,. \tag{111}$$

This is Liouville's theorem in this particular case, except Liouville had not added $-H \, dt$ to the form and, therefore, got the more inconvenient expression for the last of the integrals in (111). It is interesting to note that already in this early paper the importance of the vanishing Poisson bracket must have become clear to Liouville. For in his proof of the exactness of the form, i.e., that

$$\frac{\partial p_1}{\partial q_2} = \frac{\partial p_2}{\partial q_1} \,, \tag{112}$$

he showed that this equality is equivalent to the vanishing bracket (110). Still, it is impossible to say whether he possessed a more general version of his theorem at this time.

The above observation was a generalization of a theorem of Jacobi, which had also been discussed by Poisson. Accordingly, I shall return to the historical development of rational mechanics with the following question in mind: To what extent was Liouville's theorem perceived by others before Liouville published it in 1855? How much of this insight could have inspired Liouville? In what context did Liouville find and publish his theorem?

I shall discuss Poisson's contribution before turning to the chronology of Jacobi's publications, their impact on Liouville's work, and Jacobi's knowledge of Liouville's theorem. Finally, I shall summarize the developments in France that prompted Liouville to publish the general version of his theorem and other important ideas in rational mechanics.

Poisson on Liouville's Theorem

36.   A very interesting and often ignored link between Hamilton and the Parisians, by-passing Jacobi, was published in 1837 by Poisson in Liouville's *Journal* [Poisson 1837a]. Poisson began his paper by showing that for a system with holonomic constraints when expressed in generalized coordinates Hamilton's generating function $V = \int_0^t T \, dt$ yields the complete set of integrals (95) and (96). Hamilton, we recall, only treated free particles. Then, Poisson made the following very interesting remark:

> However, despite this extension, its applicability is very limited and almost non-existent if, as the author [Hamilton] suggests, one uses it to find all the first and second integrals of the motion and if one has to find the function V a priori, without knowing any of these integrals. When one knows a convenient number of first integrals, this method may help the integration.
> [Poisson 1837a, p. 325 (333 in the misprinted pagination)]

The method he had in mind was precisely the one used in Liouville's theorem. If so many integrals are known that one can determine the p_i's as functions of the q_i's and constants a_i and *if* $\sum p_i \, dq_i$ is exact and equal to dS, then (107) produces the complete set of integrals. At least this is a simplified version of Poisson's argument. In fact, like Hamilton in his first paper, Poisson did not use the canonical but the Lagrangean approach (q_i and $\dot{q}_i$); moreover, he assumed that half or more of the integrals are known in advance, and since he did not know of Jacobi's reformulation of Hamilton's approach, he had some trouble relating the equations $\frac{\partial V}{\partial a_i} = b_i$ to Hamilton's integrals (96), where the a_i's are necessarily equal to q_i^0 instead of being arbitrary. Moreover, he did not know of Hamilton's second paper, so he had to invent the Legendre transform S of V for himself. These

technical details are similar to some of the innovations that Jacobi simultaneously worked out in a much clearer way. Thus, in the 1837 paper, Poisson
very clearly put forward what I called (§34) the first principal component
of Liouville's theorem: If p_i can be determined so as to make $\sum p_i \, dq_i$ exact,
then the integration can be finished by quadrature. However, Poisson did
not address the second principal question; what must the given integrals
satisfy in order to make $\sum p_i \, dq_i$ exact?

Liouville viewed his theorem as an answer to this question, raised implicitly by Poisson's paper. That can be seen from his announcement of his
theorem in 1855:

> ...half of the integrals. In fact, it is by equations like $(\alpha, \beta) = 0$
> that one expresses the conditions of integrability required by Poisson
> (*Journal de Mathématique*, vol. II, 1837) in order to determine a
> function which afterward will provide the remaining integrals.
>
> [Liouville 1855c, 141, p. 136]

The Publication of Jacobi's Ideas

37.   When Poisson wrote the paper discussed above, he knew of Jacobi's
first communication to the Paris *Académie* [1836a] dealing with mechanics.
Poisson showed that the example treated in Jacobi's note was a special
case of his general formulas, but he could not know that Jacobi was on
the track of something even more far-reaching. For, Jacobi did not refer to
Hamilton and still less to his own ideas, although one year later he stated
that the example was "suitable to illustrate the spirit and the utility of the
new method" [Jacobi 1837c].

Jacobi's 1836 note dealt with the example discussed in §35, limited to
Hamiltonians that do not depend explicitly on time. Jacobi solved the
problem by way of the exactness of $\sum p_i \, dq_i$, but unlike Liouville, he did
not notice the importance of the Poisson bracket. Thus, Liouville's 1840
note [1840f] was a generalization of Jacobi's [1836a].

38.   The "new method" itself was described for the first time by Jacobi in
a letter dated November 29, 1836, to the secretary of the Berlin *Akademie
der Wissenschaften*, the astronomer Encke [1837a]. According to Jacobi,
Hamilton's work had a twofold attraction. First, it extended his own and
Pfaff's earlier works on first-order partial differential equations by reducing those of the form (102) to one system of ordinary differential equations;
second, by analogy with Lagrange's treatment of first-order partial differential equations, Jacobi could show that the knowledge of one integral to
Hamilton's equations would allow him to reduce the order not only by one,
as would be expected, but by two. However, both in this letter and in the
Note sur l'intégration des équations différentielles de la dynamique [1837c],

with which he also presented the letter to the Paris *Académie* one year later, Jacobi only described this wonderful procedure of solution in general and rather vague terms. In fact, the details of Jacobi's second method, as it is now called, were not published until after his death, when A. Clebsch brought these and other papers from Jacobi's *Nachlass* to light [cf. Jacobi 1866].

Still, Jacobi published many other aspects of his theory. Thus, at the end of 1837, Crelle's *Journal* carried his version of the Hamilton-Jacobi formalism, as explained in §33 [1837b]. This paper and the letter to Encke, which had also been published by Crelle, caught Liouville's interest. In the introduction to the 1838 volume of his *Journal*, Liouville announced that because of "the importance of the subject and the name of the author" he had asked one of his friends [L. J. 3, p. 44] to translate them for his *Journal* [Liouville 1838a], and indeed they appeared in the February and April issues. Motivated by these papers and a paper by Binet, Liouville published a short note, *Sur l'intégration d'une classe d'equations différentielle* [1838b].

39.   However, Liouville did not come up with really interesting results until Jacobi communicated a third important note to the *Académie des Sciences*. It concerned the Poisson bracket. As explained in §6, Poisson had shown that the bracket of two integrals is independent of t. However, it was left to Jacobi to unveil the importance of "Mr. Poisson's very profound discovery" [Jacobi 1840].

In this communication, provoked by an obituary notice by Humboldt about the recently deceased Poisson, Jacobi informed the Paris *Académie* that given two integrals their bracket, according to Poisson's theorem, would be another integral. That is, starting with two integrals, one can find a third, and by combining that one with one of the original integrals, one finds a fourth, "and, in general, one manages in this way to deduce from two given integrals all the integrals or equivalently the complete integration of the problem" [Jacobi 1840]. This was a very significant observation, and Jacobi's further works on the brackets turned out to be at the basis of central developments in mathematics, such as Lie group theory [Demidov 1981]. Still, as Liouville pointed out in the *Note de l'éditeur* [1840f], to the reprint of this paper in his *Journal*, Jacobi was a bit optimistic when announcing the result. For, it often happens that the Poisson brackets of two integrals is either identically zero, a constant, or a combination of the two original integrals, in which case its formation does not push the integration further. Liouville continued the note with the example I have already summarized in §35. It can be viewed as an illustration of the fact that the vanishing of the Poisson brackets, as in (110), also gives interesting information, which can be used in the integration of the system.

Although the publication of this note was prompted by Jacobi's communication about Poisson brackets, it was, as we saw in §37, a generalization of an earlier paper by Jacobi. Liouville explained:

> Anxious to include some of Mr. Jacobi's beautiful discoveries in teaching, I wrote the following Note concerning one of these theorems a long time ago. This note has served as the text of one of my Lectures.
>
> [Liouville 1840f, p. 352]

The two preceding semesters, Liouville's lectures at the *Collège de France* had dealt with other subjects [Belhoste and Lützen 1984 p 302], so the note was probably more than one year old.

Thus, despite Liouville's great enthusiasm for Jacobi's work, he had by 1840 only made one small contribution to it. His really significant work did not appear until the mid-1850s. What happened during this 15-year period? I shall first address the question of whether Jacobi discovered Liouville's theorem and then turn to the developments in France.

Jacobi on Liouville's Theorem

40.   Although Jacobi published widely on his new methods of rational mechanics and differential equations [cf. his *Werke* vol IV], it is conspicuous that he never got around to explaining the element that he had stressed with the most enthusiasm in his letter to Encke: the lowering of the order by two when one integral is known. However, his posthumous works, published during the 1860s by Clebsch, contain this method, and since Liouville's theorem is in a sense a corollary, it is not surprising that it can be found in these papers.

Indeed, in the 32nd lecture of his 1842–1843 *Vorlesungen über Dynamik* [Jacobi 1866, p. 252], Jacobi showed that $\sum p_i \, dq_i$ is exact if and only if the n known integrals are in involution. The theorem also figures as Theorem III in the undated, very long *Nova methodus, aequationes differentiales partiales primi ordinis inter numerum variabilium quemcunque propositas integrandi* [Jacobi 1862]. Here, it is characterized as a "theorema gravissimum." Thus, Jacobi knew Liouville's theorem before 1842–1843 and appreciated its importance. In Prange's reconstruction of the development of Jacobi's ideas [1933, pp. 611–615], it plays a key role.

Could Liouville have learned his theorem from Jacobi? This is certainly possible. As I have mentioned, Jacobi visited Paris in 1842, just before giving his lectures on rational mechanics at the University of Köningsberg. It is most probable that the two mathematicians discussed the contents of the lectures at this occasion, and considering the central role played by "Liouville's theorem," their discussions may very well have touched upon it.

In 1847, Liouville once more had occasion to learn about Jacobi's lectures. Indeed, during that year, Borchardt, who had attended Jacobi's lectures and had produced very careful notes that were later published, visited Paris where he was in close contact with Liouville.

If Liouville learned of Jacobi's version of Liouville's theorem on one of these occasions, I feel certain that he forgot it and rediscovered it on his own. For considering his high esteem for Jacobi and the many times he enthusiastically advertised Jacobi's new results, it seems unlikely that he would claim priority for a theorem that he knew rightly belonged to the deceased Jacobi. At any rate, Liouville's rediscovery of his theorem was closely linked with the work of Jacobi.

Parisian Developments Prompting Liouville's Publication of His Theorem

41.   However, it was events in Paris that made Liouville publish his theorem. From the early 1840s, Liouville's protégé Joseph Bertrand had written a series of papers on mechanics, some of which appeared in Liouville's *Journal*. He thereby established himself as the foremost French expert in this area, so he was a natural choice as an editor of the new edition of Lagrange's *Mécanique Analytique*. Probably in preparation, he devoted his course of 1852–1853 at the *Collège de France* to mechanics, and some of the new results taught during these lectures found their way into his edition of the *Mécanique Analytique* [Vol. I, 3rd ed., 1853, notes 6 and 7]. The 6th note concerned the Hamilton-Jacobi formalism, and the most interesting seventh note dealt with Poisson brackets (see also [Bertrand 1852]). Here, Bertrand showed that given an integral φ_0 to Hamilton's equations, one can adjoin $2n - 1$ integrals $\psi_1, \psi_2, \ldots, \psi_{2n-1}$ so as to make a complete set and such that

$$(\varphi_0, \psi_1) = 1 \quad \text{and} \quad (\varphi_0, \psi_i) = 0 \quad \text{for} \quad i \geq 2. \tag{113}$$

The integral ψ_1 is now called the conjugate of φ_0.

Edmond Bour, who had followed Bertrand's lectures, developed this idea into an explicit exposition of Jacobi's second method of integration, the first published version of this method [Bour 1855]. Upon seeing Jacobi's own procedure after Clebsch's publication of it in [Jacobi 1862], Bour wrote that he was happy to learn that his method was identical to that of the master [Bour 1862]. Given one integral of the equations of motion, he set up a partial differential equation similar to the original Hamilton-Jacobi equation, but of an order two lower, having the same integrals as the original equation, except for the given integral and its conjugate. If he knew a new integral to this lower order equation, the process could be continued. In the end, when half of the integrals were found, the rest could be determined as usual from $\sum p_i \, dq_i$. That this form is exact follows from the successive construction of integrals, which ensures that they are in involution.

42.   When Bour submitted this memoir on March 5, 1855, to the *Académie des Sciences*, Liouville, Lamé, and Chasles were appointed examiners. Liouville, who wrote the report, concluded [1855c] that "the student has shown himself worthy of the master [Bertrand]." Liouville also used the occasion to call attention to his own theorem and to the fact that Bour's method could be extended, with slight alterations, to Hamiltonians, depending on time. Bour had only treated time-independent Hamiltonians.

When the central part of Bour's paper appeared in Liouville's *Journal*, Bour explicitly referred to Liouville when using Liouville's theorem, so it is unclear in which form it had appeared in the original memoir. Moreover, Bour indicated how to carry out the extension to time-dependent Hamiltonians. To the latter remark, Liouville in turn added the small, beautiful *Note a l'occasion du mémoire précédent* [1855h], in which he showed that Hamilton's equations in the time-dependent case can be reduced à priori to the time-independent case by increasing the number of generalized coordinates by one. More precisely, let H be a Hamiltonian depending on time t. Then, Liouville introduced a new variable, $\tau = t +$ constant, to play the role of time. The variable t, on the other hand, he considered as a generalized coordinate defining its corresponding momentum u by

$$\frac{du}{d\tau} = -\frac{\partial H}{\partial t}.$$

It is now easily seen that this extended set of coordinates and momenta satisfies Hamilton's equations for the new τ-independent Hamiltonian

$$H_1 = H + u.$$

43.   Two months before publishing Bour's paper and the attached note, Liouville had reprinted his *Académie* report on Bour's paper in his *Journal*, and on that occasion, he published the note containing Liouville's theorem [1855d]. His proof never appeared, but it must have been similar to Jacobi's (see references in §40). He carefully had the secretary of the *Bureau des Longitudes* testify that the note was copied verbatim from a note in the *Procès Verbaux* of the *Bureau des Longitudes* of June 29, 1853. It was prefaced by the words:

> Mr. Liouville communicates the following Note on the integration of the equations of Dynamics. He verbally adds all the developments necessary to explain the utility and the applications.

This introduction is still preserved in the *Procès Verbaux*, but the note itself was probably removed when it was printed. I have summarized its content in §34. In contrast to Jacobi, Liouville took time-dependent integrals and Hamiltonians into account by adding the extra term $-H\,dt$ in (106).

Liouville's careful dating of the note to 1853 was clearly intended to secure him priority for the theorem over Bour and also over his earlier correspondent, William Fishburn Donkin (cf. [Neuenschwander 1984b, II, 4]), who had published it in the *Philosophical Transactions* on February 23, 1854. Liouville [1855c, p. 136] generously admitted that the "esteemed author [Donkin] does not seem to have any knowledge of the results I have obtained before him." In the same note, Liouville called attention to a "remarkable Thesis" in which Adrien Lafon "has included *my* theorem (with a reference to me and proving it in his own way)" [my italics].

Let me conclude this rather complex analysis of the genesis of Liouville's theorem by stating that Jacobi was the first to find and prove it and the first to lecture on it. Donkin was the first to publish it. Liouville's only claim to it rests on the fact that he communicated it to a learned society before Donkin discovered it.

Liouville's Lecture on Mechanics

44. Liouville drew his theorem from his lectures on *La formation et l'inté-gration des équations différentielles* given at the *Collège de France* during the second semester of 1852–1853. Possibly inspired by Bertrand's lectures the previous year (cf. §41), Liouville focused these lectures on the equations of rational mechanics. Two more of his later papers, both published in 1856, contained extracts from these lectures. In the second of these [1856g], Liouville showed how to transform an arbitrary set of first-order differential equations,

$$\frac{dx_i}{dt} = X_i(x_1, x_2, \dots, x_n), \qquad i = 1, 2, \dots, n,$$

into canonical form by doubling the number of variables. He simply defined

$$H = \sum_{i=1}^{n} p_i \, X_i,$$

where $p_i(t)$ are new variables satisfying

$$\frac{dp_i}{dt} = -\frac{\partial H}{\partial x_i}.$$

It is obvious that $x_i = q_i$, p_i satisfy Hamilton's equations, (83). Liouville pointed out that the great many results pertaining to Hamilton's equations would sometimes more than compensate the complication introduced by the new variables.

Ostrogradski had published a similar result [1850b], but Liouville does not seem to have been aware of it.

The Geometrization of the Principle of Least Action

45.   The first of the two notes, communicated to the *Académie des Sciences* during the summer of 1856, opened far greater perspectives. With his eight pages in the *Comptes Rendus* [1856k], Liouville initiated a differential geometric approach to mechanics, in particular, to the principle of least action.

He began by bringing the principle of least action into its Jacobian form. The classical form of the action is (cf. §32):

$$\int_1^2 \left(\frac{1}{2} \sum m_i \left(\frac{dx_i}{dt} \right)^2 \right) dt , \tag{114}$$

where 1 and 2 symbolize the initial and final configurations. When applying the principle of least action, t is varied, but the total energy

$$h = \frac{1}{2} \sum m_i \left(\frac{dx_i}{dt} \right)^2 + U \tag{115}$$

is kept fixed (§32). Therefore, a more appropriate expression is found by eliminating dt from (115) and inserting it into (114):

$$\text{Action} = \int_1^2 \sqrt{2(h - U) \cdot \sum m_i \, dx_i^2}. \tag{116}$$

It is this quantity that is minimized according to the principle of least action.

More precisely, assume that the physical system is described by the generalized coordinates q_i $(i = 1, 2, \ldots, n)$, which take the values q_i^1 at position (1) where the integration starts, and q_i^2 at the end point, (2). If the path of integration is parametrized by q_1, it can be represented as $q_i = f_i(q_1)$, and the integral (116) can be written in the form

$$\int_{q_1^1}^{q_1^2} \omega(q_1) \, dq_1. \tag{117}$$

The principle of least action states that the path actually followed by a physical system will minimize this integral among all those that begin at q^1 and end at q^2. The speed with which the path is traversed is determined from (115).

Jacobi had made this clarification of the principle in his 1837 communication to the Paris *Académie* [1837c]. Liouville's contribution consisted of the introduction of a "remarkable expression" of the general integral (116) or (117) which made the minimizing path stand out clearly. To this end, he remarked that $\sum m_i \, dx^2 = \sum q_{ij} \, dq_i \, dq_j$ is a positive definite quadratic differential form in the dq_i's so that it can be expressed as

$$\sum m_i \, dx_i^2 = \sum_{j=1}^{n} \left(\sum_{i=1}^{n} P_{ij} \, dq_i \right)^2 = \sum_{j=1}^{n} \ell_j^2 , \tag{118}$$

where the P_{ij}'s are functions of the q_i's and where Liouville, for brevity, has introduced the differentials

$$\ell_j = \sum_{i=1}^{n} P_{ij} \, dq_i . \tag{119}$$

Given a function $\theta(q_1, q_2, \ldots, q_n)$, Liouville wrote its total differential as a linear combination of the ℓ_j's:

$$d\theta = \sum_{j=1}^{n} n_j \ell_j . \tag{120}$$

He saw that the coefficients n_j could be written as

$$n_j = \sum_{i=1}^{n} P^{ij} \frac{\partial \theta}{\partial q_i} , \tag{121}$$

where $\{P^{ij}\}$ is defined from $\{P_{ij}\}$ by

$$\sum_{i=1}^{n} P_{ij_1} P^{ij_2} = \delta_{j_1 j_2} . \tag{122}$$

With later terminology, P^{ij} is the contravariant tensor corresponding to P_{ji}. If we assume, with Liouville, that θ satisfies the first-order partial differential equation

$$\sum_{j=1}^{n} \left(\sum_{i=1}^{n} P^{ij} \frac{\partial \theta}{\partial q_i} \right)^2 = 2(h - U) , \tag{123}$$

or equivalently,

$$\sum_{j=1}^{n} (n_j)^2 = 2(h - U) , \tag{124}$$

then by (118) and (124), the important quantity in the action integral (116) can be written

$$2(h - U) \sum m_i \, dx_i^2 = \sum_{j=1}^{n} (n_j)^2 \sum_{j=1}^{n} (\ell_j)^2 ,$$

or more conveniently,

$$2(h - U) \sum m_i \, dx_i^2 = \left(\sum_{j=1}^n n_j l_j \right)^2 + \sum_{j_1 > j_2} (n_{j_1} l_{j_2} - l_{j_2} n_{j_1})^2$$
$$= (d\theta)^2 + \sum_{j_1 > j_2} (n_{j_1} l_{j_2} - n_{j_2} l_{j_1})^2. \qquad (125)$$

Thus, the action integral can be written

$$A = \int_1^2 \sqrt{(d\theta)^2 + \sum_{j_1 > j_2} (n_{j_1} \ell_{j_2} - n_{j_2} \ell_{j_1})^2}, \qquad (126)$$

which is the "remarkable expression" Liouville was after. Why is it so remarkable? Because one can directly read off the equations of the minimizing path, namely,

$$\sum_{j_i > j_2} (n_{j_1} \ell_{j_2} - \ell_{j_2} n_{j_1})^2 = 0, \qquad (127)$$

or equivalently,

$$\frac{\ell_1}{n_1} = \frac{\ell_2}{n_2} = \cdots = \frac{\ell_n}{n_n}. \qquad (128)$$

These are independent (not shown by Liouville), first-order differential equations in the q_i's, which can be used to determine the last $n - 1$ of them as functions of q_1. Indeed, if the path γ_m from (1) to (2) is chosen so as to satisfy (128), then the action integral has the value:

$$A_m = \int_{\gamma_m} d\theta = \theta(2) - \theta(1). \qquad (129)$$

If instead we choose another path, γ, from (1) to (2), equations (128) can not all be satisfied so that the form

$$\sum_{j_1 > j_2} (n_{j_1} \ell_{j_2} - \ell_{j_2} n_{j_1})^2$$

is positive, making the action integral greater than

$$\int_\gamma d\theta = \theta(2) - \theta(1). \qquad (130)$$

Here, we have used that $d\theta$ is an exact differential, so that its integrals along the two paths are the same. Thus, (128) and the energy conservation (115) must determine the motion of the system.

46.   How does this relate to the Hamilton-Jacobi formalism? Along a trajectory, the action was determined by $\theta(2) - \theta(1)$. Thus, θ must be Hamilton-Jacobi's characteristic function V, and therefore (123) must be

the Hamilton-Jacobi equation, (93). Indeed, it is not hard to see that if the momenta

$$p_k = \frac{\partial T}{\dot{q}_k} = \frac{\partial}{\partial \dot{q}_k} \frac{1}{2} \left(\sum_{j=1}^{n} \left(\sum_{i=1}^{n} P_{ij}\dot{q}_i \right)^2 \right)$$

$$= \sum_{j=1}^{n} \left(\sum_{i=1}^{n} P_{ij}\dot{q}_i \right) P_{kj} \tag{131}$$

are introduced into T instead of $\dot{q}_i$, we get

$$h = T + U = \frac{1}{2} \sum_{j=1}^{n} \left(\sum_{i=1}^{n} P^{ij}p_i \right)^2 + U, \tag{132}$$

and so by substituting $\frac{\partial \theta}{\partial q_i}$ for p_i in the Hamilton-Jacobi equations (93) equation (123) is obtained.

Moreover, by (131), equations (128) become

$$\frac{dt \sum_{i=1}^{n} P^{i1}p_i}{\sum_{i=1}^{n} P^{i1} \frac{\partial \theta}{\partial q_i}} = \frac{dt \sum_{i=1}^{n} P^{i2}p_i}{\sum_{i=1}^{n} P^{i2} \frac{\partial \theta}{\partial q_i}} = \cdots, \tag{133}$$

from which we conclude ($\{P^{ij}\}$ being nonsingular) that

$$\frac{\partial \theta}{\partial q_i} = f(q_1, \ldots, q_n)p_i. \tag{134}$$

Inserting these values into equation (123) and comparing the result with the energy conservarion (115) we find that f is identically one. Thus equations (128) combined with the law of *force vive* yield Hamilton's integrals (95) to the laws of motion.

The formulas and arguments of this section are not contained in Liouville's paper, but the final remark in that paper reveals that he knew θ is the characteristic function:

> the function θ. ... On the other hand this function, whose importance is known today to all geometers thanks to the works of Mr. Hamilton and Jacobi, plays in my method the greatest rôle.
> [Liouville 1856k]

Moreover, Liouville mentioned that if θ is a complete solution of the Hamilton-Jacobi equations (123) containing n constants $a_1, \ldots, a_n$ (so that (126)-(130) represent the general trajectory), the equations of motion have the integrals

$$\frac{\partial \theta}{\partial a_i} = b_i, \tag{135}$$

corresponding to those of Jacobi, (100). For a proof, Liouville referred the reader to "a theorem by Jacobi," but they can also be deduced easily from the formulas in this section.

Indeed, since $(h - U)$ does not depend on a_k, we find from (123)

$$
\begin{aligned}
0 &= \frac{1}{2} \frac{\partial}{\partial a_k} \sum_{j=1}^{n} \left(\sum_{i=1}^{n} P^{ij} \frac{\partial \theta}{\partial q_1} \right) \\
&= \sum_{j=1}^{n} \left[\left(\sum_{i=1}^{n} P^{ij} \frac{\partial \theta}{\partial q_i} \right) \left(\sum_{i=1}^{n} P^{ij} \frac{\partial \theta}{\partial q_i \partial a_k} \right) \right].
\end{aligned}
\tag{136}
$$

By inserting $\frac{\partial \theta}{\partial q_i} = p_i$ as expressed in (131) and using the fundamental formulas (122) twice, this expression is reduced to

$$
0 = \sum_i \dot{q}_i \frac{\partial \theta}{\partial q_i \partial a_k} = \frac{d}{dt} \frac{\partial \theta}{\partial a_k}.
\tag{137}
$$

Hence, $\frac{\partial \theta}{\partial a_k}$ is a constant b_i.

47.   Thus, Liouville had shown that the Hamilton-Jacobi formalism could be deduced from the principle of least action, and he also indicated how to reverse the argument to show that this principle is a consequence of the Hamilton-Jacobi integrals.

This equivalence was not new, but the method was. If we disregard the equivalence of Liouville's and Jacobi's equations and focus exclusively on Liouville's own deduction (§45), one must admire its simplicity compared with Jacobi's technical calculations. The technicalities have been replaced by differential geometric considerations closely linked to Liouville's geometric researches during these years. Indeed, Liouville admitted that the central idea had been suggested to him by a purely geometrical paper:

> The idea of introducing the function θ in order to express $2(U + K) \sum m \, ds^2$ as a sum of squares of which the first term is the square of an exact differential, was suggested to me when I read (in 1847) a handwritten Memoir by Mr. Schlaefli, professor at the university of Berne, in which this geometer transformed the square of the element of length of a geodesic on an ellipsoid in the same way. I am happy to acknowledge what I owe to Mr. Schlaefli and to pay tribute to his high merit.

> [Liouville 1856k, p. 1153]

Schläfli, in turn, was led to this new expression for ds by his studies of "the beautiful works of Mr. Liouville and Mr. Chasles" [Schläfli 1847–1852, CR 25, p. 391]. Liouville read Schläfli's work in the summer of 1847 as a member of the judging commission appointed when it was presented to the *Académie des Sciences*. As we shall see in §§60–64, Liouville was at that moment very actively applying the principles of mechanics to study geodesics on various surfaces, ellipsoids in particular, by considering them as trajectories of free particles. Seeing Schläfli's work, he had the ingenious idea of reversing the process and transferring a theorem concerning geodesics to the realm of general mechanics by replacing the usual surface element ds^2 by the expression $2(U + K)\sum m_i\,dx_i^2$, which by the principle of least action plays the same role in the search for trajectories of a general system as ds^2 does for the geodesics; Liouville's θ therefore corresponds to the local coordinates determined by the length along a family of geodesics, e.g., those emanating from the initial configuration. Such coordinate systems had been studied by Gauss in his *Disquisitiones generales* [1828] and had been used by Schläfli and were well known to Liouville. The additional terms in (126) correspond to the coordinate that according to Gauss exists perpendicular to the family of geodesics [Gauss 1828 §§15,16]. There is no doubt that Liouville saw these connections.

To a modern reader, Liouville's arguments show that classical mechanics is the study of the geometry of a differential manifold with the Riemannian metric $2(h - U)\sum m_i\,dx_i^2$, or rather $2(h - U)\sum g_{ij}\,dq_i\,dq_j$, whose trajectories correspond to geodesics in this geometry. However, such an explicit formulation cannot be found in Liouville's paper, for he did not have Riemann's generalization of Gauss's ideas to higher dimensions at his disposal. Still, he had taken a decisive step in this direction by showing that by replacing ds^2 by $2(h - U)\sum m_i\,dx_i^2$ the techniques and geometric ideas are carried over from geodesics on surfaces to general trajectories in n-dimensional phase space. Here and in some of his geometric works, he came close to some of Riemann's ideas (Chapter XVII, §46–48).

Where Lagrange had proudly announced that he had freed mechanics entirely from geometry, Liouville took the first step toward a new geometrization of rational mechanics, geometry being now differential geometry (later symplectic geometry) rather than the Euclidean type of geometry abandoned by Lagrange.

48.   The geometrical consideration of the action integral was carried further by Lipschitz [1871], Thomson and Tait [1879], and Darboux [1888]. Lipschitz's arguments were rather similar to Liouville's, but because of the developments in geometry in the intervening 20 years, he expressed the geometric analogy more clearly. Moreover, he operated directly with the form $\sum g_{ij}\,dq_i\,dq_j$ instead of the sum of squares (118) obtaining thereby a simpler expression for the kinetic energy:

$$2T = \sum g^{ij}p_ip_j. \qquad (138)$$

Lipschitz does not seem to have known the work of Liouville, but Thomson and Tait included "the very novel analytical investigation of the motion of a conservative system, by Liouville," which they called "Liouville's kinetic theorem." However, they only gave it for a single free particle, loosing thereby the important geometrical implications. In the second volume of the *Leçons sur la théorie générale des surfaces* [1888], Darboux followed Lipschitz, but he also presented a new derivation of Liouville's expression, with due reference to Liouville. Darboux clarified the geometric ideas of Liouville and Lipschitz, and expressed the new approach explicitly in his introduction:

In particular, I have stressed the connection which one can find here between the methods employed by Gauss in the study of geodesics and those that Jacobi later applied to the problems of analytical mechanics. In this way, I have been able to show the great interest of Jacobi's beautiful discoveries when these are considered from a geometrical point of view.

[Darboux 1888, Preface]

The geometrization of mechanics was carried further in connection with the new ideas on differential geometry, differential forms, and tensor calculus developed toward the end of the last century by Christoffel, Levi-Civita, and Ricci and was an important background for the introduction of the general theory of relativity (cf. [Dugas 1950, Book 5, Chap. 2].

In recent years, classical mechanics has had a comeback on the mathematical stage in entirely geometrized clothing. As presented in Arnold's famous book [1978], mechanics is the study of symplectic geometry. Unfortunately, it would take us too long to follow more closely the geometrization of mechanics of which Liouville took the first step.

Liouville's Unpublished Notes; Generalized Poisson Brackets

49. Liouville's notebooks reveal that he developed his ideas in rational mechanics much further than his published papers suggest. There are scattered notes through the 1860s, but the most interesting [Ms 3618 (4), 3623 (1), and 3637 (13)] stem from the summer of 1856; at least the two first are datable from intercalated letters, and the hand as well as the content of the last notebook point to the same period, although it has been misnumbered among the notebooks from the 1870s. Their common theme is the generalization of the ideas of Lagrange, Poisson, Jacobi, Bour, and Liouville himself to more general equations than Hamilton's.

50. The note [Ms 3623 (1), pp. 47r–60r] from the early days of July 1856 seems to be the first of them. It ends with a discussion of a system of two differential equations:

$$\frac{dx_i}{dt} = X_i, \quad i = 1, 2, \tag{139}$$

where $X_1 X_2$ are homogeneous functions

$$X_i = x_1 \varphi_i \left(t, \frac{x_2}{x_1}, \ldots \right). \tag{140}$$

Liouville showed that if $f(t, x_1, x_2) = A$ is an integral of these equations, then

$$F(t, x, y) = x_1 \frac{\partial f}{\partial x_1} + x_2 \frac{\partial f}{\partial x_2} = B \tag{141}$$

is also an integral. He then transformed n equations like (139) with arbitrary X_i's to $n + 1$ equations of the same kind having homogeneous right-hand sides (140):

> Thus, [Liouville concluded], instead of n given differential eq[uati]ons, one can substitute $(n + 1)$ eq.ons of which one single integral will in general suffice to provide all the others.
> Most certainly, this statement is at least as curious as Jacobi's statement concerning the theorem of Poisson.
>
> [Ms 3623 (1), p. 60r]

51.   Soon after, he found even better general formulas by exploring the possibilities opened up by his own methods (published in September [1856g]) of transforming any system (139) into canonical form. Recall from §44 that the trick was to obtain

$$\frac{dq_i}{dt} = \frac{\partial H}{\partial p_i} \quad \text{and} \quad \frac{dp_i}{dt} = -\frac{\partial H}{\partial q_i} \tag{142}$$

by introducing $H = \sum_{i=1}^{n} p_i X_i$, $q_i = x_i$ and p_i satisfying the last of Hamilton's equations. The basic observation in Liouville's note [3618 (4), pp. 2–20] is to the effect that the function

$$\sum_{i=1}^{n} f_i(t, x_1, x_2, \ldots, x_n) p_i \tag{143}$$

is an integral of Hamilton's equations if the f_i satisfy the partial differential equations

$$\frac{\partial f_i}{\partial t} + \sum_{j=1}^{n} \frac{\partial f_i}{\partial x_j} X_j = \sum f_j \frac{\partial X_i}{\partial x_j}, \qquad i = 1, 2, \ldots, n. \tag{144}$$

Liouville now translated the theorem on the Poisson brackets into the language of (144), which only involves functions entering the original equation, and remarked:

> The above must first be presented as follows without any mention of the auxiliaries x', y', ... $[p_1, p_2, \ldots]$.
>
> [Ms 3618 (4), p. 6r]

During the translation of Jacobi's last multiplier to the general system (139), Liouville admitted that "this is difficult to explain without recourse to the canonical system" [Ms 3618 (4), p. 11]. However, he pressed the idea, further showing how the integrals of the system (139) can yield the solutions of (144) and vice versa [Ms 3618 (4), p. 19v]. In particular, he showed that if we are in possession of one solution f_i of the partial differential equation (144) and one integral a of the original system (139) we can determine a new integral a_1 by

$$\sum_{i=1}^{n} f_i \frac{\partial a}{\partial x_i} = a_1.$$

In this way, new integrals can be successively generated until we find an integral that is a function of those already found. Still, if $\sum_{i=1}^{n} f_i \frac{\partial a}{\partial x_i} = 0$, "you can lower the order of your eq.ons by one by losing the integral a but by conserving the system $p, q, \ldots [f_1, f_2, \ldots]$." This looks very much like Bour's considerations, [Bour 1855], and on the subsequent pages, Liouville generalized the rest of Bour's procedure, concluding:

> In a word, the method seems much more extensive than Bour thought.
>
> [Ms 3618 (4), p. 62r]

Liouville did argue that the system of partial differential equations (144) was more "remarkable" than the familiar partial differential equations satisfied by the multipliers F_i, making

$$\sum F_i(dx_i - X_i \, dt) = d\theta \qquad (145)$$

an exact differential, and thus determining an integral $\int d\theta = a$. Still, it may have been the rather uninviting appearance of (144) that made Liouville's notes gradually develop into a study of the canonical system [Ms 3618 (4), pp. 94v ff].

52.   A smoother generalization of the theorems of rational mechanics is found in the last manuscript mentioned above. It is by far the longest of them, covering hundreds of pages of the folio size notebook [Ms 3637 (13)], and it is also by far the most carefully worked out of Liouville's manuscripts on Jacobian mechanics. As mentioned earlier, it is probably the draft of a paper that Liouville had planned to dedicate to Poinsot. In fact, it deals with a "considerable extension... of the quantities $[a, b]$ and (a, b)." It starts out as a rather polished memoir divided into sections, but during §IX [p. 12], this plan breaks down. Having discovered that his methods could be extended much further and having sketched the start of this new version several times [p. 53 ff], he finally drew up the main ideas in 11 sections [pp. 58v–63v], remarking that "this seems complete enough, at least for a first Memoir" [Ms 3637 (13), p. 63r].

Considering that this probably represents Liouville's most far-reaching ideas in this field and that contrary to his intentions it never appeared, I shall present a rather thorough summary of it. I shall not limit the account to the most general statements, but rather illustrate Liouville's process of discovery by following the successive generalizations, which are so characteristic of his method of inquiry (cf. Chapter IX, §18).

53.   We cannot follow the earliest development of the ideas in this notebook, for when Liouville began on page one he had clearly planned the memoir on systems of first-order equations. Thus, the introduction of the theory may reflect Liouville's didactical ideas, or they may reflect the first

stage of his own discovery. At any rate, it is characteristic of Liouville that
the first three sections deal with a system of only two equations:

$$\frac{\partial V}{\partial x} = p\frac{dy}{dt}, \quad \frac{\partial V}{\partial y} = -p\frac{dx}{dt}, \tag{146}$$

or equivalently,

$$\frac{dy}{dt} = P\frac{\partial V}{\partial x}, \quad \frac{dx}{dt} = -P\frac{\partial V}{\partial y}, \tag{147}$$

where x, y are functions of t, $p = \frac{1}{P}$ is a function of x and y, and V is
a function of x, y, and t. If x, y are solutions containing two arbitrary
constants a, b (making conversely a and b functions of x, y, and t), then
Liouville showed that the generalized Lagrange and Poisson brackets,

$$p\left(\frac{\partial x}{\partial a}\frac{\partial y}{\partial b} - \frac{\partial x}{\partial b}\frac{\partial y}{\partial a}\right) = [a, b] \tag{148}$$

and

$$P\left(\frac{\partial a}{\partial x}\frac{\partial b}{\partial y} - \frac{\partial a}{\partial y}\frac{\partial b}{\partial x}\right) = (a, b), \tag{149}$$

are both functions of a and b only, i.e., constants. He even remarked that
if p contains the time t then

$$\frac{d}{dt}[a, b] = \frac{\partial p}{\partial t}\left(\frac{\partial x}{\partial a}\frac{\partial y}{\partial b} - \frac{\partial x}{\partial b}\frac{\partial y}{\partial a}\right). \tag{150}$$

Besides, this is a consequence of the theorem I have given for $\frac{dx}{dt} = X$,
$\frac{dy}{dt} = Y$, to the effect that if $u = \frac{\partial x}{\partial a}\frac{\partial y}{\partial b} - \frac{\partial x}{\partial b}\frac{\partial y}{\partial a}$ one has

$$\frac{du}{dt} = u\left(\frac{\partial X}{\partial x} + \frac{\partial Y}{\partial y}\right) \tag{151}$$

[Ms 3637 (13), p. 1v]

Here, Liouville connected his theorem from [1838e] on the "volume in phase
space" to a simple generalized version of Hamilton's equations (cf. §21).
    In section 2, Liouville added a perturbation term to (146) or (147) making

$$\frac{\partial V}{\partial x} + \frac{\partial \Omega}{\partial x} = p\frac{dy}{dt}, \quad \frac{\partial V}{\partial y} + \frac{\partial \Omega}{\partial y} = -p\frac{dx}{dt}, \tag{152}$$

or equivalently,

$$\frac{dx}{dt} = -P\frac{\partial V}{\partial y} - P\frac{\partial \Omega}{\partial y}, \quad \frac{dy}{dt} = P\frac{\partial V}{\partial x} + P\frac{\partial \Omega}{\partial x}. \tag{153}$$

As in the classical Lagrangian and Poissonian theory, Liouville assumed
the solutions of these equations to be obtained from the solutions $x(a, b, t)$

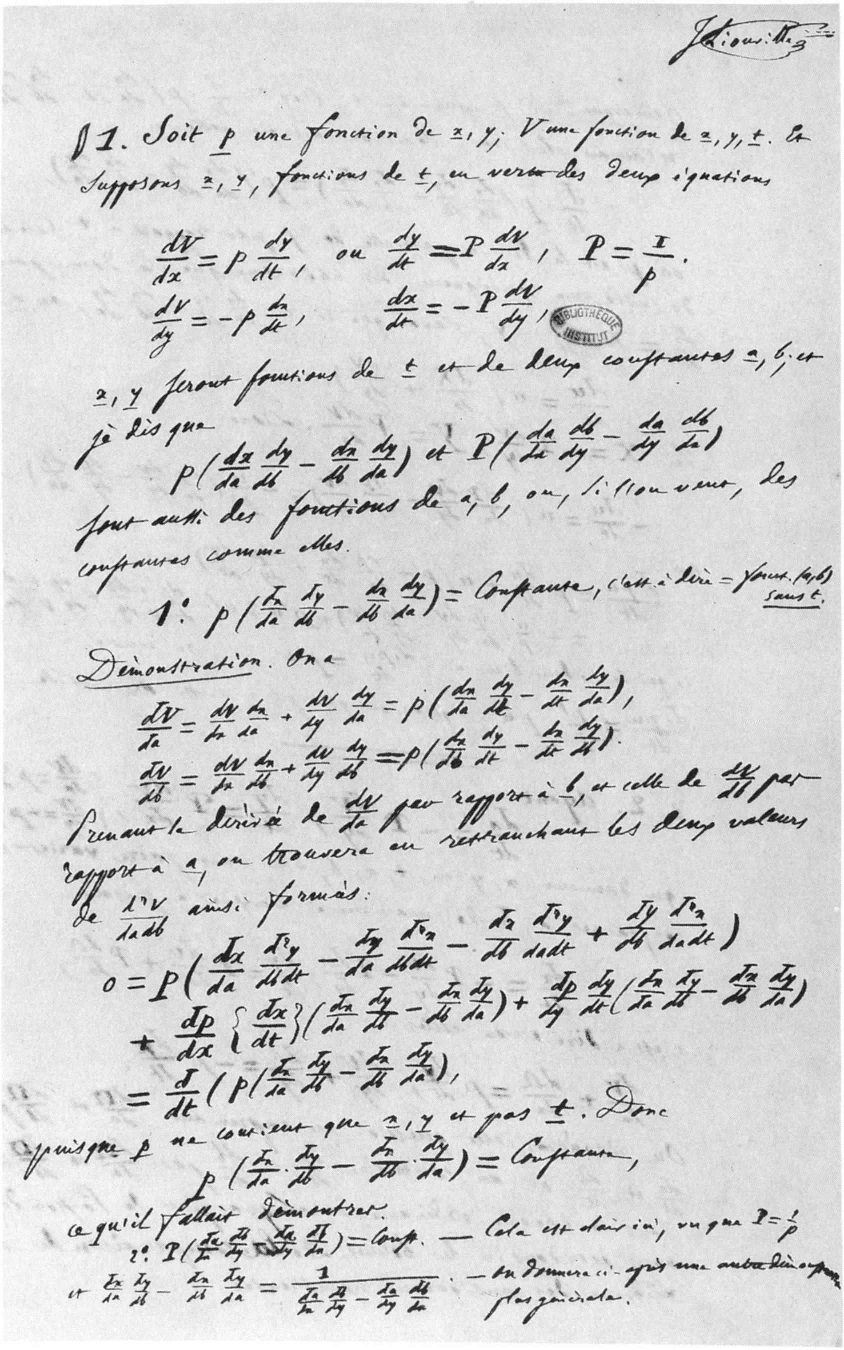

J. Liouville

§ 1. Soit p une fonction de x, y; V une fonction de x, y, t. Et supposons x, y, fonctions de t, en vertu des deux équations

$$\frac{dV}{dx} = p\,\frac{dy}{dt}, \quad ou \quad \frac{dy}{dt} = P\,\frac{dV}{dx}, \quad P = \frac{1}{p}.$$

$$\frac{dV}{dy} = -p\,\frac{dx}{dt}, \qquad \frac{dx}{dt} = -P\,\frac{dV}{dy},$$

x, y seront fonctions de t et de deux constantes a, b; et je dis que

$$p\left(\frac{dx}{da}\frac{dy}{dt} - \frac{dx}{db}\frac{dy}{da}\right) \quad et \quad P\left(\frac{da}{dx}\frac{db}{dy} - \frac{da}{dy}\frac{db}{dx}\right)$$

sont aussi des fonctions de a, b, ou, si l'on veut, des constantes comme elles.

$$1°\quad p\left(\frac{dx}{da}\frac{dy}{db} - \frac{dx}{db}\frac{dy}{da}\right) = Constante, \quad c'est à dire = fonct.(a,b) \text{ sans } t.$$

Démonstration. On a

$$\frac{dV}{da} = \frac{dV}{dx}\frac{dx}{da} + \frac{dV}{dy}\frac{dy}{da} = p\left(\frac{dx}{da}\frac{dy}{dt} - \frac{dx}{dt}\frac{dy}{da}\right),$$

$$\frac{dV}{db} = \frac{dV}{dx}\frac{dx}{db} + \frac{dV}{dy}\frac{dy}{db} = p\left(\frac{dx}{db}\frac{dy}{dt} - \frac{dx}{dt}\frac{dy}{db}\right).$$

Prenant la dérivée de $\frac{dV}{da}$ par rapport à b, et celle de $\frac{dV}{db}$ par rapport à a, on trouvera en retranchant les deux valeurs de $\frac{d^2V}{da\,db}$ ainsi formées:

$$0 = P\left(\frac{dx}{da}\frac{d^2y}{db\,dt} - \frac{dy}{da}\frac{d^2x}{db\,dt} - \frac{dx}{dt}\frac{d^2y}{da\,dt} + \frac{dy}{dt}\frac{d^2x}{da\,dt}\right)$$
$$+ \frac{dp}{dx}\left\{\frac{dx}{dt}\right\}\left(\frac{dx}{da}\frac{dy}{db} - \frac{dx}{db}\frac{dy}{da}\right) + \frac{dp}{dy}\frac{dy}{dt}\left(\frac{dx}{da}\frac{dy}{db} - \frac{dx}{db}\frac{dy}{da}\right)$$
$$= \frac{d}{dt}\left(P\left(\frac{dx}{da}\frac{dy}{db} - \frac{dx}{db}\frac{dy}{da}\right)\right),$$

puisque p ne contient que x, y et pas t. Donc

$$p\left(\frac{dx}{da}\frac{dy}{db} - \frac{dx}{db}\frac{dy}{da}\right) = Constante,$$

ce qu'il fallait démontrer.

$$2°\quad P\left(\frac{da}{dx}\frac{db}{dy} - \frac{da}{dy}\frac{db}{dx}\right) = const. \quad - \quad Cela\ est\ clair\ ici,\ vu\ que\ P = \frac{1}{p}$$

$$et\quad \frac{dx}{da}\frac{dy}{db} - \frac{dx}{db}\frac{dy}{da} = \frac{1}{\frac{da}{dx}\frac{db}{dy} - \frac{da}{dy}\frac{db}{dx}}. \quad - \quad on\ donnera\ ci-après\ une\ autre\ démonstration\ plus\ générale.$$

PLATE 33. Beginning of Liouville's generalization of the Poisson and Lagrange brackets. (Bibliothèque de l'Institut de France, Ms 3637 (13), p. 1r)

and $y(a, b, t)$ to (146), (147) by varying a and b. Thus, (146) and (147) are still supposed to hold if the t derivatives are replaced by partial derivatives so that

$$\frac{\partial \Omega}{\partial x} = p \left(\frac{\partial y}{\partial a} \frac{da}{dt} + \frac{\partial y}{\partial b} \frac{db}{dt} \right), \text{etc.}$$

From these expressions, Liouville deduced that

$$\frac{\partial \Omega}{\partial a} = [a, b] \frac{db}{dt}, \quad \frac{\partial \Omega}{\partial b} = -[a, b] \frac{da}{dt} \tag{154}$$

and

$$\frac{da}{dt} = (a, b) \frac{\partial \Omega}{\partial b}, \quad \frac{db}{dt} = -(a, b) \frac{\partial \Omega}{\partial a}, \tag{155}$$

corresponding to Laplace's and Poisson's classical results (§6). Moreover, he concluded that $(a, b) = -\frac{1}{[a, b]}$.

As a commentary on this method, which he attributed to Lagrange, Liouville wrote:

> I have treated it at the *Collège de France* since my first lectures as a substitute for Mr. Biot while imposing the ordinary theory of variation of the constants."
>
> [637 (13), p. 2v]

Thus, his early lectures on *Perturbations planétaires* may have been at the basis of the idea generalized here. Liouville also remarked that as in the usual theory

$$(a, a) = (b, b) = 0 \quad \text{and} \quad (a, b) = -(b, a), \tag{156}$$

and likewise for the generalized Lagrange brackets; §3 is devoted to the last multiplier rule. Assume one integral a of (146) or (147) to be known. First, Liouville, following Bertrand (§41), showed that there exists an integral b with $(a, b) = 1$. Indeed, we know that there exists another integral b_0 such that $(a, b_0) \neq 0$. All other integrals are functions $\varphi(a, b_0)$ of these two, and it is easy to see that

$$(a, \varphi(a, b_0)) = \frac{\partial \varphi}{\partial b_0}(a, b_0).$$

Therefore, if we choose φ such that $\frac{\partial \varphi}{\partial b_0} = \frac{1}{(a, b_0)}$, we can take $b = \varphi(a, b_0)$. (I have polished Liouville's argument here.) The problem is to determine b from the partial differential equation

$$(a, b) = P \left(\frac{\partial a}{\partial x} \frac{\partial b}{\partial y} - \frac{\partial a}{\partial y} \frac{\partial b}{\partial x} \right) = 1. \tag{157}$$

As usual, this is done by eliminating the variable y from the integral $a(x, y, t) = $ const. Using parenthesis to denote the partial derivatives when (x, a, t) are considered the independent variables, we have

$$\frac{\partial b}{\partial x} = \left(\frac{\partial b}{\partial x}\right) + \left(\frac{\partial b}{\partial a}\right)\frac{\partial a}{\partial x}, \quad \frac{\partial b}{\partial y} = \left(\frac{\partial b}{\partial a}\right)\frac{\partial a}{\partial y}.$$

Inserted into (157), this gives

$$P\left(\frac{\partial a}{\partial y}\left(\frac{\partial b}{\partial x}\right)\right) = 1,$$

so that

$$\left(\frac{\partial b}{\partial x}\right) = -\frac{1}{P\frac{\partial a}{\partial y}} = F(x, a, t). \tag{158}$$

But, being an integral, b must satisfy

$$0 = \frac{db}{dt} = \left(\frac{\partial b}{\partial t}\right) + \left(\frac{\partial b}{\partial x}\right)\frac{\partial x}{\partial t} = \left(\frac{\partial b}{\partial t}\right) + \left(\frac{\partial b}{\partial x}\right)f(x, a, t), \tag{159}$$

where $f(x, a, t)$ is the right-hand side of the last equation in (147) considered as a function of x, a, and t. Equations (158) and (159) show that

$$db = F(x, a, t)(dx - f(x, a, t)\, dt), \tag{160}$$

so that b can be found by quadrature. In other words $\frac{1}{P\frac{\partial a}{\partial y}}$ or $F(x, a, t)$ is the multiplier of the last equation of (147):

$$dx - f(x, a, t)\, dt = 0. \tag{161}$$

This is Jacobi's generalized theorem in this special case of two variables.

54.   In §4, Liouville extended these ideas to n equations similar to (146) and (147):

$$\frac{\partial V}{\partial x_1} = C(x_1, x_1)\frac{dx_1}{dt} + C(x_1, x_2)\frac{dx_2}{dt} + \cdots + C(x_1, x_n)\frac{dx_n}{dt},$$

$$\frac{\partial V}{\partial x_2} = C(x_2, x_2)\frac{dx_1}{dt} + C(x_2, x_2)\frac{dx_2}{dt} + \cdots + C(x_2, x_n)\frac{dx_n}{dt},$$

$$\vdots$$

$$\frac{\partial V}{\partial x_n} = C(x_n, x_1)\frac{dx_1}{dt} + C(x_n, x_2)\frac{dx_2}{dt} + \cdots + C(x_n, x_n)\frac{dx_n}{dt}, \tag{162}$$

and

$$\frac{dx_1}{dt} = G(x_1, x_1)\frac{\partial V}{\partial x_1} + G(x_1, x_2)\frac{\partial V}{\partial x_2} + \cdots + G(x_1, x_n)\frac{\partial V}{\partial x_n},$$

$$\frac{dx_2}{dt} = G(x_2, x_1)\frac{\partial V}{\partial x_1} + G(x_2, x_2)\frac{\partial V}{\partial x_2} + \cdots + G(x_2, x_n)\frac{\partial V}{\partial x_n},$$

$$\vdots$$

$$\frac{dx_n}{dt} = G(x_n, x_1)\frac{\partial V}{\partial x_1} + G(x_n, x_2)\frac{\partial V}{\partial x_2} + \cdots + G(x_n, x_n)\frac{\partial V}{\partial x_n}.$$

$$(163)$$

As before, the x_i's are functions of t, and V is a function of $x_1, \ldots, x_n$ and of t. Despite the notation, due to Liouville, the $C(x_i, x_j)$'s and $G(x_i, x_j)$'s are functions of $x_1, x_2, \ldots, x_n$ (but independent of t), satisfying the anti-symmetry relations:

$$C(x_i, x_i) = 0, \qquad C(x_i, x_j) = -C(x_j, x_i), \qquad (164)$$

$$G(x_i, x_i) = 0, \qquad G(x_i, x_j) = -G(x_j, x_i), \qquad (165)$$

and the cyclic relations

$$\frac{\partial C(x_i, x_j)}{\partial x_k} + \frac{\partial C(x_j, x_k)}{\partial x_i} + \frac{\partial C(x_k, x_i)}{\partial x_j} = 0. \qquad (166)$$

Initially, Liouville merely assumed that the equations of (163) were independent. If those of (162) are also independent, the $G(x_i, x_j)$ matrix is evidently the inverse of the $C(x_i, x_j)$ matrix. Liouville possessed no terminology to express this inverse relationship generally, but explicitly wrote down the formulas for $n = 3$.

Now, let $x_i(t, a_1, a_2, \ldots, a_n)$ be solutions of (163) containing n arbitrary constants $(a_1, a_2, \ldots, a_n)$. Liouville then showed that the generalized Lagrange brackets

$$[a_k, a_l] = \sum_{i<j} C(x_i, x_j) \left(\frac{\partial x_i}{\partial a_k}\frac{\partial x_j}{\partial a_l} - \frac{\partial x_i}{\partial a_l}\frac{\partial x_j}{\partial a_k} \right) \qquad (167)$$

and the generalized Poisson brackets

$$(a_k, a_l) = \sum_{i<j} G(x_i, x_j) \left(\frac{\partial a_k}{\partial x_i}\frac{\partial a_l}{\partial x_j} - \frac{\partial a_k}{\partial x_j}\frac{\partial a_l}{\partial x_i} \right) \qquad (168)$$

are independent of time, i.e., functions of $a_1, a_2, \ldots, a_n$. Lagrange's and Poisson's perturbation formulas (16) and (16a) also extend to this type of equation. That is, add to V a perturbation $\Omega(x_1, x_2, \ldots, x_n, t)$ so that (162) and (163) are changed into

$$\frac{\partial V}{\partial x_i} + \frac{\partial \Omega}{\partial x_i} = \sum_{j=1}^{n} C(x_i, x_j) \frac{\partial x_j}{\partial t}, \qquad i = 1, 2, \ldots, n, \qquad (169)$$

and

$$\frac{dx_i}{dt} = \sum_{j=1}^{n} G(x_i, x_j) \left(\frac{\partial V}{\partial x_j} + \frac{\partial \Omega}{\partial x_j} \right), \qquad i = 1, 2, \ldots, n. \qquad (170)$$

Assume that $x_i(t, a_1(t), \ldots, a_n(t))$ is a solution of the perturbed equation (169) or (170), where x_i is the solution found above for the original equation. By inserting x_i into Ω, it becomes a function of $a_1, a_2, \ldots$, and we have

$$\frac{\partial \Omega}{\partial a_i} = \sum [a_i, a_j] \frac{da_j}{dt}, \qquad (171)$$

and

$$\frac{da_i}{dt} = \sum (a_i, a_j) \frac{\partial \Omega}{\partial a_j}. \qquad (172)$$

Liouville's proofs of these theorems are simple but long. For example, in proving that $[a_i, a_j]$ is a constant, he showed by direct calculation that

$$\frac{d}{dt}[a_i, a_j] = \frac{\partial^2 V}{\partial a_j \partial a_i} - \frac{\partial^2 V}{\partial a_i \partial a_j}. \qquad (173)$$

He discussed the Lagrangian formulas (162), (164), (166), (167), (169), and (172) first [§§IV–V], stressing that they are valid even if the equations of (162) are not independent, as long as Ω satisfies the same compatibility relations as V must satisfy in order that (162) has solutions at all:

> It is very remarkable, we repeat, that these formulas [171] subsist for a function fulfilling the same conditions as V, even if our eq.ons are not sufficient to determine x, y, $\ldots [x_1, x_2, \ldots]$ in terms of t, and that it is unnecessary to deal with complementary equations.
> [Ms 3637 (13), p. 7r]

In §§VI and VII, Liouville discussed the generalized Poisson formulas (163), (165), (168), (170) and (172).

55.   The rest of the draft is devoted to a generalization of Liouville's own theorem on half the integrals in involution [1855d]. In these sections, he assumed both (162) and (163) to be linearly independent, and therefore equivalent. He first observed that the generalized Poisson bracket yields a new integral if two integrals are known:

As one can see, this is the method of integration which Jacobi has based on Poisson's theorem. It can be extended to our equations and thereby augments its importance.

However, it can fail if (a, b) reduces to a numerical constant, or to a function of a and b which only provides a combination of the already known integrals.

Suppose, for example, that the quantity (a, b) is identically zero. One cannot derive anything by Jacobi's procedure; but, then a new possibility offers itself, and we can prove that in this case where $(a, b) = 0$ it is always easy to find the two integrals that remain unknown.

[Ms 3637 (13), p. 10]

As can be seen from this quote, Liouville first discussed the special case of four equations and two known integrals. After some deleted attempts, he succeeded [p. 12v] in proving a generalization of Bertrand's result (§41) that there exist two other integrals α and β such that

$$
\begin{aligned}
(a, b) &= 0 \quad \text{(assumption)}, & (\alpha, \beta) &= 0, \\
(a, \alpha) &= 1, & (b, \alpha) &= 0, \\
(a, \beta) &= 0, & (b, \beta) &= 1.
\end{aligned}
\tag{174}
$$

Originally, Liouville had interchanged α and β, but on p. 15, he discovered the inconvenience of this notation and changed it to the one above. In order to find α, β satisfying (174), Liouville as usual eliminated x_1 and x_2, using a, b as new independent variables and found from $(a, \alpha) = 1$ that

$$
G(x_4, x_3)\left(\left(\frac{\partial \alpha}{\partial x_3}\right)\frac{\partial a}{\partial x_4} - \left(\frac{\partial \alpha}{\partial x_4}\right)\frac{\partial a}{\partial x_3}\right) = 1
\tag{175}
$$

and from $(b, \alpha) = 0$ that

$$
G(x_4, x_3)\left(\left(\frac{\partial \alpha}{\partial x_3}\right)\frac{\partial b}{\partial x_4} - \left(\frac{\partial \alpha}{\partial x_4}\right)\frac{\partial b}{\partial x_3}\right) = 0,
\tag{176}
$$

where parentheses signify that a, b, x_3, x_4 are considered the independent variables. From these two equations, one can find $\left(\frac{\partial \alpha}{\partial x_3}\right)$ and $\left(\frac{\partial \alpha}{\partial x_4}\right)$ (Liouville assumed $G(x_4, x_3) \neq 0$).

Since α is an integral, we have

$$
0 = \frac{d\alpha}{dt} = \left(\frac{\partial \alpha}{\partial t}\right) + \left(\frac{\partial \alpha}{\partial x_3}\right)\frac{dx_3}{dt} + \left(\frac{\partial \alpha}{\partial x_4}\right)\frac{dx_4}{dt},
\tag{177}
$$

which determins $\left(\frac{\partial \alpha}{\partial t}\right)$, the derivatives $\frac{dx_3}{dt}$ and $\frac{dx_4}{dt}$ being known from the original equations (163). Thus, the differential

$$
d\alpha = \left(\frac{\partial \alpha}{\partial t}\right)dt + \left(\frac{\partial \alpha}{\partial x_3}\right)dx_3 + \left(\frac{\partial \alpha}{\partial x_4}\right)dx_4
\tag{178}
$$

is known, and α can be found by quadrature. The other integral can be found in the same way. Thus, the process leading to the last multiplier for two equations (§53) can be generalized to yield Liouville's theorem for four equations.

In §IX, Liouville finally sketched how this method can generally be used to solve $2n$ equations of the form (162), (163) by quadrature when n integrals in (generalized) involution are known.

> This is both a generalization and a new proof of the theorem which I have presented to the *Bureau des longitudes* on June 29, 1853 and whose precise statement can be found in the journal.
>
> [Ms 3637 (13), p. 12r]

This conclusion shows that the original proof of Liouville's theorem was different from the one above. As I have already remarked in §43, it probably originated in the Hamilton-Jacobi formalism, whereas the proof presented here is based on a generalization of Bertrand's theorem of the existence of a conjugate system of integrals.

56.   At this point, Liouville's notes become messy, but all the mess seems to deal with "the reconstruction of the canonical form" [Ms 3637 (13)]. If a conjugate system of integrals are found, for example, by way of Liouville's theorem, then the perturbation formulas (173) take on the simple canonical form

$$\frac{\partial a_i}{\partial t} = \frac{\partial \Omega}{\partial \alpha_i}, \quad \frac{\partial \alpha_i}{\partial t} = -\frac{\partial \Omega}{\partial a_i}. \tag{179}$$

In other words, if we can determine a set of conjugate integrals of equations (162), (163) for one V, say, V_1, then the system for any other V is reduced to the canonical form (179) with $\Omega = V - V_1$. In fact, if any complete set of integrals a_i $(i = 1, 2, \ldots, 2n)$ is known for V_1, it is possible to find a conjugate system of integrals provided certain partial differential equations can be solved. For example, taking $n = 2$, if $b(a_1, a_2, a_3, a_4)$ satisfy

$$(a_1, a_2)\frac{\partial b}{\partial a_1} + (a_1, a_3)\frac{\partial b}{\partial a_2} + (a_1, a_4)\frac{\partial b}{\partial a_4} = 0,$$

then $(a, b) = 0$; α, β can be found as above.

Liouville concluded:

> It is necessary to work over the details again, but the main thing is done, the canonical form is rediscovered as an *implicit cause* of our theorems; yet our theorems have a separate existence.
>
> [Ms 3637 (13), p. 16r]

57.   At this point, where Liouville only needed some revisions in order to have a highly publishable memoir, he began to contemplate a more direct transformation of equation (162) to a manageable form. Let $x_1 =$

$f_1(a_1, x_2, \ldots, x_{2n})$ be an integral of (162) where $a_1 = a_1(t)$. If a_1 is introduced as an independent variable to replace x_1, we find a new set of equations of the form of (162). Denote the coefficients in this system C' ($C'(a_1, x_2)$etc). Then, Liouville showed that if V is independent of time, the new coefficients satisfy (164) and (166). Now, a new variable, $a_2 = a_2(t)$, is introduced to replace $x_2 = f(a_1, a_2, x_3, \ldots, x_{2n})$, etc. Let us follow Liouville, taking $n = 2$ and introducing two new variables a_1, a_2 of which $V = a_1$. If either $a_1 = $ const. and $a_2 = $ const. are integrals or $C''(a_1, a_2) = 0$, the transformed system looks like [Ms 3637 (13), p. 30v]

$$
\begin{aligned}
1 &= C''(a_1, x_3)\frac{dx_3}{dt} + C''(a_1, x_4)\frac{dx_4}{dt}\,, \\
0 &= C''(a_2, x_3)\frac{dx_3}{dt} + C''(a_2, x_4)\frac{dx_4}{dt}\,, \\
0 &= C''(x_3, x_4)\frac{dx_4}{dt}\,, \\
0 &= -C''(x_3, x_4)\frac{dx_3}{dt}\,. \qquad (180)
\end{aligned}
$$

Thus, if $C''(x_3, x_4) \neq 0$, we can determine x_4, x_3 by quadrature from the last two equations. Otherwise, (166) yields

$$
\frac{\partial C''(a_1, x_3)}{\partial x_4} - \frac{\partial C''(a_1, x_4)}{\partial x_3} = 0\,,
$$

so that

$$
C''(a_1, x_3)\, dx_3 + C''(a_1, x_4)\, dx_4
$$

is an exact differential. In that case, the first equation of (180) yields the integral

$$
t + a_3 = \int C''(a_1, x_3)\, dx_3 + C''(a_1, x_4)\, dx_4.
$$

Similarly, from the second equation of (180), we find

$$
a_4 = \int C''(a_2, x_3)\, dx_4 + C''(a_2, x_4)\, dx_3.
$$

Having obtained similar results for six equations, Liouville concluded:

> Thus here is the substance of the thing - our eq.ons $\frac{dV}{dx} = C(x, y)\frac{dx}{dt} +$ &$_c$ being given, we can substitute for x, y, ... other variables related with these by equations that are independent of t, without changing the form [of the equations] and the imposed conditions. This being stated, if half of the new variables ($2n$ in number, n variables, that was $2n + 1$ [?]) give integrals if they are set equal to constants and if the last of the new coefficients are zero, new integrations can be effectuated by quadratures (if $2n + 1$ variables, there will remain one diff. eq. to be integrated).
>
> [Ms 3637 (13), p. 33v]

Having followed this line of thought for several pages, Liouville began to experiment with the form of the equations, allowing first the C's to be functions of time and adding a new term $C(x_i, t)$ on the right-hand side of (162). That soon raised the question "However, what is the purpose of V?" [Ms 3637 (13), p. 44v], which he answered by leaving it out. The rest of the notes therefore deal with the system

$$0 = C(x_i, t) + C(x_i, x_1)\frac{dx_1}{dt} + \cdots + C(x_i, x_n)\frac{dx_n}{dt}, \qquad (181)$$

where the C's (including $C(x, t)$) satisfy (164) and (166). By successively substituting variables, Liouville transformed this system of equations into one that has all the C's equal to zero, except for one $C(x_i, x_j) = -C(x_j, x_i)$ in each line. This makes it canonical, except, of course, that $\frac{\partial V}{\partial x_i}$ is replaced by $C(x_i, t)$. I shall not describe this new approach in more detail except for pointing out the similarity with the previous one, with the coefficients C and G now playing the central role that the Lagrange and Poisson brackets played before.

After some abortive attempts Liouville then (on pp. 58v–63r) drew up a new plan of a memoir in 11 chapters, which combined the two approaches to the general system (181). As a concluding remark, he wrote: "This seems complete enough, at least for a first Memoir" [Ms 3637 (13), p. 63r]. But, instead of writing this memoir, he continued to twist and turn the ideas with the result that neither this first nor any other work on this topic ever appeared.

58.   That is regrettable, for in addition to the interesting concrete generalizations of the methods and results of rational mechanics, Liouville's projected paper contained a new vision of differential equations, in particular, Hamilton's equations and their solutions. In Liouville's generalized equations, there is no separation between generalized coordinates and generalized momenta. This indicates also that in the canonical case this separation is artificial, and thus emphasizes the role of phase space.

This idea is intimately related to the second approach to equations (162), namely, successively substituting new variables for the old ones in order to obtain canonical equations. Jacobi had briefly indicated such an approach to Hamilton's equations [1837b] by showing how to find transformations of the phase space into itself, which preserved the canonical form of the equations. Such so-called canonical transformations have since become a central tool in rational mechanics. Their history is summarized by Prange [1933 §§31–34]. Here, I shall just mention that Bour [1862] contributed to their theory before Jacobi's general ideas were published by Clebsch [1866]. Of course, Liouville did not study canonical transformations in this manuscript, for he was precisely interested in transforming noncanonical systems into canonical ones, but the systematic approach by the transformation of the entire phase space onto itself was never expressed more clearly by Jacobi than it was here.

59.   I shall conclude this summary of the great generalizations of the methods of rational mechanics found in Liouville's notebooks by quoting the projected introduction to the memoir mentioned at the end of §57. It shows
that Liouville considered it only as a first step in a "great theory" on differential equations, at least as extensive as the one we find in Jacobi's
published and unpublished works:

§1. Form of the Equations. Let the n variables x, y, z be functions
of an independent variable t and subject to satisfying n differential
equations of the 1st order which can determine the ratio between the
differentials dx, dy, dz, ... and the differential dt as functions of t, x,
y, z, ... Consequently, let

$$dx = X\,dt, \ dy = Y\,dt, \ dz = Z\,dt, \ldots,$$

and we have therefore given our system of differential equations the
simplest form.

Even when the functions X, Y, Z are arbitrary and one does
not have the intimate notions which the formation of the equations
from the problems often gives a priori, one can establish certain
general propositions which are suited to clarify the attempts that
one must make in order to integrate them. Euler's and Lagrange's
factors and Jacobi's last multiplier, the known connection between
the integrals of a system of differential equations and the integration
of a linear partial differential equation, finally the method I have
given of changing an arbitrary system

$$dx = X\,dt, \ dy = Y\,dt, \ dz = Z\,dt, \ldots$$

into another:

$$dx \ = \ \frac{dV}{dx'}\,dt, \quad dy = \frac{dV}{dy'}\,dt, \quad dz = \frac{dV}{dz'}\,dt, \ldots$$

$$dx' \ = \ -\frac{dV}{dx}\,dt, \quad dy' = -\frac{dV}{dy}\,dt, \quad dz' = -\frac{dV}{dz}\,dt, \ldots,$$

where the number of functions to determine as functions of t is, to
be sure, doubled, but which has, thanks to its canonical form some
remarkable properties; all these partial theories belong, I say, to a
great general theory on which I hope I shall soon make a communication to the Académie. My aim today is less extensive. I shall only
discuss a class of equations of a special form, but still including all
those which Lagrange discussed in the famous memoir on the theory of the variation of the constants that he read to the Bureau des
Longitudes.   [Ms 3637 (13), pp. 55v–56r]

The last remark indicates that Liouville planned to present his memoir,
or at least his results, to the Bureau des Longitudes. We do not know
whether that happened, for its Procés Verbaux from this period are lost.

The "Liouvillian" Integrable Systems

60.   We shall now turn to Liouville's three voluminous papers from 1846 and 1847 [1846i, 1847h,i], dealing with a particular class of problems for which the equations of motion can be integrated. As the note discussed in §§45–48, they were inspired by a paper on geodesics on ellipsoids, namely by Jacobi's condensed *Note von der geodätischen Linie auf einem Ellipsoid und den verschiedenen Anwendungen einer merkwürdigen analytischen Substitution*. The *merkwürdigen analytischen Substitution*, i.e., the use of ellipsoidal coordinates (cf. [Chapter XI, §22]) allowed Jacobi to set forth the equation of the geodesics on an ellipsoid thereby bypassing its second-order differential equation in Cartesian coordinates, which has "such a complicated form that one is easily discouraged from any treatment of it" [Jacobi 1839a, p. 309]. Jacobi informed both Bessel and Arago about his discovery on December 28, 1838, the day after he had made it, and Arago immediately had it published in the *Comptes Rendus* [Jacobi 1838–1839]. Neither here nor in the longer note mentioned above and published by Crelle in 1839 and translated for Liouville's *Journal* in 1841 did Jacobi include a proof. This was provided by Liouville [1844c].

As in his work on equilibrium surfaces (cf. Chapter XI), Liouville wrote the equation of the ellipsoid in the form

$$\frac{x^2}{\rho^2} + \frac{y^2}{\rho^2 - b^2} + \frac{z^2}{\rho^2 - c^2} = 1. \tag{182}$$

A point (x, y, z) on the surface is then determined by the two confocal hyperboloids

$$\frac{x^2}{\mu^2} + \frac{y^2}{\mu^2 - b^2} - \frac{z^2}{c^2 - \mu^2} = 1 \tag{183}$$

and

$$\frac{x^2}{\nu^2} - \frac{y^2}{b^2 - \nu^2} - \frac{z^2}{c^2 - \nu^2} = 1 \tag{184}$$

passing through it. The triple (ρ, μ, ν) are the ellipsoidal coordinates of the point of which μ and ν determines its position on the given ellipsoid. According to a theorem by Dupin [1813], the curves $\nu = const$ and $\mu = const$ are the lines of curvature on the surface of the ellipsoid and thus intersect orthogonally. The arc length on the surface of the ellipsoid is determined by

$$ds^2 = p\, d\mu^2 + q\, d\nu^2 , \tag{185}$$

where

$$p = \frac{(\rho^2 - \mu^2)(\mu^2 - \nu^2)}{(\mu^2 - b^2)(c^2 - \mu^2)}, \quad q = \frac{(\rho^2 - \nu^2)(\mu^2 - \nu^2)}{(b^2 - \nu^2)(c^2 - \nu^2)}. \tag{186}$$

Liouville considered the geodesic as the path of a particle moving freely on the surface. As can be seen from Jacobi's lectures on mechanics, this had also been Jacobi's idea, but it had not been explained in his published note [Jacobi 1839a]. According to (185), the kinetic energy of a unit-mass is

$$T = \frac{1}{2} d\dot{s}^2 = \frac{1}{2}(p\dot{\mu}^2 + q\dot{\nu}^2), \tag{187}$$

and thus Lagrange's equations of motion (12) have the form

$$\frac{d(p\dot{\mu})}{dt} = \frac{1}{2}\frac{\partial p}{\partial \mu}\dot{\mu}^2 + \frac{1}{2}\frac{\partial q}{\partial \mu}\dot{\nu}^2, \tag{188}$$

$$\frac{d(p\dot{\nu})}{dt} = \frac{1}{2}\frac{\partial p}{\partial \nu}\dot{\mu}^2 + \frac{1}{2}\frac{\partial q}{\partial \nu}\dot{\nu}^2. \tag{189}$$

By defining

$$m(\mu) = \frac{\rho^2 - \mu^2}{(\mu^2 - b^2)(c^2 - \mu^2)}, \quad n(\nu) = \frac{\rho^2 - \nu^2}{(b^2 - \nu^2)(c^2 - \nu^2)}, \tag{190}$$

one has

$$p = m(\mu^2 - \nu^2), \qquad q = n(\mu^2 - \nu^2). \tag{191}$$

Liouville simplified equations (188) and (189) to

$$\sqrt{m}\,\frac{d(\mu^2 - \nu^2)\sqrt{m}\dot{\mu}}{dt} = \frac{C\mu}{\mu^2 - \nu^2}, \tag{192}$$

and similarly for $\dot{\nu}$, where C is twice the constant kinetic energy. Multiplying this equation with $2(\mu^2 - \nu^2)$ and integrating with respect to μ yielded

$$m(\mu^2 - \nu^2)^2\,\dot{\mu}^2 = A + C\mu^2, \tag{193}$$

where A is an arbitrary constant. Similarly, integrating the corresponding equation for $\dot{\nu}$, Liouville got

$$n(\mu^2 - \nu^2)\dot{\nu}^2 = B - C\nu^2, \tag{194}$$

and since $T = 2C$, he concluded that $A = -B$. Liouville now divided the two equations (193) and (194) to obtain the differential equation for the geodesics:

$$\frac{m\,d\mu^2}{n\,d\nu^2} = \frac{C\mu^2 - B}{B - C\nu^2} = \frac{\mu^2 - \beta}{\beta - \nu^2}, \tag{195}$$

where $\beta = \frac{B}{C}$ is a new arbitrary constant. This is a very simple equation for the variables, which can easily be separated:

$$\frac{m}{\mu^2 - \beta}\,d\mu^2 = \frac{n}{\beta - \nu^2}\,d\nu^2. \tag{196}$$

Taking the square root and reintroducing m, n from (190), Liouville found the solution

$$\int \frac{\sqrt{\rho^2 - \mu^2}\,d\mu}{\sqrt{(\mu^2 - b^2)(c^2 - \mu^2)(\mu^2 - \beta)}} = \alpha - \int \frac{\sqrt{\rho^2 - \nu^2}\,d\nu}{\sqrt{(b^2 - \nu^2)(c^2 - \nu^2)(\beta - \nu^2)}}.$$

(197)

Jacobi's solution [1838–1839] had also been an equality of two similar Abelian integrals, but instead of μ, ν he had used the variables φ, ψ, defined by

$$\nu = b \cos\psi, \quad \mu = \sqrt{c^2 \cos^2\varphi + b^2 \sin^2\varphi}.$$

61.   Liouville's notebooks reveal that his proof of Jacobi's theorem began to grow out of his geometrical studies on July 22, 1844, with an abortive mechanical attack on the geodesic. It ended:

> It is evident that one must account for the centrifugal force, which is not normal to the ellipsoid. However, it is even simpler to start from the formulas of analytical mechanics.
>
> [Ms 3617 (5), pp. 22v–23r]

The following day, this approach turned out to yield the proof summarized above, and Liouville immediately generalized the results to other surfaces and to mechanical problems involving forces:

> One can apply it to the motion of a body attracted to two fixed centers.
>
> [Ms 3617 (5), p. 28r]

Liouville had seen that the decisive step in the above deduction is to equip the surface with orthogonal coordinates, which separate in the equation of motion (195). In his three main memoirs on mechanics, Liouville generalized this idea to mechanical systems influenced by forces. First, he treated one point mass on a surface [1846i], second, one point mass in space [1847h, results summarized in 1846h], and finally a system of point masses [1847i]. All of these levels of generality are of great interest, but in order not to overload this account with details, I shall limit it to the first step - which will show the connection to the geodesic case above - and the last step, which will show the full extent of Liouville's work.

In [1846i] Liouville considered a particle moving on a surface described by the coordinates α, β for which

$$ds^2 = \lambda(\alpha, \beta)(d\alpha^2 + d\beta^2).$$ (198)

Such a coordinate system is now called isothermal. Since there is no $d\alpha \cdot d\beta$ term, the coordinates α, β are orthogonal. The equations of motion of the particle can then be solved by quadrature if the potential U is such that

$$(C - 2U)\lambda = f(\alpha) - F(\beta),$$ (199)

where f and F are arbitrary functions and C is twice the total energy of the particle. In fact, in this case, one can separate the variables in the equations of motion and obtain

$$\frac{d\alpha}{\sqrt{f(\alpha) - A}} = \frac{d\beta}{\sqrt{A - F(\beta)}}, \tag{200}$$

corresponding to (195). Condition (199) is satisfied for all values of the total energy if and only if both λ and U can be separated in the form $f(\alpha) - F(\beta)$.

Liouville carried this method of separation furthest in §14 of [1847i], where he considered a conservative system of points with a number of holonomic constraints. He assumed that it be characterized by n generalized "orthogonal" coordinates, for which the kinetic energy could be expressed as

$$2T = \sum_{i=1}^{n} \lambda_i(q_1, \ldots, q_n)\dot{q}_i^2, \tag{201}$$

i.e., without mixed $\dot{q}_i\dot{q}_k$ terms.

Moreover, he assumed that there exist functions Π_i^j such that

$$\sum_{i=1}^{n} \frac{1}{\lambda_i} \Pi_i^j(q_i) = \begin{cases} 0 & \text{for } j = 1, 2, \ldots, n-1 \\ 1 & \text{for } j = n. \end{cases} \tag{202}$$

If the potential U is of the form

$$-U = \sum_{i=1}^{n} \frac{f_i(q_i)}{\lambda_i}, \tag{203}$$

the solution of the equations of motion can be solved by separation of the variables and quadrature. More precisely, Liouville showed that a complete solution V to Hamilton-Jacobi's equation (93) can be obtained by integrating the equations:

$$\left(\frac{\partial V}{\partial q_i}\right)^2 = \sum_{j=1}^{n-1} \alpha_j \Pi_i^j(q_i) + C\Pi_i^n(q_i) + 2f_i(q_i), \quad i = 1, 2, \ldots, n, \tag{204}$$

where the α_j's are arbitrary constants. Thus, V can be found by quadrature:

$$V = \sum_{i=1}^{n} \int \sqrt{\sum_{j=1}^{n-1} \alpha_j \Pi_i^j(q_i) + C\Pi_i^n(q_i) + 2f_i(q_i)} \, dq_i. \tag{205}$$

The integrals of the equations of motion can then, as usual, be determined by

$$\frac{\partial V}{\partial \alpha_j} = \beta_j. \tag{206}$$

In particular, if all the λ_i's are equal and of the form

$$\lambda = \sum_{i=1}^{n} F_i(q_i), \tag{207}$$

the conditions in (202) are easily seen to be satisfied by $\Pi_i^n = F_i (i = 1, 2, \ldots, n)$, $\Pi_i^i \equiv 1 (i = 1, 2, \ldots, n - 1)$, $\Pi_{i+1}^i \equiv -1 (i = 1, 2, \ldots, n - 1)$, and the rest of the Π's identically zero. Thus, in that case, the system is integrable if U is of the form

$$-U = \frac{\sum_{i=1}^{n} f_i(q_i)}{\sum_{i=1}^{n} F_i(q_i)}. \tag{208}$$

62. As indicated above, Liouville, in his first paper, used the Lagrange approach, but changed to the Hamilton-Jacobi approach in the second and third paper, although he continued to use Lagrange's equations in certain special cases. The great efforts made in the transformation from Cartesian to other coordinate systems show that Liouville was just familiarizing himself with the Hamilton-Jacobi approach in this generality.

Liouville showed that many of the integrable systems known at that time were covered by his theorem. For example, the motion of a point attracted to two fixed centers, F and F', by gravitational forces and toward their midpoint 0 with a force obeying Hooks law is of Liouvillian type. The attraction to two centers had been solved with great difficulty by Euler [1766–1767 and 1767], and Lagrange had shown how to add the third force from the midpoint [1766–1769]. Moreover, Jacobi had hinted at this application of ellipsoidal coordinates in his brief note [1839a] and spelled it out explicitly in his lectures [1866, 29th lecture]. For a particle moving in a plane, Liouville further showed [1846i p. 354] that his method allowed an additional attraction perpendicular to two axes and proportional to the inverse of the cube of the distance to the axis. The one axis must pass through the centers F and F', and the other cuts it orthogonally at the midpoint (Fig. 3). Finally, he could also allow the centers F and F' to rotate around 0 [1847h, pp. 440–441].

The rest of the three long memoirs is of a geometric nature. Orthogonal and isothermal coordinate systems, in particular, generalized and degenerate ellipsoidal systems, are discussed, and interesting theorems on geodesics are found for $U = 0$; these in turn are used to solve Abelian equations. Although it is artificial and interrupts the unity of Liouville's work, I shall postpone a discussion of the geometric aspects of these memoirs to the next chapter. Here, I shall just emphasize that they are among the richest papers in Liouville's production.

63. Liouville's investigations on integration of the equations of motion by separation of variables gained new attention around the turn of the century, when they were carried further by Stäckel [1890,1891,1893,1905], Levi-Civita [1904], Dall'Acqua [1909,1912], Burgati [1911], Hadamard [1911],

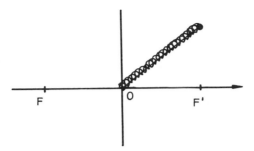

FIGURE 3

Haupt and Hilb [1924], and others. Their results are described by Prange [1933, §19a-19c], so I shall only summarize this history briefly.

Like Liouville, Stäckel focused on orthogonal coordinates and showed that Liouville's requirements for separability (202) and (203) are both necessary and sufficient. In a letter to Stäckel, Levi-Civita [1904] showed that if the potential depends on all the coordinates these coordinates must be orthogonal in order for the system to be separable. This emphasizes the importance of Liouville's work, for as Liouville had already pointed out [1847h, p. 410], the system can be simplified if a coordinate does not occur in the potential - it is cyclic in the terminology introduced by Helmholtz (see [Prange 1933, p. 572] for a discussion of the various terminologies).

As a step in his proof of this theorem, Levi-Civita also showed that if a mechanical system can be solved in this way the coordinates must be chosen so that the corresponding geodesic line (i.e., for $U = 0$) can also be determined by separating of variables. To this end, he used the geometric consideration of the action integral discussed above (§§45-48).

In a paper, *Über die Transformation Liouvillescher Mannigfaltigkeiten* [1924], Haupt and Hilb addressed the problem: given a set of coordinates with its element of arc length and a potential, decide if it is possible to find other coordinates that make the equations of motion solvable by separation of variables. Such questions became of importance in early quantum mechanics (cf. [Born 1925]).

64. This wide field was opened by Liouville's work, but although all the authors mentioned Liouville, how much of his research they were aware of varied greatly. Stäckel referred to [1846i], in which Liouville had treated only two variables. For that reason, he greatly overestimated how much he had generalized Liouville's work. Darboux [1894, Part 3, p. 9 ff] also emphasized Liouville's work on two variables, whereas Appell in his *Traité de Mécanique Rationelle* [1909-1911, Vol. 1, §305, Vol. 2, §474] mentioned Liouville's generalization to higher dimensions.

Which cases were called "Liouvillian" varied accordingly. Stäckel [1890] reserved the name "Liouvillian surfaces" for those having a line element

$$ds^2 = (F_1(q_1) + F_2(q_2))(dq_1^2 + dq_2^2).$$

In [Stäckel 1893], however, he talked about a line element of "Liouvillian form," because it is a property of the coordinate system, not of the surface. Appell, Hadamard, and others attached Liouville's name to the larger class of line elements:

$$ds^2 = (F_1(q_1) + \cdots + F_n(q_n))(g_1(q_1)\,dq_1^2 + \cdots + g_n(q_n)\,dq_n^2)$$

or the corresponding expressions for the kinetic energy. Modern textbooks, however (e.g., [Campbell 1971, p. 86]), seem to follow Whittaker [1924, p. 71] in ascribing another limited case to Liouville. When the kinetic energy is of the form

$$2T = g_1(q_1)\dot{q}_1^2 + \cdots + g_n(q_n)\dot{q}_n^2$$

corresponding to

$$ds^2 = \sum g_i(q_i)\,dq_i^2$$

and

$$U = f_1(q_1) + f_2(q_2) + \cdots + f_n(q_n),$$

they talk about a system of Liouvillian type. The summary in §44 has shown that all these "Liouvillian" forms are more restricted than the object of Liouville's own investigations.

Concluding Remarks

65.   In this section, I have tried to focus attention on Liouville's most important contributions to rational mechanics, important meaning important to Liouville, important to further development, or important in my eyes. Therefore, several of Liouville's works in the field have only been mentioned in passing or not at all. In this last category belong several unpublished notes [3615 (2)] on the motion of rigid bodies, and in particular, a voluminous paper published in 1856 but composed in the early 1840s. Its purpose was to carry over to arbitrary systems the formulas traditionally found for the motion of rigid bodies. The result was formulas of transformation between two coordinate systems moving relative to each other. The paper was written in honor of Poisson, who had suggested its content to Liouville:

> Appointed to succeed the illustrious geometer at the *Bureau des Longitudes*, I did a pious duty by developing an idea which Poisson had talked to me about many times and which he considered important. Even this indefatigable worker did not find the time.
>
>   [Liouville 1856p]

Poisson's interest may have been aroused by Coriolis' works around 1833. Compared to these works, Liouville's paper had little to offer in the 1840s, and when it was published in 1858, its content was rather pedestrian. Its main merit was that it prepared Liouville for his investigations of 1842 on rotating masses of fluids (cf. Chapter XI).

I have mentioned in passing that Liouville pointed out the importance of cyclic coordinates. This is perhaps the place to emphasize that Liouville repeated this observation [1847h, p. 410 and 1857q, p. 71], showing thereby himself as a pioneer in this respect also.

Of the discoveries that have been discussed at length in this section, I shall emphasize Liouville's theorem "on the volume in phase space," Liouville's theorem on integration by quadrature when half the integrals in involution are known, the geometrization of the action integral, and the Liouvillian integrable systems. These are all central ideas or theorems in modern mechanics, and Liouville himself knew their great value, except for the first theorem. Historically, this is a rather superficial conclusion about Liouville's work in mechanics. However, a deeper conclusion as to Liouville's originality and influence in this field is problematic.

For on the one hand, Liouville was inspired by Jacobi's work and therefore often repeated Jacobi's discoveries, but on the other hand, some of Jacobi's ideas were not published in detail until later, e.g., in the *Vorlesungen* [1866]; and on the one hand, Liouville developed a nice connected theory, both in his lectures of 1852–1853 and in the summer of 1856, but on the other hand, only fragments of it were published; finally, on the one hand, Liouville's ideas and results obtained an important place in the further development of rational mechanics, but on the other hand, most of these developments would have taken place even without Liouville's pioneering work, and often the next person to take them up did not know of Liouville's contributions (this applies to the discoveries emphasized above, except for the last one). It also complicates our evaluation that we do not know the exact content of his lectures of 1852–1853 nor his audience.

Still, it is certain that Liouville's work on mechanics stands out among his own works as a long-lasting passion, which influenced his work in other fields, such as geometry and elliptic functions. It may seem natural for a teacher who first taught mechanics at the *École Polytechnique* and later at the university to do research in this area. But, the interaction between this part of Liouville's teaching and his research should not be overestimated. In fact, Liouville never showed much interest in the rather elementary parts of mechanics that entered the curricula, and moreover, his research in mechanics was done exclusively in the period when he did not teach at either of these institutions. Indeed, it is conspicuous that he stopped this research at the very moment he was appointed Professor of Rational Mechanics at the *Faculté dés Science* in 1857.

On the short view, Liouville's most important contribution to rational mechanics may have been his introduction of Jacobi's ideas into France, first through the publication of the translated works of his friend and later through his own contributions. His lectures at the *Collège de France*, 1852–1853, were no doubt also important in this respect. To be sure, Bertrand also spread the news of Jacobi's ideas, but his interest began later (it may very well have been aroused by Liouville), and when he began to work in the field, he underestimated the work of Jacobi (as he admitted later). Bour, the leading expert in rational mechanics of the next generation in France and a student of Bertrand, cited Liouville as the greatest prophet of Jacobi [cf. Bour 1862]. In the long run, Liouville's ideas became imbedded at important points in the further development of rational mechanics, and there is little doubt that if Liouville had published the projected large volume on rational mechanics and differential equations instead of publishing only fragments in small notes his direct influence would have been much greater, perhaps comparable with that of Jacobi's.

XVII

Geometry

1. Mr. Liouville's numerous and important works are especially concerned with various branches of pure analysis, with transcendental functions, the theory of numbers and with the applications of infinitesimal analysis to astronomy and mathematical physics. However, in these works, one often finds results of interest to geometry, results that even directly concern the theory of surfaces, and views which, having given rise to new researches, have contributed to the progress of the science.

This fair evaluation opens a summary of Liouville's geometric works written by his friend Chasles in the *Rapport sur les progrès de la géometrie en France* [Chasles 1870]. It is the only connected historical survey of an aspect of Liouville's production written during his lifetime and, as such, has been of great value for the composition of this chapter. Chasles puts the development in France into perspective by referring to substantial Irish contributions, and to some English, Belgian, and Italian works, but as a picture of the general state of the art it must be used with care, for Chasles did not understand German and was therefore completely ignorant of the much more interesting development in Germany. As far as differential geometry is concerned, this one-sideness has been rectified in Karin Reich's interesting and comprehensive paper [1973]. These two papers provide an excellent general background for appreciating Liouville's work. Therefore, this chapter is limited to a localized study of Liouville's work with an emphasis on how it fits into his other research and how it is connected to his immediate predecessors and successors.

One can distinguish four periods in Liouville's geometrical research. In 1841, he published his only paper on algebraic geometry. This and a number of later papers were inspired by Chasles and Steiner. During the mid-1840s, Liouville was under the influence of Jacobi. Starting from Jacobi's integration of the equations of the geodesic curve on an ellipsoid, Liouville pursued a mechanical approach to geometry. In the third period, William Thomson's letters on the inversion in a sphere made Liouville investigate this geometric transformation. His longest paper on this subject was published in 1847, but the most important one, containing a famous theorem named after him, appeared as one of the six notes in Liouville's new edition of Monge's *Application de l'Analyse à la Géométrie* [1850]. These notes contain the most interesting results found during the fourth period, which is characterized by a strong influence from Gauss's now classical *Disquisi-*

tiones generales circa superficies curvas [1828]. This chapter is therefore divided into four sections, one for each of these periods.

Analytic versus Synthetic Geometry; Chasles's Influence

2.   As witnessed by notes from his college days, projective geometry was among Liouville's earliest interests, but this research was never pursued. In the early 1830s, while propagandizing for the fractional calculus, Liouville had given several examples of its usefulness in geometry. However, not until the early 1840s did he develop a genuine research interest in geometry, under the inspiration of Steiner (cf. Chapter III) and as a response to a remark in Chasles's *Aperçu historique sur l'origine et le développement des méthodes en géométrie...* [1837a]. Here, Chasles had shown by projective means that if one draws all the tangents to an algebraic curve (called "geometric" in those days) parallel to a given line then the barycenter (the center of mass) of the points of tangency is independent of the given line. Chasles had remarked that this theorem and a similar theorem on algebraic surfaces in space correspond to certain algebraic theorems, whose direct algebraic proof he believed to be hard.

Indeed, if $M(x, y) = 0$ is the equation of a curve, the points of tangency with tangents parallel to a line given by $y = ax$ must satisfy

$$M(x,y) = 0 \quad \text{and} \quad \frac{\partial M}{\partial x} + a\frac{\partial M}{\partial y} = 0. \tag{1}$$

Thus, Chasles's theorem says that by eliminating y from these equations, one obtains an equation of the form

$$x^i + G_i x^{i-1} + \cdots = 0, \tag{2}$$

where G_i, which represents the sum of its roots, is independent of the choice of a. The corresponding theorem in three variables is an obvious generalization.

> I had originally embarked on this research in order to obtain a direct demonstration of these theorems concerning elimination, [Liouville stated]. I must admit that this demonstration was not at all as diffi- cult as Mr. Chasles seems to have thought. In fact, it relies on very simple and very well-known principles.
>
> [Liouville 1841d, p. 347]

While pursuing this line of inquiry, Liouville discovered that his methods led to a "beautiful theorem of Jacobi" [Jacobi 1835; see also 1836b] and even beyond to the following theorem.

Let $f(x,y)$, $F(x,y)$, $\varphi(x,y)$, and $\varphi_1(x,y)$ be polynomial functions of which the degree in x and y of φ_1 is lower than that of φ. Define C and A by

$$C(x,y) = \frac{\partial f}{\partial x}\frac{\partial F}{\partial y} - \frac{\partial f}{\partial y}\frac{\partial F}{\partial x} \tag{3}$$

$$A(x,y) = \frac{\partial F}{\partial x}\frac{\partial \varphi}{\partial y} - \frac{\partial F}{\partial y}\frac{\partial \varphi}{\partial x}. \tag{4}$$

Then, we have

$$\sum \frac{\varphi_1(\alpha,\beta)}{\varphi(\alpha,\beta)} = \sum \frac{\varphi_1(\lambda,\mu)C(\lambda,\mu)}{f(\lambda,\mu)A(\lambda,\mu)}, \tag{5}$$

where the first sum ranges over all the simultaneous roots (α,β) of $f(\alpha,\beta) = 0$ and $F(\alpha,\beta) = 0$ and the second over the simultaneous roots (λ,μ) of $F(\lambda,\mu) = 0$ and $\varphi(\lambda,\mu) = 0$.

Jacobi's theorem corresponds to the special case $\varphi(x,y) = C(x,y)$. Liouville's proof is rather long, but rests on the simple idea, which he attributed to Waring [Liouville 1841d, p. 411], of solving a set of equations by eliminating variables in two different ways and equating the results. In this way, he even generalized Jacobi's theorem further [Liouville 1841d, p. 348]. In a later paper [Liouville 1841f], he presented an alternative derivation making use of the decomposition of fractions into simple fractions.

3.   In addition to Chasles's theorem, Liouville drew many new and interesting geometrical consequences of these algebraic theorems. They were presented in the original memoir and in two subsequent papers [Liouville 1844a and d]. The most important results were discussed by Chasles [1870, pp. 128–130], so here I shall just illustrate their general form by two examples. Most of the theorems are concerned with the points of tangency of an algebraic curve when all the tangents under consideration are parallel to a given line. In a letter of August 4, 1841, to Dirichlet [Neuenschwander 1984a, II, 1] he emphasized in particular that the sum of the curvatures of the curve at these points is zero, or as he put it,

$$\sum \frac{1}{\varrho} = 0. \tag{6}$$

Liouville had already presented this and other "principal results" to the *Société philomatique* at its meeting on July 10, 1841 [Liouville 1841d, pp. 345 and 365], after which Duhamel had added another consequence, namely,

$$\sum \varrho = 0, \tag{7}$$

where the sum ranges over the same points as above. When Liouville published his paper a few months later, he included Duhamel's result. It was in turn generalized by Terquem (cf. [Chasles 1870, p. 129]). Several of

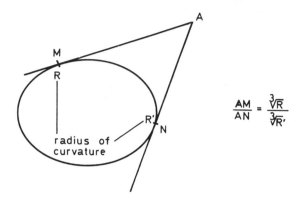

$$\frac{AM}{AN} = \frac{\sqrt[3]{R}}{\sqrt[3]{R'}}.$$

FIGURE 1

Liouville's theorems dealt with asymptotes; for example,

The sum of the cotangents of the angles under which two algebraic curves in a plane cut each other is equal to the sum of the cotangents of the angles under which their asymptotes cut each other.

In a footnote, Liouville commented on this theorem as follows:

> To these theorems correspond other elegant related theorems that are given by the principle of reciprocity or of duality, used by modern geometers.
>
> [Liouville 1841f, p. 410]

As formulated by Poncelet and Gergonne, this principle states that if a theorem on points and lines in a plane is rephrased by interchanging the words "line" and "point" the result is a valid theorem. Liouville's note continued with such a dual theorem:

> Thus, Mr. Chasles, to whom I communicated the theorem concerning $\sum \cot (\Omega - \omega)$, has in return given me the following statement:
> "Consider two geometric curves lying in the same plane and draw all their common tangents. If from an arbitrary fixed point one draws two rays to the two points of contact of each of these tangents and one considers the angle between the two rays, then the sum of the cotangents of the angles will be equal to the sum of the cotangents of the angles which the tangents drawn through the fixed point to the first curve make with the tangents drawn from the same point to the second curve."
>
> [Liouville 1841d, p. 410]

Three years after publishing [1841d], Liouville added to its last theorem a corollary [1844b] to the effect that the lengths of the two tangents drawn from a point to a conic are to each other as the cube root of the radii of curvature in the points where they touch the conic (Fig. 1).

4.   Liouville's and Chasles's attitudes illustrate the tension between the synthetic and the analytic approach to geometry. After Cartesian analytical methods had been virtually universally adopted in all geometric research for one and a half centuries, synthetic methods experienced a renaissance in the early 19th century. Monge had secured a central place for his descriptive geometry in the early curriculum at the *École Polytechnique*, and his followers, e.g., Carnot, Dupin, Chasles, and Poncelet, had continued to fight against what they considered to be a counterintuitive and unrigorous analytical machine [cf. Daston 1986]. In order to confer on the cumbersome Euclidean approach a generality and intensive power matching the undeniable analytical successes, these geometers had been forced to introduce "ideal quantities," such as double points, points at infinity, imaginary points, etc. Indeed, Liouville's theorems must be interpreted with this in mind. These innovations, which led to what was called "modern geometry" or "projective geometry," were very successful but weakened the claim of intuitive clarity and rigor and opened the door for a reaction from the analysts. In fact, Cauchy [1820] severely criticized the central principle of continuity, which Poncelet had formulated in the following way: "If one figure is derived from another by a continuous change and the latter is as general as the former, then any property of the first figure can be asserted at once for the second figure" [Poncelet 1822, introduction, transl. Kline 1972].

In this way, geometers of the early 19th century were divided into two camps who mutually denied the validity of each others' approach. The analysts, who were in the majority, ruled the Paris establishment where the synthetic geometers were considered outsiders. The hostility between the two camps only slowly diminished, and it is characteristic that the two main synthetic geometers in France after 1820, Poncelet and Chasles, were elected rather late to the *Académie* (in 1834 and 1851, respectively) and that it was not until 1846 that a special chair in *geométrie supérieure* was created for Chasles at the *Faculté des Sciences*. By that time, Germans such as Jacob Steiner (Swiss born), von Staudt, Möbius, and Plücker, had taken the lead in projective geometry.

5.   Where was Liouville in this conflict? As pointed out in the quote by Chasles heading this section, Liouville was an analyst at heart. He knew how to crank the analytical machine to obtain interesting results. The geometric papers discussed above clearly reveal his preference for analytical methods. In the concluding remark of [Liouville 1844a], Liouville stressed that the analytical method both leads to theorems a priori and provides proofs of the theorems; and earlier in the same paper, he characterized an analytic proof as being "neater and clearer" than a proof "without any calculations" [Liouville 1844a, p. 340]. However, in the introduction to the first paper on algebraic geometry, he expressed his views with great care:

There are perhaps some inconveniences in introducing the language of algebra into geometry, but in certain theories, and in particular in those we are concerned with, there are also great advantages. Here, by the way, we follow the example of Mr. Poncelet and most of the authors to whom the geometry is indebted for the immense progress it has made in recent years.

[Liouville 1841d, p. 346]

By pretending (with some justice) to follow one of the main protagonists of synthetic geometry, Liouville took the sting out of his analytical approach. His diplomatic formulation was not only meant to pacify his friend Chasles, but was sincerely felt. For although an analyst, Liouville was not an analytic chauvinist. In fact, after his early Laplacianism, Liouville was a mathematical pragmatist. He never let any philosophical ideas bar his way to new results, and even questions of rigor did not bother him very much. He enjoyed the diversity of mathematics and often stressed that different approaches to the same subject could lead to new developments in mathematics. Moreover, he was not blind to the simplicity and beauty of synthetic geometry (cf. Chapter III). Thus, after Chasles had presented a geometric approach to elliptic functions at the Académie meeting on December 9, 1844, Liouville remarked that "It is very much to be hoped that Mr. Chasles can extend his ingenious geometric considerations to transcendentals of a higher order and first to the Abelian functions of the first class," because "a detailed and profound geometric discussion will no doubt give these theorems a new and elegant form" [Liouville 1844g, pp. 1261–1262].

6.   At the same occasion, Liouville, in particular, announced that while verifying Jacobi's equation for the geodesics on an ellipsoid, which was an equation in Abelian functions (cf. Chapter XVI(60)), he had stumbled upon a first integral, namely,

$$\mu^2 \cos^2 i + \nu^2 \sin^2 i = \beta, \tag{8}$$

where μ, ν are the ellipsoidal coordinates of a point of the geodesic (cf. Chapter XI, §22) and i is the angle at that point between the geodesic and the line of curvature μ = const. Liouville stated that this integral "seems so simple that one can hope to derive it by a purely geometric method. I take the liberty to recommend this research to the talent of Mr. Chasles who is so experienced in these matters" [Liouville 1844g, p. 1262].

When Liouville published his paper on geodesics on ellipsoids later the same month [1844c], he repeated the invitation to find a geometric proof of (8) or of an equivalent theorem by Joachimsthal to the effect that $P \cdot D$ = constant, where D is the semidiameter of the ellipsoid parallel to the given geodesic at a point on it, and P is the perpendicular dropped from the center of the ellipsoid on the tangent plane at that point.

To be sure, it was commonplace to ask such questions to the other camp of geometers hoping that their approach would fail. For example, when Chasles had declared that the algebraic equivalent of the theorem quoted in §2 would be hard to prove, it was not because he longed to see a simple analytical proof, such as the one Liouville gave, but because he wanted to emphasize that synthetic methods had now surpassed the analytic ones so far that they provided shortcuts even to algebraic theorems. Liouville's invitation was not of this kind, but expressed his true interest and curiosity. This became evident when on January 19, 1846, he presented a geometric solution of his own to the *Académie* [Liouville 1846b and c]. This proof, which he characterized as "both direct and fast," was based on an idea due to Dupin.

Before Liouville's own solution became known, the Dublin mathematician Michael Roberts had sent him several results on curves on ellipsoids deduced from his equation (8). Liouville made a comprehensive summary of these "very interesting theorems" to the *Académie* [Liouville 1845i]. The entire work [Roberts 1846] appeared in the same issue of Liouville's *Journal* as Liouville's own geometric proof of (8) and another proof by Chasles [1846a]. They were followed by a number of papers by Chasles [1846c], Liouville, Roberts, and a colleague of Roberts's, Mac-Cullagh, dealing with curves on second-degree surfaces, of which some will be discussed in the next section.

7.   A genuine interest in synthetic geometry is also revealed in some of Liouville's other papers. For example, in [1842d] Liouville showed geometrically that the ellipse of smallest area passing through three points A, B, and C has its tangent at C parallel with AB. This and a similar theorem on ellipsoids, also discussed by Liouville, stemmed from Euler, who had deduced them "from analysis." Moreover, in 1846, Liouville turned to a "theorem of Joachimsthal concerning plane curves," introducing a new proof with the words:

> Mr. Joachimsthal's proof only requires a very simple calculation; yet geometric considerations lead to the goal even more easily as we shall see.
>
> > [Liouville 1846d]

As a result of the new proof, Liouville found an extension of Joachimsthal's result, without knowing that it was in fact included in a theorem published by Lancret as early as 1806 (cf. [Chasles 1870, pp. 10–11]).

But, the ultimate testimony of Liouville's love for synthetic geometry and his high esteem for Chasles's work in particular is found in his notebook [Ms 3618 (6)], which shows that he followed Chasles's geometry course at the *Faculté* during 1847–1848, the second time it was held. Liouville even went so far as to volunteer to substitute for Chasles in this course, promising to "follow in your footsteps" so that "the tradition will continue" [Neuenschwander 1984a, I, 7]. As described in Chapter V, Chasles did not accept Liouville's offer, but the instance underlines that although Liouville was an analyst he still had room for synthetic geometry in his heart and even the skill to make interesting contributions.

Relations with Mechanics and Elliptic and Abelian Functions; Jacobi's Influence

8.   Jacobi's paper [1839a] on the geodesics on an ellipsoid became the starting point of a series of works by Liouville concerned with various topics of geometry, such as geodesics on other surfaces and the study of different coordinate systems. It was probably Jacobi's announcement of the usefulness of ellipsoidal coordinates that caught Liouville's attention, for he had himself investigated their application to the problem of the equilibrium of rotating masses of fluid (cf. Chapter XI). In this connection, Liouville had given a geometric construction of the gravitational force at the surface of the equilibrium surface in [1843b], and the following year he proved Jacobi's formula for the geodesic on an ellipsoid. Joachimsthal [1843] and Minding [1840] (cf. [Jacobi 1866, p. 215]) had preceded Liouville in this respect.

As I explained in the last chapter, Liouville found the equation of the geodesic by considering it as the trajectory of a freely moving particle on the surface. The first integral

$$\mu^2 \cos^2 i + \nu^2 \sin^2 i = \beta \tag{8}$$

discussed above is a simple consequence of the formula (XV (195)). By generalizing this method, Liouville found his famous integrable mechanical systems (Chapter XVI). When the potential energy of the particle is zero, the equation of the geodesics follow as an easy corollary in the case where the line element is given by

$$ds^2 = \lambda(d\alpha^2 + d\beta^2), \tag{9}$$

where

$$\lambda = \varphi(\alpha) - \omega(\beta), \tag{10}$$

φ and ω being "arbitrary" functions. (Reich [1973, p. 311] states that Liouville succeeded in integrating the equation of the geodesic under assumption (9) alone. That is not correct.) Under assumptions (9) and (10), a geodesic satisfies a differential equation of the form

$$\frac{d\alpha}{\sqrt{\varphi(\alpha) - a}} = \frac{d\beta}{\sqrt{a - \omega(\beta)}}, \tag{11}$$

where a is an arbitrary constant. This equation can be integrated by quadrature since the variables are separated (cf. formula XVI (200)). In accordance with (8), Liouville discovered that along a geodesic satisfying (11) we have

$$\varphi(\alpha) \cos^2 i + \omega(\beta) \sin^2 i = a, \tag{12}$$

where i, as in (8), is the angle between the geodesic and the curves $\alpha =$ const [Liouville 1846i, p. 351].

9. Liouville later proved that any surface can be equipped with a coordinate system having a line element of the form (9) (so-called isothermal coordinates). It is unclear whether Liouville was aware of that when he wrote [1846i], but he did investigate what assumption (10) meant in the plane. He showed that the only coordinate system satisfying (9) and (10) is the elliptic one, i.e., the obvious restriction to two dimensions of the ellipsoidal system in space; its coordinate curves $\alpha = $ constant   and β $=$ constant consist of confocal ellipses and hyperbolas. In an appended note, Bertrand pointed out that in terms of stationary heat conduction this uniqueness theorem says that these confocal curves constitute the only system of equitemperature curves for which the radius of curvature at each point is inversely proportional to the cube of the heat flux [Bertrand 1846].

As pointed out in the preceding chapter, line elements satisfying (9) and (10) on an arbitrary surface are, in modern textbooks, called Liouvillian forms (nets, etc.) . As shown by Blaschke and Zwirner in 1927, they can be characterized by the following simple diagonal property: In every quadrangle of the coordinate net, there exist two geodesic diagonals of equal length (cf. [Blaschke and Reinhardt 1960, pp. 71–74]). The rectangular, polar, and parabolic coordinates in the plane are degenerate cases of the elliptical system and received special treatment by Liouville [1846i, §10]. He finally showed how the equation of the geodesic on an arbitrary surface of revolution and on a right helicoid can be determined in the way described above.

10. In these and his subsequent papers, Liouville gave the word "geodesic" the meaning it has today. As pointed out by Reich [1973, pp. 308–309], this term was first used by Bessel for the locally shortest lines on ellipsoids of revolution and, more generally, by Jacobi on arbitrary ellipsoids, because of the resemblance of these surfaces with the surface of the earth. On other surfaces, the term "shortest line" (linea brevissima, etc.) was universally adopted until Liouville in [1844g] generalized the term "geodesic line (the shortest line)" to an arbitrary surface:

On an arbitrary surface, the geodesic line has the property that its radius of curvature is in each point normal to the surface.
[Liouville 1844g, p. 1262]

Liouville's consistent use of "geodesic line" in this general sense was taken over by Bonnet and Dickson in 1850 and soon by the entire mathematical community (cf. [Reich 1973, pp. 308–309]).

11. In the preceding chapter, we saw that in ellipsoidal coordinates the equation of the geodesic on an ellipsoid is an equation involving Abelian functions (XVI (197)). In [1844c], Liouville analyzed this equation using the quadruply periodic inverses introduced by Jacobi and Hermite. He continued this line of thought in [1846i and 1847h] by showing that the equation of the geodesics in the plane when expressed in elliptic coordinates is an equation of elliptic functions. Since this geodesic is known to be a straight

line, he had thereby solved the elliptic equation geometrically. Moreover, he showed that other Abelian equations arose when he expressed geodesics on a sphere in conical coordinates [Liouville 1846i, p. 371], geodesics in space in ellipsoidal coordinates [Liouville 1847h, pp. 417–431], and the trajectory of a point in space attracted by a Hooke's-law force when expressed in ellipsoidal coordinates [Liouville 1847h, p. 432]. In these three cases, the known geodesics, great circles, straight lines, and conics are the solutions of the respective equations.

12.   The geodesic in space was a particularly interesting example. Here, Liouville deduced from the Hamilton-Jacobi formalism the following equations for a freely moving point in space:

$$\frac{d\rho}{\Delta(\rho^2)} + \frac{d\mu}{\Delta(\mu^2)} + \frac{d\nu}{\Delta(\nu^2)} = 0, \tag{14}$$

$$\frac{\rho^2\, d\rho}{\Delta(\rho^2)} + \frac{\mu^2\, d\mu}{\Delta(\mu^2)} + \frac{\nu^2\, d\nu}{\Delta(\nu^2)} = 0, \tag{15}$$

$$\frac{\rho^4\, d\rho}{\Delta(\rho^2)} + \frac{\mu^4\, d\mu}{\Delta(\mu^2)} + \frac{\nu^4\, d\nu}{\Delta(\nu^2)} = dt\sqrt{2C}, \tag{16}$$

where ρ, μ, ν are the ellipsoidal coordinates of the moving point defined by (XVI(182)–(184)) and where

$$\Delta(\rho^2) = \sqrt{(\rho^2 - b^2)(\rho^2 - c^2)(\rho^2 - a^2)(\rho^2 - \beta^2)}. \tag{17}$$

The last of these equations determines how the point traverses its trajectory in time so the first two must be the equations of the trajectory, that is, the equations of a straight line in ellipsoidal coordinates.

Liouville showed how to determine the position of this line from the two constants α and β. It was known that any straight line in space touches exactly two of the confocal second-degree coordinate surfaces: $\rho = \rho_0$, $\mu = \mu_0$, and $\nu = \nu_0$, where ρ_0, μ_0, ν_0 are constants. The values of α and β are equal to the constant coordinates of the two tangential coordinate surfaces. To be more precise, in order for the solution of system (14)-(16) to be real, we must have $\alpha^2 > \beta^2$, $\alpha^2 > b^2$, and $\beta^2 < c^2$. Since $0 < \nu^2 < b^2 < \mu^2 < c^2 < \rho^2$, this means that α can be a value of ρ and of μ and β can be a value of μ and of ν. The line can therefore touch the second-degree surfaces in the following ways:

one ellipsoid $\rho = \alpha$ and one hyperboloid of one sheet $\mu = \beta$,
one ellipsoid $\rho = \alpha$ and one hyperboloid of two sheets $\nu = \beta$,

two hyperboloids of one sheet $\mu = \alpha$, $\mu = \beta$,

one hyperboloid of one sheet $\mu = \alpha$ and one hyperboloid of two sheets

$\nu = \beta.$

By a geometric method, Chasles had also determined two equations for the straight line touching two homofocal second-degree surfaces [Chasles 1846b] and had remarked that they could be transformed into two differential equations in ρ, μ, and ν. Liouville added:

> However, I think it is important to add that the two differential equations obtained in this way can be reduced to a system of two equations of the type that Mr. Jacobi calls Abelian, so that the geometric considerations used by Mr. Chasles provide the integrals of such a system.

Liouville made this remark in a brief note [1847e], and soon thereafter [1847h] he supplied the details I have summarized above. In the brief note, Liouville also gave a hint of a remarkable application of these theorems on lines touching two concentric second-degree surfaces:

> By means of these straight lines, about which we have already found numerous beautiful properties, one can extend to the surfaces of second degree the known theorems on the conics concerning the maximum and minimum of the perimeter of inscribed and circumscribed polygons. I shall treat all these questions in a forthcoming issue.
>   [Liouville 1847e]

The theorems he was referring to here and in an interesting note [Liouville 1846a] in [Roberts 1846], were those he had communicated to the *Académie* earlier the same month [Liouville 1846g], at the same time as presenting a copy of [1846i]. They concern geodesic polygons (that is, "polygons" whose sides are geodesics) of a given number of sides in - and circumscribed a given line of curvature on an ellipsoid, that is, curves of the form $\mu = \mu_0$ or $\nu = \nu_0$ in the usual ellipsoidal coordinate system (Fig. 2). The theorems state:

1.   The circumscribed polygon of minimal circumference has all its corners· on one and the same line of curvature, and the first corner can be taken to lie at any point on this line of curvature.

2.   Similarly, in the inscribed polygon of maximal perimeter, all the sides touch one and the same line of curvature. These were generalizations of two theorems of Chasles's on planar and spherical ellipses [Chasles 1843].

13.   Contrary to the claim in the quote above, Liouville never treated these questions in detail. A brief allusion to the problem in [Liouville 1847h, p. 426], indicates that Liouville had originally found the theorems synthetically, using a theorem by Chasles to the effect that the tangents to the geodesic of a second-degree surface all touch the same second-degree homofocal surface [Chasles 1846a, p. 11]. But, Liouville never published this proof, nor the analytic proof that he supplied on the request of Chasles (cf.

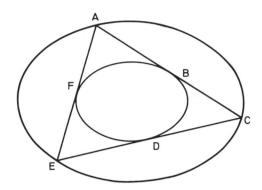

[Liouville 1846g]). The latter, however, is preserved in his notebooks [Ms 3615 (5), pp. 78r,v]. In order to understand the argument, we must return to Liouville's published work on geodesics on ellipsoids.

In accordance with the proofs in the manuscript, I shall use below β^2 instead of β, so that the equation of the geodesic is

$$\frac{\sqrt{m}}{\sqrt{\mu^2 - \beta^2}}\, d\mu = \pm \frac{\sqrt{n}}{\sqrt{\beta^2 - \nu^2}}\, d\nu\,, \tag{18}$$

instead of (XVI (196)); Liouville's first integral is

$$\mu^2 \cos^2 i + \nu^2 \sin^2 i = \beta^2\,, \tag{19}$$

instead of (8). The geometric interpretation of β was indicated by Liouville in [1846i]: If $c^2 > \beta^2 > b^2$ so that β takes on one of the values of μ, then the line of curvature $\mu = \beta$ is tangent to the geodesic; otherwise, if $\beta^2 < b^2$, the geodesic touches $\nu = \beta$. This follows easily from the fact that i is $0°$ and $90°$, respectively, in the two cases. The second fact we need to know in order to read Liouville's note is the formula for the arc length of a geodesic, which Liouville had found in [1844c]. It is of the simple form

$$ds = \pm \frac{\mu^2 \sqrt{m}}{\sqrt{\mu^2 - \beta^2}}\, d\mu \pm \frac{\nu^2 \sqrt{n}}{\sqrt{\beta^2 - \nu^2}}\, d\nu\,, \tag{20}$$

which, as Liouville pointed out, is very remakable first because the variables are separated and second because the arc lengths are found to be a combination of elliptic integrals similar to those entering the equation of the curve (18). The ambiguous sign in (18) and (20), not spelled out by Liouville, is determined such that $\pm d\mu$ and $\pm d\nu$ are always positive. In the notes, Liouville used another formula for the arc length along a geodesic, namely,

$$ds = \pm\sqrt{m}\sqrt{\mu^2 - \beta^2}\,d\mu \pm \sqrt{n}\sqrt{\beta^2 - \nu^2}\,d\nu\,, \tag{21}$$

which can be found from (20) by adding β^2 times the difference between the two sides of the equation (18).

14.   Now, we can turn to Liouville's unpublished proofs of the two theorems above. His idea is that if A, M, and B are three consecutive corners of the maximal inscribed geodesic polygon then the maximality also applies locally. Thus, while varying M and keeping A and B fixed, he must show that the variation of $AM + MB$ is zero when AM, MB both touch the same line of curvature, i.e., has the same value of β. A similar remark holds for the circumscribed polygon. Here are Liouville's proofs:

> Geodesic polygons, inscribed and circumscribed lines of curvature on the ellipsoid A and B fixed, M on a line of curvature (μ). Make $AM + BM$ maximum. (the minimum is AB when M falls on A or B).
>
> Let (μ, ν_1), (μ, ν), (μ, ν_2) be the coord. of A, M, B and for simplicity assume that ν increases all the way from A through M to B. Then
>
> $$\begin{aligned} s =\ & \int_{\nu_1}^{\nu} \sqrt{n}\,d\nu\,\sqrt{\beta^2 - \nu^2} + 2\int_{\beta}^{\mu}\sqrt{m}\,d\mu\,\sqrt{\mu^2 - \beta^2} \\ &+ \int_{\nu}^{\nu_2}\sqrt{n}\,d\nu\,\sqrt{\beta_1^2 - \nu^2} + 2\int_{\beta_1}^{\mu}\sqrt{m}\,d\mu\,\sqrt{\mu^2 - \beta_1^2}. \end{aligned}$$
>
> β and β_1 correspond to the lines of curvature that AM and BM touch. And I say that $\delta s = 0$ gives $\beta_1 = \beta$. In fact, if one differentiates the integrals $\int_{\beta}^{\mu}$ and $\int_{\beta_1}^{\mu}$ with respect to the lower limits, the variation will be zero [because the integrand vanishes for $\mu = \beta$], and those that come from β and β_1 are canceled by the corresponding variations of the integrals $\int_{\nu_1}^{\nu}$, $\int_{\nu}^{\nu_1}$, because of the Eq^{on} of the geodesics [(18)]. Thus, one has simply
>
> $$\delta s = \sqrt{n}\,d\nu\left(\sqrt{\beta^2 - \nu^2} - \sqrt{\beta_1^2 - \nu^2}\right)\,,$$
>
> $$\delta s = \frac{\sqrt{n}(\beta^2 - \beta_1^2)\,d\nu}{\sqrt{\beta^2 - \nu^2}\sqrt{\beta_1^2 - \nu^2}};$$
>
> and $\delta s = 0$ gives $\beta_1 = \beta$.
>
> Thus for a maximal polygon, β is everywhere the same. This is a beautiful generalization of Mr. Chasles's ellipses.
>
> Analogous theorem but of minimum of a circumscribed part, with fixed end points, of a geodesic polygon. In fact, let β be the parameter of the line of curvature, $A(\nu_1, \beta)$, $B(\nu_2, \beta)$ fixed and $M(\mu, \nu)$,

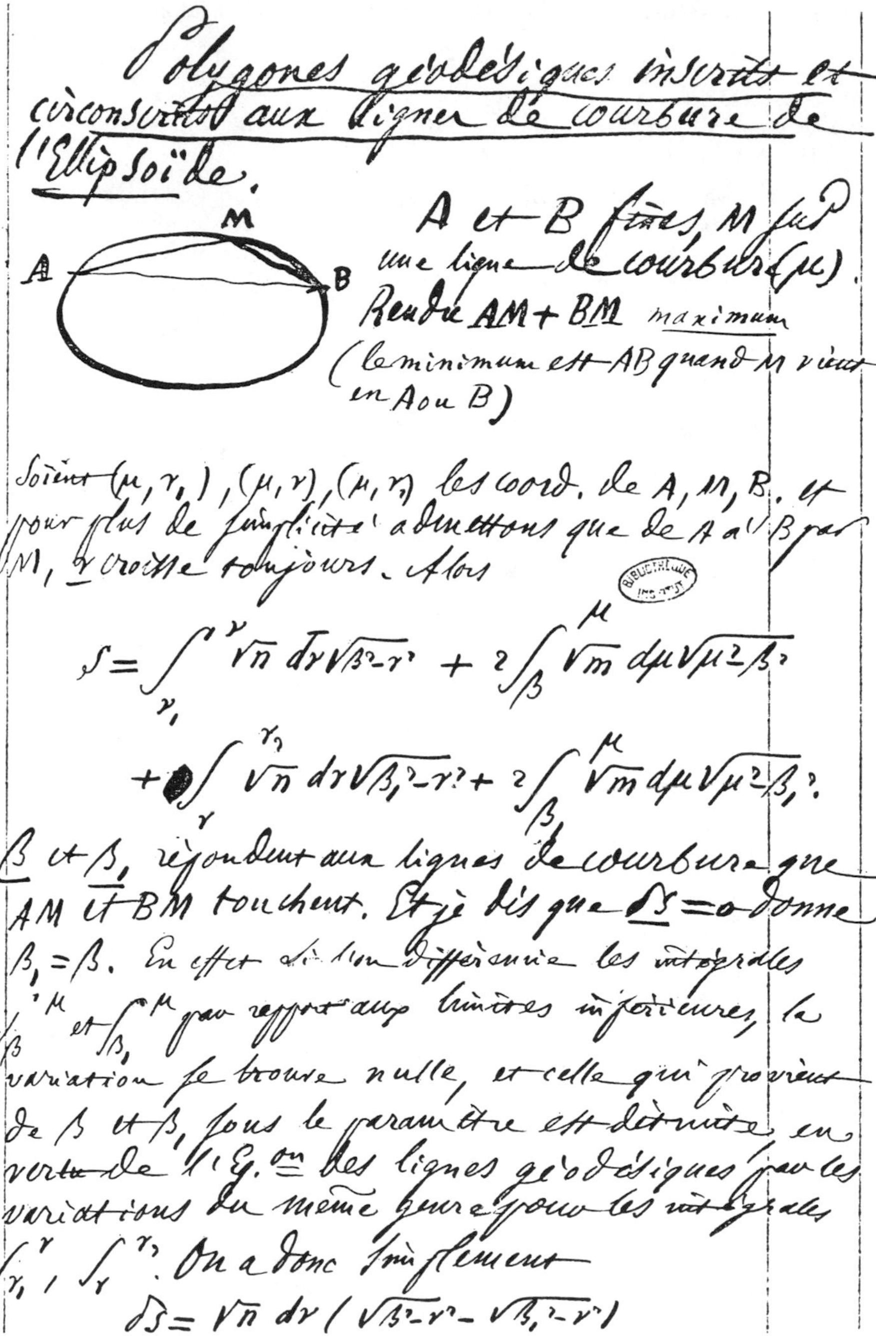

Polygones géodésiques inscrits et circonscrits aux lignes de courbure de l'Ellipsoïde.

A et B fixes, M sur une ligne de courbure (μ).

Rendre AM + BM maximum

(le minimum est AB quand M vient en A ou B)

Soient (μ, r_1), (μ, r), (μ, r_2) les coord. de A, M, B. et pour plus de simplicité admettons que de A à B par M, r croisse toujours. Alors

$$S = \int_{r_1}^{r} \sqrt{n}\, dr \sqrt{\beta^2 - r^2} + 2\int_{\beta}^{\mu} \sqrt{m}\, d\mu \sqrt{\mu^2 - \beta^2}$$

$$+ \int_{r}^{r_2} \sqrt{n}\, dr \sqrt{\beta_1^2 - r^2} + 2\int_{\beta_1}^{\mu} \sqrt{m}\, d\mu \sqrt{\mu^2 - \beta_1^2}$$

β et β₁ répondent aux lignes de courbure que AM et BM touchent. Et je dis que $\delta S = 0$ donne $\beta_1 = \beta$. En effet si l'on différencie les intégrales $\int_\beta^\mu$ et $\int_{\beta_1}^\mu$ par rapport aux limites inférieures, la variation se trouve nulle, et celle qui provient de β et β₁, sous le paramètre est détruite en vertu de l'E.ᵒⁿ des lignes géodésiques par les variations du même genre pour les intégrales $\int_{r_1}^{r}$, $\int_{r}^{r_2}$. On a donc simplement

$$\delta S = \sqrt{n}\, dr \left(\sqrt{\beta^2 - r^2} - \sqrt{\beta_1^2 - r^2} \right)$$

PLATE 34. Liouville's proof of a theorem concerning maximal geodesic polygons inscribed in a line of curvature on an ellipsoid. (Bibliothèque de l'Institut de France Ms 3615 (5) p. 78r)

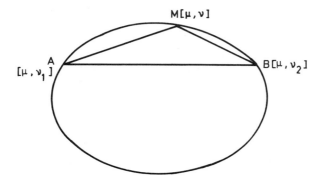

Figure 3

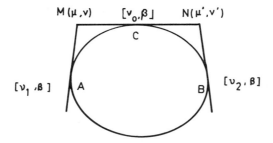

Figure 4

$N(\mu', \nu')$ variable, MN touching AB somewhere. [Liouville now sets $\mu_1 = \mu'$].

The arc$AMNB$ $= \displaystyle\int_{\nu_1}^{\nu_2} \sqrt{n}\, d\nu \sqrt{\beta^2 - \nu^2} + 2\int_{\beta}^{\mu} \sqrt{m}\, d\mu \sqrt{\mu^2 - \beta^2}$

$$+ 2\int_{\beta}^{\mu_1} \sqrt{m_1}\, d\mu_1 \sqrt{\mu_1^2 - \beta^2}.$$

Thus when μ and μ_1 varies

$$\delta s = 2\sqrt{m}\delta\mu\sqrt{\mu^2 - \beta^2} + 2\sqrt{m_1}\delta\mu_1\sqrt{\mu_1^2 - \beta^2}.$$

However by going from A to M to N, B along geodesic lines one has a relation between μ and μ_1, for let (ν_0, β) be the coord. of C: [cf. (18)]

$$\int_{\beta}^{\mu} \frac{d\mu\sqrt{m}}{\sqrt{\mu^2 - \beta^2}} = \int_{\nu_1}^{\nu} \frac{d\nu\sqrt{n}}{\sqrt{\beta^2 - \nu^2}}, \qquad (A, M),$$

$$\int_{\beta}^{\mu} \frac{d\mu\sqrt{m}}{\sqrt{\mu^2 - \beta^2}} = \int_{\nu}^{\nu_0} \frac{d\nu\sqrt{n}}{\sqrt{\beta^2 - \nu^2}}, \qquad (M, C),$$

$$\int_{\beta}^{\mu_1} \frac{d\mu_1\sqrt{m_1}}{\sqrt{\mu_1^2 - \beta^2}} = \int_{\nu_0}^{\nu} \frac{d\nu\sqrt{n}}{\sqrt{\beta^2 - \nu^2}}, \qquad (C, N)$$

$$\int_{\beta}^{\mu_1} \frac{d\mu_1\sqrt{m_1}}{\sqrt{\mu_1^2 - \beta^2}} = \int_{\nu'}^{\nu} \frac{d\nu\sqrt{n}}{\sqrt{\beta^2 - \nu^2}}, \qquad [(N, B)].$$

Summing up:

$$2\int_{\beta}^{\mu_1} \frac{d\mu_1\sqrt{m_1}}{\sqrt{\mu_1^2 - \beta^2}} + 2\int_{\beta}^{\mu} \frac{d\mu\sqrt{m}}{\sqrt{\mu^2 - \beta^2}} = \int_{\nu_1}^{\nu} \frac{\sqrt{n}\,d\nu}{\sqrt{\beta^2 - \nu^2}} = \text{const.}$$

Therefore,

$$0 = \frac{\delta\mu_1\sqrt{m_1}}{\sqrt{\mu_1^2 - \beta^2}} + \frac{\delta\nu\sqrt{m}}{\sqrt{\mu^2 - \beta^2}}, \qquad \text{and}$$

$$\delta s = 2\sqrt{m}\,d\mu\left(\sqrt{\mu^2 - \beta^2} - \frac{\mu_1^2 - \beta^2}{\sqrt{\mu^2 - \beta^2}}\right) = \frac{2\sqrt{m}\delta\mu}{\sqrt{\mu^2 - \beta^2}}(\mu^2 - \mu_1^2).$$

Thus, $\delta s = 0$ gives $\mu_1 = \mu$ &$_c$. Hurrah!!!!"

[Liouville Ms 3615 (5), p. 78 r,v]

Despite the inconsistencies in notation this proof should be easily comprehensible without further comments.

15.   The original theorems (cf. §12) also contained a statement to the effect that the first corner of the geodesic polygon can be chosen at random. More generally, Liouville [Ms 3615 (5), p. 95v] considered two given lines of curvature, say, $\mu = \mu_0$ and $\mu = \beta$, $\beta < \mu_0$. In the space between them, he drew a geodesic Poncelet trajectory: starting from a point A on the outer line of curvature $\mu = \mu_0$ (cf. Fig. 2), we draw a tangent to the inner line of curvature $\mu = \beta$, meeting the outer one in a point C; from C, we draw a new tangent to $\mu = \beta$, meeting $\mu = \mu_0$ in E, etc. If this repeated process brings us back to A after one or more tours around the inner line of curvature, we say that the Poncelet trajectory closes. In that case, we have a geodesic polygon inscribed in the outer and circumscribed around the inner line of curvature. Now, Liouville showed that if the Poncelet trajectory closes when it starts from A then it closes wherever it is started, i.e., wherever we choose the first corner of the polygon. This is a beautiful ellipsoidal analogue of Poncelet's closure theorem in the plane. (Bos et al. [1984] discuss the history of this theorem and in particular how Jacobi showed its equivalence with addition theorems for elliptic functions.)

16.   How did Liouville prove the ellipsoidal Poncelet closure theorem? That can be seen from the following passage in his notebook:

The general condition that the polygon inscribed in the line of curvature (μ) and circumscribed about (β) closes is (i is the number of turns, m the number of sides)

$$\int_{-b}^{b} \frac{\sqrt{n}\, d\nu}{\sqrt{\beta^2 - \nu^2}} = \frac{m}{i} \int_{\beta}^{\mu} \frac{\sqrt{m}\, d\mu}{\sqrt{\mu^2 - \beta^2}}$$

[Liouville Ms 3615 (5), p. 95v]

The above condition is obtained by integrating the equation of the geodesics (18) along the arcs of the geodesic polygons and adding the results. For each time we encircle the line of curvature (β) (that is, $\mu = \beta$), ν has passed once from its maximal value b to its minimal value $-b$ and back again and through each side of the polygon μ passes from μ to β and back again. Now, since the above condition is independent of the choice of the first corner, the closure is independent of its position.

Liouville generalized this theorem to a geodesic polygon having its corners on different lines of curvature. Here, he considered only polygons closing after one tour:

Consider a fixed line of curvature (β) and other lines of curvature $\mu_1, \mu_2, \ldots, \mu_p$ [i.e., $\mu = \mu_1$, etc.] on which lies the corners of a polygon circumscribed about β. The condition of possibility is

$$\int_{0}^{b} \frac{\sqrt{n}\, d\nu}{\sqrt{\beta^2 - \nu^2}} = \int_{\beta}^{\mu_1} \frac{\sqrt{m}\, d\mu}{\sqrt{\mu^2 - \beta^2}} + \int_{\beta}^{\mu_2} + \cdots + \int_{\beta}^{\mu_p},$$

and if it is satisfied, one can start at will from any point of the first line of curvature.

[Liouville Ms 3615 (5), p. 80r]

Liouville took these closure theorems as a point of departure for a study of addition and division theorems of the Abelian integrals that occur [Ms 3615 (5), pp. 80r–96v], just as Jacobi [1828] had drawn similar theorems for elliptic integrals out of the classical Poncelet closure theorem [cf. Bos et al. 1984].

17.   As I have already pointed out, Liouville only vaguely alluded to "certain questions of maxima and minima" in his most comprehensive published account of the geometric use of ellipsoidal coordinates [Liouville 1847h]. But, he continued to announce a more extensive paper:

However, at this moment we shall not add any details on this subject because we shall return to it in another memoir, as well as to the geometric developments which all the formulas of this paragraph may give rise to.

[Liouville 1847h, pp. 431–432]

In a footnote, he added:

> The work which we announce and which will deal with many different questions will carry the title: *Recherches de Géométrie analytique.* In this work, we shall also talk about properties of homofocal conics in the plane and lines of various types that one can trace on the ellipsoid, principally geodesic lines and lines of curvature, a fertile subject in which one is far from having exhausted all the details.
>
> [Liouville 1847h, p. 432]

The *Recherches de Géométrie analytique,* which was planned to include the above proofs on geodesic polygons and other interesting subjects, never appeared. But, Liouville continued to show interest in such questions for many years.

18.   Liouville's first note under the pseudonym of Besge [1842] had dealt with a geometric problem, namely, the center of gravity of a geodesic triangle. In Besge [1849b], he went on to show that the locus described by the summit of a geodesic angle of constant magnitude $2i$ whose legs touch a given line of curvature on an ellipsoid is a fourth-degree curve. He proved this generalization of a theorem proved in the plane by de la Hire by substituting rectangular coordinates in the equation

$$\mu^2 \cos^2 i + \nu^2 \sin^2 i = \beta \,,$$

which is evidently the equation of the curve in ellipsoidal coordinates. This result was confirmed shortly afterward by Roberts [1850] in a paper on curves on ellipsoids in which "I will make use of a method that Mr. Liouville has communicated to me." In this way, he corrected some results which he had earlier communicated to Liouville and which the latter had hastily published in his journal [Roberts 1847], without discovering that they were "all inexact" [Roberts 1850, p. 275].

Fifteen years later, Darboux [1865] showed that this curve lies on a second-degree surface of revolution having the same principle planes as the ellipsoid (cf. [Chasles 1870, p. 367]).

In [1851a], Liouville made his last analytical contribution to Chasles's geometry. Chasles had shown [1837, p. 392] that two homofocal second

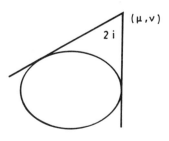

FIGURE 5

degree surfaces could be considered as the locus of the centers of curvature of a certain surface. Liouville found the equation of this surface.

Finally, Liouville devoted the first semester of 1851–1852 at the *Collège de France* to *Les coordonnés elliptiques et les transformations analytiques qui s'y rattachent* [Ms 3619 (6)]. Among the new theorems discussed here, were two new max-min properties of the tangents of an ellipse, which one of his audience, Paul Serret, later proved by synthetic means [P. Serret 1852]. However, as Liouville explained in [1856c] when he published the results, the analytic method he had used himself in the lectures had the advantage of allowing an easy extension to lines of curvature on ellipsoids.

With these published and unpublished works, Liouville contributed to the geometric study of elliptic and Abelian integrals. As I have pointed out, these works were principally inspired by Jacobi (in particular, by Jacobi [1839a]) and by Chasles.

19.   In 1845, J.A. Serret caused Liouville to make another contribution to this family of ideas. Pursuing an idea of Euler, Serret sought and found a family of plane algebraic curves whose arc length represented elliptic integrals of the first kind:

$$\int \frac{d\theta}{\sqrt{1 - k^2 \sin^2 \theta}} \, , \tag{22}$$

when k is of the form $\sqrt{\frac{n}{n+1}}$, $n \in N$. Such curves, generalizing the properties of the lemniscate and the ellipse, are of interest because they provide a geometric interpretation of addition - and division theorems. Liouville reviewed Serret's work favorably to the *Académie*, and remarked in the review that Serret's results could be generalized easily to integrals whose modulus k is of the form $\sqrt{q}$, $q \in Q$ [Liouville 1845e]. When Serret's paper was published later the same year in Liouville's *Journal* [J.A. Serret 1845a], Liouville added a note containing a proof of this extension [Liouville 1845 f], which Serret characterized as "elegant" in a subsequent paper [Serret 1845b]. Moreover, Liouville investigated some analytical aspects of Serret's formulas later the same year [1845g]. Several years later, he returned to Serret's curves in his notebooks, but since inversion in circles and spheres played a central part in these studies, I shall postpone a discussion of their content to the next section, which is devoted to Liouville's study of this transformation.

Inversion in Spheres; William Thomson's Influence

20.   In 1845 and 1846, William Thomson sent Liouville three letters concerning a transformation, which he had told him about during his stay in

Paris. He defined the image of (ξ, η, ζ) of a point (x, y, z) by

$$\xi = \frac{nx}{x^2 + y^2 + z^2}, \eta = \frac{ny}{x^2 + y^2 + z^2}, \zeta = \frac{nz}{x^2 + y^2 + z^2}. \qquad (23)$$

Geometrically speaking, the image of a point M is a point M' on the line OM having a distance from the origin determined by $OM' = \frac{n}{OM}$. It is obvious that this transformation leaves the sphere of radius $\sqrt{n}$ and center at O fixed and maps its interior onto the exterior and vice versa, whence its most common name inversion in the sphere. Thomson originally called it the principle of (electrical) images, but Liouville preferred the name "transformation par rayons vecteurs réciproque." Today, it is also called the Kelvin transform after the name Thomson took when he was raised to the peerage.

Thomson's letters were published in Liouville's *Journal* [Thomson 1845b, 1847a]. They display the power of the transformation in electrostatics. Liouville, on the other hand, was fascinated by its broad mathematical implications and added a note to the last two letters, twice as long as the letters themselves, in order to show "the great importance of the work of which the young geometer from Glasgow has given us a short excerpt" [Liouville 1847f]. Liouville's note was summarized by Terquem seven years later in the *Nouvelles Annales* [Liouville 1854e], and it was included in its entire length in the 14th chapter of Thomson's *Reprint of Papers on Electrostatics and Magnetism* [1872]. Most of Liouville's note was concerned with a formula, found by Thomson, expressing the distance between two points (x, y, z), (x', y', z') in terms of the distance between their images (ξ, η, ζ), (ξ', η', ζ'):

$$(\xi' - \xi)^2 + (\eta' - \eta)^2 + (\zeta' - \zeta)^2 = n \frac{(x' - x)^2 + (y' - y)^2 + (z' - z)^2}{(x^2 + y^2 + z^2)(x'^2 + y'^2 + z'^2)}. \qquad (24)$$

Inspired by this, Liouville posed the question:

Characterize the transformations $(x, y, z) \rightarrow (\xi, \eta, \zeta)$ *for which there exists a function*

$$p = \psi(x, y, z)$$

such that

$$(\xi' - \xi)^2 + (\eta' - \eta)^2 + (\zeta' - \zeta)^2 = \frac{(x' - x)^2 + (y' - y)^2 + (z' - z)^2}{p^2 p'^2}. \qquad (25)$$

As above, (ξ, η, ζ) and (ξ', η', ζ') are the images of (x, y, z) and (x', y', z'), respectively, and p' means $\psi(x', y', z')$.

After long calculations, Liouville found [1847f, p. 277] that the only transformations satisfying the requirements are those composed of inversions in spheres and similitudes; and what is more, such a transformation only requires one inversion.

21.   This is certainly an interesting theorem, but it was only a few years before Liouville gave it the twist that made it one of his most beautiful results. Some of the leading ideas, however, can be found in the note to Thomson's letters. Indeed, Liouville explicitly pointed out that it followed from (25) (or 24) that the transformation $(x, y, z) \rightarrow (\xi, \eta, \zeta)$ is conformal, or as he put it:

> It is a remarkable property of this type of transformation that the two triangles formed by three arbitrary infinitely close points of the primitive figure and the three corresponding points of the transformed figure are similar to each other.
>
> [Liouville 1847f, p. 280]

From this he deduced that the transformation preserves angles and remarked that the proof of the conformality "does not even require that the equation (1) [25] holds for two points at a finite distance apart. It requires only that this equation is always satisfied by two infinitely close points" [Liouville 1847f, p. 281].

In a footnote, he pointed out that such transformations were the three-dimensional analogue of the "drawing of geographic maps." The map projection problem had been discussed by Gauss in a prize essay to the Royal Danish Academy [1822–1825]. It is unclear if Liouville knew this paper in 1847, but in a footnote in the paper [Liouville 1847g, pp. 294–295] following the Thomson note, he showed that the conformal mappings of the plane $R^2 = C$ into itself are precisely the holomorphic functions and their conjugates. This result is implied by the arguments in Gauss's paper, but nowhere stated by Gauss.

One year later, Liouville wrote more explicitly about this analogue in a footnote in [Roberts 1848]. In this paper, Roberts had proved various theorems concerning a generalization of inversion in the plane. Liouville pointed out that the only property of these transformations used in the proofs was their conformality, so the theorems were in fact more general than Roberts had thought. Liouville continued:

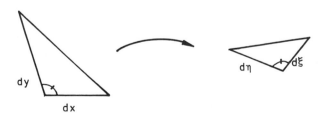

FIGURE 6

However, only the transformation by reciprocal radius vectors possesses the singular property of extending to an arbitrary number of variables and consequently of yielding, among other results, a sort of geographic transformation in three dimensions, that is, a solution of the equation

$$dx^2 + dy^2 + dz^2 = \lambda(d\alpha^2 + d\beta^2 + d\gamma^2),$$

an equation which is much more difficult to treat than the plane equation

$$dx^2 + dy^2 = \lambda(d\alpha^2 + d\beta^2),$$

from which one easily deduces:

$$x + y\sqrt{-1} = f(\alpha \pm \beta\sqrt{-1})\cdots"$$

[Liouville 1848c]

22.   Thus, by 1848, Liouville had formulated the three-dimensional analogue of Gauss's problem of geographic projection: characterize all conformal mappings of the space into itself. He had also discovered that it was a difficult problem, and as he admitted in [Liouville 1850g, p. 616] he did not know the answer then. Two years later, however, he announced this surprising theorem:

[Liouville's Theorem] The only conformal mappings $R^3 \rightarrow R^3$ are those composed of an inversion in a sphere and a similitude.

It is in fact surprising that there are so few conformal mappings $R^3 \rightarrow R^3$.

As in the quote above, Liouville characterized the conformal mappings $(x, y, z) \rightarrow (\alpha, \beta, \gamma)$ by the existence of a function $\lambda(x, y, z)$ such that

$$dx^2 + dy^2 + dz^2 = \lambda(d\alpha^2 + d\beta^2 + d\gamma^2). \tag{26}$$

That indeed is equation (25) for infinitely small distances (where $p^2 = p'^2$) to which he had alluded in the Thomson note. Having announced his theorem in his *Journal* [1850a], Liouville included its proof as note 6 in his edition of Monge's *Application de l'analyse a la géométrie*. It rested on Lamé's results on orthogonal coordinates (cf. [Liouville 1850g, p. 614]).

23.   In the Thomson note, Liouville at first thought of the transformation $(x, y, z) \rightarrow (\xi, \eta, \zeta)$ as a change of coordinates from a Cartesian system to a system whose coordinate planes are spheres. On p. 275, however, he introduced the point of view we have taken above, that is, to think of the transformation as a mapping of space onto itself, considering both x, y, z and ξ, η, ζ as being Cartesian coordinates:

However, I think it is more convenient to introduce in our research one of these transmutations of figures, which are so familiar to the geometers and which have recently contributed so much to the progress of the science.

[Liouville 1847f, p. 275]

Here, he explicitly included inversion in spheres as a new example of a transformation that could be used in synthetic geometry in order to extend theorems proven in a specific case to curves and surfaces in a more general position. In this sense, inversion was similar to central projection, which had been the essential ingredient in the development of projective geometry.

In fact, inversion in spheres had been introduced earlier by several geometers. Disregarding the ancient treatment of stereographic projection, which is a special example, the use of inversion had first been anticipated by Dandelin and Quetelet in 1822–1826 and explicitly used by Steiner in an unpublished work from 1824. It was introduced in print by Plücker and Magnus (1828–1832), and in 1836 Bellavitis gave the first complete exposition of inverse figures. He was followed by Stubbs and Ingram (1842). It will take us too far to go into any detail about the complicated problem of the origin of the principle of inversion with its many scattered and mostly independent contributors. The reader is referred to the following three investigations in which there are references to the works mentioned above: [Chasles 1870, pp. 140–146] (as usual excluding all the Germans); [Encyclopädie der Mathematischen Wissenschaften III. Band Geometrie I pp. 312, 668, 1030–31, 1042]; and [Patterson 1933].

Thomson and Liouville were the first to study inversions from a differential geometric and analytic point of view. They do not seem to have known of the earlier works on the subject, but they had a certain influence on their successors. I have already mentioned that Roberts [1848] made a generalization in the plane of Thomson's and Liouville's methods. Other French, Irish, and English contributors to the theory were aware of the works of Thomson and Liouville [Chasles 1870, pp. 142–146]; [Reich 1973, pp. 335–337], but Möbius, in whose *Theorie der Kreisverwandtschaft* [Möbius 1853, 1855] the geometric theory of inversion culminated, seems to have been aware only of the work of Magnus.

24. Let me briefly summarize some of the applications Liouville made of the inversion in synthetic geometry. From the fundamental fact that spheres are mapped into spheres (planes are considered to be spheres of infinite radius), he concluded that lines of curvature on a surface are transformed into lines of curvature on the image surface [Liouville 1847f, pp. 278–281]. (The latter theorem had been proved earlier by Stubbs (cf. [Chasles 1870, p. 142]).) Moreover, Liouville indicated how the transformation could yield elegant proofs of various geometric theorems of Serret and Dupin. The property used in these applications is that it is possible to transform three given spheres in general position into three planes by inverting in a sphere with its center at one of their points of intersection. For example, this proves a theorem of Miquel [1844] that in a triangle made up from three arcs of circles intersecting each other at one point, the sum of the angles equals two right angles, that is, the case in the transformed linear triangle, and the transformation is angle preserving.

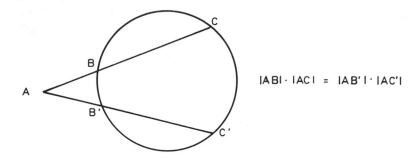

$$|ABI| \cdot |ACI| = |AB'I| \cdot |AC'I|$$

FIGURE 7

Metric relations are also easily transformed. If i denotes inversion in a unit circle with center at O, formula (24) can be written:

$$|i(A)i(B)| = \frac{|AB|}{|OA| \cdot |OB|}. \tag{27}$$

From this, Liouville concluded among other things that the well-known theorem (see Fig. 7) also holds for a figure where the lines are transformed into circles.

25.   At the end of his note [1847f], Liouville extended and clarified Thomson's applications of the transformation to mathematical physics or rather to boundary value problems of partial differential equations, in particular, Laplace's equation. These applications are based on the simple way in which the Laplacian transforms. Let $U(\xi, \eta, \zeta)$ be a solution to

$$\frac{\partial^2 U}{\partial \xi^2} + \frac{\partial^2 U}{\partial \eta^2} + \frac{\partial^2 U}{\partial \zeta^2} = 0. \tag{28}$$

Then, after the inversion in a unit sphere with center at the origin, we have

$$\frac{\partial^2 p^{-1} U}{\partial x^2} + \frac{\partial^2 p^{-1} U}{\partial y^2} + \frac{\partial^2 p^{-1} U}{\partial z^2} = 0, \tag{29}$$

where p^{-1} stands for $(x^2 + y^2 + z^2)^{\frac{1}{2}}$, and U is expressed in (x, y, z). If we can solve one of these equations, we can evidently solve the other. Therefore a boundary problem for (28) with difficult boundaries can be solved if we succeed in transforming the boundaries into a position where we know the solution. For example, it is always possible to transform two nonintersecting spheres into two concentric spheres for which the solution of Laplace's equation was well known in the nineteenth century.

These general ideas had been used in Thomson's electrostatic papers, but Liouville expressed them more clearly and generalized them to a problem of Gauss, which he had studied in [Liouville 1846e,f] (cf. Chapter XV, §6), and to the transformation of the wave equation and the Hamilton-Jacobi equation.

26. Liouville published only the three papers [1847f, 1850a and g] on inversion in spheres, but a remark in a letter from J.A. Serret to Liouville from 1853 points to a continued interest in this subject:

> Some time ago, you urged me to attempt to search for surfaces with spherical lines of curvature, using the method that I published in the *Comptes rendus* of the meeting of January 24th for the surfaces with plane lines of curvature, by adding the considerations which the transformation of reciprocal radius vectors can provide.
>
> [J.A. Serret 1853, p. 328]

When Liouville published Serret's letter in the *Comptes Rendus* of February 21, 1853, it led to a quarrel in the following issues of this journal between Serret and Ossian Bonnet, who had published similar ideas a week earlier, (cf. [Reich 1973, pp. 338–339]).

27. It may have been around the same time that Liouville took a new look at Serret's "elliptic curves" (cf. §19). He left about one hundred pages of notes [Ms 3616 (3), pp. 83v–140r] on the application of inversion in this context. An interpolated note resembling [Liouville 1853a] and a letter dated June 29, 1854, pasted into this notebook, point to this period of composition, but the dating is uncertain. The central theme in these notes is to show how one can arrive at Serret's curves by inverting in a circle a particular class of curves, whose arcs are determined by arcs of ellipses. Euler had found this class in 1781, but his discovery was not published until 1830 [Euler 1781–1830]. Liouville's approach is nicely illustrated by the following quote from his notes:

> Euler in Vol. XI (p. 95) of the *Mémoires de l'acad. imper. de St. Petersbourg* gives these formulas
>
> $$x = \frac{a}{n+1}\cos(n+1)\varphi + \frac{b}{n-1}\cos(n-1)\varphi \qquad [30]$$
>
> $$y = \frac{a}{n+1}\sin(n+1)\varphi + \frac{b}{n-1}\cos(n-1)\varphi \qquad [31]$$
>
> that can be reduced to
>
> $$x + y\sqrt{-1} = \frac{a}{n+1}e^{(n+1)\varphi\sqrt{-1}} + \frac{b}{n-1}e^{(n-1)\varphi\sqrt{-1}} \qquad [32]$$
>
> and which gives
>
> $$dx^2 + dy^2 = d\varphi^2(a^2 + b^2 + 2ab\cos\varphi) = ds^2 \qquad [33]$$
>
> so that
>
> $$ds = d\varphi\sqrt{a^2 + b^2 + 2ab\cos 2\varphi} \qquad [34]$$
>
> is the element of an arc of the ellipse. From there, algebraic curves (if $n = \frac{p}{q}$ fract. rat.) whose arc is expressed by an arc of an ellipse.

Now, let a transformation of reciprocal radius vectors be effectuated by setting

$$\xi + \eta(\sqrt{-1}) = \frac{c}{x + y\sqrt{-1}},$$
[35]

then

$$d\xi^2 + d\eta^2 = d\sigma^2 = \frac{c^2(dx^2 + dy^2)}{(x^2 + y^2)^2},$$
[36]

$$d\sigma = \frac{c\,ds}{(x^2 + y^2)},$$
[37]

but

$$x^2 + y^2 = \frac{a^2}{(n+1)^2} + \frac{b^2}{(n-1)^2} + \frac{2ab}{(n+1)(n-1)}\cos 2\varphi.$$
[38]

However, the right-hand side can be written in the form

$$x^2 + y^2 = A^2(a^2 + b^2 + 2ab\cos 2\varphi)$$
[39]

if one takes

$$\frac{a^2 + b^2}{2ab} = \frac{\frac{a^2}{(n+1)^2} + \frac{b^2}{(n-1)^2}}{\frac{2ab}{(n+1)(n-1)}}$$
[40]

that is

$$a^2 + b^2 = a^2\frac{(n-1)}{(n+1)} + b^2\frac{(n+1)}{(n-1)}$$

$$a^2\left(1 - \frac{n-1}{n+1}\right) = b^2\left(\frac{n+1}{n-1} - 1\right)$$

$$\frac{a^2}{n+1} = \frac{b^2}{n-1}, \quad \frac{a^2}{b^2} = \frac{n+1}{n-1} = \frac{p+q}{p-q}.$$
[41]

And hence (if one takes for simplicity $c = A^2$)

$$d\sigma = \frac{d\varphi}{\sqrt{a^2 + b^2 + 2ab\cos 2\varphi}}$$
[42]

such that we have algebraic curves whose arc σ represents the elliptic function of the first kind, not an arbitrary such function but one for which

$$\frac{a^2}{b^2} = \frac{p+q}{p-q},$$
[43]

p and q being integers.

———

———

This is certainly very curious and deserves a profound study.

[Liouville Ms 3616 (3), pp. 83v–84r]

Let me add a few comments on this quote. In order to see that Euler's curve given parametrically by [30] and [31] is an algebraic curve when $n = \frac{p}{q}$ is rational, one just has to find $\cos 2\varphi$ or $2\cos^2\varphi - 1$ from the expression of $x^2 + y^2$. This yields $\cos\varphi$ as an algebraic function of x and y, and since $\cos\varphi$ is a polynomial in $\cos\frac{\varphi}{q}$, the latter is also an algebraic function of x and y. By expanding $\cos(n \pm 1)\varphi = \cos(p \pm q)\frac{\varphi}{q}$ in polynomials of $\cos\frac{\varphi}{q}$ and inserting it into the expression for [30], one has the desired algebraic relation between x and y. Note also that the condition on the modulus [43] says only that $\frac{a^2}{b^2}$ must be a rational number different from 1. Finally, we see that the arc [42] of the σ curve is an elliptic integral of the form (22) because $\cos 2\varphi = 1 - -2\sin^2\varphi$.

28.   Liouville's approach was, as far as I know, entirely new. In fact, Serret had not even mentioned Euler's paper. So, indeed, it deserved to be elaborated. On the next couple of pages, Liouville convinced himself that it was Serret's curves that he had found in this way. A few pages further on he began to investigate a mechanico-geometrical generation of Euler's curves and was led to the following simple description:

> Suppose that a circle of radius B rolls on a circle of radius A and slides at the same time such that when N is the present point of contact of these circles the arcs NM, NA drawn to the two points M, A that coincided at first, have a constant ratio m to each other, it is clear that the point M describes an elongated or curtailed epicycloid, depending on the value of the ratio m; the case $m = 1$ corresponds to the ordinary epicycloid. It is known that the latter is geometrically rectifiable and that the other cannot be rectified except by means of an arc of an ellipse. However, these elongated or curtailed epicycloids (which have a double generation by interior or exterior circle rolling on A) are precisely the curves given by Euler as examples of curves whose arcs are equal to the arc of an ellipse. When there is a certain relation between the ratio A, B and the constant m, they are algebraic. If, moreover, a certain second relation holds, they lead, by way of the transformation by reciprocal radius vectors, to the elliptic curves of Mr. J. A. Serret.
>
> Eq. of our epicycloids:

$$x = OH = OK + CL = (A - B)\cos\theta + B\cos\left(\theta - \frac{a\theta}{B}m\right)$$

$$y = (A - B)\sin\theta + B\sin\left(\theta - \frac{a\theta}{B}m\right).$$

This being given, let $A - B = \frac{a}{n+1}$, $B = \frac{b}{n+1}$

$$\theta = (n+1)\varphi, \quad \theta\left(1 - \frac{A}{B}m\right) = (n-1)\varphi$$

that is,

$$(n+1)\left(1 - \frac{a+b}{b}m\right) = n - 1,$$

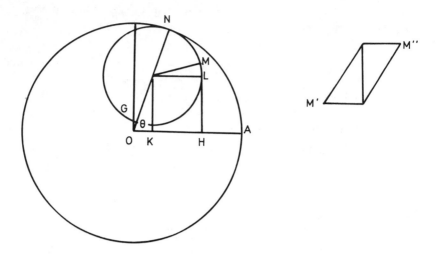

Figure 8

$$n + 1 - m\frac{(n + 1)(a + b)}{b} = n - 1$$

$$m = \frac{2b}{(n + 1)(a + b)}$$

so

$$x = \frac{a}{n + 1}\cos(n + 1)\varphi + \frac{b}{n - 1}\cos(n - 1)\varphi$$

$$y = \frac{a}{n + 1}\sin(n + 1)\varphi + \frac{b}{n - 1}\sin(n - 1)\varphi$$

And here we have Euler's equations. They give

$$dx^2 + dy^2 = ds^2 = (a^2 + b^2 + 2ab\cos 2\varphi)\,d\varphi^2,$$

so that ds is really an arc of an ellipse. For let $x = g\sin\varphi$, $y = g\cos\varphi$ [error for $x = g\cos\varphi$, $y = h\sin\varphi$], be the equations of the ellipse. One has for the ellipse

$$ds^2 = (g^2\cos^2\varphi + h^2\sin^2\varphi)\,d\varphi^2 = \left(\frac{g^2 + h^2}{2} + \frac{g^2 - h^2}{2}\cos 2\varphi\right)d\varphi^2,$$

so that one just has to make

$$\frac{g^2 + h^2}{2} = a^2 + b^2, \quad \frac{g^2 - h^2}{2} = 2ab$$

$$g^2 = (a + b)^2,\ h^2 = (a - b)^2,\ g = a + b,\ h = a - b$$

If n is rational $= P$ (I exclude $n = 1$ in order to make $\frac{b}{n-1}$ a finite quantity) the curve is algebraic. Note that

$$x^2 + y^2 = r^2 = \frac{a^2}{(n + 1)^2} + \frac{b^2}{(n + 1)^2} + \frac{2ab}{n^2 - -1}\cos 2\varphi.$$

Note also that if n is rational φ can be deduced <u>algebraically</u> or <u>geometrically</u> from θ. The normal to the epicycloid in M can easily be obtained by the method of Roberval; the rotational motion around N being accompanied by a sliding on the circle of A, the point M moves in the direction composed of MM' on the tangent and M on the circle with center N and radius NM and MM' parallel to the x axis. One easily concludes that the normal will cut ON in a point P whose distance to N remains constant.

[Liouville Ms 3616 (3), pp. 89v–90v]

This note bears witness to an interest in the kinematic generation of curves that is not revealed in Liouville's published works. Indeed, he liked Roberval's kinematic method of tangents so much that he taught it to his students at the *École Polytechnique*.

29.   Liouville's notes contain many other calculations concerning similar mechanically generated curves, their image under inversions in circles and their relation to elliptic integrals. Some of them may point to a deeper understanding of the subject, for example:

Ah! Ah! — compare with Abel, with our eq. $P^2 - Q^2 R = \&_c$ — There must be the key to many things!!

[Liouville Ms 3616 (3), p. 89r]

Here, Liouville seems to connect Serret's curves to Abel's and his own work on integration in finite terms and continued fractions (cf. Chapter IX, §11). He also linked this research with his works on ellipsoidal coordinates [Ms 3616 (3), p. 126v ff]. In several of the notes, Liouville attempted to extend his methods to all values of the modulus k of the elliptic integral (22) or equivalently to any value of $\frac{a^2+b^2}{2ab}$ in the expression for σ (42). This search apparently failed, in the sense that he did not find plane algebraic curves with the desired line element, but he did succeed in finding curves on a sphere corresponding to all elliptic integrals of the first kind. In the following final quote from these notes, it can be seen how Liouville accomplished this by inversion in a sphere having its center outside the x-y-plane in which the original Eulerian curve lies. The only thing one has to keep in mind while reading the note is that inversion in a sphere takes a plane (not through the center of the sphere) into a sphere.

It is very easy to find an arc of a spherical curve represented by the arbitrary function of the first kind. In fact consider one of our plane curves

$$x + y\sqrt{-1} = ae^{m\varphi\sqrt{-1}} + a'e^{m'\varphi\sqrt{-1}}, \qquad m' - m = 2$$

$$ds^2 = (a^2 m^2 + a'^2 m'^2 + 2aa' mm' \cos 2\varphi)\, d\varphi^2$$

$$x^2 + y^2 = a^2 + a'^2 + 2aa' \cos 2\varphi;$$

adding h^2 and taking $r^2 = x^2 + y^2 + h^2$

$$r^2 = a^2 + a'^2 + h^2 + 2aa' \cos 2\varphi$$

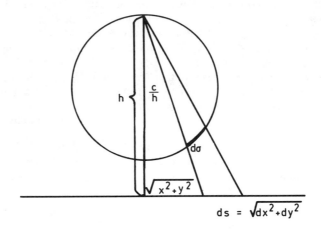

FIGURE 9

let

$$\frac{a^2 m^2 + a'^2 m'^2}{2aa'mm'} = \frac{a^2 + a'^2 + h^2}{2aa'} = k^2$$

[this "modulus" k is different from the one defined by (22)]

$$\frac{(am + a'm')^2}{(am - a'm')^2} = \frac{(a + a')^2 + h^2}{(a - a')^2 + h^2} = \frac{k^2 + 1}{k^2 - -1}$$

$$\frac{am + a'm'}{am - a'm'} = \sqrt{\frac{k^2 + 1}{k^2 - -1}} \,, \qquad a' = \mu a \,;$$

finally

$$\frac{(1 + \mu^2)a^2 + h^2}{(1 - \mu^2)a^2 + h^2} = \frac{k^2 + 1}{k^2 - -1}$$

from which h^2, and k remains arbitrary. Thus, you only have to take as the origin of a transformation by reciprocal radius vectors the point on the z-axis which lies at a height h above the xy-plane.
[Liouville Ms 3616 (3), p. 109r]

The last remark can be understood if from Figure 9 we find the element $d\sigma$ of the arc of the transformed curve. It is easily seen to be given by

$$d\sigma = \frac{c\,ds}{x^2 + y^2 + h^2} = \frac{c\,ds}{r^2} \,, \tag{44}$$

as in the plane case (42). Thus, when the constants are adjusted, as in the quote, we have

$$d\sigma = \frac{c'\,d\varphi}{\sqrt{k^2 + \cos 2\varphi}} \,, \tag{45}$$

which is the integrand of an elliptic integral of the first kind and, indeed, an arbitrary one since k can be given any value by adjusting h.

This is a remarkable result found by an equally remarkable method. However, as in so many other cases, these ideas were left unpublished in Liouville's notebooks.

Contributions to Gaussian Differential Geometry

30.   According to Karin Reich [1973, p. 291], it was principally due to Liouville that Gauss's ideas on differential geometry became known in France. To be sure, Sophie Germain had read Gauss's *Disquisitiones generales circa superficies curvas* [1828], but during the following 15 years where Lamé's theories of systems of orthogonal surfaces dominated the French scene, Gauss's work was forgotten. In 1843, in a paper in Liouville's *Journal* on this subject, Bertrand admitted that "After having written this memoir, I have learned about a memoir by Mr. Gauss entitled *Disquisitiones generales...*" [Bertrand 1843]. The following year, Bonnet also referred to Gauss. It is not impossible that it was Liouville who had called the attention of these two young talents to the *Disquisitiones*, and it is certain that when the interest in Gauss's ideas spread in France after 1847 it was due to Liouville. Karin Reich writes:

> In 1847, there arose a very keen interest in the ideas that Gauss had published almost 20 years earlier. Indeed, at this time J. Liouville gives a new proof of the above-mentioned "Theorema egregium," which in turn inspired other mathematicians to other proofs.
>
> [Reich 1973, p. 291]

Liouville repeated this proof in his edition of Monge's *Application de l'Analyse à la Géométrie*, together with several other notes chiefly concerned with Gaussian differential geometry. In fact, he explained in note III that the aim of the notes was precisely to diffuse Gauss's ideas:

> The work of Mr. Charles Dupin [*Développements de Géométrie...* [1813] in which Monge's ideas were developed] is available to everybody, and it would be quite needless to reproduce its various parts here. Naturally, the notes that we append deal with those points which this learned academician has left aside and for which Mr. Gauss has opened new ways; besides our aim is to indicate to young people the sources where they can find information rather than giving them regular lectures.
>
> [Liouville 1850d, p. 582]

Liouville not only passed on Gauss's ideas; he also made his own contributions to them. This section is mainly devoted to those contributions to the general differential geometry of surfaces.

31.   The first note in Monge's *Application...*, however, dealt with the theory of space curves. Of the greatest interest in this note is a long quotation

from a letter from Serret to Liouville. In the letter, Serret proved that the generalized helix, that is, a helix on an arbitrary cylinder, is the only curve for which the ratio between the radius of curvature and the radius of torsion is a constant. Bertrand had proved this theorem synthetically in [1848]; Serret's proof was analytical. Liouville also quoted Puiseux's proof from 1842 that the ordinary helix is the only curve for which the radius of curvature and the radius of torsion are constant separately. He felt that "Mr. Puiseux's method does not seem to lend itself to the proof of this new theorem [i.e., Bertrand's and Serret's theorem], and that is a real imperfection." This criticism was immediately met in a letter from Puiseux to Liouville, printed in Liouville's *Journal* [Puiseux 1851], in which he provided the proof Liouville had doubted would exist. (For further details see [Reich 1973, pp. 278–280]).

32.   In the letter published by Liouville, Serret had used two of the three so-called Frenet formulas, namely, those expressing the derivative of the direction cosines of the tangent and the binormal (i.e., a line normal to the osculating plane). The following year, Serret developed his results in a paper in Liouville's *Journal*, and here he also included the third Frenet formula for the normal [Serret 1851]. This was the first public appearance of these three formulas. However, the following year, Liouville printed an *Extrait d'une Thèse à la Faculté des Sciences de Toulouse, le 10 juillet 1847*, in which Frédéric Jean Frenet presented similar ideas [Frenet 1852]. At first, Frenet's paper received little attention, and when, twelve years later, Bertrand published his *Traité de calcul différentiel et de calcul intégral*, he attributed the formulas to Serret. This made Frenet take up "a small question of priority" in a letter to the editors of the rival journal, the *Nouvelles Annales de Mathematiques* [Frenet 1864].

> I repeat here an assertion that has already been recorded in the vol XII of the *Nouvelles Annales*, and that Mr. Liouville's testimony can support if necessary, namely: in the year 1847, the illustrious editor of the *Journal de Mathématique* had in his hands a manuscript by me concerning these formulas and several of their consequences. As soon as the new edition of Monge's work became known to me, and that did not happen until 1852, I sent a complaint to Mr. Liouville, who acknowledged it and inserted my work in the journal he runs.
>
> [Frenet 1864]

In the paper in the 12th volume of the Annales referred to above, Frenet had indeed mentioned "a work which I have sent to Mr. Liouville in 1847 and which he has been so kind to insert in the volume XVII of the *Journal de Mathématiques*" [Frenet 1853].

In respectful terms, meant not to hurt the powerful editor, Frenet here seems to hold Liouville responsible for having delayed his paper so much that Serret got the formulas out first. It is an example of how Liouville supported his own protégés and how the provincial mathematicians were often

neglected in the Parisian circles. It must be said in Liouville's defense, however, that, as remarked by the editors Gerono and Prouhet of the Annales,
the formulas did not appear very explicitly in Frenet's *Extrait* and "in order
for a theorem to strike the mind, it is necessary that it is stated, and it is
not sufficient that one can deduce it by comparing some formulas" [note by
Gerono and Prouhet in Frenet 1864]. Moreover, although Karin Reich calls
Frenet's formulas "the greatest discovery in the theory of curves in the period between Gauss and Riemann," the mathematicians at the time, Frenet
in particular, did not appreciate their fundamental importance. Thus, in
the complaints quoted above, Frenet only spoke about one of the formulas,
and in fact, one which Cauchy had published previously (cf. [Reich 1973,
p. 227]).

Finally, when Liouville published Frenet's paper, he did not hide the
fact that it had been written before Serret's papers, and so he can hardly
be held responsible for the fact that Serret's name was for many years
mistakenly attached to these formulas. After Frenet's complaints in 1864,
the formulas have generally been known as Frenet's formulas or the Serret-
Frenet formulas, but as pointed out by Reich [1973, pp. 282–283], this is
equally misleading. The formulas had already been published in 1831 in
Dorpat by a rather unknown mathematician, Senff.

33.   Having now recorded Liouville's not very heroic role in the theory of
space curves, I shall turn to the theory of surfaces, and first to the various
systems of intrinsic curved coordinate systems on these surfaces.

If u, v are two variables determining the position of a point on a surface,
the general line element has the form

$$ds^2 = E\,du^2 + 2F\,du\,dv + G\,dv^2. \tag{46}$$

When the curves $u = $ constant and $v = $ constant cut each other orthogonally, F is equal to zero. As mentioned by Liouville in [1847 g and 1850c],
Gauss had shown that this can be obtained on an arbitrary surface by
choosing $v = $ constant to be geodesics emanating from a given point (geodesic
polar coordinates) or by supposing them orthogonal to a given curve. The
curves $u = $ constant must be chosen as the orthogonal trajectories to this
family of geodesics. If, moreover, u is defined as the geodetic distance from
the point (u, v) to the given point or the given curve, the function E equals
1 so that

$$ds^2 = du^2 + G\,dv^2. \tag{47}$$

This is the simplest general line element in Gauss's *Disquisitiones.*

Inspired by elliptic coordinates, however, Liouville in his mechanical and
geometric work, preferred coordinate systems (α, β) with line element

$$ds^2 = \lambda(\alpha, \beta)(d\alpha^2 + d\beta^2). \tag{48}$$

(cf. §8 and Chapter XI). Since this is the most characteristic feature of
Liouville's differential geometry, it would have been appropriate to call it

a Liouvillian line element if it had not been for the fact that this term is usually reserved for the special case $\lambda(\alpha, \beta) = f(\alpha) + F(\beta)$ (cf. §§8–9 and Chapter XVI, §§61–64). Instead, coordinates satisfying (48) are called isothermal.

Already in [1846i, p. 348], Liouville observed that not only are such coordinates orthogonal, their coordinate nets also divide the surface into similar rectangles if the coordinate curves $\alpha = $ constant, $\beta = $ constant are separated by equal intervals $d\alpha$ and $d\beta$, respectively. According to an important theorem of Bertrand [1844], this is precisely the characteristic feature of the lines of curvature on a surface that is member of a family of isothermal surfaces.

34.   In [1847g, p. 293, and 1850c], Liouville proved that any surface can be equipped with a coordinate system (α, β) having an isothermal line element (48) (in the general case, of course, $\alpha = $ constant and $\beta = $ constant are no longer lines of curvature; [Reich 1973, p. 332] is misleading at this point). Liouville's proof is simple and possibly inspired by Gauss's work on map projections. Given an arbitrary coordinate system (u, v) on the surface, the line element (46) is positive definite and is therefore the product of two complex conjugate forms:

$$du\sqrt{E} + dv \, \frac{1}{\sqrt{E}}(F + i\sqrt{EG - F^2}), \tag{49}$$

and

$$du\sqrt{E} + dv \, \frac{1}{\sqrt{E}}(F - i\sqrt{EG - F^2}). \tag{50}$$

Assume that the first form becomes an exact differential if it is multiplied by $\mu + i\nu$. It is known from Euler that finding the integrating factor $\mu + i\nu$ is equivalent to solving the differential equation found by equating (49) to zero. Let $d(\alpha + i\beta)$ be the differential found by this multiplication. Then, the form (50) multiplied by $\mu - i\nu$ equals $d(\alpha - i\beta)$. Multiplying the two conjugate differentials, Liouville finally obtained

$$(\mu^2 + \nu^2)ds^2 = d\alpha^2 + d\beta^2,$$

or

$$ds^2 = \lambda(d\alpha^2 + d\beta^2), \tag{48}$$

where $\lambda = (\mu^2 + \nu^2)^{-1}$.

Thus, the determination of an isothermal system requires the solution of a first-order differential equation, and Liouville tacitly assumed the existence of such a solution.

Liouville further proved that any isothermal system (α', β') can be found from a given one (α, β) by means of the equation

$$\alpha' + i\beta' = \Pi(\alpha \pm i\beta),$$

where Π is a holomorphic function.

35.   The main aim of Liouville's paper [1847g] was to deduce the following new formula for the Gaussian curvature expressed in isothermal coordinates (48):

$$k = \frac{1}{RR'} = -\frac{1}{2\lambda} \left(\frac{\partial^2 \log \lambda}{\partial \alpha^2} + \frac{\partial^2 \log \lambda}{\partial \beta^2} \right). \tag{51}$$

It is a matter of taste whether one agrees with Liouville, who claimed that his derivation was simpler than Gauss's derivation, but it does provide an alternative proof of Gauss's important theorema egregium [Gauss 1828, §12]. This theorem states that since the curvature can be expressed in terms of the intrinsically defined E, F, and G in (46), or equivalently in terms of λ, it is not changed by flexions of the surface, i.e., by isometries.

In this way, Gauss had shown that for two surfaces to be developable on each other (i.e., to be isometric) it is necessary that the curvatures are equal at corresponding points. In general, this condition is not sufficient. From 1838, Ferdinand Minding therefore began to search for sufficient conditions, and in [1839] he succeeded. In particular, he showed that two surfaces of constant curvature are isometric if and only if they have the same curvature. Therefore, any surface of constant positive curvature k is developable on a sphere of radius $\frac{1}{\sqrt{k}}$, and those of constant negative curvature can be developed on a pseudosphere, a surface introduced by Minding.

In France, the same results were obtained first by Ossian Bonnet [1848] and then by Liouville [1850e] in the fourth Monge note. The question of the dependence or independence of these works is not clear. Although Bonnet's methods differed markedly from those of Minding and although Bonnet only referred to an earlier work of his German colleague, it is, as pointed out by Reich [1973, p. 301], not unlikely that he knew the 1839 paper as well. Liouville did not refer to either of his predecessors, nor did he explicitly claim priority even for his own novel contributions. If he knew the earlier papers, this behavior seems strange in the light of the goal of the notes (cf. quote in §30), namely, to refer the young readers to the sources. Since Bonnet's paper had not appeared in Liouville's *Journal*, it is in principle possible that he was unaware of it, and I have not located a reference to this paper in Liouville's notebooks. Still, it seems almost unthinkable that Bonnet's results were unknown to Liouville, especially since the two exchanged other unpublished material around this time (cf. §42). The question cannot be solved on the basis of the sources I know.

36.   Let me briefly summarize the content of Liouville's work on the development of one surface on another with a particular view to the important novelties he introduced.

In [1847g, pp. 301–303], Liouville had shown that if the surface has zero curvature, then it is developable on a plane. The proof consists of solving the differential equation

$$\frac{\partial^2 \log \lambda}{\partial \alpha^2} + \frac{\partial^2 \log \lambda}{\partial \beta^2} = 0, \tag{52}$$

which results from Liouville's expression of the curvature (51). Having repeated this result in [1850e], Liouville turned to the general problem. Denoting by $RR_1 = f(u,v)$ the inverse of the curvature of one surface and by $RR_1 = f'(u',v')$ the inverse curvature of another surface, we know that in order for the two surfaces to be isometric $f(u,v)$ must be equal to $f'(u',v')$ if (u,v) and (u',v') are corresponding points. First, Liouville assumed that the curvature is not constant. In that case, $f(u,v) = f'(u',v')$ can be used as a new variable on the two surfaces, replacing, say, u and u'. Thus, the line elements are of the form

$$
\begin{aligned}
ds^2 &= L\,df^2 + 2M\,df\,dv + N\,dv^2, \\
ds'^2 &= L'\,df^2 + 2M'\,df\,dv' + N'\,dv'^2.
\end{aligned}
\tag{53}
$$

The two surfaces are isomorphic if and only if there is a map $v'(f,v)$ so that $ds^2 = ds'^2$. By equating the two expressions in (53), Liouville showed that this can be decided by isolating v' as a function of v and f from the equation

$$L - \frac{M^2}{N} = L' - \frac{M'^2}{N'} \tag{54}$$

and then checking if the following integrability conditions are satisfied:

$$\frac{\partial v'}{\partial v} = \sqrt{\frac{N}{N'}} \quad \text{and} \quad \frac{\partial v'}{\partial f} = \frac{M - M'P}{N'P}. \tag{55}$$

Thus, as Minding had pointed out, no integrations are needed to solve the question. If (54) is an identity in v', Liouville proceeded in a slightly different way.

The choice of the curvature as a common coordinate on the two surfaces makes Liouville's proof more elegant than that of his predecessors. Minding, whose proof is otherwise similar to Liouville's, did mention this possibility at the end of his paper:

> The general formulas deduced above are very much simplified when one choses the measure of curvature or some function of it as the auxiliary quantities p and p'.
>
> [Minding 1839, p. 386]

Both Minding and Liouville explained that the elegance is obtained at the expense of practical utility, because the two new forms (53) are not easy to find. Did Liouville learn this method of proof from Minding, or was he honest when he said about his clever choice of coordinates: "But it abbreviates the exposition of <u>our</u> method very much." ?

37. In his treatment of surfaces of constant curvature, Liouville differed strikingly from his predecessors by using the isothermal line element. From his new expression (51) he saw that if a surface has constant curvature $\pm\frac{1}{a^2}$ then $\lambda(\alpha, \beta)$ must satisfy

$$\frac{\partial^2 \log \lambda}{\partial \alpha^2} + \frac{\partial^2 \log \lambda}{\partial \beta^2} \pm \frac{2\lambda}{a^2} = 0. \tag{56}$$

This equation is known as Liouville's differential equation (cf. [Blaschke 1923, Jordan 1887, §278]). Introducing (as he had done earlier (cf. §34)) the complex variables

$$\alpha + i\beta = u, \quad \alpha - i\beta = v,$$

he transformed this equation into

$$\frac{\partial^2 \log \lambda}{\partial u \partial v} \pm \frac{\lambda}{2a^2} = 0 \tag{57}$$

In the Monge note, he continued:

> I have found, by way of considerations whose details I shall suppress, that the complete integral, containing two arbitrary functions, $\varphi(u)$ and $\psi(u)$, of this partial differential equation is
>
> $$\lambda = \frac{4a^2 \varphi'(u)\psi'(v)e^{\varphi(u)+\psi(v)}}{[1 \pm e^{\varphi(u)+\psi(v)}]^2}. \tag{58}$$
>
> [Liouville 1850e, p. 597]

He then proved a posteriori that λ does satisfy equation (57). Three years later, he revealed that "I have been led to this result by geometric considerations, drawn naturally from the properties of the sphere" [Liouville 1853a, p. 71]. In this later note, Liouville presented an a priori proof of the solution:

> Since then another purely analytical procedure for the study of the integral of the equation (1) [(57)] has occurred to me. It is more direct and at least as simple and will be the subject of this Note.
> [Liouville 1853a, p. 71]

Liouville continued the Monge note by reintroducing α and β in φ and ψ. In order for λ to be real, $\varphi(\alpha + i\beta)$ and $\psi(\alpha - i\beta)$ must be complex conjugates, so he put

$$\varphi(\alpha + i\beta) = \zeta + i\tau \quad \text{and} \quad \psi(\alpha - i\beta) = \zeta - i\tau,$$

where ζ and τ are new functions of α, β. Taking the differential on both sides of these two equations and multiplying the results he found

$$\varphi'(u)\psi'(v)(d\alpha^2 + d\beta^2) = d\zeta^2 + d\tau^2,$$

which can be substituted into (58) to yield the line element

$$d^2s = \lambda(d\alpha^2 + d\beta^2) = \frac{4a^2 e^{2\zeta}}{(1 \pm e^{2\zeta})^2}(d\zeta^2 + d\tau^2). \tag{59}$$

Thus, for a given value of the curvature $\pm\frac{1}{a^2}$, Liouville found coordinates which give the common expression (59) for the line elements. All surfaces of the same constant curvature are therefore isometric.

38.   In the special case of a surface of rotation described by

$$\tau = b\theta \quad \text{and} \quad \frac{2abe^{\zeta}}{1 \pm e^{2\zeta}} = r, \tag{60}$$

the line element (57) reduces to

$$ds^2 = \frac{a^2 dr^2}{a^2 b^2 \mp r^2} + r^2 d\theta^2. \tag{61}$$

For $b = 1$ and the upper sign, Liouville found that the surface is a sphere, and for $b = 0$ and the lower sign, he found that the surface arose by rotation about the z axis of a curve satisfying

$$dz = dr\frac{\sqrt{a^2 - r^2}}{r}. \tag{62}$$

He observed that the tangent of this curve has the constant length (measured from the axis)

$$a = \sqrt{r^2 + \left(\frac{r\,dz}{dr}\right)^2}. \tag{63}$$

It is therefore the tractrix. In Minding's paper [1839], one finds a parametric description of the meridian curve of the pseudosphere, but Minding did not identify this curve as the tractrix.

"This brief 'Note' by Liouville attracted great attention" says Reich [1973, p. 342]. Codazzi [1857] and Beltrami [1864] discussed it at length, and Dini [1865] used Liouville's approach to show that there are three classes of surfaces of revolution with constant negative curvature. The second of these classes consisted of the one surface whose meridian curve satisfies (62). Dini called this important surface "la superficie di Liouville." Only three years later, Beltrami changed its name to the "pseudosphere," a name it has kept ever since. This happened in his celebrated work on the interpretation of the non-Euclidean geometry [Beltrami 1868], where he remarked that the geometry on a surface of constant negative curvature is non-Euclidean when geodesics are interpreted as lines.

39.   Thus, Liouville had a strong impact on the Italian school of differential geometers. Did he himself suspect the connection to non-Euclidean

geometry? This question and its answer (no) is a special case of a puzzling historical problem: Why did neither Gauss (in print) nor his successors before Beltrami discover this connection, which was so strongly suggested by Gauss's version of the Gauss-Bonnet theorem? I shall not discuss Gauss, but for those who followed him, there are at least two aspects to the answer. First, interest in the discussions about the parallel postulate was small and the "belief" in Euclidean geometry widespread before Gauss's ideas in his letters to Schumacher were published in 1860. Gauss's views and Houël's subsequent translation of Bolyai and Lobachevski's works into French spread the gospel of a new and fertile field of geometry around 1868. Second, in order to accept a geodesic as playing the role of a "line" in plane geometry, the mathematicians had to take seriously the research programme indicated in Gauss's *Disquisitiones* [1828, §13], namely, to study the intrinsic properties of surfaces and thus see that the nonzero curvature of the geodesic in <u>space</u> is irrelevant for the geometry on the surface. This idea was carried to perfection and generalized to curved spaces of higher dimension by Riemann in his *Habilitationsvortrag* (held in 1854) and published posthumously the same year as Beltrami's paper. These two factors were probably more responsible for blocking a differential geometric model of non-Euclidean geometry than the purely mathematical fact discovered by Hilbert that neither the pseudosphere nor any other regular analytic surface in space can carry the entire non-Euclidean geometry. The pseudosphere only represents a sort of non-Euclidean cylinder.

In Liouville's case, the lack of interest in non-Euclidean geometry was the most important factor, for as I shall argue later, he did appreciate the importance of the intrinsic formulation of geometry on a surface. Apparently, he did not know the work of Bolyai and Lobachevski, and the few times he ever addressed the problems of the "theory of parallels" (cf. Chapter VI) he expressed contempt for the subject. His remarks make it clear that he did not see any connection between his work in differential geometry and non-Euclidean geometry.

40.   Even less could he have foreseen that the few pages he wrote about equation (56) would cause the physicists of the 1980s to associate his name with an important part of the theory of strings in high-energy physics. This field was opened by Polyakov in [1981]. He introduced the "Liouville action"

$$S = K \iint \left(\frac{1}{2} \left[\left(\frac{\partial \varphi}{\partial x} \right)^2 \pm \left(\frac{\partial \varphi}{\partial y} \right)^2 \right] + \mu^2 e^\varphi \right) dx \, dy \,, \qquad (64)$$

corresponding to the classical equation of motion

$$\frac{\partial^2 \varphi}{\partial x^2} \pm \frac{\partial^2 \varphi}{\partial y^2} + \mu^2 e^\varphi = 0 \,,$$

which is precisely equation (56) for $\varphi = \log \lambda$. The associated quantized theory he called "quantum Liouville theory."

Polyakov's idea has been taken up by many physicists, such as Onofri and Virasoro [1982], who explicitly referred to Liouville [1853a] and used the geometric interpretation of the equation, Gervais, Neveu, Durhuus, Olesen, Petersen, and many others. For references, see Neveu's paper, *The uses of the Quantum Liouville Theory* in the Niels Bohr Centenial Volume [Neveu 1985].

41.   Note five in Liouville's Monge edition is devoted to "geographic mappings of surfaces on each other." The problem is to decide if two given surfaces can be mapped conformally onto each other, that is, so that angles are preserved, or analytically that their isothermal coordinates (α, β) and (α', β') satisfy $d\alpha'^2 + d\beta'^2 = l(d\alpha^2 + d\beta^2)$. Following Gauss (and his own earlier proof, §34) Liouville showed that $\alpha' + i\beta'$ must be a "function" (like Riemann and others, he meant a holomorphic function) of $\alpha + i\beta$. Continuing along these lines, he investigated in what cases the "scale" l is equal to one. This gave rise to a new solution to "Minding's" problem, $ds = ds'$: "The circumstances determine whether one shall choose this method or the one in Note IV or other methods, which I could add" [Liouville 1850f]. Liouville never added more than these two solutions.

In note six, Liouville generalized the question of map projection to three dimensions and thus arrived at his beautiful theorem discussed in the previous section. Liouville concluded the Monge edition with a note in which he showed that d'Alembert's original method of solving the wave equation, using exact differentials, in contrast to the more familiar Eulerian proof, could be used to solve a variety of other partial differential equations. Although inspired by Monge's geometric discussion of the problem of singularities in the solution of the vibrating string, which is important in the development in the concept of function, Liouville's note was purely analytical and contained no reflections on the foundation of analysis or on the theory of surfaces.

42.   Before closing the discussion of the Monge notes, I must mention the important introduction of the "geodesic curvature" of a curve on a surface. In [1830], Minding had studied the quantity $\cos \theta / R$, where R is the radius of curvature of the curve, and θ is the angle between its osculating plane and the tangent plane of the surface. He had shown that this quantity is the curvature of a curve found by developing the given curve on a plane. He also found an expression in terms of the coefficients E, F, G of the first fundamental form, (46).

Apparently, independently of Minding, Bonnet studied the same quantity in his great 1848 paper in which he, among other things, used it in the celebrated Gauss-Bonnet theorem:

$$\sum = A + B + C - 2\pi - \int_0^s \frac{\cos \theta}{R} \, ds \,, \tag{65}$$

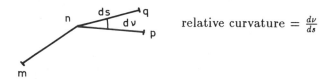

relative curvature $= \frac{d\nu}{ds}$

FIGURE 10

where A, B, C are the angles in a triangle made up by arbitrary (smooth) curves on a surface, $\sum$ is the integral curvature of the interior of the triangle, and the integral is to be taken around the perimeter of the triangle. Gauss had proved the theorem for a geodesic triangle, where the integral vanishes, and Bonnet later extended it to arbitrary polygons (see [Reich 1973, pp. 312–313]).

In this paper, Bonnet introduced the suggestive name "courbure géodésique" for the quantity $\frac{\cos\theta}{R}$, which Minding had not given a separate name. When Bonnet sent his paper to the *Académie* the following year, he admitted in the accompanying letter that this name was due to Liouville:

> This name is due to Mr. Liouville who has employed it in a Memoir which was written long ago but has not yet been published.
> [Bonnet 1849, p. 449]

In order not to give away too much priority, Bonnet at the same time pointed out that many of his important formulas for the geodesic curvature had already been presented to the *Académie* in 1844.

43.   Liouville's own ideas on geodesic curvature were published in the first and second note in the 1850 Monge edition. Here, we see that the name he introduced was motivated by an interesting new definition of the concept. He first defined the relative curvature of two curves touching each other (Fig. 10):

> Consider two curves having an element ds in common; the two elements that follow ds on these two curves form an infinitely small angle $d\nu$, which we call the relative angle of contact; and the ratio between $d\nu$ and ds measures the relative <u>curvature</u> or <u>deviation</u> of the two curves.
> [Liouville 1850b, p. 568]

Having indicated some generalizations of these notions and the mechanical notion of relative centrifugal force, Liouville concluded note 1 with the definition of geodesic curvature:

> After some reflection, the reader will see that one can make great use of these general ideas, which we shall not stress here. Let us limit ourselves to emphasizing among the relative curvatures the one of a curve drawn on a surface relative to the tangential geodesic line. I

have suggested to denote it by the expressive name <u>geodesic curvature</u>, which Mr. Bonnet has kindly adopted in a remarkable memoir.

[Liouville 1850b, p. 568]

Liouville's definition has the advantage over that of Minding and Bonnet that it is entirely intrinsic. To be sure, Minding had shown that $\cos\theta/R$, which seemingly depends on the spatial embedding of the surface, is in fact expressible in terms of E, F, and G and therefore independent of flexions of the surface. However, by insisting on an a priori intrinsic definition, Liouville seems to have grasped better than his predecessors the desirability of a purely intrinsic formulation, which had been characterized by Gauss as "a very worthy method which may be thoroughly developed by geometers." Gauss described the method as follows:

> When a surface is regarded, not as the boundary of a solid, but as a flexible, though not extensible solid, one dimension of which is supposed to vanish, then the properties of the surface depend in part upon the form to which we can suppose it reduced, and in part are absolute and remain invariable whatever may be the form into which it is bent. To these latter properties, the study of which opens to geometry a new and fertile field, belong the measure of curvature...; and the generic method of defining in a general manner the nature of the surfaces thus considered is always based on the formula $\sqrt{E\,dp^2 + 2F\,dp\,dq + G\,dq^2}$, which connects the linear element with the two indeterminates p, q.

> [Gauss 1828, §13]

As we shall see later, Liouville was consciously following this research programme of Gauss's.

44.   Liouville further supplied new, simple formulas for geodesic curvature, which were discussed later by Bonnet, Chelini, and Beltrami (cf. [Reich 1973, p. 299]). In the second Monge note, Liouville first showed that for orthogonal coordinates (u, v) the geodesic curvature is given by

$$\frac{1}{\rho} = -\frac{di}{ds} + \frac{\cos i}{\rho_v} + \frac{\sin i}{\rho_u},$$

where i is the angle the given curve makes with the curve $u = $ const. and ρ_u and ρ_v are the geodesic curvatures of the curves $u = $ const. and $v = $ const., respectively. The following year, he generalized this formula to the case where the parameter curves cut each other at an arbitrary angle ω [Liouville 1851d]. In the same note, he published three new intrinsic formulas for the Gaussian curvature:

$$\frac{D}{RR_1} = -\frac{1}{2}\frac{\partial}{\partial u}\frac{1}{D}\left(\partial v - 2\frac{\partial F}{\partial v}\right)$$
$$-\frac{1}{2}\frac{\partial}{\partial v}\frac{1}{D}\left(\frac{\partial E}{\partial v} - \frac{F}{G}\frac{\partial G}{\partial u}\right)$$

$$= \frac{\partial^2 \omega}{\partial u \partial v} + \frac{\partial}{\partial v}\left(\frac{\sqrt{E}}{\rho_v}\right) - \frac{\partial}{\partial u}\left(\frac{\sqrt{G}}{\rho_u}\right)$$

$$= \frac{\partial}{\partial v}\left(\frac{\sqrt{E}}{\rho_\omega}\right) - \frac{\partial}{\partial u}\left(\frac{\sqrt{G}}{\rho_u}\right), \tag{66}$$

where $D = \sqrt{EG - F^2}$ (cf. (46)) and ρ_ω is the geodesic curvature of a curve on the surface tangent to the given curve and cutting all the curves $u =$ constant at a constant angle.

45.   Liouville indicated that he had described these new results in the brief series of lectures he had given on "the theory of entire and homogeneous differential formulas of several variables" in the few weeks that remained of the first semester of 1850–1851 after his appointment as a professor at the *Collège de France*. The *Résumé* in his *Nachlass* [Ms 3640 (1846–1851)] shows that in addition to his published ideas Liouville included a new proof of Gauss's theorema egregium and a few more reflections on geodesic curvature. The most interesting remark in this résumé, however, is his very explicit intention of treating the theory of surfaces entirely intrinsically:

> I have been able to arrive at these various results and, in particular, at the formula
>
> $$ds^2 = E\,du^2 + 2F\,du\,dv + G\,dv^2\,,$$
>
> from which they result, without relying on the previous formula
>
> $$ds^2 = dx^2 + dy^2 + dz^2$$
>
> and without any use of coordinates outside the surface.
> [Liouville Ms 3640 (1846–1851), p. 5]

In fact, it is worth noticing that any mention of the surrounding space is absent, not only in the rest of these lectures, but also in Liouville's earlier published papers on surfaces. I have already mentioned this in connection with his definition of geodesic curvature. This indicates that among the mathematicians before Riemann, Liouville was the one who best understood the value of an intrinsic theory of surfaces.

46.   Did Liouville conceive of a generalization to higher dimensions, similar to the ideas on manifolds that Riemann put forward in his famous *Habilitationsvortrag* [1854–1867]? There are two sources indicating that he was on this track, one concerning a general theory of differential forms, the other concerning the principle of least action. The first is a published note [Liouville 1852e] in which he expanded some of the ideas and results that he had presented in his lectures at the *Collège de France*. In the lecture notes, he wrote that his aim was to study the "entire and homogeneous differential forms of several variables, such as $M\,dx + N\,dy$, $L\,dx^2 + 2M\,dx\,dy + N\,dy^2\,\&_c$ which often present themselves in pure analysis, in geometry, in mechanics" [Ms 3640 (1846–1851), p. 1]. The short semester, however, only allowed

him to "explain ...the first rudiments," namely, those related to the theory of surfaces. In the note of 1852, Liouville stated that his research into differential forms of two variables "was terminated a long time ago and could be published immediately in their entirety." This publication never materialized, so one has to guess the direction of Liouville's research from the brief indications in the note of 1852.

> I have treated this differential form $[E\,du^2 + 2F\,du\,dv + G\,dv^2]$ in itself and in all its generality, and I have tried to study it as in the theory of numbers one studies the quadratic formula
>
> $$ax^2 + 2bxy + cy^2,$$
>
> to be sure with very different methods.
>
> [Liouville 1852e, p. 478]

Following this line of thought, which bears witness to his growing interest in Dirichlet's number-theoretical research, Liouville studied the factorization of arbitrary binary differential forms into factors of the form $(l\,du + m\,dv)$.

> 47.   The forms of three or more than three variables offer much greater difficulties and there I am far from having obtained all the success that I have desired. However, I think that I have succeeded as far as the form
>
> $$E_1\,du_1^2 + E_2\,du_2^2 + E_3\,du_3^2 + 2F_1\,du_2\,du_3 + 2F_2\,du_1\,du_3 + 2F_3\,du_1\,du_2$$
> [67]
>
> is concerned when this form represents the distance of two arbitrary infinitely close points, that is, when it is reducible to the form
>
> $$dx^2 + dy^2 + dz^2."$$
>
> [Liouville 1852e, p. 479]

In particular, Liouville considered a surface given by the equation

$$\varphi(u_1, u_2, u_3) = 0$$

> by means of the value of ds^2 alone, and without knowing the relation between the curvilinear coordinates u_1, u_2, u_3 and the ordinary rectangular coordinates x, y, z, one asks for 1° the direction of the normal and of the lines of curvature in each point; 2° the size of the radius of curvature of an arbitrary normal section.
>
> [Liouville 1852e, p. 479]

In the published note, he wrote down the general formula for the radius of curvature of a normal section of the coordinate surface $u_3 = $ constant. He also mentioned other problems he could solve in terms of the u coordinates alone, for example, to find the direction of the normal and the radius of curvature of a curve given by two equations in u_1, u_2, and u_3.

These formulas are beautiful generalizations of the more specific formulas that were known for special curvilinear coordinate systems, in particular, the orthogonal ones studied by Lamé. They show that Liouville made the study of the quadratic form (67) the basis of a study of space, independent of its representation in Cartesian coordinates. This is an intrinsic point of view similar to that of Riemann. The formulas he derived in this way are in fact valid for any Riemanian metric (67), and so Liouville did formally work in an arbitrary three-dimensional Riemanian manifold. However, as opposed to Riemann, Liouville does not seem to have attached any geometric reality to his formulas, unless the form (67) is "reducible to the form $dx^2 + dy^2 + dz^2$," that is, unless the space is Euclidean. For example, having discussed his formulas for the radius of curvature of a normal section to the curve, he wrote:

> The analysis which solves this question gives in particular for the product of the principal radii of curvature a value which must reduce to the one that can be obtained directly from a known formula of Mr. Gauss, when one eliminates in the expression for ds^2 one of the quantities u_1, u_2, u_3 by way of the equation $\varphi(u_1, u_2, u_3) = 0$ of the surface. By equating the two results, one obtains a conditional equation which E_1, E_2 etc. must satisfy, for any choice of φ, if ds in fact represents the distance between two points. But, let us not be carried away by digressions which will lead us too far away.
>   [Liouville 1852e, p. 479]

Thus, Liouville had found a necessary condition that the quadratic differential form had to satisfy in order to represent the line element in Euclidean space.

48.   In the 1852 note, Liouville did not indicate how he intended to continue his theory to differential forms of more than three variables, but as we saw in the previous chapter, such forms did turn up four years later in his important reflections on the principle of least action. As stated in the quote in §46, Liouville, in his lectures, stressed the applicability of his theory of differential forms to mechanics, but in the lecture notes [Ms 3640 (1846–1851), lecture 4], he only considered the principle of least action for one point on a surface. Still, there is no doubt that Liouville must have seen the similarity between his generalization in [1856k] to many particles and his earlier programme for differential forms of any number of variables, just as he discovered the formal similarity between the determination of trajectories using the principle of least action and finding geodesics on a "surface" (of high dimension) (see Chapter XVI, §§45–48). From these considerations, Riemanian manifolds are not hard to seek. Had Liouville been interested in the theory of parallels and the possible non-Euclidean nature of space, he might have come all the way.

If Liouville did not arrive at an understanding of curved spaces, his programme for investigating differential forms and his concrete manipulation with the form in the action integral anticipated the concrete theory of

Principe de la moindre action: $\int \sqrt{(U+C)ds^2}$; si $ds^2 = \lambda(d\alpha^2 + d\beta^2)$,

cette intégrale $= \int \sqrt{\lambda(U+C)(d\alpha^2 + d\beta^2)}$, d'où une réduction au plan avec

$\lambda(U+C)$ fonction des forces et $2\lambda(U+C)$ force vive. — On pourrait ramener

à une autre surface quelconque. Lignes géodésiques ou les plus courtes;

1° décrites sur leur surface par un point sans force; 2° sur le plan avec λ

pour fonction des forces et 2λ pour force vive, on a les mêmes relations

en α, β. — Cas de $\lambda = f(\alpha) - F(\beta)$, $\dfrac{d^2\alpha}{dt^2} = f'(\alpha)$, $\dfrac{d^2\beta}{dt^2} = -F'(\beta)$,

$$\left(\frac{d\alpha}{dt}\right)^2 + \left(\frac{d\beta}{dt}\right)^2 = 2\big(f(\alpha) - F(\beta)\big), \quad \left(\frac{d\alpha}{dt}\right)^2 = 2\big(f(\alpha) - a\big), \quad \left(\frac{d\beta}{dt}\right)^2 = 2\big(a - F(\beta)\big)$$

$$\left(\frac{d\alpha}{d\beta}\right)^2 = \frac{f(\alpha) - a}{a - F(\beta)} = \tan^2 i, \quad f(\alpha)\cos^2 i + F(\beta)\sin^2 i = a.$$

$\circ$ $\sin i = \dfrac{\sqrt{f(\alpha) - a}}{\sqrt{\lambda}}$, $\cos i = \dfrac{\sqrt{a - F(\beta)}}{\sqrt{\lambda}}$, $ds = d\alpha \sin i + d\beta \cos i$

$ds = d\alpha\sqrt{f(\alpha) - a} + d\beta\sqrt{a - F(\beta)}$, $\delta ds = 0$ le long de la ligne.

— Surfaces de révolution. — Plan, sphère, ellipsoïde. —

Intégrale d'Euler pour les fonctions elliptiques. — Lignes de

courbure de l'ellipsoïde, et démonstration simple du théorème général

de C. Dupin sur les surfaces orthogonales.

PLATE 35. Liouville's notes from his lectures of 1850–1851 at the Collège de France concerning differential forms. In the fourth lecture, he drew the connection between two-dimensional mechanics and differential geometry by means of the principle of least action. This is the paragraph shown above. (Bibliothèque de l'Institut de France, Ms 3640 (1846–1851)).

differential forms in Riemann's Paris Academy Prize Essay [1858] and in the work of Christoffel and Lipschitz. As mentioned in the last chapter, Liouville's note on the principle of least action had some impact on this development, but his research programme for a general theory of differential forms does not seem to have been read by his successors.

Let me conclude that Liouville was the mathematician before Riemann who best understood the importance of a general theory of differential forms as a means to create a completely intrinsic differential geometry of surfaces (and to a certain extent of three-dimensional space) and as a means to geometrize mechanics.

Appendix I

Extract from AC AL Ms 2 on the centrality of Ampère's force law:

S'il était démontré que l'action électrodynamique est une action élémentaire, il serait par cela même évident que l'action mutuelle de deux élémens voltaïques est dirigée suivant la droite qui joint leurs milieux: Dès lors plus de difficulté; et toute la théorie en découle comme nous venons de l'exposer. Mais comme il nous semble peu philosophique d'admettre à priori une telle proposition sans la démontrer, nous avons cherché à démontrer, par l'expérience, que l'action de deux élémens de courants est en efet dirigée suivant la ligne qui les unit.

Rappellons d'abord qu'il résulte d'une expérience de Mr Oersted que dans toutes les positions l'action d'un fil voltïque sur l'aiguille aimantée est identiquement la même; d'ou il résulte que toutes les faces du fil sont parfaitement semblables, et considérons en premier lieu les courants infiniment petits mm', nn' dirigés comme on le voit sur la figure [figure A does not exist in the maniscript], désignons par leur action mutuelle $m'n'$. La direction de leur action mutuelle

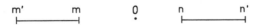

Figure A

doit être symétrique par rapport à eux. Si l'on suppose pour plus de simplicité que leurs grandeurs et leurs intensités sont égales. Cette direction est donc ou la droite $m'n'$ ou une ligne quelconque située dans un plan perpendiculaire à $m'n'$ et passant par le point O. Or par cela seul que toutes les faces du fil sont parfaitement identiques il ne peut y avoir aucune raison pour choisir une plutôt que l'autre de ces dernières droites: elles doivent donc toutes être rejettées d'ou il résulte que la droite $m'n'$ est véritablement la direction suivant laquelle les deux courants s'attirent. On sent d'ailleurs que la démonstration faite pour deux élémens égaux s'étendra par suite à deux éléments de grandeurs quelconque.

Une démonstration analogue s'applique à deux éléments mm', nn' parrallèles entr'eux et à angle droit sur la ligne qui joint leurs milieux. Supposons en effet qu'ils soient égaux et dirigés dans la

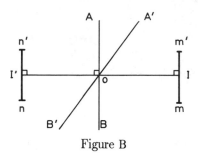

Figure B

même sens. La force qui entraine l'un d'eux mm' par exemple, sera dirigée suivant II' ou suivant une des droites OB, OA, OB', OA' passant par le milieu O de II' perpendiculaires à cette ligne et situés dans le plan des deux éléments ou dans un plan à angle droit avec celui là. Mail il n'y a aucune raison pour préférer OB à OA, OB' à OA' puisque l'on peut changer le sens des courants sans rien changer à la force produite. Ces quatre directions doivent en conséquence être rejettées et la force est dirigée suivant II'. Cette démonstration s'étendra d'ailleurs aussi bien que la précédente à des éléments de grandeurs quelconques en décomposant [aux à] en des assemblages de courants égaux à leur commune mesure.

L'égalité que nous avons supposée entre les intensités, ne présentera pas non plus de difficultés, si l'on se rappelle qu'un courant dont l'intensité serait 2 ne ferait que remplacer deux courants d'une intensité [1] et dirigés dans le même sens. D'un autre côté comme l'action qui résulte de deux courants infiniment petits quelconques est toujours susceptible d'être ramenée, comme nous l'avons fait voir, à des actions telles que celles que nous venons de considérer, il en résulte que notre démonstration est tout à fait générale et remplit parfaitement le but que nous nous étions proposé.

[AC AL Ms 2]

Appendix II

*Liouville's notes on Galois. Bibliotèque de l'Institut de France Mss 2108.
I Fol. 260–265 (consecutively paginated 1–6)*

[Fol 260] Notes sur Galois J. Liouville.

Adjonction — Expression des racines $x_0, \ldots, x_{n-1}$ par V. — $V, U \&_c$ dépendent rationnellement l'une de l'autre — C'est donc toujours en facteurs de degrés égaux que l'éq. en V se décompose par adjonction.

— Admettons qu'en adjoignant r racine d'une éq. irréductible de degré p, l'éq. en V se décompose $V^m + \&_c = f(V, r) f_1(V, r) \ldots f_{q-1}(V, r) \, f, f_1, \ldots,$ f_{q-1} de même degré s, $sq = m$ — Et aussi si $V^m + \&_c = f(V, r') f_1(V, r') \ldots$ $f_{q-1}(V, r') = f(V, r'') f_1(V, r'') \ldots$ si $r', r,'' \ldots \&_c$ sont les autres racines de $r^p + \&_c = 0$.

$$\text{Donc } (V^m + \&_c)^p = f(V, r) f(V, r') \ldots f(V, r^{(p-1)})$$
$$\times f_1(V, r) f_1(V, r') \ldots f_1(V, r^{(p-1)})$$
$$\ldots$$

Mais $f(V, r) f(V, r') \ldots f(V, r^{(p-1)})$ est rationnel avant toute adjonction, c'est donc un diviseur rationnel de $(V^m + \&)^p$ et, à cause du degré et de $m = sq$, c'est $(V^m + \&_c)^s$ [incorrect, see Fol. 282].

Mais, si $f(V, r)$ et $f(V, r')$ ont un facteur commun, ils coïncident, car soient $\varphi(V), \psi(V), \&_c$ les racines (outre V) de $f(V, r) = 0$, alors $f(\varphi V, r) = f(V, r) Q(V, r)$. Donc $f(\varphi V, r') = f(V, r') Q(V, r')$ et, si $f(V, r') = 0$, il suit $f(\varphi V, r') = 0$. Donc $\&_c$. Il y a donc s facteurs qui coïncident avec $f(V, r), \&_c$

$$-\text{Et } V^m + \&_c = f(V, r) f(V, r_1) \ldots f(V, r_{q-1})$$
$$= f(V, r) f_1(V, r) \ldots f_{q-1}(V, r).$$

De la deux genres bien distincts de décomposition. Mais, si toutes les racines $r, r_q, \ldots, r_{q-1}$ s'expriment rationnellement par une seule d'entr'elles, il est visible que le 1er genre de décomposition rentre dans le second, et que les deux genres de décomposition coïncident.

Au moment où nous adjoindrons une racine $\sqrt[p]{A}$, nous aurons le droit de supposer déjà adjointes les racines de $\alpha^p = 1$ qui dépendent de radicaux d'ordre inférieur et, dès lors, les racines de $r^p = A$ sont toutes exprimables rationnellement par une seule. D'ailleurs, si l'éq. $r^p = A$ ne restait pas irréductible et si $r^q + \&_c \ldots - H = 0$ était un de ses facteurs les racines de ce facteur donneraient par leur produit $\alpha^\mu r \cdot \alpha^\nu r = H$, $r^q = H$, en sorte que r serait une racine de degré q, non de degré p comme on le suppose. Ainsi, [Fol. 261] s'en tenant au degré p premier, les théorèmes ci-dessus s'appliqueront à ce mode d'adjonction.

Par une suite d'adjonctions de radicaux de degré 1er l'éq. $f(x) = 0$ finit par s'abaisser du degré n (que je suppose premier) à un degré moindre. Soit $r = \sqrt[p]{A}$ le radical qui produit cet abaissement, p ayant la plus petite valeur possible, en sorte que l'abaissement n'aurait pas lieu en adjoignant α, racine de $\alpha^p = 1$, et qu'après cette adjonction de α, l'éq. éq.$^{\underline{on}}$ $\xi^p = A$ reste [rait] irréductible — Soit $f(x,r)$ le plus [petit] facteur (quand au degré en x) de $f(x)$; $f(x,r')$, $f(x,\alpha r)$, $f(x,\alpha^2 r) \ldots$ en seront d'autres et seront premiers entr'eux, sans quoi le commun diviseur de deux fournirait un facteur de degré moindre. Donc $f(x,r)f(x,r') \ldots$ fait un diviseur de $f(x)$, et étant rationnel sans l'adjonction de r, $= f(x)$, donc $f(x) = f(x,r)f(x,r') \ldots f(x,r^{(p-1)})$. Donc si i est le degré de $f(x,r)$, $ip = n$; donc, pour n premier $i = 1$, $p = n$. Pour n composé p divise n. — Ainsi pour p premier l'éq. $\underline{on}$ s'abaisse en une seule fois au 1er degré.

Mais alors V s'abaisse aussi au 1er degré. Donc V était auparavant de degré n. Les adjonctions précédentes avaient déjà réduit à n son degré primitif m. Ces adjonctions peuvent se réduire à celle d'une seule quantité dépendant rationnellement des radicaux adjoints et les exprimant rationnellement à son tour.

Dans la dernière éq. en V (de degré n premier), les racines ne peuvent être que $V, \varphi V, \varphi^2 V, \ldots, \varphi^{n-1} V$ et $\varphi^n V = V$; φV est rationnelle en V dès le commencement, avant toute adjonction on a $x_0 = \lambda(V)$, et on ne peut avoir $\lambda(V) = \lambda(\varphi V)$; car il en résulterait $x_0 = \frac{1}{n}(\lambda(V) + \lambda(\varphi V) + \cdots + \lambda(\varphi^{n-1}V)) =$ quantité rationnelle avant d'adjoindre $r = \sqrt[p]{A}$. Dont $\lambda(\varphi V) = x_1, \lambda(\varphi^2 V) = x_2, \lambda(\varphi^{n-1}V) = x_{n-1}$ — on voit que le dernier groupe de l'Eq^{on} ne renferme que des permutations tournantes.

Mais, $1°$ on passe d'un groupe à l'autre par une même substitution, savoir en changeant V en V', et les substitutions sont les mêmes dans chaque groupe. Cela résulte de la double forme [fol. 262],

$$\begin{aligned}
\mathcal{F}(\mathcal{V}) &= F(V,t)F(V,t_1)F(V,t_2)\ldots \\
&= F(V,t)F_1(V,t)F_2(V,t)\ldots
\end{aligned}$$

Soit donc $x_k, x_{f(k)}$ la permutation par laquelle on passe du 1er groupe au second. Le second groupe sera

$$x_{f(0)}, x_{f(1)}, x_{f(2)}, \ldots, x_{f(k)}, \ldots$$
$$x_{f(1)}, x_{f(2)}, x_{f(3)}, \ldots, x_{f(k+1)}, \ldots$$
$$\vdots$$
$$x_{f(i)}, x_{f(i+1)}, x_{f(i+2)}, \ldots, x_{f(i+k)}, \ldots$$

Mais les substitutions doivent être celles du 1er groupe

$$x_0, x_1, x_2, \ldots, x_{n-1}$$
$$x_1, x_2, x_3, \ldots, x_0$$
$$\cdots$$

Donc le second groupe déduit par ce second moyen doit être

$$x_{f(0)}, \quad x_{f(1)}, \quad x_{f(2)}, \ldots, x_{f(k)}, \ldots$$
$$x_{f(0)+1}, \quad x_{f(1)+1}, \quad x_{f(2)+1}, \ldots, x_{f(k)+1}, \ldots$$
$$\vdots$$
$$x_{f(0)+l}, \quad \cdots \quad , x_{f(k)+l}$$

Donc $f(i+k)$ revient à $f(k) + l$; l dépend de i, non de k

$$f(i) = f(0) + l, \quad f(2i) = f(i) + l = 2f(0) + l, \quad f(3i) = 3f(0) + l$$

$f(\mu i) = \mu f(0) + l, f(k) = k\frac{f(0)}{i} + l, f(k) = ak + b$. Ainsi, il n'y a que des permutations linéaires [fol. 263].

Autrement, $x_0 = \lambda(V), x_1 = \lambda(\varphi V), x_2 = \lambda(\varphi^2 V), \ldots$ d'oú les permutations tournantes. Que fait la substitution d'une autre racine ωV du second facteur

$$F(V, t_1) = F_1(V, t) \quad \text{de} \quad F(V) \;=\; F(V, t)F(V, t_1)\ldots$$
$$=\; F(V, t)F_1(V, t)\ldots$$

$F(\varphi V, t) = F(V, t)Q(V, t)$ donne $F(\varphi V, t_1) = F(V, t_1)Q(V, t_1)$.
Donc $\omega V, \varphi \omega V, \ldots, \varphi^{n-1}\omega V$ sont les racines de $F(V, t_1) = 0$.

Mais $F_1(\omega V, t) = F(V, t)P(V, t)$; donc, changeant V en $\varphi V, \&_c$, ces racines sont encore $\omega V, \omega\varphi V, \omega\varphi^2 V, \ldots, \omega\varphi^{n-1}V$, qui doivent revenir aux précédentes.

Ainsi,

$$\varphi\omega V = \omega\varphi^b V,$$
$$\varphi^2\omega V = \varphi\omega\varphi^b V = \omega\varphi^{2b}V,$$
$$\varphi^3\omega V = \varphi\omega\varphi^{2b}V = \omega\varphi^{3b}V \; \&_c$$

soit $x_0 = \lambda(V), x_1 = \lambda(\varphi V), \&_c$. Changeant V en ωV,

x_0 se change en $\lambda(\omega V) = x_a = \lambda(\varphi^a V)$

$x_1 \qquad\qquad \lambda(\varphi \omega V) = \lambda(\omega \varphi^b V) = \lambda(\varphi^{a+b} V) = x_{a+b}$

$x_2 \qquad\qquad \lambda(\varphi^2 \omega V) = \lambda(\omega \varphi^{2b} V) = \lambda(\varphi^{a+2b} V) = x_{a+2b}$

$\vdots$

$x_{n-1} \qquad\qquad\qquad = x_{a+(n-1)b}$

En général x_k en x_{a+bk}, ou, si l'on veut, x_{ak+b} comme ci dessus — On voit qu'il y a deux marches bien différentes à suivre.

––––––––

$1°$ si tout ce qui précède a lieu, en se donnant deux des racines, on se donne visiblement toutes les autres. En effet, si on veut que $x_0 = \lambda(v)$ soit une éq. satisfaite, on ne peut prendre que x_k, x_{ak}, et alors, pour k différent de zéro, x_k change de place. Donc, si on donne en outre $x_1 = \lambda(\varphi v)$, il n'y aura qu'une [fol. 264] seule racine V qui reste possible, et $x_2, x_3, \ldots$ s'expriment par, x_0 et x_1, rationnellement — Mais la réciproque? Si les racines dépendent de deux x_0, x_1, l'éq. en V est au plus de degré $n(n-1)$. Soi t m son degré. Soit $x = \lambda(V)$ une des racines de $f(x) = 0$. L'éq. $\Pi[x - \lambda V] = (x - \lambda(V))(x - \lambda(V')) \ldots = 0$ n'a que des racines appartenant à $f(x)$. Donc $\Pi(x - \lambda V) = f(x)^\mu$ et $m = n\mu$. μ est $\leq n-1$. Mais on voit que m est divisible par n. — Il y aura μ facteurs $x - \lambda V$ pour lesquels $x_0 = \lambda V$, μ pour $x_1, x_2, \ldots$ Et, de même, si on avait pris $\lambda_1 V$ qui exprime x_1 au lieu de $\lambda(V), \&_c$. Et les racines qui donnent $x_0 = \lambda(V) = \lambda(V_1) = \ldots$ sont (à la $1^{\text{ère}}$ près) différentes de celles qui donnent $x_1 = \lambda_1(V) = \lambda_1(V') = \ldots$, car si $V' = V_1$, le changement de V en V_1 ou $V' = V_1$ laisserait deux racines à la même place, toutes par concéquent, et on aurait $V = \text{fonct}(x_0, x_1) = V_1$ ce qui est absurde.

Ainsi $\mu - 1$ racines pour lesquelles x_0 ne change pas, puis $(\mu - 1)$ pour lesquelles x_1, ne change pas, $\&_c$ en mettant ces racines à la place de V. De là, $n(\mu - 1) + 1$ permutations ou une des racines garde sa place. Mais cela n'épuise pas les racines V ou les $n\mu$ permutations du groupe. Il y a donc des permutations pour lesquelles toutes les racines changent de place.

Soit φV une de ces racines, $x_0 = \lambda(V)$ devient $\lambda(\varphi V)$ que j'appelle x_1, puis $\lambda(\varphi^2 V) = x_2, \ldots$ jusqu'à $\lambda(\varphi^{s-1} V) = x_{s-1}, \lambda(\varphi^s V) = x_0$. Mais alors $\varphi^s V = V$, sans quoi le changement de V en $\varphi' V$, laissant 2 racines x_1, x_2 à la même place, y laisserait toutes les autres et, de la, $\varphi^s V = f(\ldots x_0, x_1) = V$, puisque le changement de V en $\varphi^s V$ dans $V = f.R.(x_0, x_1)$ ne change pas les seconds membres. D'ailleurs si $\lambda(V), \lambda(\varphi V), \ldots \lambda(\varphi^{s-1} V)$ [fol. 265] n'épuisaient pas les n racines x, ces n racines se partageraient en groupes d'un nombre égal de racines, ce qui est impossible, n étant premier. Aussi, $s = n$ et $x_0 = \lambda(V), x_1 = \lambda(\varphi V), \ldots, x_{n-1} = \lambda(\varphi^{n-1}(V))$.

Maintenant soit x_k, $x_{f(k)}$ une des permutation du groupe. On peut y appliquer la permutation tournante qui vient d'être trouvée, et aussi appliquer la permutation ci-dessus sur ces permutations tournantes. De là, $x_k, x_{f(k+i)+i'}$, ou i et i' ont les n valeurs $0, 1, 2, \ldots, n$. De là, n^2 permutations qui doivent se réduire a $n\mu$ distinctes, $\mu \le n-1$. Donc, deux au moins sont égales et $f(k+i)+i' = f(k+i_0)+i'_0$ quel que soit k; $f(k+I) = f(k)+C$, C indépendant de K et $I = i-i_0$ différent de zéro sans quoi $i' = i'_0$ aussi, et absurdité. De là $f(I) = f(0)+C_1$, $f(2I) = 2f(I)+C, \ldots, f(K) = aK+b$. Ainsi nous voilà ramenés aux permutations linéaires et l'éq.$^\text{on}$ se résout.

II Fol. 278-284 (consecutively paginated 1-6).
On fol. 278, Liouville introduces the two types of factorisation of $F^p(V)$ i.e. what he called $(V^n + \&_c)^p$ on fol. 260. He then shows that for p prime they reduce to two factorisations of $F(V)$ namely

$$
\begin{aligned}
F(V) &= f(V,r)f(V,r')\ldots f(V,r^{(m-1)}) \\
&= f(V,r)f_1(V,r)\ldots f_{m-1}(V,r).
\end{aligned}
$$

They are discussed separately on the following page: [fol. 279].

Considérons les groupes qui proviennent de la décomposition

$$
\begin{aligned}
F(V) &= f(V,r)f(V,r')\ldots f(V,r^{(m-1)}) \\
f(V,r) &= 0 \quad V, \varphi V, \ \psi V, \ldots \text{indép de } r
\end{aligned}
$$

donc

$$
f(\varphi V, r') = f(V, r')Q(V, r').
$$

Donc, si $V' = \omega(V)$ est racine de $f(V', r') = 0$, $\varphi V', \psi V', \ldots$ seront les autres.

Donc, des permutations du groupe

$$
(V), \quad (\varphi V), \quad (\psi V), \ldots
$$

on passe à celles du groupe

$$
(V'), \quad (\varphi V'), \quad (\psi V') \ldots
$$

par le simple changement de V en V' effectué partout, c'est à dire à l'aide d'une même substitution opérée à la fois dans toutes les permutations du 1er groupe.

[Fol 280] On passe du 1er groupe au 2ème par une seule substitution opérée dans les permutations du 1er

$$1^{er} \text{ groupe} \left|\begin{array}{ll} x_0 = \lambda(V), & x_1 = \lambda_1(V)\ldots \\ & x_0 = \lambda_k(\varphi V) \end{array}\right.$$
$$\ldots$$

Changeons V en V' pour passer au second groupe et soit $\lambda(V') = \lambda(\omega V) = \lambda_s(V) = x_s,\ldots$ ou x_0 changé en x_s dans la première ligne. On a $x_0 = \lambda(V) = \lambda_k(\varphi V)$; donc, $\lambda(V') = \lambda_k(\varphi V')$; donc, dans la seconde ligne, $x_0 = \lambda_k(\varphi V)$ se change aussi en x_s. Donc, &$_c$.

Considérons en second lieu les groupes qui proviennent de la décomposition

$$F(V) = f(V,r)f_1(V,r)\ldots f_{m-1}(V,r)$$

et montrons que les substitutions sont les mêmes dans chaque groupe.

$$f(V,r) \ 1^{er} \text{ groupe} \left|\begin{array}{ll} x_0 = \lambda(V), & x_1 = \lambda_1(V)\ldots \\ x_k = \lambda(\varphi V) \end{array}\right.$$

soit $f_1(\omega V, r) = f(V,r)Q(V,r)$. On passe au second groupe en remplaçant V par ωV. Soit $x_0 = \lambda(\omega V)$; au dessous de x_0 dans le second groupe on aura $\lambda(\omega \varphi V)$; or $\lambda(\varphi V) = \lambda_k(V), \lambda(V) = \lambda_s(\omega V)$; de la dernière on tire $\lambda(\varphi V) = \lambda_s(\omega \varphi V) = \lambda_k(V)$; donc, $\lambda_s(\omega \varphi V) = x_k$ et au dessous de x_0 se trouve x_k dans le second comme dans le 1er groupe.

[Fol 281] Ainsi, on peut décomposer le groupe primitif de deux manières en groupes partiels 1° En groupes dont chacun se déduit d'un autre à l'aide d'une seule substitution opérée dans les permutations.

2° En groupes qui renferment tous les mêmes substitutions.

Mais, si r est $\sqrt[p]{a}$, ou si les racines $r, r',\ldots$ dépendent rationnellement l'une de l'autre, les deux genres de décomposition ne peuvent manquer de coïncider, &$_c$&$_c$

Mais, soit à présent p une nombre composé.

$$F(v)^p = f(V,r)f(V,r)\ldots = f_1(V,r)\ldots f_{n-1}(V,r)\ldots$$

alors $f(V,r)f(V,r')\ldots$ est de degré ip et peut être une puissance q de $F(V)$

$$f(V,r)f(V,r')\ldots f(V,r^{(p-1)}) = F(V)^q = V^{mq} + \&_c$$

donc $ip = mq$ et déja $m = in$; donc $ip = inq$, $p = nq$.

Dans ce cas, on a à la fois

$$F(V) = f(V,r)f_1(V,r)\ldots f_{n-1}(V,r)$$
$$F(V)^q = f(V,r)f(V,r')\ldots f(V,r^{(p-1)})$$

Il y aura donc q racines égales à V & en posant $f(V,r)f(V,r')\ldots f(V,r^{(p-1)})$? 0 Mais chaque facteur $f(V,r)$ ne peut avoir que des racines inégales. Voila ce qu'on voit d'abord mais approfondissons la chose.

[Fol 282] Soit $V, \varphi V, \psi V, \ldots$ les racines de $f(V,r) = 0$. Soit ωV une racine V' de $f(V',r') = 0$. On a donc $f(\omega V, r') = 0$ et ωV est <u>absolument</u> rationnelle. Mais

$$f(\varphi V, r) = f(V,r)Q(V,r).$$

Donc

$$f(\varphi V, r') = f(V,r')Q(V,r')$$

équation identique en V, de sorte que

$$f(\varphi V', r') = f(V',r')Q(V',r').$$

Donc, si $f(V',r') = 0$, $f(\varphi V', r') =$ aussi zéro, de là les racines $V', \varphi V', \ldots$ inégales, car si $\varphi V' = \psi V'$, comme les fonctions φ, ψ sont absolument rationnelles, $\varphi V = \psi V$ contre l'hypothèse.

Donc $f(V,r)$ et $f(V,r')$ ne peuvent avoir une racine commune sans être identiques. Il y aura donc q facteurs égaux à $f(V,r)$, q autres différentes de ceux-là et égaux à

$$f(V,r_1), \ldots \text{jusqu'à } f(V,r_n) \text{ et par suite}$$
$$F(V) = f(V,r)f(V,r_1)\ldots f(V,r_n)$$

où $r, r_1, \ldots, r_n$ sont n racines de l'Éq.^{on} irréductible de degré $p = nq$ — Ainsi deux décompositions

$$F(V) = f(V,r)f_1(V,r)\ldots f_{n-1}(V,r)$$
$$\text{et} \quad F(V) = f(V,r)f(V,r_1)\ldots f(V,r_n)$$

en n groupes. — La discussion de ces deux décompositions ressemble à celle ci-dessus.

[Fol 283] Ainsi quand il y a décomposition du groupe, ce groupe se divise en n autres appartenant soit à n facteurs relatifs à une racine r adjointe, soit à n racines $r, r_1, \ldots, r_n$. Et les autres racines $r', r,'' \ldots$ ne feraient que reproduire ces n groupes si on voulait les employer.

Dans la dernière décomposition, considérez deux des groupes partiels et vous passerez du 1er au second en opérant dans toutes les permutations du 1er une seule substitution.

Dans la première décomposition, considérez successivement les divers groupes partiels et vouz verrez que tous contiennent les mêmes substitutions.

Si les décompositions coïncident, la décomposition est dite propre et les groupes partiels jouissent à la fois des deux propriétés indiquées.

––––––––

Adjoindre toutes les racines peut se ramener à adjoindre une racine t d'Éq.^{on} irréductible dont les racines s'expriment rationnellement l'une par l'autre et dont r &$_c$ dépendent alors rationnellement. Et alors, les décompositions doivent coïncider.

[Fol 284] Soit t racine d'une Éq^{on} irréductible dont $r, r', \ldots$ dépendent rationnellement, et $t', t'', \ldots$ autres racines aussi.

Adjoignons t. De la

$$
\begin{aligned}
F(V) &= f(V,t)f_1(V,t)\ldots f_s(V,t) \\
\text{et} &= f(V,t)f(V,t_1)\ldots f(V,t_{s-1})
\end{aligned}
$$

Or, si $f(V,t)$ est le plus petit facteur rationnel possible, comme adjoindre t est aussi adjoindre $t', t'' \ldots$, voilà deux décompositions rationnelles en t et à facteurs premiers. Donc, l'égalité doit avoir lieu facteur par facteur et les deux décompositions coïncident. Donc &$_c$

––––––––

Il est donc bien vrai qu'il y a une grande différence entre adjoindre une seule racine d'une équation irréductible ou les adjoindre toutes.

Appendix III

The sketch of contents of Liouville's course on rational mechanics at the *Faculté des Sciences* 1864–1865. [Ms 3626 (1) p. 14r–21r].

After the number of each lecture, Liouville has written the day and date as follows:

$$1^{\text{ere}} \text{ Leçon (Mercredi 23, 9}^{\text{bre}} \text{ 1864)}$$

I shall leave out these dates. Otherwise I shall quote Liouville's plan verbatim.

Faculté des Sciences

1864–1865 <u>Mécanique rationnelle</u>

(Les mercredis et vendredis à 10^{h})

1er Semestre; puis 2eme le 15 Mars 1865

1. Notions préliminaires

2. Composition des forces concourantes

3. Forces parallèles

4. Suite des forces parallèles. Remarques diverses

5. Conposition des couples. Conposition générale des forces

6. Remarques générales sur la composition des forces — Cas d'une résultante unique

7. Développements sur ce qui précède

8. Centre de gravité, centre de masses, notions générales

9. Suite sur les centres de gravité

10. Suite sur les centres de gravité

11. Fin des centres de gravité — Théorème de Guldin

12. Attraction d'une couche sphérique homogène sur un point extérieur ou intérieur

$2^{\text{ème}}$ Semestre

32. Mouvement apparent sur la surface de [la] terre, $\&_c$ — nous voilà au second semestre. Passons donc à la seconde partie de notre cours et aux questions d'examen ([in margin]: Mort de Mr. Beugnot [?])

22. Nous revenons en détail sur le mouvement rectiligne — Les premiers exemples à l'ordinaire. ([in margin]: Mort de Me Blin [?] mère de la mère de Bertrand)

34. Chute des corps pesants en ayant égard à la variation de la pesanteur et mouvement ascensionnel

35. Suite et fin du sujet précédent. Mouvement dans l'intérieur de la terre supposée homogène

36. Corps attiré par deux centres fixes — Montée d'un corps en ayant égard à la résistance d'un milieu

37. Chute d'un corps pesant dans l'air — Mouvement d'un point matériel dans un tube qui tourne sur un plan horizontal ([in margin]: Mort de M. Ary [?] (de Toul), Mort de Richard Labden [?] né en 1804)

38. Oscillations du pendule dans le vide

39. Pendule simple dans un milieu résistant ([in margin]: Mort de Mr. Gillet [?] de Nancy)

40. Oscillations d'un corps pesant sur une courbe fixe — Cycloïde, tautochronisme, même dans un milieu qui résiste d'après la simple vitesse

41. Synchrones d'Euler, $\&_c$ — Pression sur la courbe pendant le mouvement, $\&_c$ — On commence le pendule sphérique.

    30 avril 1865 je reçois une lettre de faire part de la mort de Mr Edine [?] — Joseph Millot, beau père de Delaunay...

42. Suite du pendule sphérique — Projectiles dans le vide, en négligeant toujours la force centrifuge composée

43. Projectiles dans un milieu résistant — ou dans le vide en tenant compte de la force centrifuge composée

44. Déviation à l'Est des corps tombant suivant la verticale, et à l'ouest des corps lancés suivant la verticale — Rappel du problème des planètes. ([in margin]: J'apprends la mort de Marie Catalan à Liège)

45. Suite des planètes — $r = a(1 - e \cos u)$, $\tan \frac{1}{2}\theta = \sqrt{\frac{1+e}{1-e}} \tan \frac{1}{2}u$

46. Fin des planètes — Mouvement d'un corps solide autour d'un axe fixe — On commence à parler des moments d'inertie ([in margin]: Mort de mon oncle Vincent (à Toul))

47. Suite des moments d'inertie axes spontanés de rotation &$_c$. Propriétes mécaniques et géométriques

48. Dernieres remarques au sujet des moments d'inertie — On revient au pendule composé, puis on s'occupe du pendule balistique (il y aura à revenir sur le pendule composé) ([in margin]: Mort d'une des filles de mon cousin Gérard...)

49. Après quelques mots sur le pendule on commence à s'occuper des percussions ou impulsions d'un point de vue abstrait. ([in margin]: Mort de Geruser)

50. Rotation d'un corps solide autour d'un point fixe — Composition des rotations

51. On continue à s'occuper du mouvement d'un corps solide autour d'un point fixe

52. Suite du sujet

53. Éq^{ons} d'Euler — Corps solide libre — Stabilité de l'équilibre

54. Quelques développements sur la cycloïde [calculations of the Brachistochrone] — On commence l'hydrostatique

55. On continue l'hydrostatique (et on parle même de considérations plus générales) — Cas particulier des fluides pesantes

56. Suite de l'hydrostatique

57. Notions d'hydrodynamique — Fin du Cours

————

————

Examens de licence, École normale, 2ème année. Je suis convoqué pour le jeudi 6 juillet à 8^h, vendredi et samedi, avec MM Haton et Briot. Mais au lieu de Haton, Delaunay.

Notes

Notes to Chapter II

1. M. Cauchy réunit la majorité des suffrages et est nommé membre du Bureau. La nomination sera soumise à l'approbation du roi.

2. Le président dépose l'ordonnance du Roy par laquelle la nomination de Mr Liouville est confirmé. Mr Liouville sera installé apprès la formalité du serment voulu par la loi (ce sont les termes exprès de la lettre d'envoi du M^r le Ministre de l'Instruction publique).

Notes to Chapter III

1. Mr Liouville entretient le bureau des recherches qu'il a faites sur une classe de quantités qui ne sont ni rationnelles ni réductible à des rationnelles algébriques.

2. Mr Beautemps-Beaupré analyse la conversation qu'il a eu avec le ministre de l'Instruction publique, au sujet de la position où se trouve le Bureau depuis la nomination non acceptée de Mr Cauchy. Mr Beautemps-Beaupré a fait remarquer à Mr Villemain que la question doit être tranchée par le gouvernement. Le Bureau ne peut prendre à ce sujet aucune initiative.

3. Il ne lui [the president] appartient pas de rechercher les motifs qui ont empêché le gouvernement de statuer sur cette nomination, mais il accomplit un devoir en vous assurant, M^r le ministre, que l'état actuel des choses ne saurait se prolonger sans un dommage réel pour les sciences astronomique.

4. inviter le Bureau des Longitudes à procéder à une nouvelle designation (de géomètre) la première (celle de Mr Cauchy) ne pouvant être ratifiée, et demeurant sans effet d'après le refus du candidat de remplir une obligation prescrite par la loi.

5. Mr Liouville parle de la découverte qui a été faite en Allemagne des manuscrits mathématiques de Pascal que Leibniz avait fait copier et parmi lesquels se trouve, dit-on, le traité des Coniques.

6.   Mr Liouville rend un compte verbal d'un mémoire que Mr Delaunay se propose de présenter à l'académie dans la prochaine séance sur le mouvement de la terre autour de son centre de gravité.

7.   Mr Liouville parle du mémoire que Mr Jacoby [sic] a publié, en allemand, sur les inégalités séculaires. Il voudrai que le Bureau le fît traduire. On pourrait après, suivant la plus ou moins grande importance du mémoire, l'insérer en totalité ou seulement par extrait dans la *Connaissance des Temps.*

8.   Mr Faye sera invité au nom du Bureau à traduire le mémoire de Mr Jacoby sur les inégalités séculaires.

9.   Mr Leverrier signale l'intérèt du travail du géomètre allemand, tout en faisant remarquer qu'il faudra le reprendre entièrement, à cause de l'action de la nouvelle planète.

10.   On discute les propositions qui ont déjà été faites touchant le nom a donner à la planète de Mr. Leverrier et en particulier le nom de Neptune.

11.   Mr Arago met sous les yeux de Mr Leverrier placé en ce moment près de lui, mais à titre seulement de communication officieuse et individuelle, une lettre imprimé de Mr Encke à Mr Schumacher, dans laquelle l'astronome de Berlin donne le passage suivant comme tiré textuellement d'une lettre que Mr Leverrier lui aurait écrire en date du 6 octobre 1846."
     "J'ai prié mon illustre ami, Mr Arago, de se charger du soin de choisir un nom pour la planète. J'ai été eu peu confus de la décision qu'il a prise dans le sein de l'académie."
     Mr Leverrier préférant sans doute une explication officielle à une discussion particulière, lit à haute voix le passage ci dessus et déclare qu'il n'a rien pu, qu'il n'a rien du écrire de semblable. Je serais blâmable, a-t-il ajouté, si le passage rapporté était de moi; mais j'ai conservé le brouillon de ma lettre, je le chercherai et l'on verra que mes paroles n'ont pas été rapportées fidèlement.
     Mr Leverrier quitte la séance.

12.   Les directeurs des observatoires de Poulkowa et de Greenwich se sont décidé pour le nom de Neptune, après avoir reçu une lettre, en date du 1er Octobre 1846 par laquelle Mr. Leverrier leur annonçait que le Bureau des Longitudes avait choisi lui-même ce nom.
     Il est constaté par la lecture des Procès Verbaux et par le souvenir de tous les membres présents, que le Bureau n'a jamais pris aucune délibération à ce sujet...

13.   on examine comment on demandera au ministre de faire connaître sa résolution relativement à la demission de Mr Leverrier. Il est décidé qu'une commission sont chargée de ce soin.

14.   Mr. Leverrier, dans sa lettre, déclare au ministre qu'il reprend sa position de membre adjoint du Bureau des Longitudes.

15.   Le Bureau, dans sa séance du 28 juillet dernier, avait pris la résolution de donner à la planète dont l'existence a été signalée par Mr Leverrier le nom de Neptune, qui a prévalu aujourd'hui parmi les astronomes. Il avait arrêté en outre qu'une note relative à cette délibération serait insérée dans la *Connaissance des temps* pour 1850.

Mr Leverrier par l'intermédiaire de Mr Beautemps-Beaupré demande au Bureau qu'il soit ajouté à cette note les mots suivants:

Mr Leverrier n'était pas présent à la séance."

16.   Mr. Liouville rend compte d'une belle théorie de l'électricité et du magnétisme publiée par un Mr. Green. Le mérite de cette théorie paraît n'avoir pas été reconnu du vivant de l'auteur.

17.   M. Liouville parle d'un théorème sur l'attraction qu'une couche électrique en équilibre exerce sur un point extérieur placé très près de la surface de cette couche; ce théorème que Poisson dans ses mémoires sur l'électricité donne comme étant de Laplace, se trouve déjà dans les Mémoires de Coulomb (Mém. de l'acad des sciences pour 1788). Cette remarque a été communiquée à Mr. Liouville par un géomètre anglais, Mr. William Thomson.

M. Liouville annonce que des recherches étendues sur la théorie de l'électricité ont été entreprises par le même géomètre.

Notes to Chapter IV

1.   La séance n'étant pas encore ouverte, Mr. Cauchy est entré dans la salle des séances sans avoir prévenu personne, et a signé la feuille de présence.

Mr. Cauchy étant étranger au *Bureau des Longitudes*, Mr. le président l'a engagé à se retirer, et, avec l'assentiment unanimé du *Bureau*, a ragé son nom de la feuille de présence.

2.   Mr. Liouville insiste pour que dans la lettre qu'il doit adresser au Ministre Mr. le Trésorier du *Bureau* fasse resortir la différence qui existe entre la somme ordonnancée pour les traitements du *Bureau* et la somme réellement percue.

3.   À l'occasion d'un mémoire lu récemment à l'académie des sciences par Mr. Cauchy MMr. Poinsot et Liouville posent quelques considérations sur la nature des fonctions ainsi que sur celle des différentielles et des intégrales qui y sont relative.

4.   Il est question d'un mémoire ou Mr. Cauchy donne une expression analytique de la dispersion. Mr. Liouville rapelle que la condition physique de la dispersion d'après Fresnel est que le rayon de la sphère d'activité des molécules de l'ether doit être comparable à la longeur de l'ondulation, . . .

Notes to Chapter V

1.   M. Arago propose au Bureau Mr Ernest Liouville comme candidat stagiaire au titre d'Élève astronome de l'observatoire de Paris.

Cette proposition est soumise à l'approbation du *Bureau*. Il y a 8 votant, le scrutin donne 8 bulletins portant *Oui*. [P. V. Bur. Long. July 28th 1852]

2.   M. Liouville donne quelques détails sur les travaux de Mr Bienaymé relatifs aux calcul des probabilités et en particulier sur un application qu'il a fait de la méthode des moindres carrés au calcul de la probalité de l'erreur de la masse de Jupiter déterminée par MM^r Laplace et Bouvard. Mr. Bienaymé a fait voir que la formule donnée par Laplace pour cet objet était
faute defautive en ce qu'elle donnait la probabilité de la masse comme si cela était la seule inconue du problème. [P. V. Bur. Long. January 28th 1852]

3.   M. Liouville entretient le *Bureau* d'un problème relatif à la fixation la plus probable d'un point déterminée par la rencontre de trois droites menées par trois points donnés de position suivant des directions également connues. [P. V. Bur. Long. August 11th 1852]

4.   À l'occasion de la dernière opposition de Neptune, Mr Liouville insiste pour que l'on cherche si les observations de cet astre continuent à donner un grand axe d'à peu près 30 et différant ainsi de 1/6 de celui que M^r Le Verrier avait adopté hypothétiquement pour en faire les [bases ?] de ses calculs. [P. V. Bur. Long. September 11th 1852]

5.   A l'occasion du nouvel astéroïde découvert par Mr Hind, on rapelle l'hypothèse si connue d'Olbers, qui fait résulter ces petits corps de l'explosion d'une planète circulant autrefois dans la région compris entre Mars et Jupiter. MM Biot et Babinet citent à ce sujet un théorème de Mr Leverrier en vertu duquel aucun astre de masse très petite ne pourrait dans cette région conserver sous l'influence perturbatrice de Jupiter et de Saturne une orbite peu excentrique et peu inclinée à l'écliptique.

M. Liouville fait observer que l'analyse de M. Leverrier dont on parle repose sur des bases inexactes et sur une approximation mal ordonnée pour le cas tout particulier qu'on discute ici et auquel M^r Le Verrier a cru pouvoir l'étendre. C'est ce qu'on peut voir sans calcul par le résultat même, et à l'aide d'un raisonnement très simple. "Mais," ajoute M. Liouville "on pourra traiter la question par des formules rigoureuses, car je me suis assuré que les équations différentielles similaires, relatives au calcul des inclinaisons mutuelles des orbites de trois planètes qui se troublent et pour lesquelles les excentricités sont très petites et les inclinaisons quelconques, peuvent toujours s'intégrer quelles que soient les masses des trois planètes. L'intégration n'en devient de plus facile quand une des masses est insensible. C'est par ce cas simple que j'ai commencé mes recherches dont le champ s'est ensuite étendu, et j'en ai fait l'objet d'un mémoire que la discussion actuelle m'engagera à présenter au Bureau un peu plus tôt que je ne me proposais de le faire." [P. V. Bur. Long. December 22nd 1852]

Notes to Chapter VI

1.   M. Liouville communique au Bureau un théorème relatif à la théorie des nombres: le théorème est le suivant:

Soit l'équation $4m = i^2 + 3i'^2$ dans laquelle m désigne un nombre inpair donné, i et i' deux entiers indéterminés egalement impaires. Considérons l'ensemble de toutes les solutions de l'équation en question et désignons par $\sum i^2$ et $\sum i'^2$ la somme de tous les nombres i^2 et celle de tous les nombres i'^2 respectivement associés: on aura

$$\sum i^2 > \sum i'^2$$
$$\text{et} \quad \sum i^2 < 9 \sum i'^2$$

Chacune de ces inégalités peut dans certains cas tres-particuliers se changer en égalité.

M. Liouville a obtenu un grand nombre de résultats du même genre, qu'il se propose de publier prochainement. [P. V. Bur. Long. January 21st 1874]

2.   une discussion s'engage entre M. Liouville et M. Le Verrier qui tient à ce que le matériel du Bureau soit mis en harmonie avec les attributions de ses membres.

Notes to Chapter VII

1.   In fact, Ampère divided the above fact into two separate facts so that the following is called fact 2 in the lecture notes. However, I have preferred this numbering in order to facilitate a comparison with Ampère's book and Liouville's papers.

2.   Ampère gave no argument to support his claim that the dependence of r must be proportional to $\frac{1}{r^n}$ with the same n as above. In fact, Liouville [AC AL Ms 2] wrote $k\frac{ii'ds ds'}{r^k}$ at first, but when combining the two formulas he set the k in r^k equal to n.

3.   In fact, Fourier worked with the specific heat per unit mass C_m, but in order to compare his formula with Poisson's, as Liouville did, I have used the specific heat per unit volume C_v instead. They are related by the equation $C_v = C_m D$ where D is the density of the material.

4. Laplace's function φ is obtained from P by integrating P along a horizontal plane of figure 13. This explains why Laplace found the factor $\int s^2 \varphi(s)\,ds$ and Poisson the factor $k = \int s^4 P(s)\,ds$.

5. Dulong and Petit measured the heat exchange in a vacuum between a mercury-thermometer and its surrounding container. The container was kept at a constant temperature u_0 of $0°$ (or $20°, 40°, 60°, 80°$ C) while the temperature of the thermometer decreased from $300°$ C to $0°$ C. They interpreted the results as a proof that the heat exchange was proportional to $a^u - a^{u_0} = a^u - 1$. Stephan-Boltzmann's law predicts a heat exchange proportional to $T^4 - T_0^4 = T^4 - 273^4$. In the figure are shown the graphs of $T^4 - 273^4$ and $10^{10}(a^u - 1)$. If I had drawn $1.15 \cdot 10^{10}(a^u - 1)$ instead, the two curves would have been almost indistinguishable.

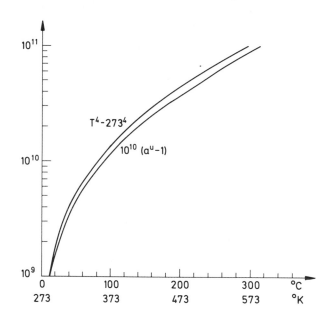

Figure

6. Poisson had earlier worked with the diffraction of light according to Fresnel's wave theory and had in 1819 predicted the famous "Poisson's spot" [Arnold 1981 p. 29].

7. In the earlier memoir [AC L Ms 4] is inserted a small note by Poisson where he states that he returns Liouville's memoir with a few comments. Thus Poisson had already read Liouville's ideas in an earlier version, but the very few comments in the margin suggest that he was not as critical the first time he saw them.

8. In [1832a], Liouville illustrated the use of fractional calculus to analyse microscopic elementary forces, but the mathematical method was the primary subject of the paper.

9.   The memoir is not in the file of the meeting and a correspondence in Liouville's personal file in the *Archive de l'Académie des Sciences* bear witness to earlier unfruitful attempts to find it.

10.   Either by inserting a power series with arbitrary coefficients into the equation or by applying the successive approximation procedure from [Liouville 1830], cf. Chapter X formula (72).

11.   This is the so-called problem of moments. Monna [1973 p. 48] writes that "the connection of this problem with the theory of probability was pointed out by the Russian mathematician Tchebycheff." As a matter of fact, Laplace had already seen this connection.

12.   This step is all right here according to Lebesgue's theorem of dominated convergence.

Notes to Chapter VIII

1.   Note présenté par M. Liouville

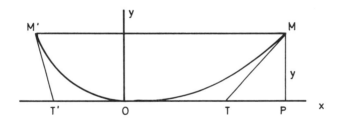

Ox horisontale, Oy verticale. - On demande que, quel que soit le point de départ M d'un mobile pesant 1° le temps du transport de M en M' soit constant 2° l'excès du temps employé à descendre sur le temps employé à montrer soit proportionnel à la différence $MT - M'T'$ des tangents extrêmes.

Solution. Soit $OM = s, OM' = s'$; on trouve

$$s = A\sqrt{y} + B \int_0^y e^{\frac{n^2}{y}} \frac{dy}{y\sqrt{y}},$$
$$s' = A\sqrt{y} - B \int_0^y e^{-\frac{n^2}{y}} \frac{dy}{y\sqrt{y}};$$

A, B, n sont des constantes quelconques.

N.B. On obtient ces formules par le calcul des différentielles à indices quelconques; le développement en séries indiqué par Mr Poisson pour le problème des <u>tautochrones</u> n'y conduirait pas. [P. V. Bur. Long. February 28th 1844].

2.   There are two more notes connecting the theory of heat and the fractional calculus. In the first [3615 (3) p. 57v] from June 1832, Liouville expressed the heat equation "supposing the law of action at a distance $= e^{-r}r^{\mu-1}$" as a fractional differential equation. The second, dated December 31st 1838, runs as follows:

Let $\dfrac{d^{\frac{1}{2}}u}{dt^{\frac{1}{2}}} = \dfrac{du}{dx}$

so that $\dfrac{du}{dt} = \dfrac{d^2u}{dx^2}.\&_c.$ application of the methods... of diff. of arbitrary index. [Ms 3616 (3) p. 5v]

This is perhaps an attempt to solve the one-dimensional heat equation by way of the fractional calculus. However, the transformation is wrong. Gregory [1841] and later Heaviside [1893–1912] (cf. [Lützen 1979]) solved this equation by operational techniques, which also use fractional calculus.

3.   This integral only converges if $\mathrm{Re}\,\nu > 0$. Otherwise one has to use the definition

$$_cD_x^\nu f(x) = \frac{1}{\Gamma(m-\nu)}\frac{d^m}{dx^m}\int_c^x (x-t)^{m-\nu-1}f(t)\,dt$$

where m is an integer greater than $\mathrm{Re}\,\nu$. It is easy to see that this expression is independent of m.

4.   A third proof of the expandability can be found in a note from April 1832 [Ms 3615 (3) p. 42v] where Liouville based it on Cauchy's integral formula

$$f(x) = \frac{1}{2\pi}\int_{-\pi}^{\pi}\frac{\overline{x}f(\overline{x})}{(\overline{x}-x)}\,dp \quad \text{where } \overline{x} = X\,e^{p\sqrt{-1}},\ |x| < X$$

To this end Liouville inserted

$$\frac{1}{\overline{x}-x} = -\int_0^\infty e^{\alpha(\overline{x}-x)}\,d\alpha \quad \text{for } \mathrm{Re}(\overline{x}-x) < 0$$

and

$$\frac{1}{\overline{x}-x} = \int_0^\infty e^{-\alpha(\overline{x}-x)}\,d\alpha \quad \text{for } \mathrm{Re}(\overline{x}-x) > 0$$

(cf. (28)) into Cauchy's formula and obtained an expression of the form

$$f(x) = \int_0^\infty \left(P(\alpha)e^{-\alpha x} + Q(\alpha)e^{\alpha x}\right)d\alpha$$

which is also an exponential "series."

Cauchy had found his integral formula the previous year and had only just sent a proof of it to the *Académie* on March 5th 1832. It was contained in the *Extrait du mémoire présenté a l'Académie de Turin le 11 octobre 1831*. [Cauchy 1824/1981]

5. I do not know if these theorems correspond to true theorems in non-standard analysis.

6. Following Liouville and Fourier I have written Fourier's formula with the p integration inside the α integration. This order of integration is inadmissible in clasical analysis but in distribution theory it is correct. As I pointed out in [Lützen 1982a, p. 112–115] Fourier intentionally used this order of integration because he understood that

$$\int_{-\infty}^{\infty} dp \cos(px - p\alpha) = 2\pi\, \delta(x - \alpha).$$

I think that Liouville followed Fourier in order to compare his results directly with Fourier's, and if he had thought of the problem he would probably have preferred to take the α integration first as Cauchy and Dirichlet had done.

7. Ross [1977 p. 77] attributes this definition to Lacroix and calls it the Lacroix-Peacock method. However, Euler explicitly stated this definition, except that he did not use the symbol Γ but wrote its integral representation.

8. Ross [1977 p. 79] suggests that Liouville might have been quite aware of the restrictions of the exponential definition and therefore formulated the second based on (32). Since Liouville explicitly preferred the original definition, this seems highly improbable to me. Moreover, talking about Liouville's two definitions, as Ross does, is slightly misleading. Liouville only applied one definition but indicated *two* others.

9. Here Liouville had $n = -2i\sqrt{\pi}$. He often returned to this equation for various values of n : $(-1, -\frac{i}{n}, 1, -n$ etc.) Since there is in principle no difference between the various equations, I shall always refer to the general equation (53) in the subsequent discussion, in order to facilitate the comparison of the results.

10. Pour intégrer toute équation linéaire à coëfficients constants ayant une second membre fonction de x on peut employer la méthode suivante qui est générale quoique je préfère la démontrer d'abord dans cette note sur un exemple particulier. Cet exemple une fois bien compris, il sera facile d'étendre notre procédé à tous les cas possibles.

Soit donc proposé d'intégrer l'équation

$$Ay + B\frac{d^{1/3}y}{dx^{1/3}} + C\frac{d^{2/3}y}{dx^{2/3}} + D\frac{dy}{dx} = f(x) \tag{1}$$

dans laquelle A, B, C, D sont des constantes et $f(x)$ une fonction de x à volonté.

Je pose $\frac{d^{1/3}y}{dx^{1/3}} = u$; $\frac{d^{2/3}y}{dx^{2/3}} = v$, et je trouve cette nouvelle forme de l'équation (1)

$$Ay + Bu + Cv + D\frac{dy}{dx} = f(x),$$

d'où je conclus par <u>deux</u> différentiations successives par rapport à x et relativement à l'indice 1/3 les <u>deux</u> égalités nouvelles:

$$A\frac{d^{1/3}y}{dx^{1/3}} + B\frac{d^{1/3}u}{dx^{1/3}} + C\frac{d^{1/3}v}{dx^{1/3}} + D\frac{d^{4/3}y}{dx^{4/3}} = \frac{d^{1/3}f(x)}{dx^{1/3}}$$

$$A\frac{d^{2/3}y}{dx^{2/3}} + B\frac{d^{2/3}u}{dx^{2/3}} + C\frac{d^{2/3}v}{dx^{2/3}} + D\frac{d^{5/3}y}{dx^{5/3}} = \frac{d^{2/3}f(x)}{dx^{2/3}}$$

lesquelles se simplifient en vertu des relations qui tient u et v à y: on a en résumé par un calcul fort simple trois équations:

$$\begin{aligned}
Ay + Bu + Cv + D\frac{dy}{dx} &= f(x) \\
Au + Bv + C\frac{dy}{dx} + D\frac{du}{dx} &= \frac{d^{1/3}f(x)}{dx^{1/3}} \\
Av + B\frac{dy}{dx} + C\frac{du}{dx} + D\frac{dv}{dx} &= \frac{d^{2/3}f(x)}{dx^{2/3}}
\end{aligned} \tag{2}$$

Or ces trois équations différentielles simultanées du premier ordre et à <u>dérivées entières</u> suffisent évidemment pour déterminer u, v et surtout y. [Ms 3615 (3) p 9v–10r]

11.   In [1832b, p. 84] Liouville found an example where $\lim \int \cdots$ is convergent whereas $\int \lim \cdots$ is divergent. He even claimed to have found another case where they are both convergent but converge towards different values. This, however, is due to the erroneous belief that

$$\int_0^\infty (e^{-zx} - 1)\frac{dz}{z} \quad \text{for } x > 0$$

is finite. In fact it diverges as $\log z$ at infinity.

Notes to Chapter IX

1.   With the possible exception of one sheet found in Abel's notebooks after his death; see Sylov and Lie's *Apercu des manuscrits d'Abel conservés jusqu'à present* in Abel's *Oeuvres* vol II p. 288.

2.   Let R be an arbitrarily given polynomial and let θ be a polynomial with the same roots as R. Define

$$P = \theta' - \frac{1}{2}\frac{R'\theta}{R}.$$

Then P is a polynomial and

$$\int \frac{P\,dx}{\sqrt{R}} = \frac{\theta}{\sqrt{R}}.$$

3. [Ms 3615 (3)] has $\frac{M+P\sqrt{1+x^2}}{Q}$ instead of $\frac{M+P\sqrt{1+x^4}}{Q}$. This is clearly a slip of the pen.

4.   Liouville in fact analysed the equation

$$A \sqrt[n]{P} + B \sqrt[m]{Q} = C$$

with $n = m$, but from the context it is clear that he had the more general equation (13) in mind.

5.   At the end of the argument Liouville found that R can be expanded:

$$R = a_2 z^{2k} + a_3 z^{3k} + a_4 z^{4k} + \cdots$$

where $z = (\alpha_1 + \beta_1 \sqrt{R})$. He claimed that this is absurd if R only has simple roots. However, since k can be a fraction, there is in fact no contradiction.

6.   In the *Procès-Verbaux* it is stated that the report was *Signé à la minute: Lacroix et Libri Rapporteur*. However, Poisson's signature also figures on the original report in the file at the Archive of the *Académie*.

7.   According to Youschkevitch [1970–1980a] the thesis existed in draft as early as 1843.

Notes to Chapter X

1.   The technique of separating the variables by searching for solutions of the general form $F(x)f(t)$ had been introduced by Fourier in [1822, §167]. However, for simple equations like (5) both Fourier and his successors knew the equation for $f(t)$ and its solution so well that they immediately wrote down the expression (9).

2.   Daniel Bernoulli also studied certain inhomogeneous chains and there he found a solution which we recognize as a first-order Bessel function.

3.   To a modern reader d'Alembert's assumption $\zeta = e^{\int p \, dx}$ for $p(x) \in \mathbf{R}$ limits the discussion to positive values of ζ in which case only the first value of λ can be found. However, this argument does not apply to d'Alembert who believed that $\log x = \log(-x)$. In the letter considered here d'Alembert [1766 p. 250] gave a new argument for this standpoint based on the equation (11) for X constant.

4.   Here Fourier has left out one multiplicative constant. Another constant has been determined from the implicit boundary value condition that $u(x)$ is regular at $x = 0$.

5.   Poisson did not use vector notation but wrote his equations in components.

6.   Sturm's memoir was presented to the *Académie* on September 28th 1833 and not on September 30th, as is stated in [Sturm 1833a] and [Prouhet 1856].

7.   The analysis of Sturm's memoir in [1833a] is written in the third person; thus it is not certain but is most probable that Sturm is the author.

8.   For $L(x) \neq 0$ in (39) Sturm found the functions of (41) to be

$$K(x) = e^{\int \frac{M(x)}{L(x)}\, dx} \quad \text{and} \quad G(x) = \frac{N(x)}{L(x)} e^{\int \frac{M(x)}{L(x)}\, dx}.$$

9.   Sturm formulated only one theorem (here called theorem E) in the whole paper. In order to make this discussion easier to grasp than Sturm's paper, I have formulated as propositions many of the properties proved by Sturm.

10.   In the 1830's the concept of differentiability had not been introduced. Following Cauchy, the rigorists differentiated only continuous functions, but even continuity of V with respect to r is never questioned by Sturm. Liouville raised this question once, when he had obtained completely inadmissible results (cf. §54). In the following presentation I take for granted sufficient smoothness in V_r as a function of r.

11.   Sturm did not argue in favour of such an ordering of the eigenvalues. However, it is a consequence of Liouville's expression for the asymptotic behaviour of the eigenfunctions (96).

12.   In the last paragraph of [Sturm 1836b, section 9], Sturm claimed that the root $x = \beta$ must be counted among the $n - 1$ roots of V_n. This mistake was corrected in the Errata.

13.   Sturm did not question the interchange of integration and summation involved in determination of the Fourier coefficients C_i.

14.   I have set all the physical constants appearing in Liouville's paper equal to 1, and have from the start assumed the interval to be of the form $[0, \beta]$. In Liouville's paper this simplification is made only in the last part.

15.   Liouville correctly deduced the orthogonality (60) [Liouville 1830, p. 179] but when he used it to derive the expression (64) for the Fourier coefficients he forgot the factor $g(x)$. [Liouville 1830 pp 179–180]

16.   In the first place Liouville's analysis shows only that U_n behaves asymptotically like $\cos \frac{m\pi}{\gamma} x$. This has $m - 1$ roots in $(0, \gamma)$, and since U_n, according to Sturm's oscillation theorem, has $n - 1$ roots in $(0, \gamma)$, Liouville concludes that $m = n$. Kline [1972 p. 1101] and Schlissel [1976/1977] counts this as one of the earliest uses of asymptotic series.

17.   Liouville wrote: "that the integral $\int_0^\gamma \sqrt{\lambda^2}\, dz$ has a finite value and can be considered as equivalent to the sum of its elements." The last remark no doubt means that the integral is not be taken in its 18th-century sense, as the opposite of the differential, but must be defined as a sum in the way Fourier and more precisely Cauchy had done. Liouville probably thought of Cauchy's extension of the integral to functions with isolated discontinuities.

18.   It is impossible to tell whether Liouville in [1836f] had forgotten the convergence problems of the successive approximation and of the Fourier series, which he had discussed in [1830], or whether he consciously postponed treatment of these problems to the following papers. As Liouville expressed himself in [1836f], he must have created the impression that he was unable to reach Cauchy's standards of rigor, which he had earlier explicitly stressed. [Chapter III]

When Sturm in [1836b p. 411] referred to Liouville's theorem he was kind enough to formulate it not as Liouville had done in [1836f] but as he ought have done:

> the sum of the series [85], if this series is convergent, cannot fail to be equal to $f(x)$, for all values of x contained between α and β.

In [Liouville 1838h §35] Liouville formulated the theorem in this way as well.

19.   Under the assumptions made in Liouville's first convergence proof [1836f] term-by-term integration is allowed since the Fourier series converges uniformly.

20.   Sturm and Liouville built their conclusion on a development of

$$\frac{V_r}{\Pi(r)}$$

in simple fractions ($\Pi(r)$ being defined by formula (4)).

21.   Liouville rejected the first deduction because it used the Taylor theorem. He explicitly referred to Cauchy's objections to this theorem.

22.   The argument works only because V_n has no double root. Liouville's careful argument in [1836e p. 275] takes this fact into account.

23.   In secondary sources it is often claimed that Liouville proved Bessel's inequality for the general Sturm-Liouville problem (1)-(3) (see e.g. [Kline 1972 p. 716–717], [Birkhoff 1973 p. 276] and [Dieudonné 1981 p. 21]). Bessel's inequality states that

$$\sum_{n=1}^{\infty} C_n^2 \leq \int_{\alpha}^{\beta} g(x) f^2(x)\, dx,$$

where C_n are the Fourier coefficients and the orthogonal system V_n has been normalized

$$\int_{\alpha}^{\beta} g(x) V_n(x) V_m(x)\, dx = \delta_{m,n}.$$

At the place referred to in the secondary sources [Liouville 1836d p 265] Liouville proves that if

$$\sigma_n(x) = \sum_{i=1}^{n} C_i V_i(x),$$
$$\varrho_n(x) = \sum_{i=n+1}^{\infty} C_i V_i(x),$$

then

$$\int_\alpha^\beta g(x)f(x)^2 \, dx = \int_\alpha^\beta g(x)(\sigma_n(x)^2 + \varrho_n(x)^2) \, dx,$$

from which

$$\int_\alpha^\beta g(x)f(x)^2 \, dx \geq \int_\alpha^\beta g(x)\sigma_n(x)^2 \, dx. \qquad (*)$$

Now it is true, but not pointed out by Liouville, that the right-hand side of $(*)$ is equal to $\sum_{i=1}^n C_n^2$, and so one obtains Bessel's inequality in the limit $n = \infty$. Liouville himself made the following comment on $(*)$:

> This last formula proves that the integral $\int_\alpha^\beta g\sigma_n^2 \, dx$, however large you make the index n, can never have a numerical value surpassing the limit $\int_\alpha^\beta gf(x)^2 \, dx$ with which it coincides when $n = \infty$.

Thus for Liouville $(*)$ was important because it is valid for a finite n, whereas Bessel's inequality has $n = \infty$. According to Liouville the two sides of $(*)$ coincide for $n = \infty$; this has made Kline [1972 p. 716–717] attribute Parseval's equality to Liouville. For Liouville, however, the equality was a simple consequence of the expansion theorem, $\sigma_\infty(x) = f(x)$, and therefore he did not consider it a theorem in its own right.

The modern mathematician evaluates the inequality $(*)$ differently from Liouville because he knows that one cannot always take the limit inside the integral sign whereas Liouville did not doubt the identity

$$\int_\alpha^\beta g(x)f(x)^2 \, dx = \int_\alpha^\beta g(x) \lim_{n\to\infty} (\sigma_n(x)^2) \, dx = \lim_{n\to\infty} \int_\alpha^\beta g(x)\sigma_n(x)^2 \, dx.$$

To ascribe Bessel's inequality, and particularly Parseval's equality to Liouville, therefore, is an overinterpretation, at least when only his work on Sturm-Liouville theory is taken into account. In Chapter XV §§36–37 I shall show that these theorems obtained a central place in his unpublished work on a certain type of integral operators arising in potential theory.

24.   In the very first note in the first notebook [Ms 3615 (1)] Liouville wrote $\frac{d^3V}{dx^3} + rV = 0$, $\frac{d^3V'}{dx^3} + r'V' = 0$, and tried without success to find an expression for $(r-r') \int VV'$. Next to this calculation he successfully carried out the corresponding calculation for $\frac{d^4V}{dx^4} + rV = 0$. Most of the notes in [Ms 3615 (1)] seem to stem from around 1830, but the above mentioned note is probably from a later date.

25.   In this note Liouville had only treated the equation

$$V^{(3)}(x) + G(r, x)V(x) = 0.$$

26.   Liouville also tried a related method in notes from February 8th-10th, March 8th, and August 5th, 1838 [Ms 3615 (5), 3616 (2)]. It consisted in studying solutions V of (138) of the form

$$V = UW' - WU',$$

where U and W are solutions to the adjoint equation with suitable boundary conditions. The investigations end without result.

27.   Liouville wrongly wrote $A_n = -V_n(\alpha)$.

Notes to Chapter XI

1.   See Todhunter [1873] for the respective merits of Legendre and Laplace in the theory of attraction.

2.   Todhunter [1870] has discussed Ivory's paper [1838] in detail and has pointed out several other weaknesses. He has even found an error in Ivory's correcting note [1839]. A list of Ivory's many articles on rotating fluids can be found in [Todhunter 1873, II §§1416–1417].

3.   Liouville's formulas include the mass and the density of the fluid, so that the dependence of these quantities can also be inferred. This is for example of importance in the discussion of the behaviour of a star when it is slowly cooling off (see [Poincaré 1885 and 1902]).

4.   See [Oppenheim 1919, §§19, 20] for the difference between static and kinetic equilibrium.

5.   Oppenheim [1919, footnote 100] attributes the formula (16) to Liouville but explicitly states that Liouville did not discover the additional term in (17). He probably learned about Liouville's formula from Liapounoff [1884, §7] and misunderstood his remark about another missing term.

6.   The exact formulation of [Ms 3617 (2) p. 78r] can be found in [Lützen 1984b] note 13.

7.   In the manuscript Liouville initially multiplied by Y_n and integrated over the sphere. However, he crossed out all the Y_n's and wrote P_n instead. Today P_n is usually considered as a function of one variable but Liouville clearly thought of a function of the two spherical coordinates θ, φ. Moreover, the following argument uses the completeness of the functions P_n. Therefore one can conclude that Liouville thought of the system consisting of $P_n(\cos\theta)$, $P_n^m(\cos\theta)\cos m\varphi$ and $P_n^m(\cos\theta)\sin m\varphi$ $(1 \leq m \leq n)$, which I have abbreviated as P_n^m. Considering this system is equivalent to considering the Y_n's since

$$Y_n(\theta, \varphi) = A_{0,n} P_n(\cos\theta) + \sum_{m=1}^{n}(A_{m,n}\cos m\varphi + B_{m,n}\sin m\varphi)P_m^n(\cos\theta).$$

An argument similar to the one leading to (29) can be found in §29.

8.   In [Ms 3616 (3) p. 43v], which must be earlier than [Ms 3617 (2) p. 78] Liouville took the variation of the quantity in (18) under the single condition (23). He concluded:

"If $g\zeta + \iint \frac{\zeta' d\omega'}{\Delta}$ is not a constant, one is not at the minimum. Thus the minimum gives (at least if there is one)

$$\iint \frac{\zeta' d\omega'}{\Delta} = g\zeta + C$$

...All this seems very tricky."

He probably could not prove the existence of a minimum and therefore introduced the condition (22).

9.   Since Liouville in dealing with the stability coefficients used two different notations, Lamé's and the notation used by Laplace and Jacobi, it is convenient to derive the translation of some of these expressions. The translations are all based on the identities (47):

$$\rho^2 = k_2^2, \quad \rho^2 - b^2 = k_1^2, \quad \rho^2 - c^2 = k^2.$$

Formula relative to $R_3 = \sqrt{\rho^2 - c^2}$:

$$2\pi k_1 k \int_0^\infty \frac{k^2\, d\alpha}{(\alpha + k^2)H}$$

$$= 2\pi\sqrt{(\rho^2 - b^2)(\rho^2 - c^2)} \int_0^\infty \frac{(\rho^2 - c^2)\, d\alpha}{(\alpha + \rho^2 - c^2)\sqrt{(\alpha + \rho^2)(\alpha + \rho^2 - b^2)(\alpha + \rho^2 - c^2)}}$$

Now substitute $\rho'^2 = \alpha + \rho^2$, $2\rho'\, d\rho' = d\alpha$. Then we get

$$2\pi\sqrt{(\rho^2 - b^2)(\rho^2 - c^2)} \int_\rho^\infty \frac{(\rho^2 - c^2)2\rho'\, d\rho'}{(\rho'^2 - c^2)\sqrt{\rho'^2(\rho'^2 - b^2)(\rho'^2 - c^2)}}$$

$$= 4\pi\sqrt{(\rho^2 - b^2)(\rho^2 - c^2)}(\rho^2 - c^2) \int_\rho^\infty \frac{d\rho'}{(\rho'^2 - c^2)\sqrt{(\rho'^2 - b^2)(\rho'^2 - c^2)}}$$

$$= \frac{4\pi R_3 S_3 \sqrt{(\rho^2 - b^2)(\rho^2 - c^2)}}{3}.$$

Thus

$$\frac{4\pi R_3 S_3 \sqrt{(\rho^2 - b^2)(\rho^2 - c^2)}}{3} = 2\pi k_1 k \int_0^\infty \frac{k^2\, d\alpha}{(\alpha + k^2)H}. \qquad (*)$$

Similarly we get:

Formula relative to $R_4 = \rho^2 - l^2$:

$$4\pi\frac{R_4 S_4 \sqrt{(\rho^2 - b^2)(\rho^2 - c^2)}}{5} = 2\pi k k_1 \int_0^\infty \frac{(k_2^2 - l^2)^2\, d\alpha}{(\alpha + k_2^2 - l^2)^2}. \qquad (**)$$

Formula relative to $R_5 = \rho(\sqrt{\rho^2 - b^2})$:

$$4\pi \frac{R_5 S_5 \sqrt{(\rho^2 - b^2)(\rho^2 - c^2)}}{5} = 2\pi k_1 k k_1^2 k_2^2 \int_0^\infty \frac{d\alpha}{(\alpha + k_1^2)(\alpha + k_2^2)H}.$$

$$(***)$$

10.   By note 9 the equilibrium condition (10) can be written in the compact form

$$\frac{R_3 S_3}{3} = \frac{R_5 S_5}{5}.$$

This was explicitly remarked by Liouville in [Ms 3617 (4) p. 29].

11.   On p. 64 of [Ms 3617 (4)] Liouville finds some of the Lamé functions of third degree, among others the critical quantity R_{15} (71), but he does not calculate $\frac{RS}{2^n+1}$, in this case.

Notes to Chapter XVI

1.   Mr. Liouville a examiné que serait le mouvement d'une lune que Mr. Laplace avait placé de telle sorte qu'elle restait toujours opposée au soleil.

D'après le résultat général que Mr. Liouville a trouvé le resultat du mouvement en question (position en ligne droite du soleil de la lune et de la terre) ne serait pas stable; la moindre perturbation apporterait aux positions relatives des dérangements considérables et fort rapides.

2.   Mr. Liouville communique les résultats d'un calcul relatif à l'augmentation d'amplitude que présenterait dans ses oscillations très petites un pendule dont la longueur irait sans cesse en diminuant. En appliquant ses formules à la Lune, dans le but de rechercher si (comme on l'a cru) la libration de cet astre pourrait à la longue devenir sensible par suite d'un changement dans les dimensions de cette astre produit par un abaissement dans la température. Mr Liouville trouve que en partant de données d'une grandeur exagérée et après des millions d'années, la libration ne serait aggrandie que d'une petite fraction de sa valeur actuelle, laquelle est insensible.

3.   Mr Liouville pense qu'un calcul des perturbations qui pourrait ne pas être très long s'il etait convenablement dirigé augmenterait beaucoup la certitude des résultats obtenues.

4.   On parle du calcul des réfractions astronomiques dans lequel les géomètres ont coutume de supposer l'atmosphère dans un état d'équilibre moyen. Mr Liouville annonce au Bureau qu'il s'occupe d'un travail que peut être [?] jeter quelque jour sur la question, dans le cas d'une atmosphère agitée.

5.   Mr. Arago parle d'un mémoire sur les réfractions astronomiques présentes a l'académie par M. Robert Lefebre. À cette occasion Mr. Liouville annonce qu'il a trouvé une démonstration nouvelle du théorème de Mr.

Biot sur les réfractions près de l'horizon et d'une équation aux différences partielles qu'il ont déja donnée et qui comprend ce théorème comme cas particulier. La demonstration nouvelle ne suppose pas qu'on ait l'expression de la réfraction. La méthode s'étend avec des modifications convenables à un atmosphère quelconque: on a même des résultats d'un genre semblable pour toutes les problèmes de Mécanique ou l'intégrale du force vives a lieu.

6.   Mr. Liouville annonce que la méthode dont il a entretenu le bureau dans une présedente séance, est plus générale encore qu'il la pensait; il n'est pas nécessaire que l'intégrale des forces vives ait lieu, pour qu'elle soit applicable. Elle s'étend à toutes les cas.

Bibliography

Unpublished Manuscripts and Other Archival Material

Académie des Sciences (Archives)

AC AL Ms 1. *Cours de physique mathématique. Collège de France. Mr Ampère prof. 1826–1827. J. Liouville É. de l'E.Po.* 122 pp. Liouville's notes from Ampère's lectures. Papiers de A.M. Ampère, Cart. 11, Chap. 11, Chem 211.

AC AL Ms 2. *Cours de physique mathématique 1826–1827.* 258 pp. Fair copy of AC AL Ms 1, with additional section on electrodynamics. Papiers de A.M. Ampère, Cart. 12, Chap. 11, Chem 213.

AC AL Ms 3. *Théorie mathématique des phénomènes électrodynamique.* 86 pp. In Liouville's hand but written as though Ampère was the author. Papiers de A.M. Ampère, Cart. 11, Chap. 10, Chem 208bis.

AC AL Ms 4. *Note sur l'Electro-dynamique.* 3 pp. In Liouville's hand. Papiers de A.M. Ampère, Cart. 10, Chap. 10, Chem 200. Published as [Liouville 1831].

AC AL Ms 5. *Action d'un courant fermé sur un élément de courant.* 4 pp. Signed by Liouville. Papiers de A.M. Ampère, Cart. 10, Chap. 10, Chem 200.

AC L Ms 2. *Sur l'électricité dynamique, et en particulier sur l'action mutuelle d'un pôle d'aimant et d'un fil conducteur.* 100 pp., by Liouville. File of meeting June 30, 1828.

AC L Ms 3a-c. *Mémoire sur le calcul aux differences partielles, $1^{\underline{er}}$–$3^{\underline{ème}}$.* 19, 23, and 24 pp., respectively, by Liouville. File of meeting December 1, 1828.

AC L Ms 4. *Recherches sur la théorie physico-mathématique de la chaleur.* Abstract 12 pp.; Mémoire 132 pp. by Liouville. File of meeting August 16, 1830.

AC L Ms 5. *Mémoire sur les questions primordiales de la théorie de la chaleur.* 2 copies of 37 and 39 pp., by Liouville. File of meeting November 2, 1830.

AC L Ms 6. *Premier Mémoire sur la détermination des intégrales dont la valeur est algébrique.* File of meeting December 17, 1832. A first version of [Liouville 1833b].

AC L Ms 7. *Premier Mémoire sur la détermination des intégrales dont la valeur est algébrique* (dated February 4, 1833). File of meeting December 17, 1832. A second version of [Liouville 1833b].

AC L Ms 8. *Second Mémoire sur la détermination des intégrales dont la valeur est algébrique.* File of February 4, 1832. Early version of [Liouville 1833c].

AC A Ms 1. *Rapport sur un mémoire de M. Liouville intitulé mémoire sur les questions primordiales de la théorie de la chaleur.* 6 pp., Ampère (rapporteur) et Navier. File of meeting April 4, 1831.

AC P Ms 1. *Observations sur le mémoire de M. Liouville.* 4 pp. in Poisson's hand. Attached to AC A Ms 1. File of meeting April 4, 1831.

1. Liouville's personal file (dossier personnelle).

2. Libri's personal file; here one finds:

3. Letter read by Sturm in the *Académie des Sciences* on June 26, 1843.

Archive Nationale

AN. F^{14} 2271^{1} Liouville's personal file.

Bibliothèque Nationale

V. 14275. Joseph Liouville. *Notes des cours à l'Ecole Centrale des Arts et Manufacteurs 1837–1838.*

Bureau des Longitudes (meeting room)

Procès Verbaux du Bureau des Longitudes. The originals from the period 1853–1876 have disappeared. Copies exist from the period 1853–1868.

Collège de France (Archives)

1. Joseph Liouville. Dossier personnelle (personal file).

2. *Procès Verbaux à l'assemblé des Professeurs.*

École Polytechnique (Archives)

1. *École Polytechnique. Conseil d'Instruction.* Handwritten minutes from the meetings.

2. *École Polytechnique. Conseil de perfectionnement.* Handwritten minutes from the meetings.

3. Report from the *Commission Mixte 1851* in Ec. Pol. Cons. Perf.

4. *École Polytechnique. Corps Enseignent de 1794 à 1854.* Handwritten lists.

5. *Programmes de l'enseignement de l'École Royale polytechnique arrêtés par le Conseil de Perfectionnement pour l'année scolaire 1825–1826.* Printed Paris 1825.

6. *École Royale Polytechnique. Registre des Notes 1825–1826.* Handwritten tables for each student.

7. *École Impériale Polytechnique, Cours de 1820–1830.* "Signalements" of students and their careers at the school.

8. *École Polytechnique. Visite de santé.* Tables of health and constitution of the students.

9. *École Polytechnique, Dosier personelle de M. Liouville.*

The following lithographed notes taken by various students in Liouville's *Cours d'Analyse*; they are all classified under the code $A^{III}a$ 88:

10. Liouville *Cours d'Analyse 1ere anné 1843–1844.*

11. Liouville *Cours d'Analyse 1ere anné 1845–1846 redigé par M. Vazeille, élève (Étienne).*

12. Liouville *Cours d'Analyse 2eme Divission, 1ere anné 1847–1848.*

13. Liouville *Calcul Intégral, 1ere anné et 2eme anné 1847–1849.*

14. Liouville *Cours d'Analyse 2eme anné et Cours de Mécani-que 2eme anné* (no year mentioned).

Institut de France (Bibliothèque)

Ms 3615–3640 The Liouville Nachlass consisting of 340 notebooks and a box of loose sheets.

Ms 2108 *Manuscrits d'Evariste Galois.* Fol. 36 and fol. 260–285 contain Liouville's notes. Fol 26.–34 are the never published proof sheets of *Mémoire sur les conditions de résolubilité des équations* by Galois intended for L. J. **8** (December 1843), 489–505.

Sorbonne (Bibliothèque)

Société Philomathique de Paris, Extraits des Procès-verb-aux des séances.

Archives départementales du Loir-et-Cher

Intégrales définies, Cours de M. Liouville. Fonds privé Saint Venant. — [Liouville/Saint Venant 1839–1840].

List of J. Liouville's Published Works

[Year] refers to the first published version of the entire paper. Within each year the ordering is only approximately chronological.

The number after the year is the number of the publication in the Royal Society's Catalogue of Scientific Papers.

From the *Comptes Rendus* notes and summaries of Liouville's verbal contributions are included if they contain any scientific information. Moreover, there are references to the reports written by Liouville (the other members of the judging committee are mentioned in paranthesis). Notes mentioning merely that Liouville gave a speach or was elected a member of some committee are not recorded. In case the note in the Compte Rendu has no title in the main text, the title in the index is usually quoted.

Joint papers with Sturm and papers written under the pseudonym Besge, are included.

ABBREVIATIONS:

C. R.: *Comptes Rendus de l'Académie des Sciences*

L. J.: *Journal de Mathématique pures et appliquées* or "Liouville's Journal."

LIST

[1829–1] *Démonstration d'un théorème d'électricité dynamique*, Ann. de Chimie et Physique **41** (1829), 415–421.

[1830–2] *Mémoire sur la théorie analytique de la chaleur*, Ann. Math. Pures Appl. **21** (1830–1831), 131–181; Summary by Sturm in Bull. Sci. Math. Phys. et Chim.

[1831] *Note sur l'électro-dynamique*, Bull. Sci. Math. Phys. et Chim. **15** (1831), 29–31.

[1832a–3] *Sur quelques questions de géometrie et de mécanique, et sur un nouveau genre de calcul pour résoudre ces questions*, Journ. Ec. Polyt. **13** (**21. cahier**) (1832), 1–69.

[1832b–4] *Mémoire sur le calcul des différentielles à indices quelconques*, Journ. Ec. Polyt. **13** (**21. cahier**) (1832), 71–162.

[1832c–5] *Sur l'intégration de l'équation* $(mx^2+nx+p)\frac{d^2y}{dx^2}+(qx+r)\frac{dy}{dx}+sy = 0$ *à l'aide des différentielles à indices quelconques*, Journ. Ec. Polyt. **13** (**21. cahier**) (1832), 163–186.

[1833a–6] *Mémoire sur l'équation de Riccati*, Journ. Ec. Polyt. **14** (**22. cahier**) (1833), 1–19.

[1833b–7] *Premier Mémoire sur la détermination des intégrales dont la valeur est algébrique*, Journ. Ec. Polyt. **14** (**22. cahier**) (1833), 124–148; Mém. Savans Etrangers Acad. Sci. Paris **5** (1838), 76–102.

[1833c–7] *Second Mémoire sur la détermination des intégrales dont la valeur est algébrique*, Journ. Ec. Polyt. **14** (**22. cahier**) (1833), 149–193; Mém. Savans Etrangers Acad. Sci. Paris **5** (1838), 103–151.

[1833d–8]   *Note sur la détermination des intégrales dont la valeur est algébrique,* Journ. Reine Angew. Math. **10** (1833), 347–359; Errata: Journ. Reine Angew. Math. **11** (1834), p. 406.

[1834a–9]   *Mémoire sur le théorème des fonctions complémentaires,* Journ. Reine Angew. Math. **11** (1834), 1–19.

[1834b–10]   *Mémoire sur une formule d'analyse,* Journ. Reine Angew. Math. **12** (1834), 273–287.

[1834c–11]   *Sur les transcendantes elliptiques de première et de seconde espèce, considérées comme fonctions de leur amplitude,* Journ. Ec. Polyt. **14** (23. cahier) (1834), 37–83.

[1834d–12]   *Note sur la figure d'une masse fluide homogène, en équilibre, et douée d'un mouvement de rotation,* Journ. Ec. Polyt. **14** (23. cahier) (1834), 289–296.

[1835a–13]   *Mémoire sur l'intégration d'une classe de fonctions transcendantes,* Journ. Reine Angew. Math. **13** (1835), 93–118.

[1835b–14]   *Mémoire sur l'usage que l'on peut faire de la formule de Fourier, dans la calcul des différentielles à indices quelconques,* Journ. Reine Angew. Math. **13** (1835), 219–232.

[1835c–15]   *Sur le changement de la variable indépendante dans le calcul des différentielles à indices quelconques,* Journ. Ec. Polyt. **15** (24. cahier) (1835), 17–54.

[1835d–16]   *Sur la détermination d'une fonction arbitraire placée sous un signe d'intégration définie,* Journ. Ec. Polyt. **15** (24. cahier), 55–60.

[1836a–17]   *Mémoire sur le développement des fonctions ou parties de fonctions en séries de sinus et de cosinus,* L. J. **1** (1836), 14–32.

[1836b–18, 48]   *Mémoire sur une question d'analyse aux différences partielles,* L. J. **1** (1836), 33–74; Mém. Savans Etranger Acad. Sci. Paris **5** (1838), 559–606.

[1836c 19]   *Note sur une manière de généraliser la formule de Fourier,* L. J. **1** (1836), 102–105.

[1836d–20]   *Note sur le calcul des inégalités périodiques du mouvement des planè-tes,* L. J. **1** (1836), 197–210; C. R. **2** (1836), p. 217.

[1836e–21]   *Démonstration d'un théorème du à M. Sturm et relatif à une classe de fonctions transcendantes,* L. J. **1** (1836), 269–277.

[1836f–22]   *Mémoire sur le développement des fonctions ou parties de fonctions en séries, dont les divers termes sont assujettis à satisfaire à une même équation différentielle du second ordre, contenant un paramètre variable,* L. J. **1** (1836), 253–265; C. R. **1** (1835), p. 418.

[1836g–23]   *Mémoire sur un nouvel usage des fonctions elliptiques dans les problèmes de mécanique céleste,* L. J. **1** (1836), 445–458; C. R. **3** (1836), 41–42.

[1836h–25]   *Lettre à M. Arago concernant la démonstration de la convergence d'une série qui se présente en analyse lorsqu'on cherche à trouver les lois du mouvement de la chaleur dans une barre hétérogène,* C. R. **3** (1836), 622–623, 653–655.

[1836i–26]   *Note ajoutée au Rapport de M. Poisson sur mon mémoire concernant une question nouvelle d'analyse,* Journ. Reine Angew. Math. **16** (1836), 41–46.

[1836j]   *Avertissement*, L. J. **1** (1836), 1–4; (chronologically this belongs before 1836a).

[1836]   Liouville and Sturm, *Démonstration d'un théorème de M. Cauchy relatif aux racines imaginaires des équations*, L. J. **1** (1836), 278–289.

[1837a–24]   *Mémoire sur l'intégration de l'équation $\frac{du}{dt} = \frac{d^3u}{dx^3}$*, Journ. Ec. Polyt. **15**(25. cahier) (1837), 85–117; C. R. **3** (1836), 572–575.

[1837b]   *Solution d'un problème d'analyse*, L. J. **2** (1837), 1–2.

[1837c–27]   *Second Mémoire sur le développement des fonctions ou parties de fonctions en séries, dont les divers termes sont assujettis à satisfaire à une même équation différentielle du second ordre, contenant un paramètre variable*, L. J. **2** (1837), 16–35.

[1837d–28]   *Mémoire sur la classification des transcendantes et sur l'impossibilité d'exprimer les racines de certaines équations en fonction finie explicite des coefficients*, L. J. **2** (1837), 56–105;, **3** (1838), 523–547.

[1837e–29]   *Sur la sommation d'une série*, L. J. **2** (1837), 107–108.

[1837f–30]   *Note sur le développement de $(1 - -2xz + z^2)^{-\frac{1}{2}}$*, L. J. **2** (1837), 135–139.

[1837g–31]   *Note sur un passage de la Mécanique Céleste, relatif à la théorie de la figure des planètes*, L. J. **2** (1837), 206–220.

[1837h–32]   *Sur une lettre de d'Alembert à Lagrange*, L. J. **2** (1837), 245–248.

[1837i–33]   *Troisième Mémoire sur le développement des fonctions ou parties de fonctions en séries, dont les divers termes sont assujettis à satisfaire à une même équation différentielle du second ordre, contenant un paramètre variable*, L. J. **2** (1837), 418–437; C. R. **5** (1837), 205–207.

[1837j–34]   *Solution nouvelle d'un problème d'analyse, relatif aux phénomènes thermo-mécaniques*, L. J. **2** (1837), 439–456; C. R. **5** (1837), 598–599.

[1837k–35]   *Sur la formule de Taylor*, L. J. **2** (1837), 483–484.

[1837l–36,39]   *Nouvelles recherches sur la détermination des intégrales dont la valeur est algébrique*, C. R. **5** (1837), 330–333; L. J. **3** (1838), 20–24.

[1837m–37]   *Mémoire sur l'intégration des équations différentielles à indices fractionnaires*, Jorn. Ec. Polyt. **15** (25. cahier) (1837), 58–84; C. R. **2** (1836), 167–168 (chronologically this belongs before 1837a).

[1837a]   Liouville and Sturm, *Extrait d'un Mémoire sur le développement des fonctions en séries dont les différents termes sont assujettis à satisfaire a une même équation différentielle linéaire, contenant un paramètre variable*, L. J. **2** (1837), 220–223; C. R. **4** (1837), 675–677.

[1837b]   Liouville and Sturm, *Note sur un théorème de M. Cauchy relatif aux racines des équations simultanées*, C. R. **4** (1837), 720–724.

[1838a–38]   *Sur les deux derniers cahiers du Journal de M. Crelle*, L. J. **3** (1838), 1–3.

[1838b–40]   *Sur l'intégration d'une classe d'équations différentielles*, L. J. **3** (1838), 31–32; slightly abbreviated as Note 6 in Navier *Résumé des Leçons d'Analyse données à l'École Polytechnique*, Paris **2** (1840), p. 345.

[1838c–41]   *Note sur la théorie des équations différentielles*, L. J. **3** (1838), 255–256.

[1838d–42]   *Sur la théorie des équations transcendantes*, L. J. **3** (1838), 337–341.

[1838e–43]   *Note sur la théorie de la variation des constantes arbitraires*, L. J. **3** (1838), 342–349.

[1838f–44]   *Observations sur un mémoire de M. Libri, relatif à la théorie de la chaleur*, L. J. **3** (1838), 350–354; C. R. **6** (1838), p. 240.

[1838g–45]   *Note sur l'intégration d'une équation aux différentielles partielles qui se présente dans, la théorie du son*, L. J. **3** (1838), 435–436; C. R. **7** (1838), 247–248.

[1838h–46]   *Premier mémoire sur la théorie des équations différentielles linéaires, et sur le développement des fonctions en séries*, L. J. **3** (1838), 561–614; C. R. **7** (1838), 1112–1116.

[1838i–47]   *Sur quelques formules de M. Poisson*, C. R. **7** (1838), 84–86; (chronologically this belongs before 1838g).

[1839a–49]   *Sur l'intégration des équations linéaires aux différentielles partielles*, L. J. **4** (1839), 1–6.

[1839b–50]   *Observations sur un mémoire de M. Ivory, sur l'équilibre des elliosoïdes homogénes*, L. J. **4** (1839), 169–174.

[1839c–51,57]   *Note sur quelques intégrales définies*, L. J. **4** (1839), 225–235; C. R. **8** (1839), p. 626.

[1839d–52]   *Note sur évaluation approchée du produit* $1.2.3\ldots x$, L. J. **4** (1839), 317–322; C. R. **9** (1839), 104–108; Note 4 in Navier *Résumé des Leçons d'Analyse données à l'École Polytechnique*, Paris **2** (1840), 332–339.

[1839e–53]   *Mémoire sur l'intégration d'une classe d'équations différentielles du second ordre en quantités finies explicites*, L. J. **4** (1839), 423–456; C. R. **9** (1839), 527–530.

[1839f–54]   *Sur les variations séculaires, des angles que forment entre elles les droites résultant de l'intersection des plans des orbites de Jupiter, Saturne et Uranus*, L. J. **4** (1839), 483–492; C. R. **8** (1839), p. 566.

[1839g–55]   *Sur la moyenne arithmétique et la moyenne géométrique de plusieurs quantités positives*, L. J. **4** (1839), 493–494.

[1839h–56]   *Sur le principe fondamental de la théorie des équations algébriques*, L. J. **4** (1839), 501–507; (see also 1840r and 1840s).

[1839i–58]   *Sur le problème des perturbations dans certains cas où l'excentricité de l'orbite de la planète troublée et son inclinaison à l'écliptique ont des valeurs quelconques*, C. R. **8** (1839), 696–699.

[1839j–59]   *Démonstration d'un théorème de M. Libri*, C. R. **8** (1839), 790–792.

[1839k–60]   *Remarques sur la critique faite par M. Libri, concernant la théorie des fonctions finies explicites*, C. R. **8** (1839), 792–798.

[1839l]   *Rapport sur des recherches analytiques de M. Ritter, concernant le problème des réfractions astronomique*, (with Savary and Sturm), C. R. **9** (1839), 650–51.

[1840a–61]   *Note sur les transcendantes elliptiques de première et de seconde espèce considérées comme fonction de leur module*, L. J. **5** (1840), 34–36; C. R. **10** (1840), 2–4.

[1840b–62]   *Sur l'irrationalité du nombre* $e = 2,718\ldots$, L. J. **5** (1840), p. 192.

[1840c–62]   *Addition a la note sur l'irrationalité du nombre* e, L. J. **5** (1840), 193–194.

[1840d–63]   *Sur la limite de* $(1+\frac{1}{m})^{m}$*, m étant un entier positif qui croit indéfiniment*, L. J. **5** (1840), p. 280.

[1840e–64]   *Sur quelques formules pour le changement de la variable innépendante*, L. J. **5** (1840), 311–312.

[1840f–65]   *Note sur un des théorèmes de Jacobi*, L. J. **5** (1840), 351–355.

[1840g–66] *Sur les conditions de convergence d'une classe générale de séries*, L. J. **5** (1840), 356–359; C. R. **11** (1840), 615–618.

[1840h–67] *Sur l'équation $Z^{2n} - Y^{2n} = 2X^n$*, L. J. **5** (1840), p. 360.

[1840i–68] *Mémoire sur les transcendantes elliptiques de première et de second espèce, considérées comme fonctions de leur module*, L. J. **5** (1840), 441–464; C. R. **10** (1840), 2–4.

[1840j] *Observations sur une note de M. Libri*, C. R. **10** (1840), 343–345.

[1840k] *Réplique de M. Liouville à M. Libri*, C. R. **10** (1840), p. 347.

[1840l–69] *Sur un théorème d'analyse indéterminée*, C. R. **10** (1840), 381–382.

[1840m–70] *Observations sur une note de M. de Pontécoulant, relative à certaines formules de la théorie analytique du "Système du Monde"*, C. R. **10** (1840), 881–884.

[1840n–71] *Sur les méthodes générales à l'aide desquelles on détermine les perturbations du mouvement des planètes*, C. R. **11** (1840), 251–255.

[1840o] *Sur la formule de Maclaurin*; Note 1 in Navier *Résumé des Leçons d'Analyse données à l'École Polytechnique*, Paris **2** (1840), 319–320.

[1840p] *Sur les fractions qui se présentent sous la forme $\frac{\infty}{\infty}$*; Note 2. in Navier *Resumé des Leçons d'Analyse données à l'École Polytechnique*, Paris **2** (1840), 321–322.

[1840q] *Sur quelques intégrales définies*; Note 3 in Navier *Résumé des Leçons d'Analyse données à l'École Polytechnique*, Paris **2** (1840), 323–331.

[1840r] *Sur une application singulière de la théorie des intégrales doubles à la démonstration d'un théorème d'algèbre*; Note 5 in Navier *Résumé des Leçons d'Analyse données à l'École Polytechnique*, Paris **2** (1840), 340–344; verbatim quote of [1839h505–508].

[1840s–56] *Addition à la Note sur le principe fondamental de la théorie des équations algé- briques*, L. J. **5** (1840), 31–34; (cf. [1839h]); (chronologically this belongs before 1840a).

[1840t] *Rapport sur un Mémoire de M. LeVerrier, concernant les variations des éléments det sept planètes principales*, (with Arago and Savary), C. R. **10** (1840), 524–527.

[1841a–72] *Remarques nouvelles sur l'équation de Riccati*, L. J. **6** (1841), 1–13; C. R. **11** (1840), p. 729.

[1841b–73] *Sur l'intégrale $\int_0^\pi \cos i(u - x \sin u)du$*, L. J. **6** (1841), p. 36.

[1841c–74] *Sur une formule de M. Jacobi*, L. J. **6** (1841), 69–73.

[1841d–75] *Mémoire sur quelques propositions générales de Géométrie et sur la théorie de l'élimination dans les équations algébriques*, L. J. **6** (1841), 345–411; C. R. **13** (1841), 412–416.

[1841e–76] *Sur une classe d'équations différentielles*, L. J. **6** (1841), p. 448.

[1841f–77] *Remarques sur un théorème de Jacobi*, C. R. **13** (1841), 467–470.

[1841g] *Rapport sur un Mémoire de M. Steiner concernant les maxima et les minima des figures géométriques*, (with Cauchy and Sturm), C. R. **12** (1841), 931–935.

[1841h] *Rapport sur une Note de M. Passot intitulée: Note sur l'inexactitude des indications du frein dynamométrique*, (with Sturm and Piobert), C. R. **13** (1841), 526–528.

[1842a–78] *Sur l'équation $\frac{d^2y}{dx^2} + f(x)\frac{dy}{dx} + F(y)(\frac{dy}{dx})^2 = 0$*, L. J. **7** (1842), 134–136.

[1842b–79] *Sur les fractions qui se présentent sous la forme indéterminée $\frac{\infty}{\infty}$,* L. J. **7** (1842), 160–162.

[1842c–80] *Sur un problème de Géométrie relatif à la théorie des maxima et minima,* L. J. **7** (1842), 163–164.

[1842d–81] *Sur l'ellipse de plus petite surface qui passe par trois points A, B, C, et sur l'ellipsoïde de plus petite volume qui passe par quatre points A, B, C, D,* L. J. **7** (1842), 190–191.

[1842e–82] *Démonstration d'un théorème de M. Biot sur les réfractions astronomiques près de l'horizon,* L. J. **7** (1842), 268–271.

[1842f] *Extrait d'une Lettre de M. Liouville à M. Arago,* C. R. **15** (1842), 425–426.

[1842g–83] *Sur un cas particulier du problème des trois corps,* C. R. **14** (1842), 503–506; L. J. **7** (1842), 110–113; Connaiss des Temps pour 1845, (1842), 3–17.

[1842h–84] *Sur la stabilité de l'éqilibre des mers,* C. R. **15** (1842), 903–907.

[1842i–97] *Note à l'occasion du mémoire de M. Chasles: Sur l'atraction des corps,* Connaiss des Temps pour 1845 (1842), 34–36.

[1842j] *Rapport sur un Mémoire de M. J. Binet relatif à la théorie de la variation des constantes,* (with Poinsot, Sturm and Coriolis), C. R. **14** (1842), 440–441.

[1842/1984] *Analyse d'un Mémoire sur la stabilité de l'équilibre des mers.* Appendix A in [Lützen 1984b].

[1842] Besge, *Sur le centre de gravité d'un triangle sphérique,* L. J. **7** (1842), p. 516.

[1843a–85] *Sur l'équation $\frac{d^2\phi}{dx^2} + \frac{d^2\phi}{dy^2} = 0$,* L. J. **8** (1843), 265–267.

[1843b–86] *Sur la loi de la pesanteur à la surface ellipsoïdale d'équilibre d'une masse liquide homogène douée d'un mouvement de rotation,* L. J. **8** (1843), p. 360.

[1843c–87] *Remarques sur un Mémoire de N. Fuss sur la solution du problème suivant: Un polygone P étant donné, de combien de manières peut-on le partager en polygones de m côtes au moyen de diagonales,* L. J. **8** (1843), 391–396.

[1843d–88] *Rapport fait à l'Académie des Sciences de l'Institut, sur un Mémoire de M. Hermite, relatif à la division des fonctions Abéliennes ou ultra-elliptiques;* (with Lamé) *suivi d'une Lettre de M. Jacobi à M. Hermite,* L. J. **8**, 502–506; C. R. **17** (1843), 292–295.

[1843e–89] *Sur les figures ellipsoïdales à trois axes inégaux, qui peuvent convenir à l'équilibre d'une masse liquide homogène, douée d'un mouvement de rotation,* C. R. **16** (1843), 216–218; Connaiss. des Temps, pour 1846 (1843), 85–96.

[1843f] *Recherches sur la stabilité de l'équilibre des fluides,* C. R. **16** (1843), p. 363.

[1843g] *Réponse aux remarques fait sur le rapport sur un mémoire de M. J. Bertrand, par M. Libri,* C. R. **17** (1843), 295–296.

[1843h] *Observations à l'occasion de la réclamation de M. Libri relativement à un passage du Rapport sur le mémoire de M. Hermite,* C. R. **17** (1843), 327–334.

[1843i] *Réplique à M. Libri,* C. R. **17** (1843), 445–449.

[1843j] *Réponse à une note de M. Libri*, C. R. **17** (1843), 552–554.

[1843k–90] *Sur la division du périmètre de la lemniscate, le diviseur étant un nombre entier réel ou complexe quelconque*, C. R. **17** (1843), 635–640; L. J. **8** (1843), 507–513.

[1843l–91] *Sur un théorème d'Abel*, C. R. **17** (1843), p. 720; L. J. **8** (1843), 513–514.

[1843m] *Rapport sur un Mémoire de M. Bertrand intitulé Développements sur quelques points de la théorie des surfaces isothermes orthogonales*, (with Lamé), C. R. **17** (1843), 290–292; (this belongs before 1843d).

[1843/1984] *Recherches sur la stabilité de l'équilibre des fluides*. Appendix B in [Lützen 1984b].

[1844a–92] *Développements sur un théorème de Géométrie*, L. J. **9** (1844), 337–349.

[1844b–93] *Sur une propriété des sections coniques*, L. J. **9** (1844), 350–353.

[1844c–94] *De la ligne géodésique sur un ellipsoïde quelconque*, L. J. **9** (1844), 401–408.

[1844d–95] *Sur les rayons de courbure des courbes géométriques*, L. J. **9** (1844), p. 435.

[1844e] *Remarques relatives 1° à des classes très-étendues de quantités dont la valeur n'est ni rationnelle ni même réducible à des irrationnelles algébri-ques; 2° à un passage du livre des Principes où Newton calcule l'action exercée par une sphère sur un point extérieur*, C. R. **18** (1844), 883–885.

[1844f–96] *Nouvelle démonstration d'un théorème sur les irrationnelles algébriques*, C. R. **18** (1844), 910–911.

[1844g] *Remarques (relatives 1° à des lignes géodésiques: 2° à des fonctions doublement périodiques) à l'occasion d'une note de M. Chasles*, C. R. **19** (1844), 1261–1263.

[1844h] *Rapport sur une Note relative à la flexion des pièces chargées debout présentée par M. E. Larmarle*, (with Poncelet), C. R. **18** (1844), 82–84.

[1844a] Besge, *Sur une équation différentielle à indices fractionnaires*, L. J. **9** (1844), p. 294.

[1844b] Besge, *Sur l'équation* $\frac{d^2u}{dx^2} = \frac{Au}{(a+bx+cx^2)^2}$, L. J. **9** (1844), p. 336.

[1845a–98] *Sur les deux Formes* $x^2 + y^2 + z^2 + t^2, x^2 + 2y^2 + 3z^2 + 6t^2$, L. J. **10** (1845), 169–170.

[1845b] *Resultats de quelques recherches concernant des questions de physique mathématique et d'analyse*, C. R. **20** (1845), 1386–1389. also 222–225 of [1845d].

[1845c–102] *Solution d'un problème relatif à l'ellipsoïde*, C. R. **20** (1845), 1609–1612. also 225–228 of [1845d].

[1845d–99] *Sur diverses questions d'analyse et de physique mathématique*, L. J. **10** (1845), 222–228; consists of 1845b and 1845c.

[1845e] *Rapport sur un Mémoire de M. Serret, sur la représentation géométrique des fonctions elliptiques et ultra-elliptiques*, (with Lamé), C. R. **21** (1845), 281–284; L. J. **10** (1845), 290–293; see also *Rapport sur la nouvelle rédaction...*, C. R. **26** (1848), 352–353.

[1845f] *Note*, (added to the report [1845e]), L. J. **10** (1845), 293–296.

[1845g–100]   *Sur un Mémoire de M. Serret, relatif à la représentation des fonctions elliptiques*, L. J. **10** (1845), 456–465; C. R. **21** (1845), 1255–1264.

[1845h–101]   *Sur une propriété générale d'une classe de fonctions*, L. J. **10** (1845), 327–328.

[1845i]   *M. Liouville donne de vive voix une idée des nouvelles recherches de M. Michael Roberts, sur les lignes géodésiques et les lignes de courbure de l'ellipsoïde*, C. R. **21** (1845), 1410–1411; L. J. **10** (1845), 466–468.

[1846a]   Note in a paper *Sur quelques propriétés des lignes géodésique et des lignes de courbure de l'ellipsoïde*, by M. Robert, L. J. **11** (1846), p. 4.

[1846b–113]   *Démonstration géométrique relative à l'équation des lignes géodésiques sur un ellipsoïde quelconque*, C. R. **22** (1846), 111–113.

[1846c–104]   *Démonstration géométrique relative à l'équation des lignes géodésiques sur les surfaces de second degré*, L. J. **11** (1846), 21–24.

[1846d–105]   *Sur un théorème de M. Joachimsthal, relatif aux lignes de courbures planes*, L. J. **11** (1846), 87–88.

[1846e–106]   *Lettres sur diverses questions d'analyse et de physique mathématique, concernant l'ellipsoïde adressées à M. P. H. Blanchet - Première Lettre*, L. J. **11** (1846), 217–236.

[1846f–107]   *Lettres sur diverses questions d'analyse et de physique mathématique concernant l'ellipsoïde, adressées à M. P. H. Blanchet - Deuxième Lettre*, L. J. **11** (1846), 261–290.

[1846g–114]   *Deux théorèmes concernant les lignes géodésiques et les lignes de courbure de l'ellipsoïde*, C. R. **22** (1846), p. 893.

[1846h–122]   *Théorème concernant l'intégration des équations du mouvement d'un point libre*, Connaiss. des Temps pour 1849 (1846), 255–256.

[1846i–108]   *Sur quelques cas particuliers où les équations di mouvement d'un point matériel peuvent s'intégrer - Premier mémoire*, L. J. **11** (1846), 345–378.

[1846j]   *Oeuvres Mathématiques d'Evariste Galois. Avertissement*, L. J. **11** (1846), 381–384; *Note*, L. J. **11** (1846), 415–416.

[1846k–109]   *Sur une transformation de l'équation* $\sin\theta\,\frac{d\sin\theta\frac{d\phi}{d\theta}}{d\theta} + \frac{d^2\phi}{d\omega^2} + n(n+1)\sin^2\theta.\phi = 0$, L. J. **11** (1846), 458–461.

[1846l–110]   *Sur la décomposition des fractions rationnelles*, L. J. **11** (1846), 462–463.

[1846m–111]   *Sur l'intégrale* $\int_0^\infty e^{-x}x^n dx$, L. J. **11** (1846), 464–465.

[1846n–112]   *Sur une classe d'équations du premier degré*, L. J. **11** (1846), 466–467.

[1846o–103]   *Sur une propriété de la couche électrique en équilibre à la surface d'un corps conducteur*, Camb. and Dubl. Math. Journ. **1** (1846), 279–281.

[1846a]   Besge, *Sur l'équation* $\frac{d^2y}{dt^2} = \frac{y}{(e^t - e^{-t})^2}$, L. J. **11** (1846), p. 96.

[1846b]   Besge, *Sur l'équation* $\frac{dy}{dx} - f(x)\sin y + F(x)\cos y + \varphi(x) = 0$, L. J. **11** (1846), p. 445.

[1847a]   *Rapport sur un Mémoire de M. Delaunay concernant la théorie analytique du mouvement de la Lune*, (with Biot and Laugier), C. R. **24** (1847), 6–11.

[1847b–115]   *Sur les équations algébriques à plusieurs inconnues*, L. J. **12** (1847), 68–72.

[1847c] *Remarques à l'occasion d'une communication de M. Lamé sur un théo-*
rème de Fermat, C. R. **24** (1847), 315–316.

[1847d–116] *Sur la loi de réciprocité dans la théorie des résidues quadratiques*,
L. J. **12** (1847), 95–96; C. R. **24** (1847), 577–578.

[1847e–117] *Note au sujet d'un Mémoire de M. Chasles sur les lignes géodésiques*
des surfaces, L. J. **12** (1847), p. 255.

[1847f–118] *Note sur deux lettres de M. Thomson relatives à l'emploi d'un sys-*
tème nouveau de coordonnées orthogonales dans quelques problèmes des
théories de la chaleur et de l'électricité, et au problème de la distribution
d'électricité sur le segment d'une couche sphérique infiniment mince, L.
J. **12** (1847), 265–290.

[1847g–119] *Sur un théorème de M. Gauss concernant le produit des deux rayons*
de courbure principaux en chaque point d'une surface, L. J. **12** (1847),
291–304; C. R. **25** (1847), p. 707.

[1847h–108] *Sur quelques cas particuliers ou les équations du mouvement d'un*
point matériel peuvent s'intégrer. Second Mémoire, L. J. **12** (1847),
410–444.

[1847i–124] *Mémoire sur l'intégration des équations différentielles du mouve-*
ment d'un nombre quelconque de points matériels, Connaiss. des Temps
pour 1850 (1847), 1–40; L. J. **14** (1849), 257–299.

[1847j–120] *Sur l'élimination par les fonctions symétriques*, Nouv. Ann. Math.
6 (1847), 295–301.

[1848a] *Note à la suite d'un article de M. Serret*, L. J. **13** (1848), 34–37.

[1848b–121] *Sur l'équation aux différences partielles qui concerne l'équilibre de*
la chaleur dans un corps hétérogène, L. J. **13** (1848), p. 72.

[1848c] *Note à la suite d'une Lettre de M. W. Roberts*, L. J. **13** (1848), p. 220.

[1848d] *Aux Electeurs du Département de la Meurthe*; Election poster . 2 pages
(1848).

[1849a–125] *Note au sujet de l'équation $x^2 + y^2 = z^2$*, C. R. **28** (1849), p. 687.

[1849b–123] *Remarques sur une classe d'équations différentielles, à l'occasion*
d'un Mémoire de M. Jacobi sur quelques séries elliptiques, L. J. **14**
(1849), 225–241.

[1849a] Besge, *Sur l'intégrale definie $\int_0^\infty \frac{\sin ax}{x} dx$*, L. J. **14** (1849), 31–32.

[1849b] Besge, *Sur un problème de géométrie*, L. J. **14** (1849), 247–248.

[1850a–126] *Théorème sur l'équation $dx^2 + dy^2 + dz^2 = \lambda(d\alpha^2 + d\beta^2 + d\gamma^2)$*, L.
J. **15** (1850), p. 103.

[1850b] *Sur les courbes à double courbure*; Note I in G. Monge *Application de*
l'Analyse à la Géométrie, 5. Ed, ed. Liouville, Paris 1850. 547–568.

[1850c] *Expressions diverses de la distance de deux points infinement voisins*
et de la courbure géodésique des lignes sur une surface; Note II in G.
Monge *Application de l'Analyse à la Géométrie, 5. Ed*, ed. Liouville,
Paris 1850. 569–576.

[1850d] *Théorème concernant l'intégration de l'équation des lignes géodésiques*;
Note III in G. Monge *Application de l'Analyse à la Géométrie, 5. Ed.*,
ed. Liouville, Paris 1850. 577–582.

[1850e] *Sur le théorème de M. Gauss concernant le produit des deux rayons*
de courbure principaux en chaque point d'une surface; Note IV in G.
Monge *Application de l'Analyse à la Géométrie, 5. Ed*, ed. Liouville,
Paris 1850. 583–600.

[1850f]   *Du tracé géographique des surfaces les unes sur les autres*; Note V in G.
Monge *Application de l'Analyse à la Géométrie, 5. Ed.*, ed. Liouville,
Paris 1850. 601–608.

[1850g]   *Extension au cas des trois dimensions de la question du tracé géographique*; Note VI in G. Monge *Application de l'Analyse à la Géométrie, 5.
Ed.*, ed. Liouville, Paris 1850. 609–616.

[1850h]   *Note à l'occasion de l'équation des cordes vibnrantes*; Note 7 in G.
Monge *Application de l'Analyse à la Géométrie, 5. Ed*, ed. Liouville,
Paris 1850. 617–638.

[1851a–127]   *Sur un théorème de M. Chasles*, L. J. **16** (1851), 6–8.

[1851b]   *Remarques à l'occation d'une communication de M. Binet sur le mouvement du pendule simple, en ayant égard à l'influence de la rotation
diurne de la terre*; [Foucault's pendulum] C. R. **32** (1851), 159–160.

[1851c]   *Réclamation de priorité relativement à quelques parties des travaux men-
tionnés dans un Rapport fait par M. Cauchy*, C. R. **32** (1851), 450–452.

[1851d–128]   *Sur la théorie générale de surfaces*, L. J. **16** (1851), 130–132; C. R.
32 (1851), 533–535.

[1851e–129]   *Sur des classes très-étendus de quantités dont la valeur n'est ni
algébrique, ni même réductible à des irrationnelles algébrique*, L. J. **16**
(1851), 133–142.

[1851f–130]   *Mémoire sur les figures ellipsoïdales à trois axes inégaux, qui peuvent convenir à l'équilibre d'une masse liquide homogène, douée d'un
mouvement de rotation*, L. J. **16** (1851), 241–254.

[1852a–131]   *Rapport sur un Mémoire de M. Jules Bienaymé, concernant la
probabilité des erreurs d'après la méthode des moindres carrés*; (with
Lamé and Chasles) C. R. **34** (1852), 90–92; L. J. **17** (1852), 31–32.

[1852b]   *Note au sujet de deux Thèses de M. O. Bonnet*, L. J. **17** (1852), p. 340.

[1852c–132]   *Théorème sur le rapport anhurmonique*, L. J. **17** (1852), 391–392.

[1852d–133]   *Sur les fonctions Gamma de Legendre*, L. J. **17** (1852), 448–453;
C. R. **35** (1852), 317–322.

[1852e–134]   *Note sur la théorie des formules différentielles*, L. J. **17** (1852),
478–480.

[1852f–146]   *Formules générales relatives à la question de la stabilité de l'équilibre
d'une masse liquide homogène douée d'un mouvement de rotation autour
d'un axe*, Connaiss. des Temps pour 1855 (1852), 26–44; L. J. **20** (1855),
164–184.

[1853a–135]   *Sur l'équation aux différences partielles* $\frac{d^2 \log \lambda}{du\,dv} \pm \frac{\lambda}{2a^2} = 0$, L. J. **18**
(1853), 71–72; C. R. **36** (1853), 371–373.

[1853a]   Besge, *Sur une transformation d'intégrales définies*, L. J. **18** (1853), p.
112.

[1853b]   Besge, *Addition à la Note sur une transformation d'intégrales définies,
insérée dans le cahier de mars*, L. J. **18** (1853), p. 168.

[1854a]   *Sur la part qu'a prise l'Académie au développement d'une branche im-
portante de la richesse agricole dans l'Algérie*, C. R. **38** (1854), p. 1038.

[1854b–136]   *Sur l'équation différentielle* $\frac{d(x-x^3)\frac{dy}{dx}}{dx} - xy = 0$, L. J. **19** (1854),
151–152.

[1854c–137]   *Expression simple du rayon de courbure géodésique d'une ligne
tracée sur un ellipsoïde*, L. J. **19** (1854), p. 368.

[1854d] *Note à l'occasion d'un Mémoire de M. Borchardt*, L. J. **19** (1854), 395–400.

[1854e–138] *Méthode métamorphique par rayons vecteurs réciproques*; (summary of [1847f]) Nouv. Ann. Math. **13** (1854), 227–245.

[1855a–139] *Note sur une formule pour les différentielles à indices quelconques, à l'occasion d'un mémoire de M. Tortolini*, L. J. **20** (1855), 115–120.

[1855b–140] *Valeur d'une intégrale définie qui se rattache aux intégrales trinômes*, L. J. **20** (1855), 133–134.

[1855c–141] *Rapport sur un Mémoire de M. Bour, concernant l'intégration des équations différentielles de la mécanique analytique*; (with Lamé and Chasles) C. R. **40** (1855), 661–662; L. J. **20** (1855), 135–136.

[1855d–142] *Note sur l'intégration des équations différentielles de la Dynamique, présentée au Bureau des Longitudes le 29 juin 1853*, L. J. **20** (1855), 137–138.

[1855e–143] *Sur l'équation différentielle du premier ordre $\frac{dy}{dx} = f(x,y)$*, L. J. **20** (1855), 143–144.

[1855f–144] *Sur un théorème relatif à l'intégrale Eulérienne de seconde espèce*, L. J. **20** (1855), 157–160.

[1855g–145] *Sur l'équation $\Gamma(t)\Gamma(t + \frac{1}{2}) = 2^{1--2t}\sqrt{\pi} \cdot \Gamma(2t)$*, L. J. **20** (1855), 161–163.

[1855h] *Note à l'occasion d'un Mémoire de M. Edmond Bour*, L. J. **20** (1855), 201–202.

[1855i–147] *Sur les fonctions elliptiques*; (contains 1844g and 1851c) L. J. **20** (1855), 203–208.

[1855j] *Discours prononcé aux funérailles de M. Sturm*; L. J. **20** (1855), 395–396; Included in [Prouhet 1856].

[1856a] *Avertissement*, L. J. (2) **1** (1856), V–VI.

[1856b–148] *Sur deux mémoires de M. Poisson. [Note historique relative à l'intégrale de l'équation sur laquelle repose essentiellement la théorie de la propagation du son dans les milieux gazeux]*, L. J. (2) **1** (1856), 1–6; C. R. **42** (1856), 465–470.

[1856c–149] *Sur des questions de minimum*, L. J. (2) **1** (1856), 7–8.

[1856d–150] *Détermination des valeurs d'une classe remarquable d'intégrales définies multiples, et démonstration nouvelle d'une célèbre formule de M. Gauss concernant les fonctions Gamma de Legendre*, L. J. (2) **1** (1856), 82–88; C. R. **42** (1856), 501–508.

[1856e–153] *Mémoire sur la réduction de classes très-étendus d'intégrales multiples*, L. J. (2) **1** (1856), 289–294; C. R. **42** (1856), 525–530; Summary in Zeitschrift f. Math. u. Phys. (Schlömilch), **1** (1856), 356–363.

[1856f–151] *Extension d'un théorème de calcul intégrale*, L. J. (2) **1** (1856), 190–191; C. R. **42** (1856), 985–990.

[1856g–157] *Sur la théorie générale des équations différentielles*, L. J. (2) **1** (1856), 345–348; C. R. **42** (1856), 1084–1088.

[1856h–152] *Sur la representation des nombres par la forme quadratique $x^2 + ay^2 + bz^2 + abt^2$*, L. J. (2) **1** (1856), p. 230; C. R. **42** (1856), 1145–1146.

[1856i–154] *Mémoire sur un cas particulier du problème des trois corps*, L. J. (2) **1** (1856), 248–264.

[1856j–155] *Note sur une équation aux différences finies partielles*, L. J. (2) **1** (1856), 295–296.

[1856k–156]   *Expression remarquable de la quantité qui, dans le mouvement d'un système de points matériels à liaisons quelconques, est un minimum en vertu du principe de la moindre action*, L. J. (2) **1** (1856), 297–304; C. R. **42** (1856), 1146–1154.

[1856l–158]   *Sur les sommes de divideurs des nombres*, L. J. (2) **1** (1856), 349–350.

[1856m–159]   *Sur l'équation* $1.2.3\ldots(p-1)+1 = p^m$, L. J. (2) **1** (1856), 351–352.

[1856n–160]   *Sur l'intégrale* $\int_0^1 \frac{t^{\mu+\frac{1}{2}}(1-t)^{\mu-\frac{1}{2}}dt}{(a+bt-ct^2)^{\mu+1}}$, L. J. (2) **1** (1856), 421–424.

[1856o–161]   *Démonstration nouvelle d'une formule de M. W. Thomson;* [donnant la valeur d'une intégrale définie multiple] L. J. (2) **1** (1856), 445–450.

[1856p–177]   *Développement sur un chapitre de la "Mécanique" de Poisson*, Connaiss des Temps pour 1859 (1856), 1–22; L. J. (2) **3** (1858), 1–25.

[1857a–162]   *Observations sur l'intégrale* $\int_0^1 \frac{t^{\mu+\frac{1}{2}}(1-t)^{\mu-\frac{1}{2}}dt}{(a+bt-ct^2)^{\mu+1}}$ *avec remarques*, L. J. (2) **2** (1857), 47–55.

[1857b–163]   *Théorème concernant les sommes de diviseurs des nombres*, L. J. (2) **2** (1857), p. 56.

[1857c–164]   *Sur l'expression* $\phi(n)$, *qui marque combien la suite* $1.2.3\ldots n$ *contient de nombres premiers à* n, L. J. (2) **2** (1857), 110–112.

[1857d–176]   *Note sur la théorie des nombres*, C. R. **44** (1857), 753–755.

[1857e–173]   *Sur un point de la théorie des équations binomes*, L. J. (2) **2** (1857), 413–423; C. R. **44** (1857), 797–801.

[1857f–165]   *Sur quelques fonctions numériques*, L. J. (2) **2** (1857), 141–144, 244–248, 377–384, 425–432.

[1857g–166]   *Sur un théorème de Dirichlet*, L. J. (2) **2** (1857), p. 184.

[1857h–167]   *Sur le produit* $m(m+1)(m+2)\cdots(m+n-1)$, L. J. (2) **2** (1857), 277–278.

[1857i–168]   *Sur l'intégrale définie* $\int_0^1 \frac{x^{p-1}(1-x)^{q-1}dx}{(1+\sqrt{1+gx})^{2p+2q}}$, L. J. (2) **2** (1857), p. 279.

[1857j–169]   *Sur le fonction* $E(x)$, *qui marque le nombre entier contenu dans* x, L. J. (2) **2** (1857), p. 280.

[1857k–170]   *Sur la décomposition d'un nombre en un produit de deux sommes de carrés*, L. J. (2) **2** (1857), 351–352.

[1857l–171]   *Généralisation d'un théorème de l'arithmétique Indienne;* [La somme $1^3+2^3+3^3+\cdots+n^3$ des cubes des nombres naturels est égale au carré de la somme des nombres eux-mêmes, c. à. d. $= (1+2+3+\cdots+n)^2$.] L. J. (2) **2** (1857), 393–396.

[1857m–172]   *Sur une relation entre deux fonctions numériques*, L. J. (2) **2** (1857), p. 408.

[1857n–172]   *Démonstration du théorème éconcé dans l'article précédent*, L. J. (2) **2** (1857), 409–412.

[1857o–174]   *Note à l'occasion d'un mémoire de M. Bouniakowsky: "Nouvelle méthode pour la théorie des formes quadratiques"*, L. J. (2) **2** (1857), p. 424.

[1857p–175]   *Sur quelques séries et produits infinis*, L. J. (2) **2** (1857), 433–440.

[1857q–179]   *Sur un problème de Mécanique*, Connaiss des Temps pour 1860 (1857), 69–72; L. J. (2) **3** (1858), 69–72.

[1858a–178]   *Genéralisation d'une formule concernant les sommes des puissances des diviseurs d'un nombre,* L. J. (2) **3** (1858), 63–68.

[1858b–180]   *Démonstration d'un théorème sur les nombres premiers de la forme* $8\mu + 3$, L. J. (2) **3** (1858), 84–88.

[1858c–181]   *Sur quelques formules générales qui peuvent être utiles dans la thé-orie des nombres;* (premier-sixième article) L. J. (2) **3** (1858), 143–152, 193–200, 201–298, 241–250, 273–288, 325–336.

[1858d–182]   *Note sur une question de la théorie des nombres,* L. J. (2) **3** (1858), 357–360.

[1858a]   Besge, *Sur deux intégrales doubles,* L. J. (2) **3** (1858), p. 324.

[1858b]   Besge, *Autre égalité d'intégrales doubles,* L. J. (2) **3** (1858), p. 416.

[1859a]   *Sur quelques formules générales qui peuvent être utiles dans la théorie des nombres (septième-onzième acticle),* L. J. (2) **4** (1859); 1–8, 73–80, 111–120, 195–204, 281–304.

[1859b–183]   *Sur la forme* $x^2 + y^2 + 5(z^2 + t^2)$, L. J. (2) **4** (1859), 47–48.

[1859c–184]   *Sur une intégrale définie multiple,* L. J. (2) **4** (1859), 155–160.

[1859d–185]   *Théorème arithmétique,* L. J. (2) **4** (1859), 271–272.

[1859e–186]   *Théorème concernant les nombres premiers de la forme* $24\mu + 7$, L. J. (2) **4** (1859), 399–400.

[1859a]   Besge, *Sur une équation différentielle,* L. J. (2) **4** (1859), p. 72.

[1859b]   Besge, *Sur les intégrales trinômes,* L. J. (2) **4** (1859), p. 194.

[1860a]   *Sur quelques formules générales qui peuvent être utiles dans la théorie des nombres (douzième article),* L. J. (2) **5** (1860), 1–9.

[1860b–187]   *Théorème concernant le double d'un nombre premier contenu dans l'une ou l'autre des deux formes linéaires* $16k + 7, 16k + 11$, L. J. (2) **5** (1860), 103–105.

[1860c–188]   *Sur le double d'un nombre premier* $4\mu + 1$, L. J. (2) **5** (1860), 119–121.

[1860d–189]   *Note à l'occasion d'un théorème de M. Kronecker,* L. J. (2) **5** (1860), 127–128.

[1860e–190]   *Théorème concernant les nombres premiers de la forme* $24k + 11$, L. J. (2) **5** (1860), 139–140.

[1860f–191]   *Théorème concernant la fonction numérique relative au nombre des réprésentation d'un entier sous le forme d'une somme de trois carrés,* L. J. (2) **5** (1860), 141–142.

[1860g–192]   *Nombre des réprésentations du double d'un entier impair sous la forme d'une somme de douze carrés,* L. J. (2) **5** (1860), 143–146.

[1860h–193]   *Sur la forme* $x^2 + y^2 + 3(z^2 + t^2)$, L. J. (2) **5** (1860), 147–152.

[1860i–194]   *Addition à la note au sujet d'un théorème de M. Kronecker, supra,* p. 127, L. J. (2) **5** (1860), 267–268.

[1860j–195]   *Sur la forme* $x^2 + y^2 + 2(z^2 + t^2)$, L. J. (2) **5** (1860), 269–272.

[1860k–196]   *Égalités entre des sommes qui dépendent de la fonction numérique* $E(x)$, L. J. (2) **5** (1860), 287–288; (addition) 455–456.

[1860l–197]   *Théorème concernant les nombres premiers de la forme* $8\mu + 5$, L. J. (2) **5** (1860), p. 300.

[1860m–198]   *Sur les nombres premiers de la forme* $16k + 7$, L. J. (2) **5** (1860), 301–302.

[1860n–199]   *Sur le produit de deux nombres premiers, l'un de la forme* $8k + 3$, *et l'autre de la forme* $8h + 5$, L. J. (2) **5** (1860), 303–304.

[1860o]   *Sur la forme $x^2 + y^2 + 4(z^2 + t^2)$*, L. J. (2) **5** (1860), 305–308.

[1860p-200]   *Nouveau théorème concernant les nombres premiers de la forme $24k + 11$*, L. J. (2) **5** (1860), 309–310.

[1860q-201]   *Théorème concernant les nombres premiers de la forme $24k + 19$*, L. J. (2) **5** (1860), 311–312.

[1860r-202]   *Théorème concernant les nombres premiers de l'une ou de l'autre des deux formes $40\mu + 11, 40\mu + 19$*, L. J. (2) **5** (1860), 387–388.

[1860s-203]   *Théorème concernant les nombres premiers de la forme $40\mu + 7$*, L. J. (2) **5** (1860), 389–390.

[1860t-204]   *Théorème concernant les nombres premiers de la forme $40\mu + 23$*, L. J. (2) **5** (1860), 391–392.

[1860u]   *Théorème concernant le triple d'un nombre premier de la forme $8\mu + 3$*, L. J. (2) **5** (1860), 475–476.

[1860]   Besge, *Somme d'une série*, L. J. (2) **5** (1860), 367–368.

[1861a-205]   *Théorèmes concernant le quadruple d'un nombre premier contenu dans l'une ou dans l'autre des deux formes $8\mu + 3, 8\mu + 5$*, L. J. (2) **6** (1861), 1–6.

[1861b-206]   *Théorème concernant les nombres premiers de la forme $16k + 13$*, L. J. (2) **6** (1861), 7–8.

[1861c-207]   *Théorèmes concernant le double d'un nombre premier de la forme $16k + 7$*, L. J. (2) **6** (1861), 28–30.

[1861d-208]   *Théorème concernant les nombres premiers de la forme $8\mu + 1$*, L. J. (2) **6** (1861), 31–32.

[1861e-209]   *Nouveau théorème concernant les nombres premiers de la forme $8\mu + 1$*, L. J. (2) **6** (1861), 55–56.

[1861f-210]   *Théorèmes concernant le quadruble d'un nombre premier de la forme $12k + 5$*, L. J. (2) **6** (1861), 93–96.

[1861g-211]   *Théorèmes concernant respectivement les nombres premiers de la forme $16k + 3$ et les nombres premiers de la forme $16k + 11$*, L. J. (2) **6** (1861), 97–100.

[1861h-212]   *Théorème concernant les nombres premiers de la forme $24k + 13$*, L. J. (2) **6** (1861), 101–102.

[1861i-213]   *Théorème concernant les nombres premiers de la forme $24k + 1$*, L.J. (2) **6** (1861), 103–104.

[1861j-214]   *Théorème concernant les nombres premiers de la forme $40\mu + 3$*, L. J. (2) **6** (1861), 105–106.

[1861k-215]   *Théorème concernant les nombres premiers de la forme $40\mu + 27$*, L. J. (2) **6** (1861), 107–108.

[1861l-216]   *Théorèmes concernant le quintuble d'un nombre premier de l'une ou de l'autre des deux formes $40\mu + 7, 40\mu + 23$*, L. J. (2) **6** (1861), 109–112.

[1861m-217]   *Sur la forme $x^2 + 3y^2 + 4z^2 + 12t^2$*, L. J. (2) **6** (1861), 135–136.

[1861n-218]   *Théorèmes concernant le quintuble d'un nombre premier de la forme $24k + 17$*, L. J. (2) **6** (1861), 147–149.

[1861o-219]   *Théorèmes concernant les nombres premiers de l'une ou de l'autre des deux formes $120k + 61, 120k + 109$*, L. J. (2) **6** (1861), 150–152.

[1861p-220]   *Théorèmes concernant le produit de deux nombres premiers égaux ou inégaux de la forme $8\mu + 3$*, L. J. (2) **6** (1861), 185–186.

[1861q-221]   *Théorème concernant le produit de deux nombres premiers, l'un de la forme $8\mu + 1$, l'autre de la forme $8\nu + 3$*, L. J. (2) **6** (1861), 187–188.

[1861r-222]   *Théorème concernant le produit de deux nombres premiers égaux ou inégaux de la forme $24\mu + 5$*, L. J. (2) **6** (1861), 189–190.

[1861s-223]   *Théorème concernant le produit de deux nombres premiers égaux ou inégaux de la forme $24\mu + 7$*, L. J. (2) **6** (1861), 191–192.

[1861t-224]   *Théorème concernant le produit de deux nombres premiers, l'un de la forme $40\mu + 3$, l'autre de la forme $40\nu + 7$*, L. J. (2) **6** (1861), 193–194.

[1861u-225]   *Théorème concernant le produit de deux nombres premiers, l'un de la forme $40\mu + 7$, l'autre de la forme $40\nu + 27$*, L. J. (2) **6** (1861), 195–196.

[1861v-226]   *Théorème concernant le produit de deux nombres premiers, l'un de la forme $40\mu + 3$, l'autre de la forme $40\nu + 23$*, L. J. (2) **6** (1861), 197–198.

[1861x-227]   *Théorème concernant le produit de deux nombres premiers, l'un de la forme $40\mu + 23$, l'autre de la forme $40\nu + 27$*, L. J. (2) **6** (1861), 199–200.

[1861y-228]   *Théorème concernant le produit de deux nombres premiers égaux ou inégaux de la forme $120\mu + 31$*, L. J. (2) **6** (1861), 201–202.

[1861z-229]   *Théorème concernant le produit de deux nombres premiers égaux ou inégaux de la forme $120\nu + 79$*, L. J. (2) **6** (1861), 203–204.

[1861α-230]   *Théorème concernant de produit de deux nombres premiers, l'un de la forme $120\mu + 31$, l'autre de la forme $120\nu + 79$*, L. J. (2) **6** (1861), 205–206.

[1861β-231]   *Théorème concernant le produit d'un nombre premier $8\mu + 3$ par le carré d'un nombre premier $8\nu + 7$*, L. J. (2) **6** (1861), 207–208.

[1861γ-232]   *Remarques nouvelles concernant les nombres premiers de la forme $24\mu + 7$*, L. J. (2) **6** (1861), 219–224.

[1861δ-233]   *Sur les deux formes quadratiques $x^2 + y^2 + z^2 + 2t^2$, $x^2 + 2(y^2 + z^2 + t^2)$*, L. J. (2) **6** (1861), 225–230.

[1861ε-234]   *Sur un certain genre de décomposition d'un entier en sommes de carrés*, L. J. (2) **6** (1861), 233–239.

[1861ζ-235]   *Sur la forme $X^2 + Y^2 + Z^2 + 8T^2$*, L. J. (2) **6** (1861), 324–328.

[1861η-236]   *Nouveaux théorèmes concernant les fonctions $N(n, p, q)$ et d'autres fonctions qui s'y rattachent*, L. J. (2) **6** (1861), 369–376.

[1861θ-237]   *Sur la forme $x^2 + 2y^2 + 4z^2 + 8t^2$*, L. J. (2) **6** (1861), 409–418.

[1861λ-238]   *Sur les deux formes $X^2 + Y^2 + Z^2 + 4T^2$, $X^2 + 4Y^2 + 4Z^2 + 4T^2$*, L. J. (2) **6** (1861), 440–448.

[1861μ]   *Réponse à une lettre de Hermite*, CR **53** (1861), 228–231; L. J. (2) vol 7 (1862), 41–44.

[1861]   Besge, *Extrait d'une lettre adressée a M. Liouville*, L. J. (2) **7** (1862), 239–240.

[1862a]   *Remarques à l'occasion d'un Mémoire de M. Bour*, CR **54** (1852), 941–942.

[1862b-239]   *Sur la forme $X^2 + 2Y^2 + 2Z^2 + 4T^2$*, L. J. (2) **7** (1862), 1–4.

[1862c-240]   *Sur la forme $X^2 + 8(Y^2 + Z^2 + T^2)$*, L. J. (2) **7** (1862), 5–8.

[1862d-241]   *Sur la forme $X^2 + 4Y^2 + 4Z^2 + 8T^2$*, L. J. (2) **7** (1862), 9–12.

[1862e-242]   *Sur la forme $X^2 + 8Y^2 + 8Z^2 + 16T^2$*, L. J. (2) **7** (1862), 13–16.

[1862f-243]   *Nouveau théorème concernant les nombres premiers de la forme $16g + 11$*, L. J. (2) **7** (1862), 17–18.

[1862g-244]   *Nouveau théorème concernant les nombres premiers de la forme $8\mu + 1$*, L. J. (2) **7** (1862), 19–20.

[1862h-245]   *Théorème concernant le produit de deux nombres premiers inégaux de la forme* $8\mu + 3$, L. J. (2) **7** (1862), 21–22.

[1862i-246]   *Théorème concernant la quatrième puissance d'un nombre premier de la forme* $8\mu + 3$, L. J. (2) **7** (1862), 23–24.

[1862j]   *Note de M. Liouville*; [to a letter by Hermite] (cf. 1861μ) L. J. (2) **7** (1862), 44–48.

[1862k-247]   *Sur la forme* $x^2 + 2y^2 + 4z^2 + 4t^2$, L. J. (2) **7** (1862), 62–64.

[1862l-248]   *Sur la forme* $x^2 + 2y^2 + 8z^2 + 8t^2$, L. J. (2) **7** (1862), 65–68.

[1862m-249]   *Sur la forme* $x^2 + 8y^2 + 16z^2 + 16t^2$, L. J. (2) **7** (1862), 69–72.

[1862n-250]   *Sur la forme* $x^2 + 4y^2 + 4z^2 + 16t^2$, L. J. (2) **7** (1862), 73–76.

[1862o-251]   *Sur la forme* $x^2 + 16(y^2 + z^2 + t^2)$, L. J. (2) **7** (1862), 77–80.

[1862p-252]   *Sur la forme* $x^2 + y^2 + 2z^2 + 4t^2$, L. J. (2) **7** (1862), 99–100.

[1862q-253]   *Sur la forme* $x^2 + y^2 + 4z^2 + 8t^2$, L. J. (2) **7** (1862), 103–104.

[1862r-254]   *Sur la forme* $x^2 + 4y^2 + 16z^2 + 16t^2$, L. J. (2) **7** (1982), 105–108.

[1862s-255]   *Sur la forme* $x^2 + y^2 + 8z^2 + 8t^2$, L. J. (2) **7** (1862), 109–112.

[1862t-256]   *Sur la forme* $x^2 + 4y^2 + 8z^2 + 8t^2$, L. J. (2) **7** (1862), 113–116.

[1862u-257]   *Sur la forme* $x^2 + y^2 + 16z^2 + 16t^2$, L. J. (2) **7** (1862), 117–120.

[1862v-258]   *Théorème concernant le double du carré d'un nombre premier* $8\mu+3$, L. J. (2) **7** (1862), p. 136.

[1862x-259]   *Sur la forme* $x^2 + 4y^2 + 8z^2 + 16t^2$, L. J. (2) **7** (1862), 143–144.

[1862y-260]   *Sur la forme* $x^2 + 2y^2 + 16z^2 + 16t^2$, L. J. (2) **7** (1862), 145–147.

[1862z-261]   *Sur la forme* $x^2 + 2y^2 + 2z^2 + 8t^2$, L. J. (2) **7** (1862), 148–149.

[1862α-262]   *Sur la forme* $x^2 + 2y^2 + 4z^2 + 16t^2$, L. J. (2) **7** (1862), 150–152.

[1862β-263]   *Sur la forme* $x^2 + 2y^2 + 8z^2 + 16t^2$, L. J. (2) **7** (1862), 153–154.

[1862γ-264]   *Sur la forme* $x^2 + y^2 + 2z^2 + 8t^2$, L. J. (2) **7** (1862), 155–156.

[1862δ-265]   *Sur la forme* $x^2 + y^2 + 4z^2 + 16t^2$, L. J. (2) **7** (1862), 157–158.

[1862ε-266]   *Sur la forme* $x^2 + 2y^2 + 2z^2 + 16t^2$, L. J. (2) **7** (1862), 161–164.

[1862ζ-267]   *Sur la forme* $x^2 + y^2 + z^2 + 16t^2$, L. J. (2) **7** (1862), 165–168.

[1862η-268]   *Sur la forme* $x^2 + y^2 + 8z^2 + 16t^2$, L. J. (2) **7** (1862), 201–204.

[1862θ-269]   *Sur la forme* $x^2 + y^2 + 2z^2 + 16t^2$, L. J. (2) **7** (1862), 205–209.

[1862λ-270]   *Sur la forme* $x^2 + 8y^2 + 8z^2 + 64t^2$, L. J. (2) **7** (1862), 246–248.

[1862μ-271]   *Sur la forme* $x^2 + 8y^2 + 16z^2 + 64t^2$, L. J. (2) **7** (1862), 249–252.

[1862ν-271]   *Extrait d'une lettre à M. Besge sur une formule concernant les sommes de diviseurs des nombres*, L. J. (2) **7** (1862), 375–376.

[1862ξ-273]   *Théorème concernant les nombres triangulaires*, L. J. (2) **7** (1862), 407–408.

[1862π]   *Sur la forme* $x^2 + 8y^2 + 64(z^2 + t^2)$, L. J. (2) **7** (1862), 421–424.

[1862]   Besge, *Extrait d'une lettre adressée à M. Liouville*, L. J. (2) **7** (1862), p. 256.

[1863a-274]   *Nouveaux théorèmes concernant les nombres triangulaires*, L. J. (2) **8** (1863), 73–84.

[1863b-275]   *Théorèmes concernant le quadruple d'un nombre premier de l'une ou de l'autre des deux formes* $20k + 3$, $20k + 7$, L. J. (2) **8** (1863), 85–88.

[1863c-276]   *Nouveau théorème comcernant le quadruple d'un nombre premier de la forme* $12k + 5$, L. J. (2) **8** (1863), 102–104.

[1863d-277]   *Sur la forme* $x^2 + y^2 + z^2 + 3t^2$, L. J. (2) **8** (1863), 105–114.

[1863e-278]   *Sur la forme* $x^2 + y^2 + 2z^2 + 2zt + 2t^2$, L. J. (2) **8** (1863), 115–119.

[1863f-279]   *Sur la forme* $x^2 + y^2 + z^2 + zt + t^2$, L. J. (2) **8** (1863), 120–123.

[1863g-280]   *Sur la forme* $x^2 + y^2 + 2z^2 + 6t^2$, L. J. (2) **8** (1863), 124–128.

[1863h-281]   *Sur la forme $x^2 + y^2 + 2z^2 + 3t^2$*, L. J. (2) **8** (1863), 129–133.

[1863i-282]   *Sur la forme $x^2 + 2y^2 + 4z^2 + 6t^2$*, L. J. (2) **8** (1863), 134–136.

[1863j-283]   *Théorème concernant les nombres premiers contenus dans une quel-conque des trois formes linéaires $168k + 43, 168k + 67, 168k + 163$*, L. J. (2) **8** (1863), 137–140.

[1863k-284]   *Sur la forme $x^2 + xy + y^2 + z^2 + zt + t^2$*, L. J. (2) **8** (1863), 141–144.

[1863l-285]   *Sur la forme $x^2 + y^2 + z^2 + 12t^2$*, L. J. (2) **8** (1863), 161–168.

[1863m-286]   *Sur la forme $x^2 + 2y^2 + 2z^2 + 12t^2$*, L. J. (2) **8** (1863), 169–172.

[1863n-287]   *Sur la forme $x^2 + y^2 + 4z^2 + 12t^2$*, L. J. (2) **8** (1863), 173–176.

[1863o-288]   *Sur la forme $x^2 + 4y^2 + 4z^2 + 12t^2$*, L. J. (2) **8** (1863), 177–178.

[1863p-289]   *Sur la forme $3x^2 + 4y^2 + 4z^2 + 4t^2$*, L. J. (2) **8** (1863), 179–181.

[1863q-290]   *Sur la forme $x^2 - y^2 + 3z^2 + 4t^2$*, L. J. (2) **8** (1863), 182–184.

[1863r-291]   *Sur la forme $x^2 + 3y^2 + 4z^2 + 4t^2$*, L. J. (2) **8** (1863), 185–188.

[1863s-292]   *Sur la forme $2x^2 + 2y^2 + 3z^2 + 4t^2$*, L. J. (2) **8** (1863), 189–192.

[1863t-293]   *Remarques nouvelles sur la forme $x^2 + y^2 + z^2 + 3t^2$*, L. J. (2) **8** (1863), 193–204.

[1863u-294]   *Sur la forme $x^2 + 4y^2 + 12z^2 + 16t^2$*, L. J. (2) **8** (1863), 205–208.

[1863v-295]   *Sur la forme $x^2 + 3y^2 + 6z^2 + 6t^2$*, L. J. (2) **8** (1863), 209–213.

[1863x-296]   *Sur la forme $2x^2 + 3y^2 + 3z^2 + 6t^2$*, L. J. (2) **8** (1863), 214–218.

[1863y-297]   *Sur la forme $x^2 + 3(y^2 + z^2 + t^2)$*, L. J. (2) **8** (1863), 219–224.

[1863z-298]   *Sur la forme $2x^2 + 2xy + 2y^2 + 3(z^2 + t^2)$*, L. J. (2) **8** (1863), 225–226.

[1863α-299]   *Sur la forme $x^2 + xy + y^2 + 3(z^2 + t^2)$*, L. J. (2) **8** (1863), 227–228.

[1863β-300]   *Sur la forme $3x^2 + 3y^2 + 3z^2 + 4t^2$*, L. J. (2) **8** (1863), 229–238.

[1863γ-301]   *Sur la forme $3x^2 + 3y^2 + 4z^2 + 12t^2$*, L. J. (2) **8** (1863), 239–240.

[1863δ-302]   *Sur la forme $3x^2 + 4y^2 + 12z^2 + 12t^2$*, L. J. (2) **8** (1863), 241–242.

[1863ε-303]   *Sur la forme $x^2 + 3y^2 + 3z^2 + 12t^2$*, L. J. (2) **8** (1863), 243–248.

[1863ζ-304]   *Sur la forme $x^2 + 3y^2 + 12z^2 + 12t^2$*, L. J. (2) **8** (1863), 249–252.

[1863η-305]   *Sur la forme $x^2 + 12y^2 + 12z^2 + 12t^2$*, L. J. (2) **8** (1863), 253–254.

[1863θ-306]   *Sur la forme $3x^2 + 4y^2 + 12z^2 + 48t^2$*, L. J. (2) **8** (1863), 255–256.

[1863λ-307]   *Remarque nouvelle sur la forme $x^2 + y^2 + 3(z^2 + t^2)$*, L. J. (2) **8** (1863), p. 296.

[1863μ-308]   *Sur la forme $x^2 + xy + y^2 + 2z^2 + 2zt + 2t^2$*, L. J. (2) **8** (1863), 308–310.

[1863ν]   *Extrait d'une lettre adressée a M. Besge*, L. J. (2) **8** (1863), 311–312.

[1863ξ-309]   *Théorèmes généraux concernant des fonctions numériques*, L. J. (2) **8** (1863), 347–352.

[1863π]   *Théorème d'Atithmétique*, L. J. (2) **8** (1863), 431–432.

[1864a-310]   *Sur la forme $x^2 + y^2 + z^2 + 5t^2$*, L. J. (2) **9** (1864), 1–12.

[1864b-311]   *Sur la forme $x^2 + y^2 + 2z^2 + 2zt + 3t^2$*, L. J. (2) **9** (1864), 13–16.

[1864c-312]   *Sur la forme $x^2 + 5(y^2 + z^2 + t^2)$*, L. J. (2) **9** (1864), 17–22.

[1864d-313]   *Sur la forme $2x^2 + 2xy + 3y^2 + 5z^2 + 5t^2$*, L. J. (2) **9** (1864), 23–24.

[1864e-314]   *Extension du théorème de Rolle aux racines imaginaires des équations*, L. J. (2) **9** (1864), 84–88.

[1864f-315]   *Sur la forme $x^2 + y^2 + z^2 + t^2 + u^2 + 3v^2$*, L. J. (2) **9** (1864), 89–104.

[1864g-316]   *Sur la forme $x^2 + 3(y^2 + z^2 + t^2 + u^3 + v^2)$*, L. J. (2) **9** (1864), 105–114.

[1864h-317]   *Sur la forme $x^2 + y^2 + z^2 + t^2 + 2u^2 + 2uv + 2v^2$*, L. J. (2) **9** (1864), 115–118.

[1864i-318]   *Sur la forme* $2x^2 + 2xy + 2y^2 + 3(z^2 + t^2 + u^2 + v^2)$, L. J. (2) **9** (1864), 119–122.

[1864j-319]   *Sur la forme* $x^2 + y^2 + 2z^2 + 2zt + 2t^2 + 3u^2 + 3v^2$, L. J. (2) **9** (1864), 123–128.

[1864k-320]   *Nouveau théorème concernant le quadruple d'un nombre premier de l'une ou de l'autre des deux formes* $20k + 3$, $20k + 7$, L. J. (2) **9** (1864), 135–136.

[1864l-321]   *Théorèmes concernant l'octuble d'un nombre premier de l'une ou de l'autre des deux formes* $20k + 3$, $20k + 7$, L. J. (2) **9** (1864), 137–144.

[1864m-322]   *Sur la forme* $x^2 + y^2 + 2yz + 2z^2 + 3t^2$, L. J. (2) **9** (1864), p. 160.

[1864n-323]   *Sur la forme* $x^2 + y^2 + z^2 + t^2 + u^2 + 2v^2$, L. J. (2) **9** (1864), 161–174.

[1864o-324]   *Sur la forme* $x^2 + 2(y^2 + z^2 + t^2 + u^2 + v^2)$, L. J. (2) **9** (1864), 175–180.

[1864p-325]   *Sur la forme* $x^2 + xy + y^2 + 6z^2 + 6zt + 6t^2$, L. J. (2) **9** (1864), 181–182.

[1864q-326]   *Sur la forme* $2x^2 + 2xy + 2y^2 + 3z^2 + 3zt + 3t^2$, L. J. (2) **9** (1864), 183–184.

[1864r-327]   *Sur la forme* $x^2 + xy + y^2 + 3z^2 + 3zt + 3t^2$, L. J. (2) **9** (1864), 223–224.

[1864s-328]   *Sur quelques formules générales qui peuvent être utiles dans la théorie des nombres;* (treizième-seizième article) L. J. (2) **9** (1864), 249–256, 281–288, 321–336, 389–400.

[1864t-329]   *Sur la forme* $x^2 + y^2 + z^2 + t^2 + 2(u^2 + v^2)$, L. J. (2) **9** (1864), 257–272.

[1864u-330]   *Sur la forme* $x^2 + y^2 + 2(z^2 + t^2 + u^2 + v^2)$, L. J. (2) **9** (1864), 273–280.

[1864v-332]   *Extrait d'une lettre adresée à M. Besge,* L. J. (2) **9** (1864), 296–298.

[1864x-331]   *Sur la forme* $x^2 + 2y^2 + 3z^2 + 6t^2$, L. J. (2) **9** (1864), 299–312.

[1864y-333]   *Sur la forme* $x^2 + 2(y^2 + z^2 + t^2 + u^2) + 4v^2$, L. J. (2) **9** (1864), 421–424.

[1865a-334]   *Sur la forme* $x^2 + y^2 + 5z^2 + 5t^2$, L. J. (2) **10** (1865), 1–8.

[1865b-335]   *Sur la forme* $2x^2 + 2xy + 3y^2 + 2z^2 + 2zt + 3t^2$, L. J. (2) **10** (1865), 9–13.

[1865c-336]   *Sur la forme* $x^2 + y^2 + 9z^2 + 9t^2$, L. J. (2) **10** (1865), 14–20.

[1865d-337]   *Sur la forme* $2z^2 + 2xy + 5y^2 + 2z^2 + 2zt + 3t^2$, L. J. (2) **10** (1865), 21–24.

[1865e-338]   *Note au sujet de la forme* $x^2 + y^2 + a(z^2 + t^2)$, L. J. (2) **10** (1865), 43–48.

[1865f-339]   *Note au sujet de la forme* $x^2 + 2y^2 + az^2 + 2at^2$, L. J. (2) **10** (1865), 49–54.

[1865g-340]   *Sur la forme* $x^2 + 4y^2 + 4z^2 + 4t^2 + 4u^2 + 4v^2$, L. J. (2) **10** (1865), 65–70.

[1865h-341]   *Sur la forme* $x^2 + y^2 + 4z^2 + 4t^2 + 4u^2 + 4v^2$, L. J. (2) **10** (1865), 71–72.

[1865i-342]   *Sur la forme* $x^2 + 2y^2 + 2z^2 + 4t^2 + 4u^2 + 4v^2$, L. J. (2) **10** (1865), 73–77.

[1865j-343]   *Sur la forme* $x^2 + y^2 + z^2 + 4t^2 + 4u^2 + 4v^2$, L. J. (2) **10** (1865), 77–80.

[1865k]   *Sur quelques formules générales qui peuvent être utiles dans la théorie des nombres;* (dixseptième-dixhuitième article) L. J. (2) **10** (1865), 169–176.

[1865l-344]   *Sur la forme $x^2 + y^2 + 2z^2 + 2t^2 + 4u^2 + 4v^2$,* L. J. (2) **10** (1865), 145–150.

[1865m-345]   *Sur la forme $x^2 + y^2 + z^2 + t^2 + 4u^2 + 4v^2$,* L. J. (2) **10** (1865), 151–154.

[1865n-346]   *Sur la forme $x^2 + y^2 + z^2 + 2t^2 + 2u^2 + 4v^2$,* L. J. (2) **10** (1865), 155–160.

[1865o-347]   *Sur la forme $x^2 + y^2 + z^2 + t^2 + u^2 + 4v^2$,* L. J. (2) **10** (1865), 161–168.

[1865p-348]   *Sur la forme $x^2 + 4y^2 + 4z^2 + 4t^2 + 4u^2 + 16v^2$,* L. J. (2) **10** (1865), 203–208.

[1865q]   *Extrait d'une lettre adressée à M. Besge,* L. J. (2) **10** (1865), p. 234.

[1865r-349]   *Théorème concernant les nombres premiers contenus dans la formule $A^2 + 20B^2$,* L. J. (2) **10** (1865), 281–284.

[1865s-350]   *Théorème concernant les nombres premiers contenus dans la formule $A^2 + 36B^2$, en y prenant B impair,* L. J. (2) **10** (1865), 285–288.

[1865t-351]   *Théorème concernant les nombres premiers contenus dans la formule $A^2 + 44B^2$, en y prenant B impair,* L. J. (2) **10** (1865), 289–292.

[1865u-352]   *Théorème concernant les nombres premiers contenus dans la formule $A^2 + 56B^2$, en y prenant B impair,* L. J. (2) **10** (1865), 293–294.

[1865v-353]   *Théorème concernant les nombres premiers contenus dans la formule $A^2 + 116B^2$, en y prenant B impair,* L. J. (2) **10** (1865), 295–296.

[1865x-354]   *Sur les deux formes $x^2 + y^2 + 6z^2 + 6t^2$, $2x^2 + 2y^2 + 3z^2 + 3t^2$,* L. J. (2) **10** (1865), 359–360.

[1866a-355]   *Nombre des représentation d'un entier quelconque sons la forme d'une somme des dix carrés,* C. R. **60** (1865), 1257–1258; L. J. (2) **11** (1866), 1–8.

[1866b-356]   *Sur les deux formes $x^2 + 2y^2 + 2yz + 2z^2 + 15t^2$, $2x^2 + 2xy + 3y^2 + 3z^2 + 3t^2$,* L. J. (2) **11** (1866), 39–40.

[1866c-357]   *Théorèmes concernant les nombres premiers centenus dans la formule $4A^2 + 5B^2$, en y prenant A impair,* L. J. (2) **11** (1866), 41–48.

[1866d-358]   *Sur les deux formes $3x^2 + 5y^2 + 10z^2 + 10zt + 10t^2$, $2x^2 + 2xy + 3y^2 + 15z^2 + 15t^2$,* L. J. (2) **11** (1866), 103–104.

[1866e-359]   *Sur les deux formes $x^2 + 2y^2 + 2yz + 2z^2 + 6t^2$, $x^2 + 2y^2 + 3z^2 + 3t^2$,* L. J. (2) **11** (1866), 131–132.

[1866f-360]   *Sur les formes quadratiques proprement primitives, dont le déterminant changé de signe est > 0 et $\equiv 3 \,(mod.\,8)$,* L. J. (2) **11** (1866), 191–192.

[1866g-361]   *Sur la forme $x^2 + 3y^2 + az^2 + 3at^2$,* L. J. (2) **11** (1966), 211–216.

[1866h-362]   *Extrait d'une lettre adressée à M. Besge,* L. J. (2) **11** (1866), 221–224.

[1866i-363]   *Sur les deux formes $2x^2 + 3y^2 + 4z^2 + 4zt + 4t^2$, $x^2 + 2y^2 + 6z^2 + 6t^2$,* L. J. (2) **11** (1866), 280–282.

[1866j-364]   *Sur la forme à cinq indéterminées $x_1x_2 + x_2x_3 + x_3x_4 + x_4x_5$,* C. R. **62** (1866), 714–715; L. J. (2) **12** (1867), 47–48.

[1866k-365]   *Nombre des représentation d'un entier quelconque sous la forme d'une somme de dix nombres triangulaires,* C. R. **62** (1866), 771–773.

[1866l-366]   *Sur la fonction numérique qui exprime pour un déterminant négatif donné le nombre des classes de formes quadratuques dont un au moins des coefficients extrêmes est impair*, C. R. **62** (1866), 1350–1354; L. J. (2) **12** (1867), 98–103.

[1866]   Besge, *Extrait d'une lettre adressée à M. Liouville*, L. J. (2) **11** (1866), p. 328.

[1868a]   *Extrait d'une lettre adressée à M. Besge*, L. J. (2) **13** (1868), 1–4.

[1868b]   *Observations au sujet d'une article de M. Allégret*, C. R. **66** (1868), 1174–1175.

[1869a-368]   *Extrait d'une lettre adressée à M. Besge. Thórème concernant la fonction $F(k)$*, L. J. (2) **14** (1869), 1–6.

[1869b-369]   *Théorème concernant les nombres entiers $\equiv 5 \,(mod.\,12)$*, L. J. (2) **14** (1869), 7–8.

[1869c-370]   *Nouveux théorème concernant la fonction numérique $F(k)$*, L. J. (2) **14** (1869), 260–262.

[1869d-371]   *Remarque au sujet de la fonction $\zeta_1(n)$, qui exprime la somme des diviseurs de n*, L. J. (2) **14** (1869), 263–264.

[1869e-372]   *Extrait d'une lettre adressée à M. Besge. Valeur de l'intégrale définie $\int_0^1 \frac{arc\,tang\,x\,dx}{1+x}$*, L. J. (2) **14** (1869), 298–301.

[1869f-373]   *Théorème concernant la fonction numérique $\rho_2(n)$*, L. J. (2) **14** (1869), 302–304.

[1869g-374]   *Sur la forme ternaire $x^2 + 2y^2 + 3z^2$*, L. J. (2) **14** (1869), 359–360.

[1870a-375]   *Extrait d'une lettre adressée à M. Besge. Sur l'intégrale $A = \int_0^\infty f\left(x + \frac{1}{x}\right) arctang\,x \frac{dx}{x}$*, L. J. (2) **15** (1870), 7–8.

[1870b-376]   *Extrait d'une lettre adressée à M. V.-A. Le Besgue. Formule nouvelle dans la théorie des nombres*, L. J. (2) **15** (1870), 133–136.

[1870c]   *Sur l'emploi de gélatine dans l'alimentation*, C. R. **71** (1870), 759–760.

[1870d]   *Protestation faite par M. Liouville, en sa qualité de Président de l'Académie, à propos de l'arrestation récente de M. P. Thenard par l'armée prussienne*, C. R. **71** (1870), p. 911.

[1871]   *Le Président sortant rend compte à l'Académie de l'etat ou se trouve l'imprssion des Recueils qu'elle publie, et des changements survenus parmi les Membres et les Correspondants pendant l'année 1870*, C. R. **72** (1871), 14–17.

[1873-377]   *Sur quelques formules générales qui se rattachent à certaines formes quadratiques*, L. J. (2) **18** (1873), 142–144.

[1873]   Besge, *Sur une équation différentielle*, L. J. (2) **18** (1873), 139–142.

[1874a-378]   *Sur une intégrale définie*, L. J. (2) **19** (1874), 55–56.

[1874b-379]   *Extrait d'une lettre adressée à M. Besge [au sujet d'un mémoire d'Euler]*, L. J. (2) **19** (1874), 189–191.

1874a Besge, *Réponse à une lettre de M. Liouville*, L. J. (2) **19** (1874), p. 192.

[1874b]   Besge, *Extrait d'une lettre adressée à M. Liouville*, L. J. (2) **19** (1874), 423–424.

[1880-380]   *Leçons sur les fonctions doublement périodiques faites en 1847*, Journ. Reine Angew. Math. **88** (1880), 277–310.

Other References

Abel, N. H.

[1881]   *Oeuvres Complètes de Niels Henrik Abel*, Edition publiée par M. M. L. Sylow et S. Lie, tome I–II, Christiania 1881.

[1823]   *Oplösning af et Par Opgaver ved Hjelp af bestemte Integraler*, Magazin for Naturvidenskaberne, ser. 1, I (1923), 11–27. French translation in Oeuvres

[1824]   *Mémoire sur les équations algébrique, ou l'on démontre l'impossibilité de la résolution de l'équation générale du cinquième degré.* Brochure, Christiania 1824; Oeuvres

[1826a]   *Sur l'intégration de la formule différentielle $\frac{\rho dx}{\sqrt{R}}$, R et ρ étant des fonctions entières*, Journ. Reine Angew. Mat. 1 (1826), 185–221; Oeuvres I, 104–144.

[1826b]   *Auflösung einer mechanischen Aufgabe*, Journ. Reine Angew. Mat. 1 (1826), 153–157; French version in Ouevres I, 97–101.

[1826c]   *Beweis der Unmöglichkeit algebraische Gleichungen von höreren Graden als dem vierten allgemein aufzulösen*, Journ. Reine Angew. Math. 1 (1826), 65–85; French version Oeuvres I, 66–87.

[1826d]   *Recherches sur la série $1 + \frac{m}{1}x + \frac{m(m-1)}{1\cdot 2}x^2 + \frac{m(m-1)(m-2)}{1\cdot 2\cdot 3}x^3 + \cdots$*, Journ. Reine Angew. Math. 1 (1826); Ouevres Complètes I, 217–262.

[1827–1828]   *Recherches sur les fonctions elliptiques*, Journ. Reine Angew. Math. 2 (1827), 101–181;, 3 (1828), 160–190; Ouevres I, 263–388.

[1829a]   *Mémoire sur une classe particulière d'équations résolubles algébriquement*, Journ. Reine Angew. Math. 4 (1829), 131–156; Ouevres I, 478–507.

[1829b]   *Précis d'une théorie des fonctions elliptiques*, Journ. Reine Angew. Math. 4 (1829), 236–277; Ouevres I, 518–617.

[1830]   *Lettre à Legendre*, written November 25, 1828, Journ. Reine Angew. Math. 6 (1830), 73–80; Oeuvres II, 271–279.

[1839]   *Théorie des transcendantes elliptique*, first publ. in Holmboe's edition of Abel's Ouevres 1839; Ouevres; (ed. Sylow and Lie), II, 87–188.

[1841]   *Mémoire sur une propriété générale d'une classe très étendu de fonctions transcendantes*, Mem. Savants Etrangèrs Acad. Sci. Paris 7 (1841), 176–264; Ouevres I, 145–211.

Agulhon, M.

[1973]   *1848 ou l'apprentissage de la république 1848-1852*, Nouvelle Histoire de la France Contemporaine 8, Paris 1973.

d'Alembert, J.

[1747]   *Recherches sur la courbe que forme une corde tenduë mise en vibration*, Hist. Acad. Roy. Sci. et Belles Let. Berlin; (1747 publ. 1749),214–219.

[1752]   *Essai d'une nouvelle théorie de la Résistance des fluides*, Paris 1752.

[1766]   *Extrait de différentes lettres de M. d'Alembert à M. de La Grange écrit pendant les années 1764 & 1765*, Miscellanea Taurenensia 3 (1766), 381–396; Hist. Acad. Roy. Sci. et Belles Lettres Berlin; (1763 publ. 1770)235–277.

[1773]   *Sur la figure de la terre*, Opuscules Mathématiques 6 (1773), 47–67.

Ampère, A.-M.

[1806]   *Recherches sur quelques points de la théorie des fonctions dérivées qui conduisent a une nouvelle démonstration de la série de Taylor, et a l'expression finie des termes qu'on néglige lorsqu'on arrête cette série à un terme quelconque*, Journ. Ec. Polyt **6**. (13. cahier) (1806) 148–181.

[1824]   *Précis de calcul différentiel et intégral*, 1 vol in 4°, 152 pages. Unfinished printed lecture note. Archive of the École Polytechnique. $A^{III}a$ 174.

[1826]   *Théorie des phénomènes électrodynamiques, uniquement déduite de l'expérience*, Paris 1826. Reproduced with few modifications in Mem. Acad. Sci. Paris; (1823, publ. 1826).

[1835]   *Note sur la chaleur et sur la lumière considérées comme resultant de mouvement vibratioires*, Ann. Chimie et Physique **58** (1835), 432–444; Bibliothèque Universelle Genova **59** (1835), 26–37; translated in Phil. Mag. (3) **7** (1835), 342–349.

Appell, P.

[1909–1911, 1932–1937]   *Traité de Mécanique Rationnelle*, 3. ed., Paris Tome 1 (1909), Tome 2 (1911), Tome 4 (2. ed), Paris 1932–1937.

Arbogast, L. F. A.

[1791]   *Mémoire sur la nature des fonctions arbitraires qui entrent dans les intégrales des équations aux différences partielles*, Sct. Petersburg 1791.

Arnold, D. H.

[1981]   *Poisson and Mechanics*, 23–37; of *Siméon Denis Poisson et la Science de son Temps*. Paris 1981

Arnold, V. I.

[1978]   *Mathematical Methods of Classical Mechanics*, Springer-Verlag New York 1978; original Russian edition: Mathematicheskie metody klassicheskoi mekkaniki. Nauka Moscow 1974.

Aucoc, L.

[1889]   *L'Institut de France Lois. Status et Réglements*. Paris 1889

Bacharac, M.

[1883]   *Abriss der Geschichte der Potentialtheorie*, Göttingen 1883.

Beer, A.

[1856]   *Allgemeine Methode zur Bestimmung der elektrischen und magnetischen Induction*, Ann. Phys. Chem **98** (1856), 137–142.

[1865]   *Elektrostatik*, Braunschweig 1865.

Belhoste, B.

[1982]   *Augustin-Louis Cauchy et la pratique des sciences exactes en France au XIXeme siècle*, Thèse de 3eme cycle 2 vols. Paris 1982. To appear from Springer-Verlag in an English translation

[1985]   *Cauchy. Un mathématicien légitimiste au XIXe siècle*. Paris 1985

Belhoste, B., and Lützen, J.

[1984]   *Joseph Liouville et le Collège de France*, Rev. Hist. Sci **37** (1984), 255–304.

Bell, E. T.
[1937]   *Men of Mathematics,.* New York 1937

Beltrami, E.
[1864]   *Intorno ad alcune proprietà delle superficie di rivoluzione,* Ann. Mat.
p. appl. **6** (1864), 271–279; Werke **1**, 199–207.
[1868]   *Saggio di interpretazione della geometria non-euclidea,* Giorn. Mat. **6**
(1868), 284–312; Werke **1**, 374–405; French translation in Ann. Sci. Ec.
Norm. Sup. **6** (1869), 251–288.

Berg, C., and Lützen, J.
[1990]   *J. Liouville's Unpublished Work on an Integral Operator in Potential
Theory. A Historical and Mathematical Analysis,* To appear in Exposi-
tiones Mathematicae.

Bernkopf, M.
[1966–1967]   *The Development of Function Spaces with particular Reference to
their Origins in Integral Equation Theory,* Arch. Hist. Exact. Sci **3**
(1966–1967), 1–96.

Bernoulli, D.
[1732–1733]   *Theoremata de oscillationibus corporum filo flexili connexorum et
catenae verticaliter suspensae,* Comm. Acad. Sci. Petrop **6** (1732–1733),
108–122.
[1734–1735]   *Demonstrationes theorematum suorum de oscillationibus...,* Comm.
Acad. Sci. Petrop. **7** (1734–1735), 162–173.

Bernoulli, J.
[1728]   *Meditationes de chordis vibrantibus,* Comm. Acad. Sci. Petrop. **3** (1728),
13–28; Opera **3**, 198–210.

Berthelot, M. P. E.
[1888]   *Notice sur les origines et sur l'Histoire de la Société Philomatique,* Mémoires publiées par la Société philomatique à l'occasion du centenaire de
sa fondation 1788–1888, I–XVII. Paris 1888

Bertrand, J.
[1841]   *Note sur la vraie valeur des fractions qui prennent la forme $\frac{\infty}{\infty}$,* L. J **6**
(1841), 14–16.
[1843]   *Démonstration de quelques théorèmes sur les surfaces orthogonales,* Journ.
Ec. Polyt. **17**, 157–173. (29. cahier) (1843)
[1844]   *Mémoire sur les surfaces isothermes orthogonales,* L. J. **9** (1844), 117–132.
[1845]   *Mémoire sur le nombre de valeurs que peut prendre une fonction quand
on y permute les lettres qu'elle renferme,* Journ. Ec. Polyt. **18** (1845),
123–140.
[1846]   *Note relative au Mémoire précédent,* L. J. **11** (1846), 379–380.
[1848]   *Sur la courbe dont les deux courbures sont constants,* L. J. **13** (1848),
423–424.
[1852]   *L'intégration des équations différentielles de la mécanique,* L. J. **17**
(1852), 393–436.

[1867]   *Rapport sur les progrès les plus récents de l'analyse mathématique*, Paris (1867), 2–5.
[1889]   *Eloges Académiques.* Paris 1889
[1902]   *Eloges Académiques, Nouvelle Série.* Paris 1902

Besge, synonym for Liouville, cf. list of Liouville's works

Betti, E.
[1851]   *Sopra la risolubilità per radicali delle equazioni algebriche irriduttibili di grado primo*, Ann. Sci. Mat. Fis. **2** (1851); Opere **1** (1903), 17–27.
[1852]   *Sulla risoluzione delle equazioni algebriche*, Ann. Sci. Mat. Fis. **3** (1852); Opere **1** (1903), 31–80.

Bienaymé, J.
[1852]   *Mémoire sur la probabilité des erreurs d'après la méthode des moindres carrées*, L. J. **17** (1852), 33–78; Mém. Acad. Sci. Paris (2) **15** (1858), 615–663.

Biermann, K. R.
[1959]   *Johann Peter Gustav Lejeune Dirichlet. Dokumente für sein Leben und Wirken.* Abh. Deutschen Akad. d. Wiss. Berlin 1959 nr. 2, Berlin 1959
[1960]   *Vorschläge zur Wahl von Mathematikern in die Berliner Akademie*, Ein Beitrag zur Gelehrten- und Mathematikgeschichte des 19. Jahrhunderts; Abh. Deutschen Akad. d. Wiss. Berlin Mat. Phys. u. Techn. Kl. 1960 nr. 3. Berlin 1960

Bigourdan
[1928–1933]   *Le bureau des longitudes, son histoire et ses travaux de l'origine (1795) à ce jour.* Annuaire du Bureau des Longitudes 1828–1833

Billy, A.
[1959]   *Le Comte-Libri, Membre de l'Acad. des Sci.*; Historia, no. 154, Sept 1959, 268–273.

Binet, J.
[1839]   *Réflexions sur le problème de déterminer le nombre de manières dont une figure rectiligne peur être partagée en triangles au moyen de ses diagonales*, L. J. (1839), 79–90.
[1843]   *Note à la suite d'un article de M. Liouville relatif à un mémoire de N. Fuss*, L. J. **8** (1843), 394–396.
[1851]   *Note sur le mouvement du pendule simple en ayant égard à l'influence de la rotation diurne de la terre*, C. R. **32** (1851), 157–159.

Biot, J. B.
[1839]   *Mémoire sur la mesure théoretique et expérimentale de la réfraction terrestre, avec son application à la détermination exacte des différences de niveau, d'après les observations des distances zénithales simples ou reciproques*; Connais. des Temps pour 1842 (1839), 3–80.
[1841]   *Réponse de M. Biot à M. Libri*, C. R. **12** (1841), p. 523.

Birkhoff, G.
[1973]   *A Source Book in Classical Analysis.* Cambridge, Mass. 1973.

Bjerknes, C. A.
[1858]   *Indberetning til Departementet for Kirke- og Undervisningsvjsnet an-
          gaaende en Reise i Udlandet for at studere den rene Matematik.* Manu-
          script communicated to me by Prof. J. E. Roos.

Bjerknes, V.
[1925]   *C. A. Bjerknes. Hans Liv og Arbeide.* Oslo, 1925.

Blanchet, M. P. H.
[1840]   *Mémoire sur la propagation et la polarisation du mouvement dans un
          milieu élastique indéfini,* L. J. 5 (1840), 1–30.

Blaschke, W.
[1923]   *Vorlesungen Über Differentialgeometri.* Berlin 1923.

Blaschke, W. and Reinhardt, H.
[1960]   *Einführung in die Differentialgeometri;* 2. ed. Berlin 1960..

Blignières, C.
[1857]   *Exposition de la Philosophie et de la Religion positive.* Paris 1857.

Blondel, C.
[1982]   *A.-M. Ampère et la création de l'électrodynamique (1820–1827);* . Paris
          1982.

Bôcher, M.
[1898–1899]   *The theorems of oscillation of Sturm and Klein,* N. Y. Bull. Amer.
               Math. Soc. 4 (1898), 295–313, 365–376;, 5 (1899), 22–43.
[1900]   *On Sturm's theorem of comparison,* N. Y. Bull. Amer. Math. Soc. 6
          (1900), 96–100.
[1899–1916]   *Randwertaufgaben bei gewöhnlichen Differentialgleichungen,* Encyk-
               lopädie der mathematischen Wissenschaften II A 7a II **1.1,** 437–463.
[1911–1912]   *The published and unpublished work of Charles Sturm on algebraic
               and differential equations,* Bull. Amer. Math. Soc. **18** (1911–1912), 1–18.
[1912]   *Boundary problems in one dimension,* Proceedings of the Fifth interna-
          tional Congress of Mathematicians. Cambridge, Mass. 1912.
[1917]   *Leçons sur les méthodes de Sturm dans la théorie des équations différen-
          tielles linéaires et leurs développements modernes.* Paris 1917.

Boltzmann, L.
[1866]   *Über die mechanische Bedeutung des zweiten Haupsatzes der Wärm-
          etheorie,* Sitzungsber. Wien Akad. **53** (1866), 195–220; Wiss. Abh. I,
          9–33.
[1871a]   *Über das Wärmegleichgewicht zwischen mehratomigen Gasmolekülen,*
           Sitzungsber. Wien Akad. **63** (1871), 397–416; Wiss. Abh. I, 237–258.
[1871b]   *Einige allgemeine Sätze über Wärmegleichgewicht,* Sitzungsber. Wien
           Akad. **63** (1871), 676–711; Wiss. Abh. I, 259–287.
[1896–1898]   *Vorlesungen über Gastheorie,* 1. Teil Leipzig 1898, 2. Teil Leipzig
               1898; Lectures on Gas Theory, transl. Stephen G. Brush, Berkeley 1969.

818   Bibliography

Bonnet, O.
[1848]   *Mémoire sur la théorie générale des surfaces*, Journ. Éc. Polyt. **19**. (32. cahier) (1848), 1–46.
[1849]   *Lettre de M. O. Bonnet concernant l'emprunt qu'il a fait à un travail inédit de M. Liouville d'une expression employée dans un travail imprimé dont il adressé un exemplaire à l'Académie*, C. R. **28** (1849), 448–449.
[1885]   *Discours de M. O. Bonnet au nom de l'Académie et de la Faculté des Sciences, au Funérailles de M. Serret Membre de l'Académie le jeudi 5. mars 1885*, C. R. **100** (1885), 677–680.

Born, M.
[1925]   *Vorlesungen über Atommechanik*. vol. 1 Berlin 1925.

Bos, H. J. M., Kers, C., Oort, F. and Raven, D. W.
[1984]   *Poncelet's Closure Theorem, its history, its modern formulation, a comparison of its modern proof with those by Poncelet and Jacobi, and some mathematical remarks inspired by these early proofs*; Preprint nr. 353 Utrecht University, Dept. of Math. (1984.).

Bottazzini, U.
[1978]   *Ricerche di P. Tardy sui differenziali di indice qualunque (1844–1868)*, Historia Mathematica **5** (1978), 411–418.
[1983]   *La matematica e le sue "utili applicazioni" nei congressi degli scienziati italiani, 1839–1847*; I congressi degli scienziati italiani nell' età del positivismo, a cura di Giuliano Pancaldi, Bologna, CLUEB 1983 11–68.
[1986]   *The Higher Calculus: A History of Real and Complex Analysis from Euler to Weierstrass*. New York 1986.

Boulanger, A.
[1897]   *Contribution à l'ètude des équations différentielles linéaires homogènes intégrables algébriquement*, These, fac. sci. Paris 1897; Journ. Ec. Polyt. Paris 2. ser, 4. cahier (1898), 1–122.

Bour, E.
[1855]   *L'intégration des équations différentielles de la mécanique analytique*, L. J. **20** (1855), 185–200; Mem. Sav. Etrang. Acad. Sci. Paris **14** (1856), 792–812.
[1862]   *Sur l'intégration des équations différentielles partielles du premier et du second ordre*, C. R. **54** (1862), 439–444, 549–554, 588–593, 645–655; Journ. Ec. Polyt. **22** (39. cahier). (1982), 149–191.

Brioschi, F.
[1877]   *La théorie des formes dans l'intégration des équations différentielles du deuxième ordre*, Math. Ann. **11** (1877), 401–411.

Briot, C.-A.-A and Bouquet, J.-C.
[1855]   *Recherches sur les fonctions doublement périodiques*, C. R. **40** (1855), 342–344..
[1859]   *Théorie des fonctions elliptiques*. Paris 1859.

References   819

Brocard, H.
[1902–1907]   *Notes biographiques sur J. Liouville*, Intermédiaire des mathématiciens **9** (1902), 215–217;, **13** (1906), 13–15; vol 14, (1907), 59–61.

Brusch, S. G.
[1970]   *The Wave Theory of Heat*, British Jrn. Hist. Sci. **15** (1970), 145–167.

Buchwald, J. Z.
[1976]   *Thomson, Sir William (Baron Kelvin of Largs)*, Dictionary of Scientific
Biography **13**, 374–388.

Burgatti, P.
[1911]   *Determinazione dell' equazioni di Hamilton Jacobi integrabili mediante
la seperazione delle variabili*, Rom. Linc. Rend. (5) **20** (1911), p. 108.

Burkhardt, H. and Meyer, W. F.
[1899–1916]   *Potentialtheorie*, Enclopädie der Math. Wiss. **II A7b** (1899–1916),
464–503.

Butzer, P.L., and Jongmans, F.
[1989]   *P.L. Chebyshev (1821–1894) and his Contacts with Western European
Scientists,,* Historia Mathematica **16** (1989), 46–68.

Butzer, P.L. and Westfall, V.
[1975]   *An Access to Fractional Differentiation via Fractional Difference Quotients*; Fractional Calculus and its Applications, ed. B. Ross, Berlin 1975,
116–145.

Campbell, R.
[1971]   *La mécanique analytique.* Paris 1971

Cantor G.
[1874]   *Über eine Eigenschaft des Inbegriffes aller reellen algebraischen Zahlen*,
Journ. Reine Angew. Math. **77** (1874), 258–262; Ges. Abhandl., 115–118.

Cantor, M.
[1898]   *Vorlesungen über Geschichte der Mathematik.* vol. 3 Leipzig, 1. ed. 1898,
2. ed. 1901, reprinted New York 1965.

Catalan, E.
[1839]   *Solution nouvelle de cette question: Un polygone étant donné, de combien
de manières peut-on le partager en triangles au moyen de diagonales*, L.
J. 4 (1839), 91–94.

Cauchy, A. L.
[1815]   *Mémoire sur le nombre des valeurs qu'une fonction peut acquérir lorsqu'-
on permute de toute les manières possibles les quantités qu'elle renferme*,
Journ. Ec. Polyt. 10 (17. cahier) (1815), 1–28; Oeuvres (2) **1**, 64–90.
[1820]   *Rapport sur un Mémoire de M. Poncelet relatif aux propriétés projectives des sections coniques*, Procès-Verbaux de l'Académie des Sciences
7 (1820–1823), p. 57; Ann. Math. Pures et Appl. 11 (1820), 69–83; Oeuvres (2) **2**, 329–342. (see also 388–396)
[1821]   *Cours d'analyse de l'École Royale Polytechnique*, 1er partie: Analyse

algébrique, Paris 1821; Oeuvres (2) **3**.

[1823]   *Résumé des leçons données à l'École Royale Polytechnique sur le calcul infinitésimal*, Tome premier Paris 1823; Oeuvres (2) **4**, 5–261.

[1824–1981]   *Equations Différentielles Ordinaires*, Paris. New York 1981. Fragment of lecture notes for the second year at the École Polytechnique. Introduction by C. Gilain.

[1826]   *Leçons sur les applications du calcul infinitésimal a la géométrie*, Paris 1826; Oeuvres (2) **5**.

[1829]   *Leçons sur le calcul différentiel*, Paris 1829; Oeuvres (2) **4**, 263–609.

[1831]   *Extrait du mémoire présenté à l'Académie de Turin de 11 octobre 1831*; Oeuvres (2) **15**, 262–441.

[1835–1840]   *Mémoire sur l'intégration des équations différentielles*, Exercices d'analyse et de physique mathématique 1 (1840) 327–384; Oeuvres (2) **11**, 399–465. A small number of lithographed versions of this paper appeared in 1835.

[1937]   *Extrait d'une lettre à M. Coriolis*, C. R. **4** (1837), p. 216; Oeuvres (1) **4**, 38–42.

[1839a]   *Mémoires sur les mouvements infiniment petits d'un système de molécules sollicitées par des forces d'attraction ou de répulsion mutuelle*, C. R. **8** (1839), 505–522, 589–597, 659–673, 767–778; Oeuvres (1) 4, 237–312; *Exercices d'analyse et de physique mathématique*, **1** (1840), 1–15; Oeuvres (2) **11**, 11–28.

[1839b]   *Mémoire sur l'intégration des équations différentielles des mouvements planétaires*, C. R. **9** (1839), 184–190; Oeuvres (1) **4**, 483–490.

[1840a]   *Théorèmes divers sur les résidues et les non-résidues quadratiques*, C. R. **10** (1840), 437–452; Oeuvres (1) **5**, 135–152.

[1840b]   *Sur quelques séries dignes de remarque, qui se présentent dans la théorie des nombres*, C. R. **10** (1840), 719–731; Oeuvres (1) **5**, 199–212.

[1840c]   *Méthodes générales pour la détermination des mouvements des planètes et de leur satellites*, C. R. **11** (1840), 179–184; Oeuvres (1) **5**, 260–266.

[1840d]   *Mémoire sur l'intégration des équations linéaires*, Exercices d'Analyse 1 (1840) 53–100; Oeuvres (2) **11**, 75–133.

[1842]   *Note sur la réflexion de la lumière à la surface des métaux*, L. J. **7** (1842), 338–344; Oeuvres (2) **2**, 338–344.

[1844a]   *Mémoire sur les fonctions continuées ou discontinuées*, C. R. **18** (1844), 116–130; Oeuvres (1) **8**, 145–160.

[1844b]   *Mémoire sur quelques propositions fondamentales du calcul des résidues, et sur la théorie des intégrales singulières*, C. R. **19** (1844), 1337–1344; Oeuvres (1) **8**, 366–375.

[1846a]   *Note sur le développement des fonctions en séries ordonnées suivant les puissances ascendantes des variables*, L. J. **11** (1846), 313–330; Oeuvres (2) **2**, 35–54.

[1846b]   *Considérations nouvelles sur les intégrales définies qui s'étend à tous les points d'une courbe fermée et sur celles qui sont prises entre des limites imaginaires*, C. R. **23** (1846), p. 689; Oeuvres (1) **10**, 153–168.

[1849]   *Mémoire sur les intégrales continues et les intégrales discontinues des équations aux dérivée partielles*, C. R. **29** (1849), 548–557; Oeuvres (1) **11**, 172–183.

[1851a] *Rapport sur un Mémoire présenté par M. Hermite et relatif aux Fonctions aux double période*, C. R. **32** (1851), 442–450; Oeuvres de Hermite **1**, 75–83.

[1851b] *Note de M. Augustin Cauchy relative aux observations présentées à l'Académie par M. Liouville*, C. R. **32** (1851), 452–454; Oeuvres (1) **11**, 373–376.

[1851c] *Sur l'influence souvent exercée par des circonstances étrangères à la science dans la solution des questions qui paraissaeint purement scientifiques, et sur le pouvoir attribué, dans une élection récente, à un billet blanc*, Arch. Ac. Sci. 13 janvier 1851. Note lithograpiée.

Chandrasekhar, S.

[1969] *Ellipsoidal Figures of Equilibrium*. New Haven 1969.

Chasles, M.

[1837a] *Aperçu Historique sur l'origine et le dévelopement des méthodes en géométrie particulièrement celles qui se rapportent a la géométrie moderne*. Bruxelles 1837.

[1837b] *Mémoire sur l'attraction d'une couche ellipsoidale infiniment mince et les rapports qui ont lieu entre cette attraction et les lois de la chaleur en mouvement dans un corps en équilibre de tempétature*, Journ. Ec. Polyt **15** (25. cahier) (1837), 266–316.

[1839] *Énoncé de deux théorèmes généraux sur l'attraction des corpts et la théorie de la chaleur*, C. R. **8** (1839), 209–211.

[1842] *Théorèmes généraux sur l'attraction des corps*, Connaiss. des Temps. pour l'an 1845 (1842), 18–33.

[1843] *Des arcs d'une section conique dont la différence est rectifiable; et des polygons de perimètre minimum circonscrits à une conique*, C. R. **17** (1843), 838–844.

[1846a] *Sur les lignes géodésiques et les lignes de courbure des surfaces du second degré*, L. J. **11** (1846), 5–20; C. R. **22** (1846), 63–72.

[1846b,c] *Nouvelle démonstration de deux équations relatives aux tangentes communes à deux surfaces du second degré homofocales; — Et propriétés des lignes géodésiques et des lignes de courbure de ces surfaces*, L. J. **11** (1846), 105–119; C. R. **22** (1846), p. 313 and 517.

[1846d] *Notes sur quelques questions de priorité, au sujet d'un Mémoire de M. Mac Cullagh*, L. J. **11** (1846), 120–123.

[1846e] *Mémoire sur l'attraction des ellipsoides*, Mem. Savants Étrang. Acad. Sci. Paris (Sci. math. phys.) **9** (1846), 629–715.

[1870] *Rapport sur les progrès de la géométrie*. Paris 1870.

Chebyshev, P. L.

[1847] *On Integration by Means of Logarithms*, Thesis in Russian; Presented 1847 to Univ. of St. Petersburg. Published 1930.

[1852] *Rapport du professeur extraordinaire de l'université de St. Pétersbourg Tchebychef sur son voyage à l'étranger*, in Russian 1852. French translation in; Oeuvres de P. L. Tchebychef **2**. VII–XVIII

[1853] *Sur l'intégration des différentielles irrationnelles*, Journ. Math. Pures Appl. **18** (1853), 87–111; Oeuvres **1**, 147–168.

[1857]   *Sur l'intégration des différentielles qui contiennent une racine carrée d'une polynome du troisième ou du quatrième degré*, L. J. (2) **2** (1857), 1–42; Mem. Acad. Imp. Sci. St. Pétersbourgh (6) **6** (1857), 203–232; Oeuvres **1**, 171–200.

[1860]   *Sur l'intégration des différentielles irrationnelles*, C. R. **60** (1860), 46–48; L. J. (2) **9** (1864), 242–247; Oeuvres **1**, 511–514.

[1861]   *Sur l'intégration de la différentielle* $\dfrac{x+A}{\sqrt{x^4+\alpha x^3+\beta x^2+\gamma x+\delta}}dx$, Bull. Acad. Imp. Sci. St. Pétersbourg **3** (1861), 1–12; L. J. (2) **9** (1864), 225–241; Oeuvres **1**, 517–530.

[1865]   *Sur l'intégration des différentielles qui contiennent une racine cubique*, in Russian, Bull. Acad. Imp. Sci. St. Pétersbourg **7** (1865 no. 5), 563–608; French translation in Oeuvres **1**.

[1867]   *Sur l'intégration des différentielles les plus simples parmi celles qui contiennent une racine cubique*, in Russian, Math. Sbornik **2** (1867), 71–78; French translation in Oeuvres **2**, 43–47.

[1951]   *Complete Collection of the Works of P. L. Chebyshev*, (in Russian); Moskva and Leningrad 1951.

Chevalier, A.
[1832]   *Nécrologie, Evariste Galois*, Révue encyclopédique **55** (September 1832), 744–754.

Chrystal, G.
[1888]   *Joseph Liouville*. Proc. Roy. Soc. Edinburgh, **14** (1888), second pagination 83–91.

Clairaut, A.-C.
[1743]   *La théorie de la figure de la terre, tirée des principes de l'hydrostatique*, Paris 1743; Ostwalds Klassiker nr. 189.

Codazzi, D.
[1857]   *Intorno alle superficie le quali hanno constante il prodotto de' due raggi di curvature*, Ann. Sci. Mat. Fis. Roma **8** (1857), 346–355.

Colladon, D. and Sturm, C.
[1827–1834]   *Mémoire sur la compression des liquides*; Mém. Savans Étrang. Acad. Sci. Paris **5** (1834), p. 267 ff; Ann. Chémie et Phys. **22**, p. 113 ff.

Condorcet, J.-M. Marquis de
[1765]   *Du calcul intégrale*, Paris 1765.

Cougny, G. and Robert, A.
[1890]   *Article on Liouville*, in, Dictionnaire des parlementaires français **2** (Paris 1890), p. 165.

Dall'Acqua, F.A.
[1909]   *Sulla integrazione delle equazioni di Hamilton-Jacobi per separazione di variabili*, Math. Ann. **66** (1909), 398–415.

[1912]   *Le equazioni di Hamilton-Jacobi, che si integrano per separazione di variabili*, Rendiconti del circolo matematico **33** (1912), 341–351.

Darboux, G.
[1865]   *Recherches sur les surfaces orthogonales*, Ann. Sci. Ec. Norm. Sup. **2**
          (1865), 55–69.
[1888]   *Leçons sur la théorie générale des surfaces*, 2.ème partie; Paris 1888 (2.
          ed. 1915).
[1894]   *Leçons sur la théorie générale des surfaces*, 3.ème partie; Paris 1894.

Daston, L. J.
[1986]   *The Physicalist Tradition in Early Nineteenth Century French Geome-
          try*, Stud. Hist. Phil. Sci. **17** (1986), 269–295.

Davenport, J. H. and Singer, M. F.
[1986]   *Elementary and Liouvillian Solutions of Linear Differential Equations*,
          Journ. Symbolic Computation. **2** (1986), 237–260.

David
[1883]   *"Liouville" Obituary*, Mem. Acad. Sci. incript et belles-lettres Toulouse
          (8) **5** (1883), 257–258.

Deakin, M. A. B.
[1981]   *The Development of the Laplace Transform 1737–1937*, 1. Euler to Spitzer
          1737–1880, Arch. Hist. Ex. Sci. **25** (1981), 343–390.

Demidov, S. S.
[1980]   *The evolution of the theory of partial differential equations of the first
          order*, (in Russian), Istoriko-Matem. Issledovaniya **25** (1980), 71–103.
[1981]   *Des parentheses de Poisson aux algèbres de Lie*; Siméon-Denis Poisson
          et la science de son temps Ec. Polyt. Palaiseau 1981.
[1982]   *The Study of Partial Differential Equations of First order in the 18th
          and the 19th Centuries*, Arch. Hist. Ex. Sci. **26** (1982), 325–350.
[1983]   *On the History of the Theory of Linear Differential Eqautions*, Arch.
          Hist. Ex. Sci. **28** (1983), 369–387.
[1987]   *Ordinary Differential Equation*, Mathematics in the 19th Century. Cheby-
          shev's works on function theory; ed. Kolmogorov and Youschkevich in
          Russian, Moskva. (1987), 80–183.

Descartes, R.
[1637]   *La Géometrie*; Leiden 1637, Dover reprint with English translation 1954.

Dini, U.
[1865]   *Sulle superficie di curvatura constante*, Giorn. Mat. **3** (1865), 241–256.
[1880]   *Serie di Fourier ed altre rappresentazioni analitiche delle funzioni di una
          variabile reale*. Pisa 1880.

Dieudonné, J.
[1978]   *Abrégé d'Histoire des Mathematiques 1700–1900*. I + II Paris 1978.
[1981]   *History of Functional Analysis*. Amsterdam 1981.

Dirichlet, P. G. Lejeune
[1829]   *Sur la convergence des séries trigonométriques qui servent à représenter
          une fonction arbitraire entre des limites données*, Journ. Reine Angew.
          Math. **4** (1829), 157–169; Werke I, 117–132.

[1835]   *Über eine neue Anwendung bestimmter Integrale auf die Summation endlicher und unendlicher Reihen*, Abh. Königl. Preuss. Akad. Wiss. (1835), 391–407; Werke **1**, 237–356.

[1836]   *Sur les intégrales eulériennes*, Journ. Reine Angew. Math. **15** (1836), 258–263; Werke **1**, 271–278.

[1837a]   *Über die Darstellung ganze willkürlicher Funktionen durch sinus und cosinus-Reihen*, Repert. Phys. **1** (1837), 157–174; Werke **1**, 133–160; Ostwalds Klassiker n° 116, 3–34.

[1837b]   *Beweis des Satzes, dass jede unbegrenzte arithmetische Progression, deren erstes Glied und Differenz ganze Zahlen ohne gemeinschaftlichen Factor sind, unendlich viele Primzahlen anthält*, Abh. Königl. Preuss. Akad. Wiss. (1837), 45–81; Werke **1**, 313–342.

[1837c]   *Sur l'usage des intégrales définies dans la sommation des séries finies ou infinies*; (French version of [Dirichlet 1835]); Journ. Reine Angew. Math. **17** (1837), 57–67; Werke **1**, 257–270.

[1837d]   *Sur les séries, dont le terme général dépend de deux angles, et qui servent à exprimer des fonctions arbitraires entre des limites données*, Journ. Reine Angew. Math. **17** (1837), 35–56; Werke **1**, 273–306.

[1838]   *Sur l'usage des séries infinies dans la théorie des nombres*, Journ. Reine Angew. Math. **18** (1838), 259–274; Werke **1**, 357–374.

[1840]   *Extrait d'une lettre de M. Lejeune-Dirichlet à M. Liouville*, C. R. **10** (1840), 285–288.

[1856]   *Sur l'équation* $t^2 + u^2 + v^2 + w^2 = 4m$, L. J. (2) **1** (1856), 210–214.

[1856–1857/1876]   *Vorlesungen über die im umgekehrten Verhältniss des Quadrats der Entfernung wirkenden Kräfte*; Herausgegeben von Dr. F. Grube, Leipzig, 1876.

[1857]   *Untersuchungen über ein Problem der Hydrodynamik*, Nachr. Königl. Gess. zu Wiss. Göttingen (1857), 205–207; Werke **2**, 215–218.

[1858–1859]   *Untersuchung über ein Problem der Hydrodynamik*, Abh. Königl. Gess. der Wiss. Göttingen **8** (1858–1859), 3–42; Math. Cl., 3–42; Journ. Reine Angew. Math. **58** (1861), 181–216; Werke **2**, 263–301.

[1862]   *Démonstration d'un théorème d'Abel*, Note de M. Lejeune Dirichlet communiquée par M. Liouville, L. J. (2) **7** (1862), 253–255; Werke **2**, 305–306.

Dolbnia, J.

[1890]   *Sur les intégrales pseudo-elliptiques d'Abel*, L. J. (4) **6** (1890), 293–311.

Donkin, W. F.

[1854]   *On a Class of Differential Equations, including those which occur in Dynamical Problems*, Part I, Phil. Trans. Roy. Soc. London. **144** (1854), 71–113.

Drach, J.

[1898]   *Essai sur une théorie générale de l'intégration et sur la classification des transcendantes*, Ann. Sci. Ec. Norm. (3) **15** (1898), 243–384.

Dugas, R.

[1950]   *Histoire de la mécanique.* Neuchatel 1950.

[1959]   *La théorie physique au sens de Boltzmann et ses prolongements moderne.* Neuchatel 1959.

Duhamel, J.-M.-C.

[1833]   *Sur la méthode générale relative au mouvement de la chaleur dans les corps solides plongés dans des milieux dont la température varie avec le temps,* Journ. Ec. Polyt. **14** (22. cahier) (1833), 20–77.

[1939a] *Note sur les surfaces isothermes dans les corps solides dont la conductibilité n'est pas la même dans tous les sens,* L. J. **4** (1839), 63–78.

[1839b] *Nouvelle règle pour la convergence des séries,* L. J. **4** (1839), 214–221.

Dulong, P. L. and Petit, A.-T.

[1820]   *Recherches sur la mesure des température et sur les lois de la communication de la chaleur,* Journ. Ec. Polyt. **11** (18. cahier) (1820), 189–294; Ann. de Chimie et Phys. **7** (1817), 113–154, 225–249, 337–367.

Dumont, C.-E.

[1843]   *Histoire de la ville et des seigneurs de Commercy.* vol. 3, Bar le-Duc 1843.

Dupin, C. F. P.

[1813]   *Développements de géométrie.* Paris 1813.

Edwards, H. M.

[1975]   *The Background of Kummer's Proof of Fermat's Last Theorem for regular primes,* Arch. Hist. Ex. Sci. **14** (1975), 219–236.

[1977]   *Postscript to "The Background of Kummer's Proof",* Arch. Hist. Ex. Sci. **17** (1977), 381–394.

[1984]   *Galois Theory.* New York 1984.

[1989]   *A Note on Galois Theory.* 10pp., to appear.

Eisenstein, G.

[1844]   *Bemerkungen über die elliptischen und Abelschen Transcendenten,* Journ. Reine Angew. Math. **27** (1844), 185–192; French translation L. J. **10** (1845), 445–450.

[1847]   *Genaue Untersuchung der unendlichen Doppelproducte aus welchen die elliptischen Funktionen als Quotienten zusammengesetzt sind,* Journ. Reine Angew. Math. **35** (1847), 153–274; Werke **1**, 357–478.

Encyclopedic Dictionary of Mathematics, by the Mathematical Society of Japan. Cambridge Mass. 1977.

Euler, L.

[1730–1731] *De progressionibus transcendentibus, seu quarum termini algebraice dari nequent,* Comment. Acad. Sci. Imp. Petropol. **5 1730–1731** (1738), 38–57; Opera Omnia (1) **14**, 1–24.

[1737]   *De fractioniobus continuis dissertatio,* Comment. Acad. Sci. Petropol. **9** (1737; 1744), 98–137; Opera Omnia (1) **14**, 187–215.

[1748]   *Introductio in alalysin infinitorum,* (2 vols); Lausanne 1748; Opera Omnia (1) **8–9**.

[1764–1766] *De motu corporis ad duo centra virium fixa atracti.* Novi Comm. Acad. Sci. Petrop. **10** (1764 publ. 1766), 207–241; and **11** (1765 publ. 1767), 152, 184; Opera Omnia (2) **6**, 209–246 and 247–273.

[1766–1767] *De motu corporis ad duo centra virium fixa atracti,* and, Novi Comm. Acad. Sci. Petrop. **11** (1764 publ. 1766), 152–184; Opera Omnia (2) **6**,

209–246 and 247–273..

[1767]   *Problème: Un corps etant attiré en raison réciproque quarrée des distances vers deux points fixes donnes trouver les cas ou la courbe decrite par ce corps sera algé- brique. Resolu par M. Euler*, Mem. Acad. Sci. Berlin **16** (1760 publ. 1767), 228–249; Opera Omnia (2) **6**, 274–293.

[1781–1830]   *De infinitis curvis algebraicis, quarum longitudo indefinita arcui elliptico aequatur*, Mém. acad. Sci. St. Pétersbourg **11** (1830), 95–99; Opera Omnia (1) **21**, 241–245.

Euler, L., and Goldbach, C.

[1965]   "Leonhard Euler und Christian Goldbach Briefwechsel 1729–1764," Ed. A. P. Juškevič and E. Winter, Akademie Verlag Berlin 1965.

Favre-Rollin, A. M.

[1836]   *Intégration de l'équation* $\frac{d^{p/q}V}{dx^{p/q}} + \frac{P d^n y}{dx^m} + \frac{Q d^n y}{dx^n} + etc. = V$ *dans laquelle on suppose p, q, m, n, etc., des nombres entiers; P, Q, etc. des coefficients constants et V une fonction quelconque de la variable indépendante x,* L. J. **1** (1836), 339–340.

Faye, H.

[1882]   *Discours de M. Faye prononcés aux funérailles de M. Liouville, au nom de l'Académie des Sciences, de la Faculté des Sciences de Paris et du Bureau des Longitudes*, C. R. **95** (1882), 468–469.

Fontaine, A.

[1764]   *Le calcul intégral*, Mémoire donnés à l'Académie Royale des Sciences non imprimés dabs leur temps par M. Fontaine de cette Académie; Paris 1764, Premier méthode 24–83 ; Seconde méthode 84–236.

Fourcy, A.

[1828]   *Histoire de l'École Polytechnique.* Paris 1828.

Fourier, J.

[1807]   *Sur la propagation de la chaleur.* Published in Grattan-Guinness 1972.

[1820]   *Sur l'usage du théorème de Descartes dans la recherche des limites des racines*; Bull. Sci. Soc. Philomatique Paris, October (1820), 156–165; December, (1820), 181–87; Oeuvres **2**, 289–309.

[1822]   *Théorie analytique de la chaleur.* Paris 1822.

[1831]   *Remarques Générales sur l'application des Principes de l'analyse algéybrique aux équations transcendentes*, Mém. Acad. Sci. **10** (1831), 119–146. [1829]

Fox, R.

[1974]   *The Rise and Fall of Laplacian Physics*, Historical Studies in the physical Sciences **4** (1974), 89–136.

Fredholm, I.

[1900]   *Sur une nouvelle méthode pour la résolution du problème de Dirichlet*, Öfversigt af Kungliga Svenska Vetenskabs-Akademiens Förhandlinger; Stockholm **57** (1900), 39–46; Oeuvres Complètes, 61–68.

[1903]   *Sur une classe d'équations fonctionnelles*, Acta. Math. **27** (1903), 365–390;

Oeuvres Complètes, 81–106.

Frenet, F. J.

[1852]   *Sur les courbes à double courbure*, L. J. **17** (1852), 437–447.

[1853]   *Théorèmes sur les courbes gauches*, Nouvelle Annales de Math. **12** (1853), 365–372.

[1864]   *Correspondance*, Nouvelle Annales de Math. (2) **3** (1864), 284–286.

Freudenthal, H.

[1972]   *Hermite, Charles,*, biographies in Dictionary of Scientific Biography (ed. Gillispie) **6**, 306–308.

Fricke, R.

[1913]   *Elliptische Funktionen*, Encyclopädie der Mathematischen Wissenschaften **II** 2, 177–348.

Fuchs, L.

[1876–1878]   *Über die linearen Differentialgleichungen zweiter Ordnung, welche algebraische Integrale besitzen, und eine neue Anwendung der Inwariantentheorie*; Journ. Reine Angew. Math. 1. Abhandlung **81** (1876), 97–147; 2. Abhandlung, **85** (1878), 1–25.

Fuss, P. H.

[1843]   *Correspondance mathématique et physique de célèbres géomètres*; Ed. P. H. Fuss I, II . Sct. Petersburg 1843.

Galois, E.

[1831–1846]   *Mémoire sur les conditions de résolubilité des équations par radicaux*, L. J. **11** (1846), 417–433; slightly revised version of a paper communicated to the Académie des Sciences in 1831.

[1832]   *Lettre à Auguste Chevalier, 29 mai 1832.* 173–185 of [Galois 1962].

[1846]   *Oeuvres Mathématique d'Évariste Galois*, L. J. **11** (1846), 381–444.

[1906]   *Manuscrits de Evariste Galois*, publié par J. Tannery Paris 1908; Bull. Sci. Math. (2) 30 and 31 (1906–1907).

[1962]   *Ecrits et Mémoires Mathématiques d'Evariste Galois*, ed. R. Bourgne and J. P. Azra, Paris 1962.

Gauss, C. F.

[1799]   *Demonstratio nova theorematis omnem functionem algebraicam rationalem integram unius variabilis in factores reales primi vel secundi gradus resolvi posse*, Dissertation, Helmsted 1799; Werke **3**, 3–20; Ostwald's Klassiker **14**; ed Netto 3–36.

[1801]   *Disquisitiones Arithmeticae*, Braunschweig 1801; Werke **1**; English translation, New Haven and London 1966.

[1811]   *Summatio quarundam serierum singularium*, Comment soc. reg. sci. Gottengensis recentiores. 1 (1811) Werke 2 9–45.

[1813]   *Disquisitiones generales circa seriem infinitam* $1 + \frac{\alpha\beta}{1\cdot\gamma}x + \frac{\alpha(\alpha+1)\beta(\beta+1)}{1\cdot2\cdot\gamma(\gamma+1)} + \frac{\alpha(\alpha+1)(\alpha+2)\beta(\beta+1)(\beta+2)}{\gamma(\gamma+1)(\gamma+2)}x^3 + etc.$; Comment. soc. reg. sci. Gottengensis recentiores **2** (1813); Werke **3**, 123–162.

[1816]   *Theorematis de resolubilitate functionum algebraicarum integrarum in factores reales demonstratio tertia*; Comment. soc. reg. sci. Gottengensis recentiores;, **3**; (ser 3) 1816; Werke **3**, 57–64; Ostwald's Klassiker, **14**; ed. Netto 61–67.

[1822–1825] *Allgemeine Auflösung der Aufgabe: Die Theile einer gegebenen Fläche auf einer andern gegebenen Fläche so abzubilden, dass die Abbildung dem Abgebildeten in den kleinsten Theilen ähnlich wird (Als Beantwortung der von der königlichen Societät der Wissenschaften in Copenhagen für 1822 afgegebenen Preisfrage)*, Astron. Abh. (1825), 1–30; Werke **4**, 189–216.

[1828] *Disquisitiones generales circa superficies curvas*; Comment. soc. reg. sci. Gottengensis recentiores **6** (1828); math. Classe, 99–146; Werke **4**, 217–258; in [Monge 1850] 505–546.

[1840] *Allgemeine Lehrsätze in Beziehung auf die im verkehrten Verhältnisse des Quadrats der Entfernung wirkenden Anziehungs- und Abstossungs-Kräfte*, Resultate aus den Beobachtungen des magnetischen Vereins im Jahre 1839 (4) Leipzig 1840; Werke **5**, 197–242; Ostwald's Klassiker nr. 2 Leipzig 1889; French transl. in L. J. **7** (1842), 273–324.

Gauss, C. F. and Schumacher, H.C.

[1861] *Letter from Gauss to Schumacher in "Carl Friedrich Gauss - H. C. Schumacher"*, Briefwehsel III; Altona 1861 p. 137.

Gilain, C.

[1988] *Condorcet et le calcul intégral.* 87–150 in "Science à l'époque de la Révolution Française, Recherches historiques." Paris 1988.

Gispert, H.

[1987] *La correspondance de G. Darboux avec J. Houël Chronique d'un rédacteur*; Cahiers du Séminaire d'Hist. des Math. **8** (1987), 67–202.

Giusti, E.

[1984] *Gli "errori" di Cauchy e i fondamenti dell'analisi*, Bolletino di Storia delle Scienze Matematiche **4** (1984), 24–54.

Grabiner, J.

[1981] *The Origins of Cauchy's Rigorous Calculus.* Cambridge Mass. 1981

Grattan-Guinness, I.

[1970] *The development of the foundations of mathematical analysis from Euler to Riemann.* Massachusetts 1970.

[1972] *Joseph Fourier.* Cambridge 1972.

[1981] *Mathematical Physics in France, 1800–1840*, Knowledge, Activity and Historiography; Annals of Science **38** (1981), 663–690.

[1988] *From the Calculus and Mechanics to Mathematical Analysis and Mathematical Physics, French Mathematicians and their institutions 1800–1840.* To appear.

Gray, J. J.

[1984] *Fuchs and the Theory of Differential Equations*, Bull. Amer. Math. Soc. **10** (1984), 1–26.

[1986] *Linear Differential Equations and Group Theory from Riemann to Poincaré.* Boston 1986.

Greatheed, S. S.

[1839]   *On General Differentiation*, Cambridge Math. Journ. **1** (1839), 11–21;
         109–117.

Green, G.

[1828]   *An Essay on the Application of Mathematical Analysis to the theories
         of Electricity and Magnetism*, Nottingham 1828; Journ. Reine Angew.
         Math. **39** (1850), 73–89;, **44** (1852), 356–374; and, **47** (1854), 161–221;
         Mathematical Papers, 3–115.

[18330–1835]   *On the determination of the Exterior and Interior Attractions
         of Ellipsoids of Variable Densities*, Trans. Cambridge Phil. Soc. (5) **3**
         (1835), 395–430; Math. Papers, 187–222.

Greenhill, Sir A. G.

[1894]   *Pseudo-elliptic integrals and their dynamical applications*, Proc. London
         Math. Soc. (1) **25** (1894), 195–304.

Gregory, D. F.

[1841]   *Examples of the Processes of the Differential and Integral Calculus*. Cam-
         bridge 1841, 2. ed. 1846.

Grunwald, A. K.

[1867]   *Ueber begrenzte Derivationen und deren Anwendungen*, Zeitschrift Math.
         Phys. **12** (1867), 441–480.

Guitard, G.

[1986]   *La querelle des infiniment petits à l'École Polytechnique au XIX^{e}e siècle*,
         Historia Scientiarum **30** (1986), 1–61.

Hadamard, J.

[1911]   *Sur les trajectoires de Liouville*, Bull. Sci. Math. **35** (1911), 106–113.

Hahn, R.

[1971]   *The Anatomy of a Scientific Institution. The Paris Academy of Sciences
         1666–1803*. Berkely 1971.

Hamilton, W. R.

[1828]   *Theory of Systems of Rays*, Trans. Roy. Irish Acad. **15** (1828), 69–174;
         Math. Papers **1**, 1–88.

[1834, 1835]   *On a general method in dynamics*, I + II; Phil. Trans. (I) (1834),
         247–308; and (II), (1835), 95–144; Math. Papers **I**, 103–161;, **II**, 162–211.

Hansen, P. A.

[1831]   *Untersuchung über die gegenseitigen Störungen des Jupiters und Sat-
         urns*. Berlin 1831.

Hardy, G. H.

[1905–1916]   *The Integration of functions of a single variable*; Cambridge 1. ed.
         1905, 2. ed. 1916.

[1910]   *Orders of Infinity*; Cambridge tracts in Mathematics No. 12, 1910..

[1912]   *Properties of logarithmico-exponential functions*, Proc. London Math.
         Soc. (2) **10** (1912), 54–90.

Haupt, O. and Hilb, E.
[1924]   *Über die Transformation Liouvillescher Manigfaltigkeiten*; Göttinger Nachrichten 1924, 77–79.

Heaviside, O.
[1893–1912]   *Electromagnetic Theory*; London **1** (1893);, **2** (1899);, **3** (1912).

Heine, E.
[1845]   *Beitrag zur Theorie der Anziehung und der Wärme*, Journ. Reine Angew. Math. **29** (1845), 185–208.
[1870]   *Über trigonometrische Reihen*, Journ. Reine Angew. Math. **71** (1870), 353–365.
[1878–1881]   *Handbuch der Kugelfunktionen, Theorie und Anwendungen*; 2. ed. 2 vols.; Berlin 1878–1881.
[1880]   *Einige Anwendungen der Residuenrechnung von Cauchy*, Journ. Reine Angew. Math. **89** (1880), 19–39.

Hellinger, E. and Toeplitz, O.
[1923–1927]   *Integralgleichungen und Gleichungen mit unendlichvielen Unbekannten*; Encycl. d. Math. Wiss. II C 13, 1335–1597.

Hermite, C.
[1846]   *Extraits de deux lettres de Ch. Hermite à C. G. Jacobi*, Journ. Reine Angew. Math. **32** (1846), 277–299; Jacobi Werke **II**, 87–114; Oeuvres de Hermite **I**, 10–37.
[1848]   *Sur la division des fonctions abéliennes ou ultralliptiques*, Mém. Sav. Etrang. Acad. Sci. Paris **10** (1848), 563–574; Oeuvres **I**, 38–48.
[1850]   *Lettres de M. Hermite à M. Jacobi sur différents objets de la Théorie des Nombres*, Opuscula Math. de Jacobi II; Journ. Reine Angew. Math. **40** (1850), 261–315; Oeuvres de Hermite **I**, 100–163.
[1861]   *Lettre de M. Hermite à M. Liouville*, C. R. **53** (1861), 214–228; L. J. (2) **7** (1862), 24–44; Oeuvres **2**, 109–124.
[1873]   *Sur la fonction exponentielle*, C. R. **77** (1873), 18–24;226–233;285–295; Oeuvres **3**, 150–181.
[1984]   *Lettres de Charles Hermite à Gösta Mittag-Leffler (1874–1883)*, ed. Dugac; Cahiers du Séminaire d'Histoire des Mathématiques **5** (1984), 49–285.

Hermite, C. and Stieltjes, T. J.
[1905]   *Correspondance d'Hermite et de Stieltjes*. 2 vols. Paris 1905.

Heyde, C. C. and Seneta, E.
[1977]   *I. J. Bienaymé. Statistical Theory Anticipated*. New York 1977.

Hilb, E. and Szász, O.
[1922]   *Allgemeine Reihenentwicklungen*; Encyclopädie der mathematischen Wissenshcaften II C 11 **II 3.2**, 1229–1276.

Hilbert, D.
[1900]   *Über das Dirichletsche Prinzip*, Jahresber. Deut. Math. Verein. **8** (1900), 184–188; Journ. Reine Angew. Math. **129** (1905), 63–67; Ges. Abhandl. **3**, 10–14.

[1904] *Über das Dirichletsche Prinzip*, Math. Ann. **59** (1904), 161–186; Ges. Abh. **3**, 15–37.

[1904–1910] *Grundzüge einer allgemeinen Theorie der linearen Integralgleichungen*, Nachrichten Königl. Ges. Wiss. Göttingen Mat.-Phys. Kl. (1904), 49–91;213–259;, (1905), 307–338;, (1906), 157–227;439–480;, (1910), 355–417;595–618. Collected as monograph Leipzig 1912. References refer to monograph.

Hirano, Y.
[1984] *La diffusion des idées de Galois et le rôle de Camille Jordan (1838–1922) - formation de la théorie des groupes et l'apport de Camille Jordan.* Thèse Paris 1984.

Holmgren, E.
[1906] *Sur la théorie des équations intégrales linéaires*, Ark. f. Mat. Astron. och Fysik;, **3** (1906 nr. 1, 24 pages); summarized in: *Sur un problème du calcul des variations*, C. R. **142** (1906), 331–333.

Holmgren, H. J.
[1863–1864] *Om Differentialkalkylen med indices af hvilken natur som helst*; Kungl. Svenska Vetenskaps-Akademiens Handlingar **5**. no. 11, (1863–1864), 83 pages.
[1867–1868] *Sur l'intégration de l'équation différentielle $(a_2 + b_2 x + c_2 x^2)\frac{d^2 y}{dx^2} + (a_1 + b_1 x)\frac{dy}{dx} + a_0 y = 0$*; Kungl. Svenska VetenskapsAkademiens Handlingar **7**,. no. 9, (1867–1868), 58 pages.

Houzel, C.
[1978] *Fonctions elliptiques et intégrales abéliennes.* in Abrégé d'histoire des mathématiques 1700–1900 **2** 1–113 ed. J. Dieudonné.

Institut de France
[1979] *Index Biographique de l'Académie des Sciences du 22 décembre 1866 au 1er octobre 1978.* Paris 1979.

Ivory, J.
[1838] *Of such ellipsoids consisting of homogeneous matter as are capable of having the resultant of the attraction of the mass upon a particle in the surface, and a centrifugal force caused by revolving about one of the axes, made perpendicular to the surface*, Phil. Trans. (1838), 57–66.
[1839] *Note of Mr. Ivory relating to the correcting of an error in a Paper printed in the Philosophical Transactions for 1838 pp. 57 &c*, Phil. Trans. (1839), 265–266.

Jacobi, C. G. J.
[1827a] *Über die integration der Partiellen Differentialgleichungen erster Ordnung*, Journ. Reine Angew. Math. **2** (1827), 317–329; Werke **IV**, 1–15.
[1827b] *Über die Pfaffsche Methode eine gewöhnliche lineare Differentialleichung zwischen 2n Variabeln durch ein System von n Gleichungen zu integriren*, Journ. Reine Angew. Math. **2** (1827), 347–357; Werke **IV**, 17–29.
[1828] *Über die Anwendung der elliptischen Transcendenten auf ein bekanntes Problem der Elementargeometrie*, Journ. Reine Angew. Math. **3** (1828),

376–389; Werke **1**, 277–293; French trans. L. J. **10** (1845), 435–444.

[1828–1834]   *De functionibus duarum variabilium quadrupliciter periodicis, quibus theoria transcendentium Abelianarum innitur*, Journ. Reine Angew. Math. **13** (1835), 55–78; Werke **2**, 23–50.

[1829]   *Fundamenta Nova Theoriae Functionum Ellipticarum*, Köningsberg 1829; Werke **1**, 49–239.

[1834]   *Über die Figur des Gleichgewichts*, Ann. Phys. u. Chem. (2) **33** (1834), 229–233; Werke **II**, 17–22.

[1835]   *Theoremata nova algebraïca circa systema duarum aequationum inter duas variabiles propositarum*, Journ. Reine Angew. Math. **14** (1835), 281–288; Werke **3**, 285–294.

[1836a]   *Sur le mouvement d'un point et sur un cas particulier du problème des trois corps*, C. R. **3** (1836), 59–61; Werke **IV**, 35–38.

[1836b]   *De relationibus quae locum habere debent inter puncta intersectionis duarum curvarum vel trium superficierum algebraicarum dati ordinis, simul cum enodatione paradoxi algebraici*, Journ. Reine Angew. Math. **15** (1836), 285–308; Werke **3**, 329–354.

[1837a]   *Zur Theorie der Variations-Rechnung und der Differential-Gleichungen*, Journ. Reine. Angew. Math. **17** (1837), 68–82; Werke **IV**, 37–55; Trans. *Sur le Calcul des Variations et sur la Théorie des Équations différentielles*, L. J. **3** (1838), 44–59.

[1837b]   *Über die Reduction der Integration der partiellen Differentialgleichungen erster Ordnung zwischen irgend einer zahl variabeln auf die Integration eines einzigen Systems gewöhnlicher Differentialgleichungen*, Journ. Reine Angew. Math. **17** (1837), 97–162; Werke **IV**, 57–127; Trans: *Sur la Réduction de l'intégration des Équations différentielles partielles du premier ordre entre un nombre quelconque de variables à l'intégration d'un seul système d'équations différentielles ordinaires*, L. J. **3** (1838), 60–96; and 161–201.

[1837c]   *Note sur l'intégration des équations différentielles de la dynamique*, C. R. **5** (1837), 61–67; Werke **IV**, 129–136.

[1838–1839]   *Lettre de M. Jacobi à M. Arago, concernant les lignes géodesiques tracées sur un ellipsoïde à trois axes*, C. R. **8** (1839), p. 284; equal to a letter in German to Bessel Werke **7**, p. 385.

[1839a]   *Note von der geodätischen Linie auf einem Ellipsoid und den verschiedenen Anwendungen einer merkwürdigen analytischen Substitution*, Journ. Reine Angew. Math. **19** (1839), 309–313; Werke **2**, 59–63; trans. as *De la ligne géodésique sur un ellipsoïde, et des différents usages d'une transformation analytique remarquable*, L. J. **6** (1841), 267–272.

[1839b]   *Über die Complexen Primzahlen, welche in der Theorie der Reste der 5ten, 8ten und 12ten Potenzen zu betrachten sind*, Monatsber. Akad. Wiss. Berlin 1839; Journ. Reine Angew. Math. **19** (1839), 314–318; french transl. in L. J. **8** (1843), 268–272.

[1840]   *Sur un théorème de Poisson*, C. R. **11** (1840), p. 529; L. J. **5** (1840), 350–351; Werke **IV**, 143–146.

[1842]   *Sur un nouveau principe de la mécanique analytique*, C. R. **15** (1842), 202–205; Werke **IV**, 289–293.

[1843]   *Extrait d'une lettre à M. Hermite*, L. J. **8** (1843), 505–506.

[1844]   *Theoria novi multiplicatoris systemati aequationum differentialium vulgarium applicandi*; Journ. Reine Angew. Math. ; and 333–376 ; Werke **IV**, 317–509.

[1848]   *Über die Reduction der quadratischen Formen auf die kleinste Anzahl Glieder*, Monatsber, Akad. Wiss. Berlin (November 1848) 414–417; Journ. Reine Angew. Math. **39** (1850), 290–292; Werke **6**, 318–321.

[1862]   *Nova methodus aequationes differentiales partiales primi ordinis inter numerum variabilium quemcunque propositas integrandi*, Posthumous manuscript publ. by Clebsch, Journ. Reine Angew. Math. **60** (1862), 1–181; Werke **5**, 1–189.

[1866]   *Vorlesungen über Dynamik gehalten an der Universität zu Köningsberg in Wintersemester 1842-1843 und nach einem von C. W. Borchardt ausgearbeiteten Hefte.* Werke Supplementband.

Jahnke, E.
[1903]   *Brief von Liouville an Jacobi 1. juin 1846*, Archif der Math. und Phys. (3) **5** (1903), p. 41.

Jardin, A. and Tudesq, A. J.
[1973]   *La France des notables; I L'évolution générale 1815-1848, II La vie de la nation 1815-1848.* Nouvelle Histoire de la France Contemporaine 6, 7 Paris 1973.

Jeans, J. H.
[1919]   *Problems of Cosmology and Stellar Dynamics.* Cambridge 1919.

Joachimsthal, F.
[1843]   *Observationes de lineis brevissimus et curvis curvaturae in superficiebus secundus gradus*, Journ. Reine Angew. Math. **26** (1843), 155–171.

Jongmans, F.
[1981]   *Quelques pièces choisies dans la correspondance d'Eugène Catalan*, Bull. Soc. Roy. Sci. Liège (50. année), 9–10;, (1981), 287–309.

[1986]   *Une election orageuse à l'Institut*, Bull. Soc. Roy. Sci. Liège **55** (1986), 581–603.

[1987]   *Les mathématiciens au XIXème siècle.* Bruxelles 1987.

Jordan, C.
[1870]   *Traité des substitutions et des équations algébriques.* Paris 1870.

[1878]   *Mémoire sur les équations différentielles linéaires à intégrales algébriques*, Journ. Reine Angew. Math. **84** (1878), 89–215.

[1887]   *Cours d'Analyse.* Paris 1887.

Kaplansky, I.
[1957]   *An Introduction to Differential Algebra.* Paris 1957, 2. ed. 1976.

Kasper, T.
[1980]   *Integration in finite terms: The Liouville Theory*, Mathematics Magazine **53** (1980), 195–201.

Kelland, P.
[1837]   *Theory of heat.* Cambridge 1837.

Kellogg, O. D.
[1929]   *Foundations of Potential Theory*. Berlin 1929.

Kiernan, B. M.
[1971–1972]   *The development of Galois theory from Lagrange to Artin*, Arch.
Hist. Exact. Sci. **8** (1971–1972), 40–154.

Kirchhoff, G.
[1879]   *Über die Transversalschwingungen eines Stabes von veränderlichem Quer-
schnitt*, Berlin Acad. Monatsber. (1879), 815–828; Ann. Phys. Chem. **10**
(1880), 501–512.
[1894]   *Vorlesungen über die Theorie der Värme*, ed. Max Planck. Berlin 1894.

Klein, F.
[1877a]   *Über lineare Differentialgleichungen*, Sitz. phys. med. Soc. Erlangen 26
Juni 1876; Math. Ann. **11** (1877), 115–118.
[1877b]   *Über lineare Differentialgleichungen*, Math. Ann. **12** (1877), 167–179.
[1881]   *Über Körper, welche von confocalen Flächen zweiten Grades begränzt
sind*, Math. Ann. **18** (1881), 410–427.
[1894]   *Riemann und seine Bedeutung für die Entwiklung der modernen Mathematik*. Amtlicher Bericht der Naturforscherversammlung zu Wien (1894)

Kline, M.
[1972]   *Mathematical Thought from Ancient to Modern Times*. New York 1972.

Kneser, A.
[1904]   *Untersuchungen über die Darstellung willkürlicher Funktionen in der
mathematischen Physik*, Math. Ann. **58** (1904), 81–147.

Knudsen, O.
[1980]   *Elektromagnetismens historie 1820–1831 og Faraday's opdagelse af in-
duktionen*. København 1980.
[1985]   *Mathematics and Physical Reality in William Thomson's Electromagnetic Theory*, in P. M. Harman (ed.) *Wranglers and Physicists. Studies
on Cambridge Physics in the 19th Century*. Manchester 1985, 149–179.

Kolchin, E. R.
[1973]   *Differential Algebra and Algebraic Groups*. New York, London 1973.

Königsberger, L.
[1887]   *Bemerkungen zu Liouville's Classificierung der Transcendenetn*, Math.
Ann. **28** (1887), 483–492.
[1889]   *Lehrbuch der Theorie der Differentialgleichungen*. Leipzig 1889.

Kummer, E. E.
[1860]   *Gadächtnissrede auf Gustav Peter Lejeune Dirichlet*, Abhandl. Königl.
Akad. Wiss. Berlin 1860; Dirichlet's Werke, **II**, 311–344.

Laboulaye, F.
[1882]   *Discours de M. Laboulaye prononcés aux funérailles de M. Liouville au
nom de collège de France*, C. R. **95** (1882), 469–71.

Lacroix, S. F.

[1797–1800]   *Traité du calcul différentiel et du calcul intégral.* 3 vols Paris 1.ed. 1797–1800, 2. ed. 1810–1816.

[1831]   *Compléments des Eléments d'Algèbre.* 6. ed. Paris 1831.

[1837]   *Traité élémentaire du calcul différentiel et du calcul intégral.* 5. ed. Paris 1837.

Lagrange, J. L.

[1766–1769]   *Recherches sur le mouvement d'un corps qui est attiré vers deux centres fixes*; Miscellanea Taurinensia 4 (1766–1769); Ouevres 2, 65–121.

[1769]   *Mémoire Sur la résolution des équations numérique*; Hist. et Mém. Acad. Roy. Sci. et Belles-Lettres de Berlin pour l'année 1767 (vol 23) publ. 1769, 311–352; Oeuvres II, 537–578.

[1770a]   *Additions au mémoire sur la résolution des équations numériques*; Hist. et Mém. Acad. Roy. Sci. et Belles-Lettres de Berlin pour l'année 1768 (vol 24) publ. 1770, 111–180; Ouevres II, 579–652.

[1770b]   *Nouvelle méthode pour résoudre les problèmes indéterminés en nombres entiers*; Mém. Acad. Roy. Sci. et Belles-Lettres de Berlin pour l'année 1768 (vol 24) publ. 1770, 181–250; Ouevres II, 653–726.

[1770–1771]   *Réflexions sur la résolution algébrique des équations.* Mém. Acad. Roy. Sci. et Belles-Lettres Berlin (1770) 134–215 (1771) 138–253; Oeuvres 3 203–421.

[1774]   *Sur l'équation séculaire de la lune*; Mém. Acad. Sci. Paris, Savants étrangers 7 (1773); Oeuvres 6, 331–399.

[1776]   *Recherches sur les suites récurrentes*; Nouveaux Mémoires de l'Acad. Roy. Sci. et Belles-Lettres pour l'année 1775 (read April and May 1776); Oeuvres 4, 150–254.

[1781–1782]   *Théorie des variations séculaires des éléments des planètes*; 1. partie Nouveaux Mémoires l'Acad. Roy. Sci. et Belles-Lettres de Berlin pour l'année 1781; Oeuvres 5, 125–207; 2. partie Nouveaux Mémoire de l'Acad. Sci. et Belles-Lettres pour l'année 1782; Oeuvres 5, 208–344.

[1788, 1811–1815]   *Mécanique Analytique*, (Analitique in the first edition) Paris 1788; 2. ed. Paris 1811–1815, 3. ed. edited and annotated by Bertrand Paris 1853, 4. ed. edited by Darboux 1888..

[1808]   *Mémoire sur la théorie des variations des éléments des planètes et en particulier des variations des grands axes de leurs orbites*; Mém. de la première Classe de l'Institut de France, 1808; Oeuvres 6, 711–768.

[1809]   *Mémoire sur la théorie générale de la variation des constantes arbitraires dans tous les problèmes de la Mécanique*; Mem. de la première Classe de l'Institut de France, 1808 [but read to the Institute on March 13th 1809]; Oeuvres 6, 769–805.

[1810]   *Seconde mémoire sur la théorie de la variation des constantes arbitraites...*; Mem. de la première Classe de l'Institut de France 1809 [but read to the Institute on February 19th 1810]; Oeuvres 6, 807–816.

Lamarle, E.

[1846]   *Note sur le théorème de M. Cauchy relatif au développement des fonctions en séries,* L. J 11 (1846), 129–141.

Lambert, J. H.

[1766]   *Forläufige Kenntnisse für die, so die Quadratur und Rectification des Circuls suchen*, Beträge zum Gebrauche der Mathematik und deren Anwendung II, Berlin 1770, 140–169; Opera Mathematica I, 194–212; reprinted in Rudio. [1892] 133–155.

[1767a]   *Mémoire sur quelques propriétés remarquables des quantités transcendantes circulaires et logarithmiques*, Mémoires de l'Académie des Sciences de Berlin **17** (1761 publ. 1768), 265–322; Opera Mathematica II, 112–159.

[1767b]   *Solution générale et absolue du problème de trois corps moyennant des suites infinies*, Hist. et Mem. Acad. Roy. Sci. et Belles-Lettres Berlin. **23** (1767 publ. 1769), 353–364.

Lamé, G.

[1833]   *Mémoire sur les surfaces isothermes dans les corps solides en équilibre de température*, Ann. Chimie, Phys. (2) **53** (1833), 190–204.

[1837]   *Sur les surfaces isothermes dans les corps solides homogènes en équilibre de température*, Mem. Savants Etrangers. Acad. Sci. Paris, Sci. Math. Phys. (2) **5** (1837), 147–183; L. J. **2** (1837), 147–183.

[1838]   *Extrait d'une Lettre de M. Lamé à M. Liouville sur cette question: Un polygone convexe étrant donné, de combien de manières peut-on le partager en triangles au moyen de diagonales?*, L. J. **3** (1838), 505–507.

[1839]   *Mémoire sur l'équilibre des températures dans un ellipsoïde à trois axes inégaux*, L. J. **4** (1839), 126–163.

[1840]   *Mémoire d'analyse indéterminée, démontrant que l'équation $x^7 + y^7 = z^7$ est impossible en nombres entiers*, L. J. **5** (1840), 195–210.

[1847]   *Démonstration générale du théorème de Fermat*, C. R. **24**. (1847), 310–315

Laplace, P.-S.

[1782–1785]   *Théorie des attractions des sphéroïdes et de la figure des planètes*, Mem. Acad. Sci. Paris (1782 publ. 1785), 113–196; Oeuvres **10**, 339–419.

[1784]   *Théorie du Mouvement et de la Figure Elliptique des Planètes*. Paris 1784.

[1799–1825]   *Traité de Mécanique Céleste*, Paris 1799–1825; Oeuvres **1–5**.

[1809]   *Mémoire sur les mouvements de la lumière dans les milieux diaphanes*, Mem. de la première Classe de l'Institut (1809), 300–342; Oeuvres **12**, 267–298.

[1812]   *Théorie analytique de probabilité*, Paris 1812 3. ed. 1820; Oeuvres **12**.

[1835]   *Exposition du Système du Monde*, 6. ed. Paris 1835; Oeuvres **6**.

Laplace, P. S. and Lavoisier, A. L.

[1780]   *Mémoire sur la chaleur*, Mem. Acad. Roy. Sci. (1780 publ. 1784), 355–408; Oeuvres de Laplace **10**, 147–200; Oeuvres de Lavoisier **2**, 283–333.

Laugwitz, D.

[1987]   *Infinitely Small Quantities in Cauchy's Textbooks*, Historia Mathematica **14** (1987), 258–274.

Laurent, H.
[1895]   *Liouville*. Livre du centenaire de l'École Polytechnique I, Paris 1895, 130–133.

Legesgue, H.
[1913]   *Sur des cas d'impossibilité du problème de Dirichlet ordinaire,*, Comp. Rend. Soc. Math. France 41 (1913), p. 17; Oeuvres, 4, p. 131.
[1922]   *Les professeurs de mathématiques du Collège de France: Humbert et Jordan, Roberval et Ramus*, Leçon d'ouverture du Cours de Mathématique pures du Collège de France, Professée le 7 janvier 1922; Revue Scientifique du 22 avril 1922, 249–62; Monographies de l'Enseignement Mathématique n° 4 *Henri Lebesgue*. *Notices d'Histoire des Mathématique*, 73–101.

Lebesgue, V. A.
[1847]   *Démonstration nouvelle et élémentaire de la loi de réciprocité de Legendre, par M. Eisenstein, précédée et suivie de Remarques sur d'autres démonstrations qui peuvent être tirées du même principe*, L. J. 12 (1847), 457–473.
[1859]   *Exercices d'analyse Numerique*. Paris 1859.

Lefébure de Fourcy, L.
[1859]   *Leçons de Géométrie analytique*. 6. ed. Paris 1859.

Legendre, A.-M.
[1784–1787]   *Recherches sur la figure des planètes*, Mem. Acad. Sci. Paris (1784 publ. 1787), 370–389.
[1794]   *Éléments de Géometrie*. Paris 1794.
[1811–1817]   *Exercices de calcul intégrale sur divers ordres de transcendantes et sur les quadratures*. Paris 1 (1811), 2 (1816) 3 (1817).
[1825–1828]   *Traité des fonctions elliptiques*. Paris 1 (1825), 2 (1826), 3 (1828).

Leibniz, G. L.
[1899]   *Der Briefwechsel von Gottfried Wilhelm Leibniz mit Mathematikern*. Ed. C. J. Gerhardt, Berlin 1899.

Le Verrier, U.J.J.
[1839]   *Sur les variations séculaires des orbites des planètes. Extrait par l'auteur;* C. R. 9 (1839), 370–372.
[1840a]   *Mémoire sur les inclinaissons respectives des orbites de Jupiter, Saturne et Uranus, et sur les mouvements des intersections de ces orbites*, L. J. 5 (1840), 95–109.
[1840b]   *Sur les variations séculaires des éléments elliptiques des sept planètes principales: Mercure, Vénus, la Terre, Mars, Jupiter, Saturne et Uranus*, L. J. 5 (1840), 220–254.

Levi-Civita, T.
[1904]   *Sulla integrazione della equazione di Hamilton-Jacobi per separazione di variabili*, Math. Ann. 59 (1904), 383–397.

Lévy, P.
[1931]   *R. Liouville (1856–1930)*, Journ. Ec. Polyt. (2) 29 (1931), 1–5.

Lhuilier, S.

[1828]   *Recherches polyèdrométriques*, Bibl. Univ. **37** (1828), 249–264.

Liapounoff, A.

[1884]   *Sur la stabilité des figures ellipsoïdales d'équilibre d'un liquide animé d'un mouvement de rotation*, Ann. Fac. Sci. Univ. Toulouse (2) **6** (1904), 5–116; French translation of Liapounoff's thesis of 1884 with few alterations.

Libri-Carucci, G.

[1827–1831]   *Mémoire sur la théorie de la chaleur*, Journ. Reine Angew. Math. **7** (1831), 116–131; earlier printed in Pisa 1827 and Firenze 1829.

[1829]   *Mémoires de mathématiques et de physique.* **1**, Florence 1829.

[1832]   *Mémoire sur la théorie des nombres*, Journ. Reine Angew. Math. **9** (1832), 54–80, 169–188, 261–276.

[1833]   *Mémoire sur les résolution des équations algébriques dont les racines ont entre elles un rapport donné et sur l'intégration des équations différentielles linéaires dont les intégrales particulières peuvent s'exprimer les unes par les autres*, Journ. Reine Angew. Math. **10** (1833), 167–194.

[1834]   *Rapport sur un mémoire de M. Liouville intitulé Sur l'intégration d'une classe de fonctions transcendantes*, Procès-Verbaux Acad. Paris (1834), 481–483.

[1836]   *Note sur les rapports qui existent entre la théorie des équations algébriques et la théorie des équations linéaires aux différentielles et aux différences*, L. J. **1** (1836), 10–13.

[1838]   *Mémoire sur la théorie des nombres*, Mem. Sav. Etrang. Acad. Sci. Paris **5** (1838), 1–75.

[1838–1841]   *Histoire des Science mathématique en Italie.* **1–4**, Paris 1838–1841.

[1839a]   *Mémoire sur la théorie générale des équations différentielles linéaires à deux variables*, C. R. **8** (1839), 732–741.

[1839b]   *Réponse de M. Libri à la Note de M. Sturm*, C. R. **8** (1839), 789–790.

[1839c]   *Réponse de M. Libri aux Observations de M. Liouville*, C. R. **8** (1839), 789–801.

[1840a]   *Note sur un théorème de M. Dirichlet*, C. R. **10** (1840), 311–314.

[1840b]   *Réponse aux observations de M. Liouville*, C. R. **10** (1840), 345–347.

[1843]   *Réponse de M. Libri à la Note insérée par M. Liouville dans le Compte rendu de la séance du 21 août*, C. R. **17** (1843), 431–445.

Lichtenstein, L.

[1918]   *Neuere Entwicklung der Potentialtheorie. Konforme Abbildung*, Encyclopädie der Math. Wiss. (1909–1921 II C 3), 177–377.

[1933]   *Gleichgewichtsfiguren rotirender Flüssigkeiten.* Berlin 1933.

Lie, S.

[1888–1893]   *Theorie der Transformationsgruppen.* in collaboration with F. Engel. 3 vols. Leipzig 1888–1893.

Lindemann, F.

[1882a]   *Ueber die Zahl π*, Math. Ann **20** (1882), 213–225.

[1882b] *Sur le rapport de la circonférence au diamètre, et sur les logarithmes népériens des nombres commensurables ou des irrationnelles algébriques Extrait d'une Lettre adressée à M. Hermite*, C. R. **95** (10 July 1882), 72–74.

Liouville, E.

[1854a] *Mémoire sur l'emploi des mires méridiennes dans le calcul de la déviation azimutale*, L. J. **19** (1854), 139–150.

[1854b] *Note sur la variation annuelle de l'inclinaison de l'axe de rotation de la lunette méridienne de Gambey*, L. J. **19** (1854), 409–412.

[1855] *De l'influence des diaphragmes sur la grandeur des diamètres aparents du Soleil et de la Lune*, L. J. **20** (1855), 105–114.

[1873] *Sur la Statistique judiciaire*, L. J. (2) **18** (1873), 145–163.

Liouville, J.

See separate list of published works.

Liouville, J., and Sturm, C.

[1836] *Démonstration d'un théorème de M. Cauchy relatif aux racines imaginaires des équations*, Journ. Math. Pures Appl. **1** (1836), 278–289.

[1837a] *Extrait d'un Mémoire sur le développement des fonctions en séries dont les différents termes sont assujettis à satisfaire à une même équation différentielle linéaire, contenant un paramètre variable*, L. J. **2** (1837), 220–233; C. R. **4** (1837), 675–677.

[1837b] *Note sur un théorème de M. Cauchy relatif aux racines des équations simultanées*, C. R. **4** (1837), 720–739.

Liouville/Lespiault

[1855–1856] *Cours de M. Liouville au Collège de France. Intégration des équations différentielles partielles*; contained in *Mathématique, cours inédits recueillis par Lespiault*, f° 46–103; Bibliothèque de la Faculté des Sciences de Bordeaux, Ms 52 (ancienne cote).

Liouville/Saint Venant

[1839–1840] *Intégrales définies, Cours de M. Liouville*. Fonds privé Saint-Venant, Archives départementales du Loir-et-Cher.

Lipschitz, R.

[1871] *Untersuchung eines Problems der Variationsrechnung, in welchen das Problem der Mechanik enthanten ist*, Journ. Reine Angew. Math. **74** (1871), 116–149.

Loria, G.

[1936] *Le Mathématicien J. Liouville et ses oeuvres*, Archeion **18** (1936), 117–139; English transl.: *J. Liouville and his Work*, Scripta mathematica **4** (1936), 147–154. (with portrait), 257–263, 301–306.

Lützen, J.

[1978] *Funktionsbegrebets udvikling fra Euler til Dirichlet*, Nordisk Mat. Tidsskr **25–26** (1978), 5–32.

[1979] *Heaviside's Operational Calculus and the Attempts to Rigorise it*, Arch. Hist. Ex. Sci. **21** (1979), 161–200.

[1982a]   *The Prehistory of the Theory of Distributions.* New York, Heidelberg, Berlin 1982.

[1982b]   *Joseph Liouville's Contribution to the Theory of Integral Equations,* Historia Mathematica **9** (1982), 373–391.

[1984a]   *Sturm and Liouville's Work on Ordinary Linear Differential Equations. The Emergence of Sturm-Liouville Theory,* Arch. Hist. Exact. Sci. **29** (1984), 309–376.

[1984b]   *Joseph Liouville's Work on the Figures of Equilibrium of a Rotating Mass of Fluid,* Arch. Hist. Ex. Sci. **30** (1984), 113–166.

Mac-Cullagh, J.

[1842]   *Mémoire sur les lois de la réflexion et la réfraction cristallines,* L. J. **7** (1842), 217–265.

Maillet, E.

[1906]   *Introduction à la théorie des nombres transcendants.* Gauthier-Villars, Paris 1906.

Maclaurin, C.

[1740]   *On the Tides,* (De causa physica fluxus et refluxus maris). Prize essay submitted to the Académie des Sciences, Paris in 1740.

[1742]   *A Treatise of Fluxions.* 2 vols Edinburgh 1742.

Mandelbaum, J.

[1980]   *La Société Philomatique de Paris de 1788 à 1835.* Thèse, 2 vols. Paris 1980.

Marielle, C.-P.

[1855]   *Répertoire de l'École Impériale Polytechnique.* Paris 1855.

Maurice, F.

[1828]   *Rapport sur un Mémoire de M. Liouville sur l'électricité dynamique et en particulier sur l'action mutuelle d'un pôle d'aimant et d'un fil conducteur,* Procès-Verbaux Acad. Sci. Paris (1828), 119–120.

[1842a]   *De l'invariabilité des grands axes et des moyens mouvements des planètes, en tenant compte de tous les ordres des forces perturbatrices,* C. R. **15** (1842), 328–343.

[1842b]   *Remarques à l'occasion d'une lettre de M. Liouville relative à la démonstration que M. Maurice a donnée du théorème de l'invariabilité des grands axes des planètes,* C. R. **15** (1842), 598–601.

[1842c]   *Nouvelles considérations sur l'invariabilité des grands axes,* C. R. **15** (1842), 853–855.

Maxwell, J. C.

[1868]   *On the dynamical theory of gases,* Phil. Mag. **35** (1868), 129–145, 185–217; Scientific Papers II, 26–78.

[1879]   *On Boltzmann's theorem on the average distribution of energy in a system of material points,* Trans. Cambridge Phil. Soc. **12** (1879), 547–570; Annal. Phys. Chem. **5** (1881), 403–417; Phil. Mag. **14** (1882), 299–312; Scientific Papers II, p. 713.

Meyer, C. O.
[1842]   *De aequilibrii formis Ellipsoidicis*, Journ. Reine Angew. Math. **24** (1842), 44–59.

Mikhlin, S. G.
[1957]   *Integral Equations and their applications to certain problems in mechanics, mathematical physics and technology.* London 1957.

Minding, F.
[1830]   *Bemerkung über die Abwicklung krummer Linien von Flächen*, Journ. Reine Angew. Math. **6** (1830), 159–161.
[1839]   *Wie sich entscheiden lässt, ob zwei gegebene krumme Flächen auf einander abwickelbar sind oder nicht; nebst Bemerkungen über die Flächen von unveränderlichem Krummungsmaasse*, Journ. Reine Angew. Math. **19** (1839), 370–387.
[1840]   *Beiträge zur Theorie der kürzste Linien auf krummen Flächen*, Journ. Reine Angew. Math. **20** (1840), 323–327.

Miquel, A.
[1844]   *Mémoire de géométrie*, L. J. **9** (1844), 20–27.

Mittag-Leffler, G.
[1925]   *Tale af Professor G. Mittag-Leffler.* Matematikerkongressen i København 1925. Beretning. Copenhagen 1915, 27–44.

Möbius, A. F.
[1853]   *Über eine neue Vervandtschaft zwischen ebenen Figuren*, Berichte Leipzig Ges. (mat-phys) **5** (1853), 14–24; Journ. Reine Angew. Math. **52** (1856), 218–228; Gesammelte Werke **2**, 206–217.
[1855]   *Die Theorie der Kreisverwandtschaft in rein geometrischer Darstellung*, Abhandl. Leipzig Gesellschaft (mat-phys) **2** (1855), 529–595; Gesammelte Werke **2**, 243–314.

Moigno, F.
[1840–1844]   *Leçons de calcul différentiel et de calcul intégral rédigé principalement d'après les méthodes de M. A.-L. Cauchy.* 2 vols. Paris 1840, 1844.

Monge, G.
[1850]   *Application de l'Analyse à la Géométrie.* 5. ed. (ed Liouville) Paris 1850.

Monna, A. F.
[1973]   *Functional analysis in historical perspective.* Utrecht 1973.
[1975]   *Dirichlet's principle. A mathematical comedy of errors and its influence on the development of analysis.* Utrecht 1975.

Morando
[1987]   *L'âge d'or de la Mécanique Céleste.* to appear in *General History of Astronomy* 2 sect. 3 part 3.

Mourey
[1828]   *Vraie théorie des quantités négatives et des quantités prétendues imaginaires.* Paris 1828.

842   Bibliography

Navier, C.L.

[1840]   *Résumé des leçons d'analyse donné à l'École Polytechnique.* 2 vols. ed. Liouville and Catalan Paris 1840.

Neuenschwander, E.

[1984a]   *Joseph Liouville (1809–1882): Correspondance inédite et documents biographiques provenant de différents archives parisiennes,* Bulletino di Storia delle Scienze Matematice **4** (1984), 55–132.

[1984b]   *Die Edition mathematischer Zeitschriften im 19. Jahrhundert und ihr Beitrag zum Wissenschaftlichen Austausch zwischen Frankreich und Deutschland.* Math. Inst. Univ. Göttingen Preprint, 4 1984.

Neumann, C.

[1870]   *Zur Theorie des Logarithmischen und des Newton'schen Potentials,* Königl. Sächsischen Ges. der Wiss. Leipzig (1870), 49–56, 264–321; Also Math. Ann. **11** (1877), 558–566.

[1877]   *Untersuchungen über das logarithmische und Newton'sche Potential* Leipzig 1877.

Neumann, F.-E.

[1835]   *Theoretische Untersuchung der Gesetze nach welchen das Licht an der Grenze zweier vollkommen durchsichtigen Medien reflectirt und gebrochen wird,* Abh. Königl. Akad. Wiss. Berlin (1835), 1–160.

Neveu, A.

[1986]   *The uses of the quantum Liouville Theory.* 117–123; in Niels Bohr centennial conferences 1985 *Recent Developments in Quantum Field Theory* ed. Ambjørn, Durhuus and Petersen, Amsterdam 1985.

Newton, I.

[1687]   *Philosophiae naturalis principia mathematica,* London 1687; English ed. F. Cajori, 3. ed. California 1946..

[1736]   *Method of Fluxions and Infinite Series.* London 1736.

Nouvelle Biographie Générale. Copenhague 1967

Onofri, E. and Virasoro, M. A.

[1982]   *On a Formulation of Polyakov's String Theory with Regular Classical Solutions,* Nucl. Phys. B **201** (1982), 159–175.

Oppenheim, S.

[1919]   *Die Theorie der Gleichgewichtsfiguren der Himmelskörper,* Encyclopädie der Math. Wiss. Bd. VI **2B** (1922–1934), 1–79.

Ore, O.

[1974]   *Niels Henrik Abel. Mathematician Extraordinary.* New York (1. ed. 1957) 2. ed. 1974.

Osgood, W. F.

[1901]   *Allgemeine Theorie der analytischen Funktionen einer und mehrer komplexen Grössen,* Encyclopädie der Math. Wiss. **II B1** (1901–1921), 1–114.

Ostrogradsky, M.
[1845]   *De l'intégration des fractions rationelles*; Bull. Acad. Imp. Sci. St. Péters-
         bourg, class phys-math. **4** (1845). 145–167, 286–300.
[1850a] *Sur les dérivées des fonctions algébriques*; Bull. Acad. Imp. Sci. St.
         Pétersbourg, class phys-math. **22** (1850), 337–342.
[1850b] *Mémoire sur les équations différentielles relatives au problème des iso-
         périmètres*, Mem. Acad. St. Pétersb. **6** (1850), 385–517.

Ostrowski, A.
[1946]   *Sur l'intégrabilité élementaire de quelques classe d'expressions*, Com-
         ment. Math. Helv. **18** (1946), 283–308.

Pappus
Collection. Pappi Alexandrini Collectiones, Ed. F. Hultsch. Berlin 1876–1878.
         Pappus d'Alexandrie, La Collection Mathématique ed. and French trans-
         lation. P. Ver Eecke. Paris-Bruges 1933.

Parseval, M.-A.
[1806]   *Mémoire sur les séries et sur l'intégration complète d'une équation aux
         différences partielles linéaires du second ordre à coéfficiens constans*,
         Mém. Inst. Sci. Lettres et Arts. Div. Sav. (Mem. Savans Étranger) **1 an
         14** (1805), 638–648.

Patterson B. C.
[1933]   *The origins of the Geometric Principle of Inversion*, Isis **19** (1933),
         154–180.

Peacock, G.
[1834]   *Report on the Recent Progress and Present State of certain Branches of
         Analysis*, Rep. 3rd Meeting. British Ass. for the Advancement of, Sci.
         London (1834), 185–352.

Peiffer, J.
[1978]   *Les premiers exposés globaux de la théorie des fonctions de Cauchy*.
         Thése, Paris 1978.
[1983]   *Joseph Liouville (1809–1882): ses contributions à la théorie des fonc-
         tions d'une variable complexe*, Rev. Hist. Sci. **36** (1983), 209–248.

Pépin, P. T.
[1863]   *Mémoire sur l'intégration sous forme finie de l'équation différentielle du
         second ordre à coefficients rationnels*, Annali di Mat. **5** (1863), 185–224.
[1878]   *Sur les équations différentielles du second ordre*, Annali di Mat. (2) **9**
         (1878–1879), 1–10.

Picard, E.
[1883]   *Sur les groupes de transformation des équations différentielles linéaires*,
         C. R. **96** (1883), 1131–1134.
[1890]   *Mémoire sur la théorie des équations aux dérivées partielles et la méthode
         des approximations successives*, Journ. Math. Pures Appl. ser. 4 **6** (1890),
         145–210.
[1908]   *Traité d'Analyse*; . Paris 1908.

Pinet

[1887]   *Histoire de l'École Polytechnique.* Paris 1887.

Plessis, A.

[1979]   *De la fête impériale au mur des fédérés (1852-1871)*, Nouvelle Histoire de la France Contemporaine 9. Paris 1979.

Poincaré, H.

[1885]   *Sur l'équilibre d'une masse fluide amenée d'un mouvement de rotation.* Acta Mathematica **7** (1885), 259–380; Oeuvres **7** 40–142.

[1890]   *Sur les équations aux dérivées partielles de la physique mathématique,* Amer. Journ. of Math **12** (1890), 211–294; Oeuvres **9**, 28–113; see also C. R. **104** (1887), p. 44.

[1892-1899]   *Les Méthodes nouvelles de la mécanique céleste.* 3 vols. Paris 1892–1899.

[1894]   *Sur les équations de la physique mathématique,* Rend. Circ. Mat. Palermo **8** (1894), 57–155; Oeuvres **9**, 123–196.

[1895]   *Sur la méthode de Neumann et le Problème de Dirichlet,* C. R. **120** (1895), 347–352; Oeuvres **9**, 197–201.

[1896]   *La méthode de Neumann et le problème de Dirichlet,* Acta Mathhematica **20** (1896–1897), 59–142; Oeuvres **9**, 202–272.

[1899]   *Théorie du Potentiel Newtonien. Leçons professées a la Sorbonne pendant le premier semestre 1894-1895.* Paris 1899.

[1902]   *Figures de l'équilibre d'une masse. fluide. Leçons professées à la Sorbonne en 1900.* Paris 1902.

Poisson, S. D.

[1809a]   *Mémoire sur les inégalités séculaires des Moyens mouvements des planètes;* Journ. Ec. Polyt. **8** (15. cahier) (1809).

[1809b]   *Mémoire sur la variation des constantes arbitraires dans les questions de la mécanique,* Journ. Ec. Polyt. **8** (1809), 266–344.

[1811-1833]   *Traité de mécanique.* Paris 1. ed. 1811 2. ed. 1833.

[1812]   *Mémoire sur la distribution de l'électricité à la surface des corps conducteurs,* Méms. de la Classe math. et phys. de l'Inst. de France **1811** (1812), 1–92; Bull. Sci. Soc. Philomatique, Paris **3**. 61, (1812), 155–157.

[1813]   *Remarques sur une équation qui se présente dans la théorie des attractions des sphéroïde,* Nouv. Bull. de la Soc. Philomatique **3** (1813), 388–392.

[1817]   *Sur la forme des intégrales des équations aux différences partielles,* Bull. Sci. Soc. Philomatique Paris (1817), 180–183.

[1822]   *Sur une nouvelle manière d'exprimer les coordonnées des planètes dans le mouvement elliptique;* Connaiss. des Temps pour l'an 1825 Paris 1822, 373–385.

[1823a]   *Mémoire sur la distribution de la chaleur dans les corps solides,* Journ. Ec. Polyt. **12** (19. cahier). (1823) 1–162, 249–403.

[1823b]   *Suite du mémoire sur les intégrales définies et sur la sommation des séries.,* Journ. Ec. Polyt. **12** (19. cahier). (1823), 404–509.

[1826]   *Note sur les racines des équations transcendantes,* Bull. Soc. Philomatique (1826), 145–148.

[1830]   *Note sur les racines des équations transcendantes*, Mem. Acad. Sci. Paris **9** (1830), 89–95.

[1831]   *Mémoire sur la propagation du mouvement dans les milieux élastiques*, Mémoires de l'Académie des Sciences **10** (1831), 549–605.

[1833a]   *Sur le développement des coordonées d'une planète dans son mouvement elliptique, et de la fonction perturbatrice de ce mouvement*, Connaiss. des Temps pour l'an 1836 (1833), 3–55.

[1833b]   *Rapport sur deux mémoires de M. Liouville ayant pour titre Détermination des intégrales dont la valeur est algébrique*, Procès-Verbaux Acad. Sci. Paris (1833), 211–213.

[1835a]   *Théorie Mathématique de la Chaleur*. Paris 1835.

[1835b]   *Rapport sur un Mémoire de M. Liouville concernant une question nouvelle d'analyse*, Procès-Verbaux Acad. Sci. Paris (1835), 643–644.

[1836]   *Rapport sur une Note de M. Liouville relative au calcul des perturbations des planètes*, C. R. **2** (1836), 394–395.

[1837a]   *Remarques sur l'intégration des équations différentielles de la Dynamique*, L. J. **2** (1837), 317–336.

[1837b]   *Note Relative à un passage de la Mécanique céleste*, L. J. **2** (1837), 312–316.

[1838]   *Note sur l'intégration des équations linéaires aux différentielles partielles*, L. J. **3** (1838), 615–623.

Polyakov
[1981]   *Quantum Geometry of Bosonic Strings*, Phys. Lett. **103 B** (1981), 207–210. see also 211–213.

Pommaret, J. F.
[1979]   *Differential Galois Theory*. 406–413 in *Geometry and Differential Geometry*. Proc. Haifa 1979.

Poncelet, J.-V.
[1822]   *Traité des propriétés projectives des figures*. Paris 1822.

Pontécoulant, P. G. compte de
[1829–1843]   *Théorie Analytique du Système du Monde*. 1–4 Paris 1829–1843
[1840]   *Lettre de M. de Pontécoulant sur la révision qu'il a faite des calculs qui avaient servi à obtenir les résultats rapportés dans le 3e vol. de sa Théorie analytique du Système du Monde*, C. R **10** (1840), 872–873.

Prost, A.
[1968]   *Histoire de l'enseignement en France 1800–1957*. Paris 1968.

Pothier, F.
[1887]   *Histoire de l'école centrale des arts et manufactures, d'après des documents authentiques et en partie inédits*. Paris 1887.

Pradalié, G.
[1979]   *Le Second Empire*. Paris 1979.

Prange, G.
[1933]   *Die allgemeinen Integrationsmethoden der analytischen Mechanik*, Encyclodädie der Mathematischen Wissenshaften IV 1, II, 505–804.

Prelle M. J. and Singer, M. F.
[1983]   *Elementary First Integrals of Differential Equations*, Trans. Amer. Math.
Soc. **279** (1983), 215–229.

Prost, A.
[1968]   *Histoire de l'enseignement en France 1800–1967.* Paris 1968.

Prouhet, E.
[1856]   *Notice sur la vie et les travaux de Ch. Sturm*, Bull. de bibliographie,
d'histoire et de biographie mathématiques **2** (1856), 72–89; repr. in
Sturm's Cours d'analyse 5 ed. **1** (1877), XV–XXIX.

Puiseux, V.
[1850]   *Recherches sur les fonctions algébriques*, L. J **15** (1850), 365–480.
[1851]   *Sur la ligne dont les deux courbures ont entre elles un rapport constant*,
L. J. **16** (1851), 208–211.

Raabe, L.
[1834]   *Note zur Theorie der Convergenz und Divergenz der Reihen*, Journ.
Reine Angew. Math. **11** (1834), 309–310; French trans. L. J. **6** (1841),
85–88.

Raffy, L.
[1885]   *Sur les quadratures algébriques et logarithmiques*, Ann. Ec. Norm. (3) **2**
(1885), 185–206.

Rayleigh, Lord (Strutt, J.W)
[1877]   *The Theory of Sound.* London 1877.

Reich, K.
[1973]   *Die Geschichte der Differentialgeometrie von Gauss bis Riemann (1828–
1868)*, Arch. Hist. Ex. Sci. **11** (1973), 273–382.

Richter, G. R.
[1977]   *Numerical Solution of Laplace's Equation as an Integral Equation of the
First Kind*, Math. and Computers in Simulation **19** (1977), 192–197.

Riemann, B.
[1847–1874]   *Versuch einer allgemeinen Auffassung der Integration und Differentiation*, Manuscript from 1847. Published in Gesammelte Mathema-
tische Werke 1876. 2. ed. 1892. Dover Reprint 1953.
[1851–1867]   *Grundlagen für eine allgemeine Theorie der Funktionen einer ver-
änderlichen complexen Grösse*, Inaugural dissertation, Göttingen 1851,
publ. Göttingen 1867; Werke, 3–45.
[1854–1867]   *Ueber die Hypothesen welche der Geometrie zu Grunde liegen*, (Habilitationsvortrag 1854), Abh. Königl. Gess. der Wiss. Göttingen Math.
Cl. **13** (1867), 1–20; Werke, 271–287.
[1860]   *Ein Beitrag zu den Untersuchungen über die Bewegung einer flüssigen
gleichartigen Ellipsoides*, Abh. Königl. Gess. der Wiss. Göttingen; Math.
Cl. **9** (1860), 3–36; Werke, 182–211.
[1861]   *Commentatio mathematica, qua respondere tenatur quaestioni ab Illma
Academia Parisiensi propositae.* Sent to the Paris Académie des Sciences

on July 1st 1861; Werke 391–404.

Risch, R. H.
[1970]   *The solution of the problem of integration in finite terms*, Bull. Amer. Math. Soc. **76** (1970), 605–608.
[1976]   *Implicitly elementary integrals*, Proc. Amer. Math. Soc. **57** (1976), 1–7.

Ritt, J. F.
[1923]   *On the integrals of elementary functions*, Trans. Amer. Math. Soc. **25** (1923), 211–222.
[1948]   *Integration in Finite Terms. Liouville's Theory of Elementary Methods.* New York 1948.
[1950]   *Differential Algebra.* New York 1950.

Ritz, W.
[1909]   *Über eine neue Methode zur Lösung gewisser Variationsprobleme der mathematischen Physik*, Journ. Reine Angew. Math. **135** (1909), 1–61; Werke, 192–250. Paris 1911.

Roberts, M.
[1846]   *Sur quelques propriétés des lignes géodésiques et des lignes de courbure de l'ellipsoïde*, L. J. **11** (1846), 1–4.
[1847]   *Extraits de deux lettres adressées à M. Liouville*, L. J. **12** (1847), 491–492.
[1848]   *Extrait d'une Lettre adressée à M. Liouville*, L. J. **12** (1848), 209–220.
[1850]   *Mémoire sur la géométrie de courbes tracées sur la surface d'un ellipsoïde*, L. J. **15** (1850), 275–295.

Rodriques, O.
[1838]   *Sur le nombre de manières de décomposer un polygone en triangles au moyen de diagonales*, L. J. **3** (1838), 547–548.

Roselicht, M.
[1968]   *Liouville's theorem on functions with elementary integrals*, Pacific Journ. Math. **24** (1968), 153–161.
[1972]   *Integration in finite terms*, Amer. Math. Monthly **79** (1972), 963–972.

Ross, B.
[1974]   *Annotated chronological bibliography on fractional calculus.* 3–15 of Oldham and Spanier *The Fractional Calculus* New York 1974.
[1975]   *A brief History and Exposition of the Fundamental Theory of Fractional Calculus.* 1–36 of *Fractional Calculus and its Applications* (ed. B. Ross). Berlin 1975.
[1977]   *The Development of Fractional Calculus 1695–1900*, Historia Mathematica **4** (1977), 75–89.

Roth, K.F.
[1962]   *Rational Approximations to Irrational Numbers.* Inaugural Lecture, London 1962.

Rothman, T.
[1982]   *Genius and Biographers: The Fictionalization of Evariste Galois*, Amer. Math. Monthly **89** (1982), 84–106.

Rudio, F.
[1892]   *Archimedes, Huygens, Lambert, Legendre. Vier Abhandlungen über die Kreismessung.* Teubner, Leipzig 1892.

Schläffli, L.
[1847–1852]   *Über das Minimum des Integrals $\int \sqrt{dx_1^2 + dx_2^2 + \cdots + dx_n^2}$ wenn die Variablen x_1, x_2, ..., x_n durch eine Gleichung zweiten Grades gegenseitig von einander abhängig sind,* Journ. Reine Angew. Math. **32** (1852), 23–36; C. R. **25** (1847), p. 391.
[1876]   *Über die Convergenz der Entwicklung einer arbiträren Funktion $f(x)$ nach den Bessel'schen Funktionen...,* Math. Ann. **10** (1876), 137–142.

Schlissel, A.
[1976–1977]   *The development of Asymptotic Solutions of Linear Ordinary Differential Equations (1817–1920),* Archive Hist. Ex. Sci. **16** (1976–1977), 307–378.

Schmidt, E.
[1907]   *Zur Theorie der linearen und nichtlinearen Integralgleichungen,* Math. Ann. **63** (1907), 433–476.
[1908]   *Über die Auflösung linearer Gleichungen mit unendlich vielen Unbekannten,* Rend. Circ. Mat. Palermo **25** (1908), 53–77.

Schneider, T.
[1957]   *Einführung in die transcendenten Zahlen.* Springer, Berlin 1957.

Schwarz, H. A.
[1870]   *Ueber die Integration der partiellen Differentialgleichung $\frac{\partial^2 u}{\partial x^2} + \frac{\partial^2 u}{\partial y^2} = 0$ unter vorgeschriebene Grenz- und Unstetigkeitsbedinungen,* Monantsber. Königl. Acad. d. Wiss. Berlin (1870), 767–795; Ges. Math. Abh. **2**, 144–171.

Serret, J. A.
[1844]   *Mémoire sur l'intégration d'une équation diff- érentielle à l'aide des différentielles à indices quenconques,* L. J. **9** (1844), 193–216.
[1845a]   *Mémoire sur la représentation géométrique des fonctions elliptiques et ultra-elliptiques,* L. J. **10** (1845), 257–285; Addition, L. J. **10** (1845), 286–289.
[1845b]   *Note sur les courbes elliptiques de la première espèce,* L. J. **10** (1845), 421–429.
[1849]   *Cours d'Algèbre supérieure, professé a la Faculté des Sciences de Paris par M. J.-A. Serret.* Paris 1. ed. 1849.
[1851]   *Sur quelques formules relatives à la théorie des courbes à double courbure,* L. J. **16** (1851), 193–207.
[1853]   *Sur les surfaces à lignes de courbure sphériques (Extrait d'une Lettre de M. J.-A. Serret à M. Liouville.),* C. R. **36** (1853), 328–334.
[1854]   *Cours d'Algèbre supérieure.* 2. ed. Paris 1854.
[1866]   *Cours d'Algèbre supérieuere.* 3. ed. Paris 1866.

Serret, P.

[1852]   *Théorème sur les Coniques,* Nouvelles Annales de mathématique 1. ser
          11 (1852), 123–126.

Sommerfeld, A.

[1900]   *Randwertaufgaben in der Theorie der partiellen Differentialgleichungen,*
          Encycl. d. Math. Wiss. **II A7c**, 504–570.

Speziali, P.

[1964]   *Charles-François Sturm (1803–1855) documents inédits.* Palais de la
          Découverte Paris 1964.

Stäckel, P.

[1890]   *Eine charakteristische Eigenschaft der Flächen deren Linienelement ds
          durch $ds^2 = (\kappa(q_1) + \lambda(q_2))(dq_1^2 + dq_2^2)$ gegeben wird,* Math. Ann. **35**
          (1890), 91–103.

[1891]   *Über die Integration der Hamilton-Jacobi Differentialgleichungen mittels
          Separation der Veränderlichen.* Habilitations-Schrift Halle 1891.

[1893]   *Über die Bewegung eines Punktes en einer n-fachen Mannigfaltigkeit,*
          Math. Ann. **42** (1893), 537–563.

[1905]   *Über die geodätischen Linien einer Klasse von Flächen deren Linienele-
          ment den Liouvilleschen Typus hat,* Journ. Reine Angew. Math. **130**
          (1905), 89–112.

Steiner, J.

[1841]   *Sur le maximum et le minimum des figures dans le plan, sur la sphère
          et dans l'espace en général,* L. J. **6** (1841), 105–170.

Steklov

[1898]   *Sur le problème de refroidissement d'une barre hétérogene,* C. R. **126**
          (1898), 215–218.

Sturm, C.

[1829a]   *Analyse d'un Mémoire sur la résolution des équations numériques,* Bull.
          Sci. Math. Astr. Phys. **11** (1829), 419–422.

[1829b]   *Extrait d'un Mémoire de M. Sturm,* Bull. Sci. Math. Astr. Phys. **11**
          (1829), 422–425.

[1829c]   *Note présenté à l'Académie par M. Ch. Sturm,* Bull. Sci. Math. Astr.
          Phys. **11** (1829), 425.

[1829d]   *Extrait d'un Mémoire sur l'intégration d'un système d'équations diff-
          érentielles linéaires,* Bull. Sci. Math. Astr. Phys. **12** (1829), 315–322.

[1829e]   *Sur la distribution de la chaleur dans un assemblage de vases,* Bull. Sci.
          Math. Astr. Phys. **12** (1829), p. 322.

[1829f]   *Nouvelle théorie relative à une classe de fonctions transcendantes que
          l'on rencontre dans la résolution des problèmes de la physique mathémati-
          que,* Bull. Sci. Math. Astr. Phys. **12** (1829), p. 322. Cauchy's review in
          the Academy on July 5, 1830 is printed in the Procès Verbeaux

[1833a]   *Analyse d'un mémoire sur les propriétés générales des fonctions qui
          dépendent d'équations différentielles linéaires du second ordre,* l'Institut,
          Journ. Acad. et Soc. Sci. nov. 9 (1833), 219–223. (summary of 1836a)

[1833b]   *Note without name.*, l'Institut, Journ. Acad. et Soc. Sci. nov. 30 (1833), 247–248. (summary of 1836b).

[1835]   *Mémoire sur la résolution des équations numériques*, Mém. Savants Étrangers, Acad. Sci. Paris **6**, 271–318.

[1836a]   *Mémoire sur les Équations différentielles linéaires du second ordre*, L. J. **1** (1836), 106–186.

[1836b]   *Mémoire sur une classe d'Équations à différences partielles*, L. J. **1** (1836), 373–444.

[1839]   *Note de M. Sturm relative au Mémoire du M. Libri inséré dans le précédent Compte Rendu*, C. R. **8** (1839), p. 788.

[1857–1859]   *Cours d'analyse de l'École Polytechnique*, (ed. Prouhet). 2 vols, Paris 1857, 1859.

[1861]   *Cours de Mécanique de l'École Polytechnique*, (ed. Prouhet). Paris 1861.

Szabó, I.
[1977]   *Geschichte der mechanischen Prinzipien*. Basel 1977.

Tannery, J.
[1910]   *Correspondance entre Lejeune Dirichlet et Liouville*, Paris 1910, Bull. Sci. Math. (2. ser) **32** (1908), 47–62, 88–95; and, **33** (1908–1909), 47–64.

Taton, R.
[1971]   *Sur les relations scientifiques d'Augustin Cauchy et d'Évariste Galois*, Révue d'Hist. des Sci. **24** (1971), 123–148.

[1973]   *Liouville, Joseph*, Bibliography in the Dictionary of Scientific Biography, ed. Gillispie,. New York 1973, **8**, 381–387

Taylor, B.
[1713]   *De motu nervi tensi*, Phil. Trans. **28** (1713), 26 32.

Tchebyschef
      See Chebyshev.

Terquem, O.
[1847]   *Sur l'éminination par les fonctions symmétriques d'après M. Liouville (Journal de mathématique t 12, p. 68, 1847)*, Nouvelles Ann. de Math. **6** (1847), 265–300.

Thompson, S. P.
[1910]   *The Life of William Thomson, Baron Kelvin of Largs.* 2 vols London 1910.

Thomson, W. (Lord Kelvin)
[1844]   *Note on the Law of Gravity at the Surface of a Revolving Homogeneous Fluid*, Combridge Math. Journ. **4** (1845), 191–192.

[1845a]   *Démonstration d'un théorème d'analyse*, L. J. **10** (1845), 137–147.

[1845b]   *Extrait d'une Lettre de M. William Thomson à M. Liouville*, L. J. **10** (1845), 364–367.

[1847a]   *Extraits de deux Lettres adressées à M. Liouville*, L. J. **12** (1847), 256–264.

[1847b]   *Note sur une équation aux différences partielles qui se présente dans plusieurs questions de Physique mathématique*, L. J. **12** (1847), 493–496;

Cambr. and Dublin Math. Journ. **3** (1848), 84–87; Math. and Phys. Papers **1**, 93–96.

[1872] *Reprint of Papers on Electrostatics and Magnetism.* London 1872.

[1882] *On the Figures of Equilibrium of a Rotating mass of Fluid,* Proc. Roy. Soc. Edinburgh **11** (1882), p. 610; Mathematical and Physical Pepers **4**, 189–192.

Thomson, W. and Tait, P. G.

[1867, 1879–1883] *Treatise on natural philosophy.* (1. ed. Oxford 1867). 2. ed. 2 vols Cambridge 1879–1883.

Todhunter, I.

[1870] *On Jacobi's Theorem respecting the relative equilibrium of a revolving Ellipsoid of Fluid, and on Ivory's discussion of the Theorem,* Proceedings of the Royal Society **19** (1870), 42–56.

[1873] *A History of the Mathematical Theories of Attraction and the Figure of the Earth from the time of Newton to that of Laplace.* 2 vols., London 1873, Dover reprint 1962.

Tortolini, B.

[1855] *Sopra gl'integrali generali di alcune equazioni a derivate parziali a coefficienti constanti,* Mem. Math. Fis. Soc. Italiana Sci. Moderna **25** (1855), 310–341.

Truesdell, C.

[1960] *The Rational Mechanics of Flexible or Elastic Bodies 1638–1788*; Euler's Opera Omnia (2) **11**. part 2. Zürich 1960.

Valson, C. A.

[1868] *La vie et les travaux du Baron Cauchy.* Paris 1868, (2. ed. 1970).

Vallée-Poussin, Ch. J. de La

[1962] *Gauss et la théorie du potentiel,* Revue des questions scientifiques **133** (1962), 314–330.

Vapereau, G.

[1880] *Liouville;* in *Dictionnaire universel des contemporains.* 5. ed. (Paris 1880) 1171–1172.

Vecten

[1824] *Démonstration d'une propriété du quadrilatère complet,* Ann. Math. Pures et Appl. **15** (1824–1825), 146–149.

Vessiot, E. P. J.

[1892] *Sur l'intégration des équations différentielles linéaires,* Thése Fac. Sci. Paris 1892; Ann. Sci. Ec. Norm. Paris **3** (1892), 197–280.

[1915] *Gewöhnliche Differentialgleichungen, Elementare Integrationsmethoden.* Encyclopädie der Mathematischen Wissenshcaften, II A 4b.

Vincent, A. J. H.

[1833–1838] *Mémoire sur la résolution des équations numériques*; Mémoires de la Societé Royale de Sciences... à Lille (1833), 1–34;, (1838), 5–24; L. J.

1 (1836), 341–372;, **3** (1838), 235–243.

Volterra, V.

[1896] *Sulla inversione degli integrali definiti*; Nota I–IV Atti della Accademia Scienze di Torino **31** (1896), 311–323, 400–408, 557–567, 693–708; Opere Matematiche **II**, 216–262.

[1913] *Leçons sur les équations intégrales et intégrodifférentielles.* Paris: Gauthier-Villars 1913.

Voss, A.

[1901] *Die Prinzipien der rationellen Mechanik.* Encyclopädie der Math. Wiss. **4** (1901–1908) IV.1 1–121.

Waldschmidt, M.

[1983] *Les débuts de la théorie des nombres transcendants (à l'occasion du centenaire de la transcendance de π)*, Cahiers du Séminaire d'Histoire des Mathématiques **4** (1983), 93–115.

Wantzel, P.L.

[1839] *Extrait d'une Lettre de M. Wantzel à M. J. Liouville*, L. J. **4** (1839), 185–188.

[1842] *Remarques à l'occasion du Mémoire de M. Maurice sur l'invariabilité des grands axes*, C. R. **15** (1842), p. 732.

Watson, G. N.

[1922] *A Treatise on the Theory of Bessel Functions.* Cambridge 1922.

Weber, H.

[1869] *Ueber die Integration der partiellen Differentialgleichung $\frac{\partial^2 u}{\partial x^2} + \frac{\partial^2 u}{\partial y^2} + k^2 u = 0$*, Math. Ann. **1** (1869), 1–36.

Weierstrass, K.

[1857] *Über die Integration algebraischer Differentiale vermittelst Logarithmen*, Monatsber. Akad. Wiss. Berlin (1857), 148–157; Werke **1**, 227–232.

[1870–1895] *Über das sogenannte Dirichletsche Princip*, (read in 1870); Werke, Berlin 1895 **2**, 49–54.

Weiss, J. H.

[1982] *The Making of Technological Man. The Social Origins of French Engineering Education.* Cambridge Mass. London 1982.

Whiteside, T.

[1976] *The Mathematical Papers of Isac Newton.* **7** Cambridge 1976.

Whittaker, E. T.

[1937] *A Treatise on the Analytical Dynamics of Particles and Rigid Bodies*, 4th ed. Cambridge Univ. Press 1937; German transl. of 3rd. ed *Analytische Dynamik der Punkte und Starren Körper.* Berlin 1924.

Wussing, H.

[1984–1969] *The Genesis of the Abstract Group Concept*, Cambridge Mass. 1984; Transl. from *Die Genesis des abstrakten Gruppenbegriffes.* Berlin 1969.

Youschkevitsch, A.P.
[1970–1980a]   *Chebychev.* Biography in Dictionary of Scientific Biography, ed. Gillispie, New York 1970–1980 **III**.
[1970–1980b]   *Ostrogradsky.* Biography in Dictionary of Scientific Biography, ed. Gillispie, New York 1970–1980 **X**.

Zolotarev, G.
[1874]   *Sur la méthode d'intégration de M. Tschebyschef,* L. J. (2) **19** (1874), 161–188.

Index

Sources in the History of
Mathematics and Physical Sciences